Patrick Moore's Practical Astronomy Series

For other titles published in this series, go to
www.springer.com/series/3192

Guidebook to the Constellations

Phil Simpson

 Springer

Phil Simpson
Raspberry Hill Rd. 73
88317 Cloudcroft, New Mexico
USA
phil73simpson@tularosa.net

ISBN 978-1-4419-6940-8   e-ISBN 978-1-4419-6941-5
DOI 10.1007/978-1-4419-6941-5
Springer New York Dordrecht Heidelberg London

Library of Congress Control Number: 2011928554

Printed on acid-free paper

Springer is part of Springer Science+Business Media (www.springer.com)

Acknowledgements

I have heard that nobody writes a book alone, and it has certainly been the case with this book. I have to thank many people, without whom I could not have completed this book.

First, the members of the Alamogordo Astronomy Club who, for years, listened to my Constellation of the Month presentations upon which this book is based. They paid close attention to each talk, asked thoughtful and pertinent questions, and made suggestions which led to improvements in the final content and format.

Next, 13 people from literally around the world contributed astrophotographs and drawings of celestial objects which made the book possible. Each of these has a short biographical sketch and snapshot in Appendix F.

Then, Lynn Rice, Co-owner of New Mexico Skies Guest Observatory and a former professional proofreader, who volunteered uncounted hours of her time to read every word of the book, correct my grammatical and typing mistakes, as well as make numerous suggestions for improved wording, changes in content and format which greatly improved the overall result.

Finally, but in many ways the most important, my wife Patty, who patiently tolerated my daily retreat to my study, who took over many of the errands and tasks which I had previously done, protected my time from numerous business and social interruptions, listened to my complaints of difficulties, celebrated my little successes along the way, and unfailingly encouraged me to keep me striving until the end.

Contents

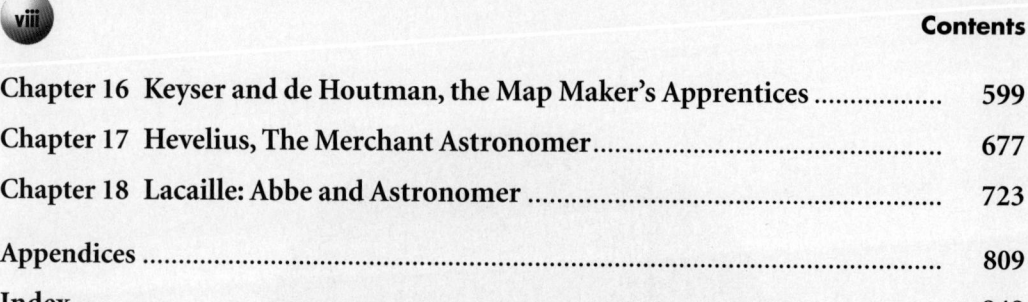

Introduction

This book actually started in 1970 when I was a member of the Valley Astronomers of San Bernardino, California. One evening, Len Farrar, our president, asked the club members to help him solve a problem. Although our membership rolls listed about 80 people, we seldom had more than 10, and usually only 5 or 6 people attending the meetings. Visitors attracted by newspaper announcements came once or twice, then failed to return. Len asked for ideas about what might be wrong, and what we could do about it.

Over the next month, I thought about the problem and came to the conclusion that we were failing to serve beginners who came with the hope of learning about astronomy and what amateur astronomers could do and see. I realized that we were not presenting anything for these potential members. At 60 miles east of Los Angeles, our club had access to astronomers at dozens of professional observatories and literally hundreds of universities, many with their own planetariums and small observatories. Len was a dedicated and advanced amateur astronomer, and was doing an admirable job of getting top-flight presenters for our monthly meetings. Many of the presenters were doing research at the frontiers of astronomy, and all had something of interest to share with us. In a way, our problem was one of riches. With so many professionals willing come to our meetings, Len was able to satisfy the most advanced members of our club. However, we had no program for beginners. They came, they saw, but they didn't understand, and they never came back.

At the next monthly meeting, I told Len about my thoughts. I got exactly the response I had expected, "Will you develop a program for the beginners?" In the process of identifying the problem, I had also thought of one solution. I proposed that we advertise a meeting time half an hour earlier than our regular time, and that I present a new constellation each month, tell the visitors how to find the constellation and recognize the pattern of the object or creature it represented, and what objects in that constellation were visible in amateur telescopes. Len agreed, and for the next two years, I gave "Constellation of the Month" talks.

These early talks were fairly simple. I took time exposure photographs of constellations which showed only the stars visible to the unaided eye and projected them on a paper so I could accurately plot their positions. I then added lines showing a simple "stick figure" demonstrating how the stars could be connected to visualize what the constellation represented. Next, I added symbols showing the location of double stars, variable stars, star clusters, nebulae and galaxies which were visible in amateur telescopes. To help beginners remember these figures and what they represented, I turned to the myths I had found so helpful when I was learning the constellations as a child. I recognized that they not only helped make the celestial figures memorable, but could often be used to present information about the constellations and their relationship to each other. Finally, I added photographs of some of the celestial objects to show beginners what they could expect to see through a telescope.

These monthly talks were a success in that, over the next two years, the attendance at our meetings gradually rose to about three or four times what they had been before. Newcomers were returning to subsequent meetings, and enough of the more experienced members were attending the pre-meeting talks that the official meeting time was moved up by half an hour. Some of the club members even suggested that I collect the presentations to form a book. I considered the idea, but didn't have the time. A few months before I was transferred to New Mexico, George Beatty, our club publications chairman became ill and had to retire from his job as instructor of physics and astronomy at Valley College in San Bernardino. The college was able to find other instructors for his daytime classes, but no one was available or willing to cover his two night classes in freshman astronomy. On the basis of my presentations, George recommended, and the college hired, me to teach the night classes.

In the classes I taught, I recognized that while most students were there because of personal interests, some were there only because they had to have at least one science class, and they thought astronomy would be preferable to alternatives of chemistry, biology or physics. I realized that these students would not remember facts like the distance to the sun, the diameter of Jupiter or similar details, the one thing that would give them a constant reminder of their astronomy class was the constellations that they could see every clear dark night for the rest of their lives. Because of this, I required all my students to observe the night sky to and learn several constellations. They also used these constellations, along with the moon, to observe the movement of celestial objects on different time scales, i.e. hours, days, weeks and months.

I continued to make constellation of the month talks to other astronomy clubs of which I was a member, and to scouting and other civic groups as well as to classes which I taught for New Mexico State University, Chapman University and Eastern New Mexico University. Over the years, these constellation presentations evolved as I used them in several ways. When computers became available with their word processing, graphics and photo processing programs, there was a quantum step in the quality of the illustrations as well as in the time it took to prepare the presentations. However, probably the greatest innovation was the recognition that by researching the various versions of the myths available, I could present a story that explained not one, but several constellations at one time.

In addition, many of the myths were the attempts of early civilizations to explain the otherwise unexplainable events and changes as acts of their gods. As such, these myths often contained information which is still useful today to help people remember important facts about the constellations. Some of these facts include the position or attitude of the figure, which constellations are adjacent or nearby, and when the constellations are visible. When presented as an interesting story or myth, these facts are easily and unconsciously absorbed.

For example, one version of the Orion myth includes the constellations of Canis Major, Corona Australis, Lepus, Ophiuchus, Sagittarius, Scorpius, Serpens and Taurus, as well as explaining why they are split into two groups with one visible in the winter and the other in the summer. It also explains why each of the constellations is located where it is in relation to the other constellations. The 48 "classical" constellations known to the ancient Greeks are organized into groups related by a single mythological story. The other 40 constellations are called "modern", and are generally grouped by their originator. In a few cases, the originator of some of the modern constellations was able to weave a new constellation into a previously related group. In this case, the modern constellation was included in the classical group.

In trying to be faithful to the origin stories, I have made the simple line drawings represent the way the constellations were visualized in Greece during its classical period. For this purpose, I used Robert Brown's *Researches into the Primitive Constellations of the Greeks, Babylonians and Phoenicians*, published in 1899. In his book, Brown included the Hipparcho-Ptolemy star catalog, with its verbal description of the stars locations in the constellations. Along with each description, Brown included the modern designation assigned by Bayer or Flamsteed. In most cases this worked well, but there were some exceptions. Most of these exceptions resulted from the action of the Intenational Astronomical Union (IAU) adopting precise definition for all constellations in 1929 (published in 1930). Until this time, the constellation boundaries had been only vaguely defined by general agreement, and each publisher of a star catalog or atlas made his own determination about which stars were included in which constellations. From this time on, stars which had previously been considered shared between two constellations, e.g. Alpheratz had been considered to be part of both Andromeda and Pegasus became Andromedae, and every other star was uniquely part of one and only one constellation. Another effect of this action was that the IAU defined boundaries were all straight lines following either constant declination or right ascension based on the 1875.0 system of coordinates. This meant that some stars which had previously been generally accepted as being in one constellation were now in a different one. Finally, two of the larger constellations were divided into smaller ones. Ophiuchus, the snake and its holder, became Ophiuchus and Serpens. Argo Navis, the ship sailed by the Argonauts, was broken into the components of Carina the Keel, Vela the Sail and Puppis the Stern.

One thing which you will notice is that there are no pictures from professional observatories in this book! The reason for this is that I wanted to show what amateurs can see and photograph with relatively modest telescopes, not what professionals can photograph with huge telescopes. Also, where possible, I have paired a photograph and a drawing. The reason for this is that many people new to amateur astronomy are disappointed when they look through a telescope and don't see the brilliant colors and fine detail which they've seen in books. Most celestial objects which appear so bright and colorful in photographs are simply too faint to activate the retina's cones (which allow daylight/color vision), so we see them with only the rods (night/black and white only), even in large telescopes. By pairing photographs and drawings, I hope to give beginners a more realistic expectation of what to expect.

CHAPTER ONE

How To Use This Book

This book is designed to help a novice or advanced amateur astronomer learn the constellations, and to provide a source of information about celestial objects visible to the eye or in amateur instruments in each of the 88 constellations. I hope it will retain its usefulness long after the beginning amateur has learned the constellations. Toward this end, I have included some objects of interest which are easily viewed by the unaided eye, many of which are visible through binoculars or small telescopes, and a few of which require larger amateur telescopes to see or photograph. It is also designed to help an experienced amateur find the fainter and less well-known constellations and, to help answer the questions which visitors frequently ask at public star parties, like "How do you find Scorpius, the Andromeda galaxy, or other celestial objects?"

How you use this book will depend on how experienced you are with the constellations and amateur astronomy in general. A beginner may not know a single constellation and may not even know anyone who can help with finding the brighter constellations. The beginner should follow the instructions below. An experienced observer will (and is encouraged to) go directly to any constellation and use the detailed chart and object list for that constellation to find objects to observe. However, even a very experienced observer probably will not be familiar with all 88 constellations, or even with all the fainter constellations visible from his home. That observer can also benefit by locating the section covering the unknown constellation(s) and following the directions for finding the fainter constellations.

Although it is practical to begin the process of learning the constellations in any season by selecting one of the bright constellations visible at the time, the most logical way for most people in the northern hemisphere to start learning is to begin with the

P. Simpson, *Guidebook to the Constellations: Telescopic Sights, Tales, and Myths*,
Patrick Moore's Practical Astronomy Series, DOI 10.1007/978-1-4419-6941-5_1,
© Springer Science+Business Media, LLC 2012

north circumpolar constellations. Not only are they familiar to most people in the form of the asterisms of The Big and Little Dippers, but they are visible to northern hemisphere observers at least part of the night all year long. The Big and Little Dippers are not constellations but asterisms which include only part of the constellations of The Big and Little Bears. The distinction is one of definition. A constellation is an area of the sky containing a group of stars which represent some figure or object and is officially recognized by the International Astronomical Union. An asterism consists of two or more stars which form some recognizable shape. An asterism may contain stars from only one constellation as does the Big Dipper, or stars from several constellations as do The Summer Triangle and The Winter Hexagon. Asterisms are in the eye of the beholder, and different people may see different asterisms in the same group of stars.

Once you are familiar with the circumpolar constellations, you can use them as guides or pointers to locate other constellations or bright stars which in turn will enable you to find surrounding constellations. No matter what time of the year you begin your study of the constellations, it is possible to find one or more bright, easily recognized constellations which can be located by extending lines from the two Dippers. In the winter there is Orion, one of the brightest and most easily recognizable constellations. During the spring, there is Leo with its first magnitude <u>Regulus</u> marking the handle of a scythe or the period at the bottom of a reversed question mark. In the summer, the three bright stars of <u>Altair</u>, <u>Deneb</u> and <u>Vega</u> form a large right triangle. Each is the brightest star in its constellation, Aquila, Cygnus, and Lyra, respectively. Autumn's choices include the constellation of Cassiopeia and the asterism called the Great Square of Pegasus to guide you. The star at the northeast corner of the Great Square lies about 30° due south of β Cassiopeiae. Although there are no first magnitude stars in Pegasus, first magnitude <u>Fomalhaut</u> lies in a relatively barren region of the sky, and thus stands out. A line drawn from Polaris, the pole star, due south to <u>Fomalhaut</u> passes along the western edge of the Great Square. The four stars of the Great Square are three rather average second magnitude stars and one third magnitude star, but they form a distinctive pattern about 15° on a side, oriented on an east/west and north/south direction. After these constellations are located, each one can be used as a guide for locating the surrounding constellations.

If you are already familiar with some of the constellations, or can find a friend or local astronomy club member willing to help you learn a few, you have a head start on learning as many as you wish because every constellation can be used as a guide to neighboring constellations, and these can be used to find even more until the entire set has been learned.

Southern hemisphere observers will need a slightly different procedure. Most of the constellations which are circumpolar for them will be very faint. These are not very helpful in finding other constellations. It will be better for inexperienced southern observers to start by finding one of the bright constellations like Crux (the Southern Cross), or an easily recognizable star or star group like <u>Canopus</u>, <u>Alpha</u> and <u>Beta Centauri</u> or <u>Achernar</u>. From those starting points, the fainter constellations can be found.

The Guidebook to the Constellations is organized so that each chapter (except Chap. 14) presents multiple constellations which are related to each other and which can be readily learned as a unit. For the 48 classical constellations, the relationship is defined by a common origin myth which explains the link between the constellations and why they are located near each other, or why they are in some particular pose or part of the sky. In the case of the constellations created by more modern astronomers, the link is usually their common creator(s). In a few cases, the creators were able to follow the ancient pattern and link the new constellation to an old myth or story, but this is usually not the case.

Although the Greek myths may at first seem to be unimportant, they have been carefully researched and selected to provide a significant amount of information about the constellations. Because these tales were old and retold by dozens or even hundreds of successive generations, many variations and elaborations arose. By carefully reviewing all the versions I could find, I was able to select the versions which include the largest number of constellations and incorporate the greatest amount of information about each group of constellations. An excellent example of this is the myth about Orion. In this case, tying together several strands of myths incorporates nine constellations into a single story. These constellations are divided into two groups, one visible in the summer and the other in the winter. This myth explains the presence of each constellation, why it is in a particular subgroup, why the two subgroups are visible only in their respective seasons, and why each constellation has a specific location relative to each of the others. It even explains why Orion is shown in a fighting stance, with his left hand holding his shield in front of him and his raised right hand holding a large club ready to strike.

Each chapter of the book begins with an alphabetical list of all the constellations included, along with their English meanings. Following this list is the myth or origin narrative for the entire group. If the constellations are located near each other, there may be one or more charts covering a large portion of the sky. These charts are called finder charts and show the constellations of interest in that particular chapter along with the surrounding constellations. These charts have lines connecting the principal stars to show how to visualize the person, animal or object for which the constellation is named. The constellations in that particular chapter are shown in dark lines, while the surrounding constellations are shown in medium gray. These charts can be used for several weeks before and after the listed date by recognizing that the star patterns rotate with the seasons, and that for each week earlier than the listed date, the stars will reach the position shown approximately 30 min later than 9:00 PM. Similarly, for each week later, they will reach the position 30 min earlier.

The finder chart is followed by subsections for each constellation. These may not be presented in alphabetical order because it may be helpful to use some constellations as guides to others, so these must be presented first, or because two or more constellations may fit on a single detailed chart, so they are presented consecutively.

Each constellation section has a specific order:

A. Three lines which consist of two columns. On the left side of the page the lines are
    The Latin nominative (naming) case of the constellation name, and its pronunciation
    The Latin genitive (possessive) case of the constellation name, and its pronunciation
    The standard three-letter abbreviation of the constellation name

    On the right side of the page, the three lines consist of
    The English meaning of the constellation name

The date on which the constellation will be on the meridian and at its highest point above the horizon at 9:00 pm

The area of the constellation in square degrees, plus its rank in size among the constellations

B. Additional elaboration or variation of the origin myth, or other explanatory material about the origin nature of the specific constellation may be included

C. Instructions for finding the constellation

D. A list of the key star names with their Bayer (Greek letter) or other designations, pronunciation, and origin

E. A list of the principal stars with their magnitudes (brightness) and their spectral type (temperature and color)

F. A list of all other stars of 5th magnitude or brighter, not included above

G. A photograph of the actual constellation (if available)

H. The same photograph with lines added to show how the constellation figure is visualized

I. A chart showing the constellation, including its figure, boundaries, the location of selected celestial objects, and the celestial coordinate system

J. A table of selected celestial objects visible to amateur astronomers

K. Astrophotographs and drawings made by amateur astronomers of some of the more prominent telescopic objects in the constellation

At the end of each chapter, there may be several notes elaborating on, or explaining, items in the text or in the object lists.

CHAPTER TWO

Long Tails of Two Bears

April–July

URSA MAJOR	URSA MINOR	DRACO
The Big Bear	The Little Bear	The Dragon

These three constellations are a good starting point since they are visible to most people in the northern hemisphere at some time of the night throughout most of the year. They also contain two asterisms familiar to many people. The brighter stars in Ursa Major and Ursa Minor form the big and little dippers, which are well known and easily recognized. In addition, the star marking the end of the Little Bear's tail is Polaris, the North Star, and two of the stars in the bowl of the big dipper always point toward Polaris. These three stars can be used to help you find your directions at night.

After the two dippers are located, it is easy to add the other stars to the big dipper to form the entire constellation of Ursa Major, the big bear. The stars forming the little dipper include almost all the easily visible stars in Ursa Minor, so the asterism and the constellation figure are virtually the same. Using pairs of stars from these constellations, we can draw imaginary lines across the sky which point to other constellations or bright stars. As additional constellations are recognized, they can then be used as guides to still more constellations or bright stars.

The ancient Greeks envisioned their gods as literally being "super-humans," with all the normal human virtues and faults. The biggest differences that distinguished the gods from humans were their immortality and their power to accomplish magical things.

P. Simpson, *Guidebook to the Constellations: Telescopic Sights, Tales, and Myths*,
Patrick Moore's Practical Astronomy Series, DOI 10.1007/978-1-4419-6941-5_2,
© Springer Science+Business Media, LLC 2012

Partly by chance and partly by choice, the Olympian gods and goddesses acquired realms of special interest in which each was supreme. Also, although each had magical powers, there were also limitations, and no deity could directly undo the acts of another. At best, one could only cast another spell to modify or lessen the effects of the earlier spell. These powers and limitations play important parts in many of the myths used by the ancient Greeks to explain the constellations.

Zeus, the king of the gods, was envisioned as being quite a woman chaser, while his wife Hera was depicted as jealous and vindictive. Zeus was supposed to have loved many mortal women and to have had many children by them, some of whom were immortal while others were mortal. These affairs were important to mortals since many of the great heroes, kings, and founders of new countries traced their ancestry back to the supposed affairs between Zeus and mortal women. Their celestial heritage both explained their extraordinary strength, skill, or ability, and at the same time, justified their positions as king, leader, or prophet.

One of the women who attracted the attention of Zeus was Callisto, the beautiful daughter of Lycaon, the king of Arcadia. Callisto bore Zeus a son named Arcas. Hera discovered Zeus' infidelity and was furious at him, and extremely jealous of the beautiful Callisto. Since she was less powerful than Zeus and could not harm him directly, Hera vented her anger by changing Callisto into the ugliest animal she could imagine – a bear with shaggy fur and long teeth and claws, unable to make any sounds except growls. Callisto tried to communicate her identity and plight, but the citizens of Arcadia saw only a wild bear and drove the unfortunate woman into the forests, forever cut off from the company of humans.

After the unexplained disappearance of Callisto, Arcas was raised by his grandparents, the king and queen. As he grew into a handsome young man, Arcas became a skilled hunter. While he was hunting one day, Callisto saw and recognized Arcas. She cried out and rushed toward him. Of course, Arcas didn't recognize his mother and to protect himself, he raised his spear and aimed it at her heart. Zeus had been watching over his son, and immediately intervened to prevent the impending tragedy. He changed Arcas into a bear cub so that the mother and son could be reunited. Then he picked up each bear by its short tail to avoid being scratched by its sharp claws. He whirled them around his head and threw them up into the heavens, where we still see them to this day. In doing this, Zeus had to whirl them so hard that their tails were stretched to great lengths, and thus the heavenly bears have much longer tails than earthly ones.

Although Callisto was no longer a threat to her, Hera was still angry and determined to have her revenge, so she persuaded the gods of the seas to forbid the bears to enter their kingdom. Thus the bears are never allowed to set beneath the seas where they could rest like other heavenly beings but must endlessly circle about the cold northern sky. Even this was not enough to satisfy Hera, so she took Ladon, the dragon which had guarded the golden apples of the Hesperides sought by Heracles in his twelfth labor, and coiled him around Arcas so that he and Callisto could never be reunited, thwarting Zeus' planned reunion of the pair.

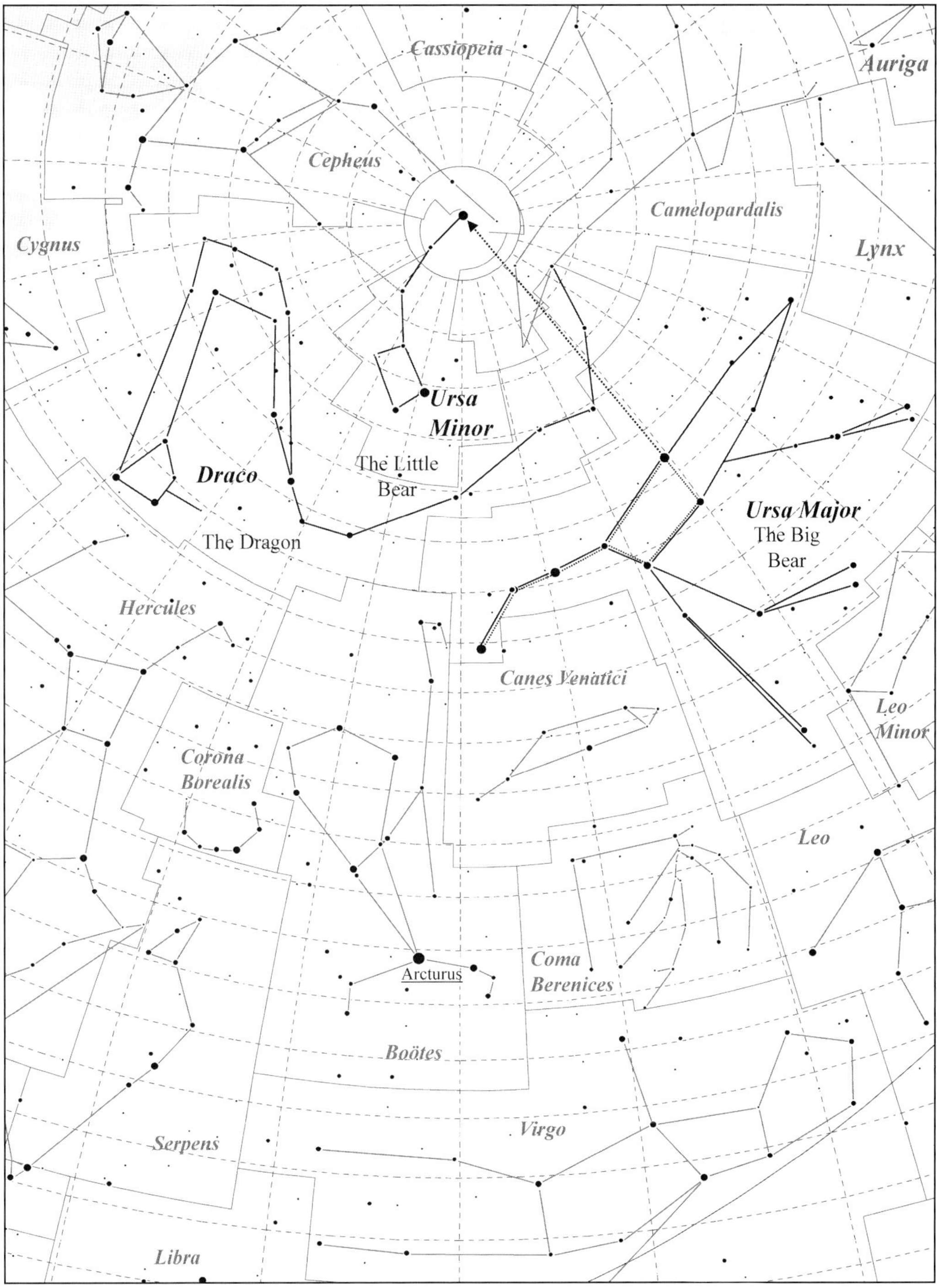

Fig. 2.1. Group finder chart. (Looking south on June 6, about 9:00 PM).

URSA MAJOR (er-sah **may**-jor)	**The Big Bear**
URSAE MAJORIS (er-sigh muh-**jor**-is)	**9:00 P.M. Culmination:** May 1
UMa	**AREA:** 1,280 sq. deg., 3rd in size

TO FIND: At 9:00 PM on May 1 (or one half hour earlier [later] for each week after [before] May 1), face north and look high in the sky, almost overhead in the Europe and the northern USA or somewhat lower from the southern USA. The seven stars of the big dipper are all fairly bright and the shape is very distinctive, so look for it first, then use the chart to trace out the rest of the constellation. The three star arc of the dipper's handle is also the large bear's tail. The four stars of the dipper's bowl form the bear's rump and flank. A line drawn through the two northern stars of the bowl, (δ, Megrez and α, Dubhe) extended 11° W ends near 23 UMa (m3.65). From there o¹, Muscida, is about 8° WSW. Muscida marks the tip of the bear's nose. To find Muscida, draw a line through the southern stars of the dipper's bowl (γ, Phecda and β, Merak) and extend it about 10° This line terminates about 1° or 2° south of υ Uma (m3.78). Muscida is about 10° WNW of υ. The two series of lines you have traced out outline the bear's body from the tail to the tip of its nose. Note that the two stars forming the top of the bowl are about 10° apart, and the two forming the outside edge of the bowl are about 5° apart. These two distances are convenient references for estimating distances between celestial objects.*¹ Using the drawing of the large bear's figure and these reference distances, you can trace out the bear's legs to complete the constellation. See Figs. 2.1–2.3.

Key Star Names:

Name	Pronunciation	Source
α Dubhe	**dub**-ee	Thahr al Dubb al Akbar (Arabic), "back of the great bear"
β Merak	**mee**-rak	al Marakh (Arabic), "loin of the bear"
γ Phecda	**fek**-dah	Fakhidh al dubb al akbar (Arabic), "the thigh of the greater bear"
δ Megrez	**mee**-grez	al Maghrez (Arabic), "the root of the tail"
ε Alioth	**al**-e-oth	probably a corruption of <u>al jaun</u> (Arabic), "the black horse or bull"
ζ Mizar	**my**-zar	al mi'zar (Arabic), "the girdle, veil, or trousers"
η Alkaid	al-**kade**	al Ka'id Banat al Na'ash (Arabic), "the leader of the daughters of the bier"
ι Talitha	**tah**-lih-thah	al-aqafza al-thalitha (Arabic), "the third leap"
λ Tania Borealis	**tah**-nih-yuh **boh**-reh-**AH**-liss	al-aqafza al-thaniya (Ind-Arabic), "the second leap" + borealis (Latin), "northern"
μ Tania Australis	**tah**-nih-yuh ous-**trah**-liss	al-aqafza al-thaniya (Ind-Arabic), "the second leap" + australis (Latin), "southern"
ν Alula Borealis	ul-**uh**-lah **boh**-reh-**AH**-liss	Al-qafza al-ula (Arabic), "the first leap" + borealis (Latin), "northern"
ξ Alula Australis	ul-**uh**-lah ous-**trah**-liss	Al-qafza al-ula (Arabic), "the first leap" + australis (Latin), "southern"
o¹ Muscida	**muh**-si-dah	musida (Latin), "muzzle"
80 Alcor	**al**-kore	(al) Khwar or Khawwar, Persian (with added "al"), "the faint one"

Magnitudes and spectral types of principal stars:

Bayer desig.	Mag. (m_v)	Spec. type	Bayer desig.	Mag. (m_v)	Spec. type	Bayer desig.	Mag. (m_v)	Spec. type
α	1.81	F7 V	β	2.34	A1 V	γ	2.41	A0 V Sb
δ	3.32	A3 V var	ε	1.76	A0p	ζ	2.23	A2 V
η	1.85	B3 V Sb	θ	2.52	F6 IV	ι	3.12	A7 IV
κ	3.57	A1 Vn	λ	3.45	A2 IV	μ	3.06	M0 III Sb
ν	3.49	K3 III Sb	ξ	3.79	G0 V	o	3.35	G4 II/III
π¹	5/63	G1.5 Vb	π²	4.59	K2 III	ρ	4.74	M3 III
σ¹	4.15	K5 III	σ²	4.80	F7 IV/V	τ	4.67	Am
υ	3.78	F0 IV	φ	4.55	A3 IV	χ	3.69	K0 III
ψ	3.00	K1 III	ω	4.66	A2 Vs	80	3.99	A5 V Sb

Fig. 2.2a. The stars of Ursa Major and adjacent constellations.

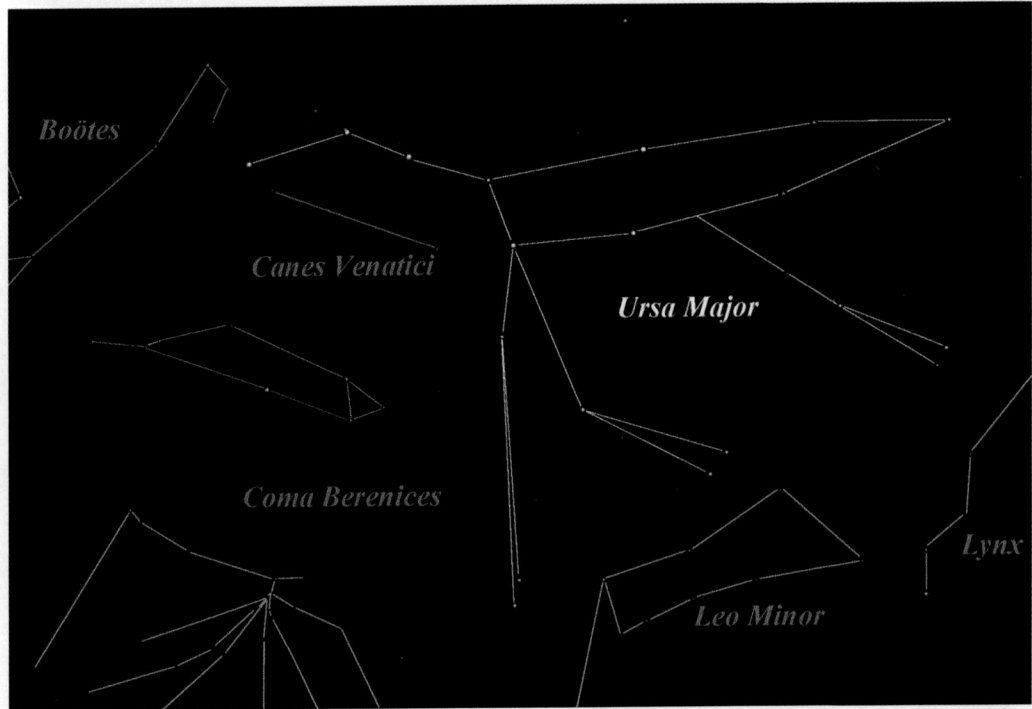

Fig. 2.2b. The figures of Ursa Major and adjacent constellations.

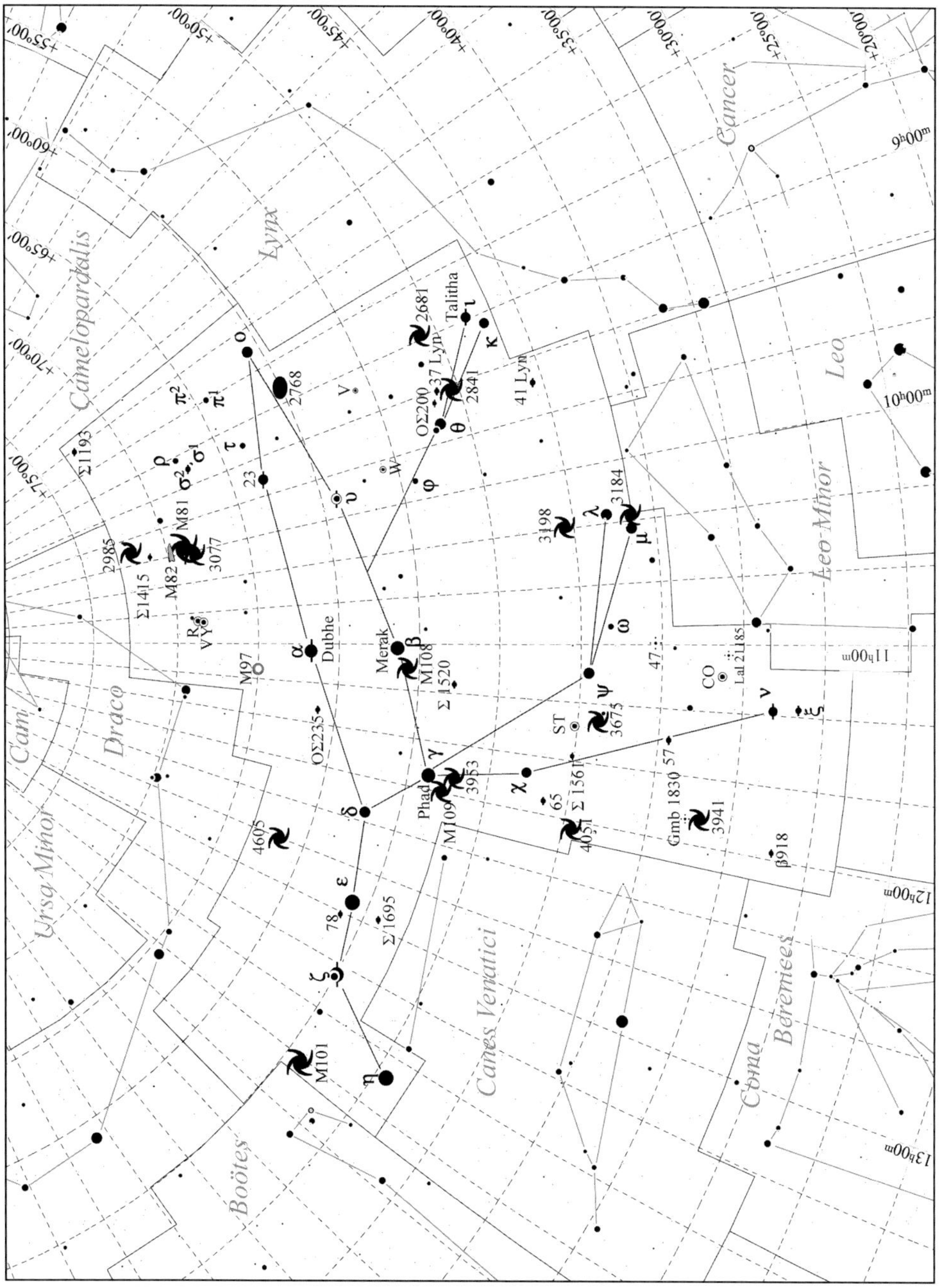

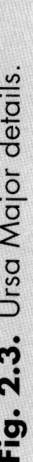

Fig. 2.3. Ursa Major details.

Table 2.1. Selected telescopic objects in Ursa Major.

Multiple stars:

Designation	R.A.	Decl.	Type	m₁	m₂	Sep. (")	PA (°)	Colors	Date/Period	Aper. Recm.	Rating	Comments
Σ1193	08h 20.7m	+72° 24'	A–B	6.2	9.7	42.7	090	dO, bGr	2001	2/3"	*****	Great at low power
9, ι Ursae Majoris	08h 59.2m	+48° 02'	A–BC	3.1	9.2	2.64	198	yW, R	**817.91**	3/4"	****	a.k.a. h2477A–BC. PA increasing
			B–C	3.1	10.1	**0.42**	232	yW, R	**39.69**	>20"	**	a.k.a. Hu 628BC. Max Sep = 0.88" in 2020 *2
13, σ² Ursae Majoris	09h 10.4m	+67° 08'	A–B	4.9	8.9	**4.15**	**350**	Y, B ?	**1,141.3**	2/3"	****	a.k.a. Σ1306AB. Slowly widening
			A–C	4.9	10.3	206.3	148	Y, ?	1999	2/3"	***	a.k.a. Σ1306AC
37 Lyncis	09h 20.7m	+51° 16'	A–B	6.2	10.0	5.6	138	W, ?	1998	3/4"	****	a.k.a. Σ199AB
			A–C	6.2	10.7	130.0	007	W, ?	1998	3/4"	***	a.k.a. OΣ199AC
OΣ 200	09h 24.9m	+51° 34'	A–B	6.5	8.6	1.3	335	Y, ?	2003	7/8"	****	Good test of seeing
41 Lyncis	09h 28.7m	+45° 36'	A–B	5.5	7.8	70.1	162	dY, grB	2005	6×30"	*****	a.k.a. S 598AB
			A–C	5.5	9.8	84.1	075	dY, ?	1999	15×80"	****	a.k.a. S 598AC
23 Ursae Majoris	09h 31.5m	+63° 04'	A–B	3.7	9.2	23.2	269	Y, brB	2003	2/3"	*****	a.k.a. Σ1351AB
29, υ Ursae Majoris	09h 51.0m	+59° 02'	A–B	3.8	11.5	11.8	296	pY, ?	1999	3/4"	***	a.k.a. OΣ521
Σ1415	10h 17.8m	+71° 04'	A–B	6.7	7.3	16.5	167	Y, W	2003	15×80"	****	
50, α Ursae Majoris	11h 03.7m	+61° 45'	A–B	2.0	5.0	**0.56**	**032**	O, ?	**44.5**	16/18"	***	a.k.a. β1077. PA increasing rapidly, opening to 0.82" in 2024
Σ1520	11h 16.1m	+52° 46'	A–B	6.5	7.8	12.3	344	O, W	2003	2/3"	*****	Showcase pair
53, ξ Ursae Majoris	11h 18.2m	+31° 32'	A–B	4.3	4.8	**1.61**	**211**	pY, oY	**59.878**	6/8"	*****	a.k.a. Σ1523. Showcase pair *3
54, ν Ursae Majoris	11h 18.5m	+33° 06'	A–B	3.5	10.1	7.4	149	pO, pB	2005	2/3"	****	a.k.a. Σ1524.
57 Ursae Majoris	11h 29.1m	+39° 20'	A–B	5.3	10.6	354	184	W, B	2003	3/4"	*****	a.k.a. Σ1543AB.
OΣ 235	11h 32.3m	+61° 05'	A–B	5.5	7.0	**0.80**	**021**	W, ?	**72.7**	12/14"	***	a.k.a. Σ235. Max separation = 1.00" in 2027
Σ1561	11h 38.7m	+45° 07'	A–B	6.5	8.2	8.96	247	dY, rW	**2050**	2/3"	****	Easy object for small scopes. Subtle but nice colors
65 Ursae Majoris	11h 55.1m	+46° 29'	AB–C	6.5	8.32	3.9	042	Y, B	2006	2/3"	****	a.k.a. A1777AB. A–B is a binary, but maximum separation is only 0.33"
β 918	11h 58.1m	+32° 16'	AB–D	6.5	6.97	63.2	114	Y, ?	2002	6×30	*****	a.k.a. Σ1579AB–C
Σ1695	12h 56.3m	+54° 06'	A–B	6.4	12.5	6.9	236	bW, ?	1999	4/5"	***	
78 Ursae Majoris	13h 00.7m	+56° 22'	A–B	6.0	7.8	3.6	281	W, gW	2006	2/3"	****	
			A–B	5.0	7.9	1.2	120	oW, grB	2005	8/10"	****	a.k.a. β1082
79, ζ Ursae Majoris	13h 23.9m	+54° 56'	A–B	2.2	3.9	14.3	153	W, W	2005	15×80	*****	a.k.a. Σ1744AB *4

Variable stars:

Designation	R.A.	Decl.	Type	Range (m_v)	Period (days)	F (f_r/f_t)	Spectral Range	Aper. Recm.	Rating	Comments
V Ursae Majoris	09^h 08.2^m	+51° 07'	SRb	9.5	207.65	–	M5–M6	4/6"	****	##
W Ursae Majoris	09^h 43.8^m	+55° 57'	EW/ KW	7.75	0.334	–	F8Vp + F8Vp	7×50	****	Two similar stars, in contact, sharing outer atmospheres Minimum is ~0.1 mag. fainter than the secondary. ##$
29, υ Ursae Majoris	09^h 51.0^m	+59° 02'	δ Sct	3.68	0.133	0.64	F2IV	Eye	****	
R Ursae Majoris	10^h 44.6^m	+68° 47'	M	6.5	301	0.39	Me3–M9e	8/10"	****	Peak luminosity ~250 suns. ##
VY Ursae Majoris	10^h 45.1^m	+67° 25'	Lb	5.7	196	–	C6,3(N0)	7×50	*****	Carbon star, extremely red. ##
CO Ursae Majoris	11^h 09.3^m	+36° 19'	Lb	5.74	–	–	M3.5IIlab	6×30	****	
ST Ursae Majoris	11^h 27.8^m	+45° 11'	SRb	6	110	–	M4–M5III	7×50	****	##$

Star clusters:

Designation	R.A.	Decl.	Type	m_v	Size (')	Brst. Star	Dist. (ly)	dia. (ly)	Aper. Recm.	Rating	Comments
None											

Nebulae:

Designation	R.A.	Decl.	Type	m_v	Size (')	Brst./ Cent. star	Dist. (ly)	dia. (ly)	Aper. Recm.	Rating	Comments
M97, NGC 3587	11^h 14.8^m	+55° 01'	Planetary	9.9	194	**16.0** star	1300	1	4/6"	*****	Owl Nebula. Requires >12" scope to see "owl's eyes" pattern

Galaxies:

Designation	R.A.	Decl.	Type	m_v	Size (')	Dist. (ly)	dia. (ly)	Lum. (suns)	Aper. Recm.	Rating	Comments
NGC 2681	08^h 53.5^m	+51° 19'	SAB	10.3	3.6×3.6	43M	46K	11G	8/10"	****	Moderately bright halo, very small brighter core
NGC 2768	09^h 11.6^m	+60° 02'	E6	9.9	6.6×3.2	77M	144K	44G	8/10"	****	Large, bright, elongated halo, slowly brightens toward center
NGC 2841	09^h 22.0^m	+50° 58'	SA	9.2	6.6×3.5	39M	65K	26G	8/10"	****	Large, bright, elongated halo, bright, round core

(continued)

Table 2.1. (continued)

Galaxies:

Designation	R.A.	Decl.	Type	m_v	Size (')	Dist. (ly)	dia. (ly)	Lum. (suns)	Aper. Recm.	Rating	Comments
NGC 2985	09^h 50.4^m	+72° 17'	SA	10.4	3.9×3.0	73M	96K	28G	8/10"	****	Large, bright, slightly elongated halo, slowly brightens toward center
M81, NGC 3031	09^h 55.6^m	+69° 04'	SB	6.9	22.1×11.1	10M	70K	11G	7×50	*****	Bright center, faint arms. 2° ESE of 24 UMa
M82, NGC 3034	09^h 55.8^m	+69° 41'	IO	8.4	11.7×4.8	17M	52K	9.1G	4/6"	*****	Very elongated, faint, mottled appearance. 0.6° NW of M18
NGC 3077	10^h 03.3^m	+68° 44'	SB	9.8	5.4×3.9	7M	10K	1.2G	8/10"	****	Fainter member of M81 group. Moderately bright, moderately large
NGC 3184	10^h 18.3^m	+41° 25'	SB	9.8	7.5×6.7	28M	61K	8.5G	8/10"	****	Pretty bright, very large, gradually brighter toward center
NGC 3198	10^h 19.9^m	+45° 33'	SB	10.3	8.8×2.9	35M	75K	11G	8/10"	****	Pretty bright, very large, gradually brighter toward center, elongated
M108, NGC 3656	11^h 11.5^m	+55° 40'	SB	10.0	8.1×2.1	42M	73K	19G	4/6"	***	Mottled edge-on galaxy, fairly bright
NGC 3675	11^h 26.1^m	+43° 35'	SB	10.2	5.8×3.0	42M	62K	13G	8/10"	****	Elongated, smooth halo, stellar nucleus
NGC 3941	11^h 52.9^m	+36° 59'	SB	10.3	3.9×2.6	62M	65K	15G	8/10"	****	Elongated halo, bright center
NGC 3953	11^h 53.8^m	+52° 20'	SB	10.1	5.9×2.9	46M	100K	28G	8/10"	****	Elongated halo, bright round center
M109, NGC 3992	11^h 57.6^m	+53° 23'	SB	9.8	7.4×3.8	55M	107K	31G	4/6"	****	Faint oval halo, small bright core
NGC 4051	12^h 03.2^m	+44° 32'	SAB	10.2	5.4×4.4	55M	84K	19G	8/10"	****	Smooth, slightly oval halo, small core, stellar nucleus
NGC 4605	12^h 40.0^m	+61° 37'	SB	10.3	5.7×2.0	13M	56K	9G	8/10"	****	Bright edge-on galaxy, long thin core
M101, NGC 5457	14^h 03.2^m	+54° 21'	SAB	7.9	23.8×23.8	18M	122K	23G	4/6"	*****	Pinwheel galaxy, gorgeous, esp. in medium/large amateur scopes

Meteor showers:

Designation	R.A.	Decl.	Period (yr)	Duration	Max Date	ZHR (max)	Comet/Asteroid	First Obs.	Vel. (mi/km/sec)	Rating	Comments
None											

Other interesting objects:

Designation	R.A.	Decl.	Type	m_v	Mass (suns)	Dist. (ly)	dia. (ly)	Lum. (suns)	Aper. Recm.	Rating	Comments
47 Ursae Majoris	10^h 59.5^m	+40° 26'	SolSys	ExoPlnt		42			Eye	****	Center of 1st extra-solar system planetary system similar to our own
Lalande 21185	11^h 03.2^m	+35° 58'	Red Dwf	7.49		8.1			7×50	****	4th nearest star, 8th largest PM = 4.78"/year, Lum. = 0.0048 Sun
Groombridge 1830	11^h 52.9^m	+37° 43'	Hi PM	6.45		28			7×50	****	Runaway star, proper motion = 7.04"/year, 3rd greatest known

Fig. 2.4a. M81 and M82, spiral galaxies in Ursa Major. Astrophoto by Gian Michele Ratto.

Fig. 2.4b. M81 and M82, astrophoto processed to simulate visual appearance in a medium sized amateur telescope.

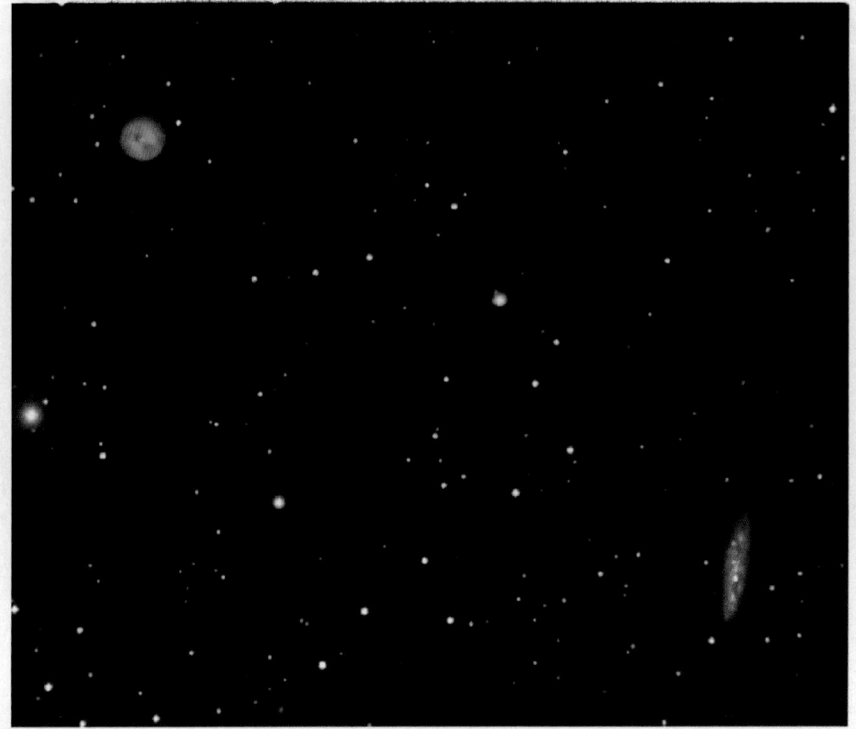

Fig. 2.5a. M97 (NGC 3587) and M108 (NGC 3556), planetary nebula and spiral galaxy in Ursa Major. Astrophoto by Gian Michele Ratto.

Fig. 2.5b. M97 and M108, astrophoto processed to simulate visual appearance in medium sized amateur telescope.

Fig. 2.6a. M101 (NGC 5457), Spiral Galaxy in Ursa Major. Astrophoto by Steve Pastor.

Fig. 2.6b. M101, astrophoto processed to simulate visual appearance in medium to large amateur telescope.

Fig. 2.7a. M109 (NGC 3992), spiral galaxy in Ursa Major. Astrophoto by Gian Michele Ratto.

Fig. 2.7b. M109 (3992), spiral galaxy in Ursa Major. Astrophoto processed to simulate visual appearance in moderate sized telescopes.

URSA MINOR (er-sah **my**-nor)	**The Little Bear**
URSAE MINORIS (er-sigh my-**nor**-iss)	**9:00 P.M. Culmination:** June 21
UMi	**AREA:** 256 sq. deg., 56th in size

TO FIND: The two westernmost stars in the bowl of the big dipper (Dubhe and Merak) point toward Polaris, the North Star. Polaris is about 30° from Dubhe and is the end star of the bear's tail (the little dipper's handle). The rest of the small bear or little dipper will be directly above Polaris, stretching out toward the end of the Big Bear's tail or the end of the big dipper's handle. Kochab and Pherkad mark the end of the little bear's body. These two stars appear to circle Polaris continuously and are sometimes called "the Guardians" (See Fig. 2.9).

KEY STAR NAMES:

Name	Pronunciation	Source
α Polaris	po-**lair**-is	polaris, Latin, "of the pole"
β Kochab	**koe**-kab	al Kaukab (Arabic), "the star"
γ Pherkad	**fur**-kad	Anwar al Farkadain, (Arabic), "dim one of the two calves"
δ Yildun	yil-**doon**	yildiz (Turkish), "the star" (originally applied to Polaris and later misapplied to δ).
11 Pherkad Minor	**fur**-kad **my**-nor	Al-farqudan (Arabic), from the asterism of "the two calves" plus minor (Latin), "smaller"

Magnitudes and spectral types of principal stars:

Bayer desig.	Mag. (m_v)	Spec. type	Bayer desig.	Mag. (m_v)	Spec. type	Bayer desig.	Mag. (m_v)	Spec. type
α	1.97	F7 Ib/IIv	β	2.07	K4 III var	γ	3.00	A3 II/III
δ	4.35	A1 Vn	ε	4.21	G5 III var	ζ	4.29	A3 Vn
η	4.95	F5 V	θ	5.00	K5 III			
π^1	6.57	G8 IV/V	λ	6.31	M1 III			
			π^2	6.89	G8 IV/V			

Fig. 2.8a. The Stars of Draco, Ursa Minor and surrounding constellations.

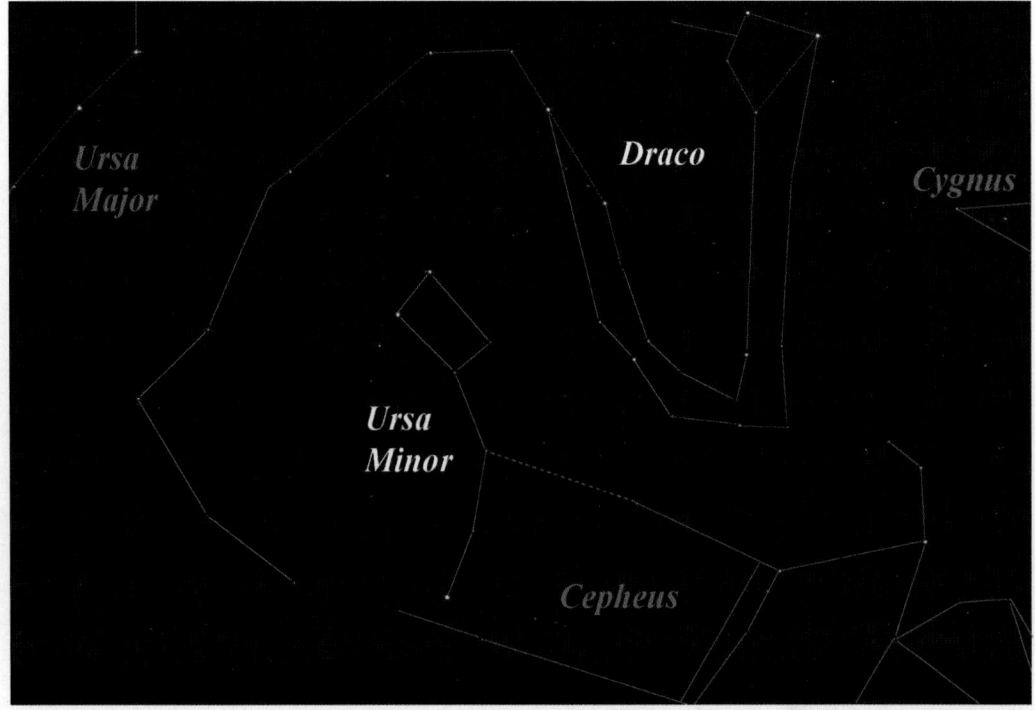

Fig. 2.8b. The figures of Draco, Ursa Minor, and surrounding constellations.

Fig. 2.9a. Rotation of the stars about the north celestial pole in a one hour period.

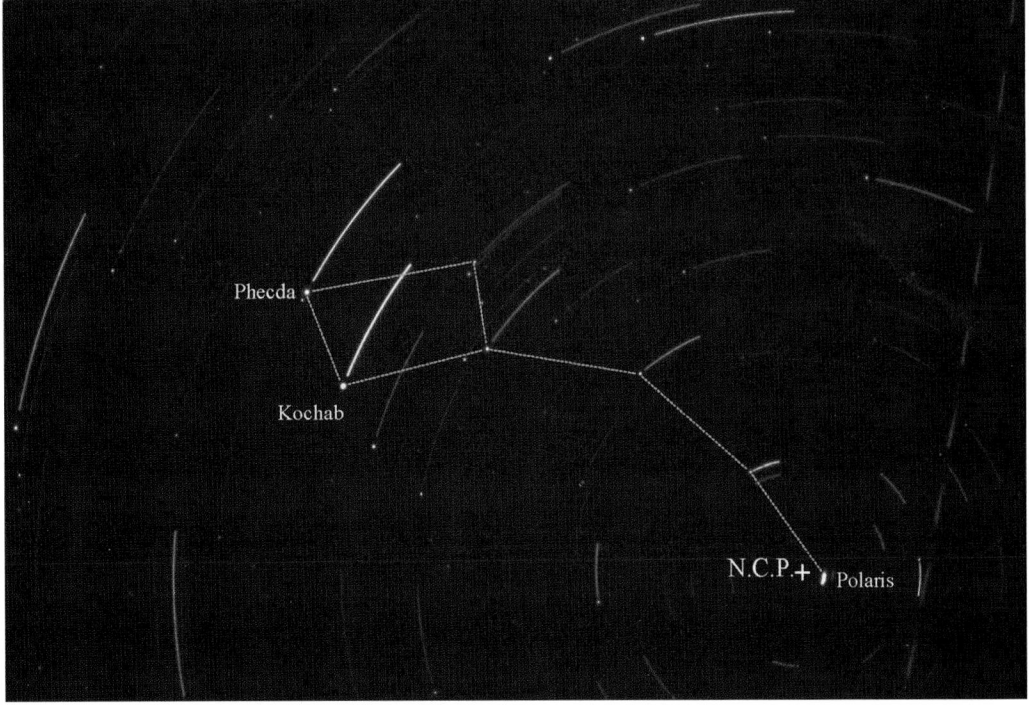

Fig. 2.9b. Same as above, with the Little Dipper and north celestial pole added.

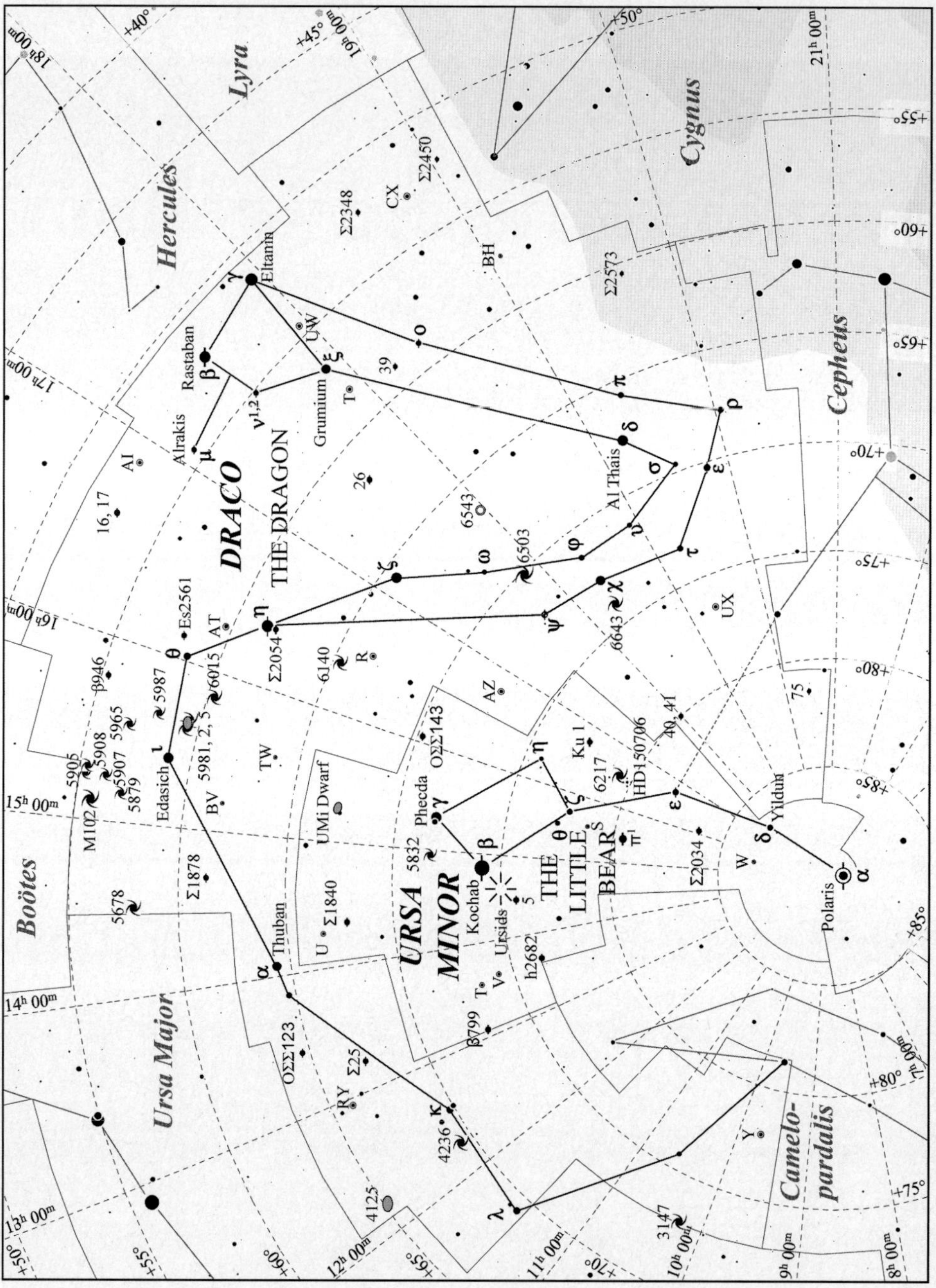

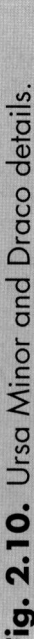

Fig. 2.10. Ursa Minor and Draco details.

Table 2.2. Selected telescopic objects in Ursa Minor.

Multiple stars:

Designation	R.A.	Decl.	Type	m_1	m_2	Sep. (")	PA (°)	Colors	Date/Period	Aper. Recm.	Rating	Comments	
1 = α Ursae Minoris	02h 31.8m	+89° 16'	A–B	2.0	9.1	18.5	233	oY, pW	2005	2/3"	****	Polaris. Showcase pair. Always visible in northern hemisphere	*5
β799	13h 04.8m	+73° 02'	A–B	6.6	8.5	1.3	265	W, ?	2004	7/8"	****		
h2682	13h 40.7m	+76° 51'	A–B	6.7	10.3	25.6	282	W, bW	2003	2/3"	****	Nice brightness contrast	
			A–C	6.7	9.0	43.8	318	W, ?	2003	7×50	****		
Σ1840	14h 19.9m	+67° 47'	A–B	7.0	10.1	27.1	221	bW, ?	2003	2/3"	****		
			A–C	7.0	10.7	144.8	239	bW, ?	1999	3/4"	****		
5 Ursae Minoris	14h 27.5m	+75° 42'	A–B	4.3	13.4	22.6	124	yO, ?	1958	7/8"	***		
			A–C	4.3	9.9	58.6	132	O, ?	1999	2/3"	****		
7 = β Ursae Minoris	14h 50.7m	+74° 09'	A–B	2.1	11.4	212.3	341	O, ?	1999	3/4"	***	Kochab	
π¹ Ursae Minoris	15h 29.2m	+80° 27'	A–B	6.6	7.3	31.1	078	yW, W	2000	10×50	*****	Very good binocular object. Very moderate brightness diff.	
Σ2034	15h 48.7m	+83° 37'	A–C	6.6	11.4	153.8	102	yW, ?	1983	3/4"	****		
			A–B	7.7	8.0	1.1	111	W, W	2001	8/10"	****	Nearly equal pair, but requires excellent seeing	
OΣΣ143	16h 04.8m	+70° 16'	A–B	6.9	8.8	46.9	084	W, ?	1999	7×50	*****	Good binocular object	
Ku 1	16h 43.1m	+77° 31'	A–B	6.1	10.2	2.6	179	yW, ?	1991	4/6"	****		
			A–C	6.0	9.8	106.6	013	O, ?	1999	2/3"	*****	Easy object for small scopes	
22 = ε Ursae Minoris	16h 46.0m	+82° 02'	A–B	4.2	11.2	77.0	003	Y, ?	1959	3/4"	****		

Variable stars:

Designation	R.A.	Decl.	Type	Range m_v		Period (days)	F (f_f/f_i)	Spectral Range	Aper. Recm.	Rating	Comments
1 = α Ursae Minoris	02h 31.8m	+89° 16'	Cδ	1.92	2.07	3.970	0.50	F7:Ib-IIv	2/3"	****	Polaris, Cepheid variable with small range
T Ursae Minoris	13h 34.7m	+73° 26'	M	7.8	15.0	301	0.45	M4e– M6e	7×50	***	2°30' WSW (PA 246°) of 3 UMi (6.43). ##
V Ursae Minoris	13h 38.7m	+74° 19'	SRb	8.8	9.9	72.0	–	M5III	2/3"	****	1°55' W (PA 265°) of 3 UMi (m6.43). #$

(continued)

Table 2.2. (continued)

Variable stars:

Designation	R.A.	Decl.	Type	Range m_v	Period (days)	F (f_r/f_t)	Spectral Range	Aper. Recm.	Rating	Comments
U Ursae Minoris	$14^h\ 17.3^m$	$+66°\ 48'$	M	7.4	12.7	330.92	0.50	M56e–M8e	6/8"	****	##
RR Ursae Minoris	$14^h\ 57.6^m$	$+66°\ 56'$	SRb	4.53	4.73	43.3	–	M5III	Eye	****	1°07' NW (PA 320°) of ζ UMi (m4.29). ##
S Ursae Minoris	$15^h\ 29.4^m$	$+78°\ 38'$	M	8.3	11.5	331.0	0.50	M7e–M9e	3/4"	****	##
W Ursae Minoris	$16^h\ 21.2^m$	$+86°\ 19'$	EA/SD	8.51	9.59	1.701	0.023	A3	7×50	****	1°22'W (PA 264°) of δ UMi (m4.34). #

Star clusters:

Designation	R.A.	Decl.	Type	m_v	Size (')	Brtst. Star	Dist. (ly)	dia. (ly)	Number of Stars	Aper. Recm.	Rating	Comments
None												

Nebulae:

Designation	R.A.	Decl.	Type	m_v	Size (')	Brtst./ **Cent.** star	Dist. (ly)	dia. (ly)	Aper. Recm.	Rating	Comments
None											

Galaxies:

Designation	R.A.	Decl.	Type	m_v	Size (')	Dist. (ly)	dia. (ly)	Lum. (suns)	Aper. Recm.	Rating	Comments
NGC 5832	$14^h\ 57.8^m$	$+71°\ 41'$	SBb	12.2	3.7×2.2	–	–	–	12/14	***	Diffuse center, faint halo. Difficult

Ursa Minor Dwarf	15h 09.1m	+67° 13'	E	11.9	30.2×19.1	0.2M	1,800	5.5K	—	**	Photographic object only	*6
NGC 6217	16h 32.6m	+74° 19'	SBbc	11.0	3.6×3.5	78M	82K	19G	8/10"	***	Faint, diffuse 1.5'×1' NNW-SSE halo, stellar nucleus	*7

Meteor showers:

Designation	R.A.	Decl.	Period (yr)	Duration	Max Date	ZHR (max)	Comet/**Asteroid**	First Obs.	Vel. (mi/**km**/sec)	Rating	Comments
Ursids	14h 38m	+75°	16.7	12/17	12/26	Nov. 14/15	10	Tuttle	~1900	22	I

*7

Other interesting objects:

Designation	R.A.	Decl.	Type	m_v	Mass (suns)	Dia. (suns)	Dist. (ly)	Planet Mass (J)	Dist. (AU)	Period (days)	Rating	Comments
HD 150706	16h 31.3m	+74° 47'	ExoPlnt	7.03	—	—	88.76	>1.0	0.82	264	**	Sun-like star (type G0), one planet

*8

DRACO (dray-koe)	**The Dragon**
DRACONIS (druh-**koe**-niss)	**9:00** P.M. **Culmination:** July 9
Dra	**AREA:** 1,083 sq. deg., 8th in size

TO FIND: α Draconis, or Thuban, is located almost precisely half-way between Mizar (in UMa) and Pherkad (in UMi). The rest of the lines of relatively faint stars delineating the dragon's tail and body can be traced out by using the chart. Although Thuban is not the brightest star in Draco, it received the designation of α because it was our pole star for a long period around 3000 BC. The tip of the dragon's tail lies at the extreme NW corner of the constellation. It is marked by an unnamed 4th magnitude star (catalog numbers SAO 1551 and HD 81817). This star lies directly between Polaris and 23 Uma (the top of the big bear's head) and about one-third of the distance from Polaris to 23 Uma. From there, a line of relatively faint stars goes SE, then ESE to a point south of the little dipper's bowl. The single line then splits into two lines forming the dragon's thicker body. These lines gradually diverge and go to the NE, then SE, and finally to the SW where they end in a distinctive group of four stars representing the dragon's head. A faint star (μ Dra) represents the tip of the dragon's tongue. μ is west of the dragon's head. Do not be fooled by a somewhat brighter star to the south. Although it looks as if it would combine with the four stars to its N to form a snake-like head, it is actually in Hercules, and represents the strongman's left ankle (See Fig. 6.3).

KEY STAR NAMES:

Name	Pronunciation	Source
α Thuban	thu-**ban** or **thuw**-ban	Raˢ al Tinnˢn (sci-Arabic), "The Serpent's Head," transliterated into thuᴸban (Arabic), "serpent"
β Rastaban	**ras**-tuh-ban	al Ras al Tinnˢn (sci-Arabic), "The Serpent's Head," originally applied to γ transliterated as above and transferred to β in error
γ Eltanin	el-tuh-**nin**	al Tinnin the medieval name for "the Serpent"
δ Al Thais	**al**-tase	Misreading of al Tinnin. (al Tais [sci-Arabic] actually means "The Goat")
ι Edasich	**eh**-duh-sik	al dhikh (Indo-Arabic) "The Male Hyena," a late Arabic name for this star
λ Giausar	jou-**zur**-or	Jauzhar (Persian), a technical term for the lunar and planetary nodes, which were called the "Head and Tail of the Dragon"
μ Alrakis	al-**rah**-kiss	al raquis (Arabic), "the Trotting Camel"
ξ Grumium	**grew**-mih-um	grunnum (late Latin), "snout" or "muzzle." According to Ptolemy, this star was on the serpent's jawbone

Magnitudes and spectral types of principal stars:

Bayer desig.	Mag. (m_v)	Spec. type	Bayer desig.	Mag. (m_v)	Spec. type	Bayer desig.	Mag. (m_v)	Spec. type
α	3.65	A0 III	β	2.79	G2 II	γ	2.23	K5 III
δ	3.07	G9 III	ε	3.83	G8 III	ζ	3.17	B6 III
η	2.74	G8 III	θ	4.01	F8 IV	ι	3.29	K2 III
κ	3.87v	B6 III	λ	3.82	M0 III var	μ	4.91	F5
ν¹	4.89	Am	ν²	4.86	Am	ξ	3.73	K2 III
ο	4.63	K0 II/IIIi	π	4.60	A2 IIIs	ρ	4.51	K3 III
σ	4.67	K0 V	τ	4.45	K3 III	υ	4.82	K0 III
φ	4.22	A0p (Si)	χ	3.55	F7 V var	ψ	4.57	F5 IV–V
ω	4.77	F5 V						

Table 2.3. Selected telescopic objects in Draco.

Multiple stars:

Designation	R.A.	Decl.	Type	m_1	m_2	Sep. (")	PA (°)	Colors	Date/**Period**	Aper. Recm.	Rating	Comments
Σ 25	13h 13.5m	+67° 17'	A–B	6.6	7.1	179.0	296	oW, W	1999	6×30	****	Easy, very wide binocular pair
			A–C	6.6	8.9	105.9	224	oW, ?	1999	7×50	****	
OΣΣ 123	13h 27.1m	+64° 44'	A–B	6.7	7.0	68.9	147	B, B	2000	6×30	*****	Striking pair, nearly equal, both unusually blue
Σ1878	14h 42.1m	+61° 16'	B–C	7.0	12.2	38.5	095	B, ?	2000	4/6"	***	Difficult, but very nice
β 946	15h 47.6m	+55° 23'	A–B	6.3	9.2	4.2	317	W, ?	2004	2/3"	****	Very difficult in small scopes
Es 2651	16h 09.0m	+57° 56'	A–B	5.9	9.5	2.2	132	bW, pB	1991	4/6"	****	Companion seems brighter than listed. ~1° ESE of θ Dra.
Σ2054	16h 23.8m	+61° 42'	A–B	6.2	12.1	1.1	353	W, pB	2005	8/10"	****	Companion seems fainter than listed. Very difficult in small scopes
14, η Draconis	16h 24.0m	+61° 31'	A–B	2.8	8.2	4.8	139	Y, ?	1996	2/3"	****	A fine yellow pair. a.k.a. Σ312
16 and 17 Draconis	16h 36.2m	+52° 55'	A–B	5.4	6.4	3.1	105	Y, Y	2006	3/4"	****	Beautiful triplet for small scopes. A–B a.k.a. Σ207
21, μ Draconis	17h 05.3m	+54° 28'	A–C	5.4	5.5	89.8	196	pY, pP	2003	6×30	****	Nearly equal binocular pair
			A–B	5.7	5.7	2.36	006	pY, W	**672**	4/6"	****	Nice nearly equal pair, but fairly difficult. B–b is binary, 0.01" sep.
β 1088AC			A–C	5.7	13.8	12.9	177	yW, yW	2006	8/10"	***	Faint companion makes this one more difficult than expected
25, $\nu^{1,2}$ Draconis	17h 32.2m	+55° 11'	A–B	4.9	4.9	62.6	312	yW, B	2005	6×30	*****	Identical twins, nice at any aperture. a.k.a. ΣA 35. Great binocular object
26 Draconis	17h 35.0m	+61° 52'	A–B	5.3	8.5	**1.05**	**317**	Y, ?	**76.1**	8/10"	****	AKA β962. Closing to 0.3" in 2017, then opening to 1.35' in 2032
31, ψ Draconis	17h 41.9m	+72° 09'	AB–C	5.2	8.1	23.9	245	Y, ?	1999	15×80	****	Nice pair for large binoculars
			A–B	4.6	5.6	**29.8**	**016**	yW, Y	**12,500**	10×50	****	Nice pair for binoculars. Orbit poorly determined. See notes
40 + 41 Draconis	18h 00.2m	+80° 00'	A–B	5.7	6.0	19.2	232	pY, pY	2006	15×80	****	Two slightly unequal pale yellow stars. a.k.a. Σ2308AB
39 Draconis	18h 23.9m	+58° 48'	A–B	5.1	8.1	3.8	349	W, bW	2005	2/3"	****	A is an extremely close binary, with maximum separation of only ~0.025". B and C, plus four other stars mag. 11–14 are within 200 arcsec, but are not physically related to A. a.k.a. Σ2303AB, AC
			A–C	5.1	8.0	90.4	020	W, W	2001	7×50	****	
Σ2348	18h 33.9m	+52° 21'	AB–C	5.5	8.7	25.3	271	pY, pGr	2002	10×50	***	A–B is binary, max sep. = 0.273"

47, o Draconis	18h 51.2m	+59° 23'	A-B	4.8	8.3	36.5	319	Y, grB	2003	6×30	****	Separated in large binoculars. a.k.a. Σ2420
Σ2450	19h 02.1m	+52° 16'	A-BC	6.5	9.5	5.2	299	yO, pB	2005	2/3"	***	Subtle, but nice colors
Σ2573	19h 40.2m	+60° 30'	A-B	6.5	8.9	18.4	025	bG, W	2006	15×80	****	Good object for large binoculars or small telescope
Draconis	19h 48.2m	+70° 16'	A-B	4.0	6.9	3.2	.019	dY, pB	2005	3/4"	****	a.k.a. Σ2603
75 Draconis	20h 28.2m	+81° 25'	A-B	5.4	11.34	109.6	011	rW, ?	2000	3/4"	***	a.k.a. BUP 211 AB
			A-C	5.4	6.7	196.6	282	rW, W	2000	6×30	****	a.k.a. STH 7AC

Variable stars:

Designation	R.A.	Decl.	Type	Range (m_v)		Period (days)	F (f_r/f_t)	Spectral Range	Aper. Recm.	Rating	Comments
Y Draconis	09h 42.4m	+77° 51'	M	6.2	15	325.79	0.45	M5e	18/20"	***	Entire cycle can be observed with a 14/16" scope. ##
RY Draconis	12h 56.4m	+66° 00'	SRb	6.0	8.0	172.5	–	C4,5J(N4p)	6×30	*****	Cool carbon star with deep red tint. Use 3–6" scope to enhance color. #
BV Draconis	15h 11.8m	+61° 51'	EW/KW	7.88	8.48	0.35007	0.45	F7V	6×30	****	Visual magnitudes would be about the same. ##$
TW Draconis	15h 33.9m	+63° 54'	EA/SD	8.0p	10.5p	2.80695	0.17	A8V + K0III	2/3"	****	Visual magnitudes would be ~ 1.5 mag brighter. ##$
AT Draconis	16h 17.3m	+59° 45'	Lb	6.8p	7.5p	–	–	M4IIIa	6×30	****	
R Draconis	16h 32.7m	+66° 45'	M	6.7	13.2	245.6	0.45	M5e– M9eIII	8/10"	****	##
AZ Draconis	16h 40.7m	+72° 40'	Lb	6.7	7.9	–	–	M2	6×30	****	##
AI Draconis	16h 56.3m	+52° 42'	EA/SD	7.05	8.09	1.98815	0.18	A0V	6×30	****	Secondary minimum can be detected in visual measurement. ##$
T Draconis	17h 56.4m	+58° 13'	M	7.2	13.5	421.62	0.44	M2e	8/10"	*****	Deep red color. ##$
UW Draconis	17h 57.5m	+54° 40'	Lb	7.0	8.2	–	–	K5p	6×30	*****	
CX Draconis	18h 46.7m	+52° 59'	γ Cas	5.68	5.99	–	–	B2.5V + F5III	6×30	****	The two stars in this system are elliptical, and undergo partial eclipses. ##
BH Draconis	19h 03.7m	+57° 27'	EA/SD	8.38	9.27	1.81724	0.11	A2V + Ap	7×50	****	#
UX Draconis	19h 21.6m	+76° 34'	SRa	5.9	7.1	168	0.5	C7,3(NO)	6×30	******	Striking crimson color. #

(continued)

Table 2.3. (continued)

Star clusters:

Designation	R.A.	Decl.	Type	m_v	Size (')	Brst. Star	Dist. (ly)	dia. (ly)	Number of Stars	Aper. Recm.	Rating	Comments
None												

Nebulae:

Designation	R.A.	Decl.	Type	m_v	Size (')	Brst. Star	Dist. (ly)	dia. (ly)	Aper. Recm.	Rating	Comments
NGC 6543	17h 58.6m	+66° 38'	Planetary	8.3	16×22"	**Cent.** star **11.3**			4/5"	*****	Cat's Eye nebula, greenish/bluish disk. Best deep sky object in Draco

Galaxies:

Designation	R.A.	Decl.	Type	m_v	Size (')	Dist. (ly)	dia. (ly)	Lum. (suns)	Aper. Recm.	Rating	Comments
NGC 3147	10h 16.9m	+73° 24'	SA	10.6	4.3×3.7	133M	58K	10G	4/6"	****	Faint oval, moderately bright core
NGC 4125	12h 08.1m	+65° 11'	E6 pec	9.7	6.1×5.1	79M	147K	54G	8/10"	****	Very elongated core. Faint companion (4121) visible in large scopes
NGC 4236	12h 16.7m	+69° 28'	SBdm	9.6	21.9×7.2	7M	12.5K	500M	10/12"	**	Large, but very low surface brightness. Difficult
NGC 5678	14h 32.1m	+57° 55'	SBb	11.4	3.3×1.4	100M	90K	22G	12/14"	***	4° SW of Edasich (ι Dra), edge on, bright even in small scopes
M102, NGC 5866	15h 06.5m	+55° 46'	SA0	9.9v	6.6×3.2	50M	90K	17G	4/6"	****	*9
NGC 5879	15h 09.8m	+57° 00'	Sbc	11.4	4.2×1.3	44M	55K	4.3G	8/10"	***	Thin, faint, slightly brighter midline
NGC 5905	15h 15.4m	+55° 31'	SB	11.7	4.7×3.6	-	-	-	6/8"	***	50'S of 5907, same FOV as 5908 and Wirtz 13, dust lane in large scopes
NGC 5907	15h 15.9m	+56° 19'	Sc	10.3	11.5×1.7	49M	112K	26G	8/10"	****	"Splinter Galaxy" edge on, very faint, almost needle-like strip
NGC 5908	15h 16.7m	+55° 25'	SA	11.8	3.2×1.6	150M	140K	28G	8/10"	***	Nearly edge-on, large scopes show nuclear bar

NGC 5965	15h 34.0m	+56° 41'	Sb	11.7	5.2×7.0	160M	325K	36G	8/10"	***	Magnitude 12.5 NGC 5963 is in same FOV
NGC 5981	15h 37.9m	+59° 23'	SC	13.0	3.1×0.6	95M	53K	6.9G	16/18"	***	Nearly edge-on, forms nearly in-line trio with 5982 and 5985 at low power
NGC 5982	15h 38.7m	+59° 21'	E3	11.1	2.5×1.8	130M	95K	49G	8/10"	***	Elliptical
NGC 5985	15h 39.6m	+59° 20'	SBb	11.1	5.5×2.9	128M	175K	58G	6/8"	***	Tilted, but arms are visible in large scopes
NGC 5987	15h 40.0m	+58° 05'	Sb	11.7	4.2×1.3	140M	171K	33G	10/12"	***	Faint, small streak
NGC 6015	15h 51.4m	+62° 19'	Sc	11.1	5.4×2.1	57M	85K	16G	6/8"	****	3.3° NE of Edasich (ι Dra). Mottled appearance
NGC 6140	16h 21.0m	+65° 23'	SBc/P	11.2	6.3×4.6	50M	91K	6.6G	12/14"	***	Large, irregular oval
NGC 6503	17h 49.4m	+70° 09'	SA	10.2	7.3×2.4	20M	28K	4.5G	6/8"	****	Nearly edge-on, inclination = 74°
NGC 6643	18h 19.8m	+74° 34'	Sc	11.0	3.7×1.8	76M	82K	18G	8/10"	***	Bright nucleus in elongated halo

Meteor showers:

Designation	R.A.	Decl.	Period (yr)	Duration	Max Date	ZHR (max)	Comet/ Asteroid	First Obs.	Vel. (mi/ km/sec)	Rating	Comments
Draconids	17h 28m	+54°	6.7	10/06 10/10	Oct. 9	*	*	1915	—	Minor	*10

Other interesting objects:

Designation	R.A.	Decl.	Type	m_v	Mass (suns)	Dist. (ly)	dia. (ly)	Lum. (suns)	Rating	Comments
α Draconis	14h 04.4m	+64° 23'	A0 III	3.7	-0.6	230			***	Thuban. Was our pole star for several hundred years around 2830 BC *11

Fig. 2.11. NGC 4236, edge-on spiral galaxy in Draco. Astrophoto by Mark Komsa.

Fig. 2.12. Spiral galaxies NGC 5908 (left) and 5905 in Draco. Astrophoto by Rob Gendler.

Fig. 2.13. Edge-on spiral galaxy NGC 5907 in Draco. Astrophoto by Rob Gendler.

Fig. 2.14. Spiral galaxies NGC 5963 and 5965 in Draco. Astrophoto by Rob Gendler.

Fig. 2.15. Galaxies NGC 5985, 5982 and 5981 in Draco. Astrophoto by Rob Gendler.

Fig. 2.16. NGC 6503, spiral galaxy in Draco. Astrophoto by Mark Komsa.

NOTES:

1. Hold your fist up at arm's length in front of the two stars at the top of the big dipper's bowl (α and δ Uma). Your knuckles should just about fit between them. These two stars are about 10° apart. Your fist can thus be used as a convenient reference for angular distances in the sky. The two stars at the end of the dipper's bowl (α and β) are approximately 5° apart. They are also convenient as a reference for distances in the sky. The latter two stars are called "the Pointers" because a line from β to α extended about 28° takes you to Polaris, the North Star.

2. The stars Flamsteed designated as 37, 39, 41, and 44 Lyncis became part of Ursa Major when the IAU defined the official boundaries of all 88 constellations. 44 Lyncis is now actually 15° from the nearest part of Lyncis! The original designations are occasionally still used in modern references.

3. ξ Ursae Majoris was the first binary star to have its orbit calculated (M. Savary, 1828). It is one of the closest binaries to the earth, and both stars are very similar to our sun in size, luminosity, and spectral type. In addition, each star is a spectroscopic binary. The A–a pair has a period of 1.832 years, and the B–b pair about 4 days. The A–a pair has a maximum separation of <0.03 and a minimum separation of <0.003. The A–B pair has the fastest binary motion which is easily visible in small telescopes.

4. Mizar has an unaided eye companion, magnitude 4.0 Alcor, 11.8' distant. Arab cultures of a few hundred years ago used Alcor as a test of satisfactory vision. There is also a 9th magnitude star slightly off the line joining Mizar and Alcor. This group is lovely at all apertures.

5. In addition to having a distant companion, Polaris is a spectroscopic binary with a maximum separation of 0.04. The Hubble Space Telescope resolved it into two separate stars.

6. The Ursa Minor Dwarf Galaxy is an extremely small elliptical satellite galaxy of the Milky Way Galaxy. Its entire light output is less than that of the single star Deneb and only about 30% more than that of Rigel.

7. Around 1900, William F. Denning noticed that every year between 18 and 22 December, he observed several meteors with a radiant point in Ursa Minor. This shower was not studied in detail until 1945, when Anton Becvár and other observers at the Skalnate Pleso Observatory in Czechoslovakia observed a shower which they estimated to average about 108/h. The following year, Becvár again observed the shower, but rates only about one-tenth as high. Since then, high rates have been observed in 1986 (110/h) and 2000 (90/h). Because Comet Tuttle is locked into a 15/13 resonance with Jupiter, it has a period of ~13.6 years, and enhanced meteor showers were also expected in 1959 and 1972. If these occurred, they apparently went unobserved. In addition to these showers at each return of Comet Tuttle, there is also a shower which appears half-way between the comet returns. This shower is due to dust ejected from the comet getting locked into a 7/6 resonant orbit. The difference in periods has delayed this dust until it is now near the earth when Comet Tuttle is at aphelion. Peter Jenniskens and Esko Lyytinen have calculated the paths of Comet Tuttle and its ejected dust streams and predicted enhanced meteor showers (~50 or more per hour) in 2016, 2017, 2018, 2020, 2028, 2030, 2032, 2044, 2047, and 2049. Of these, the ones in 2017, 2030, 2044, and 2049 occur near new moon and should offer the best opportunities for observation.

8. HD 150706 is very sun-like with a mass of 0.98 M$\odot$, and a temperature of ~5,930 K (~10,700°F) compared with ~5,830 K (~10,525°F) for our sun. Its lone planet was discovered in 2002. This planet is 0.82 AU distant from its star, has a mass of >1 M$_J$, and an orbital period of 264 days (Venus orbits the sun in 225 days). It has an orbital eccentricity of 0.38, significantly greater than Pluto, which has the most eccentric orbit in our solar system.

9. NGC 5866 is a beautiful edge-on galaxy, bright even in small telescopes, and a dust lane is visible in larger amateur instruments. NGC 5866 is sometimes used as a replacement for number 102 in Messier's famous list. His original list contained a duplicated observation of M101 with a position error.

10. The Rev. M. Davidson was one of the astronomers who made calculations of the orbit of the 1900 Comet Giacobinni–Zinner. From his results, Davidson predicted that there could be a meteor shower when the earth passed near the path of Giacobinni–Zinner in early October, 1915. Observer William F. Denning reported seeing several meteors from that shower. In 1926, the earth passed through the comet's orbit 70 days before the comet's arrival. On Oct. 9 of that year, a fireball witnessed by hundreds of people in the British Isles alerted observers who reported a shower of very slowly moving meteors peaking at rate of about 14/h on Oct. 10. Conditions were very favorable in 1933, and a meteor storm of about 6,000–10,000/h was observed in several countries. No further major showers were observed until 1946 when estimates were as

high as 10,000/h. The 1952 return was down to ~250/h, and was detected by radar during daytime, but not by visual observers. A 1985 outburst reached 700–800/h, but in 1986 the activity was less than 2/h! In 1998, an outburst of ~1,000/h occurred, and there are predictions of possible major showers in 2011 and 2018. Comet Giacobini–Zinner had close encounters with Jupiter in 1626 and 1628 in which it was deflected into an erratic orbit whose period varied between 6 and 8 years. It has now settled into a fairly stable orbit whose parameters vary slowly. Its perihelion is predicted to remain near the earth's orbit until at least 2400 AD, so periodic outbursts are likely for some time.

11. While Thuban is considerably fainter than Polaris, it was our pole star during the time when Egypt first achieved a high level of civilization and became the richest and most advanced culture in the world, but before the pyramid building period, and thus became quite famous. At its closest approach to the true celestial pole, Thuban was only 10′ away, or one-third the closest approach that Polaris will make.

CHAPTER THREE

Beauty and The Beast

September–December

Andromeda	Cassiopiea	Cepheus	Cetus
The Chained Princess	The Queen	The King	The Whale

Equuleus	Pegasus	Perseus
The Little Horse	The Winged Horse	The Hero

This group of seven-related constellations dominates the autumn sky. From the northern-most part of Cepheus to the southern-most point of Cetus, the group spans over 108°, and from the eastern-most point of Perseus to the western-most point of Equuleus, the span is 111°. Except for Equuleus, all these constellations can be remembered by a pair of intersecting myths. Equuleus is loosely connected by a few short myths.

One story line begins with King Abas of Argolis, a warrior of such renown that enemies often fled at the mere sight of his shield. The father of twin boys, Abas could not decide which one should rule Argolis after he died, so he decreed that each should rule in alternate years. The first-born Acrisius was to be followed by the second-born Proetus but, at the end of his term, Acrisius refused to relinquish the crown. Instead, he had his soldiers force Proetus to seek refuge in a neighboring kingdom.

P. Simpson, *Guidebook to the Constellations: Telescopic Sights, Tales, and Myths,*
Patrick Moore's Practical Astronomy Series, DOI 10.1007/978-1-4419-6941-5_3,
© Springer Science+Business Media, LLC 2012

After years of ruling Argolis, Acrisius, who had fathered only a daughter, worried about his legacy and the continuation of his kingdom. Acrisius sought advice about fathering a son from an oracle. The oracle only increased his anxiety when he proclaimed that not only would Acrisius never have a son, but that he was destined to die at the hands of his daughter Danaë's son. In an attempt to avoid his fate, Acrisius locked Danaë in a tower and had it constantly guarded. However, the king was foiled by Zeus, who visited Danaë in the form of a shower of gold coins which she caught in her lap. When Acrisius learned of his daughter's pregnancy, he refused to believe her fantastic story, and instead suspected Proetus of being the father.

In due time, Danaë gave birth to a son and named him Perseus. Acrisius became desperate to rid himself of the potential threat to his life, but was afraid to have Danaë and her son killed, in case Zeus really was Perseus' father. Instead, he ordered the pair locked in a covered boat and had them cast into the sea. Again he was foiled in his attempt to avoid his fate when this ark drifted ashore on the island of Seriphos. A shepherd named Dictys heard the cries of the mother and her young son and opened the locked ark. After satisfying their hunger and thirst, he took the pair to his brother, King Polydectes. Polydectes was immediately charmed by the beautiful Danaë. Plotting to make Danaë his wife, Polydectes adopted Perseus and made her the child's nursemaid. Polydectes soon began to openly court Danaë, and tried to win her hand in marriage, but she spurned his approaches.

After Perseus had grown into a handsome young man and renowned warrior, Polydectes tried to force marriage on the unwilling Danaë, and Perseus was obliged to come to her defense. Polydectes pretended to abandon his desires and seemed to be competing for the hand of Pelops' daughter Hippodameia instead. He asked Perseus to help him collect a suitable gift for Pelops. Perseus had no property and little money to contribute, but being so happy that Acrisius had apparently abandoned his pursuit of his mother, he rashly declared that he would obtain any gift that Acrisius wanted. Polydectes pretended to doubt Perseus' sincerity, and proposed a hypothetical gift of the head of the Gorgon Medusa. Perseus knew that no mortal had ever returned from an encounter with Medusa, but did not think the king could be serious, so he replied, "Even that." Seeing a chance to both rid himself of the biggest obstacle to his goal, and also a potential threat to his kingdom, Polydectes quickly accepted. Perseus immediately regretted his rashness but would not back off from his promise. He began preparing his weapons and equipment for the quest, taking his time to be exceedingly thorough. Acrisius watched Perseus' preparations carefully, and soon Perseus ran out of any reasonable excuses for delaying further.

Zeus was aware of Perseus' plight and wanted to help his son, but dared not arouse the suspicions of his jealous wife Hera. Instead of helping directly, Zeus convinced other gods to provide their assistance to Perseus. Athena accompanied Perseus to Samos where the images of all three Gorgons could be seen so Perseus could learn to distinguish the mortal Medusa from her immortal sisters, Stheno and Euryale. She then loaned Perseus her polished shield and warned Perseus that it was the act of looking directly at Medusa that was the fatal danger, so that he must never look directly at her. The only safe option for Perseus was to view Medusa's reflection in the mirror-like surface of the shield to avoid being turned into stone.

Two other Olympians also provided helpful devices. Hermes loaned Perseus a pair of winged sandals which would carry him swiftly through the air and a sword which was sharp enough to decapitate Medusa with a single blow. From Poseidon, Perseus received a helmet of invisibility and a magic bag in which he could carry the severed head without the risk of harm.

Fully armed and prepared, Perseus flew off to the west where the Gorgon sisters lived. As he approached their dwelling, he found the area littered with many stone statues of both men and animals which had been petrified by the sight of Medusa, whose head was covered with writhing snakes instead of hair. Aided by Athena, Perseus approached the Gorgons' lair by backing up while looking at the reflections in the polished shield. When Medusa heard his approach, she rushed out and charged toward the sound of the invisible Perseus, only to meet Hermes' sword, which Perseus swung over his shoulder without turning his head. Guided by the invisible hand of Athena, the sword swung straight and cleanly severed Medusa's head.

To Perseus' amazement, when Medusa's blood struck the earth, the ground sizzled, and slowly the form of the winged horse Pegasus formed and emerged from the ground. Still avoiding the direct sight of Medusa's head, Perseus quickly scooped the head into the magic bag, mounted Pegasus, and fled just before the two other Gorgons emerged. Protected by Poseidon's helmet of invisibility, Perseus evaded the Gorgons' pursuit, and eventually the sisters gave up the futile chase.

The second part of the story line begins in the Kingdom of Æthiopia, not the Ethiopia we know today, but a small kingdom located on the eastern coast of the Mediterranean, around the area where the modern city of Tel Aviv is located. King Cepheus took Cassiopeia, one of the most beautiful women of his time, to be his queen. Their first-born child, Andromeda quickly grew into a young woman whose beauty rivaled that of her mother.

Cassiopeia became very vain and began to boast of her daughter's (and indirectly her own) beauty. One day as they were walking to the market, Cassiopeia was trying to impress a group of the kingdom's leading women. Unfortunately, she went too far, and her boast that she and Andromeda were even more beautiful than the sea nymphs got her and the entire kingdom into a lot of trouble. The sea nymphs were widely regarded for their beauty, and were highly offended that any mere mortal women would even compare themselves to goddesses, much less claim that they were more beautiful than the nymphs. They took their complaint to Poseidon, god of the seas, demanding that Cassiopeia be punished.

Poseidon concurred with their complaint and dispatched Cetus, a gigantic sea monster, to attack the coast of Æthiopia. This monster roamed up and down the coast, destroying villages, devouring people and cattle until the people demanded that Cepheus get rid of the monster. Cepheus could not figure out any way to kill such a terrible creature, and feared that his entire kingdom would soon be lost. Finally, he could endure it no more, and sought help from one of Zeus' oracles. After hearing the story of Æthiopia's devastation, the oracle made a sacrifice to Zeus, meditated on the problem, and delivered his verdict. He explained that Cepheus' problem was due to his wife's boastfulness, and that the only way to appease the nymphs who had demanded the punishment was to sacrifice Andromeda to the monster. He directed Cepheus to chain Andromeda to a large rock on the shore where the monster must be allowed to devour her.

Cepheus loved his daughter and considered this the worst punishment possible. Still, he knew that the judgment of the gods could not be evaded, and sadly ordered the sacrifice to take place. Andromeda was chained to the rock, and a crowd of people gathered on a hill near the shore to watch the appeasement of the monster. The sea monster quickly detected the presence of his intended victim and headed for the shore.

Fortunately, help was on the way in the form of Perseus, whose path home took him over the coast of Æthiopia. As he traveled along the coast, Perseus passed over the tableau of the chained Andromeda and the onlookers on the hilltop. Curious, he landed to find out the meaning of this odd scene. When told of the approaching monster, Perseus offered to rescue Andromeda in return for her hand in marriage. As soon as Cepheus agreed, Perseus cautioned everyone to avoid looking at Medusa's head, remounted Pegasus, and flew out to intercept the monster Cetus. Seeing the gigantic creature surfacing, Perseus made Pegasus fly low over the monster's head. As Pegasus flew close to the terrifying creature, Perseus pulled Medusa's head out of the bag so Cetus could not avoid seeing it. The monster was immediately transformed into stone and sank from sight.

Returning to the beach, Perseus claimed his bride, and after a wedding marred by the attack of the relatives of another of Andromeda's suitors, he completed his journey to Polydectes' court with Andromeda sitting behind him. After secretly warning his friends to close their eyes, Perseus held the bag up before the king and announced his successful completion of his mission. Upon Polydectes' demand for proof of the bag's contents, Perseus closed his eyes, and pulled out the horrible head. The king and all his minions were immediately petrified.

Perseus helped Andromeda onto the back of Pegasus, seated himself behind her and flew directly back to Samos. There, he left weapons that he had been loaned to be retrieved by their owners. He also left the bag with Medusa's head for Athena, who had Hephaestus emboss the Gorgon's head into her shield as a warning to her opponents.

Afterward, Perseus and Andromeda surveyed the surrounding countryside to find an unoccupied land in which to settle. As soon as they had found a homeland, they released Pegasus and started the race of Perses, who later become the rulers of Persia.

After their deaths, all the major characters in this story were placed near each other in the sky, some in honor of their lives, but one as a reminder of the folly of human failures. Cassiopeia, the cause of the disaster, was placed near her husband, but is far outshone by the throne which she was judged unworthy of occupying. While four stars in her throne are 2nd magnitude, the brightest star in the figure of Cassiopeia's body is only 4th magnitude, and even it represents a jewel in her crown rather than part of Cassiopeia herself. As a further indignity, when they are at their highest Cepheus stands erect above the north celestial pole, while Cassiopeia and her throne are upside down.

Although all early representations of Cetus were of a serpentine monster with huge fangs, over the centuries the figure has been changed into a whale. We do not know exactly who first represented it as a whale, but the transition was certainly helped along by the seventeenth century Christians who made an atlas which showed many old constellations as figures from the bible, and created some new ones which also portray biblical characters or creatures. In this atlas, Cetus represented the whale which swallowed Jonah (Fig. 3.1).

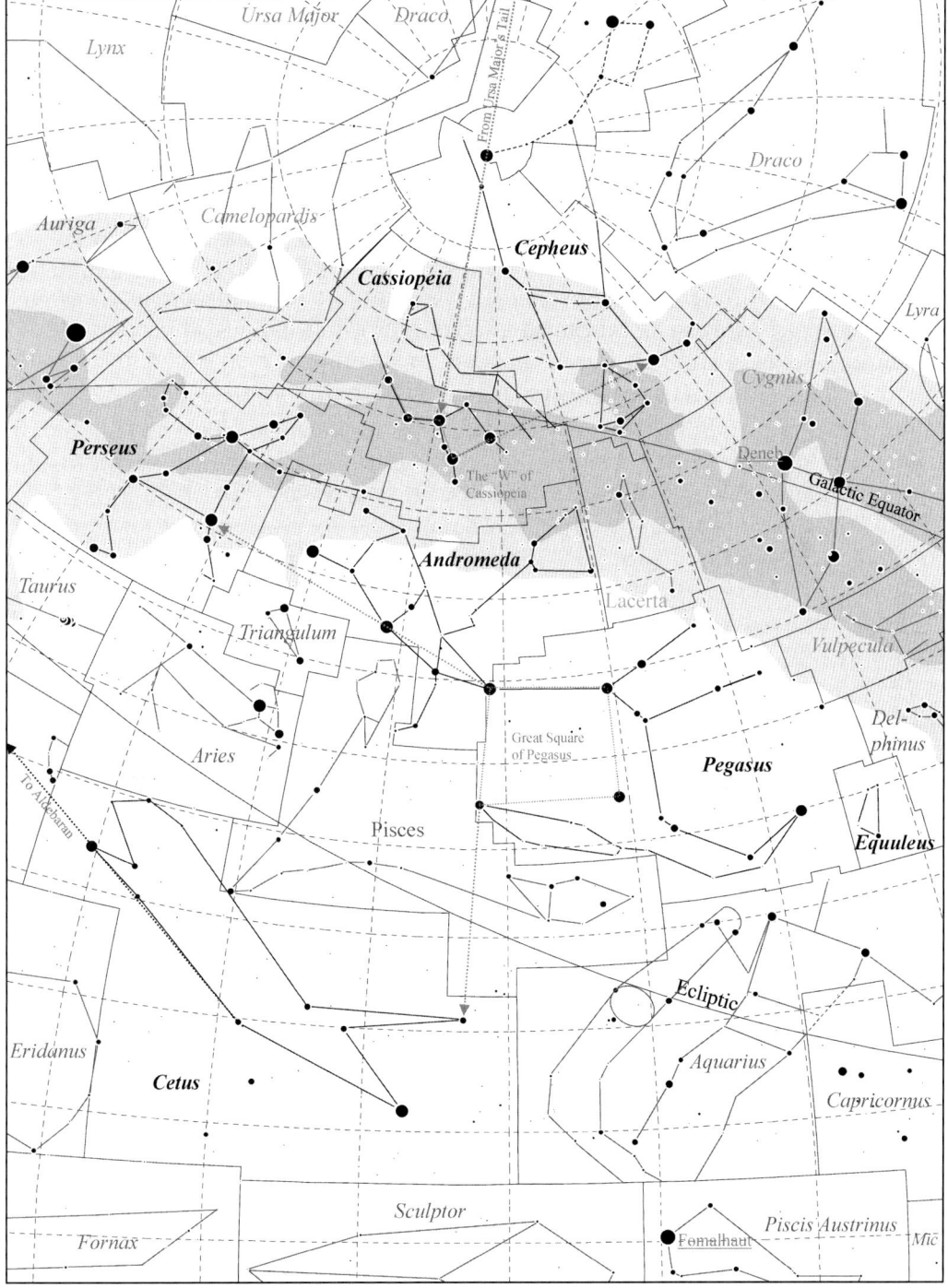

Fig. 3.1. Andromeda group finder (looking south at 9:00 PM, about November 6).

ANDROMEDA (an-**drah**-meh-duh)	The Chained Lady
ANDROMEDAE (an-**drom**-uh-die)	BEST SEEN (at 9:00 P.M.): Oct. 9
And	AREA: 722 sq. deg., 19th in size

TO FIND: If you already know how to find the asterism called "The Great Square of Pegasus," a good way to find Andromeda is to start there. The northeastern star of the Great Square is Alpheratz. Alpheratz was once considered part of both Andromeda and Pegasus, but now belongs exclusively to Andromeda. It is designated α And and represents Andromeda's head. From Alpheratz, look about 7° ENE to find δ And and then 3° N to π And. These stars represent Andromeda's shoulders. About 8° NE of δ is Mirach (β And) with the fainter μ and ν, respectively, 4° and 7° to its NE. These three stars form Andromeda's sash or waistband. Another 12.5° NE of Mirach is γ And with 51 And 8° to its NW. The latter two stars represent Andromeda's feet. Return to δ and π from which you can trace out Andromeda's arms. Her left arm is represented by ε And (1.5° S of δ) and ζ And (5° S of ε) with η And another 2.5° SSE from ζ. Andromeda's right hand is represented by σ, ρ, and θ, a small triangle of 5th magnitude stars about 5° NW of δ And from this group, a slightly larger and flatter triangle of 4th magnitude stars, ι, κ, and λ about 10° farther in the same direction, represent the top of the rock to which Andromeda is chained. Another 5° to the NW is three 5th magnitude stars and 7° to the W is ο, a 4th magnitude star. Together, these four stars represent the base of the rock (Figs. 3.2–3.7; Table 3.1).

KEY STAR NAMES:

Name	Pronunciation	Source
α Alpheratz	al-**fee**-rats	al Surrat al Faras (Arabic), "The Horse's Navel," or alpheraz id est equus (Latin), that is, the horse
β Mirach	**my**-rak	al-miʲzar (Arabic), "the girdle or waist-cloth"
γ Almach	ul-**mak**	anaq al-ard (Arabic), "the caracal" (a middle eastern black-eared lynx)
ξ Adhil	uh-**dil**	al-dhail (Arabic) "the train of a robe or dress" originally applied to χ And, mistakenly applied to ξ

Magnitudes and spectral types of principal stars:

Bayer desig.	Mag. (m$_v$)	Spec. type	Bayer desig.	Mag. (m$_v$)	Spec. type	Bayer desig.	Mag. (m$_v$)	Spec. type
α	2.06	B9 IV	β	2.06	M0 III	γ	2.18	K3 III
δ	3.27	B8 V	ε	4.37	G8 III	ζ	4.06	K1 IIIe
η	4.42	G8 IIIb	θ	4.61	A2 V	ι	4.29	B8 V
κ	4.14	B9 IVn	λ	3.82	G8 III/IV	μ	3.87	A5 V
ν	4.53	B5 V	ξ	4.88	K0 IIIb	o	3.62	B6 IIIe
π	4.36	B5 V	ρ	5.18	F5 III	σ	4.52	A5 V
τ	4.94	B8 III	υ	4.09	F8 V	φ	4.25	B7 Ve
χ	4.98	G8 III	ψ	4.95	G5 Ib	ω	4.83	F5 IV

Stars of magnitude 5.5 or brighter which have Flamsteed, but no Bayer designations:

Flamsteed	Mag. (m$_v$)	Spec. type	Flamsteed	Mag. (m$_v$)	Spec. type	Flamsteed	Mag. (m$_v$)	Spec. type
2	5.09	Ae Vn	3	4.64	K0 III	4	5.30	K4 III
6	5.31	G8 III:var	7	4.53	F0 V	8	4.82	M2 III
11	5.44	K0 III	14	5.22	K0 III	18	5.35	B9 V
22	5.01	F2 II	32	5.30	G8 III	36	5.46	K1 IV
41	5.04	A3m	49	5.27	K0 III	51	3.59	K3 III
55	5.42	K1 III	58	4.78	A5 IV–V	60	4.84	K4 III
62	5.31	A1 V	64	5.19	G8 III	65	4.73	K4 III

Stars of magnitude 5.5 or brighter which have other catalog designations:

Draper Number	Mag. (m$_v$)	Spec. type	Draper Number	Mag. (m$_v$)	Spec. type
HD 003421	5.45	G5 III	HD 010366	5.31	G8 III:var (on the And/Per border and sometimes called 6 Per)

Fig. 3.2a. The stars of Andromeda and Perseus.

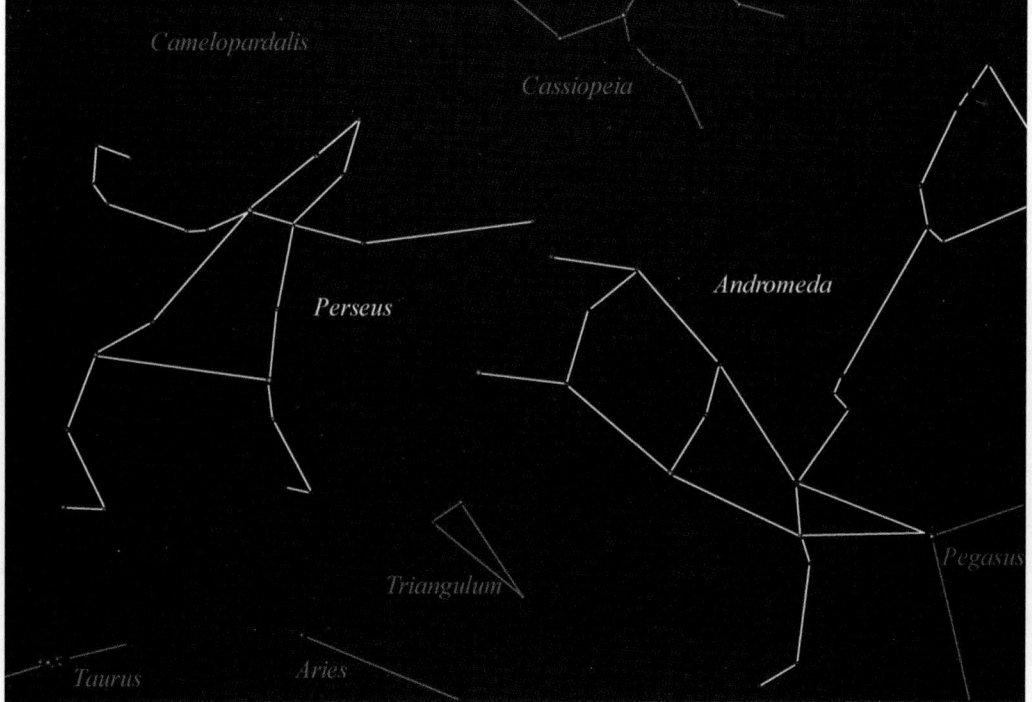

Fig. 3.2b. The figures of Andromeda and Perseus.

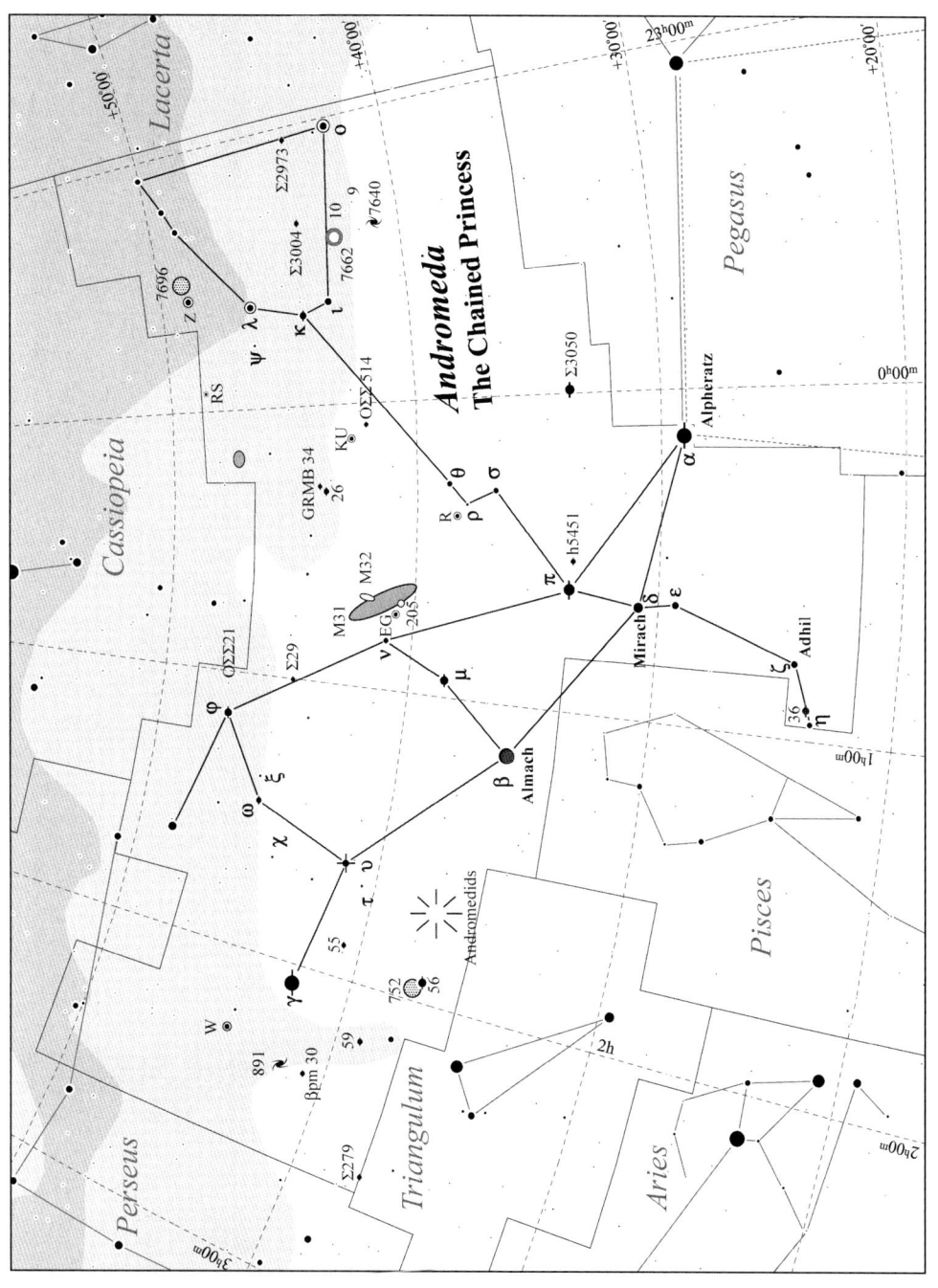

Fig. 3.3. Andromeda details.

Table 3.1. Selected telescopic objects in Andromeda.

Multiple stars:

Designation	R.A.	Decl.	Type	m_1	m_2	Sep. (")	PA (°)	Colors	Date/Period	Aper. Recm.	Rating	Comments
Σ2973	23ʰ 02.8ᵐ	+44° 04'	A–B	6.4	10.1	7.4	039	yW, ?	2003	2/3"	***	1° 18' due N (PA 001°) of 2 And
Σ3004	23ʰ 20.4ᵐ	+44° 07'	A–B	6.3	10.1	13.5	177	oY, ?	2005	2/3"	***	
κ Andromedae	23ʰ 40.5ᵐ	+44° 20'	A–B	4.1	11.3	47.8	202	bW, ?	1998	3/4"	***	A–B–C form almost a perfect right triangle, with A at the right angle
Σ3050	23ʰ 59.5ᵐ	+33° 43'	A–C	4.1	11.3	113.3	293	bW, ?	1998	3/4"	***	
			A–B	6.5	6.7	**2.11**	**336**	Y, Y	**320**	4/6"	****	Close pair of nearly identical stars
OΣΣ514	00ʰ 04.6ᵐ	+42° 06'	Aa–B	6.2	9.7	5.2	170	B, B	2002	2/3"	****	Harshaw reported the colors as yellow and red
α Andromedae	00ʰ 08.4ᵐ	+29° 05'	Aa–B	2.2	11.1	89.3	284	bW, pV	2000	2/3"	***	
Groombridge 34	00ʰ 18.2ᵐ	+44° 01'	A–B	8.1	11.0	**35.0**	**65**	R, R	**2,600**	2/3"	****	One of the nearest binaries, 11.7 ly distant
26 Andromedae	00ʰ 18.2ᵐ	+43° 47'	A–B	6.0	9.7	6.4	240	B, pB	2002	2/3"	****	
h5451	00ʰ 31.3ᵐ	+33° 35'	A–B	6.0	9.3	55.3	085	Y, pW	2005	7×50	****	
29, π Andromedae	00ʰ 36.9ᵐ	+33° 43'	Aa–B	4.3	7.1	35.7	175	W, B	2004	7×50	****	Aa is a binary with a maximum separation of 0.011", too small for amateur scopes
36 Andromedae	00ʰ 55.0ᵐ	+23° 38'	A–B	6.1	6.5	**1.06**	**323**	dY, dY	**165**	8/10"	****	Widening, maximum separation of 1.3" in 2040
μ Andromedae	00ʰ 56.8ᵐ	+38° 30'	A–B	3.87	12.9	49.8	297	W, W	1999	4/6"	***	
			A–C	3.87	11.4	29.2	147	W, W	1999	2/3"	***	
Σ79	01ʰ 00.0ᵐ	+44° 42'	A–B	6.0	6.8	7.8	193	W, pV	2004	2/3"	*****	
OΣΣ21	01ʰ 03.0ᵐ	+47° 23'	A–B	6.8	8.1	**1.11**	**176**	yW	**450**	8/10"	*****	
42, φ Andromedae	01ʰ 09.5ᵐ	+47° 15'	A–B	4.6	5.6	**0.50**	**120**	bW, ?	**370**	18/20"	***	Maximum separation will be 0.51" in 2033
ω Andromedae	01ʰ 27.7ᵐ	+45° 24'	A–B	4.8	11.7	1.8	118	W, ?	1967	4/6"	***	
			A–C	4.8	10.4	92.8	113	W, ?	2003	2/3"	***	
55 Andromedae	01ʰ 53.2ᵐ	+40° 44'	A–B	5.4	10.9	59.6	005	Y, pB	1913	2/3"	***	
56 Andromedae	01ʰ 56.2ᵐ	+37° 15'	A–a	5.8	11.9	18.3	079	O, ?	1998	3/4"	***	AB is a nearly matched pair, easy in small binoculars. A few people might be able to split AB with unaided eye
			A–B	5.8	6.1	200.5	299	O, O	2001	6 × 30	****	
Almach, γ Andromedae	02ʰ 03.9ᵐ	+42° 19'	A–BC	2.3	5.0	9.6	63	O, gB	2004	2/3"	*****	Almach, a beautiful golden and bright blue pair, one of best for small scopes
			B–C	5.1	6.3	**0.2**	**099**	B, B	**63.67**	40"	**	The BC pair requires very large aperture, excellent optics, and seeing
59 Andromedae	02ʰ 10.9ᵐ	+39° 02'	A–B	6.1	6.7	16.8	036	W, rW	2005	2/3"	****	Nice color match, slightly unmatched brightness

Double stars (continued):

Designation	R.A.	Decl.	Pair	m1	m2	Sep (")	PA (°)	Colors	Date	Aper. Recm.	Rating
βpm 30	02h 22.8m	+41° 24'	A–B	5.8	10.9	56.3	002	yW, ?	1917	2/3"	***
Σ279	02h 35.6m	+37° 19'	A–C	5.8	7.4	303.3	009	Y, ?	2002	6 × 30	****
			A–B	5.9	10.9	18.1	071	yO, ?	2002	2/3"	***
			A–C	5.9	11.7	44.9	208	yO, ?	1998	3/4"	***

Variable stars:

Designation	R.A.	Decl.	Type	Range (m_v)		Period (days)	F (f_r/f_l)	Spectral Range	Aper. Recm.	Rating	Comments	Note
o Andromedae	23h 01.9m	+42° 20'	γ Cas	3.6	3.8	–	–	B6IIIpe+A2p	Eye	*****	Quadruple system. The brightest component is variable. ##	*1
Z Andromedae	23h 33.7m	+48° 49'	Z And	8.0	12.4	–	–	M2III+B1eq	6/8"	*****	Prototype of its class. 2° 31'E (PA 084°) of λ And [m3.81].##$	*2
λ Andromedae	23h 37.6m	+46° 27'	RS	3.7	4.0	54.2	–	BG8III/IV	eye	****		
RS Andromedae	23h 55.3m	+48° 38'	SRa	8.7	10.8	136	–	M7–M10	3/4"	****	Interacting close binary. 2° 43' Ne (PA 035°) of ψ And [m4.97].#	*3
KU Andromedae	00h 06.9m	+43° 05'	M	5.8	14.9	409.33	0.38	S3, 5e–S8, 8e (M7)	18/10"	****	1° 16' at SW (PA 253°) of 26 And [m6.10]	
R Andromedae	00h 24.0m	+38° 35'	M	6.5	10.5	750	–	M10I–III	2/3"	****	Very cool star, pulsating. 1°22' # (PA 094°) of θ And [m4.61].##$	
EG Andromedae	00h 44.6m	+40° 41'	Z And	7.1	7.8	–	–	M2IIIep	6 × 30	*****	1°04' WSW (PA 248°) of η And (m4.53).##$	*4
W Andromedae	02h 17.6m	+44° 18'	M	6.7	14.6	395.93	0.42	S6, 1e–S9, 2e (M4–M10)	16/18"	***	Cool, pulsating star. 0° 47' E (PA 084°) of 60 And [m4.84].##$	

Star clusters:

Designation	R.A.	Decl.	Type	m_v	Size (')	Brst. Star	Dist. (ly)	dia. (ly)	Number of Stars	Aper. Recm.	Rating	Comments
NGC 7686	23h 30.2m	+49° 08'	Open	5.6	15'	7	3,266	14	80	7 × 50	****	Best in giant binoculars
NGC 752	01h 57.8m	+37° 41'	Open	5.7	50'	8.0	1,177	26	77	7 × 50	****	Can be glimpsed with unaided eye

Nebulae:

Designation	R.A.	Decl.	Type	m_v	Size (")	Brst./Cent. star	Dist. (ly)	dia. (ly)	Aper. Recm.	Rating	Comments
NGC 7662	23h 25.9m	+42° 33'	Planetary	8.5	0.2	2.5p	5,600	0.8	6/8"	****	"Blue Snowball," stellar appearance in 2/3", disk in 6", $T_{cs} \sim 75$ K

(continued)

Table 3.1. (continued)

Galaxies:

Designation	R.A.	Decl.	Type	Size (')	Dist. (ly)	dia. (ly)	Lum. (suns)	Aper. Recm.	Rating	Comments
NGC 7640	23h 22.1m	+40° 51'	SB(s)c III	10×2	28M	64K	9.5G	8/10"	****	Faint, edge-on, slight central brightening
M110, NGC 205	00h 41.4m	+49° 08'	E5pec	20×13	2.9M	12K	840M	4/6"	*****	Fairly faint oval, NNW-SSE, diffuse edge, companion to M31, 8–10" scope
M32, NGC 221	00h 42.7m	+40° 52'	cE2	11×7	2.9M	8K	810M	4/6"	*****	Faint oval, less elongated than M110, sharp edge, Comp to M31, 8–10" scope
M31, NGC 224	00h 42.7m	+41° 16'	SA(s)b	185×75	2.5M	220K	29G	7×50	*****	"Andromeda Galaxy," 2.2M ly dist, large spiral, ~300 billion stars, 7 × 50 binoculars *5
NGC 404	01h 09.5m	+35° 43'	E0	4.4	10M	10.5K	390M	6/8"	***	"Mirach's Ghost." 0°07' NW (PA 330°) of β And
NGC 891	02h 22.6m	+42° 21'	SB(s)c II	13×3	37M	87K	15G	6/8"	****	Edge-on, dust lane! 12–14" needed for lane. Great photo object!

Meteor showers:

Designation	R.A.	Decl.	Period (yr)	Duration	Max Date	ZHR (max)	Comet/ **Asteroid**	First Obs.	Vel. (mi/ **km**/sec)	Rating	Comments
Andromedids	01h 44m	+37°	Annual	Sep.25 Dec. 6	Nov. 14/15	2–3	Biela	1798	12/**20**	Minor	First observed by Heinrich Brandes on June 12, 1798 *6

Other interesting objects:

Designation	R.A.	Decl.	Type	m_v	Mass (suns)	Dia. (suns)	Dist. (ly)	Planet Mass (J)	Dist. (AU)	Period (days)	Eccentricity	Comments
υ Andromedae	01h 36.8m	+41° 24'	ExoPlnt	4.1	1.32	1.6	43.9	c.0.687	0.0595	69.8	0.023	First multiplanet extrasolar system discovered around a sun-like star
								b.1.98	0.832	241.23	0.262	
								d.3.95	2.54	1290.1	0.258	

Fig. 3.4. (a) M31, M32 (NGC 224 and 221), and NGC 205 in Andromeda. Astrophoto by Steven J. Goosen. **(b)** M31, M32, and NGC 205. Drawing by Jeremy Perez.

NGC 891

DEC 1, 2008 • 05:30 UT
SkyQuest XT8 – 8″ f/6 Newtonian
Pentax XW 10: 120X / 35′ FOV
Sketch by Jeremy Perez © 2008

N
W

Fig. 3.5. (**a**) NGC 891, edge-on spiral galaxy in Andromeda. Astrophoto by Rob Gendler. (**b**) NGC 891. Scale of drawing is about 55% that of astrophoto.

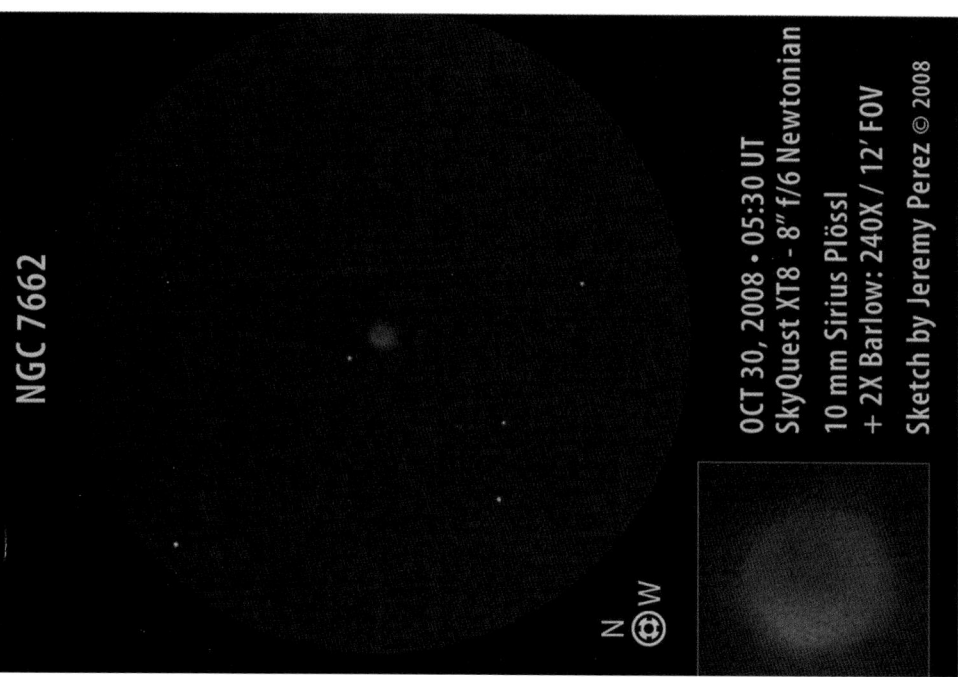

NGC 7662

OCT 30, 2008 • 05:30 UT
SkyQuest XT8 - 8" f/6 Newtonian

10 mm Sirius Plössl
+ 2X Barlow: 240X / 12' FOV

Sketch by Jeremy Perez © 2008

N ⊕ W

Fig. 3.6. NGC 7662, the Blue Snowball nebula in Andromeda.

NGC 404

NOV 24, 2008 · 08:00 UT
SkyQuest XT8 - 8" f/6 Newtonian
Pentax XW 10 + 2X Barlow:
240X / 17.5' FOV

Sketch by Jeremy Perez © 2008

N
W

Fig. 3.7. NGC 404, elliptical galaxy in Andromeda.

PERSEUS (**pur**-suse)	Perseus (The hero)
PERSEI (**pur**-see-eye)	BEST SEEN (at 9:00 **P.M.**): Dec. 23
Per	AREA: 615 sq. deg., 24th in size

TO FIND: Start from α Cas (Schedar), go 18° due east. The 3rd magnitude star at that point (η Per) is the northern point of a small triangle, about 3° long and half as wide. This group represents the Head and helmet of Perseus. The line from η Per to the brightest star in the triangle (γ Per) begins a gently curved line heading SE, and then curving due E, and finally hooking back to the north and W. This line consists of η, γ, α, ψ, δ, μ, λ, and 43 Persei. Mirfak (α Per) represents Perseus' left shoulder, while the remainder of the star row represents his left arm and the shield he is using as a mirror to avoid being petrified by the direct sight of Medusa. Stretching westward from Mirfak, there is a fairly straight line of faint stars, ι, θ, and φ Persei (with a very faint group of four stars in Andromeda lying between θ and φ Persei. This line represents the right arm and sword which Perseus is swinging behind himself to cut off the head of Medusa. Returning to η Per, trace out a slightly curved line southward through τ, ι, κ, β, π, 16, and 17 Per. The stars in this line represent the head, right shoulder, back, and right leg of Perseus. 17 Persei and 24 Persei just to its east represent Perseus' right foot. A similar line starting with Mirfak and going to the southeast through σ, ν, ε, ξ, o, and ζ represents Perseus' left shoulder, the front of his torso, and his left leg and left foot, completing the figure of the hero in the act of slaying Medusa (Figs. 3.8–3.13; Table 3.2).

NAMED STARS:

	Name	Pronunciation	Source of name
α	Mirfak or Algenib	**mir**-fuk or **mir**-fak uhl-**jeh**-nib or al-**jee**-nib	mirfik al-thurayyā (Ind-Arabic), "the elbow nearest the many little ones" Al-Janb (Arabic), "the side or flank"
β	Algol	uhl-**gol** **al**-goll	al-ghul (Arabic), "the ghoul's head"
ξ	Menkib	**men**-kib	mankib thurayyā (Ind-Arabic), "the shoulder nearest the many little ones"
o	Atik	**ah**-tik **a**-tik	ātik al-thurayyā (Ind-Arabic), "the collarbone"

Magnitudes and spectral types of principal stars:

Bayer desig.	Mag. (m$_v$)	Spec. type	Bayer desig.	Mag. (m$_v$)	Spec. type	Bayer desig.	Mag. (m$_v$)	Spec. type
α	1.79	F5 Ib	β	2.09	B8 V	γ^A	2.91	G8 III
γ^B	3.00	–	δ	3.01	B5 III	ε	2.90	B0.5 V
ζ	2.84	B1 Ib	η	3.77	K3 Ib	θ	4.10	F7 V
ι	4.05	G0 V	κ	3.79	K0 III	λ	4.25	A0 IV
μ	4.12	G0 Ib	ν	3.77	F4 IIvar	ξ	3.98	O7.5Iab
o	3.84	B1 III	π	4.70	A2 Vn	ρ	3.32	M3 IIIvar
σ	4.36	K3 III	τ^A	3.93	G4 III	τ^B	4.00	–
υ	–	–^{*7}	φ	4.01	B2 Vpe	χ	5.99	G9 III
ψ	4.32	B5 Ve	ω	4.61	K1 III			

Stars of magnitude 5.5 or brighter which have Flamsteed, but no Bayer designations:

Flams. desig.	Mag. (m$_v$)	Spec. type	Flams. desig.	Mag. (m$_v$)	Spec. type	Flams. desig.	Mag. (m$_v$)	Spec. type
4	4.99	B8 III	9	5.16	A2 Ia	12	4.91	F9 V
14	5.43	G0 Ib	16	4.22	F2 III	17	4.56	K5 III
20	5.34	F4 Vvar	21	5.10	B9pec	24	4.94	K2 III
29	5.16	B3 V	30	5.49	B8 V	31	5.05	B5 V
32	4.96	A3 V	34	4.67	B3 V	36	5.30	F4 IIIvar
40	4.97	B0.5 V	42	5.14	A3 V	43	5.28	F5 IV
48	3.96	B3 Ve	52	4.67	G5 II	53	4.80	B4 IV
54	4.93	G8 III	58	4.25	–	59	5.32	B5 V
1 Aur	4.89	K4 II						

Stars of magnitude 5.5 or brighter which have no Bayer or Flamsteed designations:

Designation	Mag. (m$_v$)	Spec. type	Designation	Mag. (m$_v$)	Spec. type
HD 026961	4.60	A2 V	HD 018970	4.77	K0 II/III
HD 020468	4.85	K2 II	HD 212780	4.99	B5 V
HD 020123	5.04	G5 II	HD 018537	5.24	B7 V
HD 027971	5.29	K1 III	HD 020809	5.32	B5 V
HD 024504	5.39	B6 V	HD 027084	5.46	A7 V
HD 018474	5.47	G4p	HD 021699	5.47	B8 IIIp
HD 024640	5.49	B1.5 V			

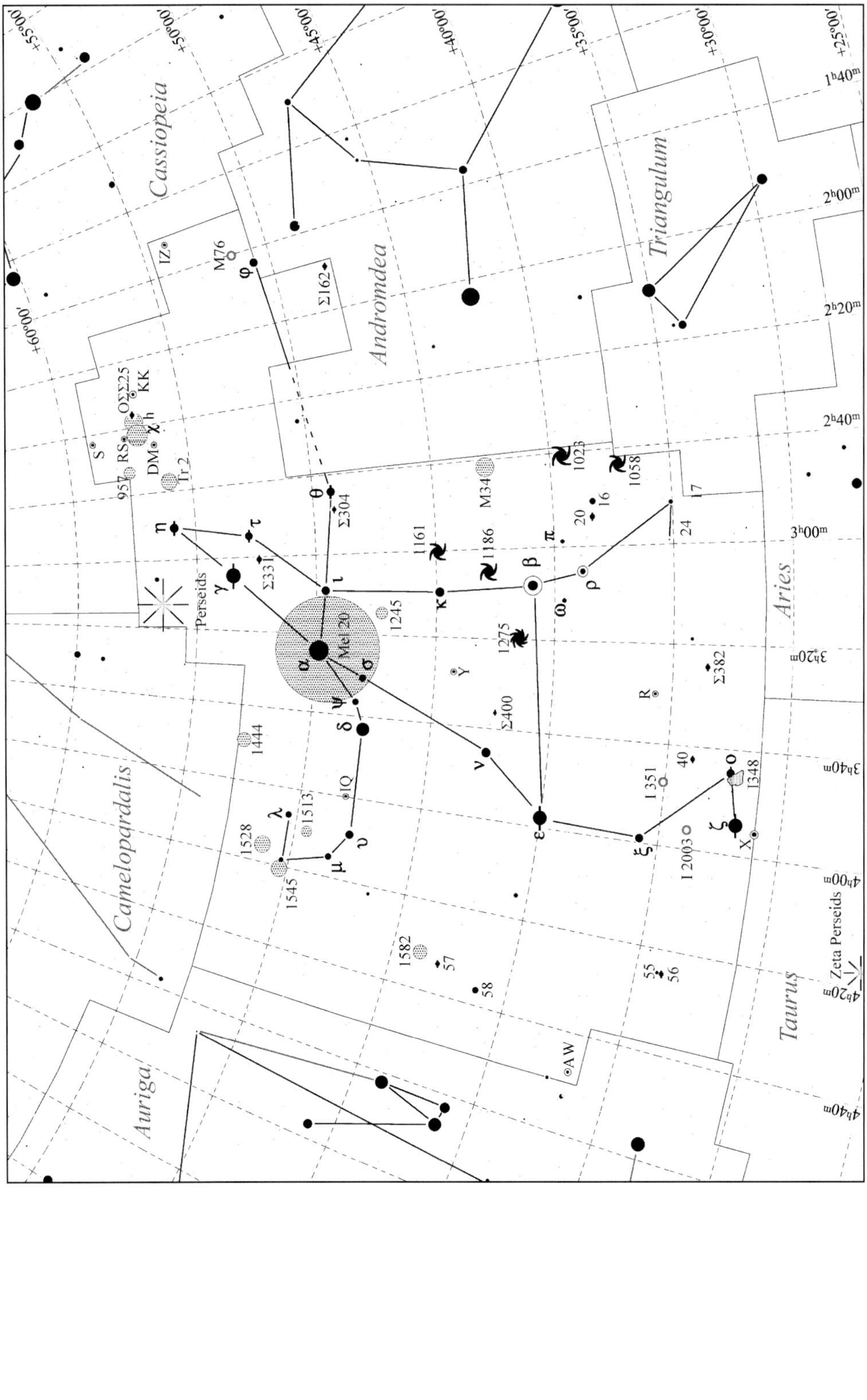

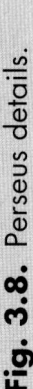

Fig. 3.8. Perseus details.

Table 3.2. Selected telescopic objects in Perseus.

Multiple stars:

Designation	R.A.	Decl.	Type	m₁	m₂	Sep. (")	PA (°)	Colors	Date/Period	Aper. Recm.	Rating	Comments
Σ162 Persei	01ʰ 49.3ᵐ	+47° 54'	Aa–B	6.5	7.2	1.9	200	Y, Y	2004	4/6"	*****	2°00' ESE (PA 110°) of 51 Persei (m3.59)
			Aa–C	6.5	9.4	20.3	178	Y, pB	2003	2/3"	****	
			Aa–D	6.5	10.0	138.3	095	Y, ?	2005	2/3"	***	
OΣΣ25 (Perseus)	02ʰ 16.9ᵐ	+57° 03'	A–B	6.5	7.41	102.9	205	bW, yW	2004	7×50	****	Easy binocular pair on WSW edge of NGC 869
13, θ Persei	02ʰ 44.2ᵐ	+49° 14'	A–B	4.1	10.0	**20.2**	**304**	Y, B	**2720**	2/3"	*****	Σ304, wide pair 40' to east
			A–C	4.1	11.0	77.2	229	Y, W	1924	3/4"	****	Very nice pair, golden yellow with faint companion nearby
Σ304 (Perseus)	02ʰ 48.8ᵐ	+49° 11'	A–B	7.5	10.7	26.10	289	W, B	1999	3/4"	****	0°46' E (PA 093°) of θ Persei (m4.20)
15, η Persei	02ʰ 50.7ᵐ	+55° 54'	A–B	3.8	8.5	28.5	301	Y, B	2002	7×50	*****	Similar to Albireo, somewhat fainter. Lovely!
			A–C	3.8	9.9	66.3	269	Y, ?	1925	11×80	****	
20 Persei	02ʰ 53.7ᵐ	+38° 20'	AB–C	5.0	9.7	13.9	236	pO, ?	1998	2/3"	****	A–B is too close for amateur instruments
18, τ Persei	02ʰ 54.3ᵐ	+52° 46'	Aa–B	4.0	12.3	47.3	107	Y, W	1999	4/6"	****	A–a is too close for amateur instruments
			Aa–C	4.0	21.7	50.7	106	Y, ?	1878	4/6"	***	
Σ331 (Perseus)	03ʰ 00.9ᵐ	52° 21'	A–B	5.2	6.2	11.9	085	yW, gB	2002	20/24"	*****	Easy showcase double. 1°05' ESE of τ Per (m3.93) *8
23, γ (Persei)	03ʰ 04.8ᵐ	+53° 30'	Aa–B	3.1	10.8	55.3	329	oY, W	1998	3/4"	***	A–a is too close for amateur instruments
Σ382 (Perseus)	03ʰ 24.5ᵐ	+33° 32'	A–B	5.8	9.3	4.8	153	yW, pB	2002	11×80	****	Very nice. Bright enough to find by unaided eye
Σ400 (Perseus)	03ʰ 36.2ᵐ	+42° 20'	A–B	8.3	9.1	**0.8**	**344**	yW, –	**153**	12/14"	*****	1°47' W (PA 281°) of υ Persei (m3.77)
38, o Persei	03ʰ 44.3ᵐ	+32° 17'	A–B	3.9	6.7	1.0	024	bW, –	2004	8/10"	*****	A challenge for larger scopes
40 Persei	03ʰ 42.4ᵐ	+33° 58'	A–B	6.0	10.0	25.6	249	bW, gB	2003	11×80	****	Lovely colors
44, ζ Persei	03ʰ 54.1ᵐ	+31° 53'	A–B	2.9	9.2	12.7	208	bW, B	2003	2/3"	****	ζ is brightest star of an association of ~20 stars, including o, ξ, 40, and 42 Persei
			A–C	2.9	11.2	33.3	287	bW, ?	2003	3/4"	***	
			A–D	2.9	10.4	98.0	196	bW, ?	2003	11×80	***	
			A–E	2.9	10.0	120.4	309	bW, ?	2003	7×50	***	
45, ε Persei	03ʰ 57.9ᵐ	+40° 01'	A–B	2.9	8.9	9.0	009	pY, dB	1998	2/3"	*****	Beautiful color contrast
56 Persei	04ʰ 24.6ᵐ	+33° 58'	A–B	5.8	9.3	4.4	017	oY, Y	2002	7×50	****	
57 Persei	04ʰ 33.4ᵐ	+43° 04'	A–B	6.1	6.8	120.5	198	bW, rO	2003	6×30	****	Easy colored pair for binoculars

Variable stars:

Designation	R.A.	Decl.	Type	Range (m_v)		Period (days)	F (f_r/f_f)	Spectral Range	Aper. Recm.	Rating	Comments
IZ Persei	01h 32.1m	+54° 01'	EA/SD	7.8	9.0	3.688	0.12	B8V	7×50	****	3°46' NNW (PA 333°) of φ Persei. #
KK Persei	02h 10.3m	+56° 34'	Lc	6.6	7.9	–	–	M1.0Iab–M3.5Iab	6×30	****	1°06' S (PA 189°) of 5 Persei (m6.38).#
RS Persei	02h 22.6m	+57° 07'	SRc	7.8	10.0	244.5	–	M4Iab	2/3"	****	0°08' SSE (PA 154°) of 64 Persei (m6.40).#$
S Persei	02h 22.9m	+58° 35'	SRc	7.9	11.5	822	–	M3eIa–M6eIa	3/4"	****	0°56' NE (PA 042°) of 8 Persei (m5.77).##$
DM Persei	02h 26.0m	+55° 06'	EA/SD	7.9	8.6	2.728	0.17	B5V	6×30	****	0°34' ENE (PA 063°) of 9 Persei (m5.16).#
ρ Persei	03h 05.2m	+38° 50'	SRb	3.3	4.0	50	–	M4IIb–IIIa	Eye	****	Also called Gorgonea Tertia, "the third gorgon."##
β Persei	03h 08.2m	+40° 57'	EA/SD	2.1	3.4	2.867	0.14	B8V	Eye	*****	Algol, prototype of well separated eclipsing binary variables.##
Y Persei	03h 27.7m	+44° 11'	M	8.1	11.3	248.6	0.48	C4,3e(R4e)	3/4"	****	1°25' NE (PA 053°) of 32 Persei (m4.96).##
R Persei	03h 30.1m	+35° 40'	M	8.1	14.8	209.89	0.049	M2e–M5e	16/18"	****	3°03' NW (PA 305°) of 40 Persei (m4.97).##
X Persei	03h 55.4m	+31° 03'	γ Cas+Xp	6.0	7.0	–	–	O9.5 (III–IV)eP	6×30	****	A high mass X-ray binary with a hot giant (~15M) and a neutron star companion; Distance estimated at 2,300–4,200 ly; L=~24,000 L. ##$ *9
IQ Persei	03h 59.7m	+48° 09'	EA/DM	7.72	8.27	1.744	0.12	B8Vp	6×30	****	2°50' E (PA 081°) of δ Persei (m2.99).#$
AW Persei	04h 47.8m	+36° 43'	δ Cep	7.04	7.89	6.464	0.25	F6–G0	6×30	****	0°55' SW (PA 272°) of 2 Persei (m4.79).#

Star clusters:

Designation	R.A.	Decl.	Type	m_v	Size (")	Brst. Star	Dist. (ly)	dia. (ly)	Number of Stars	Aper. Recm.	Rating	Comments
NGC 869, h	02h 19.0m	+57° 09'	Open	5.3	18	6.55	7,300	40	317	6×30	*****	Double cluster with χ Per, many stars visible. Beautiful field in any size binoculars

(continued)

Table 3.2. (continued)

Star clusters:

Designation	R.A.	Decl.	Type	m_v	Size (')	Brtst. Star	Dist. (ly)	dia. (ly)	Number of Stars	Aper. Recm.	Rating	Comments
NGC 884, χ	02h 22.4m	+57° 07'	Open	4.4	30	6.1	303	7,300	40	6×30	*****	Splendid binocular spectacle
NGC 957	02h 33.6m	+57° 32'	Open	7.6	11	11.0	120	6,600	21	2/3"	***	Rich, but relatively faint
Trumpler 2	02h 37.3m	+55° 59'	Open	5.9	17	7.38	110	1,800	9	6/8"	***	Blue and White stars >7 m, ~half vis. in 7×50
M34, NGC 1039	02h 42.0m	+42° 47'	Open	5.2	25	9.0	60	1,450	11	7×50	****	Most stars 7–13 m, Fine sight in binoculars
NGC 1245	03h 14.7m	+47° 15'	Open	8.4	10	10.0	200	7,300	21	4×6"	***	Some stars resolved in 4/6"
Mel 20	03h 22m	+49°	Open	2.3	300	3.0	50	535	47	10×50	****	Alpha Persei Cluster, actually an association. Nearly fills 7×50 FOV *10
IC 348	03h 44.5m	+32° 17'	Open	7.3	8	10.0	20	1,100	2	3/4"	***	Compact group of relatively faint stars with o Per on its NE edge
NGC 1444	03h 49.4m	+52° 40'	Open	6.6	4	9.6	60	3,100	4	6/8"	***	Small cluster of relatively faint stars with a bright multiple star in the middle
NGC 1513	04h 10.0m	+49° 31'	Open	8.4	12	11.0	50	2,600	9	4/6"	***	Rich cluster of faint stars
NGC 1528	04h 15.4m	+51° 14'	Open	6.4	18	10.0	165	2,400	13	4/6"	***	Only 1–2 stars visible in binoculars, 4/6" shows several with some concentration in cent
NGC 1545	04h 20.9m	+50° 15'	Open	6.2	12	9.0	65	2,500	9	4/6"	***	Only 5–6 stars visible in binoculars, including a bright triangle in the middle
NGC 1582	04h 32.0m	+43° 51'	Open	7.0	24	9.0	20	–	–	6/8"	***	Small scopes show only 8–10 stars

Nebulae:

Designation	R.A.	Decl.	Type	m_v	Size (')	Brtst./ Cent. star	Dist. (ly)	dia. (ly)	Aper. Recm.	Rating	Comments
NGC 650-1, M76	01h 42.4m	+51° 34'	Planetary	10.1	65"	15.9	3.9K	1.2	4/6"	*****	Little Dumbbell Nebula. Faintest of Messier objects
IC 351	03h 47.5m	+35° 03'	Planetary	11.9	7"	15.8	13K	0.4	8/10"	***	Nearly stellar. Central star is a Wolf-Rayet star with abundant carbon
IC 2003	03h 56.4m	+33° 52'	Planetary	11.4	7"	15.3	14K	0.5	8/10"	***	Nearly stellar. Central star is a Wolf-Rayet star with abundant carbon

Galaxies:

Designation	R.A.	Decl.	Type	m_v	Size (')	Dist. (ly)	dia. (ly)	Lum. (suns)	Aper. Recm.	Rating	Comments
NGC 1023	02h 40.4m	+39° 04'	SB	9.5	8×3	98M	65K	18G	4/6"	****	E7p, Lens-shaped, elongated E–W
NGC 1058	02h 43.5m	+37° 21'	SA	11.2	3.5×3.4	97M	32K	2.2G	12/14"	***	Faint, little or no brightening toward the center
NGC 1161	03h 01.2m	+44° 55'	SB	11.0	2.8×2.1	90M	24K	1.6G	8/10"	***	Look for NGC 1160, m12.8 just 3' N of 1161
NGC 1186	03h 05.5m	+42° 50'	SB	11.4	3.2×1.2	120M	36K	2.8G	12/14"	***	
NGC 1275	03h 19.8m	+41° 31'	Pec	11.9	2.2×1.7	230M	147K	7.3G	12/14"	***	a.k.a Perseus A, a radio source

Meteor showers:

Designation	R.A.	Decl.	Period (yr)	Duration	Max Date	ZHR (max)	Comet/ Asteroid	First Obs.	Vel. mi/ km/sec	Rating	Comments
Perseids	03h 12m	+57°	Annual	July 23 – Aug. 22	Aug. 12/13	80	Swift-Tuttle	30 AD	40 mi/s 66 km/s	Major	Most famous shower, reliable rates, first shower to be associated with a comet. Increased activity in 1862 (up to 215/h) and 1983 (~187/h), when Swift-Tuttle was near the Earth
Zeta Perseids	04h 18m	+26°	Annual	May 20 – July 05	June 13	40*	–	1947	17 mi/s 27 km/s	Day-light*	One of the most active daylight showers. Some meteors may be seen just before sunrise and after sunset about 6/13, but radio detection is more likely *11

Other interesting objects:

Designation	R.A.	Decl.	Type	m_v	Size (')	Dist. (ly)	dia. (ly)	Lum. (suns)	Aper. Recm.	Rating	Comments
NGC 1245	03h 19.8m	+41° 31'	Gal. Clstr.	11.9	3.2×2.3	235M			12/14"	****	Colliding galaxies *12
Abell 426	03h 20m	+41° 31'	Gal Clust.	12+	240×240	235M			16/18"	*****	a.k.a. Perseus I Galaxy Cluster. Over 500 galaxies, 36 NGC/IC objects

Fig. 3.9. NGC 869 and 884, the Double Cluster in Perseus.

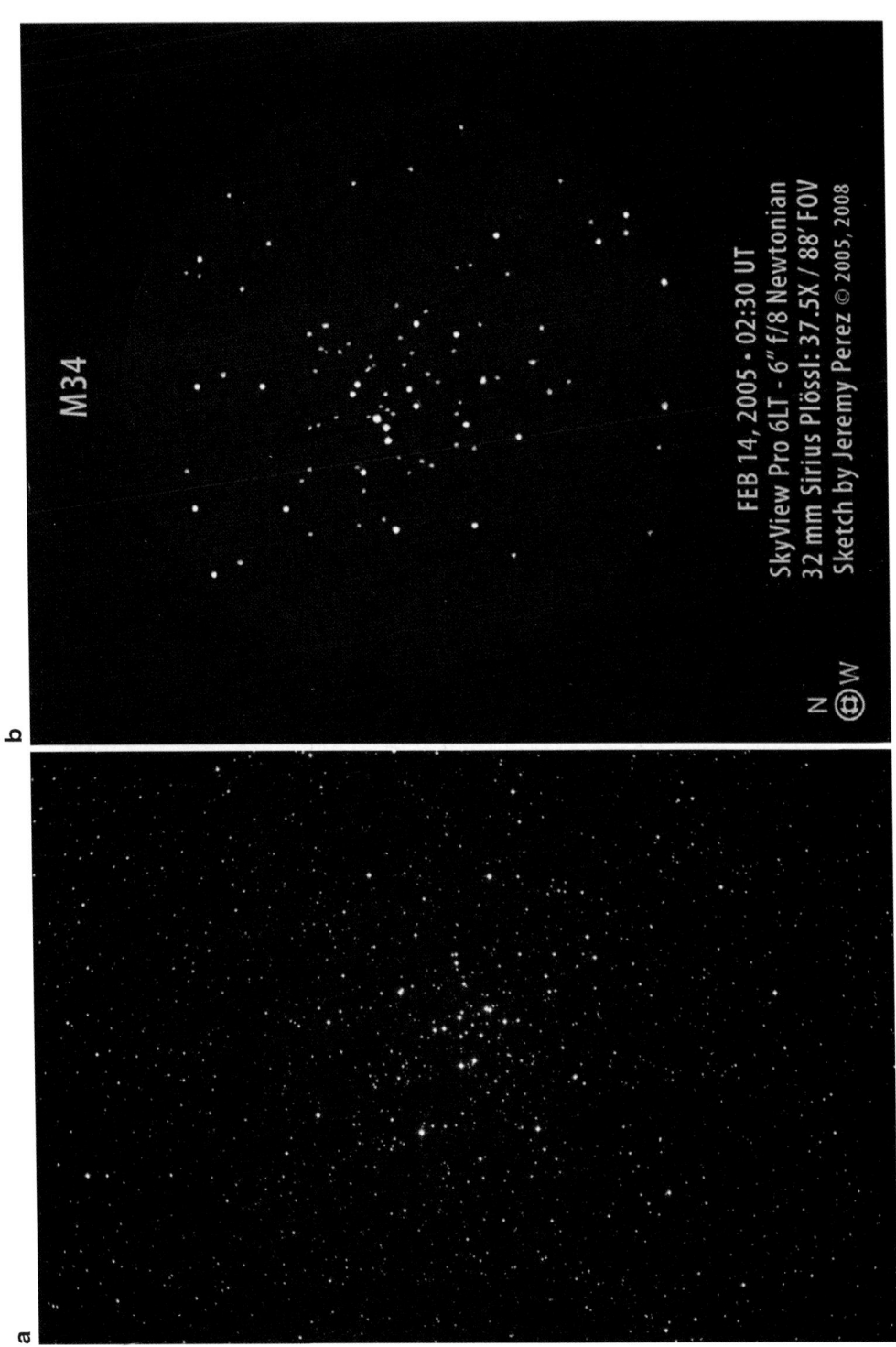

Fig. 3.10. **(a)** M34 (NGC 1039), open cluster in Perseus. Astrophoto by Mark Komsa. **(b)** M34, open cluster in Perseus.

M76

JUL 28, 2008 · 08:30 UT
SkyQuest XT8 - 8" f/6 Newtonian
10 mm Sirius Plössl: 120X / 24' FOV
Sketch by Jeremy Perez © 2008

N
W

Fig. 3.11. M76, Little Dumbbell, planetary nebula in Perseus.

NGC 1023

AUG 20, 2006 • 08:00 UT
SkyView Pro 6LT EQ - 6" f/8 Newtonian
10 mm Sirius Plössl: 120X / 24′ FOV
Sketch by Jeremy Perez © 2006

N
W

Fig. 3.12. NGC 1023, spiral galaxy in Perseus.

Fig. 3.13. NGC 1499, California Nebula in Perseus. Astrophoto by Andy Homeyer.

CASSIOPEIA (kass-ee-oh-**pee**-uh)	**Cassiopeia** (The queen)
CASSIOPEIAE (kass-ee-oh-**pee**-eye)	**BEST SEEN** (at 9:00 **P.M.**): Nov. 24
Cas	**AREA:** 598 sq. deg., 25th in size

TO FIND: Although the "W" shape formed by the six brightest stars of Cassiopeia is well known, it is also easy to find if you are not already familiar with it. This asterism lies directly opposite the stars forming the handle of the Big Dipper (or the tail of the Big Bear), with the North Celestial Pole (or Polaris) in the middle. These six stars (α, β, γ, δ, ε, and η) with κ added form the main part of Cassiopeia's throne. To get the entire throne, it is necessary to add only ι (5° NE of ε) to form the extreme top of the throne, plus ζ (2.7° S of α) and τ (3° W of β) to form the flared feet of the throne. 50 Cas lies 8.8° almost due north of ε, and represents the jewel in Cassiopeia's crown. The rest of the stars representing the unworthy Cassiopeia form a broad and irregular band to the north and west of the throne. They are all relatively faint and difficult to see except in very dark skies (Figs. 3.14 and 3.15).

KEY STAR NAMES:

	Name	Pronunciation	Source
α	Schedar	**sheh**-der	al-sadr (Arabic), "the breast" from the position description in the Arabic *Almagest*
β	Caph	kuf or kaf	al-kaff al-khadib (Arabic), "the stained hand" from the traditional use of henna in the middle east
γ	Tsih	't-seh	Tse (Chinese), "the whip," part of a Chinese constellation of Wang-liang and Tse, the Wagon and the Whip
δ	Ruchbah	**ruk**-buh	Rukbat adh-Dhat al-Kursi (Arabic), "the knee of the lady of the chair"
ε	Segin	seg-**een**	Probably an erroneous transposition of Seginus (γ Boö)
	or Navi	**nah**-vih or **nav**-ih	A modern name in honor of Virgil (Gus) Ivan Griffin*[13]
η	Achird	?	Apparently introduced by Antonin Becvar in his 1958 *Atlas Coeli*, but his source has not been found
θ	Marfak	mur-**fik** or	al-marfiq (Arabic), "the elbow," applied to both θ and μ Cas
μ	Marfak	**mahr**-fik	

Magnitudes and spectral types of principal stars:

Bayer desig.	Mag. (m$_v$)	Spec. type	Bayer desig.	Mag. (m$_v$)	Spec. type	Bayer desig.	Mag. (m$_v$)	Spec. type
α	2.24	K0 II/III	β	2.28	F2 III/IV	γ	2.15	B0 IV
δ	2.66	A5 Vv	ε	3.35	B2 var	ζ	3.69	B2 IV
η	3.46	G0 V	θ	4.34	A7 Vv	ι	4.46	A5 p
κ	4.17	B1 Ia	λ	4.74	B8 Vn	μ	5.17	G5 VIp*[14]
ν	4.90	B9 III	ξ	4.80	B2.5 V	o	4.48	B5 III
π	4.95	A5 V	ρ	4.51	F8 Ia (var)	σ	4.88	K1 III
τ	4.88	A2 V	υ¹	4.83	K2 III	υ²	4.62	G8 III/IV
φ	4.95	F0 Ia	χ	4.68	K0 III	ψ	4.72	K0 III
ω	4.97	B8 III						

Stars of magnitude 5.5 or brighter which have Flamsteed, but no Bayer designations:

Flamsteed	Mag. (m$_v$)	Spec. type	Flamsteed	Mag. (m$_v$)	Spec. type	Flamsteed	Mag. (m$_v$)	Spec. type
1	4.84	B0.5 IV	4	4.96	M1 III	6	5.43	A3 Ia
12	5.38	B9 III	23	5.42	B8 III	31	5.32	A0 Vnn
40	5.28	G8 II/IIIvar	42	5.18	B9 V	47	5.27	F0 Vn
48	4.49	A3 IV	49	5.22	G8 III	50	3.95	A2 V

Stars of magnitude 5.5 or brighter which have no Bayer or Flamsteed designations:

Designation	Mag. (m$_v$)	Spec. type	Designation	Mag. (m$_v$)	Spec. type
HD 003240	5.08	B7 III	HD 003574	5.45	K5 III
HD 004222	5.41	A2 Vs	HD 004775	5.35	A4 V
HD 005015	4.80	F8 V	HD 011946	5.29	A0 Vn
HD 012273	5.17	G8 III	HD 019275	4.85	A2 Vnn

Fig. 3.14a. The stars of Cassiopeia, Cephus and surrounding constellations.

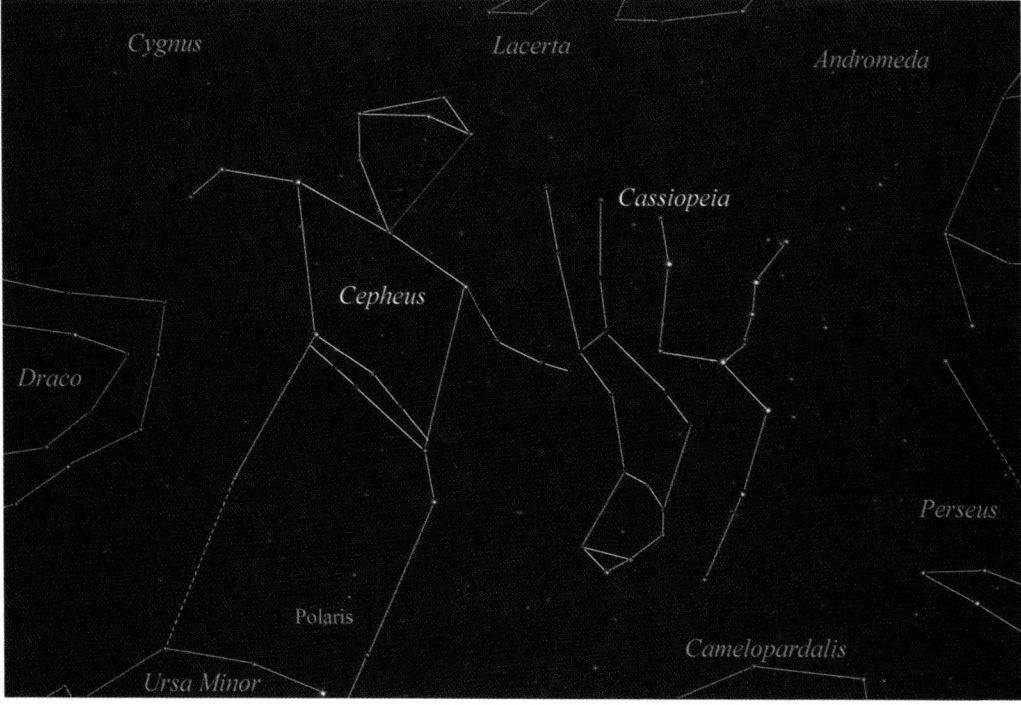

Fig. 3.14b. The figures of Cassiopeia, Cephus, and surrounding constellations.

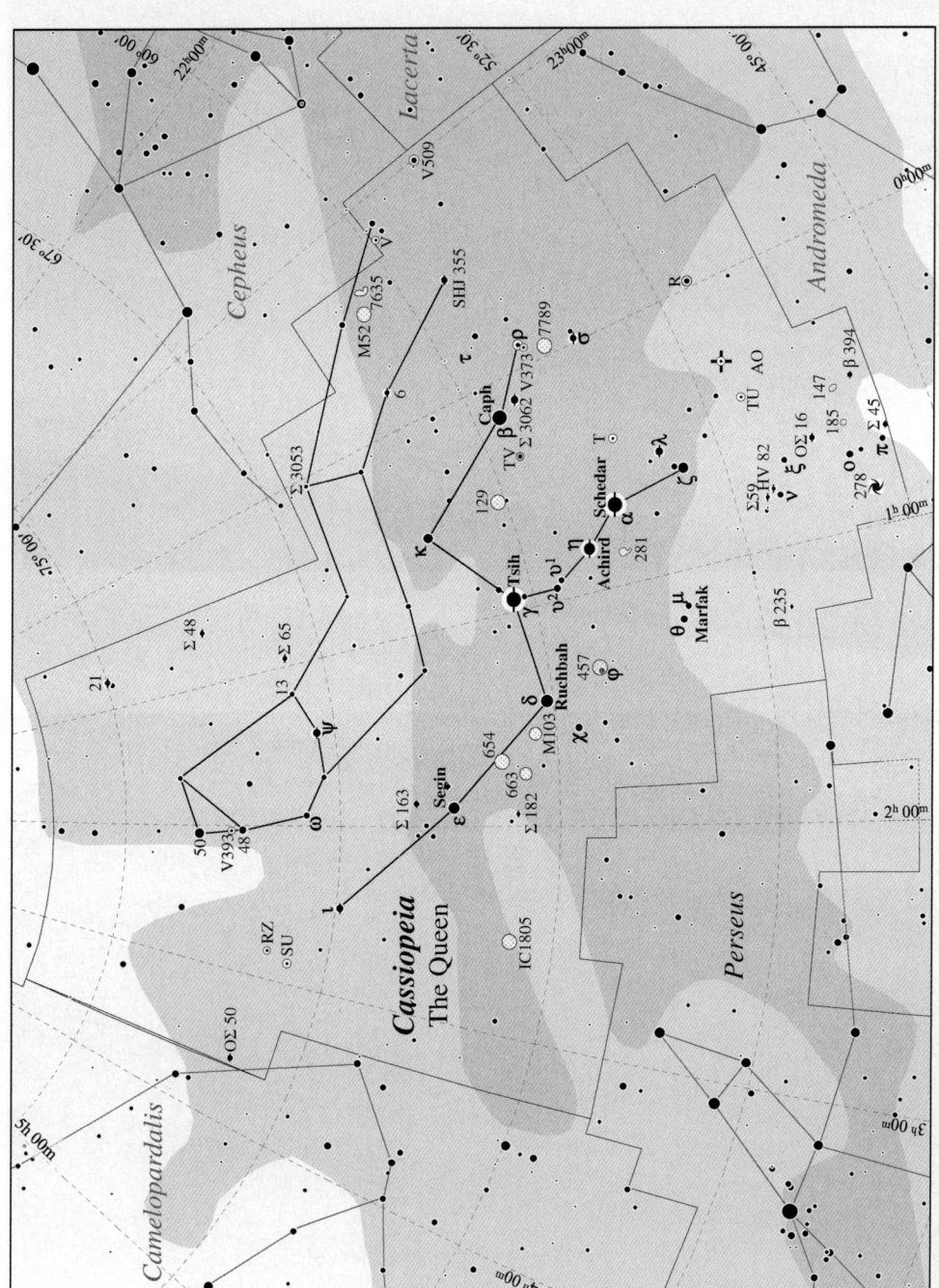

Fig. 3.15. Cassiopeia details.

Table 3.3. Selected telescopic objects in Cassiopeia.

Multiple stars:

Designation	R.A.	Decl.	Type	m_1	m_2	Sep. (")	PA (°)	Colors	Date/Period	Aper. Recm.	Rating	Comments
Schj 355	23h 30.0m	+58° 33'	A–B	4.87	9.3	0.9	335	bW, –	1970	10/12"	****	Separation and position angle slowly decreasing
			A–C	4.9	7.2	75.8	268	bW, ?	2004	6×30	****	A, C, and FG form an isosceles triangle
			C–D	7.2	9.06	1.4	214	W, ?	2005	6/8"	****	a.k.a. h1887
			F–G	8.9	9.1	10.7	073	?, ?	2005	2/3"	***	
6 Cassiopeiae	23h 48.8m	+62° 13'	A–B	5.7	8.0	1.4	197	dY, O	1994	4/6"	****	Σ508AB. Beautiful color combination
8, σ Cassiopeiae	23h 59.0m	+55° 45'	A–C	5.7	10.2	62.3	309	dY, ?	1999	2/3"	***	Σ508AC
			A–B	5.0	7.2	3.2	327	bW, Y	2004	3/4"	****	Σ3049AB. Nice star field. 10th mag. star 108" away
Σ3053	00h 02.6m	+66° 06'	A–C	5.0	10.4	106.7	066	bW, ?	2001	3/4"	***	Σ3049AC
			A–B	6.0	7.2	15.0	070	Y, W	2003	15×80	****	Easy double for small scopes
			A–C	6.0	11.0	98.5	290	Y, ?	1999	3/4"	***	
Σ3062	00h 06.3m	+58° 26'	A–B	5.9	6.9	**1.6**	**347**	Y, pB	**106.7**	6/8"	*****	A neat matched pair to test larger telescopes
β394	00h 30.8m	+47° 32'	A–B	8.5	8.8	**0.64**	**276**	Y, Y	**142.14**	12/14"	*****	
Σ45	00h 38.7m	+46° 57'	A–B	6.9	9.9	18.3	090	Y, pB	2003	2/3"	*****	Lovely colored pair
OΣ16	00h 39.2m	+49° 21'	A–B	5.7	10.2	12.8	022	yW, pB	1998	2/3"	****	
18, α Cassiopieae	00h 40.5m	+56° 32'	A–B	2.4	14.0	19.8	275	pR, ?	1934	10/12"	****	
			A–C	2.4	13.0	38.3	105	pR, ?	1908	6/8"	****	Nice color contrast
			A–D	2.4	9.0	69.5	282	pR, B	1998	7×50	****	
Σ48	00h 42.6m	+71° 22'	A–B	7.8	8.1	5.4	334	W, W	1999	2/3"	****	Nearly identical pair of white stars
21 Cassiopeiae	00h 45.8m	+74° 59'	A–B	5.66	10.57	35.9	160	W, ?	1999	2/3"	****	Nice star field. 23 Cas is in same FOV. 21 is also variable
HV82	00h 47.4m	+51° 06'	A–B	8.0	8.35	56.3	075	rW, rW	2003	2/3"	****	Faint, but easy to find HV82 and Σ59 + 25 Cas are in same low/ medium power FOV
Σ59	00h 48.0m	+51° 27'	A–B	7.24	8.06	2.2	147	Y, W	2003	4/6"	****	
24, η Cassiopeiae	00h 49.1m	+57° 49'	A–B	3.5	7.4	**13.2**	**322**	Y, B	**480**	2/3"	****	Gorgeous, sep. slowly increasing. Observers differ on color of B

(continued)

Table 3.3. (continued)

Multiple stars:

Designation	R.A.	Decl.	Type	m_1	m_2	Sep. (")	PA (°)	Colors	Date/Period	Aper. Recm.	Rating	Comments
Σ65	00h 52.7m	+68° 52'	A–B	8.0	8.0	3.0	219	rW, rW	1997	3/4"	****	Perfectly matched pair. Good test for 2/3" scopes
γ Cassiopeia	00h 56.7m	+60° 43'	A–B	2.2	10.9	2.3	252	bW, ?	1961	4/6"	****	Primary is variable. A good test for 3" telescope
β235	01h 10.6m	+51° 00'	A–B	8.5	7.5	**0.9**	**136**	yW, yW	**278**	8/10"	*****	Another neat, to test larger telescopes. Somewhat easier than β 394
ψ Cassiopeiae	01h 26.0m	+68° 07'	A–B	4.7	14.0	2.4	038	O, ?	1970	8/10"	**	a.k.a. β1101AB
			A–C	4.7	9.2	22.1	123	O, ?	1998	7×50	***	a.k.a. h583AC
			A–D	4.7	10.0	22.4	120	O, ?	1983	15×80	***	a.k.a. β1101AD
			C–D	9.4	10.0	2.6	252	O, ?	2001	3/4"	****	a.k.a. Σ117CD
Σ163	01h 51.3m	+64° 51'	A–B	6.8	9.1	33.9	036	O, B	2003	7×50	*****	Beautiful color combination
Σ182	01h 56.4m	+61° 16'	A–B	8.31	8.35	3.5	125	W, W	2002	3/4"	****	Matched pair of white stars. Test for 2" scope
ι Cassiopeiae	02h 29.1m	+67° 24'	Aa–B	4.6	6.9	**2.6**	**229**	W, Y	**620**	4/6"	*****	One of prettiest triples in the sky. Aa–B Separation is slowly increasing. C is stationary. A–a is binary, separation is always less than 0.9"
			Aa–C	4.6	9.1	7.6	115	W, B	2004	2/3"	****	
OΣ50	03h 12.1m	+71° 33'	A–B	7.6	8.5	**0.93**	**194**	yW, ?	**310.06**	10/12"	***	A challenge for larger amateur telescopes. Separation and PA are slowly decreasing

Variable stars:

Designation	R.A.	Decl.	Type	Range (m_v)		Period (days)	F (f_r/f_f)	Spectral Range	Aper. Recm.	Rating	Comments
V509 Cassiopeiae	23h 00.1m	+56° 57'	SRd	4.8	5.5	–	–	F8e–K1a+B1V	6×30	****	Photosphere pulsates, ejected shell in 1975 which is now falling back.##
V Cassiopeiae	23h 11.7m	+59° 42'	M	6.9	13.4	228.83	0.48	M5e–M8.5e	8/10"	***	##
7, ρ Cassiopeiae	23h 54.4m	+57° 30'	SRd	4.1	6.2	320	–	F8Iap–K0–Iap	6×30	****	Usual range=4.4–5.2. mp range 5.4–8.3.##$

Designation	R.A.	Decl.	Type	max	min	Period		Spectral	Aper. Recm.	Rating	Comments
V373 Cassiopeiae	23h 55.6m	+57° 25'	E/GS	5.9	6.3	13.4192	–	B0.5II+B0.5II	6×30	****	Nearly eclipsing pair, somewhat unequal double max. and min. ##
R Cassiopeiae	23h 58.4m	+51° 24'	M	4.7	13.5	430.46	0.40	M6e–M10e	8/10"	****	##
AO Cassiopeia	00h 17.7m	+51° 26'	ELL/KE	6.1	6.2	3.52349	–	O9III+O9III	6×30	****	Pearce's star, one of the most mass. and lum. binary systems in our galaxy *15
TV Cassiopeiae	00h 19.3m	+59° 08'	EA/SD	7.2	8.2	1.8126	0.18	B9V+F7IV	7×50	****	Bright SiO lines, dark ZrO bands. ##
T Cassiopeiae	00h 23.2m	+55° 48'	M	6.9	13.0	444.83	0.56	M6e–M9.0e	7/8"	****	##$
TU Cassiopeiae	00h 26.3m	+51° 17'	Cep(b)	6.9	8.2	2.1393	0.31	F3II–F5II	7×50	****	##
21, YZ Cassiopeiae	00h 45.7m	+74° 59'	EA/DM	5.2	6.1	4.46722	0.15	A2V+F2V	6×30	****	See "Binary Stars: A Pictorial Atlas" for system configuration and light curve
27, γ Cassiopeiae	00h 56.7m	+60° 43'	γ Cas	1.6	3.0	–	–	B0.5IVpe	Eye	*****	Prototype of its class. ##$ *16
V393 Cassiopeiae	02h 02.7m	+71° 18'	SRa	7.0	8.0	393	–	M0	7×50	****	#
RZ Cassiopeiae	02h 48.9m	+69° 38'	EA/SD	6.4	7.8	1.195	0.17	A3V	7×50	****	Eclipse takes ~4 h, 2 h dropping immediately followed by 2 h incr.##$
SU Cassiopeiae	02h 52.0m	+68° 53'	δ Cep	5.7	6.2	1.94932	0.40	F5 Ib-II–F7 Ib-II	6×30	****	

Star clusters:

Designation	R.A.	Decl.	m_v	Type	Size (')	Brist. Star	Dist. (ly)	dia. (ly)	Number of Stars	Aper. Recm.	Rating	Comments
NGC 129	00h 29.9m	+60° 14'	6.5	Open	12	11.0	5.3K	15	35	4/6"	***	Lg, brt, 7×50 = ~ 2 stars (121)
NGC 457	01h 19.1m	+58° 20'	6.4	Open	20	8.59	10K	60	204	7×50	****	Easily resolvable in 11×80(*), br. star is orange (121)
M103, NGC 581	01h 33.2m	+60° 42'	7.4	Open	6	6.0	8K	14	172	7×50	*****	~25 stars <10 m. req. 15×105 to resolve (**), br. star=10.6 (*), near Ruchbah (δ)
NGC 654	01h 44.1m	+61° 53'	6.5	Open	6	9.0	6.4K	11	60	7×50	****	Br. star=8.4, diff. in sm (4") scope (121)
NGC 663	01h 46.0m	+61° 15'	7.1	Open	15	9.0	7.2K	32	80	7×50	****	Conspicuous, very pretty. Br. star=8.4
IC 1805, Melotte 15	02h 32.7m	+61° 27'	6.5	Open	20	7.9	7.2K	42	40	11×80	****	Cluster surrounded by faint emission nebula IC 1805, "Heart Nebula."

(continued)

Table 3.3. (continued)

Star clusters:

Designation	R.A.	Decl.	Type	m_v	Size (')	Brtst. Star	Dist. (ly)	dia. (ly)	Number of Stars	Aper. Recm.	Rating	Comments
M52, NGC 7654	23^h 24.2^m	+61° 35'	Open	6.9	16	9.0	5.1K	24	100	7×50	*****	7×50 resolves several stars
NGC 7789	23^h 57.0^m	+56° 44'	Open	6.7	25	10.7	6.0K	44	300	7×50	****	Star Mist Cl. Rich, even distr., faint stars, m11+. Visible to unaided eye in good condition

Nebulae:

Designation	R.A.	Decl.	Type	m_v	Size (')	Brtst./Cent. star	Dist. (ly)	dia. (ly)	Aper. Recm.	Rating	Comments
NGC 7635	23^h 20.7^m	+61° 12'	Em	–	15×8	–	–	–	8/10"	**	Bubble Nebula. Large, very faint
NGC 281	00^h 52.8^m	+58° 20'	Em	–	34×29	–	–	–	8/10"	*	Very large, very faint nebula surrounding open cluster. Use UHC filter

Galaxies:

Designation	R.A.	Decl.	Type	m_v	Size (')	Dist. (ly)	dia. (ly)	Lum. (suns)	Aper. Recm.	Rating	Comments
NGC 147	00^h 33.2^m	+48° 30'	dE4	9.5	12.9×8.1	2.4M	12K	1G	4/6"	***	NGC 147 and 185 are distant satellites of M31
NGC 185	00^h 39.0^m	+48° 20'	dE0	9.2	14.5×12.5	2.2M	13K	1.4G	4/6"	***	
NGC 278	00^h 52.1^m	+47° 33'	SAb	10.8	2.5×2.5	38M	36K	25G	6/8"	****	

Meteor showers:

Designation	R.A.	Decl.	Period (yr)	Duration	Max Date	ZHR (max)	Comet/Asteroid	First Obs.	Vel. [mi/km/sec]	Rating	Comments
None											

Other interesting objects:

Designation	R.A.	Decl.	Type	m_1	m_2	Period (days)	Dist. (ly)	Mass (suns)	Lum. (suns)	Aper. Recm.	Rating	Comments
AO Cas	00^h 17.7^m	+51° 26'	SpecBin	6.05	6.3	3.5	7,000	32+30	300K	6×30	***	Pearce's star, one of most massive binary systems in our galaxy

NGC 457 and NGC 436
OCT 12, 2007 · 05:30 UT
15 X 70 Oberwerk Binoculars
Sketch by Jeremy Perez © 2007

W
N

Fig. 3.16. NGC 457 and 436, open clusters in Cassiopeia.

Fig. 3.17. NGC 654 open cluster in Cassiopeia.

Fig. 3.18. M103 (NGC 581), open cluster in Cassiopeia. Astrophoto by Mark Komsa.

Fig. 3.19. IC 1805, the Heart Nebula in Cassiopeia. Astrophoto by Warren Keller.

Fig. 3.20. M52 (NGC 7654), open cluster and NGC 7635, Bubble Nebula in Cassiopeia. Astrophoto by Mark Komsa.

Fig. 3.21. NGC 7635, Bubble Nebula in Cassiopeia. Astrophoto by Warren Keller.

CEPHEUS (see-fus)	Cepheus (The king)
CEPHEI (see-fee-ee)	BEST SEEN: Nov. 14 (at 9:00 P.M.)
Cas	AREA: 588 sq. deg., 27th in size

TO FIND: At the time listed above, face north. Above and slightly east of Polaris you will see the familiar "W" formed by the brightest stars in Cassiopeia. A line from the two southern-most stars (α and β Cas) in the "W" can be followed westward for about 25° to a point very close to Alderamin, the brightest star in Cepheus. Alderamin is the southwest corner of a fat diamond shape (almost a square) about 8° on a side. The other stars in the diamond are β, ι, and ζ Cephei. About 10° below β and ι, you will find γ Cephei, the second brightest (although only very slightly brighter than β). γ forms the peak of the roof of a tall and narrow (but upside down) house often visualized in Cepheus. Using the chart, you can trace out the figure of the king standing with his feet straddling the celestial pole (Figs. 3.14 and 3.22).

KEY STAR NAMES:

Name	Pronunciation	Source
α Alderamin	**uhl**-deh-rah-**MEN**	aldhira al yamin (Arabic), "the right arm" α will be the pole star around 7500 AD
β Alfirk	uhl-**firk**	al-firq (Arabic), "the flock"
γ Errai	err-**ray**-ee	al-rācī (Arabic), "the shepherd"
ξ Kurhah	**Koor**-hah	al-qurhā (Ind-Arabic, "the white spot on the forehead of a horse"
ρ Al Kalb al Rai	uhl **kalb** uhl **ray**-ee	al-kalbā al-rācī (Arabic), "the shepherd's dog"

Magnitudes and spectral types of principal stars:

Bayer desig.	Mag. (m$_v$)	Spec. type	Bayer desig.	Mag. (m$_v$)	Spec. type	Bayer desig.	Mag. (m$_v$)	Spec. type
α	2.45	A7 IV	β	3.23	B2 III	γ	3.21	K1 IV
δ	4.07v	G2 Ib	ε	4.18	F0 IV	ζ	3.39	K1 Ib
η	3.41	K0 IV	θ	4.21	A7 III	ι	3.50	K0 III
κ	4.38	B9 III	λ	5.05	O6 Ie	μ	4.23	M2 Ia
ν	4.25v	A2 Ia	ξ	4.29	A3 V	o	4.75	K0 III
π	4.41	G2 III	ρ	5.45	A3 V			
			υ^1	4.52	F8 IV/V			

Stars of magnitude 5.5 or brighter which have Flamsteed, but no Bayer designations:

Flam-steed	Mag. (m$_v$)	Spec. type	Flam-steed	Mag. (m$_v$)	Spec. type	Flam-steed	Mag. (m$_v$)	Spec. type
6	5.19	B3 IVe	7	5.42	B7 V	9	4.76	B2 Ib
16	5.04	F5 V	18	5.26	M5 III	19	5.07	O9.5 Ib
20	5.27	K4 III	24	4.79	G8 III	30	5.19	A3 IV
31	5.08	F3 III/IV	2 UMi	4.24	K2 II/III*[17]	78 Dra	5.18	K0 III*[17]

Stars of magnitude 5.5 or brighter which have no Bayer or Flamsteed designations:

Desig-nation	Mag. (m$_v$)	Spec. type	Desig-nation	Mag. (m$_v$)	Spec. type
HD 018438	5.49	F7 IV	HD 019978	5.44	A6 V
HD 025007	5.10	G8 III	HD 026659	5.47	G8 III
HD 026836	5.42	G6 III	HD 030338	5.09	K3 III
HD 210855	5.24	F8 V	HD 210939	5.37	K1 III
HD 212710	5.27	B9.5 Vn	HD 213022	5.47	K2 III
HD 216446	4.77	K3 III	HD 217382	4.70	K4 III
HD 218029	5.25	K3 III	HD 223274	5.05	A1 Vn

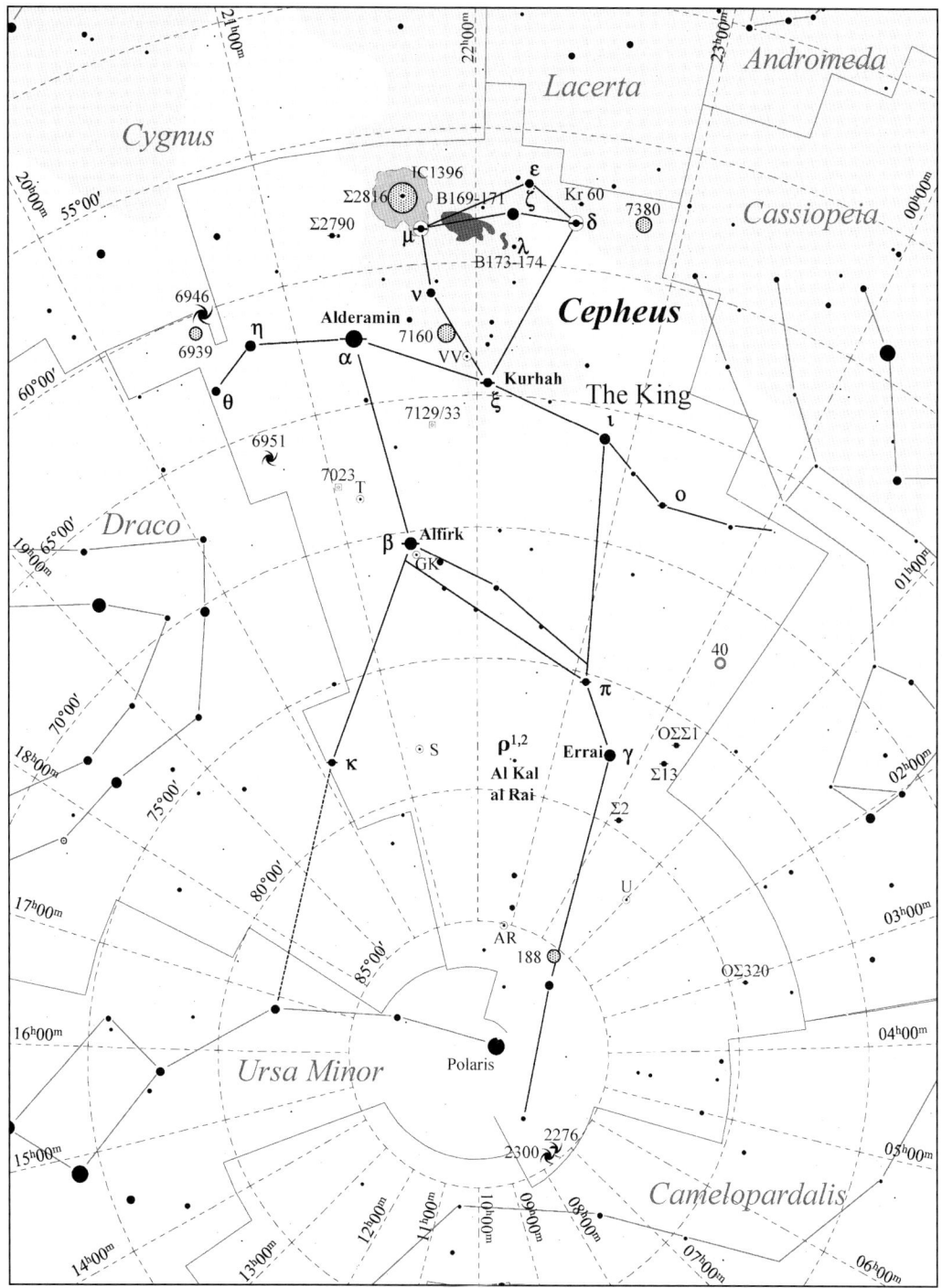

Fig. 3.22. Cepheus details.

Table 3.4. Selected telescopic objects in Cepheus.

Multiple stars:

Designation	R.A.	Decl.	Type	m₁	m₂	Sep. (")	PA (°)	Colors	Date/ Period	Aper. Recm.	Rating	Comments
1 = κ Cephei	20ʰ 08.9ᵐ	+77° 43'	A-B	4.4	8.3	7.2	120	W, B	2003	2/3"	****	Easy pair for small telescopes
5 = α Cephei	21ʰ 18.6ᵐ	+62° 35'	A-B	2.46	10.4	206.8	022	W, –	2000	4/6"	****	Alderamin. Interesting multiple star. The fainter components are chance alignments and have no real connection to the primary. α Cephei will be our pole star about 7500ᴀᴅ
			B-CD	10.4	10.4	19.9	172	–, –	2000	4/6"	***	
			C-D	10.9	11.1	2.6	104	–	2000	6/8"	**	
Σ2790	21ʰ 19.3ᵐ	+58° 37'	Aα-B	5.9	9.3	4.6	045	O, B	2002	2/3"	****	Beautiful colors
			Aα-D	5.9	10.1	74.5	352	O, –	2004	4/6"	***	
8 = β Cephei	21ʰ 28.7ᵐ	+70° 34'	A-B	3.2	8.6	13.2	248	bW, gW	1999	2/3"	****	Alfirk is a β CMa variable of small amplitude, period 0.19048 days
Σ2816	21ʰ 39.0ᵐ	+57° 29'	A-B	5.7	13.3	1.6	320	bW, –	1935	6/8"	*****	Spectacular multiple star, embedded in IC 1396 nebula. C component is also a wide double with a very faint companion. Σ2819, about 1 mag. fainter is in same LP FOV
			A-C	5.7	7.5	11.7	120	bW, –	2003	2/3"		
			A-D	5.7	7.5	19.7	339	bW, –	2003	2/3"		
17 = ξ Cephei	22ʰ 03.8ᵐ	+64° 38'	Aα-B	4.45	6.40	8.3	274	yW, Y	3,800	2/3"	****	Al Kurhah. Easy pair for small telescopes. A triple star system with a distant companion orbiting a close binary. a.k.a. Σ2863Aα-B and Aα-C
			Aα-C	4.5	12.7	95.8	097	yW, –	1925	–		
Kruger 60	22ʰ 28.0ᵐ	+57° 42'	A-B	9.8	11.3	1.9	026	R, R	44.67	6/8"	*****	A pair of red dwarf stars, only 12.89 ly away, 23rd closest star system. Secondary is a flare star
27 = δ Cephei	22ʰ 29.2ᵐ	+58° 25'	A-B	4.2	6.1	41.7	153	pO, B	2004	6/8"	****	Beautiful binocular double. In a small telescope, δ is similar to Albireo, but with much larger brightness difference
33 = π Cephei	23ʰ 07.9ᵐ	+75° 23'	A-B	4.6	6.6	1.2	359	Y, –	160	8/10"	****	The primary component is also a very close binary with a period of 567 days and a maximum separation of only 0.013". a.k.a. OΣ489AB
			AB-C	4.6	12.2	58.6	240	Y, –	2000	4/6"	***	
34 = o Cephei	23ʰ 18.6ᵐ	+68° 07'	A-B	4.9	7.1	3.3	222	O, B	1,540	3/4"	****	Lovely Albireo-like double for larger scopes. a.k.a. Σ3001AB
Σ2	00ʰ 09.3ᵐ	+79° 43'	A-B	6.01	6.13	0.86	016	–, –	540	10/12"	****	
OΣΣ1	00ʰ 14.0ᵐ	+76° 02'	A-B	7.6	7.9	73.6	103	R, –	2000	7×50	****	Easy optical pair for binoculars

Designation	R.A.	Decl.										Aper. Recm.	Rating	Comments
Σ13	00h 16.2m	+76° 57'	A–B	6.35	6.6	0.9	049	Y, W	1600			10/12"	****	
Σ320	03h 06.1m	+79° 25'	A–B	5.72	9.15	4.7	229	O, Y	–			2/3"	****	Nice pair for small telescopes

Variable stars:

Designation	R.A.	Decl.	Type	Range (m_j)		Period (days)	F (f_i/f_j)	Spectral Range	Aper. Recm.	Rating	Comments
T Cephei	21h 09.5m	68° 29'	M	5.2	11.3	388.14	0.54	M5.5e–M8.8e	3/4"	****	Circumpolar variable, visible year-round. 2°40' SW (PA 221°) of β Cep (m3.16).##
VV Cephei	21h 56.7m	63° 38'	EA+SRc	4.80	5.36	7,430	0.078	M2Ia–Iabep+B8Ve	Eye	*****	Ecl.Bin.+very long-period variable. 2°40' SE (PA 139°) of κ Cep (m4.38).#
GK Cephei	21h 31.0m	70° 49'	EB/KE	6.89	7.37	272	–	A2V+A2V	6×30	****	0°19' NE (PA 036°) of β Cep (m3.16).#
S Cephei	21h 35.2m	78° 37'	M	7.4	12.9	486.84	0.55	N8e (C7 4e)	6/8"	****	One of reddest stars visible in some telescopes. 2°31' W (PA 267°) of ρ Cep (m5.83).##
μ Cephei	21h 43.5m	53° 47'	SRc	4.43	5.10	3.76	–	M2ea	Eye	*****	Herschel's "Garnet Star," Red supergiant, luminosity may be 150K suns
27 = δ Cephei	22h 29.2m	58° 25'	Cδ	3.5	4.3	5.36	0.25	F5–G1	Eye	*****	The prototype of classical Cepheid variables.##
AR Cephei	22h 51.6m	85° 03'	SRb	7	7.9	–	–	M4III	6×30	****	#
U Cephei	01h 02.3m	81° 53'	EA/SD	6.6	9.8	2.493	0.15	B7Ve+G8III–IV	10×80	****	A large type G star occults a smaller and brighter type B8 star.##$ *18

Star clusters:

Designation	R.A.	Decl.	Type	m_v	Size (')	Brst. Star	Dist. (ly)	dia. (ly)	Number of Stars	Aper. Recm.	Rating	Comments
NGC 6939	20h 31.4m	60° 38'	Open	7.8	10.0	11.9	3,900	11	80	4/6"	****	NGC 6946 is in same FOV. Unusual cluster/galaxy combination
IC 1396	21h 39.1m	57° 30'	Open	3.5	50.0	3.82	2,600	38	50	2/3"	****	Visible to eye under good skies. Embedded in faint emission nebula
NGC 7160	21h 53.7m	62° 36'	Open	6.1	5	7.04	2,600	4	12	10×80	****	8/10" scope resolves into ~15 stars
NGC 7380	22h 47.0m	58° 06'	Open	7.2	20	10.0	9,700	57	40	12/14"	***	Triangular, embedded in faint nebulosity. A UHC filter helps greatly
NGC 188	00h 44.4m	85° 20'	Open	8.1	13	12.09	5,400	30	120	8/10"	***	Oldest known open cluster, estimates vary, *19 7–12+ billion years old!

(continued)

Table 3.4. (continued)

Nebulae:

Designation	R.A.	Decl.	Type	m$_v$	Size (')	Brtst./ Cent. star	Dist. (ly)	dia. (ly)	Aper. Recm.	Rating	Comments
C4, NGC 7023	21^h 01.2^m	68° 10'	Reflection	7.1	13	7.1	–	–	4/6"	****	One of the brightest and easiest to see refl. neb., much easier than Pleiades
IC 1396	21^h 39.1^m	57° 30'	Em+OC	3.5	50	–	–	–	2/3"	****	Extremely large, can be seen by unaided eye under *20 dark skies
NGC 7129/33	21^h 42.5^m	66° 10'	Reflection	11.5	7	–	–	–	12/14"	***	Nebula+cluster. NGC 7142 (OC) can also be seen in LP FOV
Barnard 171	22^h 03.5^m	58° 52'	Dark	–	19	–	–	–	12/14"	****	Irr., stands out well against milky way background. B169 and 170 are extensions of B171
Barnard 173	22^h 07.4^m	59° 10'	Dark	–	4	–	–	–	12/14"	****	Elongated "S" shape, mainly NE–SW orientation
IC 1470	23^h 05.2^m	60° 15'	Emission	8.9	15×1	12?	–	–	6/8"	****	Faint, may be a planetary nebula. Star at end enhances comet-like appearance
NGC 40	00^h 13.0^m	72° 32'	Planetary	12.4	37"	**11.6**	3K	0.5	12/14"	****	Spectacular in larger scopes. Conspicuous central star. Greenish oval

Galaxies:

Designation	R.A.	Decl.	Type	m$_v$	Size (')	Dist. (ly)	dia. (ly)	Lum. (suns)	Aper. Recm.	Rating	Comments
C12, NGC 6946	20^h 34.8^m	60° 09'	Sc	8.8	13.0×13.0	119M	175K	69G	4/6"	****	39' SE of NGC 6946, visible in same LP FOV. Dark sky necessary
NGC 6951	20^h 37.2^m	66° 06'	SAB	10.7	3.8×3.3	79M	125K	30G	12/14"	***	Diffuse oval halo, stellar nucleus in moderately bright core
NGC 2276	07^h 27.0^m	85° 45'	SAB	11.4	2.6×2.3	120M	98K	47G	12/14"	***	Faint round halo, little central brightening. NGC 2300 in same LP FOV
NGC 2300	07^h 32.0^m	85° 43'	SB	11.0	3.2×2.8	101M	133K	32G	12/14"	***	Small, but fairly bright, slightly elongated, stellar nucleus

Meteor showers:

Designation	R.A.	Decl.	Period (yr)	Duration	Max Date	ZHR (max)	Comet/ Asteroid	First Obs.	Vel. (mi/ km/sec)	Rating	Comments
None											

Other interesting objects:

Designation	R.A.	Decl.	Type	m$_v$	Size (')	Dist. (ly)	dia. (ly)	Lum. (suns)	Rating	Comments
None										

CETUS (see-tus)	The Whale (or Sea Monster)
CETI (set-ee)	BEST SEEN (at 9:00 P.M.): Oct. 15
Cet	AREA: 1231 sq. deg., 4th in size

TO FIND: Starting with the Great Square of Pegasus, follow the line from the northeastern corner star (Alpheratz) to the southeastern star (Algenib) southward for about 24° (little less than 1.5 times the distance between Alpheratz and Algenib) to ι Cet (m4.52, Schemali), the tip of the northern tail fluke of Cetus, the whale. Another 11° to the SE is the brighter Diphda (m2.27) which marks the tip of the southern tail fluke. 10° to the NE of Diphda is 4th magnitude η Ceti which marks the junction between the two flukes. About halfway from η Cet to α Tauri (1st magnitude Aldebaran) but slightly closer to Aldebaran lies α Ceti (Menkar) which marks the whale's lower jaw. Just 5° N of Menkar is the fainter λ Ceti which marks the whale's nose. Using these four stars to mark the ends of the whale's figure and the chart showing the location of the remaining stars, which are all 4th magnitude or fainter, the central part of the whale's figure can be traced out (Figs. 3.23 and 3.24; Table 3.5).

KEY STAR NAMES:

	Name	Pronunciation	Source
α	Menkar	**men**-car or **men**-cur	Al-minkhar (Arabic), the nostrils, wrongly applied to α
β	Deneb Kaitos	**deh**-neb **kye**-tos	dhanab qaitus al janubi (Arabic), the southern branch of the sea monster's tail
	or		
	Diphda	**dif**-duh	Difdiᶜ al-thānī (Ind-Arabic) – the Second Frog (The first frog was Fomalhaut)
ζ	Baten Kaitos	**boo**-tuhn **kye**-tos or **bay**-tuhn **kay**-tos	batn qaitus(Arabic), the sea monster's belly
ι	Deneb Kaitos al Schemali	**deh**-neb **kay**-tos al **shuh**-mal-ih	Dhanab qaitus al-Shamaliyy (Arabic), the northern branch of the sea monster's tail
o	Mira	**me**-ruh or **my**-ruh	Historiola Mirae Stella (Latin), "historical account of the wonderful star" (Hevelius 1662)

Magnitudes and spectral types of principal stars:

Bayer desig.	Mag. (m_v)	Spec. type	Bayer desig.	Mag. (m_v)	Spec. type	Bayer desig.	Mag. (m_v)	Spec. type
α	2.53	M2 III	β	2.04	K0 III	γ	3.47	A5 V
δ	4.07	B2 IV	ϵ	5.58	F8 V	ζ	3.73	K0 III
η	3.45	K1.5 III	θ	3.60	K0 IIIb	ι	3.56	K1.5 III
κ	4.99	G5 V	λ	4.70	B6 III	μ	4.27	F0 IV
ν	4.86	G8 III	ξ^1	4.37	G6 II/III	ξ^2	4.28	B9 III
o	3.04	M7 IIIe	π	4.25	B7 IV	ρ	4.89	B9.5 Vn
σ	4.75v	M3.5 III	τ	3.50	G8 Vp	υ	4.00	M0.5 III
ϕ^1	4.76	K0 III	ϕ^2	5.19	F7 IV/V	ϕ^3	5.31	K4 III
χ	4.67	F3 III	ψ	–	–	ω	–	–

Stars of magnitude 5.5 or brighter which have Flamsteed, but no Bayer designations:

Flam-steed	Mag. (m_v)	Spec. type	Flam-steed	Mag. (m_v)	Spec. type	Flam-steed	Mag. (m_v)	Spec. type
2	4.55	B9 IVn	3	4.99	K3 Ibvar	7	4.44	M1 III
13	5.20	F8 V	20	4.78	M0 III	25	5.40	K0 III/IV
37	5.14	F5 V	39	5.42	G5 III/IVe	46	4.90	K2 III
48	5.11	A0 V	50	5.41	K1 III	56	4.92	K3 III
57	5.43	M/0/1 III	60	5.41	A5 III	69	5.29	M2 III
70	5.37	K4 III	75	5.36	G3 III	94	4.07	F8 V

Stars of magnitude 5.5 or brighter which have no Bayer or Flamsteed designations:

Desig-nation	Mag. (m_v)	Spec. type	Desig-nation	Mag. (m_v)	Spec. type
HD 000787	5.29	K4 III	HD 002696	5.17	A3 V
HD 004247	5.22	V0 V	HD 004398	5.49	G8/K0 III
HD 005098	5.47	K1 III	HD 010550	4.98	K3 II/III

Fig. 3.23a. The stars of Cetus and adjacent constellations.

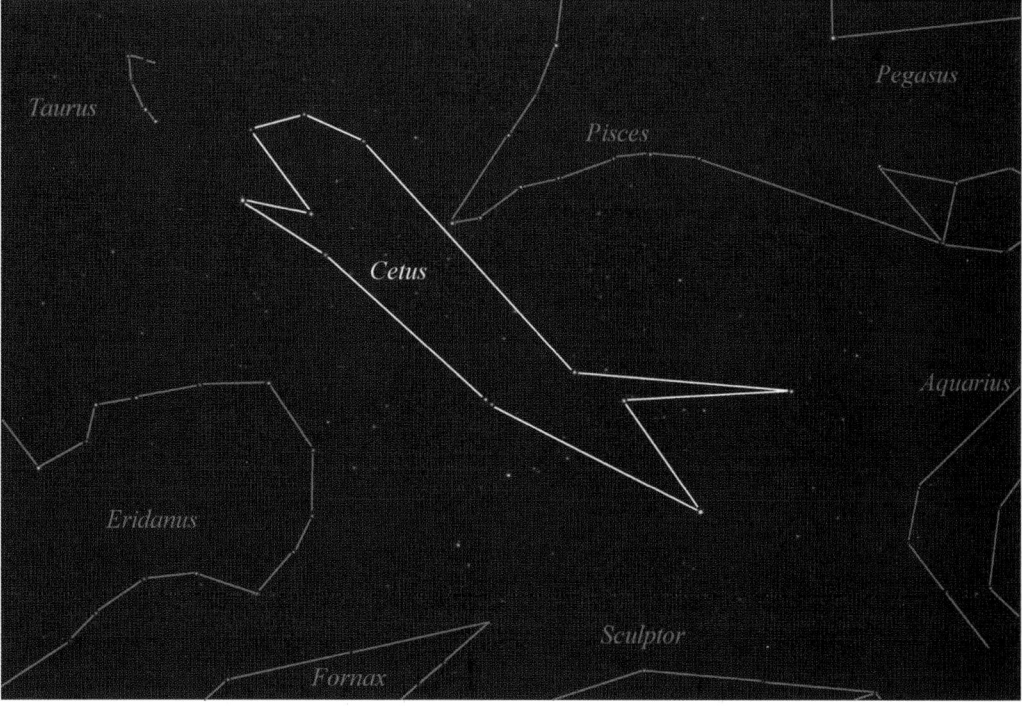

Fig. 3.23b. The figures of and adjacent constellations.

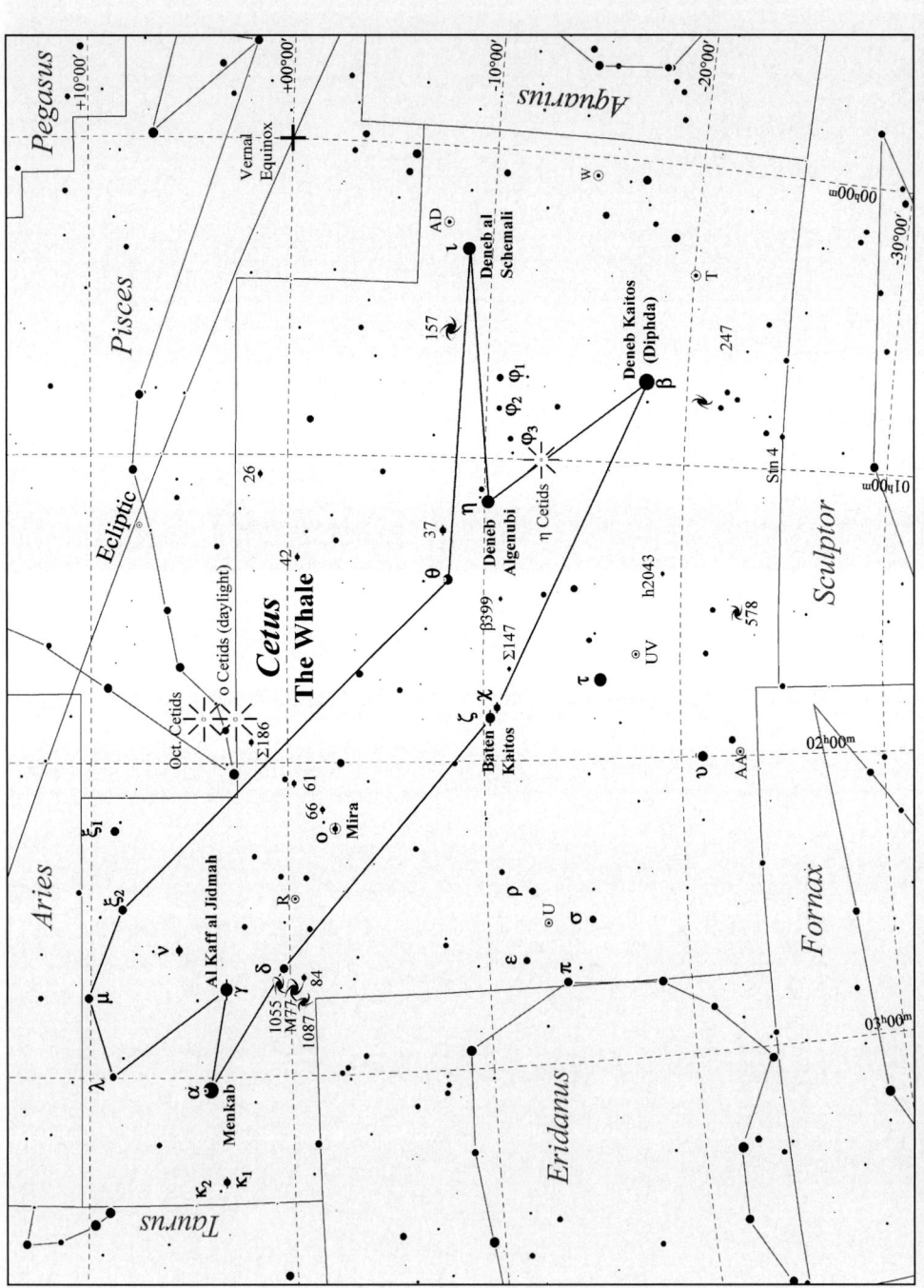

Fig. 3.24. Cetus details.

Table 3.5. Selected telescopic objects in Cetus.

Multiple stars:

Designation	R.A.	Decl.	Type	m_1	m_2	Sep. ('')	PA (°)	Colors	Date/Period	Aper. Recm.	Rating	Comments
Stone 4	00ʰ 53.2ᵐ	−24° 47'	A–B	6.6	9.2	5.4	007	yW, ?	2002	2/3''	***	
Σ84, 26 Ceti	01ʰ 03.8ᵐ	+01° 22'	A–B	6.1	9.5	16.0	253	Y, B	2003	4/6''	****	Nice color contrast
			A–C	6.1	12.7	107.2	290	Y, ?	2000	4/6''	***	
37 Ceti	01ʰ 14.4ᵐ	−07° 55'	A–B	5.2	7.9	48.4	331	yW, dY	2003	2/3''	****	Easy pair for small scopes, may be seen in 7×50 by some observers
42 Ceti	01ʰ 19.8ᵐ	−00° 31'	A–BC	6.5	7.0	1.3	331	yW, ?	2003	6/8''	****	With a separation of only 0.1'', B–C is beyond amateur scopes
h2043	01ʰ 22.5ᵐ	−19° 05'	A–B	6.5	8.7	5.2	073	yW, ?	2003	2/3''	****	Easy pair for small scopes
β399	01ʰ 27.8ᵐ	−10° 54'	A–B	6.4	8.8	1.5	303	O, ?	1991	6/8''	****	
Σ147	01ʰ 41.7ᵐ	−11° 19'	A–B	6.1	7.2	0.5	093	yW, yW	2002	18/20''	****	Separation and PA have changed, but is not known to be a binary system
53, χ Ceti	01ʰ 49.6ᵐ	−10° 41'	A–B	4.7	6.8	183.9	251	yW, ?	2001	6×30	****	
55, ζ	01ʰ 51.5ᵐ	−10° 20'	Aa–B	3.9	10.2	188.6	041	O, ?	2002	3/4''	***	With a separation of only 0.1'', A–a is beyond amateur scopes
Σ186	01ʰ 55.9ᵐ	+01° 51'	A–B	6.8	6.8	**0.8**	**070**	Y, Y	**162.0206**	12/14''	****	Close, but lovely matched pair. Binary, data calculated for 2010.0
61 Ceti	02ʰ 03.8ᵐ	−00° 20'	A–B	6.0	10.8	43.0	194	Y, ?	1998	4/6''	***	
			A–C	6.0	11.9	90.9	325	Y, ?	1998	3/4''	****	
Σ231, 66 Ceti	02ʰ 12.8ᵐ	−02° 24'	A–a	5.7	7.7	16.4	234	yW, Y	2003	15×80	****	Separable in 2/3'' scope, beautiful in 8/10''
			A–B	5.7	11.5	172.7	61	yW, ?	1908	15×80	***	
Mira, 68, o Ceti	02ʰ 19.3ᵐ	−02° 59'	A–a	Var	10.4	0.6	110	R, ?	1992	14/16''	***	Primary is a famous variable. See data in variable star section for details
			Aa–B	Var	13.0	73.1	85	R, ?	1911	6/8''	***	
			Aa–C	Var	9.6	122.0	69	R, ?	2003	3/4''	***	
			Aa–D	Var	9.3	148.5	319	R, ?	1921	2/3''	***	
78, ν Ceti	02ʰ 35.ᵐ	+05° 36'	A–B	5.0	9.1	7.9	079	Y, W	2000	2/3''	****	
84 Ceti	02ʰ 41.2ᵐ	−00° 42'	A–B	5.8	9.7	4.6	308	Y, R	1993	3/4''	****	Lovely color in 8/10'' scope
86, γ Ceti	02ʰ 43.3ᵐ	+03° 14'	A–B	3.6	6.2	2.8	294	W, Y	2002	3/4''	****	Faintest star is a red dwarf, and part of the system, but is at least 18,000 AU (0.28 ly) from the main pair
			A–C	3.6	10.2	840	?	W, R	1923	3/4''	****	
95, κ¹ Ceti	03ʰ 18.4ᵐ	−00° 56'	A–B	5.6	8.0	**1.2**	**257**	Y, ?	**282.42**	8/10''	****	Data calculated for 2010

(continued)

Table 3.5. (continued)

Variable stars:

Designation	R.A.	Decl.	Type	Range (m_v)	Period (days)	F (f_r/f_t)	Spectral Range	Aper. Recm.	Rating	Comments
W Ceti	00ʰ 02.1ᵐ	−14° 41'	M	7.1 14.8	351.31	–	S6,3e–S9.2e	12/14"	***	##
AD Ceti	00ʰ 14.5ᵐ	−07° 47'	Lb	4.9 5.2	–	–	M3III	Eye	*****	
T Ceti	00ʰ 21.8ᵐ	−20° 03'	SRc	5.0 6.9	158.9	–	M5–M6SIIe	7×50	****	~20,000 Suns Lum., 5,500 ly distant.##
UV Ceti	01ʰ 38.8ᵐ	−17° 58'	UV	7.0 13.0	–	–	dM 5.5e	8/10"	****	Prototype of dwarf flare stars with outbursts of 1–6 mag.##
AA Ceti	01ʰ 59.0ᵐ	−22° 55'	EW	6.2 **6.7**	0.54	–	F2	7×50	****	Eclipsing variable, two sun-like stars very close together.#
Mira, 68, o Ceti	02ʰ 19.3ᵐ	−02° 59'	M	2.0 10.1	331.96	0.38	M5e–M9e	3/4"	*****	"The Wonderful." Prototype of long-period pulsating variables.##$ *21
R Ceti	02ʰ 26.0ᵐ	−00° 11'	M	7.2 14.0	166.24	0.43	M4e–M9	11/12"	****	##
U Ceti	02ʰ 33.7ᵐ	−13° 09'	M	6.8 13.4	234.76	0.44	M2e–M6e	8/10"	****	##

Star clusters:

Designation	R.A.	Decl.	Type	Size (')	Brst. Star m_v	Dist. (ly)	dia. (ly)	Number of Stars	Aper. Recm.	Rating	Comments
None											

Nebulae:

Designation	R.A.	Decl.	Type	m_v	Size (')	Brst/ Cent. star	Dist. (ly)	dia. (ly)	Aper. Recm.	Rating	Comments
NGC 246	00ʰ 47.0ᵐ	−11° 53'	Planetary	8.0	4.0×3.5	**11.95**	1,300	–	8/10"	****	Can be seen as a fuzzy spot in 11×80 binoculars. Estimated age ~15,000 years

Galaxies:

Designation	R.A.	Decl.	Type	m_v	Size (')	Dist. (ly)	dia. (ly)	Lum. (suns)	Aper. Recm.	Rating	Comments
NGC 157	00ʰ 34.8ᵐ	−08° 24'	Gal, SAB	9.2	19.0×5.5	21M	56K	34G	4/6"	****	Faint, very elongated
NGC 247	00ʰ 47.1ᵐ	−20° 46'	Gal, SAB	9.0	12.5×6.5	7M	35K	2.4G	4/6"	******	Concentrated core, bright nucleus. Can be glimpsed in 7×50 binoculars

NGC 578	01ʰ 30.5ᵐ	−22° 40′	Gal, SB	11.0	3.9×2.2	76M	24K	2.5G	8/10″	****	Faint, elongated	
NGC 1055	02ʰ 42.8ᵐ	+00° 26′	Gal, SB	10.6	2.3×7.3	47M	73K	7.5G	8/10″	***	Edge-on, resembles a smaller, fainter M104 (Sombrero Galaxy)	
M77, NGC 1068	02ʰ 42.7ᵐ	−00° 01′	Gal, SB	8.9	8.2×7.3	52M	108K	56G	4/6″	****	Can be glimpsed in 15×80 binoculars as a faint spot	*22
NGC 1087	02ʰ 46.4ᵐ	−00° 30′	Gal, SB	10.9	3.7×2.8	41M	73K	20G	8/10″	****	Amorphous ellipse, stellar-like nucleus visible in larger scopes	

Meteor showers:

Designation	R.A.	Decl.	Period (yr)	Duration	Max Date	ZHR (max)	Comet/ **Asteroid**	First Obs.	Vel. (mi/**km**/sec)	Rating	Comments	
η Cetids	01ʰ 00ᵐ	−13°	–	Sep. 20 – Nov. 2	Oct. 1/7	V. Low	–	1964	–	Minor	Very few meteors, but has produced several fireballs	*23
October Cetids	01ʰ 52ᵐ	+4°	Annual	Oct. 20 – Oct. 25	Oct 22/23?	Low	–	1916	Rapid	Minor	Apparently consists of two related streams, a northern and a southern	*24
Omicron Cetids	01ʰ 52ᵐ	−3°	Annual	May 7 – June 09	May 9	–	–	1950	23/**37**	Daylight	Detected by radio measurements by A. Aspinall and Gerald Hawkins	

Other interesting objects:

Designation	R.A.	Decl.	Type	m_v	Spec. Class	Mass (suns)	Dist. (ly)	dia. (suns)	Lum. (suns)	Aper. Recm.	Rating	Comments	
L 726-8, UV Ceti	01ʰ 38.8ᵐ	−17° 58ᵐ	Flare star	11.4 / 11.9	M5.5V / M6V	0.1 / 0.1	9.0	0.14 / 0.14	0.00006 / 0.00004	3/4″	****	Sixth closest star to the sun. Very small red dwarf binary	*25
52, τ Ceti	01ʰ 44.0ᵐ	−15° 56′	Nearby	3.5	G8 Vp	0.81	11.9	0.816	0.45	Eye	*****	Seventh closest naked eye star	*26
Menkar, 86, α Ceti	03ʰ 02.3ᵐ	+04° 05′	Col. star	2.54	M2 III	3.0	220	84	384	7×50	****	A cool orange-red giant	*27

Desig-nation	Mag. (m$_v$)	Spec. type	Desig-nation	Mag. (m$_v$)	Spec. type
HD 010824	5.37	K4 III	HD 012292	5.43	M3 III
HD 014691	5.43	F0 V	HD 015694	5.27	K3 III

PEGASUS (peg-ah-sus)	**The Winged Horse**
PEGASI (peg-uh-see)	**BEST SEEN** (at 9:00 P.M.): Oct. 17
Peg	**AREA:** 1,121 sq. deg., 7th in size

TO FIND: The star α Andromedae (Alpheratz) lies almost exactly 30° due south of β Cassopeiae, the eastern-most star in the "W" of Cassiopeia's throne. Alpheratz marks the northwest corner of the Great Square of Pegasus. Although it is now formally part of Andromeda, Apheratz was formerly considered to be part of both constellations. Respectively, β Pegasi (Sheat), α Pegasi (Markab), and γ Pegasi (Algenib) form the NW, SW, and SE corners of the Great Square. This asterism is fairly regular, with the four sides ranging from 12.9° to 16.5°, with each side running approximately E–W or N–S. It also lies in a relatively sparse region of the sky, and has three 2nd magnitude and one brighter than average (γ Peg, m2.83) 3rd magnitude star, making it easy to see. The square represents the front portion of the horse's body, and two rows of 3rd and 4th magnitude stars extending westward from β Pegasi (Sheat) and μ Peg (about 4.6° SW of β) form the horse's forelegs. A similar line of stars extends first SW and then NW from α Andromedae (Alpheratz) to form the horse's neck and head. As is the case with other constellations (e.g., Taurus), early observers visualized the other part of the horse's body to be hidden behind other heavenly figures. The form of Pegasus is easier to see if you lie on your back facing north so the figure appears upright (Figs. 3.25–3.26; Table 3.6).

KEY STAR NAMES:

Name	Pronunciation	Source
α Markab	**mur**-kub	mankib al-faras (Arabic), "The horse's shoulder," for β Peg, wrongly applied to α on medieval charts
β Sheat	**Shea**-at	al-saq (Arabic), "the shin," from the Arabic *Almagest* description of its position
γ Algenib	al-**je**-nib or uhl-**je**-nib	al-janab (Arabic), "the side," originally applied to α Per and incorrectly applied to γ Peg on medieval charts
ε Enif	**eh**-nif	Probably from al-anf (Arabic) "the nose"
ζ Homam	hoe-**mam**	sa'd al-humam) (Ind-Arabic), originally applied to both ζ and ξ Peg. Origin is uncertain, but may mean "the lucky stars of the Hero"
η Matar	**mah**-tar	sa'd matar (Ind-Arabic). Origin is uncertain. Al-matar means "rain" as a common Arabic noun, but without the "al," the meaning is uncertain
θ Biham	bih-**ham**, **bye**-am	sa'd al bihām (Arabic), possibly "the lucky stars of the young beasts," originally applied to both θ and ν, but this derivation is uncertain

Magnitudes and spectral types of principal stars:

Bayer desig.	Mag. (m_v)	Spec. type	Bayer desig.	Mag. (m_v)	Spec. type	Bayer desig.	Mag. (m_v)	Spec. type
α	2.49	B9.5 III	β	2.44	M2 II/III	γ	2.83	B2 IV
δ	–	–*28	ε	2.38	K2 Ib	ζ	3.41	B8.5 V
η	2.93	G2 II/III	θ	3.52	A2 V	ι	3.77	F5 V
κ	4.14	F5 IV	λ	3.97	G8 II/III	μ	3.51	M2 III
ν	4.86	K4 III	ξ	4.20	F7 V	o	4.80	A1 V
π¹	5.58	G6 III	π²	4.28	F5 III	ρ	4.91	A1 V
σ	5.16	F7 IV	τ	4.58	A5 V	υ	4.42	F8 IV
φ	5.06	M2 III	χ	4.79	M2 III	ψ	4.63	M3 III
ω	–	–						

Stars of magnitude 5.5 or brighter which have Flamsteed, but no Bayer designations:

Flamsteed	Mag. (m_v)	Spec. type	Flamsteed	Mag. (m_v)	Spec. type	Flamsteed	Mag. (m_v)	Spec. type
1	4.08	K1 III	2	4.52	M1 III	5	5.46	F1 IV
7	5.30	M2 III	12	5.29	K0 Ib	13	5.34	F2 III/IV
14	5.07	A2 Vs	16	5.09	B3 V	30	5.37	B5 IV
31	4.82	B2 IV/V	32	4.78	B9 III	35	4.78	K0 III
51	5.45	G5 V	55	4.54	M2 III	56	4.76	K0 Iip
57	5.05	M4 Sv	58	5.39	B9 III	59	5.15	A5 Vn
64	5.35	B6 III	66	5.09	K3 III	70	4.54	G8 III
71	5.33	M4 IIIa	72	4.97	K4 III	75	5.49	A1 Vn
77	5.09	M2 III	78	4.93	K0 III	82	5.3	A4 Vn

Stars of magnitude 5.5 or brighter which have no Bayer or Flamsteed designations:

Designation	Mag. (m_v)	Spec. Type
HD 210895	5.34	K2 III

Fig. 3.25a. The stars of Pegasus and Equuleus.

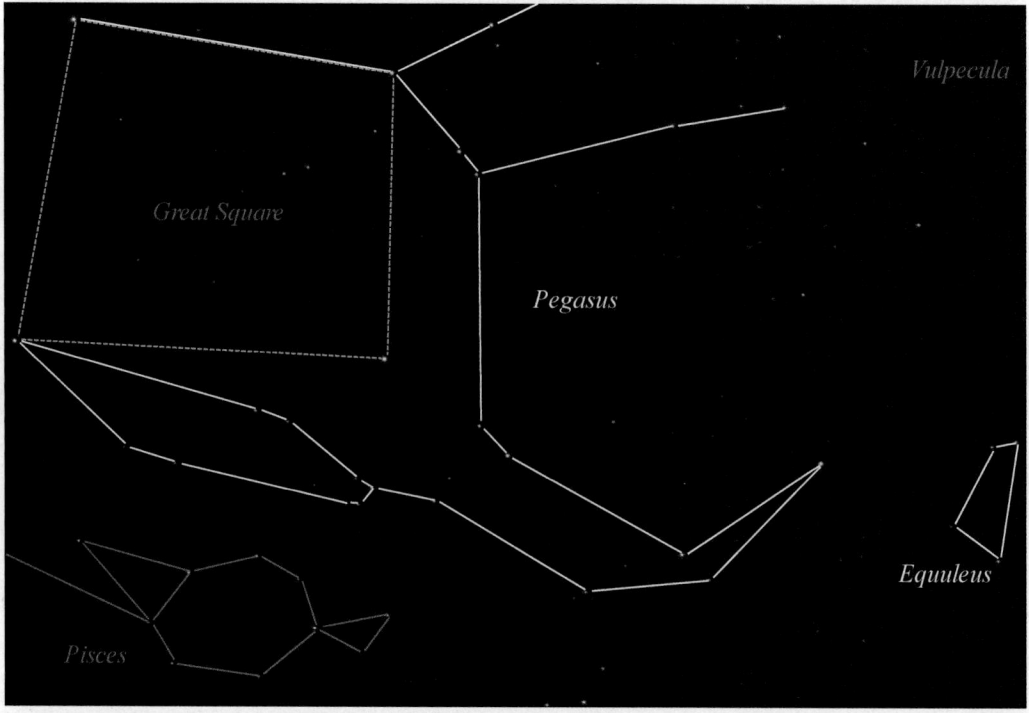

Fig. 3.25b. The figures of Pegasus and Equuleus.

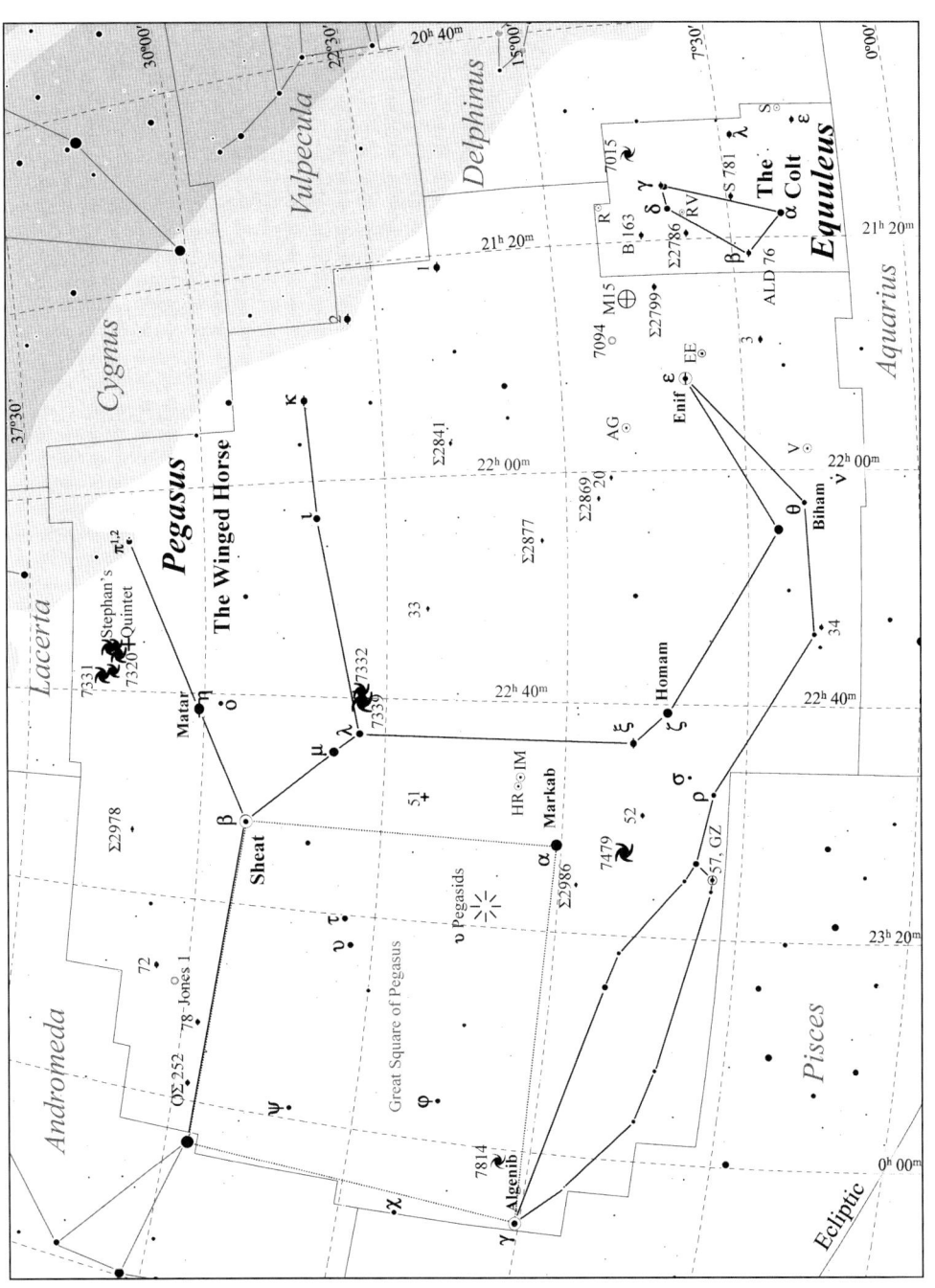

Fig. 3.26. Pegasus and Equuleus details.

Table 3.6. Selected telescopic objects in Pegasus.

Multiple stars:

Designation	R.A.	Decl.	Type	m_1	m_2	Sep. (")	PA (°)	Colors	Date/Period	Aper. Recm.	Rating	Comments
1 Pegasi	21h 22.1m	+19° 48'	A–B	4.20	7.56	36.1	312	O, B	2005	7×50	****	A 12th mag. companion lies 74.3" away at 019°
Σ2799	21h 28.9m	+11° 05'	A–B	7.37	7.44	4.7	259	oY, oY	618.89	4/6"	*****	A nearly identical pair. Lovely
2 Pegasi	21h 29.0m	+22° 11'	A–B	6.08	10.22	40.4	177	rO, B	2004	11×80	*****	Beautiful color contrast!
			A–C	6.08	10.70	40.3	129	rO, ?	1966	3/4"	****	
3 Pegasi	21h 37.7m	+06° 37'	A–B	6.18	7.50	34.7	348	yW, W	2004	7×50	****	A 3rd component is of mag. 13.5 visible in larger scopes
8, Pegasi	21h 44.2m	+09° 53'	A–B	2.53	12.66	82.6	323	pY, B	2000	4/6"	****	Color of B is hard to see because of brightness difference
			A–C	2.53	8.74	144	316	pY, B	2000	7×50	****	
10, κ Pegasi	21h 44.6m	+25° 39'	AB–C	4.13	10.8	14.5	288	dY or O, ?	2004	3/4"	****	A–B is too close for amateur instruments
Σ2841	21h 54.3m	+19° 43'	A–BC	6.45	7.99	22.3	110	oY, dB	2004	11×80	*****	Some report companion as "greenish." B–C is too close for amateur instruments
20 Pegasi	22h 01.1m	+13° 07'	A–B	5.60	11.1	59.1	323	yW, B	1999	3/4"	****	Subtle but definite color contrast
Σ2869	22h 10.4m	+14° 38'	A–B	6.33	12.4	21.3	252	pY, B	2001	3/4"	****	Color of B is hard to see because of brightness difference
Σ2877	22h 14.3m	+17° 11'	A–B	6.65	9.23	22.2	022	dY, pB	2003	11×80	****	Two more companions mag. 11 and 12 are about 100" away
33 Pegasi	22h 23.7m	+20° 51'	A–B	6.39	9.29	0.4	001	Y, ?	1995	20/24"	***	Challenge for large telescopes
			AB–CD	6.28	8.54	89.9	308	Y, pY	2003	7×50	****	CD is too close for amateur instruments
34 Pegasi	22h 26.6m	+04° 13'	A–B	5.8	12.5	1.3	039	Y, ?	420	6/8"	****	Color of secondary is hard to see because of brightness difference
44, η Pegasi	22h 43.0m	+30° 13'	Aa–BC	3.02	9.87	93.8	339	pY, pB	2001	11×80	****	A–a, B–C, and D–E pairs are too difficult for amateur instruments *29
46, ξ Pegasi	22h 46.7m	+12° 10'	A–B	4.19	12.3	11.2	097	pY, pB	1998	4/6"	****	Color of B is hard to see because of brightness difference
			A–C	4.19	11.1	145.0	33	pY, ?	1924	4/6"	****	
52 Pegasi	22h 59.2m	+11° 44'	A–B	6.11	7.27	0.34	021	pY, pY	216	28/30"	***	Closing to 0.28" in 2016, then opening. Challenging for the largest amateur scopes

Σ2978	23h 07.5m	+32° 50'	A–B	6.35	7.46	8.5	145	Y, pB	2006	2/3'	****	Lovely colors
57 Pegasi	23h 09.5m	+08° 41'	A–B	5.12	9.7	32.6	198	O, pB	2004	11×80	*****	Nice color contrast
Σ2986	23h 10.0m	+14° 26'	A–B	6.61	8.88	31.6	270	Y, pB	2004	11×80	*****	Pretty colors
72 Pegasi	23h 34.0m	+31° 20'	A–B	5.7	6.11	**0.53**	**104**	O, O	**246.17**	18/20"	****	Closing slowly
78 Pegasi	23h 44.0m	+29° 22'	A–B	5.07	8.10	**0.88**	**277**	dY, ?	**630.15**	10/12"	****	Opening slowly
OΣ252	23h 54.9m	+29° 29'	A–B	6.77	8.37	111.1	145	pY, oR	1999	7×50	*****	Lovely color contrast

Variable stars:

Designation	R.A.	Decl.	Type	Range (m$_v$)		Period (days)	F (f$_v$/f$_t$)	Spectral Range	Aper. Recm.	Rating	Comments
EE Pegasi	21h 40.0m	+09° 11'	EA/DM	6.9	7.5	2.6	0.1	A3mV+F5	6×30	****	°1 14' SW (PA 236°) of ζ Pegasi
ε, Pegasi	21h 44.5m	+09° 53'	Lc	0.7	3.5	–	–	K2Ib	Eye	*****	Enif
AG Pegasi	21h 51.0m	+12° 38'	Nc or	6.0	9.4	–	–	WN6 +M3III	7×50	*****	A symbiotic variable or slow nova *30
V Pegasi	22h 01.0m	+06° 07'	M	7.0	15	302.35	0.44	M3e–M6e	0.9	****	~20–24" may be required to make reliable estimates at some minima
IM Pegasi	22h 53.0m	+16° 50'	RS	5.6	5.85	24.44	–	K1-1, II–IIIe	6×30	****	23' WSW (PA 255°) of HR Pegasi, q.v.
HR Pegasi	22h 54.6m	+16° 57'	SRb	6.1	6.49	50	–	S5d,1 (M4)	6×30	****	3°0' NW (PA 306°) of α Pegasi (Markab)
β Pegasi	23h 04.8m	+28° 05'	Lb	2.3	2.7	–	–	M2.5II–IIIe	Eye	*****	Bright, easy to find variable
GZ Pegasi	23h 09.5m	+08° 41'	SRa	5.0	5.2	92.66	–	M4sIII	Eye	****	58'SE (PA 140°) of 55 Pegasi (m4.95)
γ Pegasi	00h 13.2m	+15° 11'	β Cep	2.8	2.9	0.152	0.5	B2IV	Eye	****	Bright, easy to find Cepheid, but small range

Star clusters:

Designation	R.A.	Decl.	Type	m$_v$	Size (')	Brtst. Star	Dist. (ly)	dia. (ly)	Number of Stars	Aper. Recm.	Rating	Comments
M15, NGC 7078	21h 30.0m	+12° 10'	Globular	6.3	18	12.6	50K	600K		7×50	*****	Showcase piece. Many stars, very bright and large *31

(continued)

Table 3.6. (continued)

Nebulae:

Designation	R.A.	Decl.	Type	m_v	Size (')	Brst./ Cent. star	Dist. (ly)	Color	Aper. Recm.	Rating	Comments
NGC7094	21ʰ 36.9ᵐ	+12° 48'	Planetary	13.4	1.6	**13.73**	—	Greenish	12/14"	**	Round ring, varying in brightness and width. 1.8° ENE (PA 070°) of M15
Jn (Jones) 1	23ʰ 35.0ᵐ	+30° 28'	Planetary	12.1	5.5	**16.1**	—	Greenish	14/16"	**	Large, low surface brightness, only a trace of a ring, very difficult object. a.k.a. PK 104-29.1

Galaxies:

Designation	R.A.	Decl.	Type	m_v	Size (')	Dist. (ly)	dia. (ly)	Lum. (suns)	Aper. Recm.	Rating	Comments
NGC 7317, 7318A	22ʰ 36ᵐ	+33° 58'	SB	11.1	0.9×0.9	—	—	—	10"	*****	Stephan's Quintet. 31' SSW of NGC 7331. Radial velocity measurements
7318B, 7319, and 7320	22ʰ 36ᵐ	+33° 58'	SB	13.5	1.7×1.2	—	—	—	10"	*****	imply different distances, but streamers seem to imply close connection
NGC 7331	22ʰ 37.1ᵐ	+34° 25'	SB	9.5	10.5×3.7	47M	120K	43G	4"	*****	Bright, highly elongated, 12" or larger scopes show 4 faint companions *32
NGC 7332	22ʰ 37.4ᵐ	+23° 48'	SB	11.1	3.7×1.0	59M	5K	10G	5"	****	Elongated NNW-SSE. 7339 is only 5' to W

	R.A.	Decl.	Type			Dia. (suns)	Dist. (ly)	Planet Mass (J)	Dist. (AU)	Period (days)	Eccentricity	Comments

| | 22ʰ 37.8ᵐ | +23° 47′ | SB | 11.1 | 3.7×1.0 | 63M | 40K | 7.1G | 6″ | | | *** | In same LP FOV as 7332 |

NGC 7339 22ʰ 37.8ᵐ +23° 47′ SB 11.1 3.7×1.0 63M 40K 7.1G 6″ *** In same LP FOV as 7332

NGC 7479 23ʰ 04.9ᵐ +12° 19′ SB 10.8 4.0×3.1 106M 117K 44G 4″ **** Very open barred spiral galaxy. Classic "S" shape

NGC 7814 00ʰ 03.3ᵐ +16° 09′ SB 10.6 6.0×2.5 49M 73K 15G 3″ **** Precisely edge-on galaxy, narrow dust lane seen in larger scopes

Meteor showers:

Designation	R.A.	Decl.	Period (yr)	Duration	Max Date	ZHR (max)	Comet/ **Asteroid**	First Obs.	Vel. (mi/ **km**/sec)	Rating	Comments
υ Pegasids	23ʰ 16ᵐ	+18°	Annual	July 25	Aug. 19 Aug. 8	2–5	–	1798	–	Minor	First observed by Heinrich Brandes on June 12, 1798 *33

Other interesting objects:

Designation	R.A.	Decl.	Type	mᵥ	Mass (suns)	Dia. (suns)	Dist. (ly)	Planet Mass (J)	Dist. (AU)	Period (days)	Eccentricity	Comments
51 Pegasi	22ʰ 57.5ᵐ	+20° 46′	G5V	5.45	1.09	1.3	50	0.472	0.0527	4.2308	0.013	First sun-like star found to have a planet. About 1.3× sun's luminosity

Fig. 3.27. M15 (NGC 7078), globular cluster in Pegasus. Astrophoto by Mark Komsa.

Fig. 3.28. Stephan's Quintet, NGC 7317, 7318A, 7318B, 7319, and 7320. Drawing by Nikolas Biver.

Fig. 3.29. NGC 7331 an edge-on galaxy in Pegasus.

EQUULEUS (e-**kwoo**-le-us)	**The Colt or Little Horse**
EQUULEI (ek-woo-**oo**-lay-ee)	**BEST SEEN** (at 9:00 **P.M.**): Sep. 23
Equ	**AREA:** 71.6 sq. deg., 87th in size

Equuleus is an extremely small constellation, larger than only Crux. It was probably first described by Hipparchus, although some researchers think that it was added by Ptolemy. Equuleus contains only a few faint stars, so it received little attention from ancient astronomers and has little detail in the few myths associated with it.

In one, Celeris was either the brother or the son of Pegasus. A pure white horse, Celeris was known for his speed (celer is Latin for swift). Celeris was a gift to Castor from Hermes, and a quite similarly named horse, Cyllarus, was given to Pollux by Hera.

TO FIND: Although faint, the quadrilateral of Equuleus is fairly easy to find, since its four main stars lie only about 7° west of ε Peg (Enif). In fact, δ Equ (m4.47) is 7°19' due west of Enif. Kitalpha (α Equ, m3.92) is 4°46' south (PA 272°) of δ Equ. β Equ (m5.16) is 2°21' from α Equ back in the direction of Enif (PA 048°). The line from Enif to δ Equ extended by only 1° ends just a few minutes south of γ Equ (m4.70). The quadrilateral marks the head of Equuleus, extending in the same direction as the head of big brother Pegasus. The rest of the little horse's neck and body are hidden behind Pegasus. The head of Equuleus is so small that the entire figure can easily be covered by your fist held at arm's length (Figs. 3.25 and 3.26; Table 3.7).

KEY STAR NAMES:

Name	Pronunciation	Source
α Kitalpha	(kih-**tal**-fah)	qit'at al-faras (Arabic), "the section of a horse"

Magnitudes and spectral types of principal stars:

Bayer desig.	Mag. (m_v)	Spec. type	Bayer desig.	Mag. (m_v)	Spec. type	Bayer desig.	Mag. (m_v)	Spec. type
α	3.92	G0 III	β	5.16	A3 V	γ	4.69	F0p
δ	4.49	F5 V	ε	5.23	F6 V	λ	6.72	F8 V[*34]

Table 3.7. Selected telescopic objects in Equuleus.

Multiple stars:

Designation	R.A.	Decl.	Type	m_1	m_2	Sep. (")	PA (°)	Colors	Date/ Period	Aper. Recm.	Rating	Comments
1, ε Equulei	20ʰ 59.0ᵐ	+04° 18'	A–B	6.0	6.3	**0.5**	**284**	pY, pY	**101.485**	18/20"	***	Presently closing, will begin opening again in 2019. See orbital plot for details
			AB–C	5.4	7.1	10.9	067	pY, B	2005	2/3"	****	Beautiful quartet with nice range of separations
			A–D	5.4	12.4	78.6	286	pY, ?	2000	4/6"	****	630 to 1 brightness difference significantly increases difficulty
2, λ Equulei	21ʰ 02.2ᵐ	+07° 11'	A–C	7.4	7.6	2.8	216	Y, Y	2005	3/4"	*****	Nearly matched pair. Ideal for test of small aperture scopes. Beautiful
5, γ Equulei	21ʰ 10.3ᵐ	+10° 08'	A–B	4.7	8.7	1.5	264	Y, W	1994	6/8"	*****	Good test of atmospheric conditions
S 781	21ʰ 13.5ᵐ	+07° 13'	A–B	7.4	9.4	**0.5**	**284**	W, W	**113**	18/20"	****	Another quartet with wide range of difficulties
			AB–C	7.3	14.0	31.9	029	W, ?	2005	8/10"	****	500 to 1 brightness difference significantly increases difficulty
			AB–D	7.3	7.2	184.0	172	W, ?	2002	6×30	*****	Nearly matched pair. Ideal for test of small aperture scopes
β 163 (Equulei)	21ʰ 18.6ᵐ	+11° 34'	A–B	7.4	8.9	**0.8**	**078**	Y, oY	**78.5**	10/12"	****	Opening to max in 2029. Observe periodically to test your resolution limit
Σ2786 Equulei	21ʰ 19.7ᵐ	+09° 31'	A–B	7.5	8.2	2.8	188	W, W	1997	3/4"	****	Easy, attractive, only slightly mismatched in brightness
10, β Equulei	21ʰ 22.9ᵐ	+06° 49'	A–B	5.2	13.6	38.5	266	W, B	2002	8/10"	****	Interesting quintuplet with one bright stars accompanied by 4 faint ones
			A–C	5.2	11.6	72.0	302	W, Y	2002	3/4"	****	
			A–E	5.2	12.1	93.6	275	W, ?	2000	3/4"	****	
			C–D	11.6	12.6	2.0	188	Y, ?	2002	4/6"	****	
Alden 76 (Equuleus)	21ʰ 22.9ᵐ	+06° 44'	A–B	11.4	11.4	3.0	63	?, ?	2002	3/4"	***	Faint, matched double only 5' S (PA 180°) of β Equ and in same mod/high power FOV

(continued)

Table 3.7. (continued)

Variable stars:

Designation	R.A.	Decl.	Type	Range (m$_v$)		Period (days)	F (f$_r$/f$_f$)	Spectral Range	Aper. Recm.	Rating	Comments
S Equulei	20h 12.8m	+05° 05'	EA/SD	8.0	10.1	3.4361	0.13	B9V+F9III-IV	2/3"	***	0°55'NW (PA 329°) of ε Equ (m5.16). #$
R Equulei	21h 13.2m	+12° 36'	M	8.7	15.0	260.76	0.44	M3e-M4e	16/18"	***	2°46'NNE (PA 015°) of γ Equ (4.70). ##
RV Equulei	21h 14.5m	+08° 47'	SRb	9.1	9.4	80	–	K0	2/3"	***	1°01'S (PA 174°) of δ Equ (m4.47). #

Star clusters:

Designation	R.A.	Decl.	Type	m$_v$	Size (')	Brtst. Star	Dist. (ly)	dia. (ly)	Number of Stars	Aper. Recm.	Rating	Comments
None												

Nebulae:

Designation	R.A.	Decl.	Type	m$_v$	Size (')	Brtst./Cent. star	Dist. (ly)	dia. (ly)	Aper. Recm.	Rating	Comments
None											

Galaxies:

Designation	R.A.	Decl.	Type	m$_v$	Size (')	Dist. (ly)	dia. (ly)	Lum. (suns)	Aper. Recm.	Rating	Comments
NGC 7015	21h 05.7m	+11° 25'	Sbc	11.5	1.9×1.6	–	–	–	12/14"	***	Elongated NNE/SSW. 1°45' NW (PA 045°) of γ Equ (m4.7) *35

Meteor showers:

Designation	R.A.	Decl.	Type	Period (yr)	Max Date	Duration	ZHR (max)	Comet/**Asteroid**	First Obs.	Vel. (mi/**km**/sec)	Rating	Comments
None												

Other interesting objects:

Designation	R.A.	Decl.	Type	m$_v$	Size (')	Dist. (ly)	dia. (ly)	Lum. (suns)	Rating	Comments
None										

NOTES:

1. ο Andromedae is actually a quadruple system consisting of two spectroscopic binaries. Each pair orbits the center of mass of the entire system. Visually, it appears to be an ordinary binary star with a period of 68.6 years, and a separation varying between 0.353" (1971) and 0.035" (2012).

2. Z Andromedae is the prototype of variables consisting of a red giant with a hot companion star. In this type of variable, the red giant is so distended that some of its outer atmosphere escapes and goes into orbit about the pair. Z Andromedae variables have nova-like outbursts, and often show signs of multiple periods with amplitudes smaller than that of the outbursts. Z And itself exhibits periodic outbursts at intervals of 10–20 years and periods of a few days and a longer one averaging 632 days.

3. RS Andromeda is an interacting close binary. Large sunspots on the individual components result in significant differences between one period and the next.

4. EG Andromedae consists of a variable red giant and a white dwarf which is eclipsed by the red giant every 480 days.

5. NGC 404 is located so close to Mirach (m2.06) that it is a difficult object to observe, and when it is seen, it is sometimes mistaken for an internal reflection in the telescope, hence the name. It is the closest dwarf spheroidal galaxy, and lies just beyond the boundary of the Local Group. Its nearest neighbor galaxy is 3.6 million light years distant, and it does not appear to be gravitationally bound to the Local Group, making it the nearest isolated (not a member of any galaxy cluster) galaxy.

6. On Dec. 6, 1798, Heinrich Brandes of Göttingen, Germany noticed a large number of meteors, and counted them for several hours. Corrected for the portion of the sky and the altitude of the radiant, his counts indicated a ZHR of ~100/h for 4 h followed by a sharp decline. This shower produced several thousand meteors per hour on Nov. 27 in both 1872 and 1885. Its orbit was subsequently deflected by several close passages near Jupiter, and such meteor storms have not been seen since.

7. Bayer gave the υ Persei designation to a star which Ptolemy had originally placed in Andromeda. In the 1930 official definition of constellation boundaries, the IAU returned this star to its original constellation as 51 Andromedae.

8. Σ331 forms a small triangle (~1–1.5° on each side) with γ and τ Persei. All three doubles can be seen in a single low-to-medium power FOV.

9. X Persei has an orbital period of 250 days and a separation of ~2 AU. Rotation of the neutron star gives an X-ray pulse every 835 s. X Per is 0°53' SSE (PA 162°) of ζ Persei.

10. Mel 20, the α Persei Cluster, is actually an association rather than a true cluster. Although the stars in this group are moving in approximately the same direction with about the same speed, the gravitational force is not sufficient to keep the group from slowly dispersing and eventually dissolving.

11. Although this shower was originally thought to have its radiant in Perseus, the radiant is actually in northern Taurus, about 8° SE of ζ Perseus. The best chance of detecting the ζ Perseids is through UHF/VHF radio signals reflected off the ionized trails of the meteors. Be careful not to confuse these meteors with the Arietid shower, also near maximum near the same date.

12. NGC 1245 was originally listed by Karl Seyfert as having the same characteristics as his other peculiar spectrum galaxies. Later research showed this object to be a collision between a spiral and an elliptical galaxy rather than a single galaxy with an active nucleus. NGC 1245 is the brightest member of the Abell 426 galaxy cluster. It is also an X-ray and radio source second only to Centaurus A. There is an AAVSO comparison chart available for tracking brightness variations.

13. Virgil Ivan (Gus) Grissom, Edward White II, and Roger B. Chafee were, respectively, the Command Pilot, Senior Pilot, and Pilot of the Apollo 1 mission. All three were killed in a fire in the capsule during a training session on Jan. 27, 1967. Earlier, as a joke, Grissom had surreptitiously introduced the names Navi (Ivan reversed), Dnoces (second reversed), and Regor (Roger reversed) on training documents for ε Cas, ι Uma, and γ Vel, stars which were to be used in calibrating equipment during their flight. After their deaths, many other people started using these names, even though all three stars already had traditional names.

14. μ Cas is only 28' SW of θ Cas, and the pair make an easy unaided eye double and a nice binocular sight with θ nearly pure white and μ yellow.

15. Pearce's star is a close binary whose components are nearly touching. Their orbit is inclined 51.2° to our line-of-sight, so the system undergoes only small partial eclipses. The A and B components, respectively, have masses of 32 and 30 solar masses. Their total luminosity is ~300,000 suns. The high luminosity makes it visible at the extraordinary distance of 7,000 ly.

16. γ Cassiopieae variables are hot blue giant stars of spectral class B. Their spectra show emission lines which originate in an extend atmosphere, shell, or disk of gas.

17. 2 Ursa Minoris and 78 Draconis were moved to Cepheus when the IAU defined the constellation boundaries. They kept the original names to avoid confusion in the existing astronomical literature.

18. Herschell's "Garnet Star" (μ Cephei) is one of the reddest stars visible to the unaided eye or in binoculars.

19. NGC 188 contains over 1,000 stars, of which only ~30–40 are seen in 4–20" scopes. NGC 188 lies along a line from Polaris toward the middle of Cassiopiea and 1° past the star SAO 181 (m4.24). Because of uncertainty in the amount of interstellar adsorption, estimates of both the distance (4,000–9,000 ly) and the age (5–9 billion years) of NGC 188 vary greatly.

20. IC 1396 is an extremely large nebula containing both emission and dark nebulae. Among the dark nebulae are B 160, 161, and 162. In addition, the doubles Σ2816 and 2819 are located in IC 1396.

21. ο Ceti, or Mira, is the prototype of the long-period pulsating red giants. These stars can be called both long-period variables (LPVs) and Mira-type variables (M). Mira is about 250 times as bright to the eye at maximum as at minimum. When at maximum in 1779, Mira was nearly as bright as Aldebaran.

22. M77 is the prototype of Seyfert galaxies. Named after Carl K. Seyfert, who first studied and described them as a class, these galaxies have active nuclei which vary in brightness. The nuclei appear to contain massive black holes which are emitting strong, broadband radiation as the result of gas, dust, and even entire stars which fall into them. They may be relatively nearby, low-power versions of quasars.

23. An Oct. 9, 1969 meteor from the η Cetid shower was measured at magnitude −20 (>900 times the brightness of the full moon)!

24. The Northern October Cetids have an orbit similar to that of the asteroid Hermes. This asteroid was discovered on Oct. 28, 1937 by Karl Reinmuth, who noted its high apparent motion across the sky of over 5°/h. At its closest, it was only twice the distance of the moon. Reinmuth lost track of it after only 5 days, when it was lost in the sun's glare, and it was not rediscovered until Oct. 15, 2003. Hermes' orbit is chaotic because its path takes it near the orbits of the Earth, Venus, and Mars twice in the 777 days of its orbital period. The gravitational effects of these planets makes it extremely hard to predict the future path of Hermes, especially when it passes close both the Earth and Venus in a single revolution, as it did 1954. The orbital motion of the Northern October Cetids is similarly affected by these gravitational forces.

25. UV Ceti (a.k.a. L 726-8) is a red dwarf binary consisting of two of the smallest and faintest stars known. The fainter companion is also a flare star which can brighten one or two magnitudes in less than a minute. The two form a binary star with a period of 26.52 years, and apparent separation varying from 0.45" to 2.19". Resolving these two stars requires a large telescope.

26. τ Ceti is a somewhat smaller, cooler, and fainter version of our sun. It has been the subject of several searches for exoplanets, but so far none have been found. Any planets of the "hot Jupiter" type have been ruled out, unless their orbital plane is nearly perpendicular to our line of sight. The density of dust surrounding τ Ceti is about ten times that around our sun, possibly indicating that even smaller planets (which would have swept up most of the dust) also are not present.

27. α Ceti is a cool orange-red giant. Compare its color with the nearby white γ Ceti (m3.47) 4.8° W and blue-white δ Ceti (m4.08) 6.6° SW. All three stars can be seen in the same 7×50 binocular FOV.

28. The former δ Pegasi is now 21 Andromedae.

29. η Pegasi A–a has a separation of <0.05", B–C has a separation of ~0.2", and the D and E component are magnitudes 14.0 and 16.0. All three pairs are too difficult for most amateur telescopes.

30. AG Pegasi is a binary system consisting of a red giant and an extremely hot subdwarf ($T > 100,000$ K). The two stars are close enough that the hot subdwarf creates a hot spot on the red giant's near side. As this spot rotates into and out of view, the overall brightness changes.

31. M15 is very bright and large, containing many stars. It can be seen with unaided eye under good conditions, resolved in 10" or larger telescopes.

32. NGC 7331 has dust lanes which can be resolved in 16/18" telescopes.

33. The υ Pegasid shower produced several thousand meteors per hour in 1872 and 1885. The orbital path was subsequently deflected by a close passage near Jupiter, and no meteor storms have been observed since.

34. Bayer did not use any letters in Equuleus beyond δ. Bode added several letters beginning with ε, but only it and λ are still in use.

35. NGC 7015 has a faint halo, which slowly brightens toward the center. Astrophotos or very large telescopes show multiple branching arms.

CHAPTER FOUR

The Hunter and the Hunted

January–March

ORION	CANIS MAJOR	LEPUS	TAURUS
The Hunter	The Larger Dog	The Hare	The Bull

The group of constellations related by the story of Orion contains some of the brightest constellations in the sky. Three of the constellations in this section are very prominent in the winter sky and, Lepus, though relatively faint, is easy to recognize.

Orion is one of the oldest recognized constellations, with some references dating back to at least 4000 BC. It is also one of the best known for two reasons. First, it is one of the brightest and most distinctive of the constellations. Second, it is very easy to view the brighter stars of Orion as representing a human figure. Many ancient cultures from widely separated areas of the world viewed this group of stars as representing a leader, hunter, warrior, or local hero of some sort.

Like Orion, Taurus is also one of the oldest recognized constellations, with some references also dating back to at least 4000 BC. Although there are many different myths which refer to bulls, and different cultures have used different myths to explain the heavenly bull, I have chosen the myth of Orion because of the large number of constellations explained by this one myth. In addition to the four winter constellations, the version of the Orion myth related here connects five additional constellations, all in the summer sky. One of the summer constellations, Scorpius, is also one of the brightest in the sky.

Although references to Sirius, the dog star, also appear very early in history, the constellation as we know it is of much later origin. Originally, the constellation of

P. Simpson, *Guidebook to the Constellations: Telescopic Sights, Tales, and Myths*,
Patrick Moore's Practical Astronomy Series, DOI 10.1007/978-1-4419-6941-5_4,
© Springer Science+Business Media, LLC 2012

the dog often consisted of only the star Sirius. The Egyptians thought Sirius to be so important in affecting the weather and the flooding of the Nile that they oriented temples toward its rising point as early as 2700 BC. Although not as old as Orion and Taurus, Canis Major had taken on its present configuration well before the period of classical Greek culture.

In one version of the Greek myth about Orion, a total of nine (or ten if the small dog is included) constellations are connected and explained. This particular version also has the advantage of being the best to describe the positions of the various constellations and even the time of the year when they can best be seen.

In this version, Orion was the son of Poseidon, the Greek god of the seas. Orion grew up to become a handsome young man of huge stature and great skill as a hunter. He was so skilled as a hunter that he was invited by Artemis, the goddess of the hunt, to become a member of her hunting party. He was the only male in the group, and quickly became one of her favorite companions because of his great skill. In spite of being one of her favorites, Orion angered Artemis by chasing the maidens in her hunting band and by boasting that he would prove that he was the world's greatest hunter by killing every wild animal on earth.

Not content with boasting, Orion actually tried to carry out his threat. Accompanied by his hunting dog, Orion set out into the forest. He first encountered one of the most timid and harmless of wild creatures, the hare. Orion immediately prepared to take his first victim. This action so infuriated the gods that Hera sent a giant scorpion to deliver a fatal sting to his heel. Knowing that the scorpion alone would be unlikely to defeat the great hunter, Hera also sent an enormous bull to distract Orion's attention. While Orion was occupied in fighting the bull, the scorpion approached him from behind and delivered the fatal sting to his heel. The scorpion, in turn, was killed by an arrow from the bow of Chiron, the centaur who had been Orion's boyhood teacher and friend.

After Orion's death, Chiron called for help from his most famous pupil, Asclepius. The son of Apollo, the Greek god of medicine, Asclepius had learned everything about medicine the great Chiron had to teach and surpassed him, becoming the greatest physician of all time. Asclepius' resume included the resurrection of several people, including Lycurgus, Capaneus, and Tyndareus. However, his attempt to revive Orion drove Hades to complain that his subterranean kingdom of the dead would soon be depopulated by Asclepius' actions, and Zeus ended this mortal's impertinence with one of his thunderbolts.

Orion, his dog, the scorpion, the bull, the hare, Chiron, and Asclepius, were all placed in the skies as constellations. In this case, they were separated into two groups. Orion still faces the charge of the bull (Taurus) with his left hand holding his shield up to fend off the bull's horns, and his right hand holding an enormous club raised to deliver a crushing blow to the bull's head. His dog (Canis Major) stands behind him, while the hare (Lepus) crouches forgotten beneath Orion's feet. The constellations Orion, Taurus, Canis Major, and Lepus retell the story of Orion every clear night of the winter. The placement of these constellations is very appropriate to their characters, since the hunter was visible during the hunting season (he would have been visible in the fall when the constellation was first formed about 4000 BC).

In spite of all the trouble Orion had caused, Artemis regretted the loss of her companion, and requested that the scorpion be moved to the opposite part of the heavens, so he could never again harm Orion. Poseidon demanded that Chiron also be moved to the summer sky to prevent any further attempts at resurrection of Orion.

Chiron was also charged with guarding the scorpion to ensure that he never left the summer sky to again attack Orion. Asclepius was originally shown as holding a large snake, one of the symbols of physicians. In later times, the snake held by Asclepius was made a separate constellation, Serpens.

Orion contains two 1st magnitude stars, <u>Betelgeuse</u>[1] and <u>Rigel</u>.[2] Orion's belt, with its row of three second-magnitude stars and the somewhat fainter sword stand out clearly, while both shoulders and feet are marked by 1st or 2nd magnitude stars. The central figure of the winter group, Orion is shown with his body facing the viewer while his head is turned to his left to face the charging Taurus. His left hand holds up his shield to fend off the bull's horns, while his right hand holds a giant club ready to strike a crushing blow to the bull's head. The bull is charging from our right, with his enormous horns extending over Orion's head, and his face (represented by the "V" shaped Hyades cluster) is approaching the shield. Beneath Orion's feet Lepus (the hare) crouches in terror.

To Orion's left, Canis Major backs up his master. Like Orion, Canis Major also contains two bright stars, one of which (<u>Sirius</u>) is the brightest in the entire sky. Adhara (ε CMa) is the second brightest star in Canis Major and almost bright enough to be a first magnitude star, but its brightness is just at the borderline between first and magnitude stars, and it is usually considered a second magnitude star. The shape of the larger dog is easily found SE of Orion, and the profile of a standing dog is easily recognized.

Taurus contains the bright orange <u>Aldebaran</u>, and the small, but bright and well-known group of stars known as the Pleiades, or the seven sisters. Lepus, a minor and passive character in this tale is relatively faint, but a small oval of stars which represent the Hare's body is not hard to recognize.

In the following list, the "best seen" date is that on which the constellation crosses the meridian and is highest in the sky at 9:00 PM. Of course, the constellation can be seen equally well on other dates if you look 30 min earlier for each week after the given. Similarly, you should look 30 min later for each week before the given date. The pronunciation of the constellation and star names is given by using common words or syllables in a way that I hope is obvious and easy to understand. The names of the 1st magnitude stars are underlined. In one case (Adhara or ε Canis Majoris), recent measurements have shown it to have a magnitude of 1.49, which by today's standards would qualify it as a 1st magnitude star, even though it was not included in any of the classical lists, and it is not usually included in today's lists. It likely was not considered a 1st magnitude star in ancient times because of two factors. First, it suffered by comparison to nearby Sirius, the brightest star in the sky, and second, it was very low in the sky from Greece and the middle eastern countries where early astronomers lived. Both factors made Adhara seem less bright than it otherwise would (Fig. 4.1).

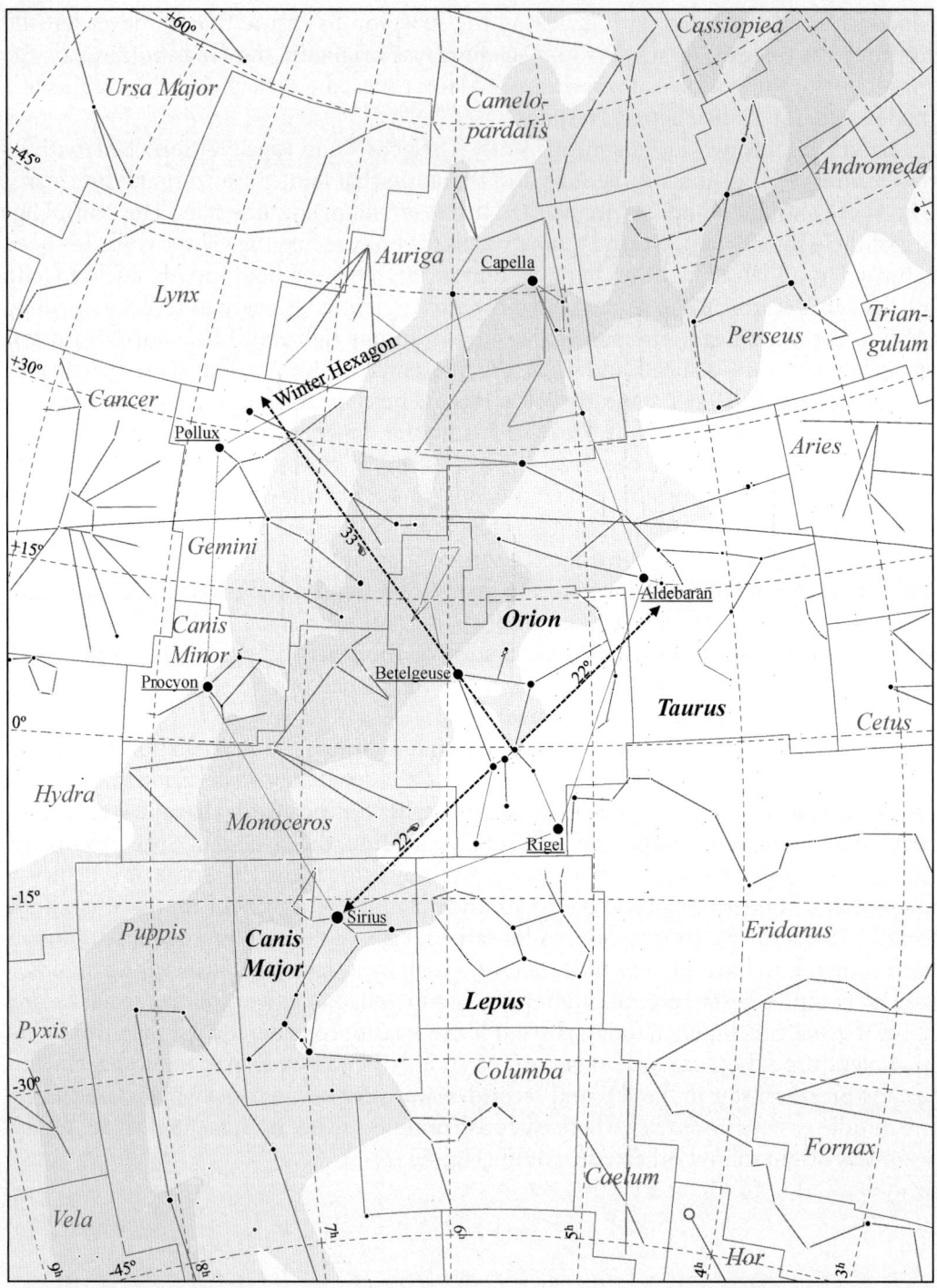

Fig. 4.1. Orion Group Finder (Looking south at 9:00 PM about Feb. 5).

ORION (oar-**eye**-uhn)	**The Hunter**
ORIONIS (oar-ee-**oh**-niss)	**9:00 P.M. Culmination:** Feb. 1
Ori	**AREA:** 594 sq. deg., 26th in size

TO FIND: Look due south at about 9:00 PM a week or two before or after Feb. 1, and about one-half to two-thirds the distance from the horizon to the zenith, depending on your latitude. With its evenly spaced row of three nearly equally bright stars, Orion's belt is easy to spot. Since Mintaka, the right-hand star of Orion's belt is almost exactly on the celestial equator, it will be the same number of degrees from the zenith as your distance from the equator (your latitude in degrees). Hence, it will be above the southern horizon by 90° minus your latitude in degrees. Reddish Betelgeuse is about 10° NE of the belt, while blue-white Rigel is nearly the same distance SW of the belt. Bellatrix and Saiph are, respectively, about 8° NNW and SSE of the belt. Betelgeuse, Bellatrix, Rigel, and Saiph outline Orion's body and form a nearly rectangular box around the belt (Figs. 4.2 and 4.3).

KEY STAR NAMES:

	Name	Pronunciation	Source
α	Betelgeuse	**bet**-el-juze	ibt al jauzah, (Arabic), "armpit of the central one," a miscopying of yad al-jauzah "hand of the central one"
β	Rigel	**rye**-juhl	rigl jauzah al Yusra (Arabic), "left leg of the central one"
γ	Bellatrix	bel-**lay**-tricks	bellatrix (Latin) "female warrior or amazon"
δ	Mintaka	**min**-tah-kuh	al mintaka (Arabic), "the belt"
ε	Alnilam	uhl-nih-lam or al-**nye**-lam	al nizam (Arabic), "the string of pearls"
ζ	Alnitak	uhl-nih-**tak** or al-**nye**-tak	al nitak (Arabic), "the belt or girdle"
η	Saif al Jabbar or just Saif	safe al jabbar	saif al jabbar (Arabic), "sword"
ι	Hatsya	(?)	The source and pronunciation of this name are uncertain
κ	Saiph	safe	saif al jabbar (Arabic), "sword" (eta has similar name)
λ	Meissa	**ma**-sah or **mye**-sah	al maisan (Arabic), "the proudly marching one," apparently a mistaken duplication of the name for γ Geminorum
	or Heka	**hek**-uh	al hakah (Arabic), "the white spot"
π³	Thabit	**thah**-bit	al thabit (Arabic), "the fixed or enduring one"
	or Tabit	**tah**-bit	

Magnitudes and spectral types of principal stars:

Bayer desig.	Mag. (m$_v$)	Spec. type	Bayer desig.	Mag. (m$_v$)	Spec. type	Bayer desig.	Mag. (m$_v$)	Spec. type
α	0.45	M2 Ib (var)	β	0.18	B8 Ia	γ	1.64	B2 III
δ	2.25	O9.5 II	ε	1.69	B0 Ia	ζ	1.74	O9.5 Ib
η	3.35	B1 V	θ^1	4.98	O7	θ^2	4.98	O9.5 Vpec
ι	2.75	O9 III	κ	2.07	B0.5 Ia(var)	λ	3.39	O8 III
μ	4.12	Am	ν	4.42	B3 IV	ξ	4.45	B0 IV
o^1	4.71	M3 Sv	o^2	4.06	K2 III	π^1	4.64	A0 V
π^2	4.35	A1 Vn	π^3	3.19	F6 V	π^4	3.68	B2 III
π^5	3.71	B2 III	π^6	4.47	K2 II(var)	ρ	4.46	K3 III
σ	3.77	O9.5 V	τ	3.59	B5 V	υ	4.62	B0 V
φ^1	4.39	B0 IV	φ^2	4.09	B8 III–IV	χ^1	4.39	B0 V
χ^2	4.64	B2 Ia (var)	ψ	4.59	B2 IV	ω	4.50	B3 IIIe

Stars of magnitude 5.5 or brighter which have flamsteed but no Bayer designations:

Flam-steed	Mag. (m$_v$)	Spec. type	Flam-steed	Mag. (m$_v$)	Spec. type	Flam-steed	Mag. (m$_v$)	Spec. type
5	5.33	M1 III	6	5.18	A3 V	11	4.65	A0p Si
14	5.33	Am	15	4.81	F2 IV	16	5.43	A2m
21	5.34	F5 Ivar	23	4.99	B1 V	25	4.89	B1 V:pe
27	5.07	K0 III	29	4.13	G8 III	31	4.71	K5 III
32	4.20	B5 V	33	5.46	B1.5 V	38	5.32	A2 V
42	4.58	B2 III	45	5.24	F0 III	49	4.77	A4 V
52	5.26	A5 V	55	5.36	B2 IV/V	56	4.76	K2 IIvar
60	5.21	A1 Vs	64	5.14	B8 V	69	4.95	B5 Vn
71	5.20	F6 V	72	5.34	B7 V	74	5.04	F1 IV/V
75	5.39	A2 V						

Stars of magnitude 5.5 or brighter which have no Bayer or flamsteed designations:

Desig-nation	Mag. (m$_v$)	Spec. type	Desig-nation	Mag. (m$_v$)	Spec. type
HD 030034	5.39	F0 V	HD 030210	5.35	Am
HD 031296	5.33	K1 III	HD 033554	5/18	L5 III
HD 034043	5.50	K4 III	HD 036591	5.34	B1 IV
HD 036960	4.78	B0.5 V	HD 037756	4.95	B2 IV/V
HD 040657	4.53	K2 IIIvar	HD 044131	4.91	M1 III

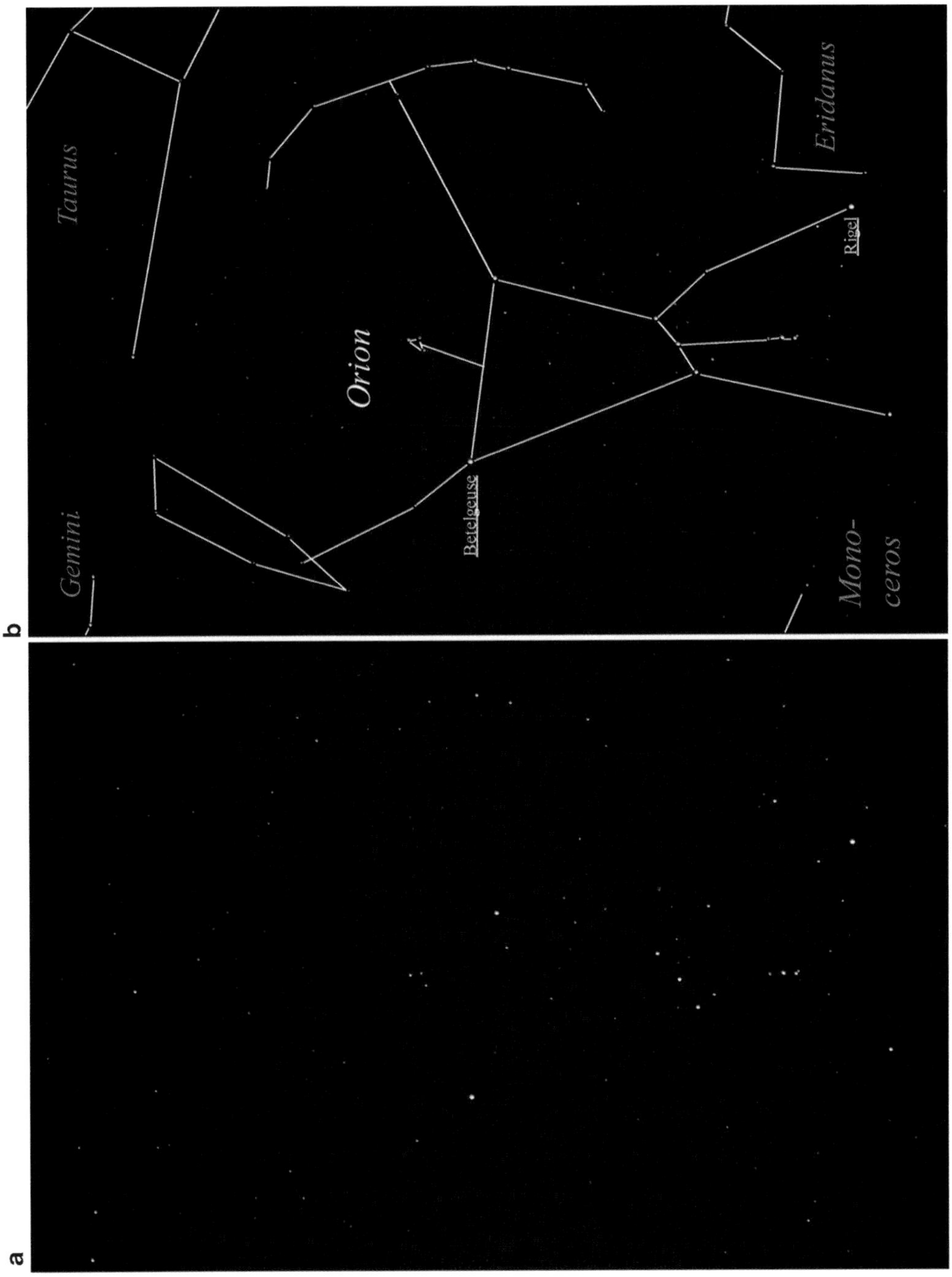

Fig. 4.2. (**a**) The stars of Orion and adjacent constellations. (**b**) The figures of Orion and adjacent constellations.

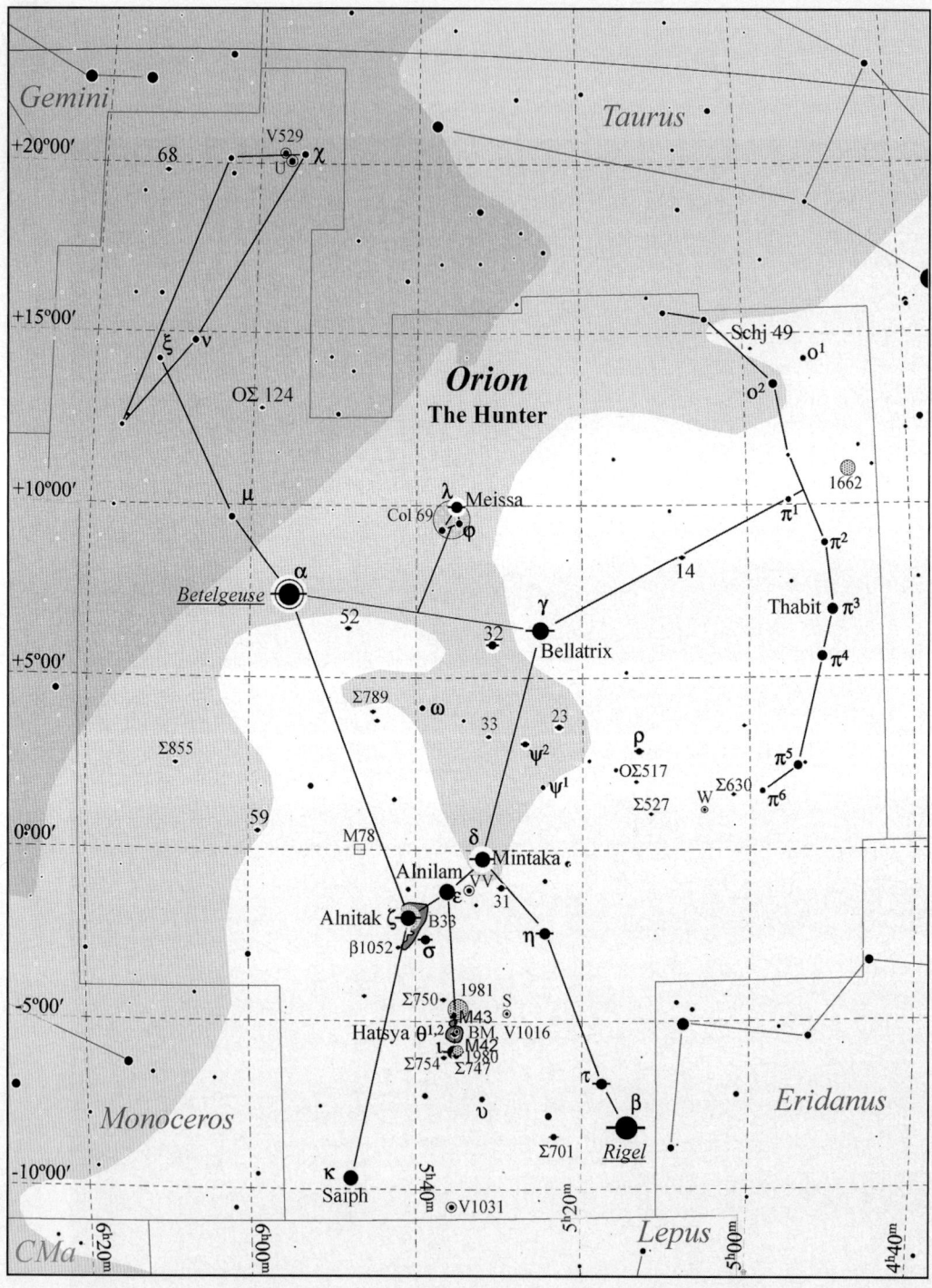

Fig. 4.3. Orion details.

Table 4.1. Selected telescopic objects in Orion.

Double/multiple stars:

Designation	R.A.	Decl.	Type	m_1	m_2	Sep. (")	PA (°)	Colors	Date/ Period	Aper. Recm.	Rating	Comments
Schj 49	04h 59.0m	+14° 32'	A–B	6.1	7.4	40.3	305	W, –	2006	7×50	****	Two faint companions flanking brighter primary in a boomerang formation
Σ627	05h 00.6m	+03° 37'	A–B	6.6	9.6	53.3	089	W, –	2002	15×80	***	
					7.0	21.8	260	W, W	2006	15×80	****	Nice pair of near twins
Σ630	05h 02.0m	+01° 36'	A–BC	6.5	7.7	14.8	050	W, pB	2006	2/3"	****	B–C is too close for amateur scopes
			A–D	6.5	10.4	131.3	101	W, pB	2002	15×80	****	
14 Orionis	05h 07.9m	+08° 30'	A–B	5.8	6.7	0.8	305	dY, pY	2006	12/14"	****	Challenging: a.k.a. OΣ 98
Σ527	05h 11.8m	+01° 02'	A–B	6.3	7.4	1.6	180	Y, dY	2003	6/7"	***	
17, ρ Orionis	05h 13.3m	+02° 52'	A–B	4.6	8.5	6.9	064	O, B	2004	2/3"	*****	Lovely color contrast for small scopes. Some see Y and O
OΣ 517	05h 13.5m	+01° 58'	A–C	4.6	11.4	182.3	157	O, –	2000	2/3"	**	
			A–B	6.1	6.7	0.7	241	yW, –	530	14/15"	***	Passed min. 0.06" in 1924, slowly opening to max. 0.99" in 2190
19, β Orionis, Rigel	05h 14.5m	–08° 12'	A–B	0.1	6.8	9.4	204	bW, B	2005	4/5"	*****	Beautiful, more difficult than separation would suggest. a.k.a. Σ 668AB *1
20, τ Orionis	05h 17.6m	–06° 51'	A–B	3.6	11.0	31.3	251	pO, B	1998	3/4"	****	Bright orange star flanked by nearly twin faint bluish stars
23 Orionis	05h 22.8m	+03° 33'	A–C	3.6	10.9	35.9	060	pO, pV	1998	3/4"	****	
			A–B	5.0	6.8	32.1	028	W, BW	2005	10×50	****	Nice group in 4/6" scopes
Σ701	05h 23.3m	–08° 25'	A–B	6.1	8.1	6.0	139	bW, B	2003	2/3"	******	
28, η Orionis	05h 24.5m	–00° 18'	Aa–B	3.4	4.9	1.7	078	W, pB	2003	5/6"	****	Test for 4", difficult in 6" unless seeing is very good
ψ² Orionis	05h 26.8m	+03° 06'	Aa–C	3.4	9.4	115.1	051	W, B?	1998	15×80	**	
			A–B	4.6	8.6	2.8	327	Y, B	1991	3/4"	***	
			A–C	4.6	13.9	83.4	197	Y, –	2000.0	8/9"	**	
31 Orionis	05h 29.7m	–01° 06'	A–B	4.7	9.7	12.8	087	O, pB	2001	2/3"	****	Nice color but requires large aperture to see color in secondary
32 Orionis	05h 30.8m	+05° 57'	A–B	4.5	5.8	1.3	045	Y, Y	613.69	7/8"	****	Binary, challenging test for 4/6" a.k.a. Σ729AB
33 Orionis	05h 31.2m	+03° 18'	A–B	5.7	6.7	1.8	028	W, bW	2006	5/6"	****	
34, δ Orionis, Mintaka	05h 32.0m	–00° 18'	Aa–B	2.2	14.0	33.2	228	bW, B	1998	9/10"	****	Easily identified test for 7×50 binoculars. 10×50 may be needed for many people

(continued)

Table 4.1. (continued)

Double/multiple stars:

Designation	R.A.	Decl.	Type	m_1	m_2	Sep. (")	PA (°)	Colors	Date/Period	Aper. Recm.	Rating	Comments
β 1048	05h 32.7m	−01° 35′	Aa–C	2.2	6.8	53.3	001	yW, bW	2004	6×30	***	
Σ747	05h 35.0m	−06° 00′	A–B	5.3	9.8	2.0	355	W, W	1953.0	4/5"	***	In same FOV as ι Orionis
HDS 742			A–B	4.7	5.5	36.2	224	W, W	2005	6×30	*****	
			Ba–Bb	5.7	8.8	0.6	120	W, –	1991	16/18"	****	
39, λ Orionis	05h 35.1m	+09° 56′	A–B	3.5	5.5	4.3	044	bW, bW	2005	2/3"	****	Brightest member of its namesake cluster, a 1° diameter group of ~20 stars Σ738
			A–C	3.5	10.7	28.5	184	bW, pB	2005	15×80		
			A–D	3.5	9.6	78.3	271	bW, pB	2005	15×80		
41, θ¹ Orionis	05h 35.3m	−05° 23′	A–B	6.7	7.9	8.8	031	bW, bW	1975	2/3"	*****	The famous "Trapezium" in the Orion Nebula. The four brightest stars are easy in small scopes. The E and F stars are more difficult but can be seen in 12/14" scopes
			A–C	6.7	5.1	12.8	132	bW, B	1975	2/3"		
			A–D	6.7	6.7	21.5	096	bW, ?	1975	15×80		
			A–E	6.7	11.1	4.1	351	bW, ?	1934	3/4"		
			C–F	5.1	11.5	4.0	122	–, –	1957	3/4"		
42, θ² Orionis	05h 35.4m	−04° 50′	A–B	4.6	7.5	1.1	205	yW, –	1995	8/9"	***	Σ16
44, ι Orionis	05h 35.4m	−05° 55′	A–B	2.9	7.0	11.3	141	W, grW	2002	2/3"	*****	Nice triple, with Σ747 only 8" to SW and visible in same low to medium power views
			A–C	2.9	9.7	49.4	103	W, pR	2002	15×80		
Σ750	05h 36.6m	−04° 22′	A–B	6.4	8.4	4.2	061	W, W	2003	2/3"	***	
Σ754	05h 36.6m	−06° 04′	A–B	5.7	9.2	5.3	288	bW, –	2002	2/3"	***	
48, σ Orionis	05h 38.7m	−02° 36′	A–B	4.0	6.0	**0.2**	**086**	W, W	**155.3**	–	****	Lovely multiple system, all visible in small scopes, except AB pair which is too close for most amateur instruments to resolve. AB is binary, period 15.3 years
			AB–C	3.8	8.8	11.5	238	W, B	2002	2/3"		
			AB–D	3.8	6.6	12.7	084	W, B	2002	2/3"		
			AB–E	3.8	6.3	41.5	062	W, B	2003	7×50		
50, ζ Orionis, Alnitak	05h 40.8m	−01° 57′	A–B	1.9	4.0	**2.2**	**166**	W, W	**1508.6**	4/5"	****	Bright white pair
			A–C	1.9	9.6	57.3	010	W, ?	2003.0	15×80		

Name	R.A.	Decl.	Pair	m₁	m₂	Sep	PA	Colors	Year	Aper. Recm.	Rating	Comments
β 1052	05h 41.7m	−02° 53'	A-B	6.4	7.6	**0.7**	**188**	W, −	**109.39**	14/15"	***	Slight opening from present 0.65" to 0.69" in 2023, closing to 0.05" in 2074
Σ788	05h 45.1m	+03° 50'	A-B	7.6	10.1	7.5	089	W, −	1999	2/3"	****	In same FOV with Σ789
			A-C	7.6	10.4	35.6	148	W, −	2003	15×80		
Σ789	05h 45.0m	+04° 00'	A-B	6.1	10.2	14.3	149	Y, −	1999	2/3"	****	Forms a double-triple with Σ788
				10.2	12.2	1.1	112	−, −	1958	9/10"		
Σ790	05h 46.1m	−04° 15'	A-B	6.4	9.0	7.1	089	rY, B	2003	2/3"	****	Somewhat faint, but colors show up well in larger scopes
52 Orionis	05h 48.0m	+06° 27'	A-B	6.0	6.0	1.1	216	yW, yW	2003	8/9"	****	Nearly perfect test object for a 4" telescope. Σ795
58, α Orionis, Betelgeuse	05h 55.2m	+07° 24'	A-B	var	11.0	176.4	154	dO, bW	2000	3/4"	****	Seven other stars ranging from magnitude 11.2 to 14.5 lie within 5' of Betelgeuse *2
59 Orionis	05h 58.4m	+01° 50'	A-B	5.9	10.4	36.5	206	W, B	1999	15×80	***	a.k.a. H5 100AB
0Σ124	05h 58.9m	+12° 48'	A-B	5.7	6.8	**0.4**	**303**	Y, W	**140**	−	***	Passed max. of 0.46" in 1986, slowly closing to 0.03" in 2042
Σ855	06h 09.0m	+02° 30'	A-B	5.7	6.7	29.1	114	yW, W	2005	10×50	****	a.k.a H6 72
			A-C	5.7	9.7	118.5	107	yW, −	2002	15×80	****	
68 Orionis	06h 12.0m	+19° 47'	Aa-B	5.7	9.4	89.9	205	W, −	2002	15×80	****	

Variable stars:

Designation	R.A.	Decl.	Type	Range (m)		Period (days)	F (f_1/f_2)	Spectral Range	Aper. Recm.	Rating	Comments
W Orionis	05h 05.4m	+01° 11'	SRb	8.2	11.1	212	−	N5(C5)	3/4"	***	A carbon star and very red. $
S Orionis	05h 29.0m	−04° 42'	M	7.2	14.0	414.3	0.48	M6.5e–M9.5e	12/14"	****	1.6° W (PA 275°) of 42 Orionis
VV Orionis	05h 33.5m	−01° 09'	EA/KE	5.3	5.7	1.485	−	B1V–B7V	6×30	****	W UMa type, stars almost in contact. 40' W (PA 274°) of Orionis (Alnilam)
BM Orionis	05h 35.3m	−05° 23'	EA	7.9	8.7	6.471	0.12	B2V+A7IV	7×50	****	Northernmost star in trapezium.
V1016 Orionis	05h 35.3m	−05° 23'	EA	6.7	7.7	65.432	−	B0/5Vp	6×30	****	6" WNW of q¹ Orionis C, the brightest star in the trapezium *3
V1031 Orionis	05h 35.4m	−10° 32'	EA/DM	6.0	6.4	3.406	−	A4V	6×30	****	32' S (PA 185°) from κ Orionis (Saiph)
Betelgeuse	05h 55.2m	+07° 24'	SRc	0.0	1.3	2120	−	M1–M2, Ia–Iab	Eye	****	The brightest variable star in the sky
U Orionis	05h 55.8m	+20° 10'	M	4.8	13.0	368.3	0.38	M6e–M9.5e	8/10"	***	21' ENE (PA 106°) of c¹ Orionis
V529 Orionis	05h 58.4m	+20° 15'	Nr?	6.0	>11.0	−	−	M2.5III	3/4"?	****	May have been first observed as Nova Orionis 1667 *4

(continued)

Table 4.1. (continued)

Star clusters:

Designation	R.A.	Decl.	Type	m_v	Size (')	Brtst. Star	Dist. (ly)	dia. (ly)	Number of Stars	Aper. Recm.	Rating	Comments
NGC 1662	04ʰ 48.5ᵐ	+10° 56′	Open	6.4	12	8.34	1,230	4.5	35	6/8″	***	Loose, but obvious cluster, several dozen faint stars
Collinder 69	05ʰ 35.1ᵐ	+09° 56′	Open	2.8	70	–	1,650	33	20	7×50	****	The three stars forming Orion's head, plus the surrounding fainter stars
NGC 1981	05ʰ 35.2ᵐ	–04° 26′	Open	4.2	28	10	1,300	11	20	4/6″	****	1° N of M42, 3 mag. six stars on E. edge may also be part of the cluster
NGC 1980	05ʰ 35.4ᵐ	–05° 54′	OC and Em	2.5	15	–	–	–	30	8/10″	****	Triple star i Ori lies on the SE edge of the cluster

Nebulae:

Designation	R.A.	Decl.	Type	m_v	Size (')	Brtst./Cent. star	Dist. (ly)	dia. (ly)	Aper. Recm.	Rating	Comments
M42, Orion Nebula	05ʰ 35.4ᵐ	–05° 27′	Em	–	65×60	–	1,500	33	eye+	*****	Great Orion Nebula, beautiful with any aperture
NGC 1980	05ʰ 35.4ᵐ	–05° 54′	Em and OC	–	14×14	–	–	–	6/8″	****	Located just SE of M42 and NW of i Orionis
M43	05ʰ 35.6ᵐ	–05° 16′	Em and Refl.	–	20×15	–	1,500	10	15×80	****	Just N of M42 *5
Barnard 33	05ʰ 40.9ᵐ	–02° 28′	Dark	–	6×4	–	–	–	10/12″	****	Horsehead Nebula. Spectacular in photos but difficult visually
M78	05ʰ 46.7ᵐ	+00° 03′	Em and Refl.	–	8×6	–	–	–	8/10″	****	Bright fan-shaped neb w 2 10th mag. stars imbedded

Galaxies:

Designation	R.A.	Decl.	Type	Size (')	Dist. (ly)	dia. (ly)	Lum. (suns)	Aper. Recm.	Rating	Comments
None										

Meteor showers:

Designation	R.A.	Decl.	Period (yr)	Duration	Max Date	ZHR (max)	Parent body	First Obs.	Vel. (mi/ **km**/sec)	Rating	Comments
Orionids	06ʰ 20ᵐ	+16°	Annual	10/11	10/21	23	Halley	~1833	41/66	1	Major shower, reliable, occasional outbursts (2–3× normal in 1996) *6

Other interesting objects:

Designation	R.A.	Decl.	Type	m_v	Size (')	Number of Stars	Dist. (ly)	dia. (ly)	Aper. Recm.	Rating	Comments
Orion's Belt	05ʰ 35.6ᵐ	–01° 05′	Asterism	140	2.5	100	1,400	60	7×50	****	Belt and surrounding stars are sometimes called Collinder 70 *7

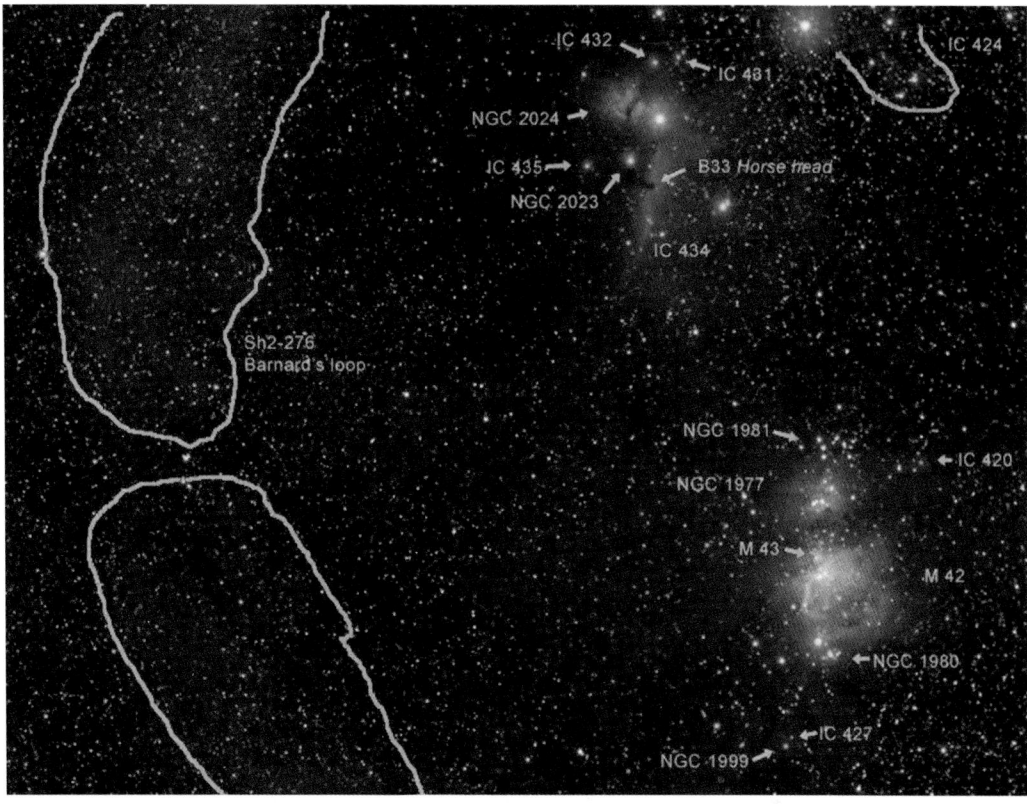

Fig. 4.4. Central Orion, showing prominent nebulae. Astrophoto and labelling courtesy of Gian Michele Ratto.

M42 and M43

OCT 22, 2006 • 07:30 UT
SkyView Pro 6LT EQ - 6" f/8 Newtonian
32 mm Sirius Plössl: 37.5X / 88' FOV
Sketch by Jeremy Perez © 2006

N
W

Fig. 4.5. M42 & M43, the Orion Nebula and its companion (NGC 1973, 1975, &1977). Astrophoto by Warren Keller.

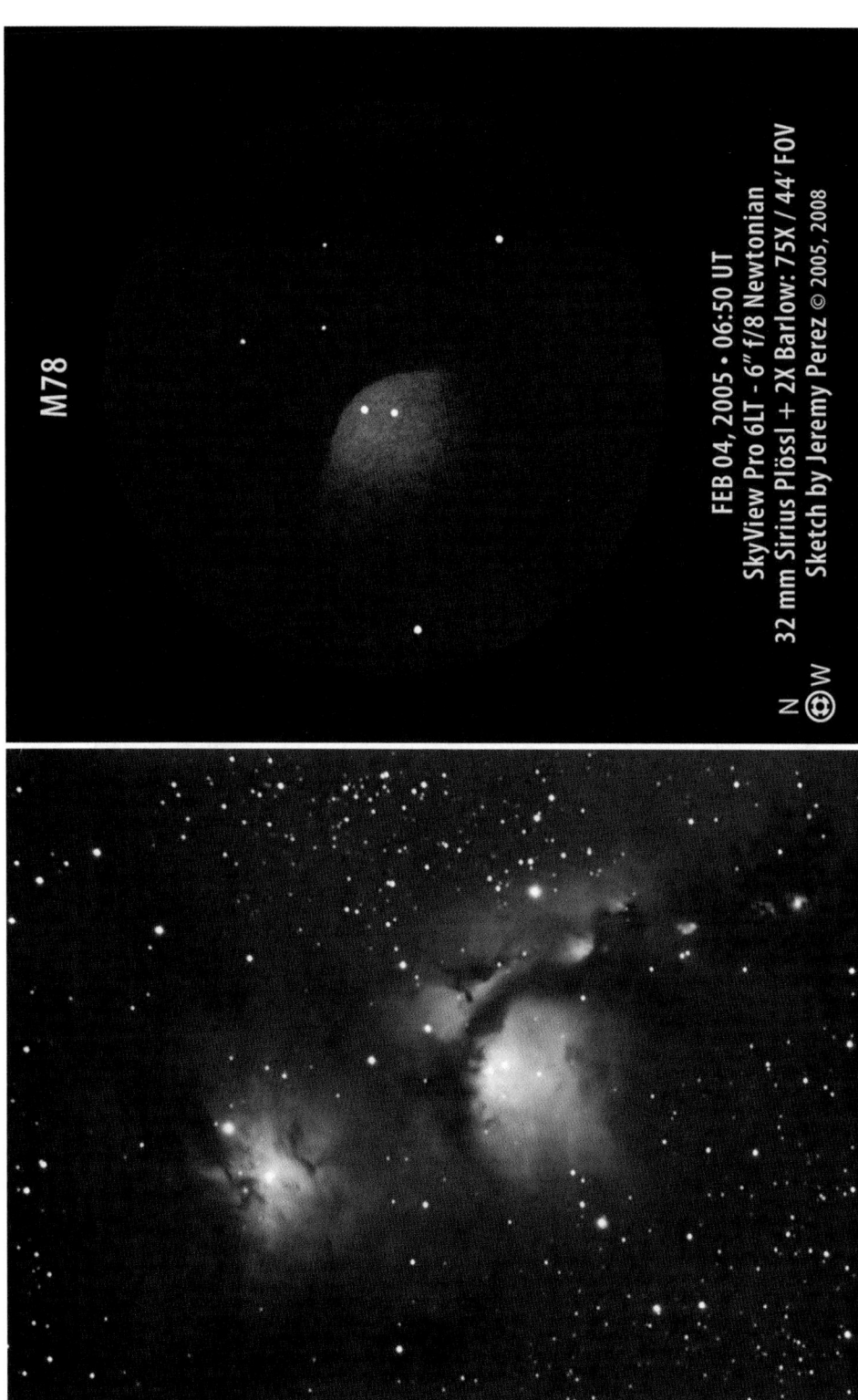

Fig. 4.6. M78 (NGC 2068), reflection nebula in Orion. Astrophoto by Warren Keller.

Barnard 33, IC 434, NGC 2023

JAN 22, 2006 · 04:00 UT
SkyView Pro 6LT - 6″ f/8 Newtonian
32 mm Sirius Plössl + Ultrablock: 37.5X / 88′ FOV
Sketch by Jeremy Perez © 2005, 2008

N
W

Fig. 4.7. B33, the Horsehead Nebula in Orion. Astrophoto by Stuart Goosen.

CANIS MAJOR (**kay**-nis **may**-jer)	The Large Dog
CANIS MAJORIS (**kay**-nis may-**jor**-iss)	BEST SEEN: March 1 (at 9:00 P.M.)
CMa	AREA: 380 sq. deg., 43rd in size

TO FIND: Extend the line of Orion's belt to the southeast about 25° to Sirius, the brightest star in the sky. Sirius marks the dog's shoulder, while η Canis Majoris marks his tail. A row of fainter stars forms his back, while β Canis Majoris and Adhara (ε) represent the front and hind feet, while a small triangle of faint stars to Sirius' north east represent the dog's head (Figs. 4.8 and 4.9).

KEY STAR NAMES:

	Name	Pronunciation	Source
α	Sirius	**see**-ri-us	Seirioz (Greek), "the scorching one"
β	Murzim or Mirzam	**mir**-zum	al Mirzam (Arabic), "the roarer" or "the announcer or proclaimer"
γ	Muliphen	muh-lih-**fan**	al muhlif n (Arabic), "two things causing dispute" from two stars (named hadari and al wazn) whose identification was in doubt
δ	Wezen	**weh**-zahn	al wazn (Arabic), "the weight"
ε	Adhara	ah-**day**-rah	al ʿadhārā (Arabic), "the virgins"
ζ	Furud	fuh-**rude** or foo-**rude**	al Qurud (Arabic), "the male apes"
η	Aludra	uhl-**ud**-ruh	al 'Udhra (Arabic), "the virgin of Orion"

Magnitudes and spectral types of principal stars:

Bayer desig.	Mag. (m_v)	Spec. type	Bayer desig.	Mag. (m_v)	Spec. type	Bayer desig.	Mag. (m_v)	Spec. type
α	−1.44	A0 m	β	1.98	B1 II/III	γ	4.11	B8 II
δ	1.83	F9 Ia	ε	1.50	B2 II	ζ	3.02	B2.5 V
η	2.45	B5 Ia	θ	4.08	K4 III	ι	4.36	B3 Ib/II
κ	3.50	B1.5 Vne	λ	4.42	B4 V	μ	5.00	B9.5 V
ν¹	5.71	G8/K0 III	ν²	3.95	K1 III	ν³	4.42	K0 II/III
ξ¹	4.34	B1 III	ξ¹	4.54	A0 III	o¹	3.89	K3 Iab
o²	3.02	B3 Ia	π	4.66	F2 IV/V	ρ	−	−*8
σ	3.49	K4 III	τ	4.37	O9 Ib	ω	4.01	B2 IV/V

Stars of magnitude 5.5 or brighter which have flamsteed but no Bayer designations:

Flam-steed	Mag. (m$_v$)	Spec. type	Flam-steed	Mag. (m$_v$)	Spec. type	Flam-steed	Mag. (m$_v$)	Spec. type
10	5.23	B2 V	11	5.28	B8/9 III	15	4.82	B1 IV
27	4.42	B3 IIIe	29	4.88	O7 Ia			

Stars of magnitude 5.5 or brighter which have no Bayer or flamsteed designations:

Desig-nation	Mag. (m$_v$)	Spec. type	Desig-nation	Mag. (m$_v$)	Spec. type
HD 043445	5.00	B9 V	HD 043827	5.15	K1 III
HD 044951	5.21	K3 III	HD 046184	5.16	K1 III
HD 047536	5.25	K1 III	HD 047667	4.82	K2 III
HD 049048	5.30	A1 IV/V	HD 049662	5.39	B7 IV
HD 051283	5.29	B2/3 III	HD 051733	5.45	F3 V
HD 053974	5.41	B0.5 IV	HD 055070	5.46	G8 III
HD 056342	5.36	B2 V	HD 056405	5.46	A1 V
HD 056618	4.66	M2 III	HD 057146	5.29	G2 II
HD 057821	4.94	B5 II/III	HD 058155	5.40	B3 V
HD 058215	5.37	K4 III	HD 058286	5.41	B2/3 II/III
HD 058343	5.18	B2 Vne	HD 058535	5.35	K1 III

Fig. 4.8a. The stars of Canis Major and Lepus.

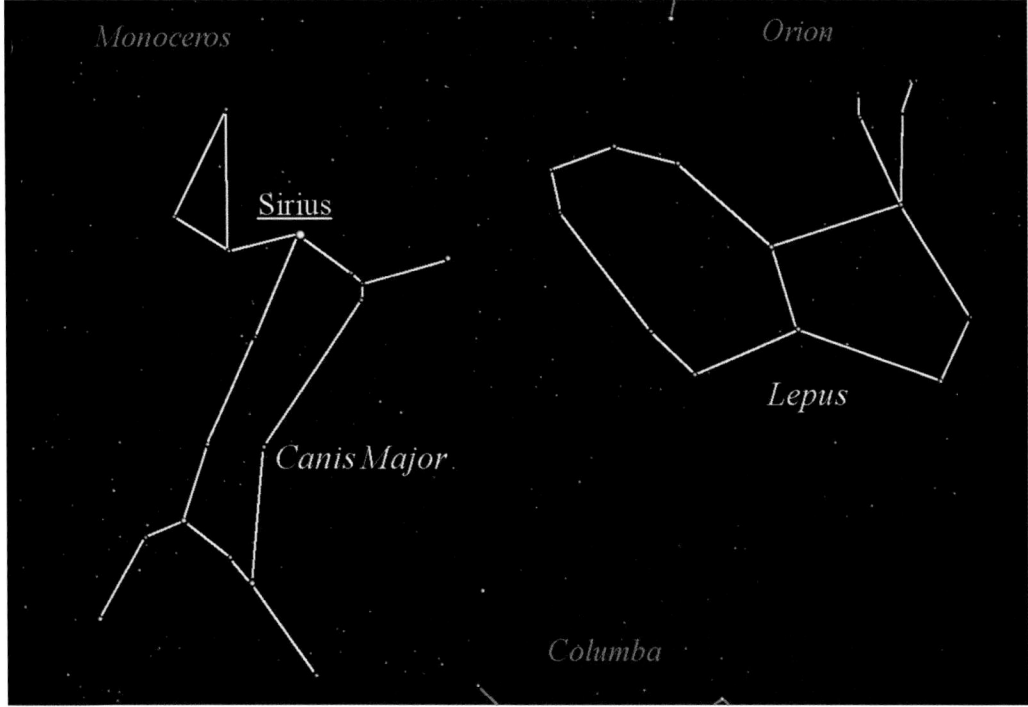

Fig. 4.8b. The figures of Canis Major and Lepus.

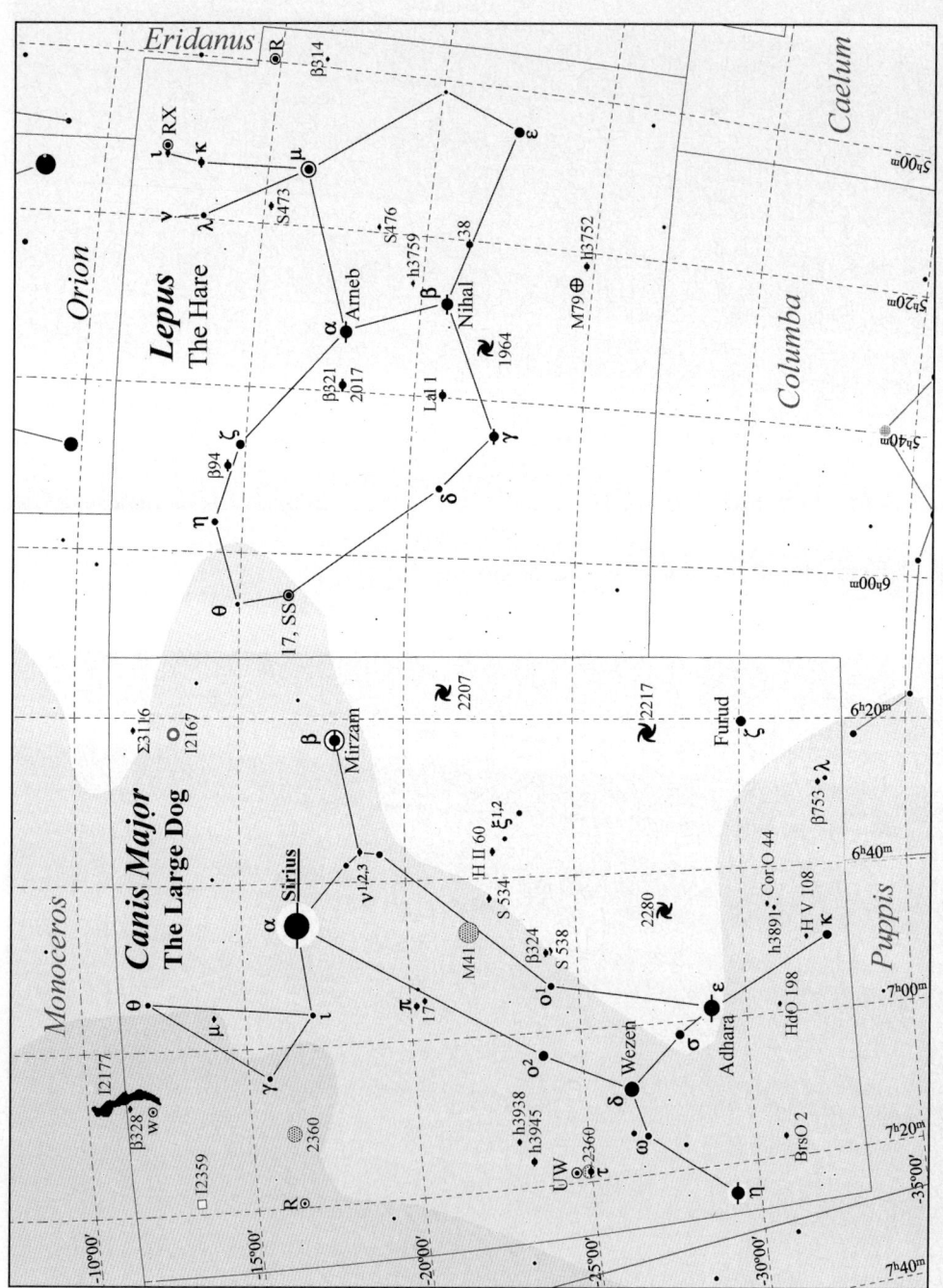

Fig. 4.9. Canis Major and Lepus details.

Table 4.2. Selected telescopic objects in Canis Major.

Multiple stars:

Designation	R.A.	Decl.	Type	m_1	m_2	Sep. (")	PA (°)	Colors	Date/Period	Aper. Recm.	Rating	Comments
Σ3116	06ʰ 21.4ᵐ	−11° 46′	A-B	5.6	9.7	3.8	021	pY, W	1999	3/4″	***	Bright oR star in same FOV
			A-C		11.62	56.6	274	pY, −	1999	15×80	***	
Adhara, β Canis Majoris	06ʰ 22.7ᵐ	−17° 57′	A-B	1.9	10.49	186.8	340	bW, −	1999	15×80	***	Also see variable star listing. a.k.a. 88
β 753	06ʰ 28.7ᵐ	−32° 22′	A-B	5.9	7.6	1.2	043	W, ?	1999	8/9″	****	
6, ν¹ Canis Majoris	06ʰ 36.4ᵐ	−18° 40′	A-B	5.8	7.4	17.8	264	Y, Y	2002	15×80	****	Easy object for small telescopes. ν² is in same low power field. a.k.a. Shj 73
H II 60	06ʰ 36.7ᵐ	−22° 37′	A-B	6.4	9.3	9.2	327	brY, bGr	1999	2/3″	****	
			A-C		12.0	35.5	239	brY, −	1999	3/4″	***	
S 534	06ʰ 42.8ᵐ	−22° 27′	A-B	6.3	8.3	18.2	144	yW, −	1999	15×80	****	
CorO 44	06ʰ 45.0ᵐ	−30° 35′	A-B	6.5	9.8	4.4	221	bY, ?	1999	3/4″	****	
9, α Canis Majoris	06ʰ 45.1ᵐ	−16° 43′	A-B	−1.5	8.5	**6.0**	**118**	W, W	**49.98**	2/3″	****	Separation and PA are changing rapidly *9
h 3891	06ʰ 45.5ᵐ	−30° 57′	A-B	5.7	8.2	4.9	223	pY, ?	1999	2/3″	****	
S 538	06ʰ 49.6ᵐ	−24° 09′	A-B	7.2	8.2	27.0	004	W, pB	1999	10×50	****	In same FOV as S 538, making a lovely double-triple group
β 324	06ʰ 49.7ᵐ	−24° 05′	A-B	6.6	7.9	1.8	209	W, W	1991	5/6″	****	AKA S 537AC
			A-C		8.3	30.1	283	W, B	1999	10×50	***	AKA S 537AD
			A-D		13.0	30.7	003	W, −	1999	6/7″	**	
H V 108	06ʰ 50.4ᵐ	−31° 42′	A-B	5.8	7.7	42.7	066	pY, pB	1999	7×50	***	Similar to Albireo but somewhat fainter and less intensely colored
17 Canis Majoris	06ʰ 55.0ᵐ	−20° 24′	A-B	5.8	8.7	44.9	148	W, O	2002	7×50	*****	Lovely! A, B, C form right triangle, with D farther out and nearly along line of the hypotenuse. a.k.a. H 5 65
			A-C		9.0	49.4	185	W, O	2002	7×50	****	
			A-D		9.7	128.4	187	W, ?	2002	15×80	**	
19, π Canis Majoris	06ʰ 55.6ᵐ	−20° 08′	A-B	4.6	9.6	11.7	018	yW, B	1999	2/3″	****	SW of 17 CMa and in same low power field. a.k.a. H 6 123

(continued)

Table 4.2. (continued)

Multiple stars:

Designation	R.A.	Decl.	Type	m_1	m_2	Sep. (")	PA (°)	Colors	Date/ Period	Aper. Recm.	Rating	Comments
18, μ Canis Majoris	06h 56.1m	−14° 03'	A–B	5.3	7.1	3.2	345	O, B	2004	3/4"	****	Beautiful contrasting colors
			A–C		10.3	85.9	289	O, –	2004	15×80	***	
			A–D		10.6	103.8	061	O, –	2002	15×80	***	
21, ε Canis Majoris	06h 58.6m	−14° 03'	A–B	1.5	7.5	7.0	161	bW, W	2000	2/3"	***	Adhara, a.k.a. Cpo 7
HdO 198	06h 58.7m	−31° 00'	A–B	6.4	10.2	40.1	338	bW, ?	1999	2/3"	**	
			A–C	6.4	11.4	70.0	320	bW, ?	1987	3/4"	**	
β 328	07h 06.7m	−11° 18'	A–B	5.7	6.9	0.6	111	bW, ?	2003	16/18"	***	
			AB–C	5.4	9.1	17.5	350	bW, ?	1999	15×80	***	
h3938	07h 13.8m	−22° 54'	A–B	6.3	9.1	19.4	251	bW, W	1999	15×80	***	
h3945	07h 16.6m	−23° 19'	A–B	5.0	5.8	26.7	052	dO, B	2002	10×50	******	Beautiful, similar to Albireo but 1.7 m (4.8x) fainter
BrsO 2	07h 17.0m	−30° 54'	A–B	6.3	7.8	37.7	183	W, W	1999	7×50	****	Nice multiple, in NGC 2362. Rich background. a.k.a. Fin 313. A–a separation ~0.1"
30, τ Canis Majoris	07h 18.7m	−24° 57'	Aa–B	4.4	10.2	8.6	093	bW, B	2002	2/3"	****	
			Aa–C		11.2	14.2	087	bW, ?	2002	3/4"	***	a.k.a. Opened box cluster
			Aa–D		8.2	84.8	077	bW, ?	2002	7×50	***	a.k.a. h 3948 Aa–B, Aa–C, and Aa–D
31, η Canis Majoris	07h 24.1	−29° 18'	A–B	2.5	6.8	179.0	286	W, B	1999	6×30	****	Beautiful, very similar to Albireo but 1.7 m (4.8x) fainter. a.k.a. Smy 2

Variable stars:

Designation	R.A.	Decl.	Type	Range (m_v)		Period (days)	F (f_r/f_t)	Spectral Range	Aper. Recm.	Rating	Comments
2, β Canis Majoris	06h 22.7m	−17° 57'	β CMa	1.9	2.0	0.25	–	B1II–III	Eye	****	Prototype of rare class of giant B stars. Note CMa B
W Canis Majoris	07h 08.1m	−11° 55'	Ib	6.4	7.9	–	–	C6, 3(N)	7×50	*****	Carbon type star, deep red. Note CMa C
UW Canis Majoris	07h 18.7m	−24° 34'	EB/ KE?	4.8	5.3	4.393	–	O7Ia + Ob	Eye	****	Massive and luminous binary, masses ~23 and ~19 times sun, luminosity ~16,000 times sun
R Canis Majoris	07h 19.5m	−16° 24'	EA/ SD	5.7	6.3	1.136	0.15	F1V	7×50	***	Algol type, stars semi-detached

Star clusters:

Designation	R.A.	Decl.	Type	m_v	Size (')	Brtst. Star	Dist. (ly)	dia. (ly)	Number of Stars	Aper. Recm.	Rating	Comments
M41, NGC 2287	06ʰ 46.1ᵐ	−20° 46'	Open	4.5	38	6.9	2,350	24	80	4/6"	*****	Large, Bright. 4° S of Sirius. Visible to unaided eye in dark skies
NGC 2360	07ʰ 17.8ᵐ	−15° 37'	Open	7.2	12	10.4	5,300	0.4	80	6/8"	***	a.k.a. Opened box cluster
NGC 2362	07ʰ 18.8ᵐ	−24° 57'	Open	4.1	8	4.4	5,100	9	60	2/3"	*****	Bright, beautiful, at ~25 million years old, it is one of the youngest open clusters

Nebulae:

Designation	R.A.	Decl.	Type	m_v	Size (')	Brtst./Cent. star	Dist. (ly)	dia. (ly)	Aper. Recm.	Rating	Comments
IC 2165	06ʰ 21.7ᵐ	−12° 59'	Planetary	10.5	9	17.9	10K	0.4	6/8	**	
IC 2177	07ʰ 05.3ᵐ	−10° 38'	Em/Dark	–	120×40	–	–	–	2/3	***	Can be glimpsed in 7×50 binoculars on clear, dark night
Gum 4, NGC 2359	07ʰ 18.6ᵐ	−13° 12'	Em	–	7×6	–	4.0K	8	8/10	**	a.k.a. Thor's Helmet and Duck Nebula. Faint, ring-shaped

Galaxies:

Designation	R.A.	Decl.	Type	m_v	Size (')	Dist. (ly)	dia. (ly)	Lum. (suns)	Aper. Recm.	Rating	Comments
NGC 2207	06ʰ 16.4ᵐ	−21° 22'	SB	10.8	4.6 × 2.8	110M	80K	93G	8/10	***	Pretty faint, pretty large, elongated
NGC 2217	06ʰ 21.7ᵐ	−27° 14'	SBa	10.2	4.8	54M	75K	19G	6/8	***	Small, very bright, round. Very bright, diffuse nucleus
NGC 2280	06ʰ 44.8ᵐ	−27° 38'	SA	10.5	6.8×2.4	75Mly	150K	40G	12/14	***	Pretty faint, pretty large, slightly elongated. Several arms

Meteor showers:

Designation	R.A.	Decl.	Type	Period (yr)	Duration	Max Date	ZHR (max)	Comet/Asteroid	First Obs.	Vel. (mi/km/sec)	Rating	Comments
None												

Other interesting objects:

Designation	R.A.	Decl.	Type	m_v	Size (')	Dist. (ly)	dia. (ly)	Lum. (suns)	Aper. Recm.	Rating	Comments
None												

M41

JAN 16, 2005 • 06:30 UT
SkyView Pro 6LT - 6" f/8 Newtonian
32 mm Sirius Plössl: 37.5X / 88' FOV
Sketch by Jeremy Perez © 2005, 2008

N
W

b

a

Fig. 4.10. (a) M41 (NGC 2287), open cluster in Canis Major. Astrophoto by Rob Gendler. (b) M41 in Canis Major.

Fig. 4.11. NGC 2359, Thor's Helmet in Canis Major. Astrophoto by Rob Gendler.

LEPUS (lee-pus)	**The Hare**
LEPORIS (lee-**por**-iss)	**9:00 P.M. Culmination:** Feb. 1
Lep	**AREA:** 290 sq. deg., 51st in size

TO FIND: Although it is relatively faint, with its brightest star being slightly fainter than the 2nd magnitude limit, Lepus is not difficult to find. The 10° long oval pattern of stars forming the Hare's body is easily identified about 8° south of Saiph (η Orionis). The stars forming the hare's head lie about 11° south of Rigel β Orionis), with the long ears extending to less than 4° from Rigel. Also, once you have seen the form of a crouching hare, it is easily recognized again (see Figs. 4.8 and 4.9).

KEY STAR NAMES:

Name	Pronunciation	Source
α Arneb	**uhr**-neb, **are**-neb	al-Arnab (Arabic), "the hare"
β Nihal	nih-**hal** or **nigh**-ahl	al-Nihal (Arabic), "the camels quenching their thirst"

Magnitudes and spectral types of principal stars:

Bayer desig.	Mag. (m_v)	Spec. type	Bayer desig.	Mag. (m_v)	Spec. type	Bayer desig.	Mag. (m_v)	Spec. type
α	2.58	F0 Ib	β	2.81	B5 II	γ	4.29	F7 V
δ	3.76	G8 III/IV	ε	3.19	K4 III	ζ	3.55	A2 V
η	3.71	F1 V	θ	4.67	A0 V	ι	4.45	B8 V
κ	4.36	B7 V	λ	4.29	B0.5 IV	μ	3.29	B9 IV:
ν	5.29	B7/B8 V						HgMn

Table 4.3. Selected telescopic objects in Lepus.

Multiple stars:

Designation	R.A.	Decl.	Type	m₁	m₂	Sep. (")	PA (°)	Colors	Date/Period	Aper. Recm.	Rating	Comments
β 314	04ʰ 59.0ᵐ	−16° 23′	A–B	5.9	7.5	**0.8**	**320**	oY, ?	**55**	11/12″	****	Max. sep. was 0.83″ in 2006, min. will be 0.042 in 2033
			AB–C	5.9	1.0	53.3	053	oY, ?	1998	6×30	****	
ι Leporis	05ʰ 12.3ᵐ	−11° 52′	A–B	4.5	9.9	12.0	337	W, pV	1998	2/3″	****	a.k.a. STF 655
4, κ Leporis	05ʰ 13.2ᵐ	−12° 56′	A–B	4.4	6.8	2.0	357	W, W	2004	4/5″	****	a.k.a. STF 661
S473	05ʰ 17.6ᵐ	−15° 13′	A–B	7.0	8.5	20.6	306	W, B	2002	15×80	****	Suitable for small telescopes
S476 (Lepus)	05ʰ 19.3ᵐ	−18° 31′	A–B	6.3	6.5	39.3	018	W, W	2006	7×50	****	Nearly perfectly matched pair for small instruments
38 Leporis	05ʰ 20.4ᵐ	−21° 14′	A–C	6.3	9.6	128.1	021	W, ?	2002	2/3″	***	Called "a most beautiful double" by discoverer William Herschel. a.k.a. h3750
			A–B	4.7	8.5	4.1	277	yW, B	2002	2/3″	****	
h3752	05ʰ 21.8ᵐ	−24° 46′	A–B	5.4	6.6	3.4	097	oY, bW	2002	2/3″	****	Nice pair for small scopes
			AB–C	5.4	9.2	62.7	104	Y, ?	2002	7×50	****	
h 3759	05ʰ 26.0ᵐ	−19° 42′	A–B	5.9	7.3	26.7	318	oY, brB	2002	2/3″	*****	aka BU 320AB
β Leporis	05ʰ 28.2ᵐ	−20° 46′	A–B	3.0	7.5	2.2	339	dY, ?	1993	4/5″	****	aka HJ 3761AC
			A–C	3.0	12.0	58.5	140	?, ?	2000	3/4″	****	
α Leporis	05ʰ 32.7ᵐ	−17° 51′	A–B	2.6	11.2	35.6	157	yW, ?	1999	3/4″	***	a.k.a. HJ 3766AB
			A–C	2.6	11.9	91.2	186	?, ?	1999	3/4″	***	a.k.a. HJ 3766AC
β 321	05ʰ 39.3ᵐ	−17° 51′	A–B	6.7	7.8	0.4	153	W, W	2002	24″	****	This multiple star is a.k.a. NGC 2017. There are at least 6 stars in this group, gravitationally bound into a miniature cluster
				6.7	8.9	89.9	138	?, ?	2002	7×50	***	
				6.7	7.9	76.0	080	?, ?	2002	7×50	***	
				6.7	8.3	132.3	299	?, ?	2002	2/3″	***	
				8.9	9.6	1.5	352	?, ?	2006	6/7″	***	
Lal 1	05ʰ 39.7ᵐ	−20° 21′	A–B	6.9	7.9	10.9	124	W, P	2002	2/3″	***	
			A–C	6.9	11.3	32.7	084	?, ?	1999	3/4″	***	
13, γ Leporis	05ʰ 44.5ᵐ	−22° 27′	A–B	3.6	6.3	97.1	350	Y, rO	2002	6×30	****	Easy pair
			A–C	3.6	11.0	112.5	008	?, ?	1999	3/4″	***	Background star, separation is widening
β 94	05ʰ 49.6ᵐ	−14° 29′	A–B	5.7	8.2	2.3	165	dY, W	2001	4/5″	****	

(continued)

Table 4.3. (continued)

Variable stars:

Designation	R.A.	Decl.	Type	Range (m$_v$)	Period (days)	F (f$_1$/f$_2$)	Spectral Range	Aper. Recm.	Rating	Comments
R Leporis	04^h 59.6^m	−14° 48′	M	5.5	427.07	0.55	N6e(C7e)	4/6″	*****	Hind's Crimson Star. 2° 16′ S (PA 182°) of 64 Eri, m4.8
RX Leporis	05^h 11.4^m	−11° 51′	SRb	4.9	–	–	M6.2III	7×50	****	Lep B
μ Leporis	05^h 13.2^m	−16° 12′	αCVn	3.0	2	–	B9IIIp(Hg-Mn)	Eye	*****	13.5′ W (PA 275°) of m5.0 of ι Lep
SS Leporis	06^h 05.0^m	−16° 29′	Z And	4.8	–	–	A0Veq + gM 1 III	Eye	****	aka 17 Lep

Star clusters:

Designation	R.A.	Decl.	Type	m$_v$	Size (′)	Brtst. Star	Dist. (ly)	dia. (ly)	Number of Stars	Aper. Recm.	Rating	Comments
M79, NGC 1904	05^h 24.2^m	−24° 31′	Globular	7.7	9.6	13.1	42K	118	–	4/6″	****	About 1/2° ENE of h3752
NGC 2017	05^h 39.4^m	−17° 51′	Open?	–	10	8	–	–	8	8/10″	****	See comment for β 321 above

Nebulae:

Designation	R.A.	Decl.	Type	m$_v$	Size (′)	Brtst./Cent. star	Dist. (ly)	dia. (ly)	Aper. Recm.	Rating	Comments
IC 418	05^h 27.5^m	−12° 42′	Planetary	9.3	12″	10.2	10.2	–	12/14″	***	Tiny but bright disk, greenish color

Galaxies:

Designation	R.A.	Decl.	Type	m$_v$	Size (′)	Dist. (ly)	dia. (ly)	Lum. (suns)	Aper. Recm.	Rating	Comments
NGC 1964	05^h 33.4^m	−21° 57′	SB	10.7	5.0×2.1	65M	97K	23G	8/10″	****	NNE–SSW elongation, stellar nucleus

Meteor showers:

Designation	R.A.	Decl.	Period (yr)	Duration	Max Date	ZHR (max)	Comet/Asteroid	First Obs.	Vel. (mi/ km/ sec)	Aper. Recm.	Rating	Comments
None												

Other interesting objects:

Designation	R.A.	Decl.	Type	m$_v$	Size (′)	Dist. (ly)	dia (ly)	Lum. (suns)	Aper. Recm.	Rating	Comments
None											

Fig. 4.12. M79 in Lepus. Astrophoto by Mark Komsa.

TAURUS (**taw**-rus or **tor**-us)	**The Bull**	
TAURI (**taw**-ree or **tor**-ee)	**BEST SEEN:** January 16 (AT 9:00 **P.M.**)	
Tau	**AREA:** 797 sq. deg., 17th in size	

TO FIND: Extend the line of Orion's belt to the right (NW) about 25°, and <u>Aldebaran</u>, the red–orange eye of the bull will be just above the end of the line. The V shape of the Hyades cluster forms the bull's face, the Pleiades cluster marks the top of the bull's shoulder, and β and ζ Tauri, directly above Orion's head, mark the tips of the bull's horns (Figs. 4.13 and 4.14).

KEY STAR NAMES:

Name	Pronunciation	Source
α Aldebaran	al-**deb**-a-ran	Nair Al Dabaran "Bright one of The Follower" (the Hyades rise after the Pleiades)
β El Nath	el-nut or **el**-nath)	Al Natih "The butting one"
γ Hyadum I	**high**-a-dum eye or **high**-a-dum one	First leading one of the Hyades Cluster
δ¹ Hyadum II	**high**-a-dum eye eye or **high**-a-dum two	Second leading one of the Hyades Cluster
η Alcyone	al-**sigh**-o-ne	Mother of Hyrieus by Poseidon (a.k.a. 25 Tau, a Pleiad)
or Al Jauz	(?)	"the Walnut," Arabic
ε Ain	ayn	ain al-thaur, Arabic "the bull's eye"
16 - Celaeno	ke-**lye**-no or seh-**lee**-no)	Lost Pleiad said to be struck by lightning
17 - Electra	eh-**lek**-truh	Pleiad who covered her eyes at destruction of Troy
19 - Taygete	tah-**ih**-geh-tuh or tay-**ih**-jee-tuh)	Pleiad, mother of Lacedaemon by Zeus
20 - Maia	**mah**-yuh or **may**-yeh)	First-born and most beautiful Pleiad
21 - Sterope I	**steh**-row-peh eye	a.k.a. Asterope [like Haydum I and II, the Roman]
22 - Sterope II	**steh**-row-peh eye eye	a.k.a. Asterope [numerals may be pronounced "one" and "two"]
23 - Merope	**meh**-row-peh)	The Pleiad who hid face because she was ashamed to be the only sister to marry a mortal
27 - Atlas	**ut**-lus or **at**-las	Father of the Pleiades
28 - Pleione	plee-**ih**-oh-nee or **plee**-yo-nee	Mother of the Pleiades

Magnitudes and spectral types of principal stars:

Bayer desig.	Mag. (m_v)	Spec. type	Bayer desig.	Mag. (m_v)	Spec. type	Bayer desig.	Mag. (m_v)	Spec. type
α	0.87	K III	β	1.65	B7 III	γ	3.65	G8 III
δ^1	3.77	G8 III	δ^2	4.80	A5 Vn	δ^3	4.30	A2 IV
ε	3.53	K0 III	ζ	2.97	B4 IIIp	η	2.85	B7 III
θ^1	3.84	G7 III	θ^2	3.40	A7 III	ι	4.62	A7 V
κ^1	4.21	A7 IV/V	κ^2	5.27	A7 V	λ	3.41	B3 V
μ	4.27	B3 IV	ν	3.91	A1 V	ξ	3.73	B6 III
o	3.61	G8 III	π	4.69	G8 III[*8]	ρ	4.65	A8 V[*8]
σ^1	5.08	A4 m	σ^2	4.67	A5 Vn	τ	4.27	B3 V
υ	4.28	A8 Vn[*8]	ϕ	4.97	K1 III[*8]	χ	5.38	B9 V[*8]
ψ	5.21	F1 V[*8]	ω	4.93	A3 m			

THE PLEIADES:

Bayer desig.	Mag. (m_v)	Spec. type	Bayer desig.	Mag. (m_v)	Spec. type	Bayer desig.	Mag. (m_v)	Spec. type
16	5.40	B5 V	17	3.72	B6 III	18	5.66	B8 V
19	4.30	B6 V	20	3.87	B8 III	21	5.76	B8 V
22	6.43	A0 Vn	23	4.14	B6 IV	25	2.85	B7 III (η Tau)
27	3.62	B8 III	28	5.05	B7 p			

Fig. 4.13a. The stars of Taurus and adjacent constellations.

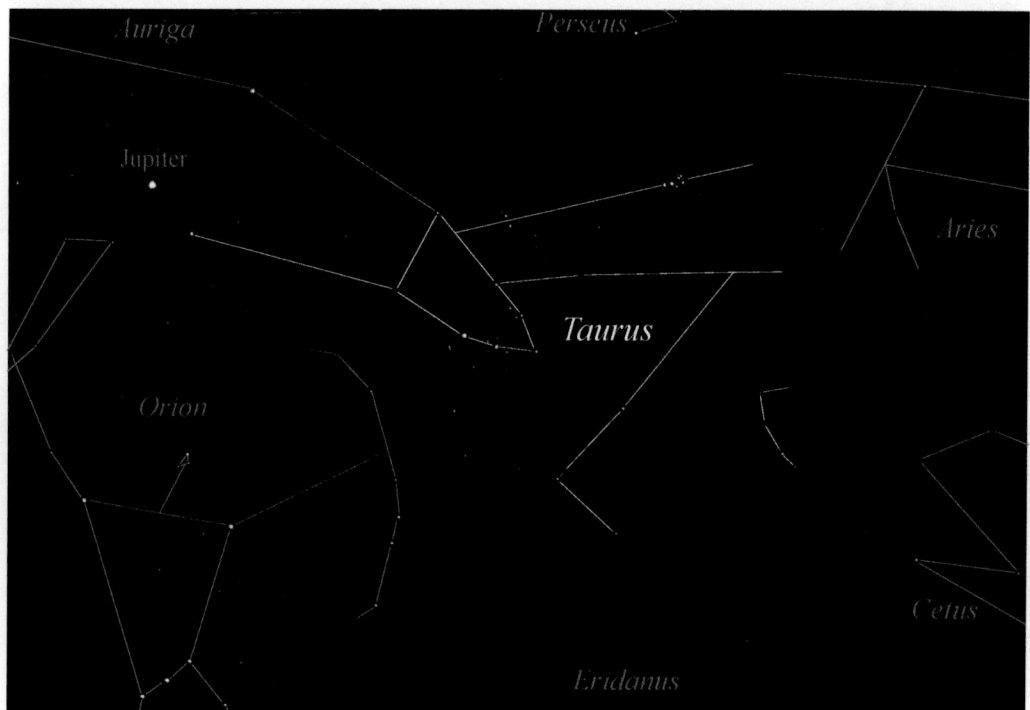

Fig. 4.13b. The figures of Taurus and adjacent constellations.

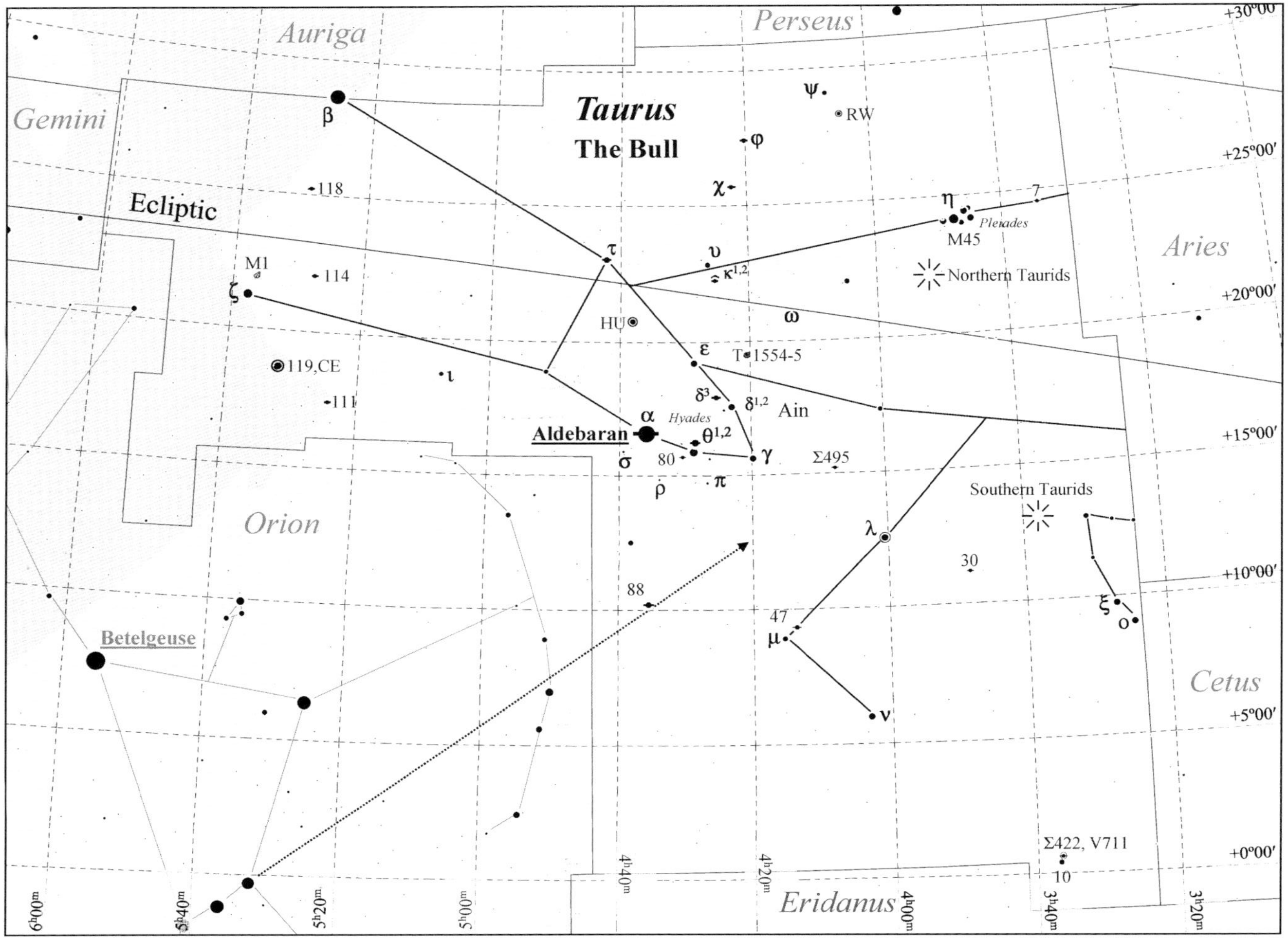

Fig. 4.14. Taurus details.

Table 4.4. Selected telescopic objects in Taurus.

Multiple stars:

Designation	R.A.	Decl.	Type	m_1	m_2	Sep. (")	PA (°)	Colors	Date/Period	Aper. Recm.	Rating	Comments
7 Tauri	03h 34.3m	+24° 28'	A-B	6.6	6.9	**0.7**	**353**	W, W	**522.16**	13/14"	****	Nearly perfect test object for resolving power of 6 or 7" telescope. a.k.a. ΣΔ12
Σ422, Taurus	03h 36.8m	+00° 35'	AB-C	6.0	9.9	22.9	055	W, –	2006	15×80	****	Pale but nice colors. Binary, period 2,102 years. a.k.a. h 3251AB
			A-B	5.8	8.7	**6.7**	**272**	Y, bW	**2101**	2/3"		
19 Tauri, Taygeta	03h 45.2m	+24° 28'	A-B	4.4	8.1	68.9	329	W, V	1925	7×50	****	In Pleiades. 10' N of 10 Tauri
21 and 22 Tauri	03h 45.9m	+24° 33'	A-B	5.8	6.4	168.0	130	bW, W	2000	6×30	****	In Pleiades
25, η Tauri, Alcyone	03h 47.5m	+24° 06'	A-B	2.8	6.3	117.5	290	bW, W	2003	6×30	****	In Pleiades
30 Tauri	03h 48.3m	+11° 09'	A-C	2.8	8.2	180.8	313	bW, W	2002	7×50	****	Very nice colors
27 and 28 Tauri	03h 49.2m	+24° 03'	A-B	5.1	9.8	9.2	059	bW, pR	2000	2/3"	****	In Pleiades. Atlas and Pleione. Pleione is also variable. a.k.a. Σ 453
			A-B	3.6	5.1	300.0	180	bW, bW	1929	6×30		
Σ495, Taurus	04h 07.7m	+15° 10'	A-B	6.1	8.8	3.8	223	Y, Y	2006	2/3"	****	May be long period binary. Orbit not determined. a.k.a. Σ 495
47 Tauri	04h 13.9m	+09° 16'	A-B	5.1	7.3	1.2	341	dpY, dpY	2001	8/9"	****	
52, φ Tauri	04h 20.4m	+27° 21'	AB-C	4.8	13.2	28.7	228	dpY, –	1999	6/7"	****	Very nice colors but large brightness difference
			A-B	5.1	7.5	49.2	256	dpY, B	1999	7×50		
59, χ Tauri	04h 22.6m	+25° 38'	A-B	5.1	12.0	118.4	025	dY, –	1997	3/4"	****	
			A-B	5.4	8.5	19.1	025	W, B	2004	15×80		
65 and 67, κ¹ and κ² Tauri	04h 25.4m	+22° 18'	A-B	4.2	5.3	339.7	174	pY, pY	2002	6×30	****	Naked eye pair for some people. An outlying member of the Hyades
68, δ³ Tauri	04h 25.5m	+17° 56'	A-CD	4.2	9.0	184.3	170	pY, –	1873	7×50	****	a.k.a. Σ 541CD
			C-D	9.5	9.8	5.4	329	–, –	2000	2/3"		a.k.a. Kui 17AB
			A-B	4.3	7.9	1.9	340	W, –	1999	5/6"		a.k.a. H 6 101AC
			A-C	4.3	8.7	76.8	234	–, –	1997	7×50		
θ¹, θ² Tauri	04h 28.7m	+15° 52'	A-B	3.4	3.9	336.7	348	W, pY	2002	Eye	*****	Easy unaided eye double
80 Tauri	04h 30.1m	+15° 38'	A-B	5.7	8.1	**1.6**	**016**	W, –	**180**	5/6"	****	Passed max. of 1.75" in 1985, opening to max. of 0.08" in 2065. a.k.a. Σ 554
88 Tauri	04h 35.7m	+10° 10'	A-B	4.4	6.6	0.3	150	W, –	1999	–	***	Widening. ~0.01"/year
			A-B	4.3	7.84	69.1	300	W, –	2003	7×50		a.k.a. Chr 18Aa

87 Tauri, Aldebaran	04h 35.9m	+16° 31'	A-B	0.9	13.6	30.8	111	O, –	1962	7/8"	****	Difficult because of great brightness difference (100,000 times). a.k.a. β 550AB
			A-C	0.9	11.3	133.2	031	O, –	1997	3/4"		
			A-E	0.9	12	31	331	O, –	1899	3/4"		
			C-D	11.3	13.7	1.3	271	–, –	1962	8/9"		
94, τ Tauri	04h 42.2m	+22° 58'	A-a	4.3	7.0	0.3	049	bW, W	2003	–	***	a.k.a. Mca 16Aa
				4.2	7.0	63	214	bW, W	1999	6×30		
111 Tauri	05h 24.4m	+17° 23'	A-B	5.0	9	102.7	271	Y, O	2002	7×50	****	Nice colors. a.k.a S 478
114 Tauri	05h 27.6m	+21° 56'	A-B	4.88	10.9	38.1	348	W, –	1997	3/4"	***	a.k.a h 365
			A-C	4.88	10.4	58.7	194	W, –	1997	2/3"		
			A-D	4.88	11.6	74.1	280	W, –	1997	3/4"		
118 Tauri	05h 29.3m	+25° 09'	A-B	5.83	6.7	4.7	208	W, Y	2006	2/3"	****	
			AB-C	5.4	11.6	140.6	099	W, –	1997	15×80		

Variable stars:

Designation	R.A.	Decl.	Type	Range (m$_v$)		Period (days)	F (f$_r$/f$_t$)	Spectral Range	Aper. Recm.	Rating	Comments	
V711 Tauri	03h 36.8m	+00° 35'	RS	5.7	5.9	2.841	–	G5IV/Vea+K1IVea	6×30	****	11' N (PA 353°) of 10 Tau (m4.29). #	
Pleione, 28, BU Tauri	03h 49.2m	+24° 08'	γ Cas	4.8	5.5	Irr	–	B8Vne	eye	****	Hot, rapidly rotating B type star, losing mass at equator. ##	
35, λ Tauri	04h 00.7m	+12° 29'	EA/DM	3.4	3.9	3.953	0.15	B3V+A4IV	Eye	****	The point star in the "V" of the Hyades	*10
RW Tauri	04h 03.9m	28° 08'	EA/SD	8.0	11.6	2.769	0.14	B8Ve+K0IV	4/6"	*****	One of greatest ranges (4.49 m, 62.5×) known for eclipsing binary.$	*11
T Tauri	04h 22.0m	+19° 32'	T Tau	9.3	13.5	–	–	F8V- K1IVVe	10/12"	*****	See Text.##	*12
HU Tauri	04h 38.3m	+20° 41'	EA/SD	5.9	6.7	2.056	0.15	B8V	6×30	****	#$	
119, CE Tauri	05h 32.2m	+18° 36'	SRc	4.2	4.5	165	–	M2Iab-Ib	Eye	*****	a.k.a. "Ruby Star."##	*13

(continued)

Table 4.4. (continued)

Star clusters:

Designation	R.A.	Decl.	Type	m_v	Size (')	Brst. Star	Dist. (ly)	dia. (ly)	Number of Stars	Aper. Recm.	Rating	Comments	
Pleiades, M45	03h 47.0m	+27° 07'	Open	1.2	110	2.87	410	7	100	7×50	*****	Beautiful in any aperture. Note Tau F. Lum. 4,800 suns	*14
Hyades	04h 26.9m	+15° 52'	Open	0.5	330	3.4	150	14	380	7×50	****	Best in 7×50 binoculars. Fills entire FOV. Luminosity = 1,150 suns	*15

Nebulae:

Designation	R.A.	Decl.	Type	m_v	Size (')	Brst./Cent. star	Dist. (ly)	dia. (ly)	Aper. Recm.	Rating	Comments	
NGC 1554–5	04h 21.8m	+19° 32'	Reflection	var	var, 1–7	star	–	–		*	"Hind's Variable Nebula" Changes in size, brightness and shape	*16
Crab Nebula, M1	05h 34.5m	+22° 01'	SN Rem	8.2	6×4		6,300	11		***	Expanding supernova remnant. Seen in Europe and America in 1054 AD	*17

Galaxies:

Designation	R.A.	Decl.	Type	m_v	Size (')	Dist. (ly)	dia. (ly)	Lum. (suns)	Aper. Recm.	Rating	Comments
None											

Meteor showers:

Designation	R.A.	Decl.	Type	Period (yr)	Duration	Max Date	ZHR (max)	Dist. (ly)	Comet/Asteroid	First Obs.	Vel. (mi/km/sec)	Rating	Comments
Southern Taurids	03h 38m	+13°	Annual	10/01	11/25	11/05	5		2004 TG	1869	17	II	Giuseppe Zezioli (Bergamo, Italy)
Northern Taurids	03h 52m	+22°	Annual	10/01	11/25	11/05	5		2P Enke	1869	18	II	T. W. Backhouse (Sunderland, England)

Other interesting objects:

Designation	R.A.	Decl.	Type	m_v	Mass (suns)	Dia. (suns)	Dist. (ly)	Planet. Mass (J)	Dist. (AU)	Period (days)	Eccentricity	Comments
Ain, Tauri	05h 34.5 m	+22° 01'	ExoPlnt	3.5	2.7	13	155	7.55	1.93	645.6	0.152	Ain is the most massive star known to have a planetary system

Fig. 4.15a. M45, The Pleiades, an open cluster in Taurus. Astrophot by Warren Keller.

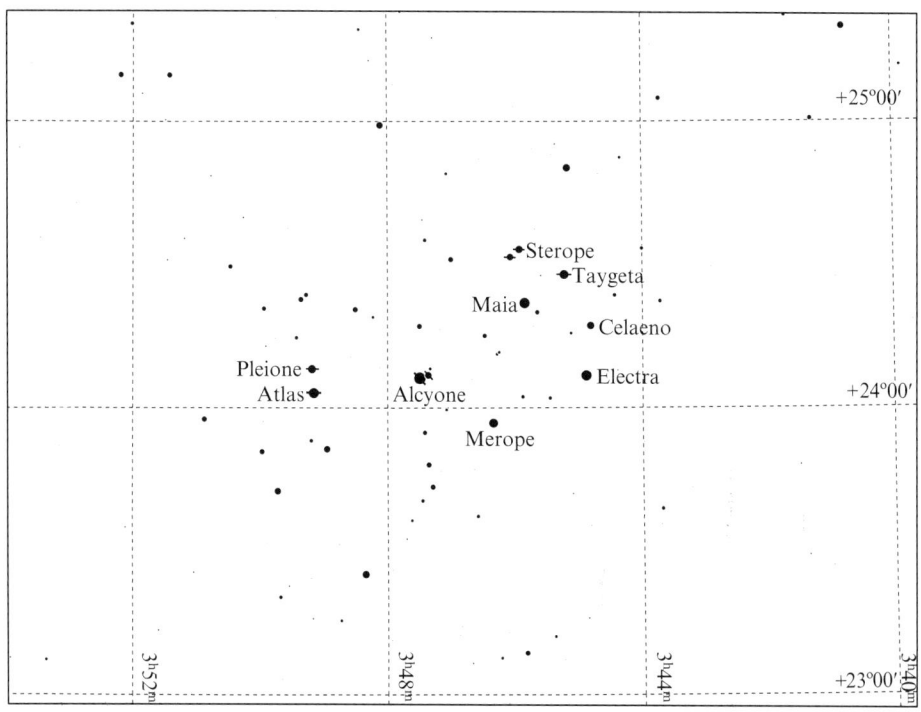

Fig. 4.15b. M45, The Pleiades detail chart.

Fig. 4.16a. M45, The Hyades, an open cluster in Taurus.

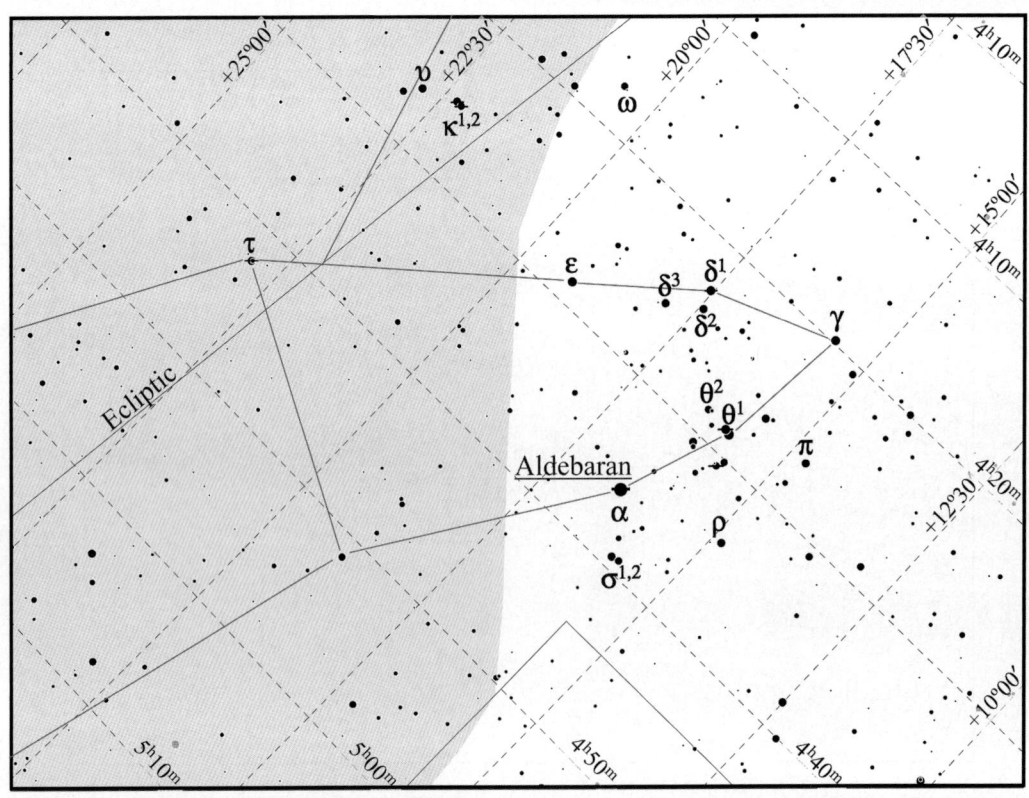

Fig. 4.16b. The Hyades detail chart.

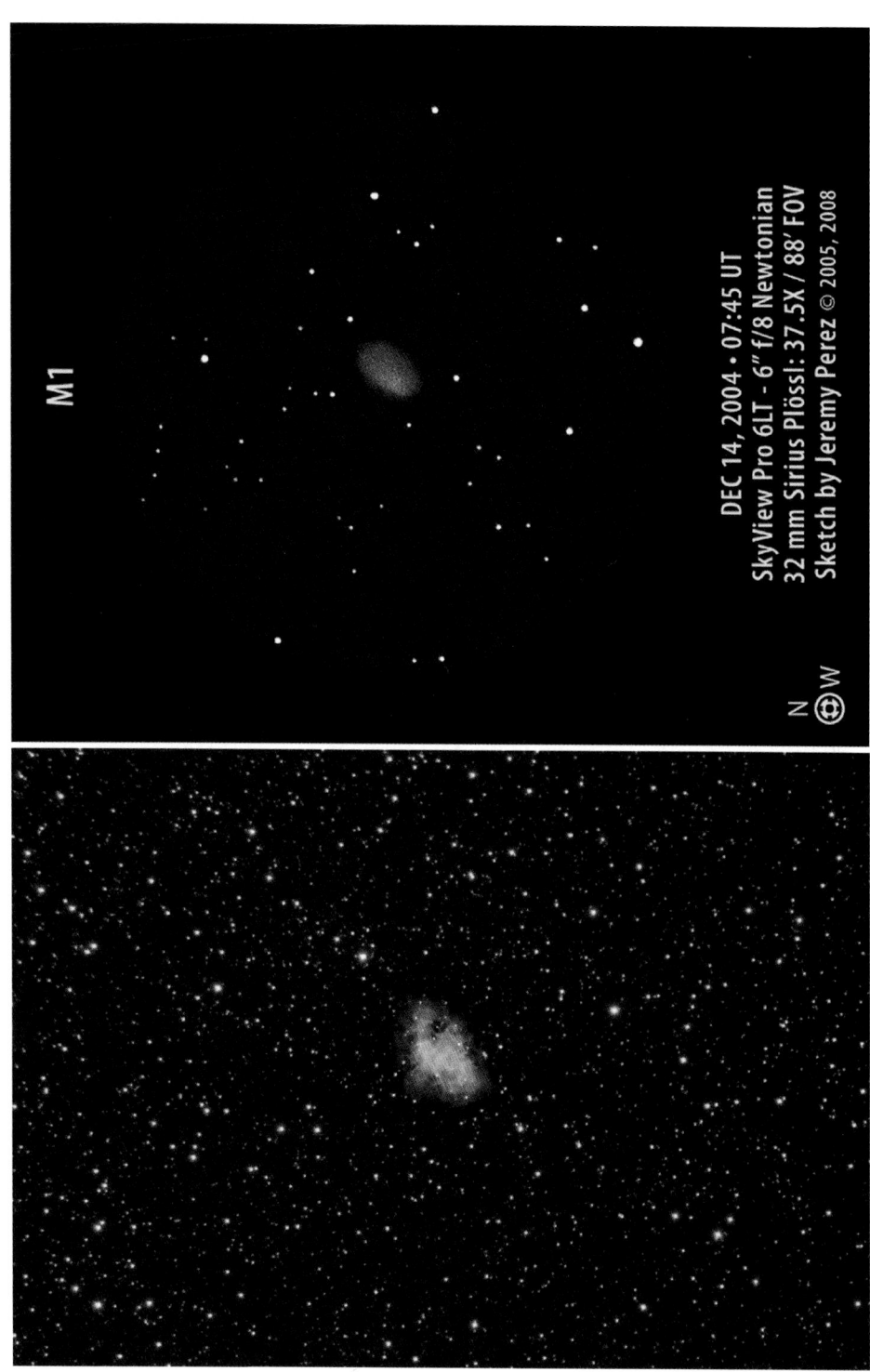

M1

DEC 14, 2004 • 07:45 UT
SkyView Pro 6LT - 6" f/8 Newtonian
32 mm Sirius Plössl: 37.5X / 88' FOV
Sketch by Jeremy Perez © 2005, 2008

N ⊕ W

Fig. 4.17. M1, the Crab Nebula, a supernova remnant in Taurus. Astrophoto by Warren Keller.

Notes:

1. Rigel (β Orionis) is the seventh brightest star in the sky, with a visual magnitude of 0.18. Except for occasional times when variable Betelgeuse (α Orionis) becomes exceptionally bright, Rigel is the brightest star in Orion. Spectroscopic measurements indicate that it is a blue supergiant of class B8 Iab with an estimated distance 700–900 light years. Measurements from Hipparchos yield an estimate of 773 light years, near the center of the spectroscopic range. The Hipparchos distance indicates an absolute magnitude of −6.75, or a luminosity of ~43,000 suns. Rigel's companion, faint by comparison, is a spectroscopic binary with a total luminosity about 100 times more than our sun. Rigel has a mass of ~17 suns and a diameter of ~70 suns.

2. Betelgeuse (α Orionis) is the brightest variable star in the sky. When at its brightest, it is the 5th brightest star, and at its faintest, the 20th. It pulsates between about 480 and 800 million miles in diameter. The normal range of variation is magnitude 0.4–1.3, but it sometimes rises to about 0.0 magnitude or brighter. Betelgeuse is a very slow variable, taking almost 6 years to complete each cycle. Betelgeuse has an average visible luminosity of ~10,000 suns, but a total luminosity (including invisible infra-red and ultra-violet rays) of ~63,000 suns.

3. V1016 Orionis is a very hot, bluish variable.

4. A second outburst of V529 Orionis occurred in 2004 ($m < 8.3$), and possibly others have occurred. This star deserves monitoring for possible future outbursts.

5. M43 is actually part of the same nebula as M42, but a dark lane makes them appear to be two separate objects, with M43 just north of M42.

6. The Orionid meteor shower is very reliable but occasionally has outbursts, such as the one in 1996 when the count was two to three times normal. The η Aquarids also originate from Comet Halley.

7. Although Mintaka (δ Orionis) is one of the belt stars, it may not be a member of the Collinder 70 cluster or association, since it appears to be considerably more distant than ε or ζ and the other stars in this area. Collinder 70 may not even be a true cluster, because its diameter of 57 light years would be exceptionally large for a true open cluster.

8. Bayer did not use π, ρ, υ, φ, χ, or ψ in Canis Major. All these letters were added by later astronomers.

9. Sirius is not only the brightest star in the night sky, but one of the nearest, and a binary star with an orbital period of only 50 years.

10. 35 (λ) Tauri is the point star in the "V" shape of the Hyades cluster. It exhibits deep primary and shallow secondary eclipses with a continuous variation throughout its entire cycle.

11. RW Tauri has a bright B8 primary which is totally eclipsed by the much less bright K class subgiant companion. This eclipse lasts 9 h, of which 84 min is total. It has one of the greatest ranges known for an eclipsing binary (4.49 magnitudes, or a maximum to minimum brightness ratio of 62.5 times).

12. T Tauri was discovered in 1852 by the well-known asteroid hunter, John Russell Hind, who also discovered R Leporis (a.k.a. Hind's Crimson Star). NGC 1555 (Hind's Variable Nebula) is just west of T Tauri. T Tauri is the prototype of a group of low mass pre-main sequence stars which are just beginning to shine by fusing hydrogen. These stars often exhibit relatively strong lines of lithium (Li), which is usually absent from more advanced stars because it is quickly consumed in the fusion reactions taking place in the star's center.

13. CE (119) Tauri is a deep red star which was given the name "Ruby Star" by Abdul Ahad when he identified it as the second reddest star (after Herschel's Garnet Star) visible to the unaided eye in April 2004. Its color is most obvious in binoculars.

14. M45, the Pleiades, is beautiful at any aperture. About 300 stars are known to be members of the cluster with some outliers as far as 20 light years from the center.

15. The Hyades is another nearby cluster. It appears best in 7 × 50 binoculars, where it fills the entire field-of-view. Aldebaran is not a member of the cluster, but a foreground star only 65 light years distant.

16. Hind's variable nebula is illuminated by T Tauri and has been alternately visible and invisible to telescopic observers several times since its discovery in 1852. It has been steadily brightening since about 1930 but can still be a challenge for visual observers.

17. M1, the Crab Nebula, is one of the most famous and most studied objects in the sky. It is the remnant of a supernova explosion which was observed in 1054 AD. The explosion was observed widely in Europe and in the Americas, especially the USA. Southwest. Chinese records tell of a "Guest Star" appearing near ζ Tauri on July 4, 1054 AD (if computed by our present calendar), being visible in daylight for 23 days, and finally fading from sight almost 2 years later. Native American petroglyphs in the western USA show a bright star near the rising crescent moon, as it would have appeared on the morning of July 5, 1054 AD.

CHAPTER FIVE

The Assassin and His Associates

July and August

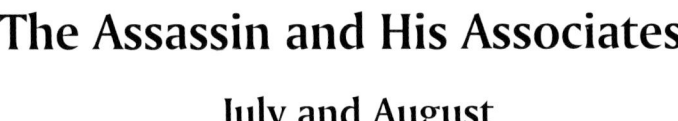

CORONA AUSTRALIS	OPHIUCHUS	SAGITTARIUS	SCORPIUS	SERPENS
The Southern Crown	The Serpent Holder	The Archer	The Scorpion	The Snake

In Chap. 4, we related the story of Orion and his death at the hands of the gods. After Orion's death and an attempt to revive him (described in the next chapter), the gods divided the characters into two groups and separated them to prevent any further violence. Orion, his hunting dog, the hare which they hunted and the bull which distracted Orion so that the giant scorpion could carry out his deadly task were placed in a contiguous group in the winter sky. The remainder of the characters in this story was placed in a second contiguous group in the summer sky.

Two of the latter characters, Sagittarius and Ophiuchus, have their own stories which preceded their involvement with Orion, and we relate them here, along with short descriptions of the origins of the other three.

The stars of Scorpius have apparently been recognized as representing the stinging desert arachnid since 5000 BC. The people of the Tigris-Euphrates area drew zodiacs with six constellations or signs, including the scorpion, and it was recognized in Akkadia, Persia, Turkey, Egypt, India, and other Middle Eastern areas.

When the gods decided to punish Orion for his hubris, Hera created a giant scorpion from the desert sand. She directed this scorpion to attack Orion from behind while the

P. Simpson, *Guidebook to the Constellations: Telescopic Sights, Tales, and Myths*,
Patrick Moore's Practical Astronomy Series, DOI 10.1007/978-1-4419-6941-5_5,
© Springer Science+Business Media, LLC 2012

bull made a frontal attack. The scorpion was able to fatally sting Orion only because the great hunter was distracted by the charging bull. After Orion's death, the gods placed all the characters in the sky to remind humans of the dangers of excessive pride and boastfulness (see Chap. 4 for additional details).

The scorpion was placed in the summer sky, opposite Orion. When Orion rises, Scorpius sets. When the scorpion reappears, the hunter takes his leave. They can never again come into contact and conflict. After its placement in the sky, the giant scorpion also played a part in the death of Apollo's son, Phaeton. When the inexperienced youth was granted the privilege of driving the sun chariot as proof of his heavenly origin, he allowed the chariot to stray from its regular path. When the chariot passed too close to the scorpion, the arachnid lashed out with its potent stinger. The horses spooked and randomly went so high that the earth froze, and so low that it burned. Seeing the situation out of control, Zeus slew Phaeton with a thunderbolt.

The original heavenly scorpion was an even more formidable beast than today's Scorpius. Most of the stars we now know as the constellation of Libra formed the "Northern Claw" and the "Southern Claw" of the giant scorpion. In 26 BC, Julius Caesar directed Roman astronomers to cut off the scorpion's claws to form Libra. The reason for this change was probably the fact that by then, precession had moved the autumnal equinox into the region where the stars of Libra are located. Since the autumnal equinox is where sun is located when the daylight and nighttime hours were equal, or balanced in the fall, a pair of scales (a balance) is an appropriate symbol. Two of the stars in Libra still carry names which indicate their former positions in the constellation Scorpius.

The stars of Ophiuchus represent the mythical healer Asclepius, the son of Apollo and Coronis, a nymph. The story of Asclepius's birth and how he came to be trained by Chiron is told in Chap. 8. Chiron taught Asclepius about many things, but Asclepius became especially adept at surgery and the medical use of natural plants to cure the ills of all living creatures. One day, a pair of snakes entered Asclepius's home. Unthinkingly, Asclepius killed one of the pair. The other fled, only to return soon afterward with a leafy branch in its mouth. Asclepius watched in amazement as the second snake rubbed the branch all over the body of its mate, and the dead snake came back to life. Asclepius was amazed by this, and became determined to add this herb to his stock of medicinal plants. He had never seen this plant before, so he followed the two snakes until they led him to the plant's source.

With this plant added to his already powerful collection of remedies, Asclepius saved the lives of several heroes, and became recognized as the founder of medicine among humans. He also raised several important characters, including Lycurgus, Campaneus, and Tyndareus, from the dead. With these skills, Asclepius became so famous that ancient Greek medical schools were called Asclepions, and the priest–physicians Asclepiadae. Asclepius's symbol became a staff in the form of a rough-hewn branch with a snake coiled around it. This symbol eventually merged with Hermes's symbol, a rod with two ribbons coiled around it and a pair of wings at the top. In the merger, the ribbons were transformed into snakes, and the resulting caduceus is the symbol of modern physicians in the USA and several other countries. Physicians in many other countries use a symbol with a single snake, which more closely resembles Asclepius's original staff.

Although humans and many of the gods admired Asclepius, his raising of the dead angered Hades, the ruler of the underworld. Hades feared that Asclepius's ministrations would soon depopulate his kingdom of the dead. Hades wanted to kill Asclepius, but he was loath to harm Apollo's son. Finally, Asclepius went too far, and tried to raise Orion from his death by the sting of a giant scorpion. Hades then took his complaints to the council of Olympians. The council judged his complaint valid, and Zeus slew Asclepius with one of his thunderbolts.

At Apollo's request, Asclepius was placed in the heavens to honor his contributions to medicine. Like the scorpion and the centaur, he was placed in the summer sky, far away from the dead hero Orion. Asclepius is shown holding one of the snakes which led him to the life-giving herb and adjacent to both Scorpius and Sagittarius. The snake holder and the snake were originally a single constellation, but were eventually separated into two more conveniently sized constellations called Ophiuchus (Asclepius) and Serpens (the snake). In the case of Serpens, the constellation is divided where it crosses Ophiuchus. The two parts are informally known as Serpens Caput (the head) and Serpens Cauda (the tail). From head to tail, the constellation of Serpens twists and turns across over 80° of the summer sky.

The constellation Sagittarius is another with a long and varied history. It first appeared in Sumerian clay tablets as representing their war god, Nergal. Nergal evolved through many guises, as a sun god whose summer heat brought disease and destruction, a war god, a storm god, the god of the planet Mars, the ruler of the underworld and possibly others. Sagittarius has followed a similar pattern. At various times, it has represented Nergal, Enkidu, Crotus, Chiron, and Pholus. It has been shown as half man, half goat (both with two legs and four legs), and as half man, and half horse. There has also been much confusion about which centaur was represented by Sagittarius and which by Centaurus. Sagittarius is included in both the Heracles (Hercules) and Orion myths. Since the name Sagittarius is identical with the Roman word for arrow or archer, we have chosen here to have it represent Chiron (sometimes spelled Cheiron) as portrayed in both the Heracles and Orion myths.

Chiron was the offspring of Chronos (who had assumed the form of a horse to avoid detection by his wife, Rhea) and Philyra, his niece. This strange coupling resulted in a half-man half-horse, the first of the centaurs. As the child of two immortals, Chiron was himself immortal. Although the other centaurs had a different origin and were mortal, they shared the same shape with Chiron. In spite of their differences, or perhaps because of them, Chiron was the king of the centaurs, and one of the few who was not rowdy and occasionally very violent. He was so gentle, wise, and learned that he became the tutor and guardian of many of the sons of the gods and of great Greek heroes, including Achilles, Asclepius, Jason, and Heracles. He taught them medicine, archery, and all the other human skills they needed to become great men.

All of Chiron's charges loved him, but none more than Heracles. However, Heracles was eventually the cause of his death. During his quest to capture the Erymanthian Boar, Heracles accidentally wounded Chiron with one of his poisoned arrows. This wound was incurable, and so excruciatingly painful that Chiron relinquished his immortality to escape it. (For more details, read Heracles fourth labor in Chap. 6).

After the death of Orion, and Asclepius's attempted revival, Chiron was placed in the sky at Apollo's request in honor of his teaching great heroes and other humans the knowledge to improve their lives. His skill at archery got him the position as the guardian of the scorpion. Sagittarius' arrow eternally points directly at Antares, the heart of the scorpion. This is to prevent the scorpion from leaving its place in the summer sky to again attack Orion. In the sky, Chiron is shown facing the scorpion, with his bow and arrow in one hand and his martial cape blowing behind his head and over the back of the equine part of his body.

Ophiuchus is also a very old constellation, and at one time was included in the zodiac as the 13th sign or constellation. Because rapid changes in the appearance of the moon and its apparent motion relative to the stars are much more obvious than the slow movement of the sun through the constellations, many early cultures based their calendars on the moon's movements. This led to years of 13 months (or "moonths") rather than the modern 12. These cultures usually also had zodiacs of 13 signs. Although Ophiuchus is not part of the modern zodiac, the sun does pass through it, and actually spends 2½ times as much time within its borders as it does in Scorpius, which is in the zodiac.

About 10° south of the brightest stars of Sagittarius lies the small and relatively faint constellation of Corona Australis, the Southern Crown. Although it played no direct part in the story of Chiron, it has long been associated with Sagittarius, even being called "Corona Sagittarii" by some Greeks and the "Golden Crown of Sagittarius" by some Romans. In spite of its small size and dim stars, Corona Australis has long been recognized by many cultures as a separate constellation, although often representing different objects. It has also been known as the "Southern Wreath," the "Little Crown," the "Southern Coil," and the "Crown of Eternal Life." The Arabs saw it as a tortoise, and as the Nest of the nearby constellation of the ostrich. They also sometimes called it the Woman's Tent.

Corona Australis was one of Ptolomy's original 48 constellations, and has probably been recognized as a separate constellation associated with Sagittarius for at least 4,000 years (Fig. 5.1).

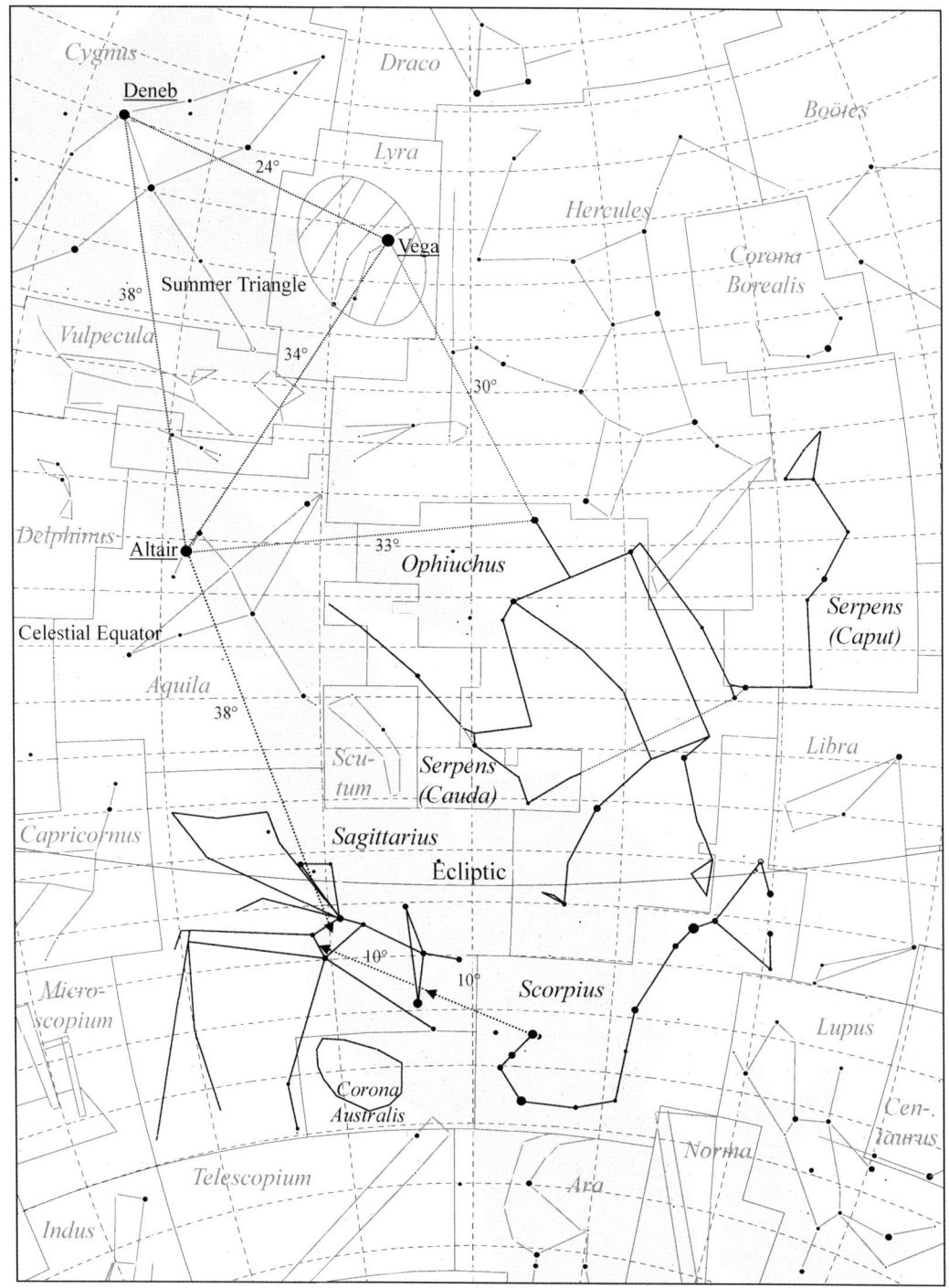

Fig. 5.1. Assassin's group finder chart.

SCORPIUS (**skor**-pee-us)	The Scorpion
SCORPII (**skor**-pee-ee)	9:00 P.M. **Culmination:** July 19
Sco	**AREA:** 497 sq. deg. (33rd in size)

KEY STAR NAMES:

Name	Pronunciation	Source
α Antares	an-**tah**-reez	Ἀντάρης (Greek) "like Ares" from its red color and the comparison when Aries (Mars) was nearby
β Acrab	uhk-**rab**	al-ʿaqrab (Arabic) "the scorpion"
δ Dschubba	**jub**-buh	jabhat al-ʿaqrab (Arabic) "the scorpion's forehead"
θ Girtab	gir-**tab**	GIR.TAB (Babylonian) "the scorpion"
λ Shaula	**shou**-lah	al-shaula (Arabic), "the Scorpion's Stinger"
σ Al Niyat	uhl-nih-**yat**	al-niyāt (Arabic) with τ,"The Arteries" because Antares, the heart of the scorpion, was between them. Later, the name applied to σ only
υ Lesath	**les**-uht	al-latkha,(Arabic), "the Spot," corrupted in the middle ages to alascha, then al-latkha and Lesath

TO FIND: Scorpius is bright enough and distinctive enough that it is practical to find it without reference to other constellations. Scorpius culminates at 9:00 PM on July 19. On that date, or others adjusted by +4 min/day earlier or −4 min/day later, look directly south. Antares, the brightest star in Scorpius, will be at an altitude of 90°- (26°+ your latitude in degrees, with North being positive). Antares is easily recognized by its brightness and distinct orange–red color. There are no other first magnitude stars within 39° of Antares (45° if you are north of +30° latitude) to cause confusion. Antares has another distinctive feature; it is flanked by two relative bright third magnitude stars, one about 2° to either side forming a gently curved arc. From Antares, the curve of the Scorpion's body and tail can easily be traced to the south and east until it curves back north to end in a fairly bright pair of stars only slightly farther apart than the width of the full moon. This pair is λ and υ Sco, and is often called the "Cat's Eyes." The head and claws can be traced the opposite direction from Antares. An alternate way is to reverse the directions for finding Sagittarius and draw a line from the center of the teapot's handle through the center of the spout and extend it 10° to arrive at the Cat's Eyes, then trace the figure of the Scorpion from the tail to Antares and the claws (Fig. 5.2).

Magnitudes and spectral types of principal stars:

Bayer desig.	Mag. (m$_v$)	Spec. type	Bayer desig.	Mag. (m$_v$)	Spec. type	Bayer desig.	Mag. (m$_v$)	Spec. type
α	1.06	M1 Ib	β	2.56	B0.5 V	γ	0.00	O0 0*[1]
δ	2.29	B0.2 IV	ε	2.29	K2 IIIb	ζ^1	4.73	B1 I
ζ^2	3.62	K4 III	η	3.32	F3p	θ	1.86	F1 II
ι^1	2.99	F3 Ia	ι^2	4.78	A6 Ib	κ	2.39	B1.5 III
λ	1.62	B1.5 IV	μ^1	3.00	B1.5 V	μ^2	3.56	B2 IV
ν	4.00	B2 IV	ξ	5.07	F5 IV	o	4.55	A4 II/III
π	2.89	B1 V	ρ	3.87	B2 IV/V	σ	2.90	B1 III
τ	2.82	B0 V	υ	2.70	B2 IV	φ	–	–*[2]
χ	5.24	K3 III	ψ	4.94	A3 IV	ω^1	3.93	B1 V
ω^2	4.31	G3 II						

Stars of magnitude 5.5 or brighter which have other catalog designations:

Draper Number	Mag. (m$_v$)	Spec. type	Draper Number	Mag. (m$_v$)	Spec. type
HD 164840	4.79	B8 Ib/II	HD 163145	4.85	K2 III
HD 163376	4.88	M0 III	HD 145570	4.93	A3 IV
HD 143787	4.96	K3 III	HD 153613	5.03	B8 V
HD 154948	5.06	G8/K0 III+	HD 145250	5.09	K0 III
HD 157243	5.10	B7 III	HD 151804	5.23	O9e
HD 148688	5.31	B1 Ia	HD 144690	5.35	M2 III
HD 142165	5.38	B5 V	HD 147722	5.40	G0
HD 142184	5.41	B2 V	HD 147628	5.42	B8 V
HD 142990	5.43	B5 V	HD 1249404	5.46	O9 Ia
HD 152234	5.46	B0.5 Ia	HD 151078	5.48	K0 III
HD 144987	5.50	B8 V			

Fig. 5.2a. The stars of Sagittarius, Corona Australis and Scorpius.

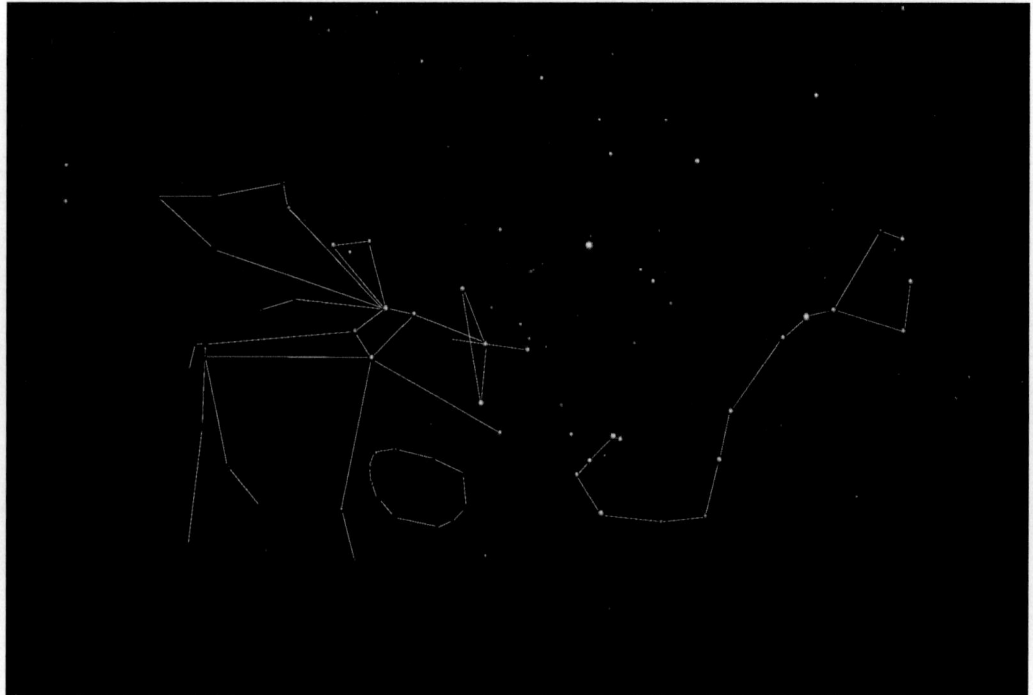

Fig. 5.2b. The figures of Sagittarius, Corona Australis and Scorpius.

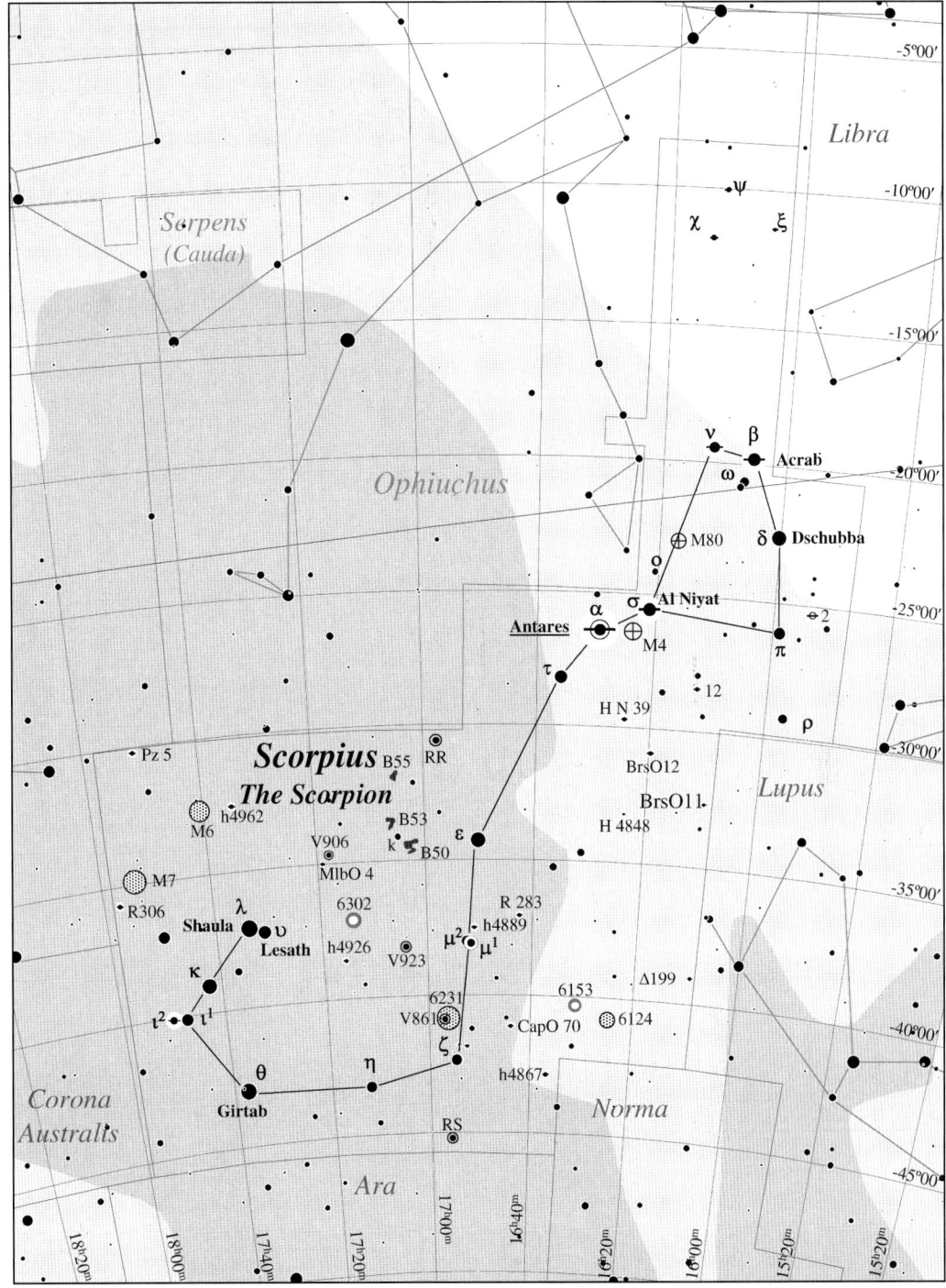

Fig. 5.3. Scorpius details.

Table 5.1. Selected telescopic objects in Scorpius.

Multiple stars:

Designation	R.A.	Decl.	Type	m_1	m_2	Sep. (")	PA (°)	Colors	Date/ Period	Aper. Recm.	Rating	Comments
2 Scorpii	15h 53.6m	−25° 20′	A–B	4.7	7.0	2.1	269	Y, yW	1991	4/5″	****	Double-Double. Separating AB is a good test for 10″ scopes. a.k.a. STF 1998
ξ Scorpii	16h 04.4m	−11° 22′	A–B	5.2	4.9	**1.0**	**355**	Y, B	**45.68**	10/12″	****	
			A–C	4.9	7.3	7.6	046	rO, rO	2005	2/3″	****	C is easy even in small scopes. Not as symmetrical as 1yr, but beautiful
8, β¹ Scorpii	16h 05.4m	−19° 48′	A–C	2.6	4.5	13.1	024	bW, pB	2005	2/3″	****	Both A and C have close companions at 0.3″ and 0.1″. a.k.a. H 37
Δ 199	16h 08.6m	−39° 06′	A–C	6.6	7.1	44.1	184	W, O	2000	6×30	****	Wide, easy in small scopes
BrsO 11	16h 09.5m	−32° 39′	A–B	6.7	7.2	7.8	083	Y, Y	1998	2/3″	****	Easy in small scopes
14, ν Sco	16h 12.0m	−19° 28′	Aa–B	4.4	5.3	1.3	002	W, W	2003	7/8″	****	a.k.a.β 120
			Aa–C	4.2	6.6	41.3	338		2005	6×30	****	a.k.a. H 5 6
			C–D	6.6	7.2	2.4	054	Y, Y	2003	4/5″	****	a.k.a. NTK 2
12 Scorpii	16h 12.3m	−28° 25′	A–B	5.8	8.1	3.8	071	B, B	1999	2/3″	*****	Nicely separated by small scopes.a.k.a. h4839
BrsO 12	16h 19.5m	−30° 54′	A–B	5.6	6.9	23.9	318	Y, Y	2006	10×50	*****	a.k.a. BrsO 12
20, σ Sco	16h 21.2m	−25° 36′	Aa–B	2.9	8.4	20.0	273	W, prB	1999	15×80	*****	a.k.a. Blm 4
h4848	16h 23.9m	−33° 12′	A–B	6.9	7.3	6.2	153	W, W	1998	2/3″	****	
HN 39	16h 24.7m	−29° 42′	A–B	5.9	6.6	4.4	358	pY, pY	1998	2/3″	****	
21, α Sco	16h 29.4m	−26° 26′	A–B	1.0	5.4	**2.1**	**277**	rO, Gr	**900**	4/5″	****	Beautiful in 16/18″ scopes on good nights, a.k.a. GNT 1 *3
h4867	16h 38.4m	−43° 24′	A–B	5.8	9.4	16.8	294	bW, ?	1999	15×80	***	
R 283	16h 42.5m	−37° 05′	A–B	7.0	7.8	**0.6**	**245**	?, ?	**201**	17/18″	*****	Closing to 0.29″ in late 2036, then opening to.924″ in early 2104
CapO 70	16h 43.9m	−41° 07′	A–B	6.1	6.2	97.2	258	W, W	2000	6×30	*****	Lovely matched pair
h4889	16h 51.0m	−37° 31′	A–B	6.2	7.8	6.8	004	W, W	1999	2/3″	****	
h4926	17h 14.5m	−39° 46′	A–B	6.7	10.1	14.1	334	oR, W	1999	2/3″	***	
			A–C	6.6	11.3	16.9	209	oR, W	1999	3/4″	***	
MlbO 4	17h 19.0m	−34° 59′	A–B	5.4	7.4	**1.4**	**189**	yO, brY	**42.15**	6/7″	****	Closing to min. of 0.479″ in mid 2018, then opening to max. of 2.158″ in late 2031

h4935	17h 19.0m	−34° 59'	AB–C	6.0	10.8	32.5	138	?, ?	1998	15×80	***	
h4962	17h 34.7m	−32° 35'	A–B	5.7	10.5	5.1	101	Y, Y	1998	2/3"	***	
i² Sco	17h 50.2m	−40° 05'	A–B	4.8	11.0	31.6	036	W, W	2000	3/4"	*****	a.k.a. HdO 279
Pz 5	17h 51.2m	−30° 30'	A–B	6.7	8.1	10.1	189	W, W	1998	2/3"	****	
R 306	17h 57.9m	−36° 00'	A–B	6.8	9.5	3.4	015	W, ?	1999	2/3"	***	

Variable stars:

Designation	R.A.	Decl.	Type	Range (mᵥ)	Period (days)	F (fᵣ/fₜ)	Spectral Range	Aper. Recm.	Rating	Comments
Antares, α Scorpii	16h 29.6m	−26° 26'	SRa	0.9 1.8	–	–	M1.5.Iab-Ib	Eye	*****	Supergiant star, 700 times diameter of our sun
μ¹ Scorpii	16h 51.9m	−38° 03'	EA/SD	2.9 3.2	1.446	–	B1.5V + B6.5V	Eye	****	39' NW (PA 320°) of ζ² Scorpii (m3.62)
V919 Scorpii	16h 52.3m	−41° 51'	WR	6.5 6.6	2	–	WN7 + O-B1I	7×30	*****	*4
RS Scorpii	16h 55.6m	−45° 06'	M	6.2 13.0	319.91	0.50	M5e-M8e	8/10"	****	1°35' NNE (PA 14°) of ζ² Scorpii (m3.62)
RR Scorpii	16h 56.6m	−30° 35'	M	5.0 13.4	281.45	0.18	M6II-IIIe–M8IIe	10/12"	*****	2°16' E (PA 94°) of μ² Scorpii (m3.56)
V861 Scorpii	16h 56.6m	−40° 49'	EB	6.1 6.4	7.848	0.47	B0.5Iae	7×30	****	40' NW (PA 320°) of ζ² Scorpii (m3.62)
V923 Scorpii	17h 03.8m	−38° 09'	EA/DM	5.9 6.2	34.827		F3IV + F3V	7×30	****	
V906 Scorpii	17h 53.9m	−34° 45'	EA/DM	6.0 6.2	2.785	–	B9V + B9V	7×30	****	

Star clusters:

Designation	R.A.	Decl.	Type	mᵥ	Size (')	Brtst. Star	Dist. (ly)	dia. (ly)	Number of Stars	Aper. Recm.	Rating	Comments
M80, NGC 6093	16h 17.0m	−22° 59'	Globular	7.3	8.9	12.5	2,8000	72	–	3/4"	*****	Easily found, 1.25° E of Antares
M4, NGC 6121	16h 23.6m	−26° 32'	Globular	5.8	26.3	10.8	6,500	59	–	2/3"	*****	Nearest globular, but very small for globular, Lum. only 44K suns
Cr 301, NGC 6124	16h 25.6m	−40° 40'	Open	5.8	29	9.0	1,600	13	100	2/3"	****	8/10", ~75 stars. 12/14", >100 stars
Cr 315, NGC 6231	16h 54.0m	−41° 48'	Open	2.6	14	6.0	6,200	25	120	2/3"	*****	Very beautiful cluster, Lum. = 1M suns! About equal to ω Centauri

(continued)

Table 5.1. (continued)

Star clusters:

Designation	R.A.	Decl.	Type	m_v	Size (')	Brtst. Star	Dist. (ly)	dia. (ly)	Number of Stars	Aper. Recm.	Rating	Comments
M6, NGC 6405	17ʰ 40.1ᵐ	−32° 13'	Open	4.2	33	5.5	1,500	14	80	2/3"	*****	Visible to unaided eye! Lum. = 8,300 suns
M7, NGC 6475	17ʰ 53.9ᵐ	−34° 49'	Open	3.3	80	7.0	820	19	80	2/3"	*****	Visible to unaided eye! Lum. = 2,500 suns

Nebulae:

Designation	R.A.	Decl.	Type	m_v	Size (')	Brtst./ **Cent.** star	Dist. (ly)	dia. (ly)	Aper. Recm.	Rating	Comments
NGC 6153	16ʰ 31.5ᵐ	−40° 15'	Planetary	10.9	25"	15.4	–	–	4/6"	****	Readily seen in 8/10", appears greenish with darker center in 12/14"
Barnard 50	17ʰ 02.9ᵐ	−34° 24'	Dark	–	15'	–	–	–	8/10"	****	Irregular dark area, readily seen in 8/10" scope on dark nights
Barnard 53	17ʰ 06.1ᵐ	−33° 56'	Dark	–	30×10'	–	–	–	10/12"	****	4.5° NE of ε Sco, not quite as dark as Barnard 50
Barnard 55	17ʰ 06.6ᵐ	−32° 00'	Dark	–	30×10'	–	–	–	10/12	****	4.5° NE of ε Sco, not quite as dark as Barnard 50
NGC 6302	17ʰ 13.7ᵐ	−37° 06'	Planetary	9.6	50"	16	2.0K	0.1	2/3"	****	"Bug Nebula" Bug shape apparent in 12/14" scopes

Galaxies:

Designation	R.A.	Decl.	Type	m_v	Size (')	Dist. (ly)	dia. (ly)	Lum. (suns)	Aper. Recm.	Rating	Comments
None											

Other interesting objects:

Designation	R.A.	Decl.	Type	m_v	Size (')	Dist. (ly)	dia. (ly)	Lum. (suns)	Aper. Recm.	Rating	Comments
None											

Fig. 5.4. M4, globular cluster in Scorpius. Astrophoto by Gian Michelle Ratto.

Fig. 5.5. M4, open cluster in Scorpius. Astrophoto by Mark Komsa.

Fig. 5.6. M7, open cluster in Scorpius. Astrophoto by Mark Komsa.

Fig. 5.7. M80, globular cluster in Scorpius. Astrophoto by Mark Komsa.

SAGITTARIUS (sa-jih-**tare**-ee-us)	**The Archer**
SAGITTARII (sa-jit-**air**-ee-ee)	**9:00 P.M. Culmination:** Aug. 22
Sgr	**AREA:** 867 sq. deg., 15th in size

TO FIND: Around 9:00 PM on Aug. evenings, the brilliant white star <u>Vega</u> will be nearly directly overhead for mid-latitude northern hemisphere observers. <u>Vega</u> is the western-most star in the Summer Triangle, which also includes the first magnitude stars <u>Deneb</u> and <u>Altair</u>. The Summer Triangle is a convenient starting place to locate several summer constellations, including Sagittarius. A line from <u>Deneb</u> (the easternmost point of the triangle) to <u>Altair</u> (the southernmost) is ~38° long. That line extended another 38° ends near the center of a compact group of two 2nd and two 3rd magnitude stars. These stars are ζ, σ, τ, and φ Sgr. They form the "handle" of a well-known and easily recognized asterism called The Teapot. About 10° WSW of this group lie three (one 2nd and two 3rd magnitude) stars which form the spout of the teapot. Between and north of these groups lies 2nd magnitude λ Sgr which designates the top of the teapot's lid. Some people even see the bright spot in Milky Way just above the teapot's spout as "steam" exiting the spout. Most of the other stars in Sagittarius are relatively faint, and recognizing the entire figure of the centaur requires careful comparison of the constellation drawing with the night sky.

If you have previously found the constellation of Scorpius, an alternative method of finding Sagittarius is even easier. The two stars (the Cat's Eyes) at the end of the scorpion's tail are λ and υ Sco. They are fairly bright (m1.62 and 2.70) and close together (only 36', or slightly more than the apparent diameter of the full moon apart), so they are easy to recognize. A line from υ to λ extended by 10° ends near the center of the teapot's spout, and if extended another 10° ends near the center of the teapot's handle.

Another easily recognized, but less well-known, asterism in Sagittarius is the "Milk Dipper." The same four stars which form the Teapot's handle can be seen as the bowl of the Milk Dipper. The stars λ (the teapot's lid) and μ Sgr form the dipper's handle and complete the asterism. This asterism is appropriately named, since it appears to be about to dip up the brightest part of the Milky Way (Figs. 5.1 and 5.8).

KEY STAR NAMES:

Name	Pronunciation	Source
α Rukbat	**ruhk**-bat, **ruek**-but	rukbat al-rami (Sci-Arabic), The archer's knee
β Arkab	**are**-cab	ʿurq_b al-r_m_ (Sci-Arabic), the archer's Achilles' tendon
γ Alnasl	uhl-**nusl**	al-nasl (Arabic), the point {of the arrow}
δ Kaus Media	**kous me**-dih-uh	al-qaus (Arabic), the bow + media (Latin), middle
ε Kaus Australis	**kous** ous-**tray**-liss	al-qaus (Arabic), the bow + australis (Latin), southern
ζ Ascella	uh-**sell**-uh	a medieval version of axilla (Latin), armpit

Name	Pronunciation	Source
λ Kaus Borealis	kous **bo**-reh-ah-lis	al-qaus (Arabic), the bow+borealis (Latin), northern
σ Nunki	**nun**-key	NUN[ki] (Babylonian), untranslatable proper name

Magnitudes and spectral types of principal stars:

Bayer desig.	Mag. (m$_v$)	Spec. type	Bayer desig.	Mag. (m$_v$)	Spec. type	Bayer desig.	Mag. (m$_v$)	Spec. type
α	3.96	B8 V	β^1	3.96	B9 V	β^2	4.27	F2 III
γ^1	4.66	G0 Ib/II	γ^2	2.98	K0 III	δ	2.72	K3 III
ε	1.79	B9.5 III	ζ	2.60	A3 IV	η	3.12	M2 III
θ^1	4.37	B2.5 IV	θ^2	5.30	A4/A5 IV	ι	4.12	K0 III
κ^1	5.60	A0 V	κ^2	5.64	A3 V	λ	2.82	K1 IIIb
μ	3.84	B2 III	ν^1	4.86	K1 II	ν^2	5.00	K1 Ib/II
ξ^1	5.02	B9.5 Ib	ξ^2	3.52	G8/K0 II/III	o	3.76	K0 III
π	2.88	F2 II/III	ρ^1	3.92	F0 III/IV	ρ^2	5.84	K0 III
σ	2.05	B2.5V	τ	3.32	K1/K2 III	υ	4.52	B2 Vpe
φ	3.17	B8.5 III	χ^1	5.02	A4 IV/V	χ^2	7.27	B7 IV
χ^3	5.45	K3 III	ψ	4.86	K0/K1 III	ω	4.70	G3/G5 III

Stars of magnitude 5.5 or brighter which have other catalog designations:

Draper Number	Mag. (m$_v$)	Spec. type	Draper Number	Mag. (m$_v$)	Spec. type
HD 165634	4.55	K0 III	HD 167818	4.66	K3 III
HD 189831	4.77	K4 III	HD 172910	4.86	B2 V
HD 190056	4.99	K1 III/IV	HD 163755	5.00	K5/M0 III
HD 170680	5.12	B9/9.5 V	HD 171034	6.28	B2 III/IV
HD 187474	5.32	A0 p	HD 168733	5.33	B7 Ib/II
HD 183275	5.46	K1/2 III	HD 184985	5.46	F7 V
HD 177074	5.49	A0 V	HD 186185	5.49	F5 V

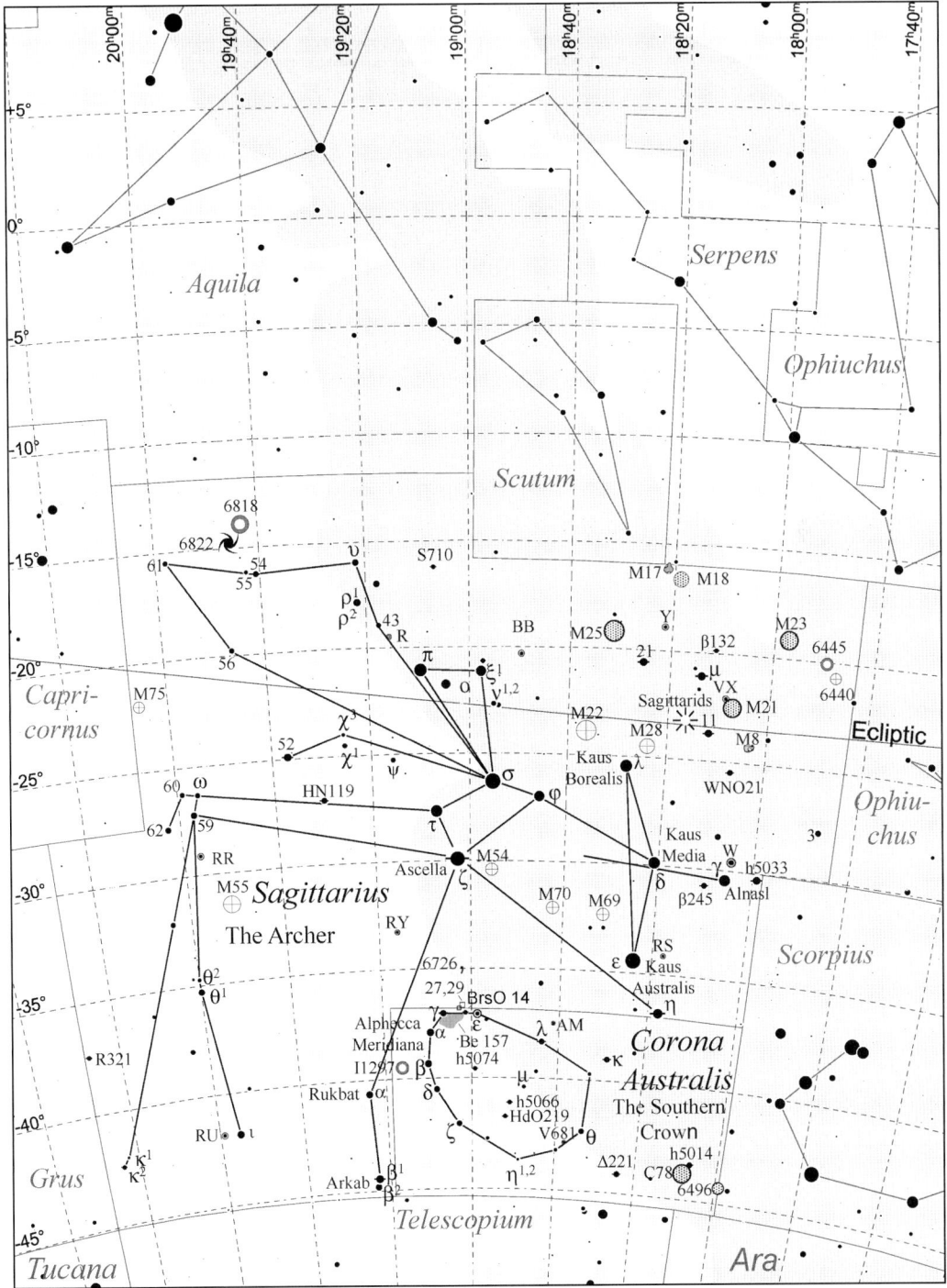

Fig. 5.8. Sagittarius and Corona Austsralis details.

Table 5.2. Selected telescopic objects in Sagittarius.

Multiple stars:

Designation	R.A.	Decl.	Type	m_1	m_2	Sep. (")	PA (°)	Colors	Date/Period	Aper. Recm.	Rating	Comments
h5003	17h 59.1m	−30° 15'	A–B	5.4	7.0	4.3	104	rO, rO	2000	2/3"	*****	Beautiful pair, in nice MW field. A is slightly redder than B. a.k.a. PZ 6
WNO 21	18h 08.9m	−25° 28'	A–C	5.2	13.1	26.0	239	rO, ?	1919	6/7"	***	
β245	18h 10.1m	−30° 44'	A–B	6.8	8.8	13.5	065	Y, ?	1999	2/3"	****	
			A–B	5.8	8.0	3.9	353	O, Y	1994	2/3"	*****	Rich starfield, some fainter pairs in FOV
β132	18h 11.2m	−19° 51'	A–B	7.0	7.1	1.3	190	W, W	2001	7/8"	****	a.k.a. HJ 5030
11 Sagittarii	18h 11.7m	−23° 42'	A–B	5.0	11.5	43.0	287	O, ?	1998	2/3"	****	B and D components are easy. a.k.a. H7AB
13, μ Sagittarii	18h 13.8m	−21° 04'	A–B	3.9	10.5	16.0	261	pY, B	2002	2/3"	****	
			A–C	3.9	13.5	25.2	112	pY, pR	2002	7/8"	****	a.k.a. BU 292AC
			A–D	3.9	10.0	47.7	312	pY, pR	2002	2/3"	****	a.k.a. HJ 2822AD
			A–E	3.9	9.2	50.4	114	pY, ?	2002	7×50	****	a.k.a. HJ 2028AE
η Sagittarii	18h 17.6m	−36° 46'	A–B	3.1	7.8	3.6	107	W, rO	1988	2/3"	****	Arkab. a.k.a. BU 760AB
			A–C	3.2	13.0	33.3	321	W, ?	2002	6/7"	***	a.k.a. BU 760AC
			A–D	3.2	10.0	92.9	319	W, ?	2002	2/3"	****	a.k.a. BU 760AD
21 Sagittarii	18h 25.4m	−20° 32'	A–D	5.0	7.4	1.5	275	O, bG	1997	6/7"	****	Beautiful color, difficult to resolve, but sometimes appears as orange spot with blue or green spike on one side. a.k.a JC 6
38, ζ Sagittarii	19h 02.6m	−29° 53'	AB–C	2.6	10.6	72.0	302	W, ?	2002	2/3"	***	Ascella. A–B is a 21-year binary, max. sep. 0.59" in 199. a.k.a. HdO 150AB
β¹ Sagittarii	19h 22.6m	−44° 28'	A–B	4.0	7.2	28.6	076	W, pY	1999	2/3"	****	a.k.a. DUN 226
HN 119	19h 29.9m	−26° 59'	A–B	5.6	8.8	7.8	142	O, B	2006	2/3"	******	Nice color, comparable to Albireo if larger telescope is used
52 Sagittarii	19h 36.7m	−24° 53'	A–B	4.7	9.2	2.4	173	W, ?	1999	4/5"	****	a.k.a. BU 654
54 Sagittarii	19h 40.7m	−16° 18'	A–B	5.3	12.6	38.1	273	yO, pB	1905	5/6"	*****	Nice triple for small telescopes. a.k.a. HJ 599
			A–C	5.3	7.7	44.7	042	dY, pY	2003	7×50	****	

	R.A.	Decl.							Aper. Recm.	Rating		Comments
κ² Sagittarii	20ʰ 23.9ᵐ	−42° 25'	A–B	5.9	7.3	0.3	330.75	W, W	700	—	****	Min. sep. 0.16″ in 2009, then widening to 0.5″ in 2024 and 1.0″ in 2055. a.k.a. BU 763
R 321	20ʰ 26.9ᵐ	−37° 24'	A–B	6.6	8.1	1.5	126.71	yO, yO	177.5	6/7″	*****	

Variable stars:

Designation	R.A.	Decl.	Type	Range (mᵥ)		Period (days)	F (fₑ/fₚ)	Spectral Range	Aper. Recm.	Rating	Comments
3, X Sagittarii	17ʰ 47.6ᵐ	−27° 50'	δ Cep	4.2	4.9	7.013	0.36	F5–G2II	Eye	***	#
W Sagittarii	18ʰ 05.0ᵐ	−29° 35'	δ Cep	4.3	5.1	7.595	0.32	F4–G2Ib	Eye	****	#
VX Sagittarii	18ʰ 08.1ᵐ	−22° 13'	M	6.5	14.0	732	–	M4ela–M10ela	12/14″	*****	Longest period Mira variable known.##
RS Sagittarii	18ʰ 17.6ᵐ	−34° 06'	EA/SD	6.0	7.0	2.416	0.17	B3IV–V + A	6×30	****	Algol type, stars are not gravitationally distorted. $F = T_{ecl}/T_p$.#
Y Sagittarii	18ʰ 21.4ᵐ	−18° 52'	δ Cep	5.3	6.2	5.773	0.34	F5–G0Ib–II	6×30	***	1°56' NNW (PA 331°) of 25 Sgr (m4.80).#
BB Sagittarii	18ʰ 51.0ᵐ	−20° 18'	δ Cep	6.6	7.3	6.637	0.34	F6–G1	6×30	***	#
RY Sagittarii	19ʰ 16.5ᵐ	−33° 31'	RCrb	5.8	14.0	–	–	G0Iaep (C1.0)	12/14″	*****	## RY Sgr also exhibits other modes of variation *5
R Sagittarii	19ʰ 16.7ᵐ	−19° 18'	M	6.7	12.8	269.84	0.46	M4e–M6e	6/8″	****	##
RR Sagittarii	19ʰ 55.9ᵐ	−29° 11'	M	5.4	14.0	336.33	0.43	M4e–M9e	12/14″	****	##
RU Sagittarii	19ʰ 58.7ᵐ	−41° 51'	M	6.0	13.8	240.49	0.43	M3e–M6e	10/12″	****	39'E (PA 89°) of ι Sgr (m4.12).##

Star clusters:

Designation	R.A.	Decl.	Type	mᵥ	Size (')	Brtst. Star	Dist. (ly)	dia. (ly)	Number of Stars	Aper. Recm.	Rating	Comments
NGC 6440	17ʰ 48.9ᵐ	−20° 22'	Globular	9.3	4.4	16.7	2,300	30	–	6/8″	***	22' S of NGC 6445, both in same LP FOV
M23, NGC 6494	17ʰ 56.8ᵐ	−19° 01'	Open	5.5	25	10.0v	2,100	15	150	7×50	*****	Stunning. Nearly as large as the full moon. >100 stars vis in 8/10″
NGC 6530	18ʰ 04.8ᵐ	−24° 20'	Open	4.6	15	–	5,200	23	100	Eye	*****	Hot O-type stars in this cluster supply the energy of the eastern part of M8
M21, NGC 6531	18ʰ 04.6ᵐ	−22° 30'	Open	5.9	16	8.0v	4,000	19	70	7×50	***	Sometimes called "Webb's Cross" because of its shape

(continued)

Table 5.2. (continued)

Star clusters:

Designation	R.A.	Decl.	Type	m_v	Size (')	Brst. Star	Dist. (ly)	dia. (ly)	Number of Stars	Aper. Recm.	Rating	Comments
M18, NGC 6613	18h 19.9m	−17° 08'	Open	6.9	7	8.65v	3,900	8	20	7×50	****	a.k.a. "Black Swan" because of the shape of the dark area outlined by cluster stars
M17, NGC 6618	18h 20.8m	−16° 11'	Open	6.0	27	9.28v	4,900	38	40	7×50	*****	Part of Omega or Swan Nebula
M28, NGC 6626	18h 24.5m	−24° 52'	Globular	6.9	13.8	12.0v	19K	77	–	7×50	*****	Some stars resolved in 4/6", impressive in 8" or larger scopes
M69, NGC 6637	18h 31.4m	−32° 21'	Globular	7.7	9.8	13.7	33K	94	–	7×50	*****	One of fainter Messier objects
M25, IC 4725	18h 31.6m	−19° 15'	Open	4.6	26	8.0v	2,300	18	30	Eye	*****	Visible to unaided eye under good conditions
M22, NGC 6656	18h 36.4m	−23° 54'	Globular	5.2	32	10.7	9,800	91	70K	Eye	*****	One of closest globular clusters, first to be recognized as such
M70, NGC 6681	18h 42.2m	−32° 18'	Globular	7.8	8	13.0v	30K	70	–	7×50	***	Fainter than average globular cluster (lum. = ~180K suns)
M54, NGC 6715	18h 55.1m	−30° 29'	Globular	7.7	12	15.2	69K	240	–	7×50	***	Brighter than average globular cluster
M55, NGC 6809	19h 40.0m	−30° 58'	Globular	6.3	19	11.2	16K	85	–	7×50	***	Nice, but loose globular, easily resolved
M75, NGC 6864	20h 06.1m	−21° 55'	Globular	8.6	6.8	14.6	60K	120	–	11×80	***	Barren area, somewhat hard to find. 1° S and 14m E of π Sagittarii

Nebulae:

Designation	R.A.	Decl.	Type	m_v	Size (')	Brst./Cent. star	Dist. (ly)	dia. (ly)	Aper. Recm.	Rating	Comments
NGC 6445	17h 49.2m	−20° 01'	Planetary	11.2	33"	19v	3,200	2	6/8"	****	22' N of NGC 6440, both in same LP FOV
M20, NGC 6514	18h 02.3m	−23° 02'	Em and Rfl	–	20'×20'	–	5,000	30	7×50	*****	The "Trifid" or "Cloverleaf" nebula

Designation	R.A.	Decl.	Type	m_v	Size (')	Dist. (ly)	dia. (ly)	Lum. (suns)	Aper. Recm.	Rating	Comments
Barnard 86	18h 03.0m	-27° 53'	Dark	NA'	5'	-	-	-	10×50	*****	The "Ink Spot." Open cluster NGC 6520 lies on the eastern edge
M8, NGC 6523	18h 03.8m	-24° 23'	Emission	3	90'×40'	-	5,200	135×60	Eye	*****	Lagoon Nebula" Open cluster NGC 6530 lies just to the NE
Barnard 90	18h 10.2m	-28° 19'	Dark	NA	3'×2'	-	-	-	4/6"	****	Dust superimposed on "Great Sagittarius Star Cloud"
M17, NGC 6618	18h 20.8m	-16° 11'	Emission	6	40'×30'	-	4,900	55×40	7×50	*****	Omega or Swan Nebula; includes open cluster of ~40 stars [*6]
NGC 6818	19h 44.0m	-14° 09'	Planetary	9.3	22'×15'	>15	4,900	0.5	4/6"	****	The "Little Gem." Bright oval bluish disk, slightly darker center

Galaxies:

Designation	R.A.	Decl.	Type	m_v	Size (')	Dist. (ly)	dia. (ly)	Lum. (suns)	Aper. Recm.	Rating	Comments
M54, NGC 6715	18h 55.1m	-30° 29'	Elliptical	7.7	12	90K	60	16K	7×50	***	The globular cluster M54 appears to be the core of the Sagittarius dwarf elliptical galaxy
NGC 6822	19h 44.9m	-14° 48'	IB(s)m	8.8	20×14.9	1.6M	7,500	50M	6/8"	***	Disc. E.E. Barnard, 1884. Very faint, use short FL and low magnification

Meteor showers:

Designation	R.A.	Decl.	Period (yr)	Duration	Max Date	ZHR (max)	Comet/Asteroid	First Obs.	Vel. (mi/km/sec)	Rating	Comments
φ Sagittariids	18h 16m	-23°	N/A	06/01 07/15	06/19	5	Unkn	1917	19	Minor	Disc. William F. Denning. [*7]

Other interesting objects:

Designation	R.A.	Decl.	Type	m_v	Size (')	Dist. (ly)	dia. (ly)	Lum. (suns)	Aper. Recm.	Rating	Comments
Galactic Center	17h 45.6m	-28° 56'	Blk Hole	-	-	30K	-	-	Various	****	3 ly dia. sphere contains 3M solar masses
M24	18h 16.5m	-18° 50'	StarCloud	2.5	95×35	9.4K	260	-	Eye	*****	"Small Sagittarius Star Cloud." Stunning sight at low power

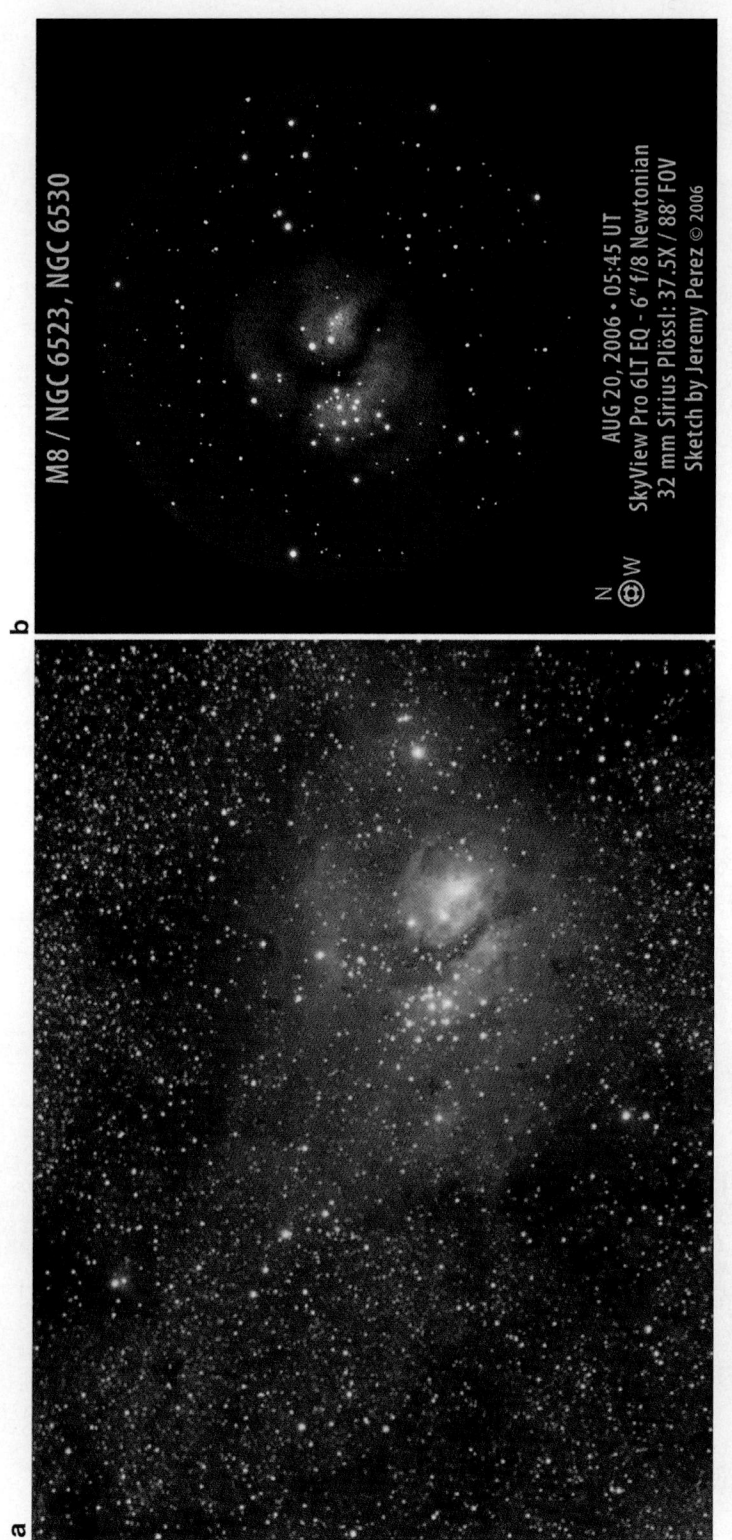

M8 / NGC 6523, NGC 6530

AUG 20, 2006 · 05:45 UT
SkyView Pro 6LT EQ - 6" f/8 Newtonian
32 mm Sirius Plössl: 37.5X / 88' FOV
Sketch by Jeremy Perez © 2006

N
W

Fig. 5.9. (a) M8 (NGC 6523 & 6530), emission nebula and open cluster in Scorpius. Astrophoto by Warren Keller. **(b)** M8 (NGC 6523 & 6530).

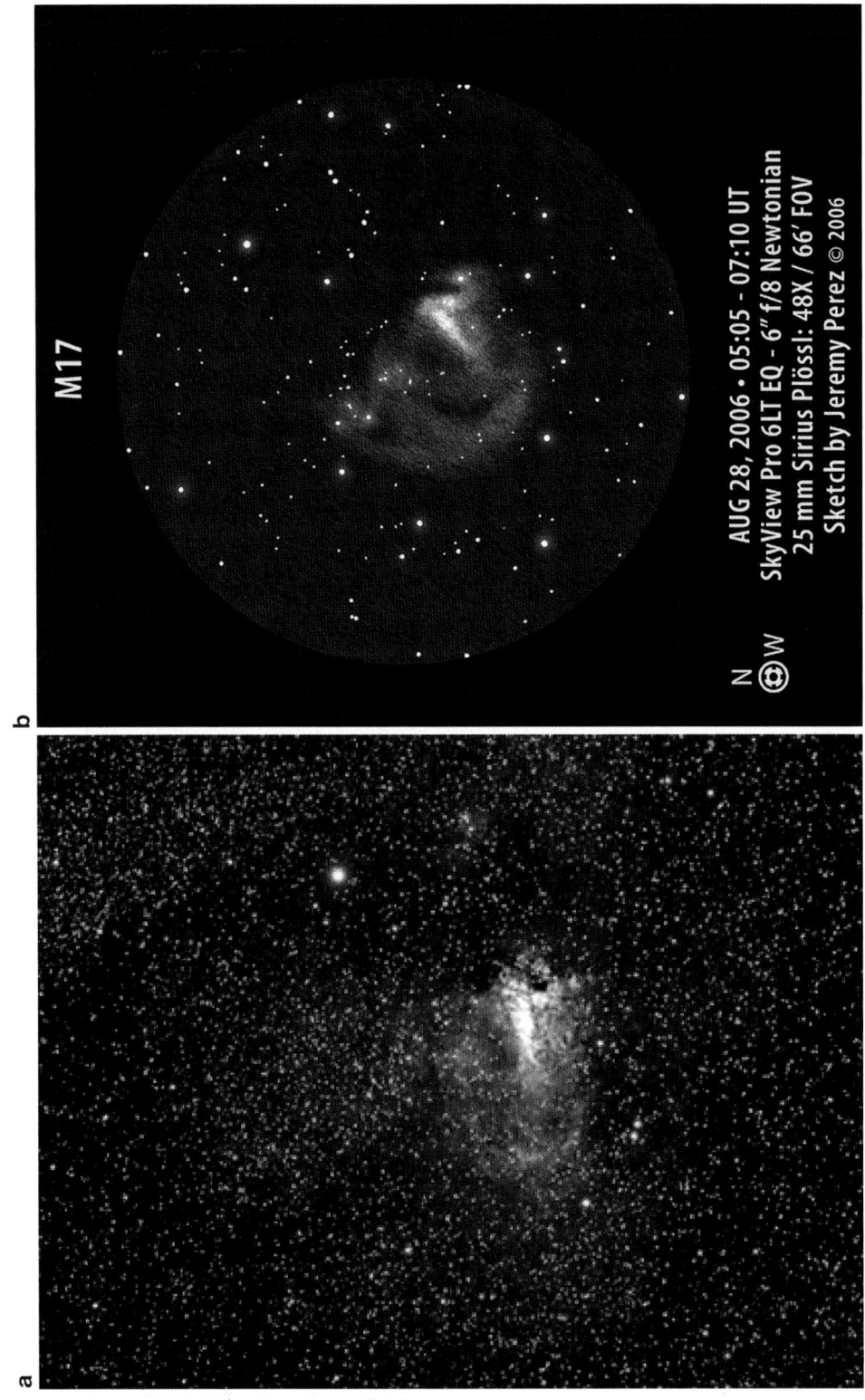

Fig. 5.10. (a) M17, emission nebula in Scorpius. Astrophoto by Steve Pastor. (b) M17.

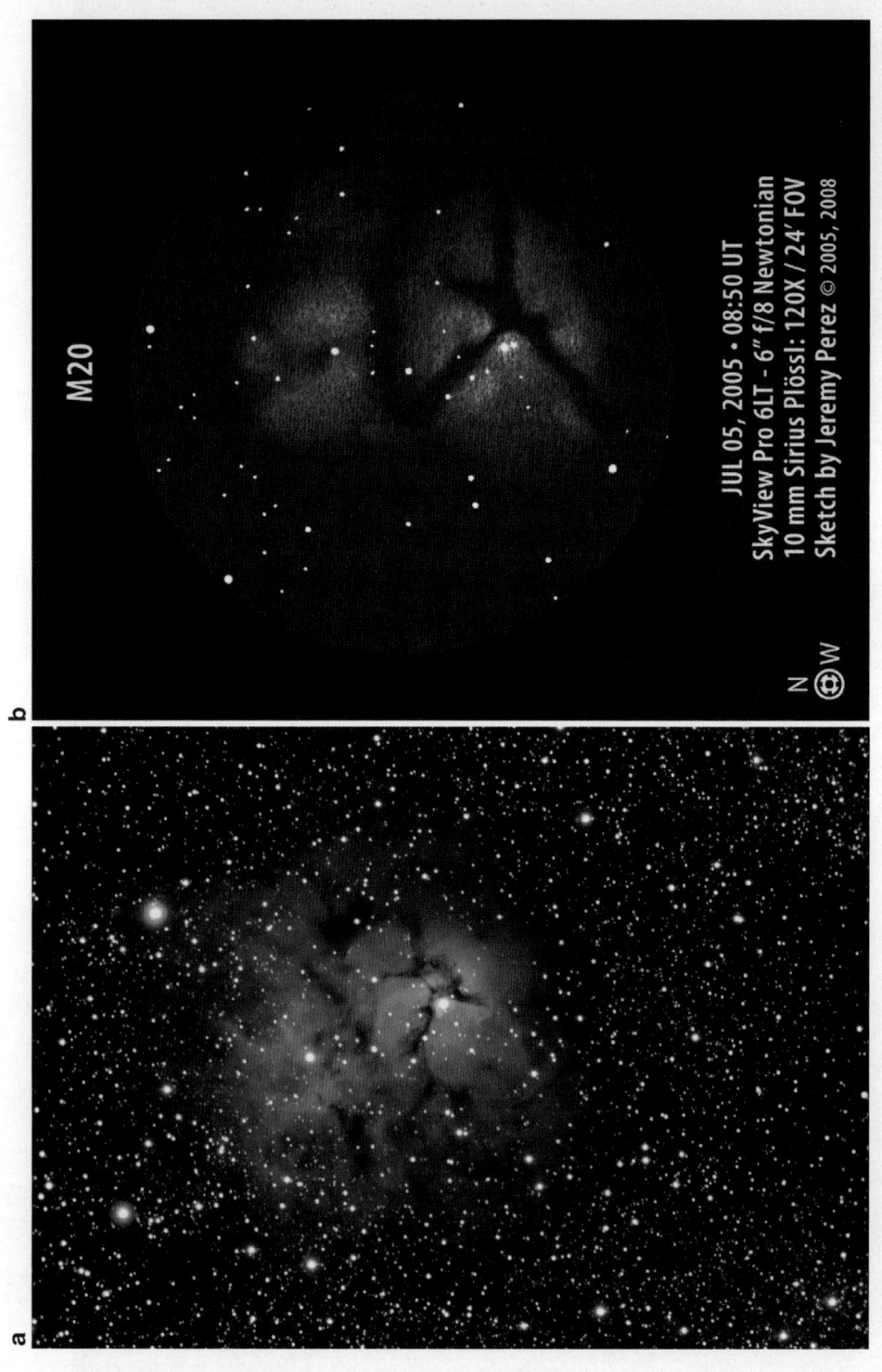

M20

JUL 05, 2005 · 08:50 UT
SkyView Pro 6LT - 6" f/8 Newtonian
10 mm Sirius Plössl: 120X / 24' FOV
Sketch by Jeremy Perez © 2005, 2008

N
W

Fig. 5.11. (a) M20, emissiion nebula in Sagittarius. Astrophoto by Stuart Goosen. Photo has been cropped and rotated to approximately match 5.11b. (b) M20, emission nebula in Sagittarius.

Fig. 5.12. M23 (NGC 6994), open cluster in Sagittarius. Astrophoto by Mark Komsa.

Fig. 5.13. M27 (NGC 6853), planetary nebula in Sagittarius. Astrophoto by Steve Pastor.

CORONA AUSTRALIS (kor-**oh**-nuh os-**tral**-is)	The Southern Crown
CORONAE AUSTRALIS (kor-**oh**-nye os-**tral**-is)	9:00 P.M. Culmination: Sep. 10
CrA	AREA: 128 sq. deg., 80th in size

TO FIND:The human eye is very good at finding apparent lines, either straight or curved, in patterns of dots or stars. Corona Australis is easily found by looking for a faint oval centered about 10° south of the group of bright stars forming the archer's torso. This group of stars is also known as the "Teapot" asterism and is described under Sagittarius above. The eastern side of the oval is considerably brighter than the western one, and the crown is often depicted as open on the western side. Once you are familiar with the Teapot asterism, you might think of the oval of Corona Australis as a burner on a stove, where the pot of freshly brewed tea was heated. The stars and constellation of Corona Australis are included in Fig. 5.2a, b, while the detailed chart for Corona Australis is included in the Fig. 5.8, Sagittarius details.

KEY STAR NAMES:

Name	Pronunciation	Source
α Alphecca Meridiana	(uhl-**feck**-uh meh-**RID**-ih-**ahn**-uh)	Al-fakka (Arabic) Meridianus (Latin), the southern Alphecca (see α Coronae Borealis)

Magnitudes and spectral types of principal stars:

Bayer desig.	Mag. (m_v)	Spec. type	Bayer desig.	Mag. (m_v)	Spec. type	Bayer desig.	Mag. (m_v)	Spec. type
α	4.11	A0/1 V	β	4.10	K0II/III	γ	4.23	F7 IV/V
δ	4.57	K1 III	ε	4.83	F3 IV/V	ζ	4.74	A0 Vn
η¹	5.46	A2 Vn	η²	5.60	B9 IV*8	θ	4.62	G5 III
ι	–	–*9	κ¹	6.31	B8	κ²	5.71	B9.5
λ	5.11	A0/A1 V	μ	5.20	G5/G6 V			

Stars of magnitude 5.5 or brighter which have other catalog designations:

Draper Number	Mag. (m_v)	Spec. type	Draper Number	Mag (m_v)	Spec. type
HD 1165189	4.92	AS5 V	HD 168592	5.09	K4/K5 III
HD 170642	5.16	AS3 V	HD 168905	5.24	B2.5 Vn
HD 175219	5.35	G5 III/IV	HD 125362	6.36	B3 V
HD 172991	5.40	K1/K2 III+	HD 171967	5.42	M2 III
HD 167096	5.45	G8/K0 III	HD 166596	5.47	B2.5 III

Table 5.3. Selected telescopic objects in Corona Australis.

Multiple stars:

Designation	R.A.	Decl.	Type	m_1	m_2	Sep. (″)	PA (°)	Colors	Date/Period	Aper. Recm.	Rating	Comments
h5014	18^h 06.8^m	−43° 26′	A–B	5.7	5.7	**1.7**	**002**	W,W	**450**	14/16″	****	Nicely matched pair, opening to max. of 2.15″ in 2171
Δ221	18^h 24.3^m	−44° 07′	A–B	5.2	10.1	74.2	164	bW, ?	1999	10×80	****	Easy in small scopes. Common proper motion, probably associated a.k.a. Δ 222. Stabilized binoculars help, easy in small scopes
κ Coronae Australis	18^h 33.4^m	−38° 44′	A–B	5.6	6.2	20.8	359	bW,W	2002	10×80	*****	
λ Coronae Australis	18^h 43.8^m	−38° 19′	A–B	5.1	10.0	28.5	215	W, O	2002	10×80	****	Very nice color contrast, easy in small scopes
h 5066	18^h 51.0^m	−41° 04′	A–C	5.1	–	42.9	052	W, ?	2002	–	**	
Hdo 291	18^h 52.2^m	−41° 43′	A–B	6.5	9.2	10.1	084	bW, –	1999	2/3″	****	
			A–BC	6.5	10.0	36.7	342	yO, –	1999	15×80	***	
I 1385BC			B–C	10.2	12.2	0.6	292	–, –	1926	16/17″	**	a.k.a. I 1385BC
h 5074	18^h 59.2^m	−39° 32′	A–B	6.5	11.8	16.0	246	W, –	1999	3/4″	****	
BrsO 14	19^h 01.1^m	−37° 04′	A–B	6.3	6.6	13.0	281	W, W	2002	2/3″	****	Easy in small scopes. Common proper motion, probably associated
γ Coronae Australis	19^h 06.4^m	−37° 04′	A–B	4.84	5.08	**1.3**	**013**	yW, yW	**121.76**	7/8″	****	Nearly matched, min. of 1.26″ in 1994, max. of 2.50″ in 2185. a.k.a. h 5084

Variable stars:

Designation	R.A.	Decl.	Type	Range (m_v)		Period (days)	F (f_r/f_l)	Spectral Range	Aper. Recm.	Rating	Comments
V681 Coronae Australis	18^h 37.7^m	−42° 57′	EA/DM	**7.6**	**8.1**	2.164	–	B9.5V	7×50	****	Eclipsing, Algol type, normal stars, well separated, double minimum
AM Coronae Australis	18^h 41.2^m	−37° 29′	SR	**8.6**	**12.7**	187.5	–	M3e	6/8″	****	Me type red giant, variations in range and period
Coronae Australis	18^h 58.7^m	−37° 06′	EW	4.7	5.0	0.591	–	F2V	eye	****	Eclipsing, W UMa type, distorted dwarf stars almost in contact

Star clusters:

Designation	R.A.	Decl.	Type	m_v	Size (')	Brtst. Star	Dist. (ly)	dia. (ly)	Number of Stars	Aper. Recm.	Rating	Comments
NGC 6496	17h 59.0m	−44° 16'	Globular	8.6	5.6	143	10K	30		6/8"	***	
NGC 6541, C78	18h 08.0m	−43° 42'	Globular	6.1	12.1	15	22K	95		7×50	****	Mod. conc. core, resolved in 12/14". Same LP FOV as h5014. a.k.a. C74

Nebulae:

Designation	R.A.	Decl.	Type	m_v	Size (')	Brtst./Cent. star	Dist. (ly)	dia. (ly)	Aper. Recm.	Rating	Comments
NGC 6726-27 (CrA)	19h 01.7m	−36° 53'	Reflection	–	9×7'	–	–	–	20×100	***	50' W of γ CrA. Mod. brt, 2 lobes, SW and NW
NGC 6729 (CrA)	19h 01.9m	−36° 57'	Reflection	–	Variable	–	–	–	12/14"	***	Comet shaped. Illuminated by irregular variable R CrA
Bernes 157 (CrA)	19h 02.9m	−37° 08'	Dark	–	55×18'	–	–	–	7×50	****	Large, very dark, irregular, between CrA and NGC 6726
IC 1297 (CrA)	19h 17.4m	−39° 37'	Planetary	10.7	7"	14.22	–	–	12/14"	***	

Galaxies:

Designation	R.A.	Decl.	Type	m_v	Size (')	Dist. (ly)	dia. (ly)	Lum. (suns)	Aper. Recm.	Rating	Comments
None											

Other interesting objects:

Designation	R.A.	Decl.	Type	m_v	Size (')	Dist. (ly)	dia. (ly)	Lum. (suns)	Aper. Recm.	Rating	Comments
None											

OPHIUCHUS (oaf-ee-**youk**-us)	**The Serpent Holder**
OPHIUCHI (oaf-ee-**youk**-ee)	**9:00 P.M. Culmination:** July 27
Oph	**AREA**: 948 sq. deg., 11th in size

The name Ophiuchus describes the constellation since it is derived from two Greek words, "ophis" (ὄφις), meaning serpent and "choulnter" (χόουλντερ), meaning holder.

TO FIND: The stars of Ophiuchus are not very bright, and there are no convenient bright stars or pointer stars nearby to help find it. However, you can start with the "keystone" of Hercules, which most people can find and recognize (see the directions in the section for Hercules). Start with π Her (m3.16), the star on the NE corner of the Keystone, and go 12° due south to δ Her (m2.74), then continue another 11° to α Her (Rasalgethi, m3.20). α Oph (Rasalhague, m2.08) is only 5° ESE of Rasalgethi. From Rasalhague, you can easily find the two pairs of stars marking the Serpent Holder's shoulders about 8° ESE and 10° WSW. From there, you can use the illustration as a guide to trace out the remainder of the figure. An alternative way to find Ophiuchus is to start with Antares (α Sco) and look about 16° due north to find ζ Oph (m2.54) which represents Ophichus' left knee. Slightly less that 10° SE ζ of is ε use (m5.43) representing his right knee. About 3° north of the line joining η and ζ is another, almost parallel, line 24° long defined by 3rd magnitude δ and ε Oph 8° NW of ζ Oph and 3rd magnitude ξ Ser and an unlettered 4th magnitude star about 5° NW of ξ. Again, use the chart to find the rest of the figure (Figs. 5.14–5.18; Tables 5.4).

KEY STAR NAMES:

Name	Pronunciation	Source
α Rasalhague	**ras**-ul-huh-**WEE** or ras-al-**HAY**-gwee	ra's al-hawwa, (Arabic) "the head of the serpent collector"
β Cebalrai	**seb**-al-**RAH**-ee or se-bal-**RAY**-ee	kalb al-r_'i, (Arabic) "the shepherd's dog"
γ Muliphen	muh-le-**fan**	muhlifan (Arabic), "two (things) causing a dispute," originally because it was uncertain which two stars the names al-wazn and hadari applied to. The name Muliphen is usually applied to γ CMa, but it is sometimes also used for γ Ophiuchi
δ Yed Prior	yed **pry**-er	al-yad (Arabic), "the hand," plus prior (Latin)
ε Yed Posterior	yed pos-**teh**-rih-or	al-yad (Arabic), "the hand," plus posterior (Latin)
η Sabik	**say**-bik or **sah**-bik	al sabiq ath-Thani (Arabic), "the first winner" or "conqueror," referring to conquering the serpent
λ Marfik	mur-**fik** or **mahr**-fik	al-marfiq (arabic), "the elbow"

Magnitudes and spectral types of principal stars:

Bayer desig.	Mag. (m_v)	Spec. type	Bayer desig.	Mag. (m_v)	Spec. type	Bayer desig.	Mag. (m_v)	Spec. type
α	2.08	A5 III	β	2.76	K2 III	γ	3.75	A0 V
δ	2.73	M1 III	ε	3.23	G8 III	ζ	2.54	O0.5 V
η	2.43	A2.5 V	θ	3.27	B2 IV	ι	4.39	B8 V
κ	3.19	K2 IIIvar	λ	3.82	A1 V	μ	4.58	B8 II/III
ν	3.32	K0 III	ξ	4.39	F2/F3 V	o	5.14	K0 II/III
π	–	–*10	ρ	4.57	B2 V	σ	4.34	K3 IIvar
τ	4.77	F5 V	υ	4.62	A3m	φ	4.29	G8/K0 III
χ	4.22	B2 Vne	ψ	4.48	K0 III	ω	4.45	Ap

Stars of magnitude 5.5 or brighter which have other catalog designations:

Designation	Mag. (m_v)	Spec. type	Designation	Mag. (m_v)	Spec. type
HD 165634	4.55	K0 IIICNp	HD 167818	4.66	K3 III
HD 189831	4.77	K4 III	HD 172910	4.86	B2 V
HD 190056	4.99	K1 III/IV	HD 163755	5.00	K5/M0 III
HD 170680	5.12	B9/9.5 B	HD 171034	5.28	B2 III/IV
HD 187474	5.32	A0p	HD 168733	5.33	B7 Ib/II
HD 170479	5.37	A5 V	HD 183275	5.46	K1/2 III
HD 184985	5.46	F7 V	HD 177074	5.49	A0 V
HD 186185	5.49	F5 V			

Fig. 5.14a. The stars of Ophiuchus and adjacent constellations.

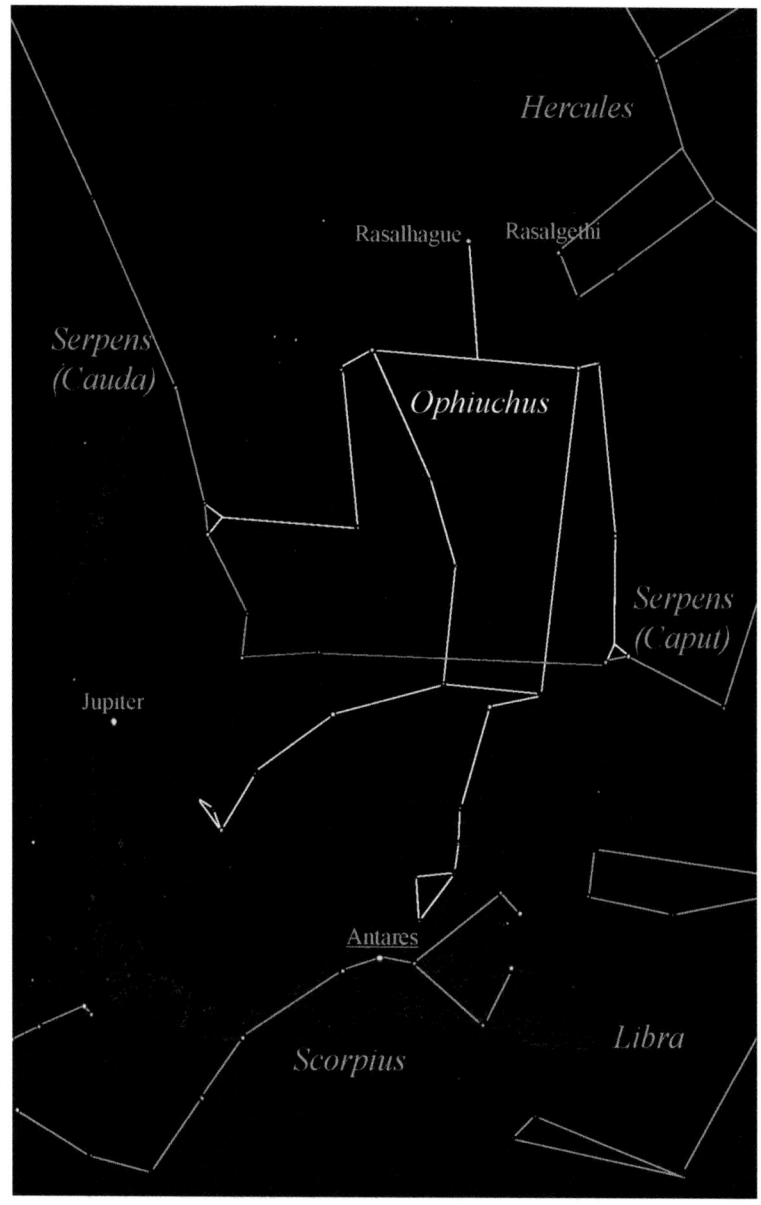

Fig. 5.14b. The figures of Ophiuchus and adjacent constellations.

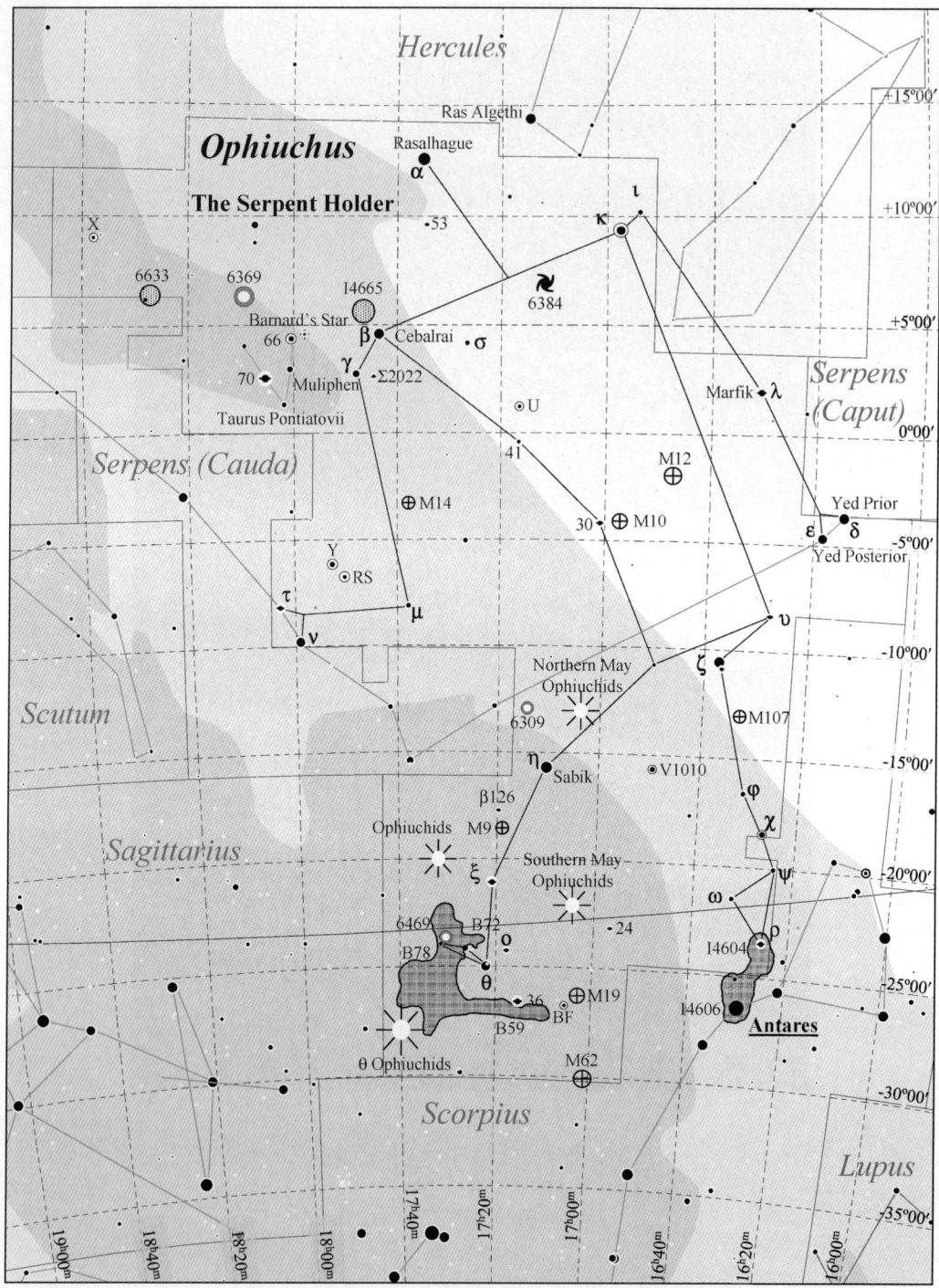

Fig. 5.15. Ophiuchus details.

Table 5.4. Selected telescopic objects in Ophiuchus.

Multiple stars:

Designation	R.A.	Decl.	Type	m₁	m₂	Sep. (")	PA (°)	Colors	Date/Period	Aper. Recm.	Rating	Comments
5, ρ Ophiuchi	16ʰ 25.6ᵐ	−23° 27'	A–B	5.1	5.7	2.9	340	B, B	2001	3/4"	*****	a.k.a H 2AB. 3.25° NNW of Antares. S. Haas reports color as amber yellow for both stars. a.k.a. H 2AC and AD,
			A–C	5.1	7.3	150.0	000	B, B	1999	6×30	****	a.k.a. H 2AC
			A–DE	5.1	6.8	156.4	252	B, B	1999	6×30	****	a.k.a. Bu 1115DE
			D–E	7.0	8.7	0.4	318	B, ?	1991	24"	**	
3, υ Ophiuchi	16ʰ 27.8ᵐ	−08° 22'	A–B	4.7	8.8	0.5	315	W, ?	1999	18/20"	****	a.k.a. RST 3949
10, λ Ophiuchi	16ʰ 30.9ᵐ	+01° 59'	A–B	4.2	5.2	**1.5**	**037**	W, pY	**129.0**	6/7"	******	a.k.a. STF 2055 AB. Calculated values for 2010
			AB–C	3.8	11.0	119.0	170	W, ?	2000.0	3/4"	***	a.k.a. STF 2055 AB–C.
			A–D	3.8	9.8	308.1	247	W, ?	2002.0	2/3"	****	a.k.a. STF 2055 AD.
24 Ophiuchi	16ʰ 56.8ᵐ	−23° 09'	A–B	6.3	6.3	1.0	300	W, pY	1995	9/10"	****	a.k.a. BU 1117
30 Ophiuchi	17ʰ 01.1ᵐ	−04° 13'	A–B	5.0	9.7	102.4	068	yO, ?	2005	7×50	****	Easy pair for small scopes or large binoculars. a.k.a. ENG 59AB
				5.0	8.8	220.2	070	yO, ?	2002	7×50	****	a.k.a. ARN 16AC
36 Ophiuchi	17ʰ 15.3ᵐ	−26° 36'	A–B	5.1	5.1	**5.0**	**143**	O, O	**470.9**	2/3"	******	a.k.a. SHJ 243. A and B are a binary pair, others are background stars. See Note C
			A–C	5.1	6.5	733.0	074	O, O	1991.0	6×30	***	Very pretty, distinct orange tints for primary pair
			A–D	5.1	7.8	267.4	337	O, ?	1991.0	2/3"	***	
			A–a	5.1	13.5	38.4	312	O, ?	1998.0	7/8"	***	
39, o Ophiuchi	17ʰ 18.0ᵐ	−24° 17'	A–B	5.2	6.6	10.3	352	Y, B	2006	2/3"	******	Lovely colors. Reports vary, some give O and Y, some O and W. a.k.a. H 3 25
β 126 (Ophiuchi)	17ʰ 19.9ᵐ	−17° 45'	A–B	6.3	7.6	2.4	263	W, W	2006	4/5"	*****	Very nice all-white triplet
				6.3	11.3	12.0	138	W, W	1998	3/4"	***	
ξ Ophiuchi	17ʰ 21.0ᵐ	−21° 07'	A–B	4.4	8.9	3.7	050	pY, ?	1989	2/3"	*****	a.k.a. DON 832. Brightness difference makes secondary color difficult to see
53 Ophiuchi	17ʰ 34.6ᵐ	−09° 35'	A–B	5.8	7.5	41.5	190	W, pB	2004	7×50	****	a.k.a. SFTA 34
			A–C	5.8	11.9	92.7	344	W, ?	2000	3/4"	****	One of the distant companions is reported to be orange, the color of the other is unknown
			B–D	7.8	10.8	91.6	223	pB, ?	1989	3/4"	****	
Σ 2022	17ʰ 44.6ᵐ	+02° 35'	A–B	6.1	6.5	21.3	093	W, W	2006	15×80	****	
69, τ Ophiuchi	18ʰ 03.1ᵐ	−08° 11'	A–B	5.2	5.9	**1.6**	**285**	yO, yO	**257.0**	6/7"	******	The distance between the A–B pair is closing (decreasing). a.k.a. STF 2262 closing (decreasing)
			A–C	5.2	9.3	100.3	127	yO, ?	1999.0	7×50	******	
70 Ophiuchi	18ʰ 05.5ᵐ	+02° 30'	A–B	4.2	6.0	**5.7**	**132**	yO, R	**88.38**	2/3"	****	Nice colors. Color reports vary from W to pO for both stars. a.k.a. 2272VX

(continued)

Table 5.4. (continued)

Variable stars:

Designation	R.A.	Decl.	Type	Range (m_v)	Period (days)	F (f_r/f_l)	Spectral Range	Aper. Recm.	Rating	Comments	
7, χ Ophiuchi	16h 27.0m	-18° 27'	γ Cas	4.2 / 5.0	334	-	B1.5IV-Vpe	Eye	****	5.0° due W (PA 270°) of η Ophiuchi (m2.43)	*11
V1010 Ophiuchi	16h 49.5m	-15° 40'	EB/KE	6.1 / 7.0	0.661	-	A5V	7×50	****	Components are gravitationally distorted. Light curve varies continuously	
κ Ophiuchi	16h 57.7m	+09° 22'	Lb	4.1 / 5.0	-	-	K2III	Eye	****	Population II Cepheid. ~2m fainter than δ (classical) Cepheid of same period	*12
BF Ophiuchi	17h 06.1m	-26° 35'	W Vir	6.9 / 7.6	4.06	0.30	F6-G2	6×30	****		
U Ophiuchi	17h 16.5m	+01° 13'	Algol	5.8 / 6.5	1.677	0.17	B5V + B5V	6×30	****	1° 29' due N (PA 359°) of 41 Oph. (M4.72) *12	
RS Ophiuchi	17h 50.2m	-06° 43'	Nr	5.3 / 12.3	-	-	Ocp + M2ep	6/8"	****	Recurrent nova, eruptions in 1898, 1933, 1958, 1967	
Y Ophiuchi	17h 52.6m	-06° 09'	δ Cep	5.9 / 6.5	17.124	0.44	F8Ib-G3Ib	6×30	****	δ Cephei type. Used for distance estimates of nearby galaxies	
66, V2048 Ophiuchi	18h 00.3m	+04° 22'	γ Cas/UV	4.6 / 4.9	-	-	B2-B6IV-Vnev	eye	******	Two modes of variation create a very complex light curve	
X Ophiuchi	18h 38.3m	+08° 50'	M	5.9 / 9.2	328.85	0.53	M5e-M9e	7×50	****	Mira type pulsating red giant. $	

Star clusters:

Designation	R.A.	Decl.	Type	m_v	Size (')	Brtst. Star	Dist. (ly)	dia. (ly)	Number of Stars	Aper. Recm.	Rating	Comments
M107, NGC 6171	16h 32.5m	-13° 03'	Globular	7.8v	13	13.0	21K	80	-	7×50	****	Lum. = 26K suns
M12, NGC 6218	16h 47.2m	-01° 57'	Globular	6.1v	16	12.0	18K	85	-	7×50	****	Lum. = 95K suns
M10, NGC 6254	16h 57.1m	-04° 06'	Globular	6.6v	20	12.0	14K	82	-	7×50	*****	Lum. = 36K suns
M62, NGC 6266	17h 01.2m	-30° 07'	Globular	6.4v	15	13.2	18k	78	-	7×50	*****	Lum. = 71K suns
M19, NGC 6273	17h 02.6m	-26° 16'	Globular	6.8v	17	~14	35k	174	-	7×50	*****	Lum. = 190K suns
M9, NGC 6333	17h 19.2m	-18° 31'	Globular	7.8v	20	13.5	25K	85	-	2/3"	****	Lum. = 36K suns *13
M14, NGC 6402	17h 37.6m	-03° 15'	Globular	7.6v	11	14.0	33K	105	-	2/3"	****	Lum. = 79K suns
IC 4665	17h 46.3m	+05° 43'	Open	4.2v	70	6.0	1.1K	23	60	7×50	****	Very loose cluster. Size estimates vary from 40 to 70'. Lum. = 2K suns
NGC 6633	18h 27.7m	+06° 34'	Open	4.6v	20	8.0	1.0K	6	160	7×50	****	Lum. = 1.7K suns

Nebulae:

Designation	R.A.	Decl.	Type	m_v	Size (')	Brtst./ **Cent.** star	Dist. (ly)	dia. (ly)	Aper. Recm.	Rating	Comments
IC 4604	16ʰ 25.6ᵐ	–23° 27'	Diffuse	–	60×25	4.6	–	–	7×50	****	ρ Ophiuchi nebula. Extr lge and faint, reflection and absorption
NGC 6309	17ʰ 14.1ᵐ	–12° 55'	Planetary	11.5	16"	13.0	8.5K	0.7	8/10"	****	The "Box Nebula," rectangular shape. Only 25" S of a m 11
Barnard 59	17ʰ 21ᵐ	–27° 23'	Dark	–	300×60	–	–	–	7×50	****	Pipe Nebula, stem
Barnard 78	17ʰ 33ᵐ	–26° 30'	Dark	–	200×140	–	–	–	7×50	****	Pipe Nebula, bowl
Barnard 72	17ʰ 23.5ᵐ	–23° 38'	Dark	–	4	–	–	–	7×50	*****	"Snake Nebula." Small, but very dark. S-shaped, extension of B78 between θ and ξ Oph.
NGC 6369	17ʰ 29.3ᵐ	–23° 46'	Planetary	11.4	30"	15.6	3.9K	0.6	6/8"	****	Little Ghost Nebula"
NGC 6572	18ʰ 12.1ᵐ	+06° 51'	Planetary	8.1	8"	12.9	2.0K	0.1	11×70	****	Small very green disk. Easy to mistake as a star at low power

Galaxies:

Designation	R.A.	Decl.	Type	m_v	Size (')	Dist. (ly)	dia. (ly)	Lum. (suns)	Aper. Recm.	Rating	Comments
NGC 6384	17ʰ 32.4ᵐ	+07° 04'	Sbc	10.6	6.3×4.0	87M	157K	52G	6/8"	****	Pr brt, sm, sl elong. 4 fil arms. Sm, brt nucl.

(continued)

Table 5.4. (continued)

Meteor showers:

Designation	R.A.	Decl.	Period (yr)	Duration	Max Date	ZHR (max)	Comet/**Asteroid**	First Obs.	Vel. (mi/**km**/sec)	Rating	Comments
Northern May Ophiuchids	17h 04m	−13°	Annual	5/8 – 6/18	May 18–19	2–3	–	1896	Slow	Minor	Discoverer (H. Corder) reported meteors as "slow." Reliable. See text
Southern May Ophiuchids	17h 04m	−22°	Annual	5/5 – 6/4	May 13–18	2–3	–	1926	Slow	Minor	First noted by Ronald A. McIntosh of Aukland, NZ. Reliable. See text
Ophiuchids	17h 32m	−20°	Annual	5/19 – 7/2	June 20	6	–	1887	v. slow	Minor	Disc. Wm. F. Dunning. Very reliable, meteors reported nearly every year
Theta Ophiuchids	17h 40m	−28°	annual	5/21 – 6/16	June 7–12	10	–	~1948	slow	Minor	Disc. Cuno Hoffmeister. Reliable, meteors reported in many years '14 *14

Other interesting objects:

Designation	R.A.	Decl.	Type	m$_v$	Size (')	Dist. (ly)	dia. (ly)	Lum. (suns)	Aper. Recm.	Rating	Comments
Barnard's Runaway Star	17h 58.1m	+04° 41'	High PM	9.5		5.95			2/3"	*****	Largest proper motion of any star, 10.31"/year
Taurus Poniatovii	18h 12m	+02° 50'	Asterism	–					Eye	****	A "V" of stars (66, 67, 68, 70, and 73 Oph) resembling a faint Hyades

Fig. 5.16. M9 (NGC 6333), globular cluster in Ophiuchus. Astrophoto by Mark Komsa.

Fig. 5.17. M10 (NGC 6254), globular cluster in Ophiuchus. Astrophoto by Mark Komsa.

Fig. 5.18. M12 (NGC 6218), globular cluster in Ophiuchus. Astrophoto by Mark Komsa.

SERPENS (**sir**-pens)	**The Serpent 9:00 P.M. Culmina-**
SERPENTIS (sir-pen-tiss)	**tion:** July 22
Ser	**AREA:** 637 sq. deg., 23rd in size

TO FIND SERPENS: Serpens is the only constellation which is divided into two separate parts which have no point of contact with each other. The serpent's head is relatively easy to find, being about 10° south of Corona Borealis. Look for three 3rd magnitude stars in a triangle about 3° on a side. Between 2 and 3° north of this triangle are three 4th magnitude stars. These seven stars form a somewhat lopsided "X," with κ Serpentis being the point where the arms cross. This group forms the head. Eight degrees SW of κ is δ Ser which is at the northwestern end of a 35° line of 2nd, 3rd, and 4th magnitude stars which form the western part of the serpent's body. Only the northernmost three of these stars are in Serpens. The other four are in Ophiuchus, but still indicate where the serpent's body passes in front of the serpent holder. The first two in Ophiuchus represent his left hand grasping the serpent. From η Oph at the southeastern end of the chain, the serpent's body jogs NE 4° to ν Serpentis, SE 5° to ξ Ser and NE 3° to o Ser before crossing a corner of Ophiuchus, where ν and τ Ophiuchui show where Asclepius's right hand holds the serpent's body. The serpent's tail then continues NE 7° to η Ser, and finally, 11° NE to θ Ser (passing directly over the 5th magnitude star 4 Aquilae) at the end of the tail. From head to tail, Serpens measure 48°, but the serpent's body has wound through over 80° getting there (Figs. 5.14 and 5.15).

KEY STAR NAMES:

Name	Pronunciation	Source
α Unukalhai	**uh**-nuk-uhl-HIGH)	unuq al-hayya (Arabic), "the serpent's neck"
θ Alya	**uhl**-you	alya (Arabic), the sheep's tail, erroneously applied to θ Ser in recent times

Magnitudes and spectral types of principal stars:

Bayer desig.	Mag. (m_v)	Spec. type	Bayer desig.	Mag. (m_v)	Spec. type	Bayer desig.	Mag. (m_v)	Spec. type
α	2.63	K2 III	β	3.65	A3 V	γ	3.85	F6 V
δ	3.80	F0 IV	ε	3.71	A2 m	ζ	4.62	F3 V
η	3.23	K0 III/IV	θ	4.03	A5 V	ι	4.51	A1 V
κ	4.09	M1 III	λ	4.42	G0 Vvar	μ	3.54	A0 V
ν	4.32	A0/A1 V	ξ	3.54	F0 IIIp	o	4.24	A2 V
π	4.82	A3 V	ρ	4.74	K5 III	σ	4.82	F0 V
τ^1	5.16	M1 III	τ^2	6.22	B9 V	τ^3	6.10	G8 III
τ^4	6.51	M5 II/III	τ^5	5.93	F3 V	τ^6	5.93	G8 V
τ^7	5.80	A2m	τ^8	6.15	A0 V	υ	5.71	A3 V
φ	5.54	K1 IV	χ	5.34	A0p	ψ	5.86	G5 V
ω	5.21	G8 III						

Stars of magnitude 5.5 or brighter which have other catalog designations:

Draper Number	Mag. (m_v)	Spec. type	Draper Number	Mag. (m_v)	Spec. type
HD 145206	5.39	K4 III	HD 168415	5.39	K3 III
HD 142574	5.45	M0 III			

Fig. 5.19a. The stars of Serpens Caput and adjacent constellations.

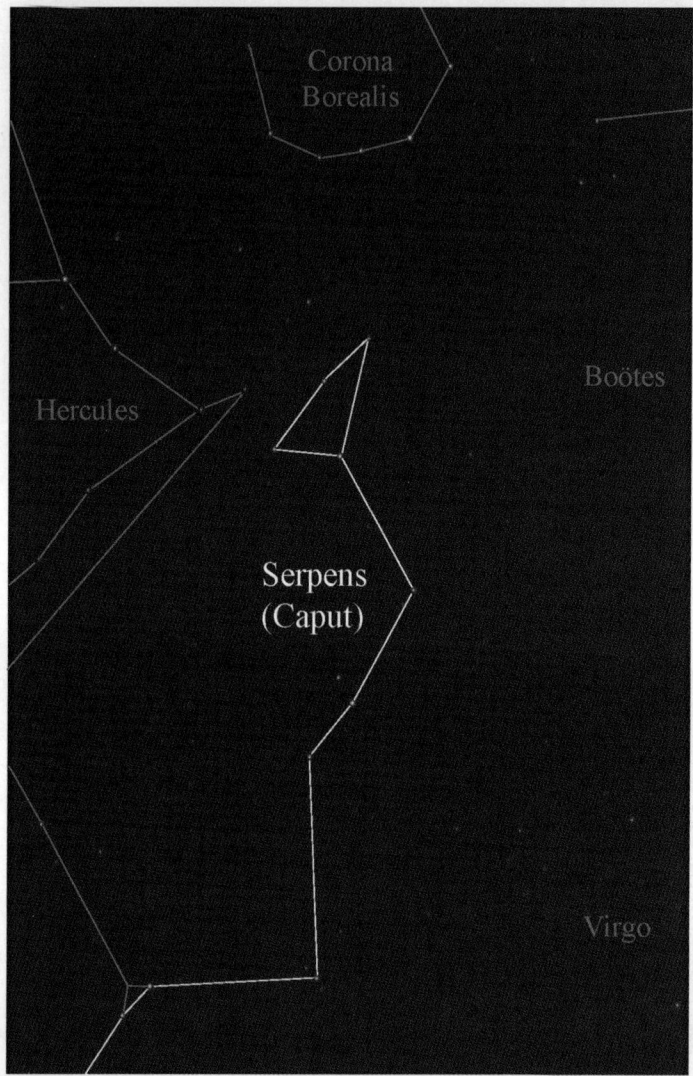

Fig. 5.19b. The figures of Serpens Caput and adjacent constellations.

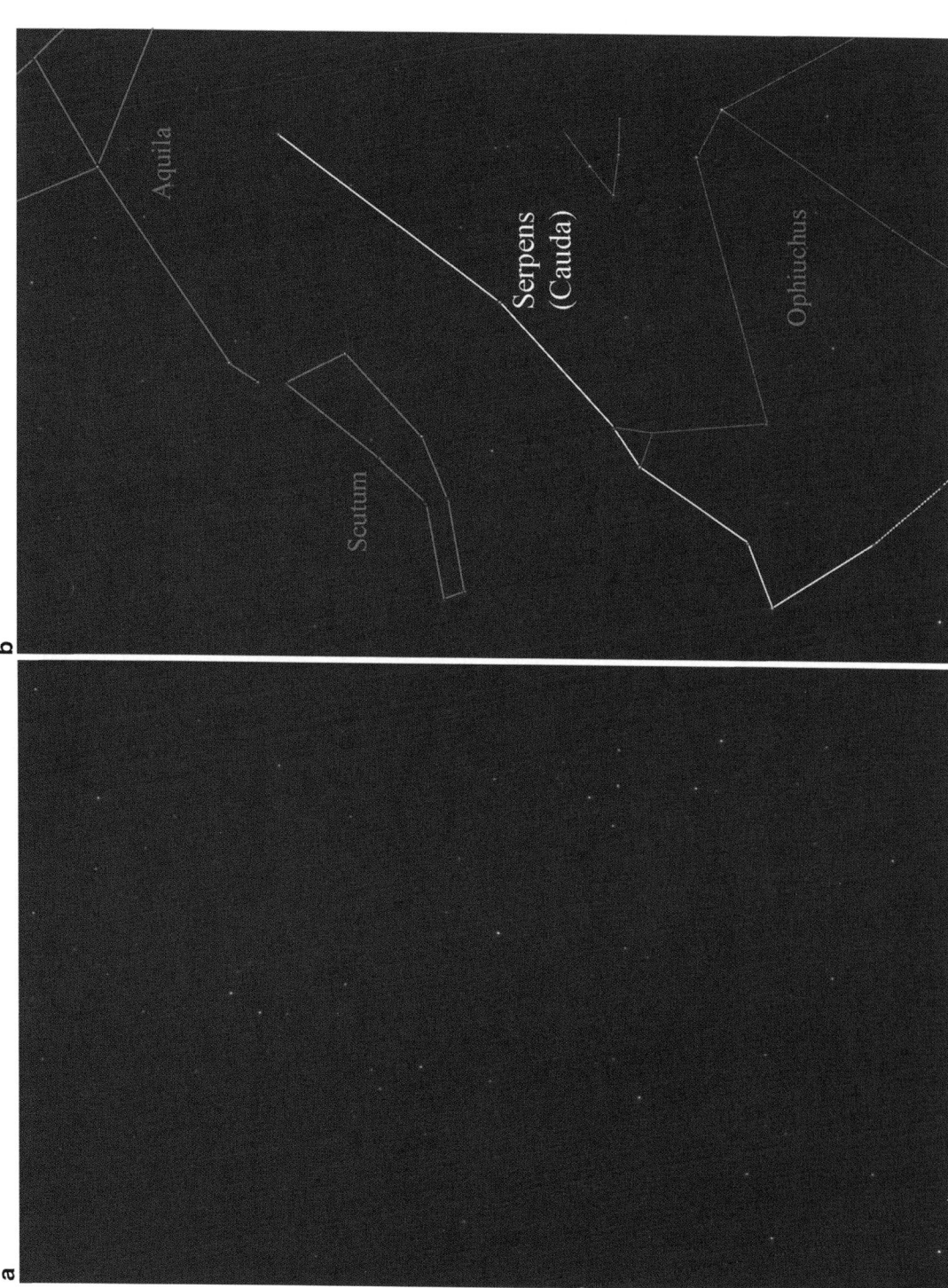

Fig. 5.20. (a) The stars of Serpens Cauda and adjacent constellations. (b) The figures of Serpens Cauda and adjacent constellations.

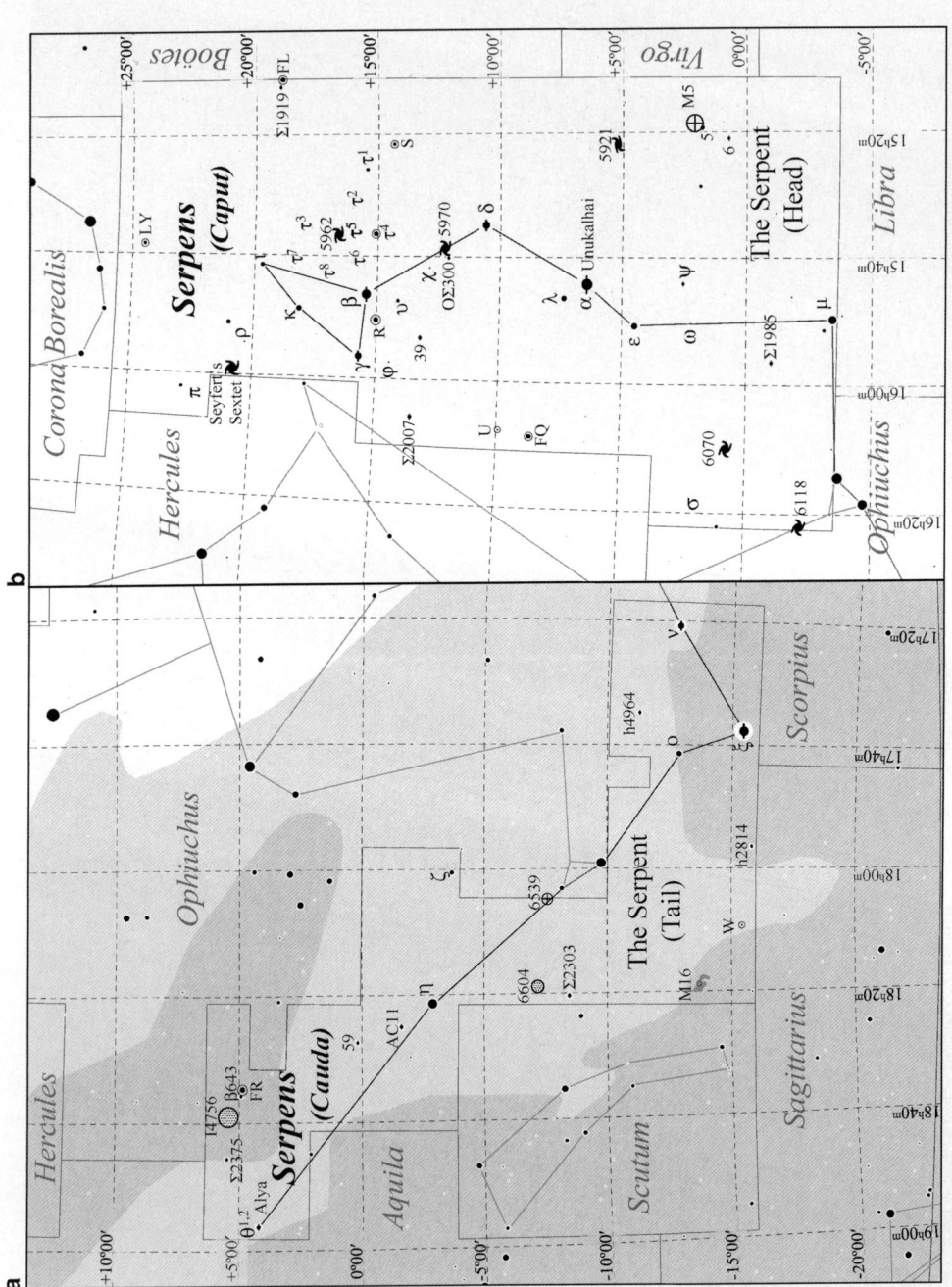

Fig. 5.21. Serpens details.

Table 5.5. Selected telescopic objects in Serpens.

Multiple Stars:

Designation	R.A.	Decl.	Type	m_1	m_2	Sep. (")	PA (°)	Colors	Date/Period	Aper. Recm.	Rating	Comments
Σ1919	15h 12.7m	+19° 17'	A–B	6.7	7.4	23.7	011	Y, dY	2005	10×50	****	Nice LP double
5 Serpentis	15h 19.3m	+01° 46'	A–B	5.1	10.1	11.4	035	Y, W	2000	2/3	***	a.k.a. Σ1930. Beautiful double and field. M5 is in NE edge of LP FOV
6 Serpentis	15h 21.0m	+00° 43'	A–C	5.1	9.2	127.2	040	W, –	1924	7×50	***	
			A–B	5.5	8.8	3.0	0.25	O, Y	2000	3/4"	****	a.k.a. β 32. Somewhat difficult, but lovely colors
13, δ Serpentis	15h 34.8m	+10° 32'	A–B	4.2	5.2	**4.4**	**175**	yW, yW	**3,168**	2/3"	*****	Showcase pair. a.k.a. Σ1954
OΣ 300	15h 40.2m	+12° 03'	A–B	6.3	10.1	15.0	260	rW, –	2003	15×80	****	
ψ Serpentis	15h 44.0m	+02° 31'	A–B	6.0	12	4.4	037	yW, –	1974	3/4"	***	a.k.a. A 2230
			A–C	6.0	9.2	195.5	208	yW, –	2002	7×50	***	
24, α Serpentis	15h 44.3m	+06° 26'	A–C	2.7	11.8	57.9	349	O, –	1911	3/4"	***	a.k.a. h1277
28, β Serpentis	15h 46.2m	+15° 25'	A–B	3.7	10.0	30.9	264	yW, pV	1999	15×80	****	a.k.a. Σ1970. Some observers call colors Y and B
39 Serpentis	15h 53.2m	+13° 12'	A–B	6.1	11.8	100.7	088	Y, –	1993	3/4"	***	a.k.a. OΣ 583
Σ1985	15h 55.9m	-02° 09'	A–B	7.0	8.7	**6.3**	**354**	yW, –	**2,331.2**	2/3"	****	
41, γ Serpentis)	15h 56.4m	+15° 41'	A–B	3.9	10.5	220.3	318	Y, –	1924	15×80	***	a.k.a. OΣ 584AB
			A–C	3.9	10.9	177.1	165	Y, –	1998	15×80	***	a.k.a. OΣ 584AC
Σ2007	16h 06.0m	+13° 19'	A–B	6.9	8.0	37.8	322	Y, R	2006	7×50	****	Unusual color combination
53, ν Serpentis	17h 20.8m	-12° 51'	A–B	4.3	9.4	45.9	026	W, R	2002	7×50	****	a.k.a. Shj 247. Nice colors
h 4964	17h 34.8m	-11° 15'	A–B	5.5	9.9	54.3	225	W, B	2002	11×70	****	a.k.a. Rst 5090
55, ξ Serpentis	17h 37.6m	-15° 24'	A–B	3.5	13.0	24.4	079	W, –	1998	6/7"	***	
h 2814	17h 56.3m	-15° 49'	A–B	5.9	9.2	21.0	157	W, rW	2006	11×70	***	
Σ 2303	18h 20.1m	-07° 59'	A–B	6.6	9.3	1.6	240	Y, –	1991	6/7"	***	
AC 11	18h 25.0m	-01° 35'	A–B	6.2	6.6	**0.8**	**355**	Y, Y	**248**	11/12"	****	Max. sep. 0.867" in 1992, min. 0.034" in 2083. Almost matched pair
59 Serpentis	18h 27.2m	+00° 12'	Aa–B	5.2	7.6	3.9	320	Y, dB	2003	2/3"	****	a.k.a. Σ2316. Very nice colors
β 643	18h 35.6m	+04° 56'	A–B	6.5	12.5	10.5	330	W, –	1924	4/5"	***	
Σ2342	18h 35.6m	+04° 56'	A–C	6.5	9.6	34.1	359	Y, –	2006	11×70	***	
Σ2375	18h 45.5m	+05° 30'	Aa–Bb	6.3	6.7	2.6	119	W, W	2006	3/4"	****	Attractively close and fairly closely matched
63, θ Serpentis	18h 56.2m	+04° 12'	A–B	4.6	4.9	23.0	104	yW, yW	2006	11×70	****	a.k.a. Σ2417. Another fairly similar pair

(continued)

Table 5.5. (continued)

Variable stars:

Designation	R.A.	Decl.	Type	Range (m_v)	Period (days)	F (f_r/f_f)	Spectral Range	Aper. Recm.	Rating	Comments
FL Serpentis	15^h 12.1^m	+18° 59'	Lb	6.0 / 5.8	–	–	M4IIIab	6×30	****	1.° 20' SW (PA 222°) of τ¹ Serpens (m5.16). ##
S Serpentis	15^h 21.7^m	+14° 19'	M	14.1 / 7.0	371.84	0.43	M5e–M6e	12/14"	****	##
17, τ⁴ Serpentis	15^h 36.5^m	+15° 06'	SRb	7.1 / 5.9	100	–	M5IIb–IIIa	6×30	****	
LY Serpentis	15^h 38.3^m	+24° 31'	Lb	7.0 / 6.7	–	–	M4III	6×30	****	
R Serpentis	15^h 50.7^m	+15° 03'	M	14.4 / 5.2	356.41	0.41	M5IIIe–M9e	14/16"	*****	1° 7' E (PA 104°) of β Serpens (m3.65). Very red at min.##
U Serpentis	16^h 07.3^m	+09° 56'	M	14.7 / 7.8	237.50	0.28	M3e–M6e	16/18"	****	6' WNW (PA 296°) of 45 Serpens (m5.63). ##
FQ Serpentis	16^h 08.6^m	+08° 37'	Lb	6.6 / 6.3	–	–	M4III	6×30	****	5' NNE (PA 25°) of 47 Serpens (m5.68) #
W Serpentis	18^h 09.8^m	–15° 33'	EA/GS	10.2 / 8.4	14.155	0.20	F5elb(Shell)	10×80	*****	
FR Serpentis	18^h 35.6^m	+04° 56'	α CVn	6.5 / 6.3	2.144	–	A2p(Sr-Cr)	6×30	****	

Star clusters:

Designation	R.A.	Decl.	Type	m_v	Size (')	Brtst. Star	Dist. (ly)	dia. (ly)	Number of Stars	Aper. Recm.	Rating	Comments
M5, NGC 5904	15^h 18.6^m	+02° 05'	Globular	5.7	17.4	12.2	24K	120	>100K	7×50	*****	Almost equal to M13. Concentrated core, resolved at edges
NGC 6539	18^h 04.8^m	–07° 35'	Globular	8.9	7	15.9	7.5K	30	–	4/5"	***	V sm, faint, low concentration
NGC 6604	18^h 18.1^m	–12° 14'	Open	6.5	4	7.5	2.3K	12	105.0	6/8"	***	~30 stars, visible in medium-sized telescopes, well separated from background
M16, NGC 6611	18^h 18.8^m	–13° 47'	Open	6.0	21	8.2	5.9K	21	543.0	7×50	****	In Eagle Nebula
IC 4756	18^h 39.0^m	+05° 27'	Open	4.6	52	4.6	1.3K	15	466.0	3/4"	***	~80 stars visible in small telescopes, separated, little concentrated toward center

Nebulae:

Designation	R.A.	Decl.	Type	m_v	Size (')	Brst./Cent. star	Dist. (ly)	Colors	Aper. Recm.	Rating	Comments
M16, NGC 6611	18h 18.8m	–13° 47'	Emission	–	120×25	–	5.9K	Red star	7×50	****	a.k.a Eagle or Star Queen Nebula. *Color shows only in photographs

Galaxies:

Designation	R.A.	Decl.	Type	m_v	Size (')	Dist. (ly)	dia. (ly)	Lum. (suns)	Aper. Recm.	Rating	Comments
NGC 5921	15h 21.9m	+05° 04'	S(B)b	10.8	4.8×4.0	65M	91K	16G	3/4"	****	Brt core, mod brt halo, filamentary arms. Dark lanes in larger scopes
NGC 5962	15h 36.5m	+16° 37'	S(B)c	11.3	2.6×1.8	90M	68K	20G	4/6"	***	Lge, fairly faint, elongated PA 110°, filamentary arms
NGC 5970	15h 38.5m	+12° 11'	Sc	11.5	2.7×18	90M	71K	16G	6/8"	***	Mod faint, mod lge, round, sm brt nucl.
NGC 6070	16h 10.0m	+00° 43'	Sc	11.8	3.6×1.9	90M	92K	12G	6/8"	***	Lge, faint, slightly elong PA °062. a.k.a. Holmberg 729a
NGC 6118	16h 21.8m	–02° 17'	Sb	11.7	4.6×1.9	70M	94K	8.2G	8/10"	***	V lge, v faint, elong PA 058°, faint nucl., 3 filamentary arms

Other interesting objects:

Designation	R.A.	Decl.	Type	m_v	Size (')	Dist. (ly)	dia. (ly)	Lum. (suns)	Aper. Recm.	Rating	Comments
Seyfert's Sextet, NGC 6027A–E	15h 59.2m	+20° 45'	Gal. Grp.	11.3–13.7	2	190M			16/18"	**	Compact group of four galaxies, a fragment gravitationally torn from one, and one background galaxy. Originally thought to be six galaxies at some distance

Fig. 5.22. M5.

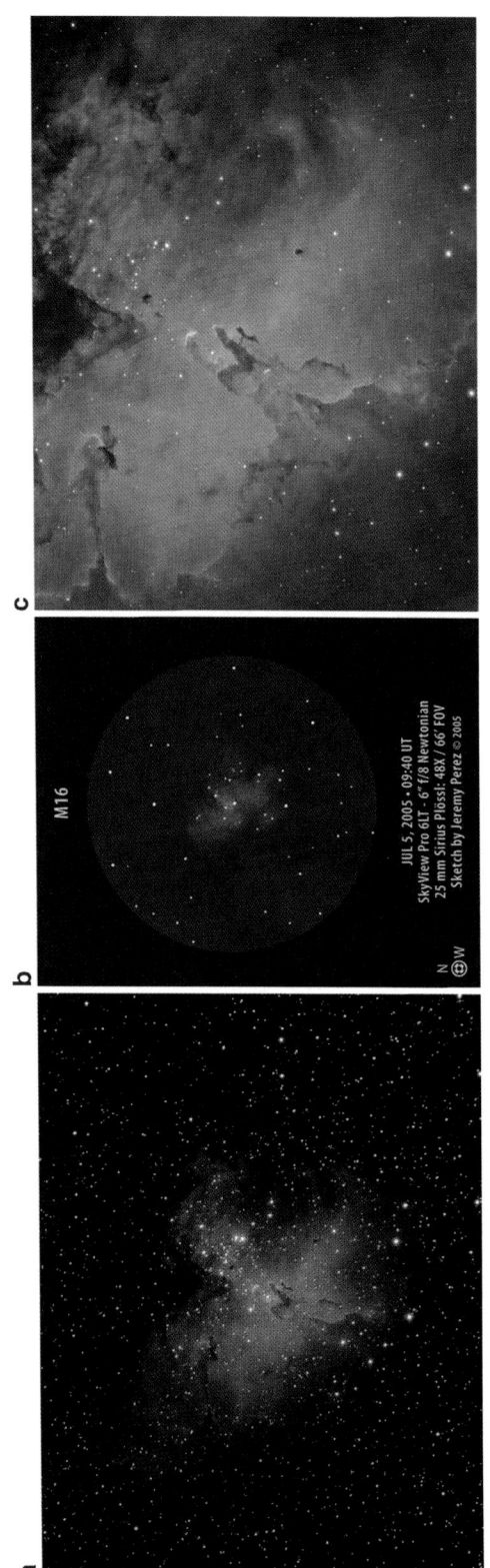

Fig. 5.23. (**a**) M16 (NGC 6611), emission nebula in Serpens. Astrophoto by Warren Keller. (**b**) M16 Drawing. (**c**) M16, recreation of Hubble telescope "Pillars of Creation" by Stuart Goosen, using a Hα (α represents Greek letter alpha) to highlight the back-lighted pillars.

NOTES:

1. The former γ Scorpii is now 20, or σ Librae.
2. Bayer listed φ Scorpii as a fifth magnitude star in the southern claw, but no such star can be identified.
3. Antares has a fainter blue giant companion which orbits in about 900 years, but the orbit is poorly known. Calculations using the 900 year period indicate that the apparent separation between the two stars was greatest (3.21″) in 1889. Resolving this pair was fairly easy in the late nineteenth and early twentieth centuries, but is becoming more difficult. The pair continues to get closer together until 2114 when the separation is only 0.28″, and no amateur scopes are able to resolve it. Antares A is slightly variable, ranging from magnitude 0.86 to 1.06.
4. V919 Scorpii is one of only a few known Wolf–Rayet eclipsing binaries. Wolf–Rayet stars are extremely hot (25,000–100,000 K) and extremely luminous. They are also rare, with only a few hundred being known in the Milky Way galaxy. Of these, only a very small number are eclipsing binaries.
5. When near maximum, RY Sagittarii also exhibits moderate range (~0.5 mag.) variations with a period of ~39 days. These are similar to δ Cepheus. The period also appears to be shortening by a noticeable amount over an interval of 10 years.
6. NGC 6822 (Sgr) was thought to be a globular cluster in the Milky Way Galaxy until 1994, when new distance measurements showed it to be outside our galaxy and a member of the Local Group of galaxies. It is shown in two places because some references still differ on whether it is a small galaxy or a very large cluster.
7. Denning also claimed to have seen meteors from this part of the sky (Sgr) in 1887, and several observations of meteors from the Scorpius–Sagittarius area were recorded by various observers between 1908 and 1938. This shower seems to be from a fragment of the θ Ophiuchid stream. It is also sometimes called the Scorpiids–Sagittariids.
8. Lacaille assigned the designation "η" Coronae Australis to an apparently single star. Later, astronomers discovered this star to consist of two components, and added the "1" and "2" superscripts. This is an anomaly, since normally uppercase letters were used to designate the components of double or multiple stars.
9. Most of Corona Australis was too far south for Bayer to observe, Lacaille was the first to assign Greek letters to the stars, and he did not assign the letter "ι" to any star in Corona Australis.
10. Bayer gave the designation of "π" Ophiuchi to a star listed in Tycho Brahe's second catalog. However, Brahe's catalog had several errors, and later astronomers could not make an identification of which star, if any, Bayer intended.
11. χ Ophiuchi (a.k.a. 7 Ophiuchi) is a star which rotates so rapidly that it is losing matter at its equator. Its variations are irregular.
12. U Ophiuchi is an Algol type eclipsing binary whose components are nearly equal in brightness. This makes the two minima very nearly equal, and the period appears to be half the actual period.
13. M9 (NGC 6333) is accompanied by two fainter globular clusters, NGC 6342 and 6356, which can be seen in the same low power field of view.
14. Discovery of the θ Ophiuchids has also been attributed to Wm. F. Dunning. It is a very reliable shower, with meteors having been reported nearly every year of the twentieth century.

CHAPTER SIX

A Strongman's Struggles

March–May, July, December

ARA	CANCER	CENTAURUS	COMA BERENICES
The Altar	The Crab	The Centaur	Berenices' Hair

HERCULES	LEO	LUPUS
The Strongman	The Lion	The Wolf

The constellation Hercules (the Roman name for Heracles) represents the mythological Theban hero celebrated for performing 12 labors which required prodigious strength and heroism. In addition to Hercules, 12 other constellations represent creatures or objects related to Heracles and his labors.

After Zeus and the other Olympians won the battle against the Titans to secure their place as the ruling gods, Zeus was warned by an oracle that an even greater enemy would try to displace them. This enemy would be the Giants, and their power would be so great that the Olympians could not defeat them alone but would need to have the help of the greatest mortal who had ever lived.

Zeus long pondered this augury in his mind. He concluded that only he had the power to father this hero, and that a very special mortal woman would be required as the mother. He saw this special woman in Alcmena, the wife of a distinguished general

P. Simpson, *Guidebook to the Constellations: Telescopic Sights, Tales, and Myths*, Patrick Moore's Practical Astronomy Series, DOI 10.1007/978-1-4419-6941-5_6, © Springer Science+Business Media, LLC 2012

named Amphitryon. When Amphitryon went on a campaign to avenge the death of Alcmena's brothers, Zeus impersonated Amphitryon and fooled Alcmena into conceiving a child by him. Amphitryon returned the next day, and Alcmena also conceived a child by him.

When the twins were born they were named Alcides and Iphicles. Zeus' wife Hera learned that Alcides was actually Zeus' son and, for the rest of his life, did all she could to destroy or demean him. When the twins were only 8 months old, she sent two giant serpents into the nursery late one night to kill the pair. Iphicles screamed so loudly that Amphitryon and Alcmena awoke and rushed into the room to rescue the twins. To their amazement, Alcides was sitting up in the crib laughing and holding a strangled serpent in each hand. They knew at once that this was no ordinary baby.

Still, Alcides was raised as if he were actually a son of Amphitryon. He had many adventures and accomplished many exploits, even as a child and young man. However, it was also evident that some powerful being was trying to thwart Alcides and harm him if possible. Amphitryon visited an oracle and learned that this being was Hera and was advised to change Alcides name to Heracles (glory of Hera) to placate her. This was done, but it did not accomplish its purpose.

After Heracles had grown up, he challenged a minor river god over the love of a beautiful princess, Megara, and won her hand in marriage. After several years of a happy marriage, Hera caused Heracles to suffer a temporary madness in which he killed his wife and the three sons they had by that time. When Heracles recovered and was shown what he had done, he was devastated by his actions. Although he was not punished for the actions committed while he was under the spell of Hera, Heracles imposed a permanent exile upon himself and wandered throughout the Greek world.

In the course of his travels, Heracles visited the oracle at Delphi (the most revered in the country), and learned that he could redeem himself by temporarily becoming the slave of his cousin Eurystheus, who was the King of Mycenae. To achieve redemption, Heracles had to work for him for 1 year, and perform any ten tasks demanded by Eurystheus. The oracle knew that Eurystheus was a cowardly and vindictive man, and would assign Heracles the most difficult and dangerous tasks conceivable. Heracles submitted to this punishment, even though he despised his cowardly cousin. In the end, Heracles had to perform a total of 12 tasks because Eurystheus refused to accept 2 of the tasks.

The cowardly Eurystheus feared Heracles and tried to make each labor as lengthy and dangerous as he could. Each time he sent Heracles out on a mission, he hoped that Heracles would not return alive. Each time, he was disappointed and became more fearful of his heroic cousin.

The first of these tasks, or labors as they came to be known, was to slay the man-eating lion which was terrorizing the countryside around Nemea. This lion was larger than any beast on earth, had claws which could cut through metal, and a skin which could not be penetrated by any spear or arrow. Heracles had no trouble finding the lion, since all he had to do was follow the trail of devastation and bodies left by the lion. When he caught up with the lion, Heracles first attempted to kill it with spears and arrows, only to see them bounce harmlessly off. The lion retreated into his den, a cave with two entrances. Heracles rolled a large stone into one opening to seal it, then entered the other. When the lion attacked, Heracles struck him a mighty blow to the head with his massive club. This blow would have crushed the skull of an elephant but only stunned the lion. The force of the impact broke the huge club, and Heracles was left weaponless. Undaunted, Heracles leaped upon the back of the momentarily dazed lion and wrapped his muscular arms around the lion's neck. Holding on with his legs wrapped around the lion's body, Heracles tightened his choke hold on the lion and slowly strangled it.

Heracles then used one of the lion's own claws to cut through the almost impenetrable hide and fashioned a protective cape from it. Dressed in this cape, with the head of the lion draped over his own head, Heracles returned to Eurystheus' home. The cowardly king fled at the terrible sight and from then on, refused to see Heracles in person. Each time Heracles returned, Eurystheus sent the new task by a messenger. Zeus later put the Nemean lion in the sky as the constellation Leo to honor Heracle's victory. The associated constellation of Leo Minor (the smaller lion) was added by Johannes Hevelius in 1690.

For his second labor, Heracles was ordered to slay the Lernaean Hydra. The Hydra inhabited a swamp near Lerna, a town near the northern end of the Gulf of Argos, about 65 miles WSW of Athens. The Hydra was an enormous beast with a serpentine body and nine snake-like heads, one of which was immortal. The Hydra had been terrorizing the entire area by destroying crops and devouring livestock along with an occasional human.

The goddess Athena helped Heracles by pointing out the Hydra's lair, and advising him that the best way to drive the Hydra out into the open, where he was more vulnerable, was to pepper him with burning arrows. When the Hydra had been flushed out, Heracles leaped on it, wrapped one arm around its trunk, and began smashing the heads one by one with his club.

Hera, always alert to opportunities to destroy Zeus' illegitimate son, tried to distract Heracles by sending a giant crab to pinch his heels. Heracles felt the crab's claws and quickly crushed it with a single backward step. When Heracles returned his attention to the Hydra, he was horrified to see that each of the smashed heads was being replaced by two new and rapidly growing heads. Heracles called to his cousin and chariot driver Iolaos to bring fire-brands and cauterize each stump as the heads were destroyed so they could not regrow. When the eight mortal heads were disposed of, Heracles severed the immortal one, dug a deep grave and buried the head with a huge rock holding it down.

The blood of the Hydra was highly poisonous so Heracles dipped the heads of his arrows in it. Thereafter, any injury with one of his arrows was fatal. When Heracles returned to Eurystheus' court, Eurystheus had heard of the assistance provided by Iolas and refused to recognize the feat, raising the total number of labors to 11.

Another of Heracles' labors was to capture the Erymanthian boar. This was a fierce and gigantic beast that lived on the slopes of Mount Erymanthus and ravaged the surrounding country. On his way, Heracles passed through an area called Pholoë and was entertained by his friend, a Centaur named Pholus. In his cave, Pholus had a wine bottle left by Dionysus for ceremonial occasions. Heracles persuaded Pholus to open the bottle, and all the centaurs in the vicinity smelled the wine. Driven into a frenzy by the aroma, they charged toward the cave hurling rocks and other objects. Heracles drove them off, first with firebrands and then with his poisoned arrows.

One of only two centaurs who had not taken part in the battle was Pholus, who had stayed in his cave, yelling for his companions to stop. When the fight was over, all the attackers were dead or dying from their wounds. Pholus was amazed at the deadliness of Heracles' arrows and picked up one to examine it. He accidently dropped it on his foot, causing a small but fatal injury. Pholus had been the last of the mortal centaurs, and was placed in the sky as the constellation of Centaurus.

During the melee, one of the arrows had struck an attacking centaur but passed through his upper arm and lodged in the knee of Chiron, the king of the centaurs. Chiron was the only other centaur who had not participated in the attack but merely watched. He was also the only one who was the child of two gods and thus immortal. Even so, the poison was so potent, and the pain so great that after suffering for many days, Chiron begged Zeus to allow him to give away his immortality to escape the torture. Zeus agreed and, Prometheus became immortal in Chiron's place. Chiron was later placed in the heavens as the constellation Sagittarius, and his crown became Corona Australis.

Later, when Heracles cleaned out the stables of Augeias for his fifth labor, he tried to turn the task to his advantage. Knowing that Augeias had the largest herd in Greece and, that the waste had accumulated to a depth of several feet, Heracles went to him and offered to do the job in 1 day for the price of one-tenth of the cattle in Augeias' herd. Since the stables had never been cleaned and Augeias had five hundred bulls, and many times that number of cows and calves, he laughed and quickly accepted the offer of a seemingly impossible performance.

Heracles quickly dammed and diverted two rivers so that their waters merged and flowed through the stables and back into the river beds and the stables were quickly washed clean. He then redirected the rivers into their original channels. By the time Heracles returned to collect his pay, Augeias had learned that Heracles had been ordered to perform this task for Eurystheus and refused to pay Heracles. When he returned to Mycenae, King Eurystheus heaped more misfortune on Heracles by refusing to recognize this labor, saying that Heracles was working for Augeias. This increased the Labors of Heracles from the original 10 to 12.

After his death, Zeus carried Heracles to Olympus where he was reconciled with Hera, became immortal, and married the goddess Hebe. Zeus also placed figures representing Heracles and many of the creatures and objects related to his 12 labors in the skies as constellations.

The Romans accepted many of the Greek myths but usually changed the names. In this case, they renamed the hero as Hercules, and we still know him and his constellation by that name today.

Related Constellations:

Name	Representing
Ara (**air**-uh)	altar upon which Chiron sacrificed a wolf
Aquila (**ack**-wi-lah)	eagle tormenting Prometheus (see Chap. 9 for details).
Cancer (**kan**-ser)	crab sent by Hera
Centaurus (sen-**taw**-russ)	Chiron (sometimes spelled Cheiron)
Corona Australis (koh-**ro**-nah **aws**-tray-liss)	crown or wreath of Pholus
Draco (**dray**-koe)	guardian of golden apples in garden of Hesperides (see Chap. 2)
Hydra (**high**-drah)	nine-headed monster of Lerna
Leo (**lee**-oh)	lion of Nemea
Leo Minor (**lee**-oh **my**-nor)	small lion (see Chap. 17)
Lupus (**loo**-pus)	The Wolf sacrificed by Chiron
Sagitta (suh-**jit**-a)	arrow shot at the eagle, 11th labor (see Chap. 9)
Sagittarius (**saj**-i-**TAY**-ree-us)	Pholus, 4th labor

HERCULES (**her**-cu-leez)	**The Strongman**
HERCULIS (**her**-cu-lis)	**9:00 P.M. CULMINATION:** July 29
Her	**AREA**: 1,225 sq. deg., 5th in size

TO FIND: In late July, Hercules is at its highest point in the sky around 9:00 PM. To find it, first draw an imaginary line between the two brightest stars in the northern sky – Vega and Arcturus. The point on this line which is one-third the distance from Vega to Arcturus lies inside the group of four stars called the "Keystone." This group (ξ, η, π, and ε Herculi) forms the lower torso of the figure. Adding the stars β and δ Herculi to the Keystone creates a slightly asymmetric pattern some people see as the "Butterfly." The legs will be to the north of the Keystone, and to the south β and δ Herculi mark the shoulders. Going southwest from β you can see a club in Heracles' right hand and east–northeast from δ are the neck and heads of the Hydra in the left hand.[*1] The figure is easier to see if you recline or lie on your back with your head to the south (Figs. 6.1–6.3).

KEY STAR NAMES:

Name	Pronunciation	Source
α Rasalgethi	**ras**-uhl-**JEH**-thee	raʾs al jāthī (Sci-Arabic), "the Kneeler's Head"
β Kornephoros	kor-**neh**-foh-ros	χορυνηφόρος, koruneforoz (Greek) "club-bearer"
λ Maasym	**muh**-uh-sim	al-miʾ ṣam (Arabic), "the wrist" as used in the Almagest for o Herculi and erroneously applied to λ during the Renaissance
χ Marsic	mur-**sik** or **mahr**-sik	al-marfiq (Arabic), "the elbow"
ω Cujam	**cue**-yahm	caiam [a misuse of caia] (Latin), "club"

Magnitudes and spectral types of principal stars:

Bayer desig.	Mag. (m$_v$)	Spec. type	Bayer desig.	Mag. (m$_v$)	Spec. type	Bayer desig.	Mag. (m$_v$)	Spec. type
α¹	3.31	A3 IV	β	2.78	G8 III	γ	3.74	A9 III
δ	3.12	A3 IV	ε	3.92	A0 V	ζ	2.81	F9 IV
η	3.48	G8 III/IV	θ	3.86	K1 IIvar	ι	3.82	B3 V
κ	6.25	K1 IIIAus	λ	4.41	K3 IIIvar	μ	3.42	G5 IV
ν	4.41	F2 II	ξ	3.70	K0 III	ο	3.84	B9.5 V
π	3.16	K3 IIvar	ρ	4.15	B9.5 V	σ	4.20	B9 Vvar
τ	3.91	B5 IV	υ	4.72	B9 III	φ	4.23	B9 pmn
χ	4.60	F9 V	ψ¹	5.04	K3 V*²	ψ²	4.98	K5 V*²
ω	4.57	B9 pCr						

Stars of magnitude 5.5 or brighter which have Flamsteed, but no Bayer designations:

Flamsteed	Mag. (m$_v$)	Spec. type	Flamsteed	Mag. (m$_v$)	Spec. type	Flamsteed	Mag. (m$_v$)	Spec. type
2	5.35	M3 III	5	5.10	G8 III	9	5.46	K5 III
29	4.84	K4 III	30	4.83	M6 III:var	42	4.86	M2.5 III
43	5.15	K5 III	45	5.22	B9p (Cr)	47	5.48	A3m −
51	5.03	K2 II/III	52	4.82	A2p	53	5.34	F0 V
54	5.35	K4 III	59	5.27	A3 IV	60	4.89	A4 IV
68	4.80	B1.5 Vp	69	4.64	A2 V	70	5.13	A2 V
72	5.38	G0 V	82	5.35	K1 III	87	5.09	K2 III
89	5.47	F2 Ia:var	90	5.17	K3 III	93	4.67	K0 II/III
95	4.26	G5	96	5.25	B3 IV	98	4.96	M3 IIIa
99	5.05	F7 V	101	5.10	A8 III	102	4.34	A5 III
104	4.96	M3 III	105	5.30	K4 II	106	4.92	M1 III
107	5.12	A7 V	110	4.19	F6 V	113	4.41	F2 II
32 Oph	4.97	M3 III						

Stars of magnitude 5.5 or brighter which have no Bayer or Flamsteed designations:

Designation	Mag. (m$_v$)	Spec. type	Designation	Mag. (m$_v$)	Spec. type
HD 147365	5.48	F3 IV/V	HD 148897	5.24	G8p
HD 152815	5.39	G8 III	HD 155103	5.41	A5m
HD 157049	5.01	M2 III	HD 155410	5.07	K3 III
HD 166208	5.00	G8 III	HD 157087	5.36	A3 III
HD 169110	5.41	K5 III	HD 166229	5.49	K2 III
			HD 169191	5.25	K3 III

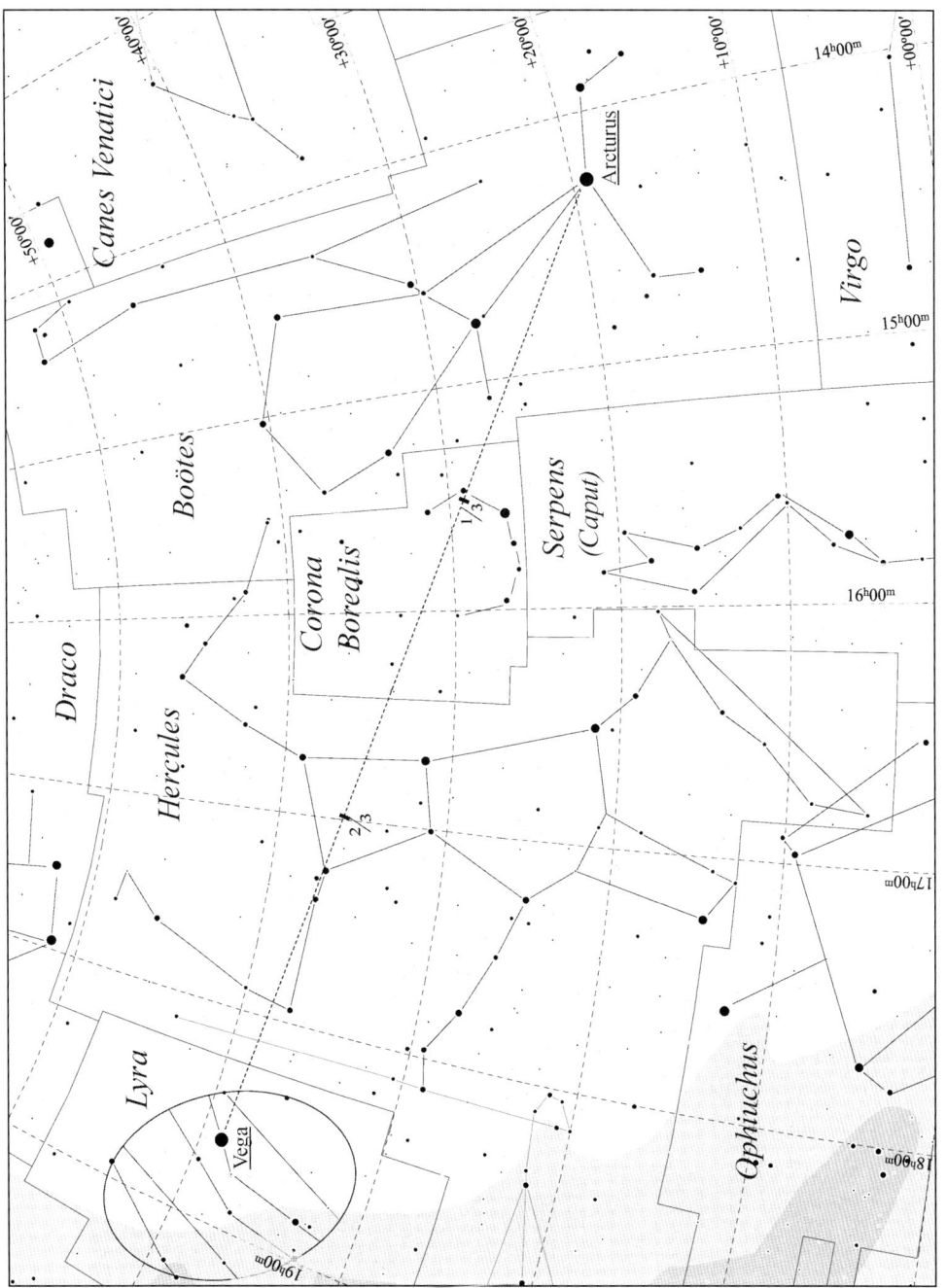

Fig. 6.1. Finder chart for Hercules and Corona Borealis.

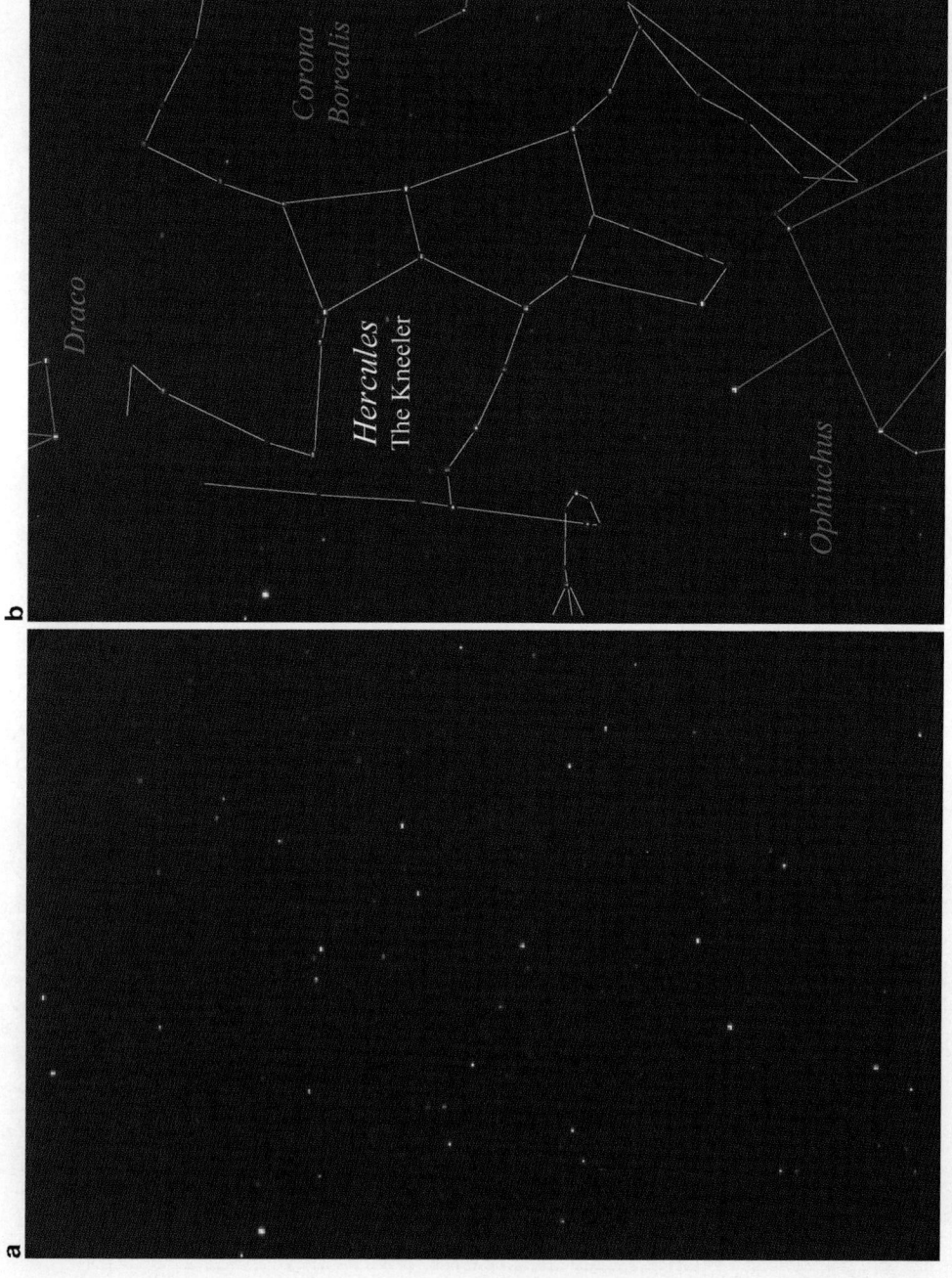

Fig. 6.2. (a, b) Stars and constellation of Hercules.

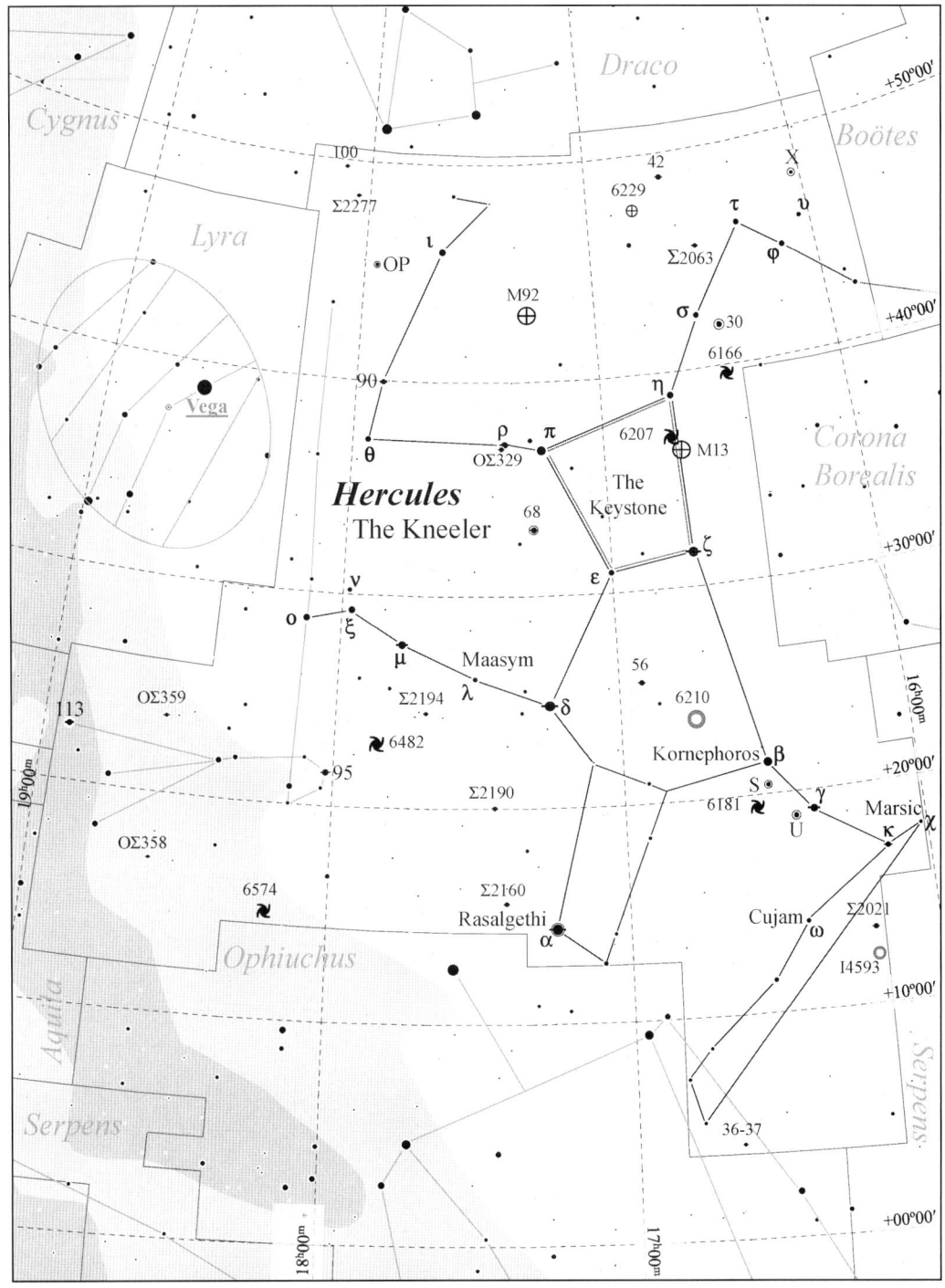

Fig. 6.3. Hercules details.

Table 6.1. Selected telescopic objects in Hercules.

Multiple stars:

Designation	R.A.	Decl.	Type	m₁	m₂	Sep. (")	PA (°)	Colors	Date/ Period	Aper. Recm.	Rating	Comments
7, κ Herculis	16ʰ 08.1ᵐ	+17° 03'	A–B	5.1	6.2	87.7	013	yO, wR	2006	6×30	****	Easily split in small scopes, C star may be difficult to see. a.k.a. ΣΣ2010
			A–C	5.1	13.6	62.6	212	wR, –	1998	7/8"	**	
Σ 2021	16ʰ 13.3ᵐ	+13° 32'	Aa–B	6.7	6.8	**4.1**	**356**	dY, –	**1,354**	2/3"	*****	Neat, nearly equal pair. A–a is too close for amateur instruments
20, γ Herculis	16ʰ 21.9ᵐ	+19° 09'	A–B	3.8	10.1	43	228	Y, P	1999	2/3"	****	Good color contrast. a.k.a. Shj 227
			B–C	10.1	12.2	84.7	298	Y, –	1910	4/5"	***	
Σ 2063	16ʰ 31.8ᵐ	+45° 36'	A–B	5.7	8.7	16.3	196	Y, pR	2003	15×80	****	Subtle color difference
42 Herculis	16ʰ 38.7ᵐ	+48° 56'	A–B	4.9	11.8	27.2	092	O, B	1996	3/4"	****	a.k.a. Σ2082. Nice color but requires larger instruments to see
36–37 Herculis	16ʰ 40.6ᵐ	+04° 13'	Aa–B	5.8	6.9	69.1	229	yW, rW	2004	6×30	***	a.k.a. Σ31. A–a is too close for amateur instruments
			B–C	6.9	11.4	25.2	316	rW, –	2002	3/4"	***	
40, ζ Herculis	16ʰ 41.3ᵐ	+31° 36'	A–B	2.8	5.5	**1.1**	**178**	Y, G	**34.45**	8/9"	****	a.k.a. Σ2084. Nice color
56 Herculis	16ʰ 55.0ᵐ	+25° 44'	A–B	6.1	10.8	18.0	091	O, B	2000	3/4"	****	Beautiful color contrast. a.k.a. Σ2110. A is variable
64, α Herculis	17ʰ 14.6ᵐ	+14° 23'	A–B	3.5	5.4	**4.6**	**104**	O, bG	**3,600**	2/3"	*****	a.k.a. Σ2140 and Ras Algethi. Beautiful color contrast.
65, δ Herculis	17ʰ 15.0ᵐ	+24° 50'	A–B	3.1	8.3	11.8	283	W, bP	2005	2/3"	*****	AB – Good color contrast. LP binary, changes are very slow. a.k.a. Σ3127
68 Herculis	17ʰ 17.3ᵐ	+33° 06'	A–B	4.8	10.2	4.2	059	W, –	2001	3/4"	***	
75, ρ Herculis	17ʰ 23.7ᵐ	+37° 09'	A–B	4.5	5.4	4.1	319	yW, yW	2006	2/3"	*****	Fine yellow–white pair. a.k.a. Σ2161
OΣ 329	17ʰ 24.5ᵐ	+36° 57'	A–B	6.4	9.9	33.2	012	dY, yW	2003	2/3"	***	
Σ 2160	17ʰ 24.6ᵐ	+15° 36'	A–B	6.4	9.3	3.7	065	pW, pR	2003	2/3"	***	
Σ 2190	17ʰ 36.0ᵐ	+21° 00'	A–B	6.1	9.5	10.3	021	Y, –	2006	2/3"	***	
			A–C	6.1	12.1	70	050	Y, –	1944	2/3"	**	
Σ 2194	17ʰ 41.1ᵐ	+24° 31'	A–B	6.5	9.3	16.6	005	oY, B	2006	4/5"	***	
86, μ Herculis	17ʰ 46.5ᵐ	+27° 43'	A–a	3.4	12.7	**0.2**	**094**	Y, – Y, R	**65**	6/7"	**	Passed max of 0.26" in 2002, min 0.07" in 2015. Orbit poorly known
			Aa–BC	3.4	9.8	34.9	249	R, R	2000	15×80	***	a.k.a. Σ2220
			B–C	10.2	10.7	0.7	187		2002	13/14"	***	binary red dwarfs, 43.02-year period
90 Herculis	17ʰ 53.3ᵐ	+40° 00'	A–B	5.3	8.8	12.5	116	yO, B	1999	2/3"	****	a.k.a. β 130. Subtle but pretty colors
95 Herculis	18ʰ 01.5ᵐ	+21° 36'	A–B	4.9	5.2	6.5	257	pGr, rO	2006	2/3"	****	a.k.a. Σ2264

Designation	R.A.	Decl.	Pair							Aper. Recm.	Rating	Comments
Σ 2277	18h 03.1m	+48° 28'	A–B	6.3	8.9	27.1	128	brY, bW	2006	10×50	****	Almost perfectly matched pair
100 Herculis	18h 07.8m	+26° 06'	A–C	6.3	10.2	97.8	297	BrY, –	2003	2/3"	***	
			Aa–B	5.8	5.8	14.5	183	W, W	2006	2/3"	****	
			Aa–C	5.8	11.0	76	127	W, –	2000	4/6"	***	
OΣ 359	18h 35.5m	+23° 36'	A–B	5.6	5.8	0.7	004	dY, –	210.9	13/14"	****	Max 0.75" in 2028, min. 0.07" in 2136
OΣ 358	18h 35.9m	+16° 58'	A–B	6.2	6.3	1.5	149	–, –	380	6/7"	****	Max. 1.94" in 1916, max. 0.39" in 2199
113 Herculis	18h 54.7m	+22° 39'	A–B	4.6	10.4	37.7	023	Y, –	025	15×80	***	a.k.a. β646
			B–C	11.1	11.1	8.7	317	–, –	317	3/4"	***	

Variable stars:

Designation	R.A.	Decl.	Type	Range (mv)		Period (days)	F (fr/ff)	Spectral Range	Aper. Recm.	Rating	Comments
X Herculis	16h 02.7m	+47° 14'	SRb	7.5	8.6	95.0	0.50	M6a	4/5"	****	$
U Herculis	16h 25.8m	+18° 54'	M	6.5	13.4	406.10	0.40	M6.5e–M9.5e	10/12"	****	May be visible to the unaided eye at some maxima
30, g Herculis	16h 28.6m	+41° 53'	SRb	4.3	6.3	89.2	–	M6III	6×30	****	$
S Herculis	16h 51.9m	+14° 56'	M	6.4	13.8	307.28	–	M4Se–M7.5Se	10/12"	****	Forms a wide double with 49 Her, separation 299" PA 235°
64, α Herculis	17h 14.6m	+14° 23'	SRc	2.7	4.0	50	–	M5Ib–II	Eye	****	Ras Algethi; one of the largest known red giants, dia=350M miles
u, 68 Herculis	17h 17.3m	+33° 06'	EB	4.6	5.3	2.05	–	B1.5Vp + B5III	Eye	****	Spectral class B giants, so close that they are egg shaped
OP Herculis	17h 56.8m	+45° 21'	SRb	5.9	6.7	120.5	–	M5IIb–IIIa(S)	6×30	****	

Star clusters:

Designation	R.A.	Decl.	Type	mv	Size (')	Brst. Star	Dist. (ly)	Number of Stars	Aper. Recm.	Rating	Comments
M13, NGC 6205	16h 41.7m	+36° 28'	Globular	5.8	20.0	11.9	25K	500K	7×50	*****	Lum=250K suns, M13 is closer, larger, and brighter than average globular
NGC 6229	16h 47.0m	+47° 32'	Globular	9.4	4.5	–	102K	–	2/3"	***	
M92, NGC 6341	17h 17.1m	+43° 08'	Globular	6.4	11.2	12.1	25K	–	7×50	*****	Lum=150K suns

(continued)

Table 6.1. (continued)

Nebulae:

Designation	R.A.	Decl.	Type	m_v	Size (')	Brst./Cent. star	Dist. (ly)	dia. (ly)	Aper. Recm.	Rating	Comments
IC 4593	16ʰ 12.2ᵐ	+12° 04'	Planetary	10.7	13"	**11.2**	6.5K	0.4	3/4"	***	Appears stellar at low magnification, higher magnification reveals nature
NGC 6210	16ʰ 44.5ᵐ	+23° 49'	Planetary	8.8	16"	**13.7**	3.6K	0.3	10×80	****	3/4" required to show disk

Galaxies:

Designation	R.A.	Decl.	Type	m_v	Size (')	Dist. (ly)	dia. (ly)	Lum. (suns)	Aper. Recm.	Rating	Comments
NGC 6166	16ʰ 28.6ᵐ	+39° 33'	SB	12.0	2.4×1.8	~500M	>440K	330G	8/10"	***	Supergiant spiral, brightest member of Abell 2199 galaxy cluster
NGC 6181	16ʰ 32.3ᵐ	+19° 56'	SB	11.9	2.6×1.3	120M	77K	36G	6/8"	***	Bright halo, elongated N–S, concentrated core
NGC 6207	16ʰ 41.3ᵐ	+36° 28'	SB	11.6	3.0×1.4	57M	43K	10G	6/8"	***	40' NNE of M13. Halo elongated NNE–SSW, stellar nucleus
NGC 6482	17ʰ 51.8ᵐ	+23° 04'	SB	11.3	2.3×2.0	–	–	–	6/8"	***	Fairly faint halo, elongated E–W, bright stellar nucleus
NGC 6574	18ʰ 11.9ᵐ	+14° 59'	SB	12.0	1.4×1.1	114M	50K	32G	8/10"	***	Halo elongated NNW–SSE, slightly brighter center

Meteor showers:

Designation	R.A.	Decl.	Period (yr)	Duration	m_v	Max Date	ZHR (max)	Comet/**Asteroid**	Dist. (ly)	dia. (ly)	First Obs.	Vel. (mi/**km**/sec)	Rating	Comments
None														

Other interesting objects:

Designation	R.A.	Decl.	Type	m_v	Size (')	Dist. (ly)	dia. (ly)	Lum. (suns)	Aper. Recm.	Rating	Comments
Harrington 7	16ʰ 18ᵐ	+13°	Asterism	–		–			7×50	****	17 m7-9 stars in a zigzag pattern, half-way between ω Herculi and IC 4593
The Keystone	16ʰ 55ᵐ	+34° 45'	Asterism	–		–			Eye	***	η, ζ, ε, and π Herculi form a quadrilateral with 3 sides ~7° and one ~4° in length

a

Fig. 6.4. (**a**) M13 (NGC6205) in Hercules. Astrophoto by Stuart Goosen. (**b**) M13 in Hercules. Drawing by Richard Weatherston. Not to same scale as (a).

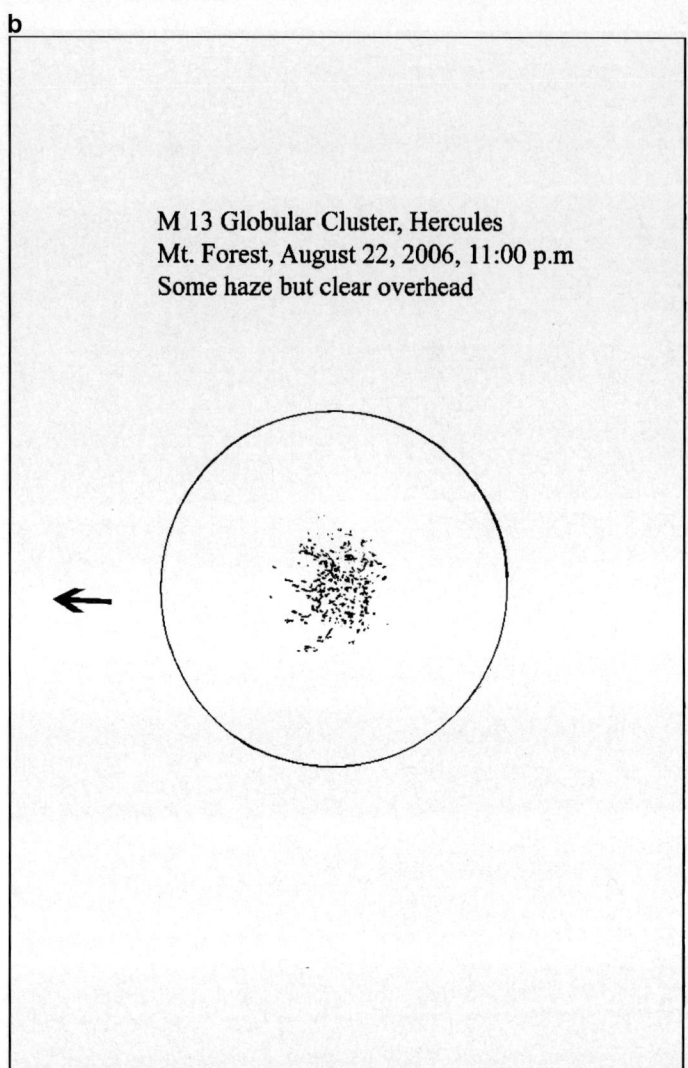

b

M 13 Globular Cluster, Hercules
Mt. Forest, August 22, 2006, 11:00 p.m
Some haze but clear overhead

Fig. 6.4. (continued)

a

Fig. 6.5. (a) M92 (NGC 6341) in Hercules. Astrophoto by Mark Komsa. (b) M92 in Hercules. Drawing by Richard Weatherston. Not to same scale as (a).

b

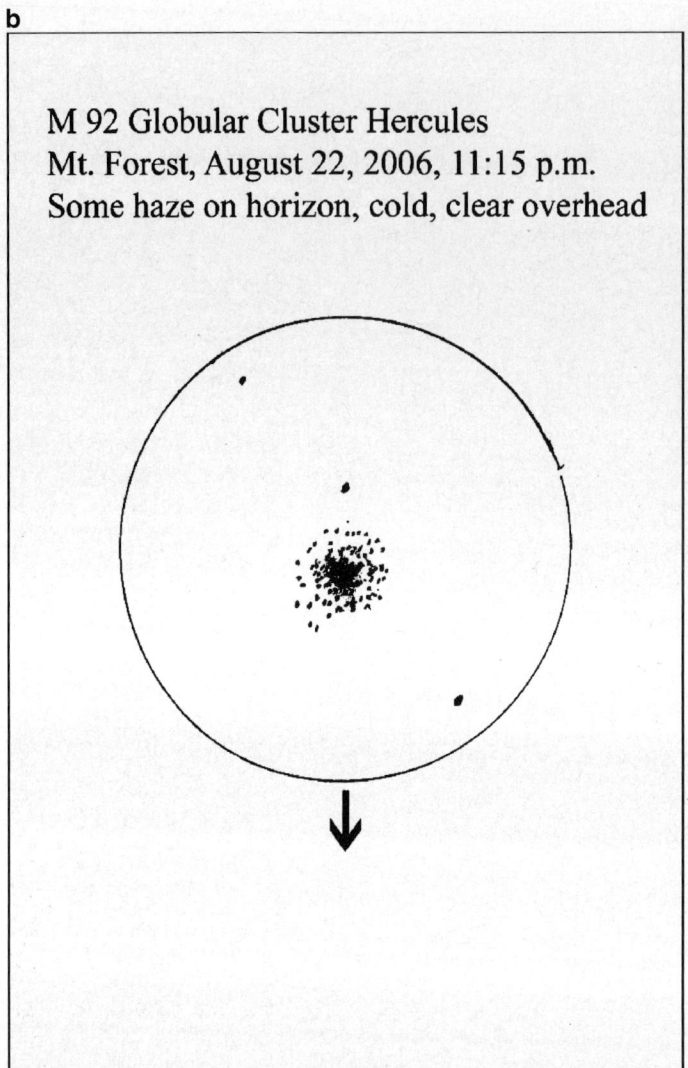

M 92 Globular Cluster Hercules
Mt. Forest, August 22, 2006, 11:15 p.m.
Some haze on horizon, cold, clear overhead

Fig. 6.5. (continued)

LEO (lee-owe)	**The Lion**	
LEONIS (lee-**owe**-niss)	**9:00 P.M. CULMINATION:** April 16	
Leo	**AREA:** 945 sq. deg. (12th in size)	

TO FIND: Leo is at its highest at 9:00 PM (ST) or 10:00 PM (DST) about April 16th. One way to find Leo is to look for the "sickle" or reversed "question mark" about 75° south of Polaris at that time. The sickle is fairly distinctive and easily recognized. Regulus, a 1st magnitude star and the 21st brightest in the sky marks the handle of the sickle, or the dot at the bottom of the reversed question mark. An alternative way is to use the bowl of the big dipper. The two stars where the handle joins the bowl are Alkaid (δ UMa) and Phecda (γ UMa). A line drawn between them and extended WSW about 40° terminates near the center of the hook of the sickle or curved top of the reversed question mark. Starting with the other side of the dipper's bowl, a line drawn from Dubhe (α UMa) through Merak (β Uma) and extended about 40° S ends just W of the triangle consisting of Denebola (β Leonis), Zosma (δ Leonis), and Chertan (θ Leonis). Denebola marks the present end of the lion's tail, and Zosma and Chertan mark the hindquarters of the lion. The lion's forelegs extend a short distance (~2°) S of Regulus to the faint 31 Leonis (m4.37), then about 7° W to slightly brighter o Leonis (m3.52) marking the forepaws. The hindlegs extend S and SW from Chertan in two irregular curved lines of 4th and 5th magnitude stars (Figs. 6.6–6.8).

KEY STAR NAMES:

Name	Pronunciation	Source
α Regulus	**reg**-u-lus	rex (Latin) "king," combined with the last syllable of $\beta\alpha\sigma\iota\lambda\iota\sigma\chi$ [basilisx] (Greek), "little king"
β Denebola	deh-**neb**-boh-lah	dhanab al asad (Arabic), "tail of the lion"
γ Algieba	al-**jee**-bah or ul-**jeh**-boo	al-jabha (Arabic), "the forehead"
δ Zosma	**zohs**-muh	$\zeta\omega\sigma\mu\alpha$[zosma] (Greek), "girdle or loin cloth"
ζ Adhafera	uh-duh-**fee**-ruh or uh-**day**-feh-ruh	al-dafira (Arabic), "the lock of hair" (Coma Berenices was originally part of Leo)
η Al Jabhah	al-**job**-hah	al-jabha (Arabic), "the forehead"
θ Chertan	**kehr**-tan or **cher**-tan	al khurtan (Arabic), "the two small ribs"
ι Al Minliar	(?)	Origin and pronunciation uncertain
λ Alterf	ul-**terf** or **al**-terf	al-tarf (Arabic), "the glance" from its location in the lion's eyes
μ Rasalas	**rue**-sah-las	ra's al-asad (sci-Arabic), "northern (part) of the lion's head"
o Subra	**zuhb**-rue, or **sue**-brah	al-zubra (indo-Arabic), "the mane, or shoulder"
ρ Shir, Ser	(?)	Origin and pronunciation uncertain

Magnitudes and spectral types of principal stars:

Bayer desig.	Mag. (m_v)	Spec. type	Bayer desig.	Mag. (m_v)	Spec. type	Bayer desig.	Mag. (m_v)	Spec. type
α	1.36	B7 V	β	2.14	A3 Vvar	γ	2.01	K0 III
δ	2.56	A4 V	ε	2.97	G0 II	ζ	3.43	F0 III
η	3.48	A0 Ib	θ	3.33	A2 V	ι	4.00	F2 IV (Sb)
κ	4.47	K2 III	λ	4.32	K5 IIIvar	μ	3.88	K0 III
ν	5.26	B9 IV	ξ	4.99	K0 IIIvar	o	3.52	A7 V
π	4.68	M2 III	ρ	3.84	B1 Ib (Sb)	σ	4.05	B9.5 V
τ	4.95	G8 II/III	υ	4.30	G9 III	φ	4.45	A7 IVn
χ	4.62	F2 III/IV	ψ	5.36	M2 III	ω	5.40	F9 V

Stars of magnitude 5.5 or brighter which have Flamsteed, but no Bayer designations:

Flamsteed	Mag. (m_v)	Spec. type	Flamsteed	Mag. (m_v)	Spec. type	Flamsteed	Mag. (m_v)	Spec. type
6	5.07	K3 III	10	5.00	K1 III:var	22	5.29	A5 IV
31	4.39	K4 III	37	5.42	M1 III	40	47.8	F6 IV
46	5.43	M2 III	48	5.07	G8 II/III	51	5.50	K3 III
52	5.49	G4 III	53	5.32	A2 V	54	4.30	A1 V
58	4.84	K1 III	59	4.98	A5 III	60	4.42	A1m
61	4.73	K3 III	69	5.40	A0 V	72	4.56	M3 III
73	5.31	K3 III	75	5.18	M0 III	79	5.39	G8 III
87	4.77	K4 III	92	5.26	K1 III	93	4.50	A7 V

Stars of magnitude 5.5 or brighter which have no Bayer or Flamsteed designations:

Designation	Mag. (m_v)	Spec. type
HD 094402	5.45	G8 III

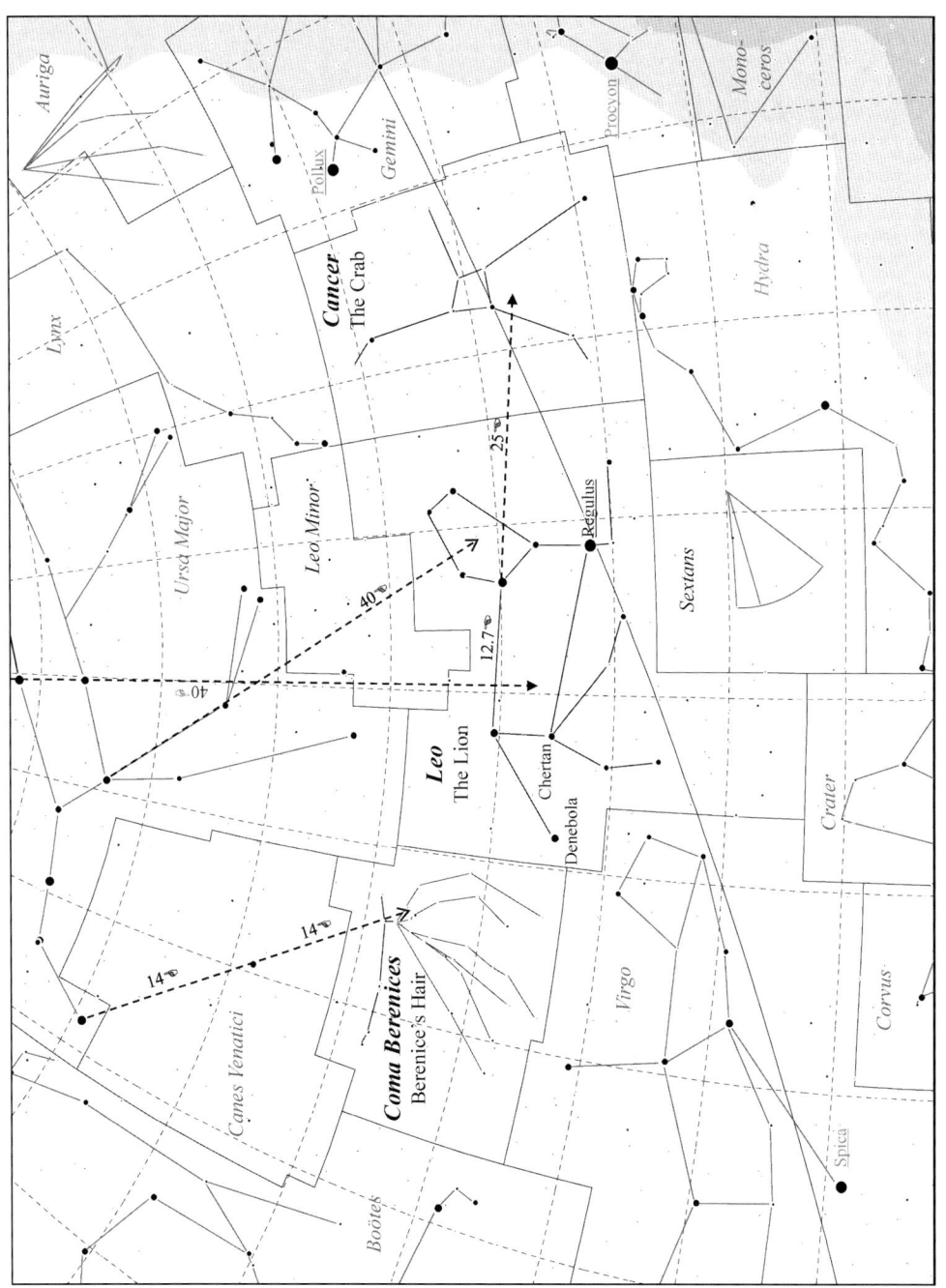

Fig. 6.6. Leo, Cancer, and Coma Berenices finder chart.

a

b

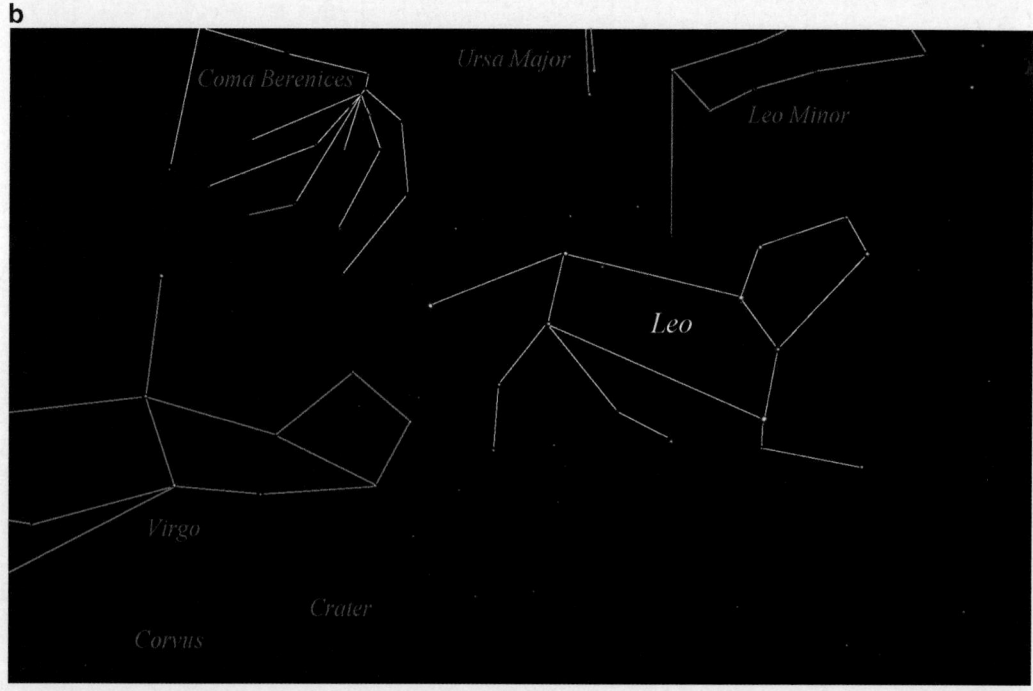

Fig. 6.7. (**a**, **b**) The stars and constellation of Leo.

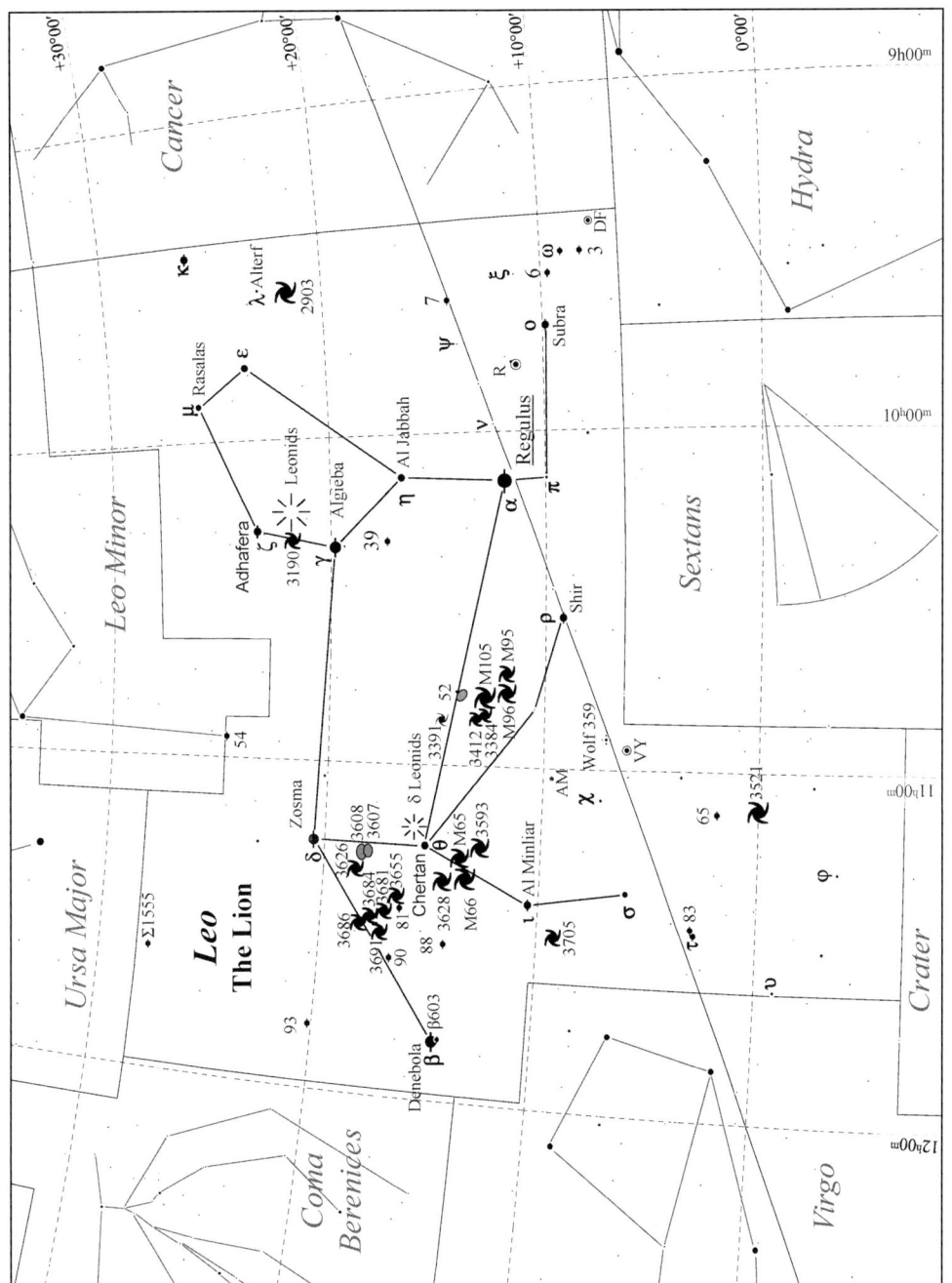

Fig. 6.8. Leo details.

Table 6.2. Selected telescopic objects in Leo.

Multiple stars:

Designation	R.A.	Decl.	Type	m₁	m₂	Sep. (")	PA (°)	Colors	Date/Period	Aper. Recm.	Rating	Comments
κ Leonis	09ʰ 24.7ᵐ	+26° 11'	A–B	4.5	9.7	2.4	211	Y, B	1975	4/6"	****	a.k.a. β105 AB
2, ω Leonis	09ʰ 28.5ᵐ	+09° 03'	A–B	5.7	7.3	**0.7**	**103**	pY, gY	**118.227**	12/14"	***	a.k.a. Σ1356. Passed sec. min. of 0.447" in 1985, pri max. of 1.069" in 2041
3 Leonis	09ʰ 28.5ᵐ	+08° 11'	A–B	5.7	11.1	25.8	079	pY, gY	2000	3/4"	****	a.k.a. H4 47
6 Leonis	09ʰ 32.0ᵐ	+09° 43'	A–B	5.2	9.3	37.5	075	R, Gr	2006	2/3"	*****	a.k.a. Schj 107
7 Leonis	09ʰ 35.9ᵐ	+14° 23'	A–B	6.3	9.4	41.0	080	bW, bGr	2003	2/3"	****	a.k.a. H5 58
14, o Leonis	09ʰ 41.2ᵐ	+09° 54'	Aa–B	3.5	10.8	95.0	0.47	Y, B	2000	3/4"	****	a.k.a. H6 76Aa–B. 3 close companions are too close or faint for amateur scopes
32, α Leonis	10ʰ 08.4ᵐ	+11° 58'	A–B	1.4	8.2	176.0	308	BW, Y	2000	7×50	****	Regulus.
			B–C	8.1	13.1	2.5	086	Y, ?	1943	6/8"	****	a.k.a. HDO 127 BC
36, ζ Leonis	10ʰ 16.7ᵐ	+23° 25'	A–B	3.46	6.0	333.8	338	W, Y	2002	6×30	****	
39 Leonis	10ʰ 17.2ᵐ	+17° 44'	A–B	5.8	11.4	7.7	299	Y, ?	2005	3/4"	***	
41, γ Leonis	10ʰ 20.0ᵐ	+19° 51'	A–B	2.4	3.6	**4.4**	**126**	dY, pY	**618.56**	2/3"	******	Beautiful! Opening slightly to 4.546" in 2063. a.k.a Σ1424AB
49 Leonis	10ʰ 35.0ᵐ	+08° 39'	A–B	5.8	7.9	2.2	154	yW, pB	2006	4/6"	****	
54 Leonis	10ʰ 55.6ᵐ	+24° 45'	Dbl	4.5	6.3	6.6	113	Y, B	2005	7×50	*****	Good color contrast
65 Leonis	11ʰ 06.9ᵐ	+01° 57'	A–B	5.7	9.7	2.7	104	dY, ?	1991	3/4"	****	
δ Leonis	11ʰ 14.1ᵐ	+20° 31'	A–B	2.6	8.6	203.9	342	Y, W	1999	7×50	****	Zosma
			A–a	2.6	12.1	95.4	041	Y, ?	1911	3/4"	***	
78, ι Leonis	11ʰ 23.9ᵐ	+10° 32'	Dbl	4.1	6.7	**2.0**	**100**	Y, W	**186**	5/6"	****	Will be at max. of 2.714" in 2059. a.k.a. Σ1536
81 Leonis	11ʰ 25.6ᵐ	+16° 27'	A–B	5.6	10.8	55.1	002	Y, R	1998	3/4"	****	
83 Leonis	11ʰ 26.8ᵐ	+03° 00'	A–B	6.6	7.5	**28.7**	**147**	W, pR	**32,000**	10×50	****	Max. sep. of 46.7" will not occur until 27361 ADI a.k.a Σ1540AB.
			A–C	6.6	9.9	90.5	187	W, ?	1937	2/3"	****	AB is 20' WNW (PA °298) of τ Leo, forming an easy and pretty double-double
τ Leonis	11ʰ 27.9ᵐ	+02° 51'	A–B	5.1	7.5	88.9	181	Y, B	2004	7×50	****	Two more companions, C and D are too faint or distant to be of interest
88 Leonis	11ʰ 31.7ᵐ	+14° 22'	A–B	6.3	9.1	15.4	331	W, ?	2004	2/3"	****	A is double, but separation is only 0.1"
90 Leonis	11ʰ 34.7ᵐ	+16° 48'	A–B	6.3	7.31	3.5	209	W, prW	2006	2/3"	*****	Nice group in 4/6" scopes
			A–C	6.3	8.8	62.3	236	B, pR	2004	7×50	****	
Σ1555AB	11ʰ 36.3ᵐ	+27° 46'	A–B	5.8	6.0	**0.9**	**147**	yW, ?	**2,651.43**	10/12"	****	Very slowly widening (~0.01"/year), use hi pwr (>250x). Max of 3.141" in 2376
h503AB–C			AB–C	5.8	11.2	22.2	155	yW, ?	2000	3/4"	****	
93 Leonis	11ʰ 48.0ᵐ	+20° 13'	A–B	4.6	9.0	74.1	357	Y, ?	2002	7×50	******	Lovely contrast. A has an extremely close companion

β603	$11^h 48.6^m$	+14° 17′	A–B	6.0	8.5	**1.2**	**336**	?, ?	**134**	8/10″	****	In same FOV as β Leo, 19′ distant. Max. 1.172″ in 2004, closing to 0.205″ in 2077
β Leonis	$11^h 49.1^m$	+14° 34′	A–C	2.1	13.2	80.3	359	bW, pR	1918	6/8″	***	At m15.7, B is too faint for most amateur scopes

Variable stars:

Designation	R.A.	Decl.	Type	Range (m_v)		Period. (days)	F (f_r/f_i)	Spectral Range	Aper. Recm.	Rating	Comments
DF Leonis	$09^h 23.8^m$	+07° 42′	SRb	6.7	7.0	70	–	gM4	6×30	***	
R Leonis	$09^h 47.6^m$	+11° 26′	M	4.4	11.3	309.95	0.43	M6e–M9.5e	3/4″	*****	Very red, especially near minimum
VY Leonis	$10^h 56.3^m$	+06° 10′	Lb	5.7	6.0	–	–	M5.5IIIv	6×30	****	
AM Leonis	$11^h 02.2^m$	+09° 54′	EW	9.3	9.8	0.3658	–	F8	10×80	****	W UMa type, stars are almost in contact

Star clusters:

Designation	R.A.	Decl.	Type	m_v	Size (′)	Brtst. Star	Dist. (ly)	dia. (ly)	Number of Stars	Aper. Recm.	Rating	Comments
None												

Nebulae:

Designation	R.A.	Decl.	Type	m_v	Size (′)	Brtst./ **Cent.** star	Dist. (ly)	dia. (ly)	Aper. Recm.	Rating	Comments
None											

Galaxies:

Designation	R.A.	Decl.	Type	m_v	Size (′)	Dist. (ly)	dia. (ly)	Lum. (suns)	Aper. Recm.	Rating	Comments
NGC 2903	$09^h 32.2^m$	+21° 30′	SAB(rs)b	9.0	12.0×5.6	25M	105K	19G	2/3″	****	One of the brightest galaxies not in Messier or Caldwell lists
NGC 3190	$10^h 18.1^m$	+21° 50′	SA(s)ap	11.2	4.1×1.6	52M	60K	7.1G	6/8″	****	
M95, NGC 3351	$10^h 44.0^m$	+11° 42′	SB(r)b II	9.7	7.3×4.7	29M	65K	7.9G	4/6″	*****	Same LP FOV as M96
M96, NGC 3368	$10^h 46.8^m$	+11° 49′	SB(rs)c	9.2	7.4×5.0	26M	71K	16G	3/4″	*****	Same LP FOV as M95
NGC 3377	$10^h 47.7^m$	+13° 59′	E5/6	10.4	4.3×2.6	31M	50K	5.2G	6/8″	****	
M105, NGC 3379	$10^h 48.7^m$	+12° 23′	SB	9.3	4.8×3.9	26M	42K	8.9G	2/3″	*****	0.8°NE of M96

(continued)

Table 6.2. (continued)

Galaxies:

Designation	R.A.	Decl.	Type	m_v	Size (')	Dist. (ly)	dia. (ly)	Lum. (suns)	Aper. Recm.	Rating	Comments
NGC 3384	10h 48.3m	+12° 38'	SB(s)0	10.0	5.3×2.6	31M	50K	8.4G	6/8"	****	
NGC 3412	10h 50.9m	+13° 25'	SB(s)0	10.5	3.6×2.2	33M	35K	5.5G	4/6"	****	Elongated NNW-SSE, concentrated core, bright nucleus
NGC 3521	11h 05.8m	+00° 02'	SBbc	9.1	10.4×5.3	26M	80K	12G	2/3"	*****	
NGC 3593	11h 14.6m	+12° 49'	SA(s)0	10.6	5.2×2.9	45M	70K	9.3G	8/10"	****	
NGC 3607	11h 16.9m	+18° 03'	E2	9.8	4.9×4.4	54M	77K	28G	6/8"	****	
NGC 3608	11h 17.0m	+18° 09'	E3	10.8	3.4×2.7	60M	60K	14G	6/8"	****	
M65, NGC 3623	11h 18.9m	+13° 05'	SAB(s)b	8.9	8.7×4.0	28M	78K	21G	2/3"	*****	
NGC 3626	11h 20.1m	+18° 21'	(R)SA(rs)0	10.9	2.9×2.1	59M	50K	12G	6/8"	***	
M66, NGC 3627	11h 20.2m	+12° 59'	Sb	9.0	8.8×4.1	22M	43K	11G	2/3"	*****	
NGC 3628	11h 20.3m	+13° 36'	Sb pec sp	9.5	13.0×3.5	34M	130K	15G	6/8"	****	Beautiful trio w M65 and M66 in small scopes, central dust lane
NGC 3655	11h 22.9m	+16° 35'	SA(s)c	11.7	1.5×1.0	54M	30K	4.9	6/8"	***	
NGC 3681	11h 26.5m	+16° 52'	SAB(r)bc	11.2	2.0×2.0	53M	31K	7.4G	6/8"	***	
NGC 3684	11h 27.2m	+17° 02'	SA(rs)bc	11.4	3.2×2.2	58M	54K	7.4G	6/8"	***	
NGC 3686	11h 27.7m	+17° 13'	Sb	11.3	3.1×2.4	41M	40K	4G	6/8"	***	
NGC 3691	11h 28.2m	+16° 55'	SBb?	11.8	1.2×0.9	39M	15K	2.3G	10/12"	***	
NGC 3705	11h 30.1m	+09° 17'	SaB(r)ab	11.1	4.7×2.0	39M	53K	4.4G	8/10"	***	

Meteor showers:

Designation	R.A.	Decl.	Period (yr)	Duration	Max Date	ZHR (max)	Comet/ **Asteroid**	First Obs.	Vel. (mi/ **km/sec**)	Rating	Comments
Leonids	10h 12m	+22°	33	11/10 – 11/23	11/19	>1,000	Tempel-Tuttle	902 AD	44	I	First observed in Egypt and Italy. Meteor storms were observed in 902, 1799, 1833, 1866, 1966. Some 1966 estimates were as high as 100,000/hl
δ Leonids	11h 12m	+16°	Annual	02/15 – 03/10	02/24	2	1999 RD?	1911	14	II	Discovered in analysis of William F. Denning's records

Other interesting objects:

Designation	R.A.	Decl.	Type	m_v	Mv	Lum. (suns)	Dist. (ly)	Eff. temp (K)	Pr. Mo. ("/year)	Aper. Recm.	Rating	Comments
Wolf 359	10h 56.0m	+07° 01'	Red dwf	13.4	16.55	0.00002	7.78	2,800	4.71	8/10"	**	Third closest star after α Centauri and Barnard's Star (in Ophiuchus)

Fig. 6.9. Leo Galaxy Trio, *clockwise from upper left*: M66, M65, and NGC 3628. Astrophoto by Stuart Goosen.

Fig. 6.10. M96 (NGC 3368), spiral galaxy in Leo. Astrophoto by Rob Gendler.

Fig. 6.11. (**a**) Another Galaxy Trio in Leo. Left to right: NGC 3389, 3384, and 3379 (M105). Astrophoto by Mark Komsa. (**b**) Another Galaxy Trio in Leo. Drawing by Eiji Kato.

b

S
→E

Fig. 6.11. (continued)

a

S
→E

Fig. 6.12. (a) M95 (NGC 3351), spiral galaxy in Leo. Astrophoto by Rob Gendler. (b) M95 (NGC 3351), spiral galaxy in Leo. Drawing by Eiji Kato.

b

Fig. 6.12. (continued)

a

S
E

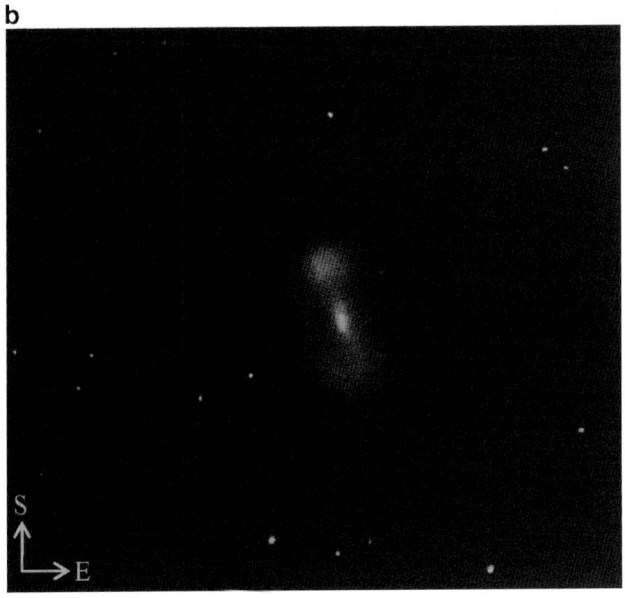

b

S
E

Fig. 6.13. (a) NGC 2903, spiral galaxy in Leo. Astrophoto by Rob Gendler. Astrophoto has been cropped to allow scale matching. (b) NGC 2903, spiral galaxy in Leo. Drawing by Eiji Kato.

CANCER (kan-sir)	The Crab
CANCRI (kan-kri)	9:00 P.M. CULMINATION: March 15
Cnc	AREA: 506 sq. deg. (31st in size)

TO FIND: Locate <u>Regulus</u>, the bright star at the bottom of the "sickle" in Leo, and Castor and <u>Pollux</u>, the two brightest stars in Gemini, the twins. Almost 2/3 of the distance from <u>Regulus</u> to <u>Pollux</u> and slightly below the line joining them, you can see a fuzzy spot (because of distortion on the chart, the straight line would have to be shown in segments to accurately show how it passes through Cancer). This spot is M44, the Praesepe (manger or crib), or Beehive cluster, which lies in the heart of Cancer. Just to its left, with one above and one below M44, are two faint stars, Asellus Borealis and Ascellus Australis. These mark the front part of the crab's body, while ζ (Tegmen) marks the rear of the crab's shell. Acubens (α) represents the end of one of the claws, and ι (unnamed) represents the other claw (Figs. 6.6–6.8).

KEY STAR NAMES:

	Name	Pronunciation	Source
α	Acubens	**ah**-cue-benz	Al Zubanah, (Arabic) "the claw"
	or Sertan	sir-**tan**	al-saraṭān (Sci-Arabic), "the Crab"
β	Tarf or Al Tarf	al tarf	Al Tarf, (Arabic) "the end" (of the southern foot)
γ	Asellus Borealis	uh-**sell**-luhs	asellus (Latin) "little ass (or donkey)" +
		boh-reh-**ah**-lis	borealis (Latin) "northern" *6
δ	Asellus Australis	uh-**sell**-luhs	asellus (Latin) "little ass (or donkey)" +
		ous-**tray**-lis	australis (Latin), "southern" *6
ε	Praesepe	**pree**-sep-ee	(Latin) Manger or crib, name also applied to M44 as a whole*7
ζ	Tegmine	**teg**-mih-neh	tegimen (Latin) "covering" or "shell"
	or Tegmen	**teg**-min	

Magnitudes and spectral types of principal stars:

Bayer desig.	Mag. (m_v)	Spec. type	Bayer desig.	Mag. (m_v)	Spec. type	Bayer desig.	Mag. (m_v)	Spec. type
α	4.26	A5 m	β	3.53	K4 III	γ	4.66	A1 IV
δ	3.94	K0 III	ε	6.29	Am	ζ^1	4.67	G0 V
ζ^2	6.01	G5 V	η	5.33	K3 III	θ	5.33	K5 III
ι	4.03	G8 Iab	κ	5.23	B8 IIIp	λ	5.92	B9.5 V
μ^1	5.99	M3 III	μ^2	5.30	G2 IV	ν	5.45	A0 pSi
ξ	5.16	K0 III	o^1	5.20	A5 III	o^2	5.68	F0 IV
π	5.36	K1 III	ρ^1	5.95	G8 V	ρ^2	5.23	G8 II/III
σ^1	5.67	A8 Vms	σ^2	5.68	F0 IV	σ^3	5.23	G9 III
τ	5.42	G8 III	υ^1	5.71	F0 IIIn	υ^2	6.35	G9 III
φ^1	5.58	K5 III	φ^2	6.30	A3 V	χ	5.13	F6 V
ψ	5.73	G8 IV	ω	5.87	G8 III			

Stars of magnitude 5.5 or brighter which have Flamsteed but no Bayer designations:

Flamsteed	Mag. (m_v)	Spec. type	Flamsteed	Mag. (m_v)	Spec. type	Flamsteed	Mag. (m_v)	Spec. type
8	5.14	A1 V	57	5.40	G7 III	60	5.44	K5 III

Stars of magnitude 5.5 or brighter which have no Bayer or Flamsteed designations:

Designation	Mag. (m_v)	Spec. type	Designation	Mag. (m_v)	Spec. type
HD 071115	5.13	G8 II			

a

b

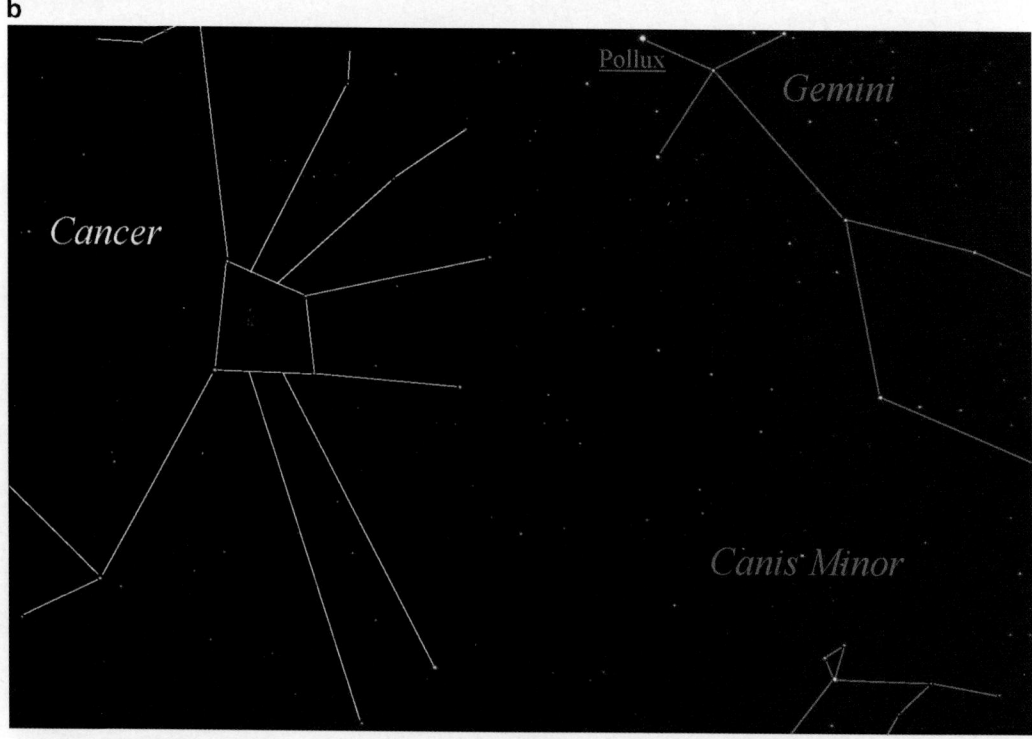

Fig. 6.14. **(a)** The stars of Cancer and adjacent constellations. **(b)** Cancer and adjacent constellations.

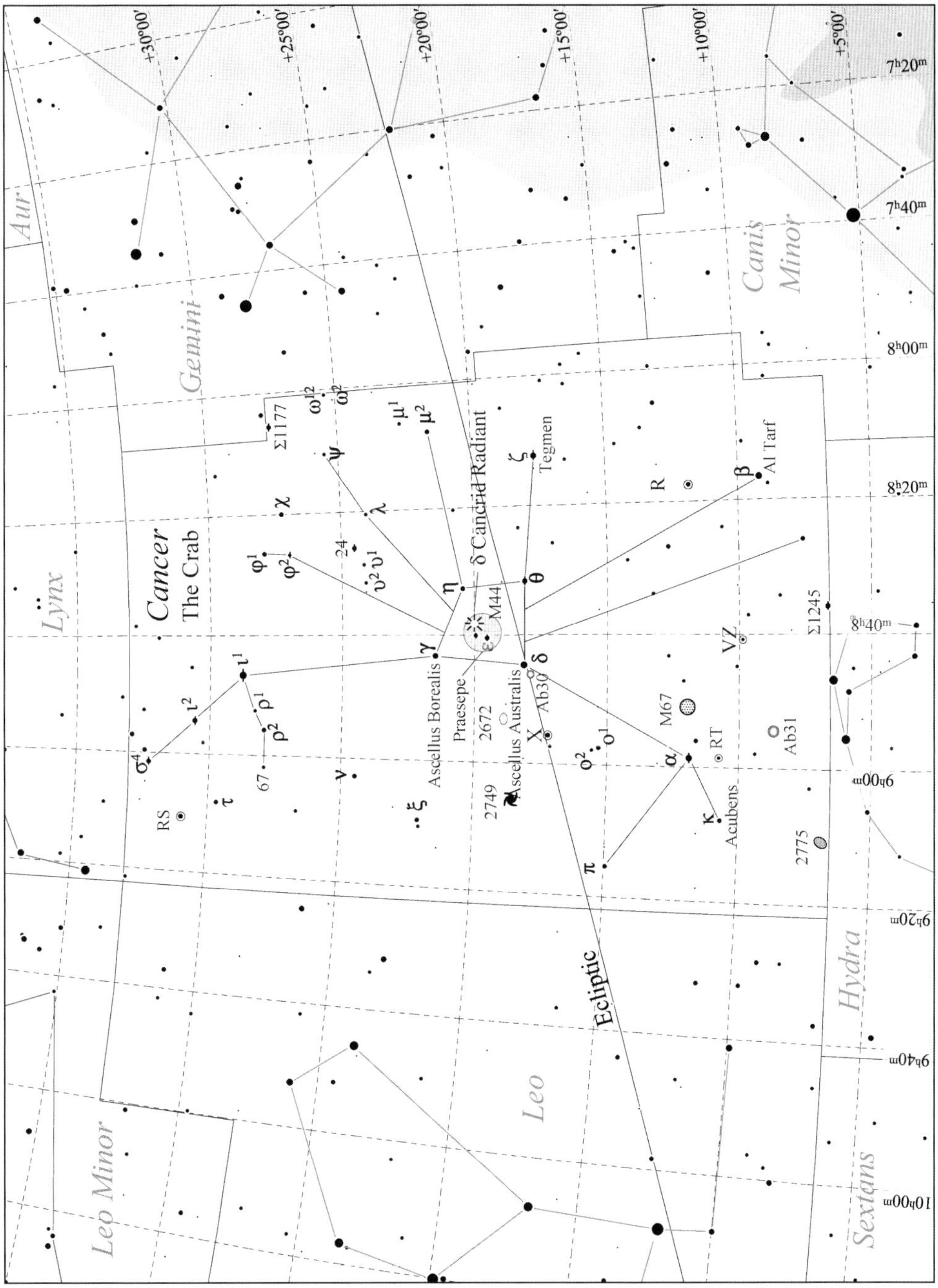

Fig. 6.15. Cancer details.

Table 6.3. Selected telescopic objects in Cancer.

Multiple stars:

Designation	R.A.	Decl.	Type	m_1	m_2	Sep. (")	PA (°)	Colors	Date/Period	Aper. Recm.	Rating	Comments
Σ1177	08h 05.6m	+27° 32'	A–B	6.7	7.4	3.3	350	yW, bW	2005	2/3"	*****	
ζ Cancri	08h 12.2m	+17° 39'	A–B	5.6	6.0	1.1	036	Y, Y	**59.56**	9/10"	*****	Tegmine, difficult but lovely triple – 3 yellow stars almost in a straight line. AB AB separation is increasing. C is a spectroscopic double. Data calculated for 2010
			AB–C	5.1	6.2	5.9	068	O, –	**1,115**	2/3"	****	
24 Cancri	08h 26.7m	+24° 32'	A–BC	6.9	7.5	5.7	051	dY, W	2003	2/3"	****	BC is a binary with maximum separation of 0.16"
23, φ² Cancri	08h 26.8m	+26° 56'	A–B	6.2	6.2	5.0	219	yW, yW	2005	2/3"	*****	Showcase pair of matched stars
θ Cancri	08h 31.6m	+18° 06'	A–B	5.4	10.0	70.4	062	Y, W	1988	2/3"	****	
Σ1245 (Cancer)	08h 35.8m	+06° 37'	A–B	6.0	7.2	10.0	025	oY, B	2004	2/3"	****	Observers differ on color of secondary, it has been called purple, rose, pale red, and greenish yellow
			A–C	6.0	10.7	97.5	112	oY, –	1991	3/4"	***	
			A–D	6.0	12.0	110.0	290	oY, –	2000	3/4"	***	
			A–E	6.0	9.6	115.2	207	oY, –	2002	15×80	***	
39–40 Cancri	08h 40.1m	+20° 00'	A–Bb	6.5	6.6	150.1	153	O, O	2002	6×30	******	A–B is an easy binocular pair of matched stars. Larger binoculars make this appear as a triple, quadruple, or quintuple system, and large telescopes or telescopes with spectragraphs reveal at least 20 components!
			A–C	6.5	9.0	134.2	309	O, –	2002	7×50	****	
			A–D	6.5	8.8	135.3	111	O, –	2002	7×50	****	
			Bb–R	6.5	10.4	147.0	147	O, –	1921	2/3"	***	
			C–S	9.0	10.1	137.1	340	O, –	2002	2/3"	***	
Σ1254	08h 40.4m	+19° 40'	A–B	6.4	10.4	20.7	054	Y, B	2003	2/3"	*****	
			A–C	6.5	7.6	63.2	343	Y, –	2002	7×50	***	
			A–D	6.5	9.2	82.6	044	Y, –	2002	7×50	***	
48, ι¹ Cancri	08h 46.7m	+28° 46'	A–B	4.1	6.0	30.7	308	Y, B	2003	10×50	*****	Lovely gold and blue pair, ~1 mag. fainter version of Albireo
57, ι² Cancri	08h 54.2m	+30° 35'	A–B	6.1	6.4	1,555.8	311	Y, Y	2004	6/7"	****	
			AB–C	5.5	9.2		203	Y, B	1988	7×50	****	
64, α Cancri	08h 58.5m	+11° 51'	A–B	4.3	11.8	10.6	321	Gld, ?	2000	3/4"	*****	Lovely gold primary with faint jewel nearby
66, σ⁴ Cancri	09h 01.4m	+32° 15'	A–B	6.0	8.6	4.5	134	rW, B	2003	2/3"	*****	Very nice color contrast
67 Cancri	09h 01.8m	+27° 54'	A–B	6.1	9.2	103.7	330	dY, pR	2005	2/3"	****	Very wide with nice colors

Variable stars:

Designation	R.A.	Decl.	Type	Range (m$_v$)	Period (days)	F (f$_r$/f$_f$)	Spectral Range	Aper. Recm.	Rating	Comments
R Cancri	08^h 16.6^m	+11° 44'	Mira	6.1 11.8	361.6	0.47	M6e–M9e	4/5"	****	Cycle nearly synchronized with year. Full rotation through seasons takes ~100 years
VZ Cancri	08^h 40.9^m	+09° 49'	δ Sct	7.18 7.9	0.1784	0.26	A7III–F2III	7×50	****	Nearly doubles brightness in 1h 7m and complete cycle is only 4h 17m
X Cancri	08^h 55.4^m	+17° 14'	SRb	5.6 7.5	195	–	C5,4(N3)	6×30	*****	Lovely orange-red star
RT Cancri	08^h 58.3^m	+10° 51'	SRb	7.12 8.6	60	–	M5III	6×30	****	Very cool star, pulsating. 1° 22' # (PA 094°) of θ and (m4.61). ##$
RS Cancri	09^h 10.6^m	+30° 58'	SRc	**6.2** 7.7	120	–	M6eIb–II(S)	6×30	****	

Star clusters:

Designation	R.A.	Decl.	Type	m$_v$	Size (')	Brtst. Star.	Dist. (ly)	dia. (ly)	Number of Stars	Aper. Recm.	Rating	Comments
M44, NGC 2632	08^h 40.4^m	+19° 40'	Open	3.1	70	6.3	561	17	100	Eye	*****	a.k.a. "Beehive Cluster," "Praesepe," or "The Manger" '7
M67, NGC 2682	08^h 50.4^m	+11° 49'	Open	6.9	25	9.7	2.6K	24	500	4/6"	*****	One of oldest open clusters, ~5 billion years '8

Nebulae:

Designation	R.A.	Decl.	Type	m$_v$	Size (')	Brtst./ **Cent.** star	Dist. (ly)	dia. (ly)	Aper. Recm.	Rating	Comments
Abell 30	08^h 46.8^m	+17° 53'	Planetary	15.5	0.2'	**14.3**	7,200	0.68	16/18"	*	Large, extremely faint, use OIII filter. 0° 54' ESE (PA 118°) from δ Cnc.
Abell 31	08^h 54.2^m	+08° 55'	Planetary	12.2	15.6	**15.5**	–	–	10/12"	**	Very large, use OIII filter and LP, can be seen in 8", but very difficult

(continued)

Table 6.3. (continued)

Galaxies:

Designation	R.A.	Decl.	Type	m$_v$	Size (')	Dist. (ly)	dia. (ly)	Lum. (suns)	Aper. Recm.	Rating	Comments
NGC 2672	08^h 49.4^m	+19° 04'	E1	11.7	3.0×2.8	180M	157K	54G	12/14"	***	Mod bright halo, slightly brighter core. Fainter NGC 2673 nearly in contact, 45" ESE
NGC 2749	09^h 05.4^m	+18° 19'	SB	11.8	1.9×1.7	180M	100K	50G	12/14"	***	Three other m13/14 galaxies are within 15' of NGC 2749
NGC 2775	09^h 10.3^m	+07° 02'	SA(r)ab	10.1	4.6×3.7	55M	73K	17G	8/10"	****	Bright core, fairly bright halo

Meteor showers:

Designation	R.A.	Decl.	Period (yr)	Duration	Max Date	ZHR (max)	Comet/ Asteroid	First Obs.	Vel. (mi/ km/sec)	Rating	Comments
δ Cancrids	08^h 40^m	+20°	annual	01/01 01/24	01/17	4	1991AQ?	1872	17	Minor	Bright, fairly slow. First detected by members of Italian Meteoric Association

Other interesting objects:

Designation	R.A.	Decl.	Type	m$_v$	Mass (suns)	Dia. (suns)	Dist. (ly)	Planetary Mass (J)	Dist. (AU)	Period (days)	Eccentricity	Comments
55, ρ¹ Cancri	08^h 52.6^m	+28° 20'	ExoPlnt	5.95	0.95	1.1	40.9	>0.045	0.038	2.81	0.174	Disc. 1997. 55 Cancri is on list of 100 top candidates for NASA Terrestrial Planet Finder
								>0.784	0.115	14.67	0.0197	Discovered 2004.
								>0.217	0.24	43.93	0.44	Existence inferred 2002, confirmed 2004. Distance ~0.6 distance of Mercury
								>3.92	5.257	4,517.4	0.327	Discovered 2002. Distance and period very similar to Jupiter's

a

b

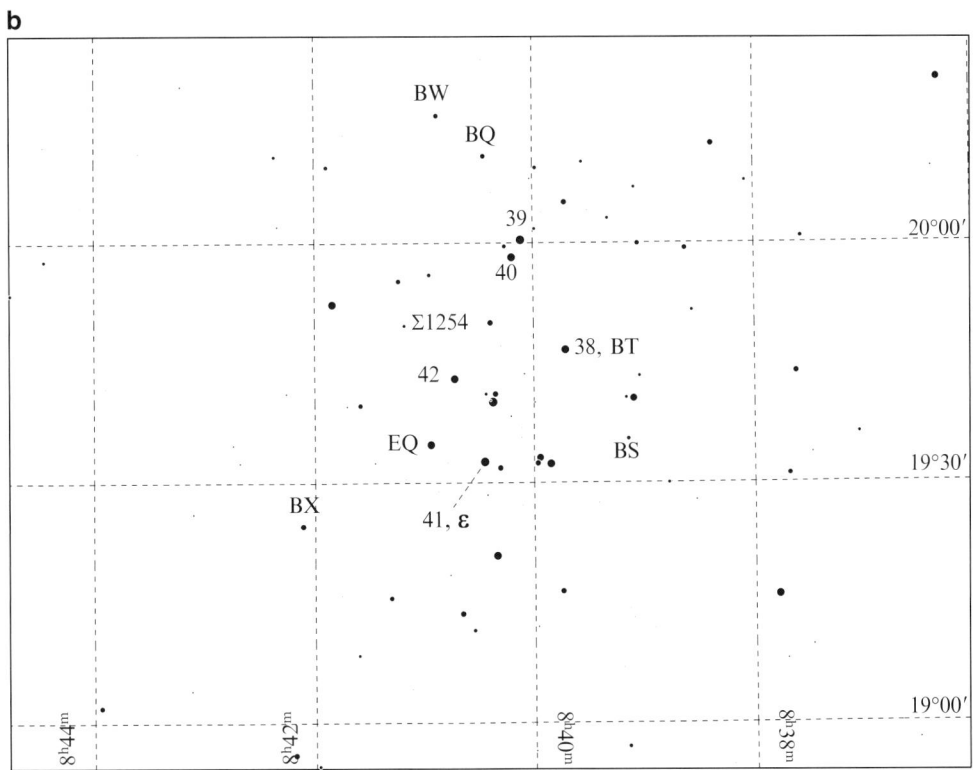

Fig. 6.16. (a) M44 (NGC 2632), open cluster in Cancer. Also known as the Praesepe, Manger, or Beehive Cluster. Astrophoto by Mark Komsa. **(b)** M44 details. Scales are somewhat different, but both fields-of-view are encompassed in large binoculars or small telescopes.

Fig. 6.17. M67 (NGC 2682) open cluster in Cancer. Astrophoto by Mark Komsa.

COMA BERENICES (**koe**-muh bare-uh-**nye**-seez) Berenice's Hair	
COMAE BERENICES (**koe**-my burr-uh-nye-seez)	9:00 P.M. CULMINATION: May 16
Com	AREA: 386 sq. deg., 42nd in size

Coma Berenices was the first of what are generally considered the "Modern Constellations." Actually it is both relatively new and relatively old. The Coma Cluster of stars was recognized by the ancient Arabs, Greeks, and other people as a fuzzy spot in the sky. It was first called Berenice's Hair about 240 BC; however, it was generally considered the tuft of hair at the end of The Lion's (Leo's) tail until fairly modern times.

When Alexander the Great entered Syrian-occupied Egypt in 332 BC, the Egyptian people welcomed him as a liberator, and the Syrian Governor, who had almost no army, surrendered without a fight. In return, Alexander allowed him to continue to continue to rule, along with Cleomenes, one of Alexander's generals. After Alexander's death in 323 BC, one of his bodyguards called Ptolemy, son of Lagus, ousted the corrupt Cleomenes. Ptolemy ruled so wisely that he was asked by the people and their leaders to become their King or Pharoah. His immediate successors, including his son Ptolemy Philadelphus who established the famous Library at Alexandria, also ruled wisely, and their dynasty lasted almost 300 years before ending with Ptolemy XVI.

Ptolemy III, also known as Ptolemy Euergetes, married Berenice, the beautiful daughter of King Magas of Cyrene. In 243 BC, Ptolemy III set out on a military campaign to punish the Assyrians for the murder of his sister. Berenice knew that this was to be a long and extremely dangerous expedition, and that her husband might well not return alive. To win favor for her husband's battles, Berenice went to the temple of Aphrodite at Zaphyrion and promised to sacrifice her long golden tresses to the goddess if Aphrodite would help her husband return safely.

Ptolemy III did return safely, and Berenice quickly honored her pledge. The priests placed her shorn locks on the alter before Aphrodite's statue. At some time during the following celebrations, Berenice's locks disappeared. Berenice and Ptolemy were furious at the sacrilege, and Ptolemy vowed to execute the thief or, if the thief could not be found, the priests charged with guarding the temple. Neither the missing locks nor the thief were found, and the king was about to order the executions when one of his advisors saved the priests' lives.

Conon of Samos, a mathematician and astronomer in Ptolemy's court, stopped the executions by declaring that Berenice's beautiful hair had so pleased Aphrodite that she had not only accepted them but had placed them in the heavens for all the world to see. He proved his assertion by taking the group out under the dark Egyptian skies and pointing to a hazy spot in the sky. This hazy spot was the group of faint stars that we now call the Coma Cluster or Melotte 111. The cluster had long been known to dedicated sky watchers as the tuft at the tip of the lion's tail, but apparently was not known to everyone, most importantly Ptolemy Eugertes!

Although Conan's new constellation saved the lives of several priests, it was not formally recognized as a constellation for over 1,800 years. Around 60 BC, the Roman Poet Catullus referred to Aphrodite's having placed Berenice's golden tresses in the sky, but over 200 years later, Ptolemy's Almagest still included these stars in the constellation Leo. In the western world, Ptolemy was the final authority on almost everything until the European Renaissance, so Conan's new constellation was almost forgotten.

It was not until the sixteenth century that astronomers started recognizing Coma Berenices as a distinct constellation. In 1515, Johann Schöner published a set of gores, or lens-shaped sections, which could be assembled on a celestial globe with this area labeled "Trica" or hair, which comes from a Greek word of similar meaning. In 1536, Casper Vopel published a true celestial globe which showed the figure of a small woman enveloped in a mass of long hair, and labeled it "Coma Berenices." The following year, Gerard Mercator, the famous map maker, published a celestial globe which not only gave the modern name but showed a drawing of only a lock of hair. The ultimate approval came with Tycho Brahe's 1602 atlas, followed in 1603 by Johannes Bayer's famous atlas, and Coma Berenice was firmly established as a widely recognized constellation.

TO FIND: It seems appropriate that this group of stars which used to mark the end of a tail is found almost midway between the ends of two other tails. Start with Denebola (the new end of the lion's tail) and draw an imaginary line to Alkaid (the end of the great bear's tail). γ Comae, the third brightest star in Coma Berenices lies a little more than ⅓ of the distance from Denebola to Alkaid (because of distortion on the chart, the straight line would have to be shown in segments to accurately show how it passes through Coma Berenices). γ Com also marks the northern tip of the roughly triangular group of faint stars which mark the point where the lock of hair is attached to an inverted "L"-shaped support. α Comae Berenices is similarly found near the middle of an imaginary line from Denebola to Arcturus. β Comae Berenices is located 10° north of α. The three stars α, β, and γ form the inverted "L" support from which the lock of hair hangs (Figs. 6.7, 6.17 and 6.18).

KEY STAR NAMES: None.

Magnitudes and spectral types of principal stars:

Bayer desig.	Mag. (m_v)	Spec. type	Bayer desig.	Mag. (m_v)	Spec. type	Bayer desig.	Mag. (m_v)	Spec. type
α	4.32	F5 V	β	4.23	G0 V	γ	4.35	K2 III CN

Stars of magnitude 5.5 or brighter which have Flamsteed but no Bayer designations:

Flamsteed	Mag. (m_v)	Spec. type	Flamsteed	Mag. (m_v)	Spec. type	Flamsteed	Mag. (m_v)	Spec. type
6	5.09	A3 V	7	4.93	K0 III	11	4.72	G8 III
13	5.17	A3 V	12	4.78	F8:p	16	4.98	A4 V
14	4.92	F0:p	18	5.47	F5 III	21	5.47	A2p:var
23	4.80	A0 IV	24	5.03	K2 III	26	5.49	G9 III
27	5.12	K3 III	31	4.93	G0 III	35	4.89	G8 III
36	4.76	M0 III	37	4.88	K1 IIIp	41	4.80	K5 III

Stars of magnitude 5.5 or brighter which have no Bayer or Flamsteed designations:

Designation	Mag. (m_v)	Spec. type
HD 106760	4.99	K1 III

a

b

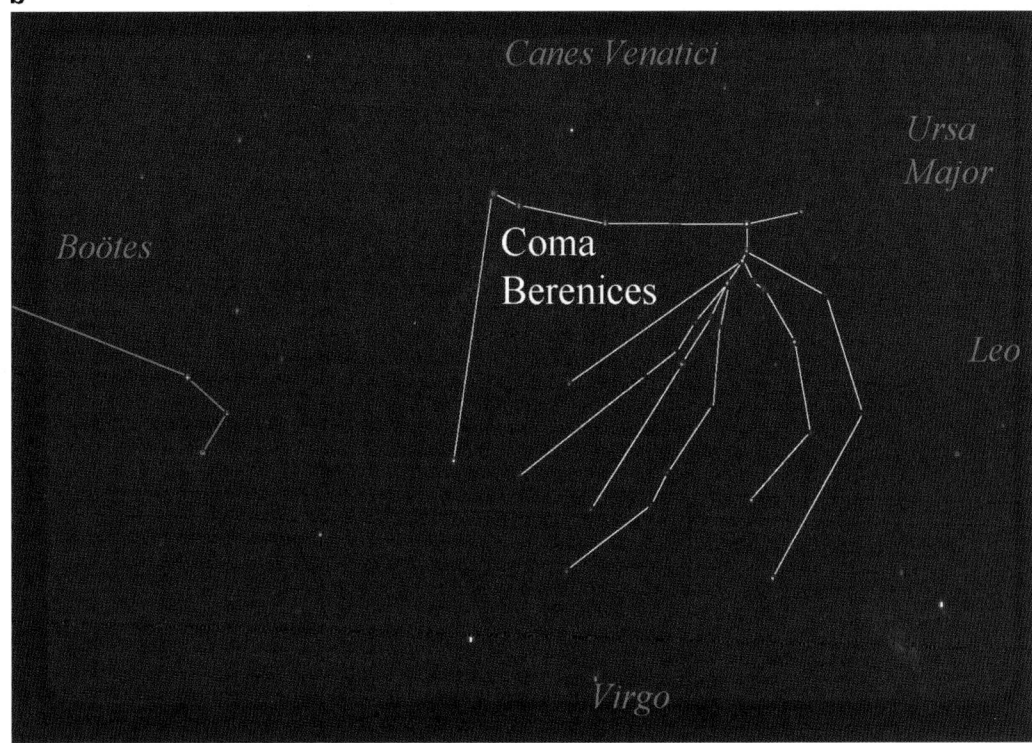

Fig. 6.18. (**a**, **b**) Stars and constellation of Coma Berenices.

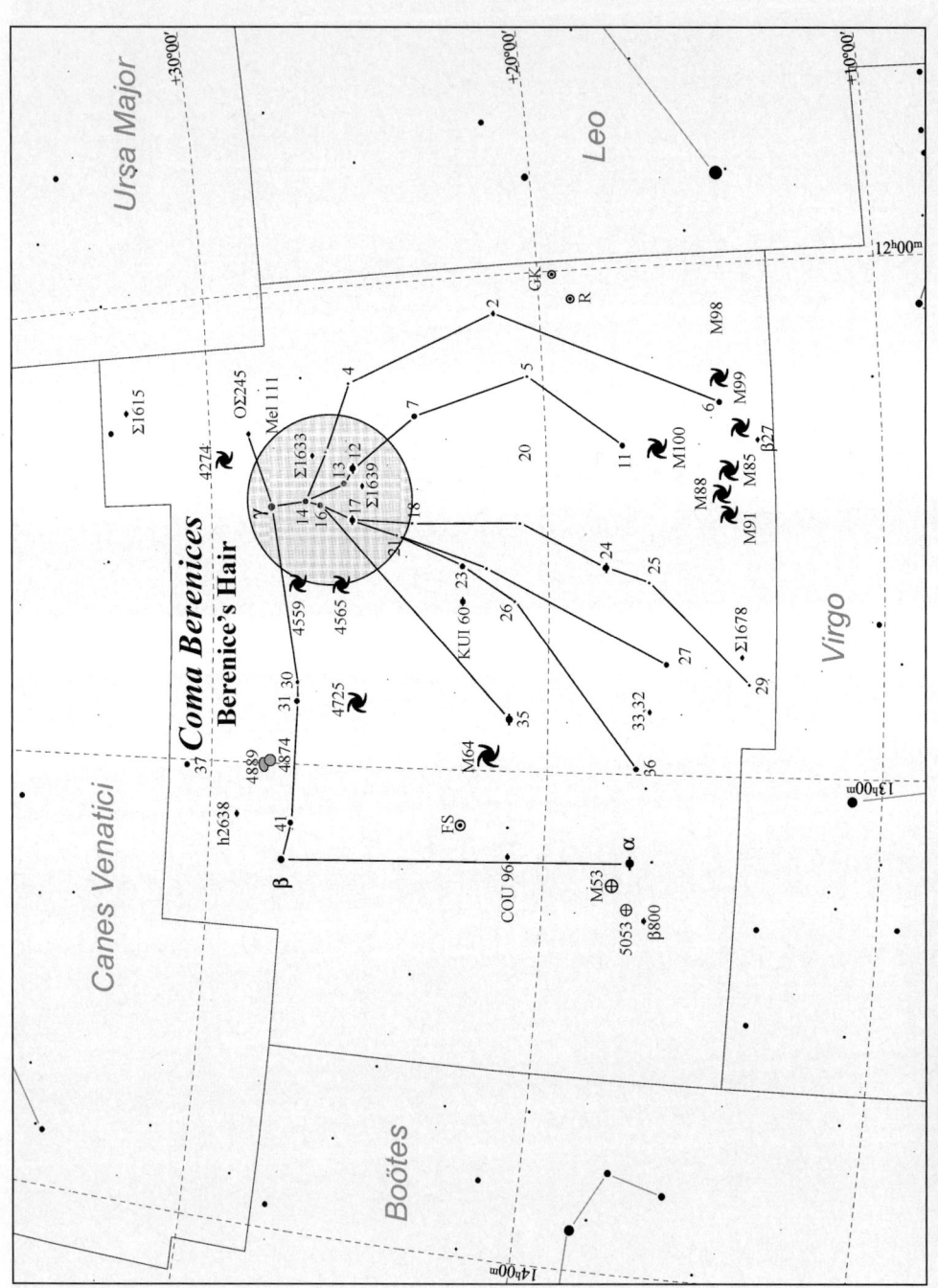

Fig. 6.19. Coma Berenices details.

Table 6.4. Selected telescopic objects in Coma Berenices.

Multiple stars:

Designation	R.A.	Decl.	Type	m₁	m₂	Sep. (")	PA (°)	Colors	Date/ Period	Aper. Recm.	Rating	Comments
2 Comae Berenices	12ʰ 04.3ᵐ	+21° 28'	A–B	6.2	7.5	3.7	236	yW, bW	2006	2/3"	*****	Beautiful. Very subtle colors
Σ1615	12ʰ 14.1ᵐ	+32° 47'	A–B	7.0	8.6	26.8	088	oY, ?	2004	2/3"	****	a.k.a. A 2058 ʙᴄ
			B–C	8.8	14.0	2.7	278	?, ?	1935	9/10"	***	
OΣ245	12ʰ 17.5ᵐ	+28° 56'	A–B	5.7	10.2	8.3	218	W, ?	1999	2/3"	****	Subtle color contrast. Some people see white and lilac
β27	12ʰ 20.1ᵐ	+13° 51'	A–B	7.1	10.5	3.6	108	oY, ?	2000	3/4"	***	Large brightness ratio ~700:1
Σ1633	12ʰ 20.6ᵐ	+27° 03'	A–B	7.0	7.1	8.9	245	yW, ?	2004	2/3"	******	A springtime Albireo. Optical pair, not related. Nearly equal brightness
12 Comae Berenices	12ʰ 22.5ᵐ	+25° 51'	A–B	4.8	11.8	36.8	057	pB, ?	2000	5.6"	****	Somewhat faint, but well worth seeking out, A and B yellowish. a.k.a. Schj 143.
			A–B	4.9	8.9	65.2	168	Y, pB	2002	2/3"	******	C is one of few stars to appear purple, probably due to contrast
Σ1639	12ʰ 24.4ᵐ	+25° 35'	A–B	6.7	7.8	**1.8**	**324**	W, grW	**574.44**	5/6"	***	Very rapid binary, at max. separation 2010.0, orbit is edge-on
17 Comae Berenices	12ʰ 28.9ᵐ	+25° 55'	A–BC	5.2	6.6	146.2	251	bW, W	2002	6×30	***	Lovely orange and reddish pair, with distant companion
			B–C	6.6	13.7	1.8	156	W, ?	1998	8/9"	*****	
24 Comae Berenices	12ʰ 35.1ᵐ	+18° 23'	A–B	5.1	6.3	20.2	271	yO, B	2006	15×80	****	a.k.a. Σ1657
Kui 60	12ʰ 39.0ᵐ	+22° 40'	A–B	6.4	12.5	33.4	226	oY, ?	1997	4/5"	****	
Σ1678	12ʰ 45.4ᵐ	+14° 22'	A–B	7.2	7.7	36.5	172	W, ?	2006	7×50	****	
30 Comae Berenices	12ʰ 49.3ᵐ	+27° 33'	Aa–B	5.8	13.5	42.4	013	W, ?	1999	7/8"	***	Aa is a close double, sep ~0.3". a.k.a. CHR 179-Aa
32-33 Comae Berenices	12ʰ 52.2ᵐ	+17° 04'	A–B	6.5	7.0	195.6	051	YO, W	2000	6×30	****	a.k.a. ΣA 23-AB
35 Comae Berenices	12ʰ 53.3ᵐ	+21° 15'	A–B	5.2	7.1	**1.03**	**195**	O, Y	**359**	9/10"	****	a.k.a. Σ1687-AB.
			AB–C	5.0	9.8	27.9	127	O, P	2004	2/3"	****	a.k.a. Σ1687-AC
h2638	13ʰ 06.2ᵐ	+29° 02'	A–B	6.6	11.1	6.2	222	W, ?	2000	3/4"	***	
42, α Comae Berenices	13ʰ 10.0ᵐ	+17° 32'	A–B	4.9	5.5	**0.65**	**192**	W, W	**25.804**	14/15"	***	a.k.a. Σ1728-AB.
			AB–C	4.4	10.2	88.7	345	W, ?	1998	2/3"	****	a.k.a. Σ1728-AB-C
Cou 96	13ʰ 10.9ᵐ	+21° 14'	A–B	6.8	10.8	10.6	310	yW, ?	2003	3/4"	****	
β800	13ʰ 16.8ᵐ	+17° 01'	A–B	6.6	9.5	**7.67**	**105**	O, R	**770**	2/3"	****	
			A–C	6.6	10.5	120.6	339	O, ?	1999	2/3"	****	
			A–D	6.6	10.6	50.8	088	O, ?	1990	3/4"	****	

(continued)

Table 6.4. (continued)

Variable stars:

Designation	R.A.	Decl.	Type	Range (m_v)	Period (days)	F (f_r/f_t)	Spectral Range	Aper. Recm.	Rating	Comments
GK Comae Berenices	12h 00.1m	+19° 25'	SRb	6.8 / 7.1	50	–	M4III	6×30	****	##
R Comae Berenices	12h 04.0m	+18° 47'	M	7.1 / 14.6	362.82	0.38	M5e-M8ep	14/16"	****	Varies from easy binocular object to difficult 10/12" object. ##$
Al Comae Berenices	12h 28.9m	+25° 55'	α CVn + δ Sct	5.2 / 5.4	5.063	–	A0p(Cr-Eu-Sr)	Eye	****	Small variation, but detectable without optical aid
40, FS Comae Berenices	13h 06.4m	+22° 37'	SRb	5.3 / 6.1	58	–	M5III	6×30	****	##

Star clusters:

Designation	R.A.	Decl.	Type	m_v	Size (')	Brst. Star	Dist. (ly)	dia. (ly)	Number of Stars	Aper. Recm.	Rating	Comments
Melotte 111	12h 25m	+26° 06'	Open	1.8	300	5.0	260	22	273	Eye	*****	3rd closest cluster, very loose, ideal for binoculars
M53, NGC 5024	13h 12.9m	+18° 10'	Globular	7.7	13.0	13.8	60K	225	–	7×50	*****	Total Lum. = 200,000 suns
NGC 5053	13h 16.5m	+17° 42'	Globular	9.0	10.0	13.8	52K	150	–	6/8"	***	Total Lum. = 55,000 suns, low for globular cluster

Nebulae:

Designation	R.A.	Decl.	Type	m_v	Size (')	Brst./Cent. star	Dist. (ly)	dia. (ly)	Aper. Recm.	Rating	Comments
None											

Galaxies:

Designation	R.A.	Decl.	Type	m_v	Size (')	dia. (ly)	Dist. (ly)	Lum. (suns)	Aper. Recm.	Rating	Comments
M98, NGC 4192	12h 13.8m	+14° 54'	SAB	10.1	9.1×2.1	185K	70M	85G	4/6"	*****	Bright, mottled core and irregular halo
M99, NGC 4254	12h 18.8m	+14° 25'	SA	9.9	4.6×4.3	95K	70M	53G	4/6"	*****	Large bright core, arms may be glimpsed in 8/10" scope
NGC 4274	12h 19.8m	+29° 37'	SB	10.4	6.7×2.5	51K	32M	9.9G	6/8"	*****	E-W with gradual brightening toward center
M100, NGC 4321	12h 22.9m	+15° 47'	SAB	9.3	6.2×5.3	130K	70M	83G	4/6"	*****	Spiral structure first detected by Lord Rosse with 72" reflector

Designation	R.A.	Decl.	Type	m_v	Size (')	Dist. (ly)	dia. (ly)	Number of Gal	Aper. Recm.	Rating	Comments
M85, NGC 4382	12h 25.4m	+18° 11'	SA	9.1	7.5×5.7	70M	150K	83G	4/6"	*****	NGC 4394 (m10.9) is in same FOV, only 7.5' to east
M88, NGC 4501	12h 32.0m	+14° 25'	SB	9.6	6.1×2.8	70M	125K	48G	4/6"	*****	Broad oval core, spiral structure may be detected in 12/14"
M91, NGC 4548	12h 35.4m	+14° 30'	SB	10.2	5.0×4.1	65M	100K	28G	4/6"	****	Inconspicuous nucleus, nearly circular core, faint halo
NGC 4559	12h 36.0m	+27° 58'	SB	10.0	12.0×4.9	32M	89K	17G	6/8"	*****	Mottled, NW–SE elongated core, faint NNW–SSE halo, 2° N of 4665
NGC 4565	12h 36.3m	+25° 59'	SB	9.6	14.0×1.8	31M	106K	21G	4/6"	*****	Possibly the finest edge-on galaxy, prominent dust lane
NGC 4725	12h 50.4m	+25° 30'	SAB	9.4	11.0×8.3	40M	118K	28G	4/6"	*****	NE–SW, Faint central bar can be seen in 12/14" scopes
M64, NGC 4826	12h 56.7m	+21° 41'	SB	8.5	9.2×4.6	13M	27K	7.1G	4/6"	*****	1° NE of 35 Com. "Black Eye Galaxy"
NGC 4874	12h 59.6m	+27° 58'	E0	11.7	2.8×2.5	310M	250K	31G	6/8"	***	NGC 4874 and 4889 are the two giant ellipticals which dominate the Coma
NGC 4889	13h 00.1m	+27° 59'	E4	11.5	3.2×2.2	280M	260K	25G	6/8"	***	Cluster, one of the three closest rich galaxy clusters

Meteor showers:

Designation	R.A.	Decl.	Period (yr)	Duration	Max Date	ZHR (max)	Comet/Asteroid	First Obs.	Vel. (mi/km/sec)	Rating	Comments
Coma Berenicids	11h 40m	+25°	Annual	12/12 01/23	12/20	5	Lowe *	1952	40/65	II	Richard E. McCrosky/Annette Posen (Harvard Meteor Project) USA. 1931

Other interesting objects:

Designation	R.A.	Decl.	Type	m_v	Size (')	Dist. (ly)	dia. (ly)	Number of Gal	Aper. Recm.	Rating	Comments
Abell 1656	12h 59.7m	+27° 59'	GalClust	–	240	321M	13M	>1000	–	*****	a.k.a. Coma Cluster of Galaxies

a

b

Fig. 6.20. (**a**) NGC 4546 (C38), edge-on spiral galaxy in Coma Berenices. Astrophoto by Rob Gendler. (**b**) NGC 4546. Drawing by Eiji Kato. The scale is twice that of Fig. 6.22a.

Fig. 6.21. M53 (NGC 5024), globular cluster in Coma Berenices. Astro-photo by Gian Michelle Ratto.

Fig. 6.22. M88 (NGC 4501) in Coma Berenices. Astrophoto by Rob Gendler.

Fig. 6.23. M99 (NGC 4554) in Coma Berenices. Astrophoto by Rob Gendler.

Fig. 6.24. M100 (NGC 4321), NGC 4322 and 4328 in Coma Berenices. Astrophoto by Gian Michelle Ratto.

Fig. 6.25. NGC 4725 in Coma Berenices. Astrophoto by Gian Michaell Ratto.

CENTAURUS (sen-**tawr**-us)	**The Centaur**
CENTAURI (sen-**tawr**-ee)	**9:00 P.M. CULMINATION:** May 15
Cen	**AREA**: 1,060 sq. deg., 9th in size

TO FIND: α and β Centauri are so bright and so close together that they are difficult to miss, especially since they are so near to the bright and distinctive pattern of Crux, the Southern Cross. If you are observing from south of latitude 29° north, you should be able to see α and β lying almost parallel to the southern horizon and 1° above it for each degree that you are south of 29° latitude. This pair of stars represent the forefeet of the Centaur. ε Centauri lies slightly less than 8° to their NNW, and represents the point where the forelegs join the body. ζ Centauri is about 6°40' (⅔ the distance across your knuckles at arm's length) NNE of ε Centauri, and represents the junction between the human and the horse parts of the centaur. From ζ northward, two diverging lines of stars trace the left and right sides of the human torso which is turned to nearly face the observer. From ζ NNE, υ¹, χ, and θ delineate right side of the centaur's human body from the waist to the shoulder. On the left (NNW) side, μ and ι delineate the left side, with an unlettered 4th magnitude about halfway between filling in the gap. About

3° above the line joining the shoulders is a group of four 4th and 5th magnitude stars represent the centaur's head. Returning to ζ Centauri, a line of stars going west, then southwest (f, τ, δ, and π) outline the back of the horse's torso. A similar row of stars (M, K, ρ, and A) ending just east of π trace the horse's underside. The hind legs go southward from π to λ, and from A to a pair of unlettered stars about 1.5° east of λ. From π, a line to C² through an unlettered star to B Centauri (about 6° NNW of δ) can be seen as the horse's tail. In many older atlases, the centaur is shown holding a spear on which the wolf is impaled and being borne to the altar as a sacrifice to the gods. The point of the spear is just east of κ, with the rest of the shaft marked by η, φ, ν, n, and I. Northern hemisphere observers above about 30° north latitude will have to be content to locate 2nd magnitude θ Centauri (Menkent) about 30° WSW of Antares and tracing out the human part of the centaur as described above. The remainder of the centaur never rises above their horizon (Figs. 6.26 and 6.37).

KEY STAR NAMES:

Name	Pronunciation	Source
α Rigel Kentaurus,	**rye**-jil ken-**tawr**-us	rijl qantūris (Sci-Arabic), "the centaur's foot"
Rigel Kent,	**rye**-jil kent	
or Toliman	toh-lee-**man**	al-zulman (Arabic), "the ostriches," an Arabic constellation, but attribution is uncertain
β Hadar	huh-**dahr**	hadāri (Arabic), untranslatable proper name
or Agena	uh-**gee**-nah	A recent name, possibly from a mistaken designation as "alpha" plus genu (Latin), "knee" from its position as described by Ptolemy
γ Muhlifain	mu-lih-**fane**	Erroneously transferred from γ CMa with a slight spelling change
ζ Alnair	uhl-**nahr**	(al) nayyir badan qantūris (Arabic), "the bright one in the body of the centaur"
θ Hadar	muh-le-**fan**	muhlifan (Arabic), "two (things) causing a dispute," because at one time it was uncertain as to which two stars the names al-wazn and hadari applied. The name Muliphen is usually applied to γ CMa, but it is sometimes also used for γ Ophiuchi
ε Birdun	(?)	Origin and pronunciation uncertain
– Proxima	**prok**-si-muh	proxima (Latin), "nearest"

Magnitudes and spectral types of principal stars:*9

Bayer desig.	Mag. (m_v)	Spec. type	Bayer desig.	Mag. (m_v)	Spec. type	Bayer desig.	Mag. (m_v)	Spec. type
α	−0.01	G2 V	β	0.61	B1 III	γ	2.20	A1 IV
δ	2.58	B2 IVne	ε	2.29	B1 III	ζ	2.55	B2.5 IV
η	2.33	B1 Vn	θ	2.06	K0 IIIb	ι	2.75	A2 V
κ	3.13	B2 IV	λ	3.11	B9 II	μ	3.47	B2 IV/V
ν	3.41	B2 IV	ξ^1	4.83	A0 V	ξ^2	4.27	B1.5 V
o^1	5.07	G0 Ia	o^2	5.12	A3 IV	π	3.90	B5 Vn
ρ	3.97	B3 V	σ	3.91	B3 V	τ	3.85	A2 V
υ^1	3.87	B2 IV/V	υ^2	4.27	B1.5 V	φ	3.83	B2 IV
χ	4.36	B2 V	ψ	4.05	A0 IV	ω	5.33	−*10
a	4.41	B2 V	b	4.01	B2.5 V	c^1	4.06	K3 III
c^2	4.92	A0 V	e	4.33	K3/4 III	f	4.71	B5 V
g	4.19	M5 III	h	4.75	B4 IV	i	4.23	F3 V
j	4.30	B3 V	k	4.32	B5 −	l	4.63	B8 II/III
m	4.52	G5 III/IV	n	4.25	A4 IV	p	4.90	B9 V
r	5.10	K0 III	u	5.45	B8/9 V	v	4.30	B6 Ib
w	4.66	K0 III	x^2	5.32	B8/9 V	z	5.15	A0 V
A	4.62	B9 V	B	4.47	K4 III	C^2	5.26	A7m0
C^3	5.46	K2 III	D	5.31	K3 III	E	5.34	A1 V
F	5.01	M1 III	G	4.82	B3 Vn	H	5.17	B8 V
J	4.52	B3 V	K	5.04	A0 V	M	4.64	G8/K0 III
N	5.26	B9 Vn				Q	4.99	B8 Vn

Stars of magnitude 5.5 or brighter which have no Bayer or Flamsteed designations:

Desig-nation	Mag. (m_v)	Spec. type	Desig-nation	Mag. (m_v)	Spec. type
HD 096616	5.15	Ap SrCrEu	HD 097495	5.37	A2 III
HD 098993	5.00	K4 III	HD 099322	5.21	K0 III
HD 099453	5.18	F7 V	HD 099556	5.22	B3 IV
HD 099803	5.14	B9 V	HD 100493	5.39	A2 IV/V
HD 100708	5.50	K1 III/IV	HD 101021	5.14	K1 III
HD 101189	5.15	B9 IV	HD 101947	5.00	F9 Ia
HD 102232	5.28	B6 III	HD 102350	4.11	G0 II
HD 102365	4.89	G3/5 V	HD 102461	5.42	K5 III
HD 102776	4.30	B3 V	HD 109536	5.12	A7 III
HD 110458	4.66	K0 III	HD 110730	5.63	B8 II/III
HD 111597	4.90	B9 V	HD 111915	4.33	K3/4 III
HD 111968	4.25	A4 IV	HD 112213	5.46	M0 III
HD 113703	4.71	B5 V	HD 114474	5.24	K1/2 III
HD 115310	5.10	K0 III	HD 115823	5.47	B6 V
HD 116243	4.52	G5 III/IV	HD 116457	5.32	F2 III
HD 116713	5.11	Kp	HD 117440	3.90	G8 II/III
HD 118978	5.38	B9 IV	HD 119971	5.46	K2 III
HD 124367	5.03	B4 Vne	HD 125158	5.22	Am
HD 125288	4.30	B6 Ib	HD 129116	4.01	B2.5 V
HD 129456	4.06	K3 III	HD 129685	4.92	A0 V
HD 131120	5.02	B7 II/III	HD 131625	5.32	A0 V

Fig. 6.26a. The stars of Centaurus, Crux and Lupus.

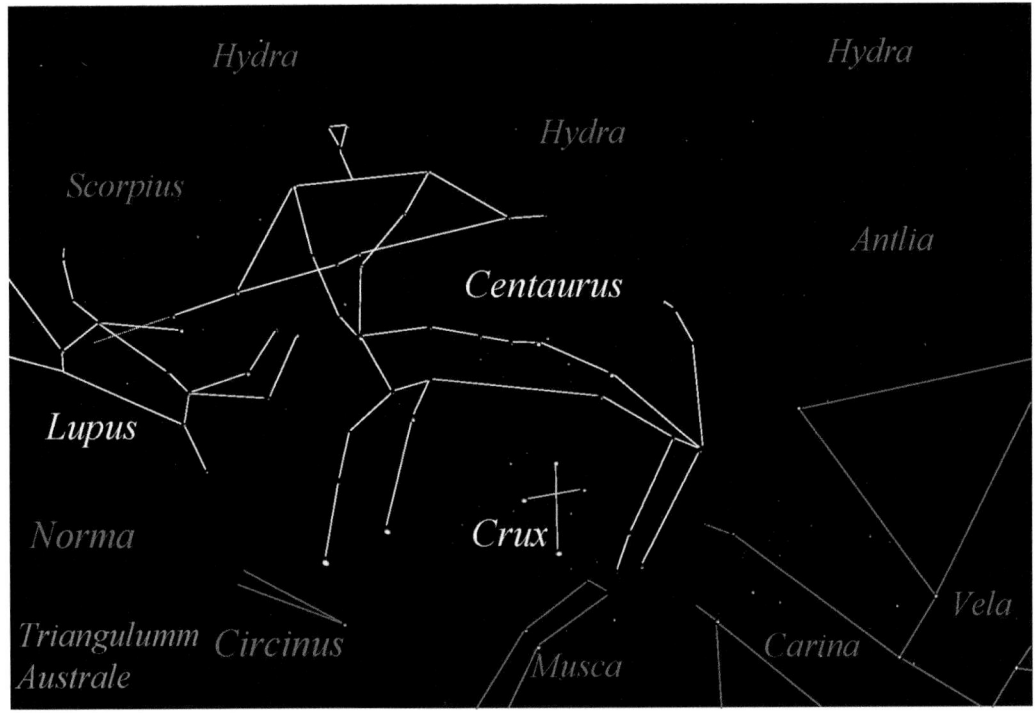

Fig. 6.26b. The figures of Centaurus, Crux and Lupus.

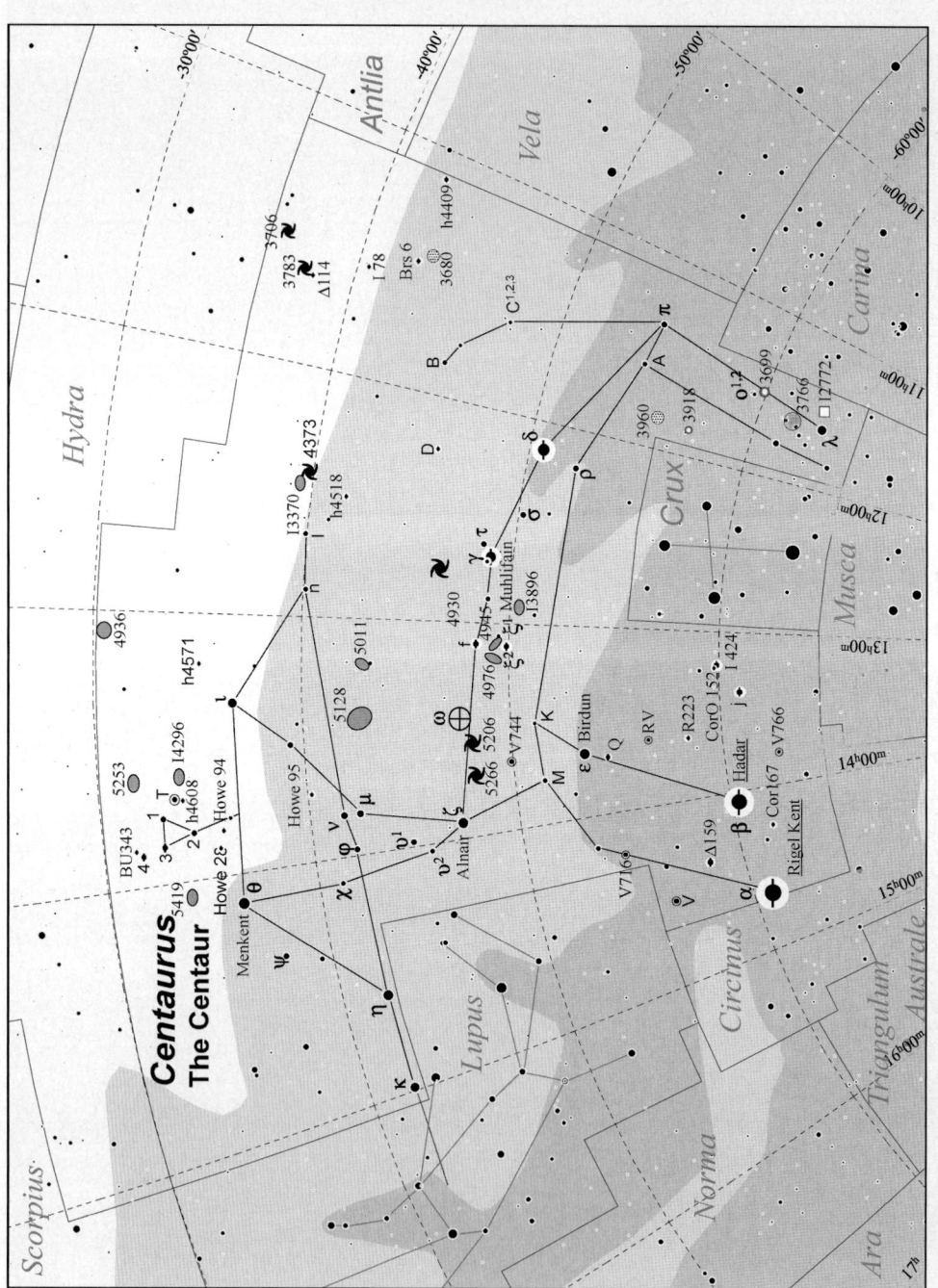

Fig. 6.27. Centaurus details.

Table 6.5. Selected telescopic objects in Centaurus.

Multiple stars:

Designation	R.A.	Decl.	Type	m_1	m_2	Sep. ["]	PA [°]	Colors	Date/ **Period**	Aper. Recm.	Rating	Comments
h 4409	11h 07.3m	−42° 38'	A–B	5.2	7.73	1.3	252	W, ?	1995	7/8"	****	Easy pair for small telescope
BrsO 6	11h 28.6m	−42° 40'	A–B	5.1	7.4	13.1	168	pY, Y	1999	2/3"	****	Almost perfect critical test for 5" scope
I 78	11h 33.6m	−40° 35'	A–B	6.1	6.2	0.7	098	W, W	1998	13/14"	***	
Δ 114	11h 40.0m	−38° 07'	A–B	6.7	8.0	17.1	095	Y, ?	1999	18×80	****	a.k.a. JC 2
δ Centauri	12h 08.4m	−50° 43'	A–B	2.5	4.4	268.9	325	B, W	1992	eye	****	
			A–C	2.5	6.4	216.9	227	B, bGr	1999	6×30	****	
D Centauri	12h 14.0m	−45° 43'	A–B	5.8	7.0	2.8	243	oR, W	1995	3/4"	*****	Lovely pair. a.k.a. Rmk 14
h 4518	12h 24.7m	−41° 23'	A–B	6.5	8.4	10.0	208	O, ?	1999	2/3"	***	
γ Centauri	12h 41.5m	−48° 58'	A–B	2.8	2.9	**0.4**	**324**	W, W	**84.494**	26"	****	Nearly identical pair. Closing, minimum of 0.145" in 2013. Max. of 1.667" in 2056
f Centauri	13h 06.3m	−48° 28'	A–B	4.7	10.8	11.5	078	bW, ?	1999	3/4"	****	4° SW of ω Cen.
ξ2 Centauri	13h 06.9m	−49° 54'	A–B	4.2	10.1	25.0	099	W, ?	1999	2/3"	****	Nice surrounding field
h4571	13h 11.5m	−35° 08'	A–B	6.8	9.1	23.7	266	W, O	1999	10×50	****	In same FOV as CorO 152. A–B has a maximum separation of only 0.25"
I 424	13h 12.3m	−59° 55'	AB–C	4.8	8.4	1.9	007	?, ?	1991	5/6"	****	
CorO 152	13h 12.9m	−59° 49'	AB–C	6.3	9.4	25.0	146	Y, ?	2000	18×80	****	Unequal double-double with I4e24. A–B is a close (0.2") double
i Centauri	13h 22.6m	−60° 59'	AB–C	4.5	6.2	60.7	346	W, Y	2000	6×30	*****	Easy, beautiful. A–B is a close (0.2") double
R 223	13h 38.1m	−58° 25'	A–B	6.6	9.9	2.6	013	O, W	1991	3/4"	*****	Nice orange, profuse FOV
Q Centauri	13h 41.7m	−54° 34'	A–B	5.2	6.5	5.5	163	yW, pB	2000	2/3"	*****	Showcase pair. a.k.a. Δ 141. 1° S of ε Cen.
h 4608	13h 42.3m	−33° 59'	A–B	7.4	7.5	4.2	008	yW, yW	2000	2/3"	****	Almost a classic test for a 1" refractor with perfect optics
Howe 95	13h 43.8m	−40° 11'	A–B	7.5	9.9	1.1	186	pY, pY	1991	8/9"	****	Slightly mismatched pair but lovely
Howe 94	13h 48.9m	−35° 42'	A–B	6.6	10.2	11.6	358	dY, oR	1998	2/3"	****	Nice color combination
3 Centauri	13h 51.8m	−33° 00'	A–B	4.5	6.0	7.7	102	gY, gY	2002	2/3"	*****	Showcase pair. NE corner of nearly equilateral triangle with 1 and 2 Cen
BU 343	13h 52.0m	−31° 37'	A–B	6.5	7.8	**0.5**	**179**	?, ?	**161.6**	18/20"	****	Min. separation was 0.244" in 2000, max. will be 1.154" in 2064
N Centauri	13h 52.0m	−52° 49'	A–B	5.2	7.5	18.1	289	bW, W	1999	18×80	*****	Showcase pair. 1.5° ENE of ε Cen. a.k.a. Rmk 18
4 Centauri	13h 53.2m	−31° 56'	A–B	4.7	8.5	14.8	185	pY, W	1998	2/3"	*****	Showcase pair
Howe 28	13h 53.5m	−35° 39'	A–B	5.5	5.6	**1.0**	**316**	YyW, ?	**292.11**	10/12"	****	Closing to a secondary min. of 0.696" in 2089, then max. of 1.076" in 2196

(continued)

Table 6.5. (continued)

Multiple stars:

Designation	R.A.	Decl.	Type	m_1	m_2	Sep. (")	PA (°)	Colors	Date/Period	Aper. Recm.	Rating	Comments
β Centauri	14h 03.8m	−60° 22'	Aa–B	0.9	4.0	1.1	234	bW, ?	1991	8/10"	****	Challenging because it is close and unequal. A–a is a very close pair (<0.1")
CorO 167	14h 15.0m	−61° 42'	A–B	6.9	8.7	2.7	158	B, ?	1991	4/5"	***	
Δ 159	14h 22.6m	−58° 28'	A–B	5.0	7.6	8.9	159	bY, Y	1994	2/3"	*****	Showcase pair
α Centauri	14h 39.6m	−60° 50'	A–B	0.14	1.24	**6.8**	**245**	Y, O	**79.914**	2/3"	*****	Closest star system to our solar system *11

Variable stars:

Designation	R.A.	Decl.	Type	Range (m_v)		Period (days)	F (f_r/f_l)	Spectral Range	Aper. Recm.	Rating	Comments
RV Centauri	13h 37.6m	−56° 29'	M	7.0	10.8	446	0.56	N3e	2/3"	****	3° 02' S (PA 186°) of ε Cen (m2.29)
V744 Centauri	13h 40.0m	−49° 57'	SRb	5.1	6.7	90	–	M8III	6×30	****	3° 31' NNE (PA 020°) of ε Cen. (m2.29)
T Centauri	13h 41.8m	−33° 36'	SRa	5.5	9.0	90.44	0.47	K0e-M4IIe	6×30	*****	0° 59' WSW (PA 236°) of 1 Cen. (m5.50)
V766 Centauri	13h 47.2m	−62° 35'	S Dor	6.2	7.5	–	–	G8Ia0	6×30	****	2° 58' SW (PA 220°) of β Cen. (m0.61)
V716 Centauri	14h 13.7m	−54° 38'	EB/KE	6.0	6.5	1.4901	–	B5V	6×30	***	2° 00' NNW (PA 331288°) of HD 125288 Cen. (m4.30)
V Centauri	14h 32.5m	−56° 53'	Cδ	6.4	7.2	5.494	0.26	F5Ib-II-G0	6×30	*****	1° 45' WNW (PA 285°) of HD 125288 Cen. (m4.30)

Star clusters:

Designation	R.A.	Decl.	Type	m_v	Size (')	Brst. Star	Dist. (ly)	dia. (ly)	Number of Stars	Aper. Recm.	Rating	Comments
NGC 3680	11h 25.7m	−43° 15'	Open	7.6	7.0	10.0	2.6K	5.3	66	4/6"	***	Small, bright core, a few stars resolved at edge. Total luminosity ~500 suns
NGC 3766	11h 36.1m	−61° 37'	Open	5.3	15.0	8.0	5.6K	24	137	2/3"	****	Small, bright core, slightly larger halo than NGC 5824, not resolved. Lum. ~19K suns
NGC 3960	11h 50.9m	−55° 42'	Open	8.3	7.0	11.5	5.5K	1.1	317	4/6"	****	Moderately concentrated, edge well resolved. Luminosity ~1.1K suns
ω Centauri, NGC 5139	13h 26.8m	−47° 29'	Globular	3.9	55.0	11.5	17K	270	>1.0M	7×50	*****	Brightest Globular in the sky. A remarkable object!! *12

Nebulae:

Designation	R.A.	Decl.	Type	Size (')	Brtst. / **Cent.** star	Dist. (ly)	dia. (ly)	Aper. Recm.	Rating	Comments
NGC 3699	11h 28.0m	−59° 57'	Planetary	45"	–	–	–	5/6"	****	Narrow NE–SW dark lane divides nebula
IC 2872	11h 29.0m	−62° 57'	Emission	4.0 × 0.3	–	–	–	4/6"	****	NNE–SSW. Primary illumination is a magnitude 9.3 star, several other stars involved
NGC 3918	11h 50.3m	−57° 11'	Planetary	12"	**13.24**	2.6K	0.2	3/4"	***	a.k.a. the Blue Planetary. Uniform brightness circle. Expansion vel. 12 mi/s

Galaxies:

Designation	R.A.	Decl.	Type	m_v	Size (')	Dist. (ly)	dia. (ly)	Lum. (suns)	Aper. Recm.	Rating	Comments
NGC 3706	11h 29.7m	−36° 24'	SA(rs)0–	11.3	3.0 × 1.8	120M	100K	35G	6/8"	****	Pretty bright, small, round
NGC 3783	11h 39.0m	−37° 44'	SB(r)ab	11.4	1.9 × 1.4	120M	66K	32G	6/8"	****	Quite bright. Seyfert Galaxy with very small, very bright nucleus
NGC 4373	12h 25.3m	−39° 46'	SAB(rs)0–	10.9	3.5 × 2.8	140M	140K	68G	4/6"	****	Pretty bright, small, moderately bright nucleus. Fainter IC 3290 is 37' NE
IC 3370	12h 27.6m	−39° 20'	E2-3	11.0	3.1 × 2.5	120M	110K	46G	6/8"	****	Pretty bright, pretty large, moderately bright, patchy arms, bright rectangular nucleus
NGC 4696	12h 48.8m	−41° 19'	E1 pec	10.4	4.5 × 3.2	130M	170K	93G	6/8"	****	Pretty bright, large, round. Brightest member of Centaurus Galaxy Cluster
NGC 4706	12h 49.9m	−41° 17'	SAB(s)	12.9	1.4 × 0.6	126M	5.1K	8.8G	12/14"	***	Dwarf member of Centaurus Galaxy Cluster
IC 3896	12h 56.7m	−50° 21'	E1	11.2	2.2 × 1.8	86M	55K	20G	6/8"	***	Bright middle. 13th magnitude NGC 3896A is 20' to NW
NGC 4936	13h 04.3m	−30° 31'	E3	10.7	2.7 × 2.3	130M	100K	71G	4/6"	****	Pretty bright, small, moderately bright, diffuse nucleus
NGC 4945	13h 05.4m	−49° 28'	SB(s)cd	8.6	20 × 4.4	16M	93K	7.4G	2/3"	*****	Edge-on spiral, bright, very large, extremely elongated
NGC 4976	13h 08.6m	−49° 30'	E4p	10.0	5.0 × 2.9	49M	71K	19G	4/6"	****	Bright, pretty large, moderately bright nucleus
NGC 5011	13h 12.9m	−43° 06'	E1-2	11.2	2.4 × 2.0	120M	84K	38G	6/8"	****	Pretty bright, quite small, very bright nucleus
NGC 5128	13h 25.5m	−43° 01'	S0 pec	6.7	25 × 20	14M	100K	33G	2/3"	*****	One of brightest and oddest galaxies in the sky *13

(continued)

Table 6.5. (continued)

Galaxies:

Designation	R.A.	Decl.	Type	m_v	Size (')	Dist. (ly)	dia. (ly)	Lum. (suns)	Aper. Recm.	Rating	Comments
NGC 5206	13h 33.7m	−48° 09'	SB(rs)0	10.7	3.6 × 3.1	18M	19K	1.4G	4/6"	***	Faint, pretty large, moderately bright nucleus
IC 4296	13h 36.6m	−33° 58'	E0	10.5	3.0	150M	130K	11G	4/6"	***	Pretty faint, pretty small, Seyfert Galaxy
NGC 5253	13h 39.9m	−31° 39'	E5 pec?	10.2	4.8 × 1.8	11M	15K	800M	4/6"	****	Bright, pretty large, elongated, very bright center *14
NGC 5266	13h 43.0m	−48° 10'	SA0−	10.6	3.2 × 2.1	120M	110K	66G	4/6"	****	Bright, pretty large, slightly elongated; small bright nucleus
NGC 5419	14h 03.6m	−33° 59'	E2	10.8	4.0 × 3.2	180M	210K	124G	4/6"	****	Pretty bright, pretty large

Meteor showers:

Designation	R.A.	Decl.	Period (yr)	Duration	Max Date	ZHR (max)	Comet/ Asteroid	First Obs.	Vel. (mi/ **km**/sec)	Rating	Comments
α Centaurids	14h 00m	−59°	Annual?	01/28 02/21	02/08	6	Unknown	1980	35/**56**	II	Michael Buhagair (Western Australia) published observations of 1969 to 1980

Other interesting objects:

Designation	R.A.	Decl.	Type	m_v	Size (')	Dist. (ly)	dia. (ly)	Lum. (suns)	Aper. Recm.	Rating	Comments
None											

Fig. 6.28. Omega Centauri (NGC 5139), globular cluster in Centaurus. Astrophoto by Rob Gendler.

Fig. 6.29. Collinder 249, open cluster in IC 2948 (Running Chicken Nebula), and NGC 3766, open cluster in Centaurus. Astrophoto by Christopher J. Picking.

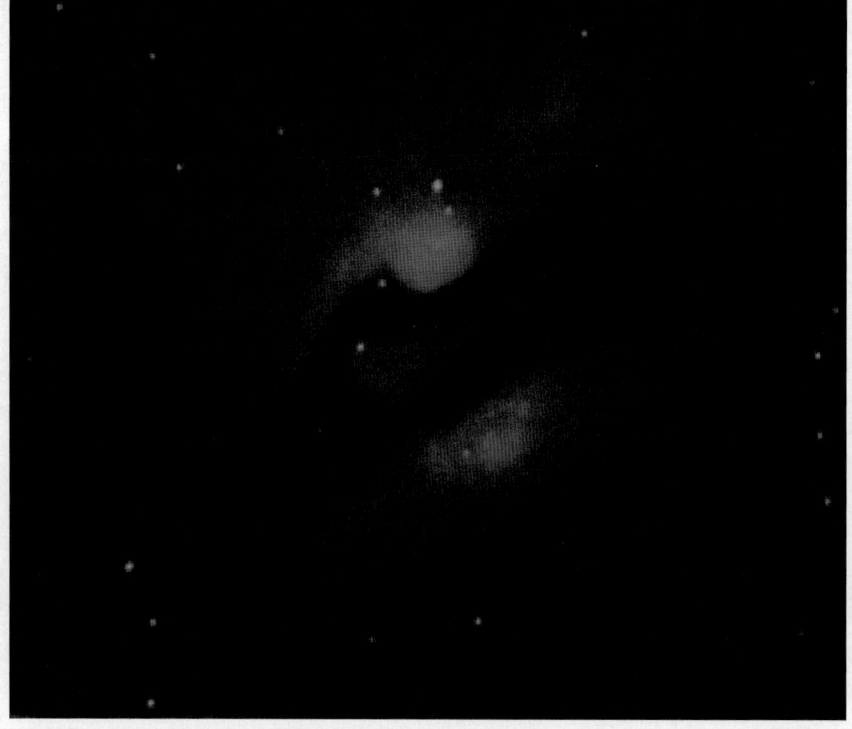

Fig. 6.30. (**a**) NGC 5128, peculiar galaxy in Centaurus. Astrophoto by Mark Komsa. (**b**) NGC 5128 in Centaurus.

Fig. 6.31. NGC 3699, planetary nebula in Centaurus. Drawing by Eiji Kato.

Fig. 6.32. NGC 4696, elliptical galaxy in Centaurus. Drawing by Eiji Karo.

Fig. 6.33. NGC 4706, spiral galaxy in Centaurus. Drawing by Eiji Kato.

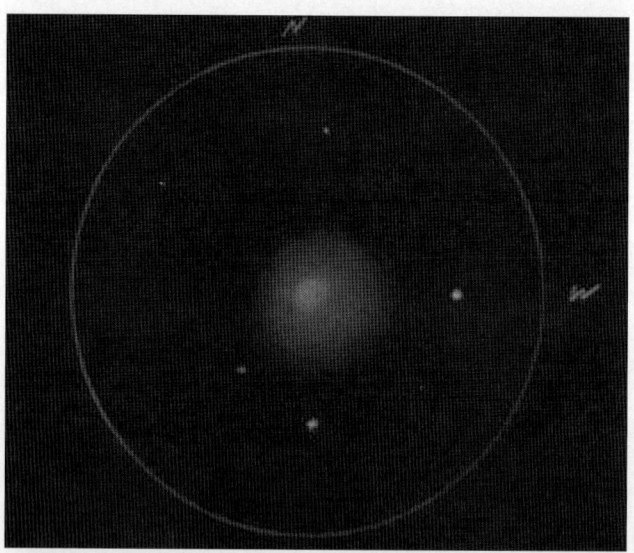

Fig. 6.34. NGC 4709, elliptical galaxy in Centaurus. Drawing by Eiji Kato.

Fig. 6.35. NGC 4945, edge-on spiral galaxy in Centaurus. Drawing by Eiji Kato.

Fig. 6.36. NGC 5253, elliptical galaxy in Centaurus. Drawing by Eiji Kato.

LUPUS (loo-pus)	**The Wolf**
LUPI (loo-pee)	**9:00 P.M. CULMINATION** July 29
LUP	**AREA:** 334 sq. deg., 46th in size

The group of stars now known as Lupus, the Wolf, is a fairly old constellation, being known to the ancient Akkadians. To them, it was simply "Urbat," a beast associated with death. To the ancient Arabs, it was known as "Al Asadha," the Lioness. The ancient Turks, Greeks, and Romans knew this group of stars as the "wild animal" or just "the beast." At some later time, a pre-existing Greek story about King Lyacon became associated with this constellation, and it has since been a wolf.

King Lyacon of Arcadia was infamous throughout the ancient world for his cruelty and impiety, and his 50 sons were little better. When Zeus learned about King Lyacon, he could scarcely believe the tales of his wickedness, and decided to learn about this first hand. To do so, he descended from Olympus to the earth disguised as a common laborer.

When he appeared at Lyacon's palace, Zeus was received with mock hospitality. He was entertained and invited to a banquet, only to be served human flesh secretly mixed in with the other meats. Infuriated by this sacrilege, Zeus overturned the table and began hurling thunderbolts right and left, destroying the King's palace and sending him and his 50 sons fleeing for their lives. Screaming as he ran, Lyacon soon found that he was unable to utter any sounds except a howl, and as he continued to run, he was transformed into a wild beast with long fangs and shaggy hair. Falling to all fours, Lyacon bared his fangs and attacked his own sheep.

Although the Greeks were not very specific about the type of animal, they did visualize this constellation as some kind of wild animal which had been speared by Centaurus and being carried, still impaled, to the altar for a sacrifice. By the seventeenth century, this group of stars had become widely accepted as representing a wolf, and the publication of Johannes Bayer's catalog of constellations cemented this form.

TO FIND: Most observers in the northern hemisphere can see at least part of Lupus. All those below latitude 60°north can see at least some of the constellation, those south of 35°north can see all of it, but the southern parts will be very low for those north of 25°N. Northern hemisphere observers can find Lupus by starting at Scorpius. Second magnitude δ Scorpii is the brightest star in the north–south arc 7° west of Antares. From δ, 3rd magnitude π Sco is about 3.5° south (PA 188°), and somewhat fainter ρ Sco is about 3° southwest (PA 196) of π. Continuing the gentle curve to the southwest for 4.5° brings you to nearly equally bright χ Lupi, the tip of the wolf's nose. Fourth magnitude η Lupi, marking the top of the wolf's head lies 5° southeast of χ. From η Lupi you can trace the back of the wolf's neck by going about 6° southwest to ω Lupi, then along the wolf's back another 8° in almost the same direction to κ Lupi. The tail then extends a little over 3° directly south to end at ζ Lupi. Returning to the top of the wolf's head, the ears extend from η about 2° northeast to fourth magnitude φ Lupi. To trace out the underside of the wolf, start again with χ Lupi, the tip of the nose. From there, just 6° to the west–southwest, the bottom of the wolf's jaw is marked by a pair of 5th magnitude stars, ψ¹ and ψ². From there, the throat is defined by a line to γ Lupi,

7°to the south–southwest. 2.6° west–northwest of γ, δ Lupi defines the wolf's shoulder, with one paw ending at φ¹ 4.4° to its north, and the other at β, 5° west–southwest of δ. The wolf's underside is defined by a line from δ 5° south–southwest to λ, and another 2° in almost the same direction to π Lupi. The wolf's hind legs extend west from π Lupi 4° to α and 3.5°to τ¹, and 5° west–southwest to ρ and then 4.6° to ι Lupi (Figs. 6.37–6.38).

KEY STAR NAMES:

Name	Pronunciation	Source
α Kakkab	**kak**-kab	kakkab (Arabic), "guide star"
γ Thusia	thu-**see**-uh	thusía theríou, θυσια θησου (Greek), "sacrificial animal"
δ Hilasmus	hi-las-**mus**	ilasmoz, ιλαμός (Greek) "the propitiation or appeasement"

Magnitudes and spectral types of principal stars:

Bayer desig.	Mag. (m$_v$)	Spec. type	Bayer desig.	Mag. (m$_v$)	Spec. type	Bayer desig.	Mag. (m$_v$)	Spec. type
α	2.30	B1.5 III	β	2.68	B2 III	γ	2.80	B2 IV
δ	3.22	B1.5 IV	ε	3.37	B1.5 V	ζ	3.41	G8 III
η	3.42	B2.5 V	θ	4.22	B2.5 V	ι	3.55	B2.5 IV
κ¹	3.88	B9 V	κ²	5.70	A 3 IV	λ	4.07	B3 V
μ	4.27	B8 V	ν¹	4.99	F8 V	ν²	4.83	G2 V
ξ¹	5.14	A3 V	ξ²	5.59	B9 V	o	4.32	B5 IV
π	3.91	B5 IV/V	ρ	4.05	B5 V	σ	4.44	B2 III
τ¹	4.56	B2 IV	τ²	4.33	F7 C	υ	5.336	A0:pSi
φ¹	3.57	K5 III	φ²	4.54	B4 V	χ	3.97	B9.5 III/IV
ψ¹	4.66	G8 III	ψ²	4.75	B5 V	ω	4.34	K4.5 III
a	5.39	K0 III	b	5.22	G6 III	c	5.38	A2 III
d	4.55	B3 IV	e	4.83	Bs3 IV	g	4.64	F5 IV/V
h	5.23	G8/K0 III	k	4.60	A0 V			

Stars of magnitude 5.5 or brighter which have no Bayer or Flamsteed designations:

Designation	Mag. (m$_v$)	Spec. type	Designation	Mag. (m$_v$)	Spec. type
HD 125442	4.78	F0 IV	HD 126983	5.38	A1 V
HD 132955	5.45	B3 V	HD 133340	5.13	G8 III
HD 134270	5.45	G2 Ib/II	HD 135345	5.15	G5 Ia
HD 137432	5.46	B5 V	HD 137709	5.26	K4 III
HD 138186	5.44	M0 III	HD 143009	4.99	G8 III
HD 143699	4.90	B6 III/IV			

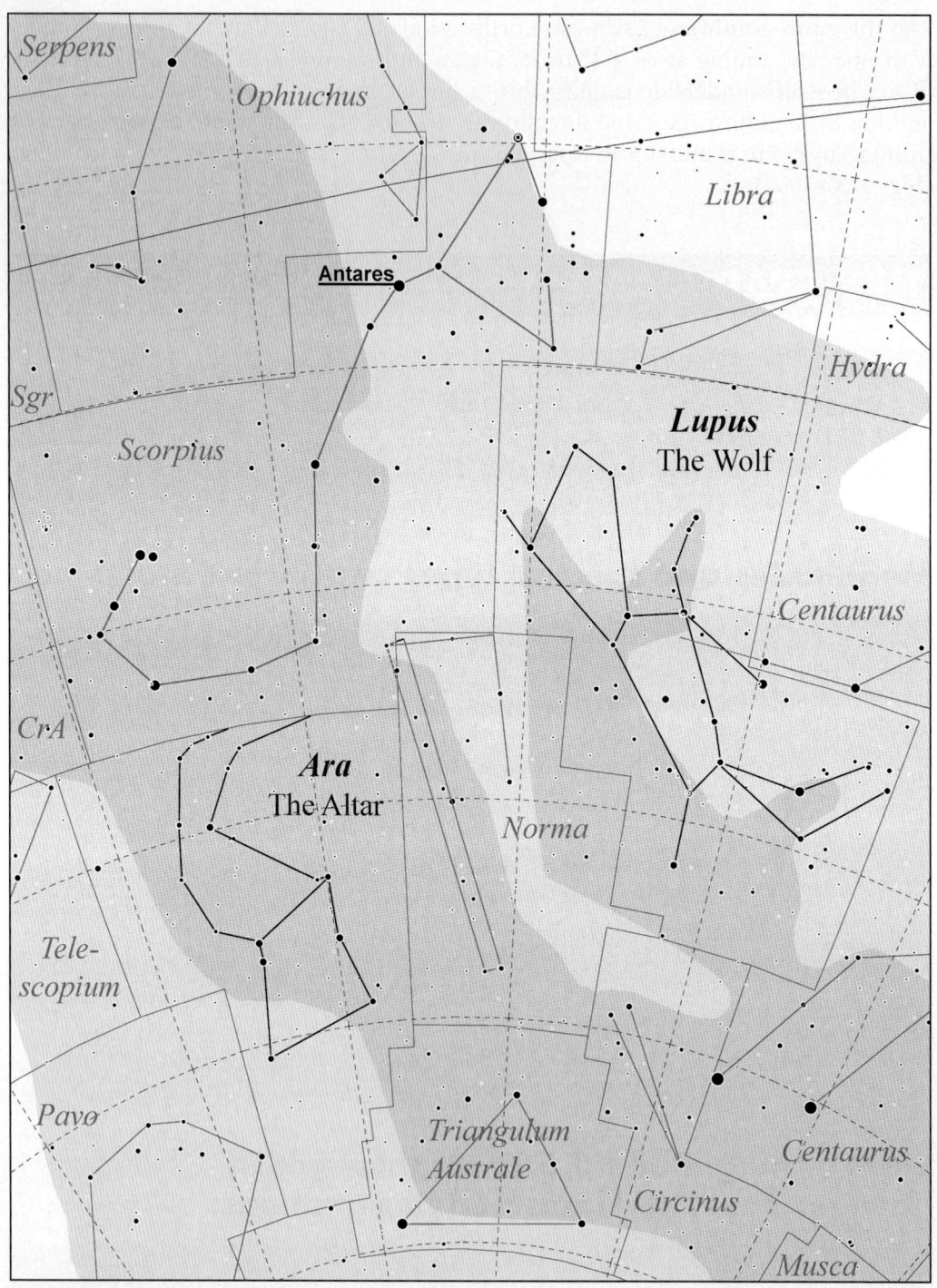

Fig. 6.37. Lupus and Ara finder chart.

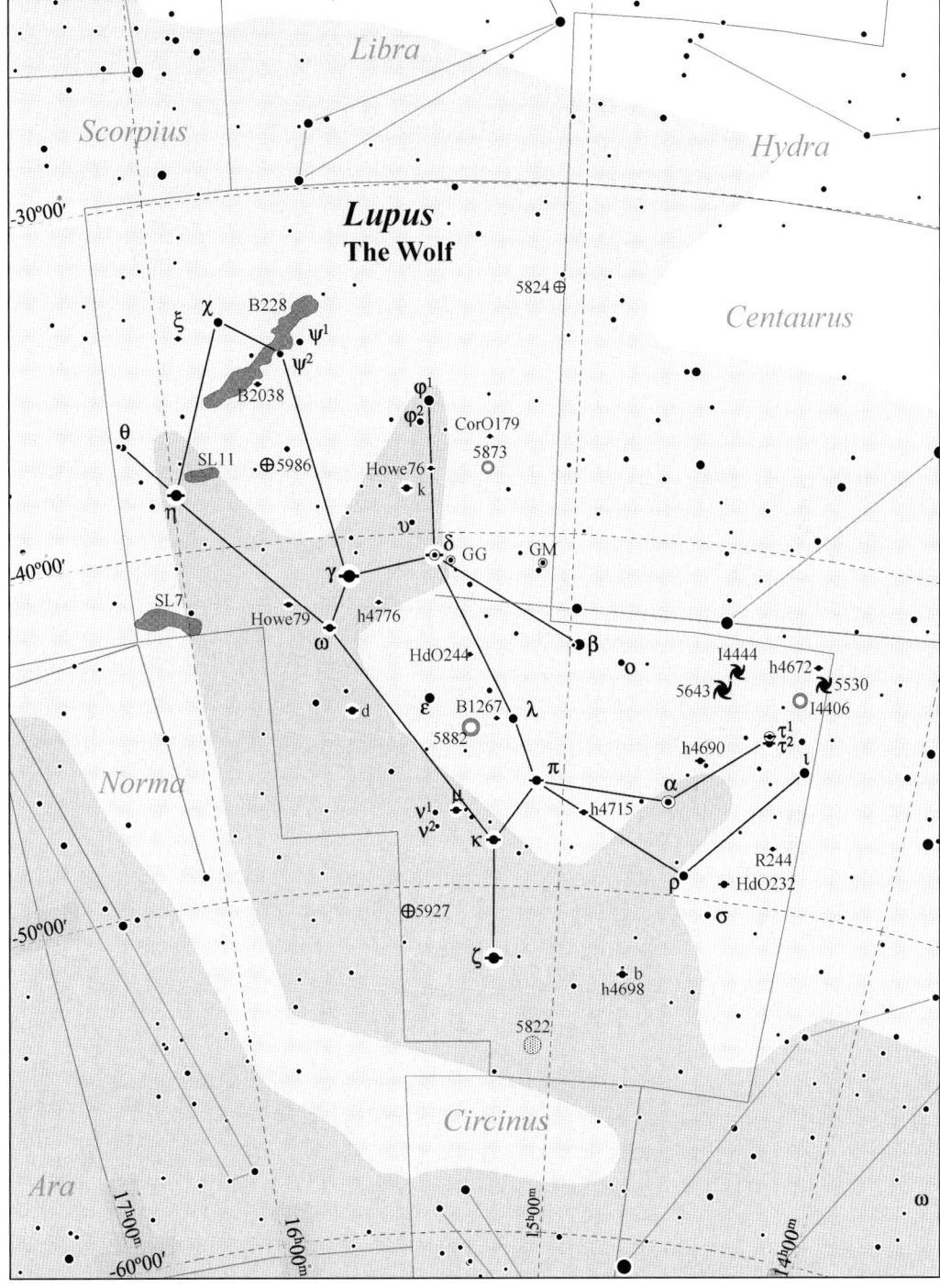

Fig. 6.38. Lupus details.

Table 6.6. Selected telescopic objects in Lupus.

Multiple stars:

Designation	R.A.	Decl.	Type	m₁	m₂	Sep (")	PA (°)	Colors	Date/Period	Aper. Recm.	Rating	Comments
h4672	14ʰ 20.2ᵐ	−43° 04'	A–B	5.8	7.9	3.5	301	Y, ?	1991	2/3"	****	Nice, easy pair for any binoculars and those with excellent eyesight.
R244	14ʰ 22.6ᵐ	−48° 19'	A–B	6.1	9.5	4.5	122	bW, W	1991	2/3"	****	τ¹ is also a small range variable. a.k.a. Dun 160.
τ¹ and ² Lupi	14ʰ 26.1ᵐ	−45° 13'	AB–CD	4.6	5.0	569.0	177	W, W	2002	eye	****	
τ¹	14ʰ 26.1ᵐ	−45° 13'	A–B	4.5	8.9	157.0	205	bW, W	**26.2**	7×50	****	
τ²	14ʰ 26.2ᵐ	−45° 23'	C–D	4.9	5.6	**0.2**	**183**	W, yW	–	–	****	Rapid orbital motion. Extremes: 0.24" in 2008, 0.008" in 2021. a.k.a. I 402
HdO232	14ʰ 30.3ᵐ	−49° 31'	A–B	5.5	12.0	24.0	019	W, bW	2000	3/4"	***	
h4690	14ʰ 37.3ᵐ	−46° 08'	A–B	5.6	7.7	19.6	026	O, bW	2002	15×80	****	Lovely colors. Aa has a separation of ~0.2" a.k.a. h 4698
b Lupi	14ʰ 47.0ᵐ	−52° 23'	A–B	5.2	13.0	8.9	260	dY, W	2000	6/7"	***	
h4715	14ʰ 56.5ᵐ	−47° 53'	A–B	6.0	6.8	2.2	278	W, ?	1991	4/5"	****	
π Lupi	15ʰ 05.1ᵐ	−47° 03'	A–B	4.6	4.6	1.7	066	bW, bW	2002	5/6"	****	Beautiful matched pair, requires medium power, ~200x. a.k.a. h 4728
B 1267 Δ 178	15ʰ 11.6ᵐ	−45° 17'	A–B	6.5	9.6	1.2	314	yO, ?	1991	8/9"	***	
			A–C	6.5	7.3	30.9	258	yO, W	1999	10×50	****	
κ Lupi	15ʰ 11.9ᵐ	−48° 44'	A–B	3.8	5.5	28.9	143	yO, W	2002	10×50	*****	Showcase pair. a.k.a. Δ 177
ζ Lupi	15ʰ 12.3ᵐ	−52° 06'	A–B	3.5	6.7	71.8	249	dY, rY	1999	6×30	*****	a.k.a. Δ 176
CorO 179	15ʰ 13.0ᵐ	−37° 15'	A–B	8.0	8.1	6.4	228	W, W	2000	2/3"	******	Nearly identical pair
HdO 244	15ʰ 15.3ᵐ	−44° 09'	A–B	6.7	9.6	13.9	040	W, ?	1999	15×80	****	
μ Lupi Δ 180	15ʰ 18.5ᵐ	−47° 53'	A–B	5.0	4.9	1.0	323	bW, bW	1999	9/10"	****	a.k.a. h 4753. AB is loosely matched pair, very similar to π Lupi but slightly closer and fainter. AB–C is easy in small telescopes
			A–C	5.0	6.3	22.7	127	bW, Y	2002	10×50	****	
δ Lupi	15ʰ 21.3ᵐ	−40° 39'	A–B	3.2	6.2	396.0	158	bW, W	–	6×30	****	Some with excellent eyesight may see with unaided eye
Howe 76	15ʰ 21.5ᵐ	−38° 13'	A–B	6.6	9.3	5.7	121	W, W	1999	2/3"	****	
h 4776	15ʰ 30.4ᵐ	−41° 55'	A–B	6.3	8.4	5.6	229	W, ?	1999	2/3"	***	
γ Lupi	15ʰ 35.1ᵐ	−41° 10'	A–B	3.0	4.5	**0.8**	**277**	bW, W	**190**	11/12"	****	Nearing max. of 0.827" in 2013, then min. of 0.031" in 2068. a.k.a. h 4786
d Lupi	15ʰ 35.9ᵐ	−44° 58'	A–B	4.7	6.5	2.1	009	W, pY	1005	4/5"	****	aka h 4788
ω Lupi	15ʰ 38.1ᵐ	−42° 34'	A–B	4.3	11.2	12.7	031	yO, ?	2002	3/4"	***	aka Hdo 250
Howe79	15ʰ 44.3ᵐ	−41° 49'	A–B	6.1	7.9	3.2	338	Y, Y	1999	3/4"	****	
B 2038 Δ192	15ʰ 47.1ᵐ	−35° 31'	A–B	7.0	8.9	0.6	063	W, ?	1991	16"	****	
			AB–C	7.0	7.3	34.7	143	W, ?	2998	7×50	***	

ξ Lupi	15h 56.9m	−33° 58′	A–B	5.1	5.6	11.7	049	pY, pGr	2002	2/3″	****	Nice, easy pair for small scope. Colors are very subtle. a.k.a. Pz 4
η Lupi	16h 00.1m	−38° 24′	A–B	3.4	7.5	15.5	019	bW, ?	2002	15×80	****	Nice, easy group for small telescopes. a.k.a. Rmk 21AB
			A–C	3.4	9.3	115.0	249	bW, ?	2002	7×50	***	

Variable stars:

Designation	R.A.	Decl.	Type	Range (mv)		Period (days)	F (fₑ/fᵢ)	Spectral Range	Aper. Recm.	Rating	Comments
τ¹ Lupi	14h 26.5m	−45° 15′	Cb	4.5	4.6	0.177	–	B2IV	Photom	***	Bright Cepheid with rapid, variations too small for eye measurements
α Lupi	14h 42.3m	−47° 25′	Cb	2.3	2.3	0.260	–	B1.5III	Photom	***	Bright Cepheid with rapid, variations too small for eye measurements
GM Lupi	15h 05.1m	−40° 53′	Lb	6.4	6.7	–	–	M6III	6×30	****	14′ NNW (PA 331°) of m5.1 SAO 225435
GG Lupi	15h 18.9m	−40° 47′	Eb/DM	5.5	6.0	2.164	–	B7V	6×30	****	Comp. very close and gravitationally distorted. 29′ WSW (PA 253°) of δ Lupi
δ Lupi	15h 21.7m	−40° 40′	Cb	3.2	3.2	0.165	0.50	B1.5IV	Photom	***	Bright Cepheid with rapid, small variations, photometric object only

Star clusters:

Designation	R.A.	Decl.	Type	mv	Size (′)	Brtst. Star	Dist. (ly)	dia. (ly)	Aper. Recm.	Rating	Comments
NGC 5824	15h 04.4m	−33° 04′	Globular	9.1	7.4	15.5	106K	225	12/14″	***	Small, bright core, a few stars resolved at edge
NGC 5822	15h 05.2m	−54° 21′	Open	6.5	35	9.1	2375	5	7×50	****	
NGC 5927	15h 28.0m	−50° 40′	Globular	8.0	6.0	14.5	26K	45	7×50	****	Small, bright core, slightly larger halo than NGC 5824, not resolved
NGC 5986	15h 46.1m	−37° 47′	Globular	7.6	9.6	13.2	34K	95	7×50	****	Moderately concentrated, edge well resolved in 12/14″

(continued)

Table 6.6. (continued)

Nebulae:

Designation	R.A.	Decl.	Type	m_v	Size (')	Brst./ Cent. star	Dist. (ly)	dia. (ly)	Aper. Recm.	Rating	Comments
IC 4406	14^h 22.4^m	–44° 09'	Planet	10.2	28"	**14.7**	4.9K	0.7	8/10"	****	Slightly elongated N–S, irregular brightness
NGC 5873	15^h 12.8^m	–38° 08'	Planet	11.0	3"	**15.6**	16K	0.2	8/10"	***	Nearly stellar appearance
NGC 5882	15^h 16.8^m	–45° 39'	Planet	9.4	7"	**13.6**	6.2K	0.5	8/10"	***	Very small, round
Bernard 228	15^h 45.5^m	–37° 47'	Dark	–	240×20	–	–	–	7×50	****	NW–SE dust lane, near ψ^1 Lupi
SL 11	15^h 57.0^m	–37° 48'	Dark	–	40×5	–	–	–	7×50	***	
SL 7	16^h 01.8^m	–41° 52'	Dark	–	160×10	–	–	–	7×50	***	

Galaxies:

Designation	R.A.	Decl.	Type	m_v	Size (')	Dist. (ly)	dia. (ly)	Lum. (suns)	Aper. Recm.	Rating	Comments
NGC 5530	14^h 18.5^m	–43° 23'	Sb	11.2	4.2×1.9	8M	98K	170M	6/8"	***	Faint, elongated. Patchy arms
IC 4444	14^h 31.7^m	–43° 25'	S(B)b	11.4	1.7×1.3	77M	38K	13G	6/8"	***	Faint, small. Thick arms. Stellar nucleus
NGC 5643	14^h 32.7^m	–44° 10'	S(B)c	10.2	4.6×4.0	42M	93K	57G	4/6"	****	Diffuse halo, gradually brighter toward center

Meteor showers:

Designation	R.A.	Decl.	Period (yr)	Duration	Max Date	ZHR (max)	**Comet/ Asteroid**	First Obs.	Vel. (mi/ **km**/sec)	Rating	Comments
None											

Other interesting objects:

Designation	R.A.	Decl.	Type	m_v	Size (')	Dist. (ly)	dia. (ly)	Lum. (suns)	Aper. Recm.	Rating	Comments
None											

Fig. 6.39. NGC 5643, spiral galaxy in Lupus. Drawing by Eiji Kato, 18.5" telescope.

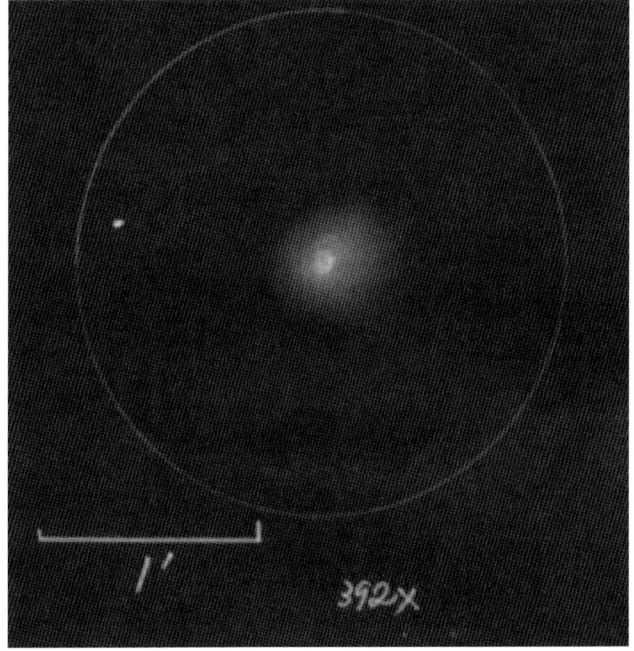

Fig. 6.40. NGC 5824, open cluster in Lupus. Drawing by Eiji Kato.

ARA (air-uh)	**The Altar**
ARAE (air-eye)	**9:00 P.M. CULMINATION**: 26 July
Ara	**AREA**: 237 sq. deg., 63rd in size

The constellation Ara represents the altar upon which the 12 Olympian gods and goddesses (Zeus and his brothers and sisters) swore allegiance to Zeus as leader and to the struggle to overthrow Chronus and his brothers and sisters (the 12 Titans). After a 10-year war with no clear advantage to either side, Gaea (mother earth) instructed Zeus to free the Hecatonchieires (hundred-handed Titans) and the Cyclopes (one-eyed Titans) who had been imprisoned in Tartarus by Chronus.

Grateful to the Olympians, the Cyclopes made weapons for the Olympians. They built a forge on which they fabricated the devastating thunderbolts for Zeus, the helmet of invisibility for Hades, and the trident for Poseidon. Over the forge, they built a cover to hide the activity from Chronus.

The Hecatonchieires then helped the Olympians fight the other Titans, while the Cyclopes continued to turn out the thunderbolts with which Zeus terrified his enemies. Their help turned the tide, and the Olympians defeated the Titans and ruled the universe. To the Greeks, this victory symbolized the triumph of intellect over brawn.

After their foes were defeated, the gods divided up the regions of influence. By a game of chance, Zeus won control of the heavens, Poseidon control of the seas, and Hades control of the underworld. The surface of the earth was neutral territory. The other gods and goddesses received influence over various aspects of nature and human activities. As leader of the rebellion, Zeus became the overall leader. It was he who placed the altar in the heavens (without the protective cover), both in gratitude for their win, and as a reminder to humans that they needed to make their sacrifices to the gods before important undertakings in order to insure their help.

The Greeks visualized the altar as standing above the southern horizon, with the smoky stream of the summer Milky Way rising from the sacrifice burning on top of the altar. Johannes Bayer inverted the Altar in his 1603 atlas, and many subsequent atlases have followed suit. Here, the altar is shown in its original position.

Like some other constellations in the area, parts of Ara were not visible to Bayer, and so many of his star positions and letter assignments were erroneous. Consequently, Lacaille reassigned the Greek letters based on newer and more accurate charts.

TO FIND: The two southernmost lettered stars in Scorpius are ζ and η Scorpii. They lie just under 5°apart, in a nearly east–west line. α Arae lies about 7° below this pair. α Arae is the brightest star in the smoke rising from the altar as well as the whole constellation. Another 6° south lie the pair β and γ Arae. They are less than 1° apart and represent the top left of the altar as viewed from the northern hemisphere. ε Arae is about 4° nearly due west of this pair, and represents the top right of the altar. δ Arae lies about 4° south of β and γ, and η lies about 3° south of ε. δ and η Arae form the bottom of the altar. Observers above 40° north latitude will not be able to see any of Ara's brighter stars, including α (Fig. 6.41).

KEY STAR NAMES: None

Magnitudes and spectral types of principal stars:

Bayer desig.	Mag. (m_v)	Spec. type	Bayer desig.	Mag. (m_v)	Spec. type	Bayer desig.	Mag. (m_v)	Spec. type
α	2.84	B2 Vne	β	3.84	K3 Ib-II	γ	3.31	B1 Ib
δ	3.60	B8 V	ε¹	4.06	K4 III	ε²	5.27	F6 V
ζ	3.12	K5 III	η	3.77	K5 III	θ	3.65	B2 Ib
ι	5.21	B2 IIIne	κ	5.19	K1 III	λ	4.76	F3 IV
μ	5.12	G3 IV/V	ν¹	6.58	B2 V	ν²	6.09	B9.5 III/IV
ξ	–	–*15	o	–	–*15	π	5.25	A7 V
ρ	6.30	B3 Vnpe	σ	4.56	A0 V			

Stars of magnitude 5.5 or brighter which have Flamsteed but no Bayer designations:

Flamsteed	Mag. (m_v)	Spec. type
41	5.47	G8 V

Stars of magnitude 5.5 or brighter which have no Bayer or Flamsteed designations:

Designation	Mag. (m_v)	Spec. type
HD 157661	5.28	B8 V

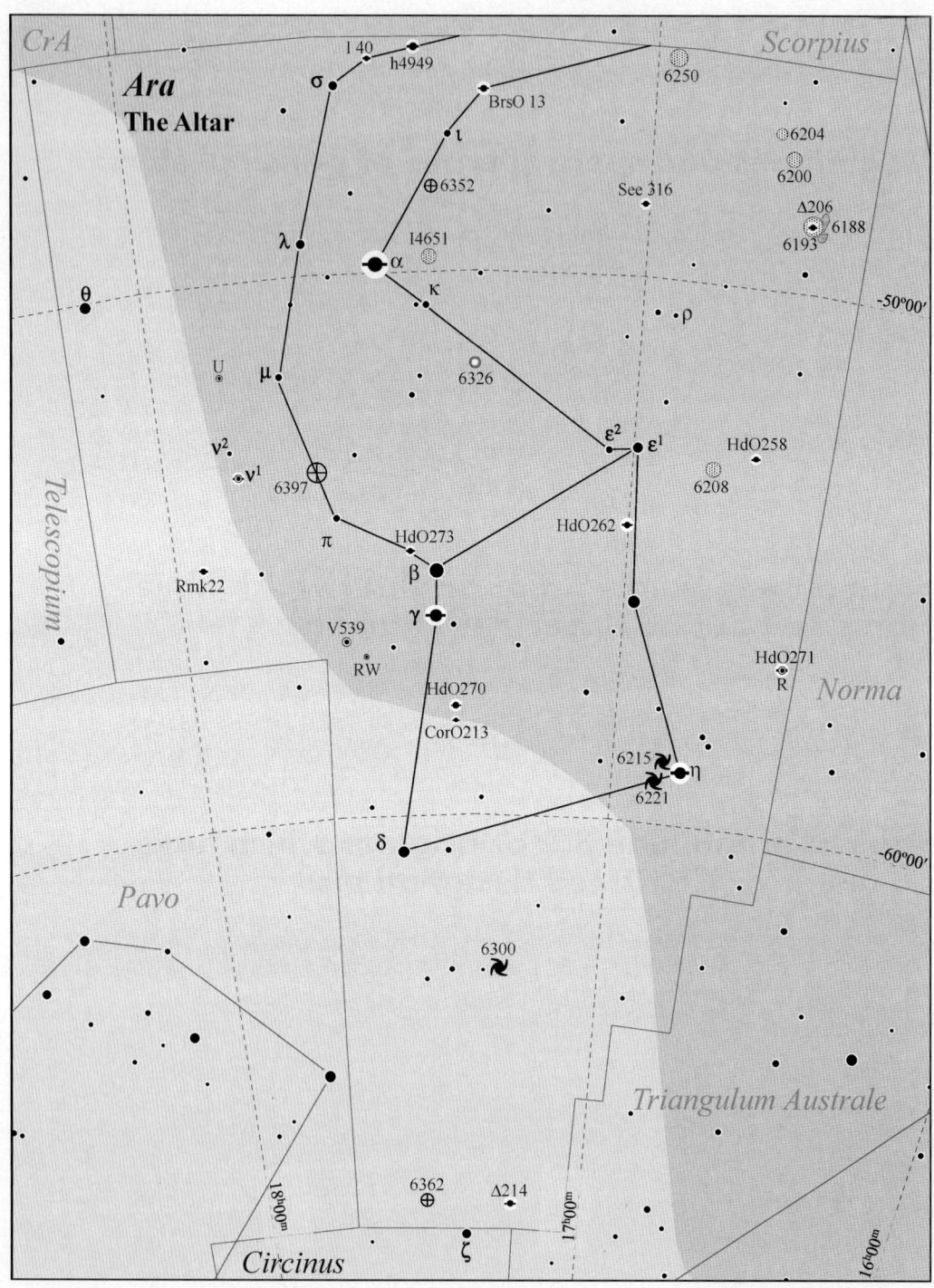

Fig. 6.41. Ara details.

Table 6.7. Selected telescopic objects in Ara.

Multiple stars:

Designation	R.A.	Decl.	Type	m_1	m_2	Sep (")	PA (°)	Colors	Date/ Period	Aper. Recm.	Rating	Comments
R Arae	16ʰ 39.7ᵐ	−57° 00'	A–B	7.2	7.8	3.3	120	Y, Y	1999	2/3"	****	a.k.a. Hdo 271A is the eclipsing variable R Ara
Δ 206	16ʰ 41.3ᵐ	−48° 46'	A–B	5.8	8.5	1.6	015	yW,	1938	6/7"	*****	Remarkable multiple. Almost a small cluster. a.k.a. BSO 13-AB
			A–C		6.8	9.5	265	pB	2002	2/3"	****	a.k.a. BSO 13-AC R Arae is embedded in NGC
			A–D		10.5	13.3	160	yW, ?	1999	2/3"	***	2193, with NGC 2188 just W of
			A–E		11.4	14.0	015	yW, ?	1938	3/4"	***	a.k.a. HJ 4876-AD
			A–F		12.5	30.0	190	yW, ?	1999	4/5"	***	a.k.a. HJ 4876-AE
								yW, ?				a.k.a. I 96-AF
HdO 258	16ʰ 44.7ᵐ	−53° 09'	A–B	6.0	12.0	40.0	045	oY, ?	1900	3/4"	***	
Rst 5067	16ʰ 49.8ᵐ	−59° 02'	A–B	3.8	13.5	23.6	119	O, ?	1967	7/8"	***	Difficult because of brightness difference ~ 9.7 mag. or 7,500×!
HdO 262	17ʰ 00.1ᵐ	−54° 36'	A–B	5.7	12.0	25.0	071	W, ?	1900	3/4"	***	
See 316	17ʰ 00.4ᵐ	−48° 39'	A–B	6.3	7.7	1.0	172	dY, ?	1996	9/10"	****	
Δ 214	17ʰ 13.3ᵐ	−67° 12'	A–B	6.0	8.8	37.4	014	yO, ?	2000	7×50	****	~ 1° NW of ζ Apodis
BrsO 13	17ʰ 19.1ᵐ	−46° 38'	A–B	5.6	8.9	**9.9**	**256**	O, ?	**693.24**	2/3"	****	*16
			A–C	5.6	13.3	41.8	279	O, ?	1900	7/8"	****	
			A–D	5.6	14.0	47.0	030	O, ?	1900	9/10"	***	
CorO 213	17ʰ 22.9ᵐ	−58° 28'	A–B	6.9	9.34	9.2	284	W, ?	2000	2/3"	****	
HdO 270	17ʰ 22.9ᵐ	−58° 01'	A–B	6.0	9.41	1.8	183	yO, ?	1991	5/6"	****	
ι Arae	17ʰ 23.3ᵐ	−47° 28'	A–B	5.5	10.8	45.5	015	bW, ?	2002	3/4"	****	A is the eclipsing variable V539 Ara. aka HdO 271
γ Arae	17ʰ 25.4ᵐ	−56° 23'	A–B	3.3	10.2	18.0	328	bW, ?	2002	2/3"	****	
			A–C		12.2	42.9	065	bW, ?	2002	4/5"	***	
h 4949 Δ 216	17ʰ 26.9ᵐ	−45° 51'	A–B	5.6	6.46	2.1	253	W, ?	1991	4/5"	****	a.k.a. Δ 216
			AB–C	5.2	7.12	102.4	312	W, ?	1999	6×30	****	
HdO 273	17ʰ 28.7ᵐ	−55° 10'	A–B	5.9	12.5	37.5	148	oY, ?	2000	4/5"	***	
I 40	17ʰ 31.8ᵐ	−46° 02'	A–B	6.0	10.5	20.1	209	Y, ?	1999	2/3"	****	
α Arae	17ʰ 31.8ᵐ	−49° 52'	A–B	2.8	11.1	49.7	168	W, ?	2000	3/4"	****	
v¹ Arae	17ʰ 50.5ᵐ	−53° 37'	A–B	5.7	9.2	12.1	268	bW, ?	2000	2/3"	****	a.k.a. h4978
Rmk 22	17ʰ 57.2ᵐ	−55° 23'	A–B	7.0	7.9	2.4	094	yW, ?	1991	4/5"	****	

(continued)

Table 6.7. (continued)

Variable stars:

Designation	R.A.	Decl.	Type	Range (m_v)		Period (days)	F (f_r/f_f)	Spectral Range	Aper. Recm.	Rating	Comments
R Arae	16h 39.7m	−57° 00'	EA/DM	6.0p	6.9p	4.425	0.09	B9IV–V	7×50	****	R is 2° 11' SSW (PA 326°) of η. #
RW Ara	17h 34.8m	−57° 09'	EA/SD	8.9	11.45	4.367	0.13	A1IV+K3III	3/4"	***	1° 30' ESE (PA 122°) of γ Arae (m3.31). #
V535 Ara	17h 38.0m	−56° 49'	EW/DW	7.2	7.75	0.629	–	A8V	6×30	*****	1° 48' ESE (PA 106°) of γ (m3.31). #) Entire cycle can be observed in one winter night
V539 Ara	17h 50.5m	−53° 37'	ED/AM	5.66	6.18	3.169	0.14	B2V+B3V	7×50	****	Equidistant from π Arae [2° 01' SSE (PA 152°, PA 152)] and μ [2° 01' ENE (PA 065°, m5.12)]
U Arae	17h 53.6m	−51° 41'	M	8.4	13.6	225.21	0.44	M3e M5e	10/12	***	Much of period can be observed in binoculars. 0.5°E (PA 85°) of μ Arae. ##

*17

Star clusters:

Designation	R.A.	Decl.	Type	m_v	Size (')	Brst. Star	Dist. (ly)	dia. (ly)	Number of Stars	Aper. Recm.	Rating	Comments
NGC 6193	16h 41.3m	−48° 47'	Open	5.2	20	5.7	4,300	25	30	7×50	****	Detached, weak conc. Lum. ~12K suns. Assoc. w NGC 6188, a faint mixed nebula
NGC 6200	16h 44.2m	−47° 29'	Open	7.4	15	9.2	7,400	32	40	4/6"	***	Luminosity, ~5K suns
NGC 6204	16h 46.5m	−47° 01'	Open	8.2	6	9.7	3.9K	6.8	45	4/6"	***	
NGC 6208	16h 49.5m	−53° 49'	Open	7.2	16	10.0	3.3K	15	60	4/6"	***	Just barely resolved in 7×50 binoculars. Lge scope, long exp. show >300 stars
NGC 6250	16h 58.0m	−45° 48'	Open	5.9	16	7.6	3.3K	15	15	4/6"	***	~15 stars, sparse, but beautiful binocular/RFT background
IC 4651	17h 24.6m	−49° 56'	Open	6.9	10	10.0	2.4K	7	80	7×50	***	Detached, weak concentration. Resolvable in large binoculars. Age ~2.4G years
NGC 6352	17h 25.5m	−48° 25'	Globular	7.8	9	13.4	18K	47	–	4/6"	***	Luminosity ~24K suns
NGC 6362	17h 31.9m	−67° 03'	Globular	8.1	15	12.7	23K	100	–	4/6"	***	Luminosity ~31K suns
NGC 6397	17h 40.7m	−53° 40'	Globular	5.3	30	10.0	7.2K	63	–	7×50	****	Resolved in 3" scope, probably closest globular cluster. Larger and slightly brighter than M13 but too far south for most northern hemisphere observers

Nebulae:

Designation	R.A.	Decl.	Type	m_v	Size (')	Brtst./ Cent. star	Dist. (ly)	dia. (ly)	Aper. Recm.	Rating	Comments
NGC 6188	16ʰ 40.5ᵐ	–48° 47'	Em + Refl.	11	20×12	–	–	–	7×50	****	Bright and dark nebulae, similar to Horsehead; difficult
NGC 6326	17ʰ 20.8ᵐ	–51° 45'	Planet	12	14"	15.5	7.8K	0.5	8/10"	***	Small but pretty bright

Galaxies:

Designation	R.A.	Decl.	Type	m_v	Size (')	Dist. (ly)	dia. (ly)	Lum. (suns)	Aper. Recm.	Rating	Comments
NGC 6215	16ʰ 46.8ᵐ	–58° 54'	SB(s)c	10.9	1.9×1.6	84M	46K	25G	8/10"	***	~0.4° E of η Arae (m3.76), which may make 6215 harder to see
NGC 6221	16ʰ 48.4ᵐ	–59° 08'	Sc	10.6p	3.9×2.2	83M	94K	32G	8/10"	***	~0.4° ESE of η Arae (m3.76), which may make 6221 harder to see
NGC 6300	17ʰ 12.3ᵐ	–62° 46'	Scp	10.5p	5.2×2.6	47M	71K	11G	8/10"	***	~2.5° SW of δ Arae

Meteor showers:

Designation	R.A.	Decl.	Period (yr)	Duration	Max Date	ZHR (max)	Comet/ Asteroid	First Obs.	Vel. (mi/ km/sec)	Rating	Comments
None											

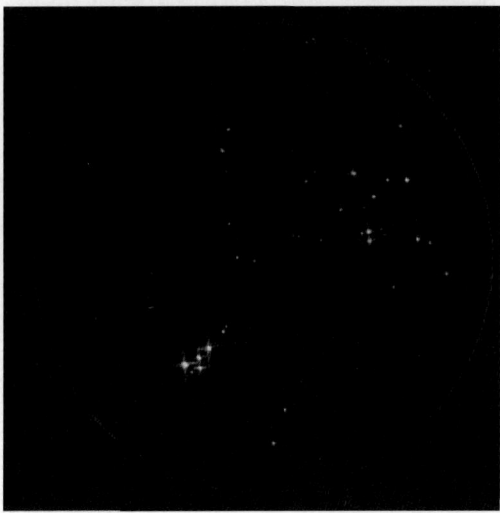

Fig. 6.42. NGC 6204, open cluster in Ara. Drawing by Eiji Kato.

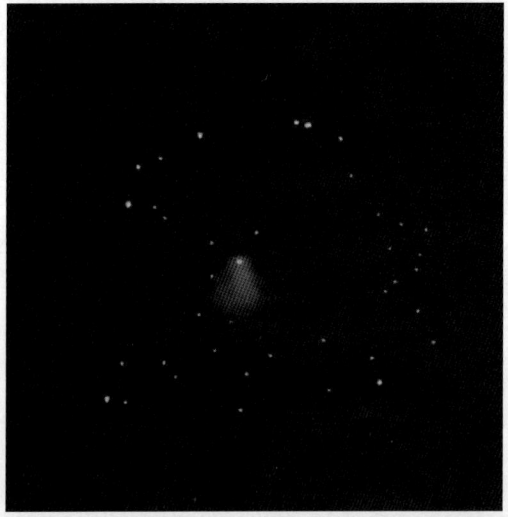

Fig. 6.43. NGC 6215, spiral galaxy in Ara. Drawing by Eiji Kato.

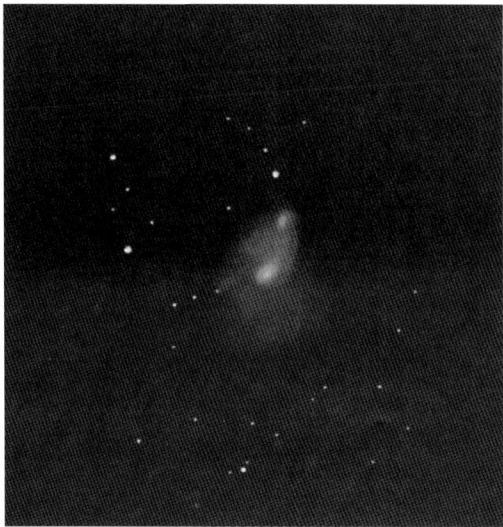

Fig. 6.44. NGC 6221, spiral galaxy in Ara. Drawing by Eiji Kato.

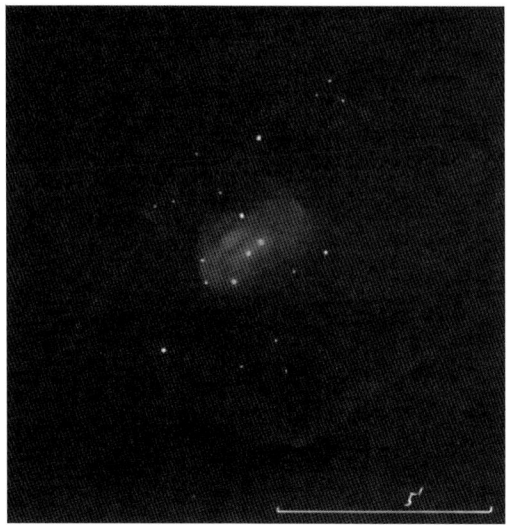

Fig. 6.45. NGC 6300, spiral galaxy in Ara. Drawing by Eiji Kato.

NOTES:

1. The Hipparcho–Ptolemy catalog description of Heracles' right arm ends at the elbow. However, several later atlases show him holding a large club, so it is included on the chart but in fainter lines than the rest of the figure. Similarly, the catalog description of his left arm ends at the wrist but again, later atlases show him holding various things, e.g., a cluster of the golden apples of the Hesperides or an animal associated with his labors. These charts show him holding the multiheaded Hydra, again in fainter lines than those of the main figure. In addition, this drawing differs from the Hipparcho–Ptolemy constellation in that it lacks a right foot. The stars forming his missing foot (originally ψ^1 and ψ^2 Herculi) were originally considered part of two constellations and became v^1 and v^2 Boötes when the IAU officially defined the constellations.
2. Bayer assigned a single letter to these two stars. See Note 1.
3. ζ Her is one of the most rapidly changing binaries which can be resolved in amateur telescopes. It passed its minimum separation of 0.50″ in mid-2001, and will reach its maximum of 1.53″ in mid-2025. Contrast this with nearby α Her (Ras Algethi), whose period is over 100 times longer.
4. NGC 6166 is a supergiant elliptical galaxy. It is the central and largest galaxy in the galaxy cluster Abell 2199, and one of the largest known in the universe. The distance to Abell 2199 is estimated to be 440 to 600 million light-years, which give the calculated diameter of NGC 6166 between 1.5 and 2.0 million light-years. NGC 6166 has three distinct cores, indicating that it is probably the result of the merger of several smaller galaxies. NGC 6166 has a central black hole estimated to be 800 million to 1 billion solar masses and jets of relativistic particles emerging from the galaxy center.
5. Wolf 359 is near the lower end of the enormous range of luminosity found among the stars. Although it is the third closest star (if the α Centauri system is counted as one) to our solar system, it is visible only in an 8″ or larger telescope. From a planet at the distance of the earth from our sun, Wolf 359 would appear only about ten times brighter than our full moon. Wolf 359 has a mass of $\sim 0.10 \pm 0.03$, and a diameter of $\sim 0.165 \pm 0.015$ times that of our sun. Wolf 359's high proper motion carries it across the sky at the rate of one moon diameter every 400 years. Wolf 359 lies 50′ N (PA 008°) of 56 Leo (m5.91) but is so faint that a detailed finder chart or very accurate settings are needed to find it. Like many red dwarf stars, Wolf 359 is subject to outbursts known as flares and is also known by the variable star designation of CN Leonis.
6. Ascellus Borealis and Acellus Australis, as well as Praesepe, are derived from an episode wherein Dionysus (the Greek god of grapes and wine) encountered two donkeys grazing near a swamp. He caught and rode one of them across the swamp to avoid getting wet. The second followed along. Dionysus then placed them in the sky along with a manger with a plentiful of food as a reward for their service.
7. The designation ε Cancri is now applied to either M44 as a whole as designated by Bayer, or to the brightest star in M44 which Flamsteed called 41 Cancri.
8. The stars in M67 are about the same age and chemical composition as our sun, consequently astronomers have extensively studied its ~100 stars similar in mass to our sun. NGC 188 in Cepheus is the only open cluster which appears to be older than M67.
9. Most of Centaurus was not visible to Bayer, so his letter assignments were largely based on inaccurate charts provided by untrained sailors. Lacaille reassigned the letters using the newer and more accurate charts available to him.
10. This globular cluster is clearly visible to the unaided eye as a fuzzy "star" and was assigned the designation of ω in spite of the fact that it could clearly be seen as nonstellar in binoculars or small telescopes.
11. The alpha Centauri system is the closest star system to our sun. It consists of α Cen A, α Cen B, and α Cen C, a faint red dwarf 2°11′ SSW (PA 213°) of the main pair. The AB pair is the brightest overall double and binary in the sky. Together, the pair has an apparent magnitude of −0.17 and rank as the 3rd brightest star in the sky. Individually, α Centauri A would be the 7th brightest, and α Centauri B 20th brightest star in the sky. Because of its short period, the position angle of this binary changes rapidly, and because of its closeness, the apparent separation varies rapidly. α Centauri C is also known as Proxima Centauri because it is slightly closer than the other two stars. It has a proper motion very similar to that of the AB pair and is considered to be a physical part of the α Centauri system, but orbital motion has not been established. The distance to the AB pair is 4.365 light-years, while that to Proxima is only 4.22 light-years. The 0.145 light-year difference is equivalent to over 9,000 astronomical units. If Proxima is orbiting about the AB pair, the orbital period must be millions of years.
12. Bayer considered the modern Centaurus and Crux to be one constellation and lettered the stars from α to ω, and then from A to Z and a to q. However, most of Centaurus and all of Crux was invisible from Europe,

and Bayer's letters did not match the actual configuration of stars. Lacaille re-lettered them from α to ω, and then from a to z and A to Q. Since this is the longest lettering of stars, when variable stars were lettered, the sequence started with R to avoid confusion with previously lettered nonvariable stars.

13. NGC 5128 is shaped like an elliptical galaxy but has a well-defined and unusual dust lane across its center. The dust lane is narrow in the middle and widens near the edges of the galaxy. Studies done with the Hubble Space Telescope indicate that NGC 5128 is a giant elliptical galaxy undergoing a collision with a smaller spiral galaxy. In addition, NGC 5128 has two relativistic jets along a line perpendicular to the dust lane. These jets apparently originate from an approximately one billion solar mass black hole at the center of the elliptical galaxy. The jets also give rise to a huge amount of radio and X-ray radiation. NGC 5128 is the 5th brightest galaxy in the sky and the nearest one with an active nucleus.

14. NGC 5253 has been classified as spiral, elliptical, irregular, and peculiar, and there is some justification for each of these categories. However recent research has identified a new class of relatively small galaxies called "Blue Dwarf Galaxies." NGC 5253 is one of the closest of these galaxies. It is composed of mostly gas clouds with relatively little dust, similar to what we believe is the composition of the material which formed the earliest galaxies in the universe. In spite of the apparent age of its material, NGC 5253 is undergoing a vigorous episode of star formation. Many of the star-forming regions appear large enough to create globular clusters similar to those formed in the Milky Way Galaxy billions of years ago.

15. Lacaille did not assign the letters ξ or o to any stars in Ara.

16. The binary star BSO 13AB has been observed for only a small fraction of its orbital period. Because of the limited observation time and measurement errors, finding an orbit which exactly fits the motion is difficult to do precisely, and different researchers have calculated different orbits. In addition to the 693.24-year orbital period used here, an orbit using slightly different inclinations and other parameters with a period of 2,204.98 years fits the present observations almost equally well. These two orbits give nearly the same results for 2010, 9.89″ separation and a position angle of 256.23° for the shorter period orbit and 10.48″ and 255.06° for the longer one. Although the predicted positions for the near term are similar, they gradually diverge with time. The first set of parameters leads to a prediction of maximum separation of 17.9″ in 2275, and the second to 43.0″ in 3081. The shorter period is taken for the purposes of this book.

17. R Arae is a binary pair exhibiting extra-stellar gas and mass transfer. R is west of, and forms a nearly equilateral triangle with, ζ and η Arae.

18. NGC 6215 and 6221 are connected by a wide, two-stranded bridge of matter over 100 kpc long. This mass of mostly neutral hydrogen was apparently torn out of the two galaxies when they collided about 500 million years ago. The bridge contains at least 300 million times the mass of the sun. In addition, there are three dwarf galaxies in this system, with the smallest one of the three almost directly between the two major galaxies, and probably formed from the matter ripped out of them during and after the collision.

CHAPTER SEVEN

Hospitality and Homicide

February, May–July

BOÖTES	CANIS MINOR	VIRGO
The Herdsman	The Small Dog	The Maiden

CANES VENATICI	CORONA BOREALIS	LIBRA
The Hunting Dogs	The Northern Crown	The Scales

The six constellations in this group are related by Greek origin myths, but not by a single myth. Boötes, Virgo, and Canis Minor are related by a single myth, but Libra is related to Virgo by an entirely different story, and Corona Borealis is related to Boötes by an independent myth which involves the same Greek god. A seventh constellation, Triangulum, is also related to Virgo by another myth, but is covered in Chap. 10 because it is also related to the voyage of the Argonauts.

When Dionysus was a young god trying to establish his position as the god of wine and vineyards, he went on a world tour to recruit followers. The origin of the constellations Boötes, Virgo, and Canis Minor are explained in the following tragic tale of too much wine.

As Dionysus was passing through the countryside near Athens, he came upon the cottage of Icarius, a herdsman and vineyard owner. Although they did not know their visitor was a god, Icarius, his wife Phanothea, and daughter Erigone welcomed and entertained him. After they had eaten and drunk grape juice, Icarius showed Dionysus his grape vines. Dionysus was so pleased by his welcome, and impressed by how carefully Icarius

P. Simpson, *Guidebook to the Constellations: Telescopic Sights, Tales, and Myths*,
Patrick Moore's Practical Astronomy Series, DOI 10.1007/978-1-4419-6941-5_7,
© Springer Science+Business Media, LLC 2012

had tended his vineyard, that he taught Icarius the art of wine making. Prior this time, no human had discovered how to make wine, so Icarius became the first vintner.

When his first batch of wine had matured, Icarius showed off his new skill by giving some of his wine to his fellow herdsmen. They were pleased by the new taste, and the relaxed and happy feeling that ensued. Since none of them had tasted wine before, they did not know what to expect, and kept drinking until they were so drunk that they all fell asleep. When they awakened, the men all had hangovers and thought that Icarius had tried to poison them. In a rage, they attacked and killed Icarius. Afterwards, realizing that they had committed murder, the men buried Icarius under a tree near his vineyard. When they returned to their homes, they claimed that a lion had killed Icarius and dragged his body away.

Since lions were common in that part of Greece at the time, nobody doubted them. However, they were undone by Maera, the faithful sheepdog of Icarius. Maera had been with the group of drinkers when the murder and burial had taken place. She led the grieving Erigone to the burial site and began digging where the body of Icarius was hidden.

When Erigone saw her father's body, she was so distraught that she wandered about for some time before hanging herself from a branch of the tree over the spot where Icarius was buried. Dionysus observed the events and was so angry that he sent a plague of madness on the young women of Athens which caused them to wander about in a daze for a time and then hang themselves in same manner as Erigone.

The people of Athens were frightened by the sudden rash of suicides, and consulted an oracle to find out the cause. When they were told the cause, they found and punished the murderers, and the plague stopped. In remembrance of the events, the Athenians initiated an annual festival in which the men had a silent drinking contest on the first day, and the second day, the young women would swing from tree branches singing a song called "Aletis" in honor of Erigone. Aletis was a descriptive surname of Erigone, and probably related to a very similar classical Greek word (ἀλητς) meaning "wanderer." This festival was called the Aletides.

Dionysus also memorialized the events by placing the trio of Icarius, Erigone, and Maera in the sky as the constellations of Boötes, Virgo, and Canis Minor. Although the tragic story of Icarius and Erigone can explain the presence of the constellations Boötes, Virgo, and Canis Minor, some details of these constellations require other stories to explain.

For example, the ancient Greeks sometimes called Boötes the "Bear Watcher," or "Bear Guard" because he holds the leashes of the Hunting Dogs (Canes Venatici) and seems to chase Ursa Major and Ursa Minor (the Large and Small Bears), around the celestial pole. Also, the name of Boötes' primary star, Arcturus, literally means "Bear Guard." Some older star atlases show Boötes holding a spear instead of the shepherd's crook, and some show him holding the shepherd's crook in his left hand and a spear in his right.

Virgo is traditionally shown holding a spike (an ear) of wheat in her left hand. This comes from the Greek myth of Demeter, the goddess of agriculture whose daughter was kidnapped by Hades to be his wife and rule over his underground kingdom of the dead (see Chap. 12). In this myth, Demeter refused to allow crops to grow until a compromise was reached so that Persephone could spend 6 months of each year with her mother in Olympus. Demeter then allowed the crops to grow and ripen during that portion of the year. This story also explains why Virgo is visible only half the year, and below the earth the other half. A somewhat similar story explains the part-time visibility of Gemini.

In her right hand, Virgo is traditionally shown holding an olive branch. This stems from the tradition of Virgo representing Astraea, the goddess of justice and peace. At first all humans lived in the Golden age where there was no strife or illness. But this did not last, and the people gradually became more corrupt. The Golden age was succeeded by Silver, Bronze, and Iron ages with living conditions progressively deteriorating and the gods abandoning human society to dwell in Olympus. Astraea remained on the earth among men after all the other gods and goddesses had fled to Olympus. She stayed, using her scales of justice and olive branch, to promote justice and harmony until the race of men became so corrupt that even she fled to the heavens. The olive branch as a symbol of peace and the nearby constellation of Libra, the scales, come from this myth. Other myths and stories are also associated with Virgo, since several cultures used this constellation to represent one or more of their goddesses.

Canis Minor is the only constellation in this group which is not visible in the summer sky. It is located near both Orion and Canis Major instead of close to Boötes and Virgo and is visible in the winter, so it is often included in the story of Orion as one of his hunting dogs. However, few myths of Orion mention more than one dog (Fig. 7.1).

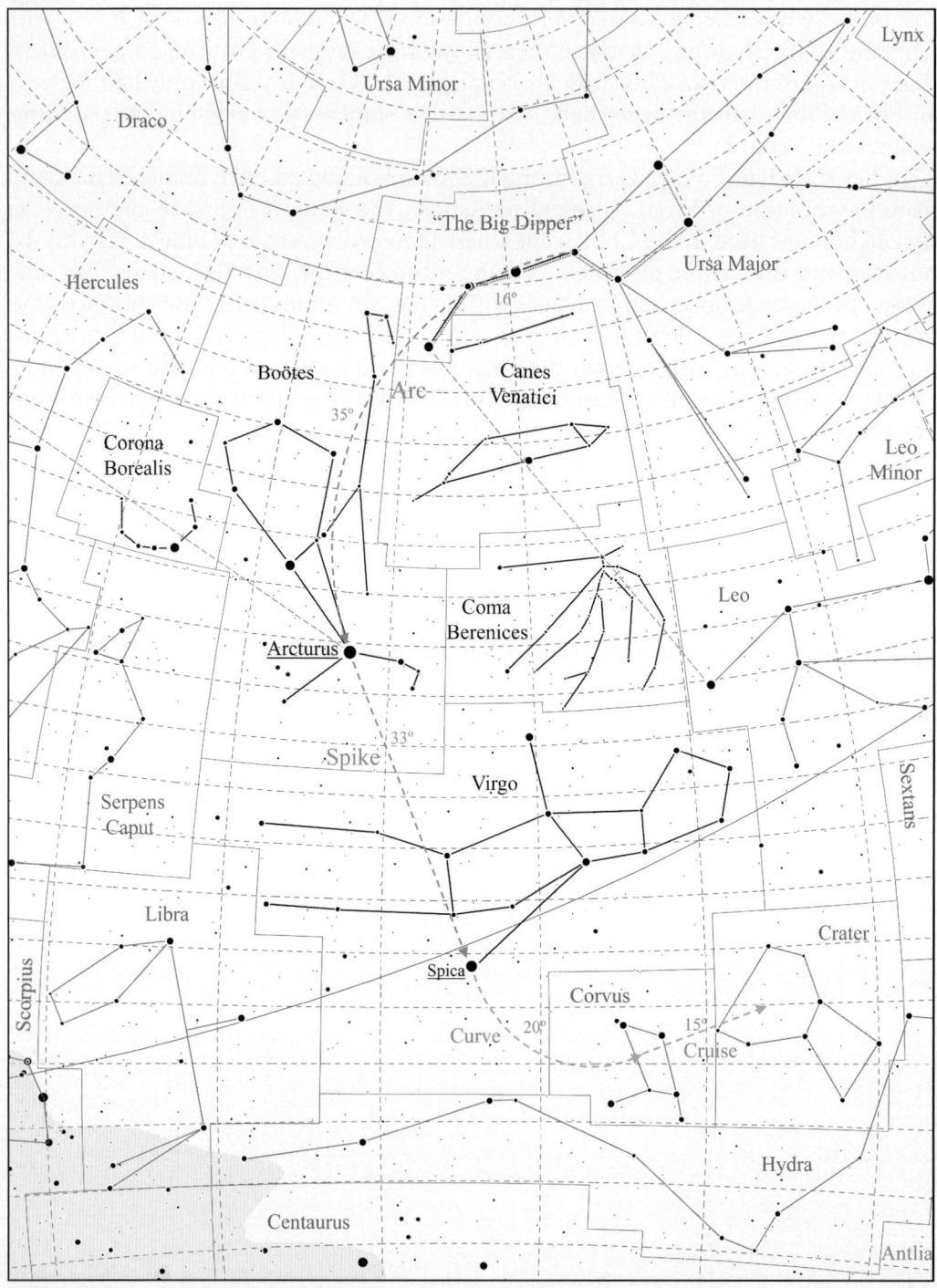

Fig. 7.1. Boötes Group Finder.

BOÖTES (bow-**owe**-teez)	**The Herdsman**
BOÖTIS (bow-**owe**-tiss)	**9:00 P.M. Culmination:** June 17
Boo	**AREA:** 907 sq. deg., 13th in size

TO FIND: Although Boötes has only two stars brighter than 3rd magnitude, it is easy to find because one of the two is orange <u>Arcturus</u>, the 4th brightest star in the entire sky, and the second brightest visible from mid-northern latitudes. If you are not already familiar with Arcturus, it is easily located using the mnemonic "arc to Arcturus, spike to Spica." This saying refers to starting with the tail of Ursa Major (the handle of the big dipper) and following the curve formed by these stars across the sky for about 35° from the end of the tail (or handle). This curve or arc will lead you to <u>Arcturus</u>. 10° NE of <u>Arcturus</u> is Izar, the second brightest star of Boötes, and 5° WSW of <u>Arcturus</u> is Muphrid, the third brightest star. Izar is at the SE point of an irregular hexagon which is 14° high (N–S) and 10° wide (E–W). This hexagon forms the upper part of Boötes body. Muphrid marks one of Boötes' knees, while his foot and other leg are marked by much fainter stars. 4° NE of Izar is 4th magnitude ρ which lies in Boötes' left arm which reaches out to his shepherd's crook. The business end of the crook is formed by five 4th and 5th magnitude stars about 5° E of the end of the bear's tail or the dipper's handle. Venabulum (δ Boö) lies 9° NE of Izar and in some atlases it is in the spear Boötes holds with his left hand (Figs. 7.2 and 7.3).

KEY STAR NAMES:

Name	Pronunciation	Source
α Arcturus	ark-**too**-russ	Αρχτουρυζσ (Greek), "bear watcher" or "bear guardian"
β Nekkar	neck-**car**	A misreading of the name al-baqqār (Sci-Arabic) for the constellation of the "ox-driver"
γ Seginus	seh-**jeye**-nus	Uncertain, but perhaps from "βοωτηζ" [Boötes] (Greek) translated into Arabic, then corrupted and applied to γ as ceginus, now seginus
δ Venabulum	win-ah-**buh**-lum	venabulum (Latin), "The hunting spear"
ε Izar	**ih**-zar, **eye**-zar	izar (Arabic), "girdle" or "loin cloth"
Pulcherrima	pul-ker-**rih**-mah	pulcherrima (Latin), "The most beautiful" as described by Wilhem Struve
η Muphrid	**mooff**-rid	mufrad-al-rāmih (Ind-Arabic), "the isolated single one," erroneously applied to η
μ Alkalurops	**al**-kah-**LU**-rops	χολλόροβοζ (Greek) , "Club" or "Staff," adapted by Arabic astronomers and retranslated into Latin, preceded by al (Arabic), "the"
h Merga	**mehr**-guh	merga (Latin), "a reaping hook" from the spear or hook sometimes shown by the right hand

Magnitudes and spectral types of principal stars:

Bayer desig.	Mag. (m$_v$)	Spec. type	Bayer desig.	Mag. (m$_v$)	Spec. type	Bayer desig.	Mag. (m$_v$)	Spec. type
α	−0.05	K2 IIIp	β	3.49	G8 III	γ	3.04	A7 IIIvar
δ	3.46	G8 III	ε	2.35	K0 II	ζ	3.78	A3 IVn
η	2.68	G0 IV	θ	4.04	F7 V	ι	4.75	A9 V
κ¹	6.62	F1 V	κ²	4.53	A8 IV	λ	4.18	A0 sh
μ¹	4.31	F0 V	μ²	6.51	G1 V	ν¹	5.04	K5 III
ν²	4.98	A5 V	ξ	4.54	G8 V	o	4.60	K0 III
π¹	4.49	B9p	π²	5.88	A6 V	ρ	3.57	K3 III
σ	4.47	F3 Vvar	τ	4.50	F7 V	υ	4.05	K5 IIIvar
φ	5.25	G8 III/IV	χ	5.28	A2 V	ψ	4.52	K2 III
ω	4.80	K4 III						

Stars of magnitude 5.5 or brighter which have Flamsteed, but no Bayer designations:

Flamsteed	Mag. (m$_v$)	Spec. type	Flamsteed	Mag. (m$_v$)	Spec. type	Flamsteed	Mag. (m$_v$)	Spec. type
6	4.92	K4 III	9	5.02	K3 III:var	12	4.82	F9 IVw
13	5.26	M2 III:var	15	5.29	K1 III	18	5.41	F5 IV
20	4.84	K3 III	22	54.0	F0m	31	4.86	G8 III:var
33	5.39	A1 V	34	4.80	M3 III	44	4.30	G2 V
45	4.93	F5 V	50	5.38	B9 Vn			

Stars of magnitude 5.5 or brighter which have no Bayer or Flamsteed designations:

Designation	Mag. (m$_v$)	Spec. type	Designation	Mag. (m$_v$)	Spec. type
HD 123657	5.13	M4.2 III	HD 125351	4.80	K1 III
HD 126128	4.86	A0 V	HD 131111	5.47	K0 III/IV
HD 134190	5.24	G8 III	HD 137 704	5.46	K4 III
HD 140728	5.48	B9p			

Fig. 7.2a. The stars of Boötes, Corona Borealis and adjacent constellations.

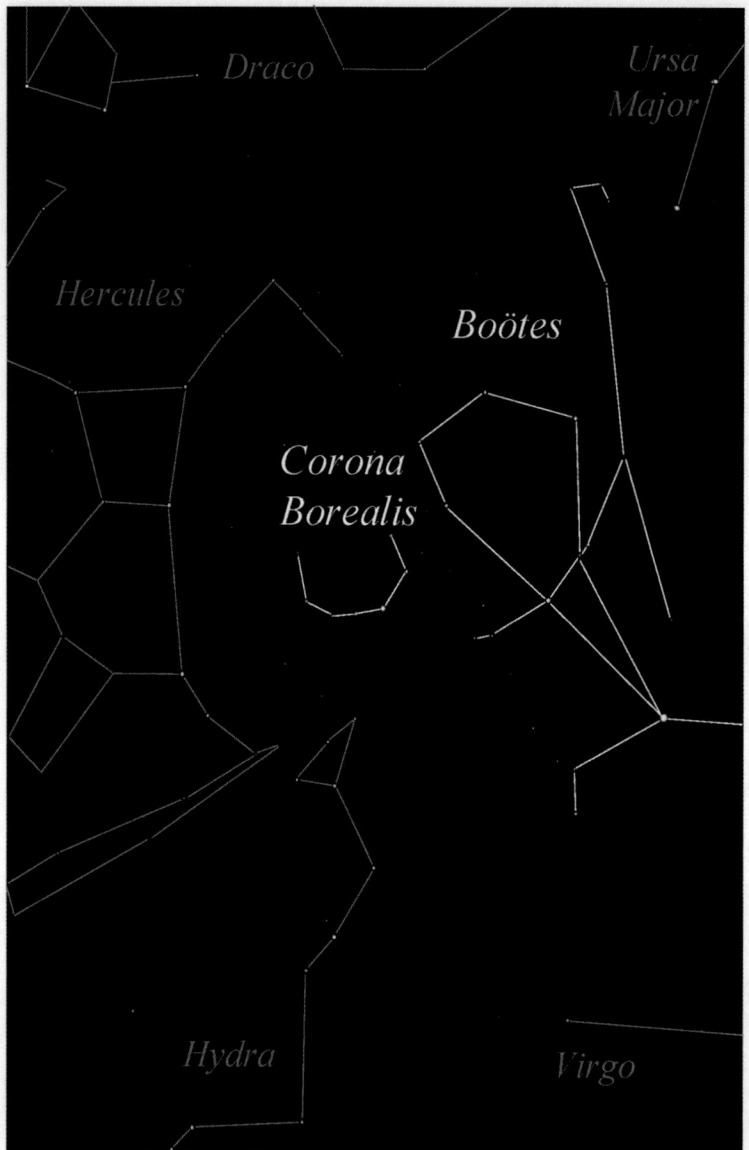

Fig. 7.2b. The figures of Boötes, Corona Borealis and adjacent constellations.

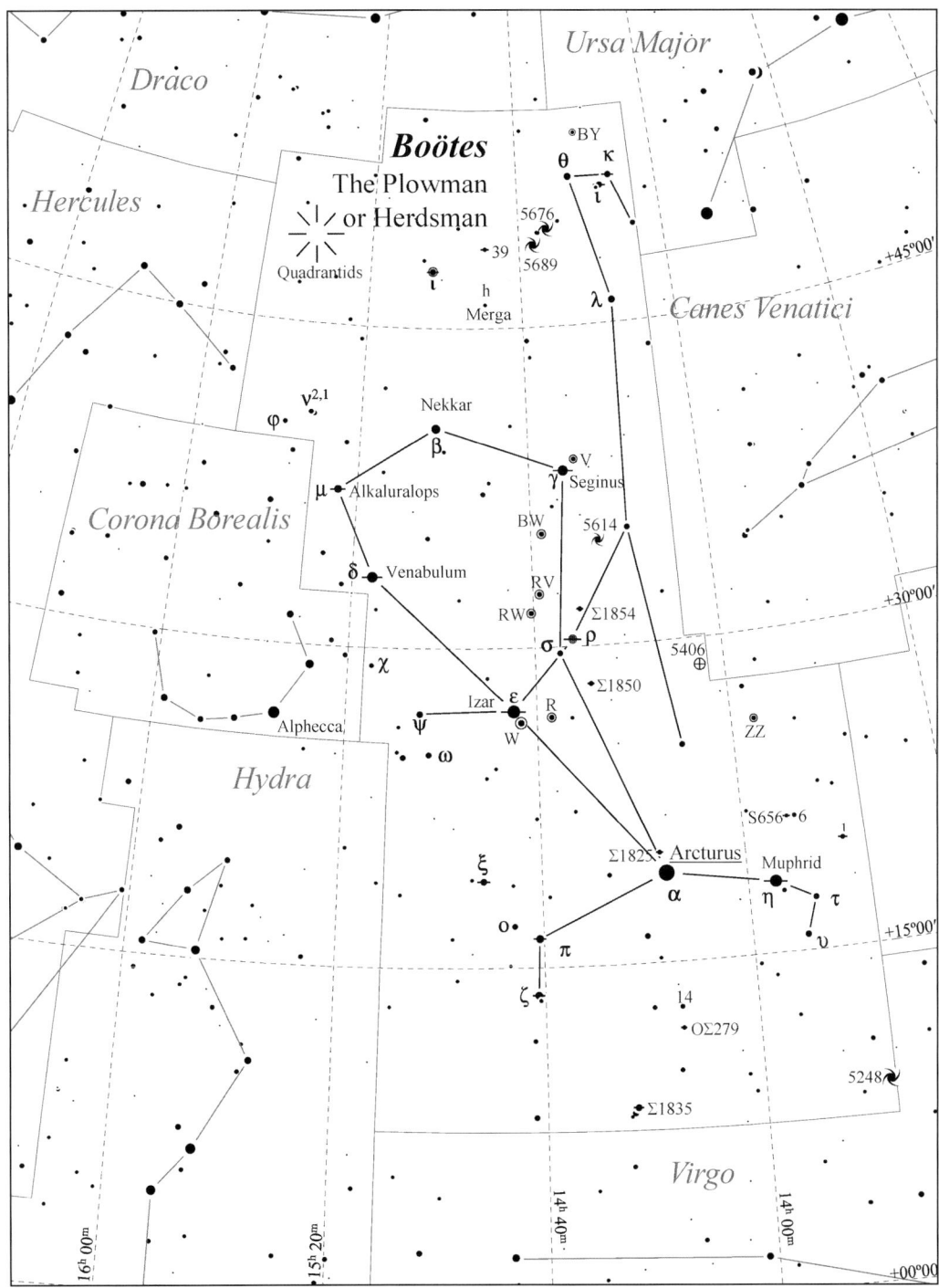

Fig. 7.3. Boötes details.

Table 7.1. Selected telescopic objects in Boötes.

Multiple stars:

Designation	R.A.	Decl.	Type	m₁	m₂	Sep. (")	PA (°)	Colors	Date/ Period	Aper. Recm.	Rating	Comments
1 Boötis	13ʰ 40.7ᵐ	+19° 17'	A–B	5.8	9.6	4.50	133°	W, ?	2000	3/4"	★★★★	a.k.a. Σ1772AB
4, τ Boötis	13ʰ 47.3ᵐ	+17° 27'	A–B	4.5	11.1	2.8	033°	yW, ?	2000	4/5"	★★★★	a.k.a. OΣ270
S 656 (Boötes)	13ʰ 50.4ᵐ	+21° 17'	A–B	6.9	7.4	88.5	208°	Y, Y	2003	6×30	★★★★	Nice, similar pair. Easy binocular double. 9'30" E (PA 085°) of 6 Boo
17, κ Boötis	14ʰ 13.5ᵐ	+51° 47'	A–B	4.5	6.6	13.5	236°	W, pB	2004	2/3"	★★★★	Slight color difference. a.k.a. Σ1821
OΣΣ279	14ʰ 13.9ᵐ	+11° 59'	A–B	6.6	8.9	2.2	257°	W, O	2004	4/6"	★★★★	0° 58' S (PA 183°) of 14 Boo (m5.53)
21, ι Boötis	14ʰ 16.2ᵐ	+51° 22'	A–B	4.8	7.4	39.7	034°	Y, B	2004	6×30	★★★★	Nice color contrast, a difficult object in binoculars. a.k.a. Σ26AB
			A–C	4.9	12.6	85.7	187°	Y, ?	2000	5/6"	★★★	
Σ1825	14ʰ 16.5ᵐ	+20° 07'	A–B	6.5	8.4	4.4	156°	yW, ?	2003	2/3"	★★★★	0° 58' NNE (PA 013°) of Arcturus (m−0.05)
Σ1835A-BC	14ʰ 23.4ᵐ	+08° 26'	A–BC	5.0	6.8	6.2	194°	W, yW	2006	2/3"	★★★★	BC is too close for amateur telescopes. 2° 42' SE (PA 128) of 15 Boo (m5.50)
Σ1850	14ʰ 28.6ᵐ	+28° 17'	A–B	7.1	7.6	25.7	262°	W, W	2006	10×50	★★★	2° 12' SSW (PA 199°) of ρ Boo (m3.57)
Σ1854	14ʰ 29.8ᵐ	+31° 47'	A–B	6.1	10.6	25.8	255°	W, ?	2000	3/4"	★★★	1° 29' NNW (PA 344°) of ρ Boo (m3.57)
ρ Boötis	14ʰ 31.8ᵐ	+30° 22'	A–B	3.6	11.5	35.7	339°	oY, ?	2000	3/4"	★★★	a.k.a. h2728
27, γ Boötis	14ʰ 32.1ᵐ	+38° 18'	Aa–B	3.0	12.7	33.4	111°	W, ?	2925	6/6"	★★★	a.k.a. β616AaB. Aa is too close for amateur instruments
29, π Boötis	14ʰ 40.7ᵐ	+16° 25'	A–B	4.9	5.8	5.5	111°	W, W	2006	2/3"	★★★★	a.k.a. Σ1864AB
30, ζ Boötis	14ʰ 41.2ᵐ	+13° 44'	A–B	4.5	4.6	**0.55**	**295°**	W, W	**123.44**	16/18"	★★★★★★	Binary, per ~125 year, separation decreasing. Test for 6/8". a.k.a. Σ1865AB
36, ε Boötis	14ʰ 45.0ᵐ	+27° 04'	A–B	2.6	4.8	2.9	343°	Gld, pB	2005	3/4"	★★★★★★	a.k.a. Σ1487 and Mirak or Izar. Excellent color contrast, best double in Boötes *1
39 Boötis	14ʰ 49.7ᵐ	+48° 43'	A–B	6.3	6.7	2.7	046°	pY, pY	–	3/4"	★★★★	Two very similar stars. Nice in most scopes. a.k.a. Σ1890
37, ξ Boötis	14ʰ 51.4ᵐ	+19° 06'	A–B	4.8	7.0	**6.03**	**308°**	Y, rO	**151.6**	2/3"	★★★★	Moderate color contrast. a.k.a. Σ1888AB
44, ι Boötis	15ʰ 03.8ᵐ	+47° 39'	A–B	5.2	6.1	**1.61**	**060°**	Y, Y	**206**	6/8"	★★★★	Maximum separation was 2.03" in 2001. By 2025, the separation will be only 0.3" *2
49, δ Boötis	15ʰ 15.5ᵐ	+33° 19'	A–B	3.6	7.9	102.4	079°	pY, pB	2004	7×50	★★★★	Binocular double, subtle, but nice color contrast. ΣA27
51, μ Boötis	15ʰ 24.5ᵐ	+37° 23'	Aa–BC	4.3	7.1	107.1	170°	Y, oY	2,002	6×30	★★★★★	A–BC, easy binocular double. a.k.a. Σ1938
			B–C	7.1	7.6	**2.25**	**005°**	Y, O	**257**	4/6"	★★★★★	BC requires more power. a.k.a. Σ1938BC

Variable stars:

Designation	R.A.	Decl.	Type	Range (m$_v$)	Period (days)	F (f$_r$/f$_f$)	Spectral Range	Aper. Recm.	Rating	Comments
ZZ Boötis	13^h 56.2^m	+25° 55′	EA/DM	6.79	4.992	0.06	F2 + F2V		****	1° 32′ S (PA 183°) of 9 Boo.##$
BY Boötis	14^h 07.9^m	+42° 51′	Lb	5.0	–	–	M4–M4.5III	Eye	*****	2° 41′ SW (PA 214°) of 19, λ Boo. AKA BD 123657, SAO 44904
S Boötis	14^h 22.9^m	+53° 49′	M	7.8	270.73	0.44	M3e–M6e	12/14	****	1° 29′ N (PA 350°) of θ Boo.##
V Boötis	14^h 29.6^m	+38° 52′	SRa	7.0	258.01	0.49	M6e	6/8″	****	The amplitude of variation has been decreasing since 1900.##$ *3
BW Boötis	14^h 37.1^m	+35° 56′	EA/DM	7.13	3.333	0.06	F0V	6×30	****	##
R Boötis	14^h 37.2^m	+26° 44′	M	6.2	223.4	0.46	M3e–M8E	08/10	****	Pulsating red giant. Requires 8/10″ to view minima. 1° 46′ WSW (PA 259°) of ε Boo *4
RV Boötis	14^h 39.3^m	+32° 32′	SRb	7.9	137	0.50	M5e–M7e	10×80	****	2.7 NNE (PA 036°) of 25 Boo.#$
RW Boötis	14^h 41.2^m	+31° 34′	SRb	8.0	209	–	M5	10×80	****	2.3 NE (PA 059°) of 25 Boo.#$ *5
W Boötis	14^h 43.4^m	+26° 32′	SRb	4.7	450	–	M2III–M4III	Eye	****	39′ SW (PA 213°) of ε Boo.##
44, ι Boötis	15^h 03.8^m	+47° 39′	EW/KW	5.8	0.268	–	G2V + G2V	6×30	****	Triple star system. Complex variations.# *2

Star clusters:

Designation	R.A.	Decl.	Type	m$_v$	Size (′)	Brtst. Star	Dist. (ly)	dia. (ly)	Number of Stars	Aper. Recm.	Rating	Comments
NGC 5466	14^h 05.5^m	+28° 32′	Globular	9.2	9.0	13.8	51.5K	140		8/10″	***	Visible in 11×80 binoculars under good circumstances, but is moderately faint in 8/10″ scopes

Nebulae:

Designation	R.A.	Decl.	Type	m$_v$	Size (′)	Brtst./ Cent. star	Dist. (ly)	dia. (ly)	Aper. Recm.	Rating	Comments
None											

(continued)

Table 7.1. (continued)

Galaxies:

Designation	R.A.	Decl.	Type	m$_v$	Size (')	Dist. (ly)	dia. (ly)	Lum. (suns)	Aper. Recm.	Rating	Comments
NGC 5248	13^h 37.5^m	+08° 53'	SAB	10.3	1.6×1.0	74M	123K	42G	6/8"	****	1° 56' SSW (PA 195°) of HD 178889 (m5.57, on Norton maps 9 and 11)
NGC 5614	14^h 24.1^m	+34° 52'	SA	11.7	2.6×2.0	–	–	–	12/14"	**	Interacting with NGC 5613, 2' to NNW, faint in 12/14". 1° 24' ESE (PA 116°) of (m4.80)
NGC 5676	14^h 32.8^m	+49° 28'	SA	11.2	3.7×1.6	112M	112K	59G	6/8"	****	Fairly bright, elongated NE–SW, sl. brighter cent. 0° 47' ESE (PA 119°) of 24 Boo (m5.58)
NGC 5689	14^h 35.5^m	+48° 45'	SB	11.9	3.7×1.0	116M	102K	16G	6/8"	****	

Meteor showers:

Designation	R.A.	Decl.	Period (yr)	Duration	Max Date	ZHR (max)	Comet/ Asteroid	First Obs.	Vel. (mi/ km/sec)	Rating	Comments
Quadrantids	15^h 16^m	+49°	Annual	Dec. 12 to Jan. 7	Jan. 3/4	45–20	2003 EH1	1825	–	Major	Named for the obsolete constellation of "Quadrans Muralis," the Wall Quadrant *6

Other interesting objects:

Designation	R.A.	Decl.	Type	m$_v$	Size (')	Dist. (ly)	dia. (ly)	Lum. (suns)	Aper. Recm.	Rating	Comments
None											

Canes Venatici (**kay**-neez ven-ah-**tee**-see)	The Hunting Dogs
Canum Venaticorum (**kay**-num ven-at-ih-**kor**-um)	9:00 P.M. Culmination: May 21
CVn	AREA: 465 sq. deg., 38th in size

Like many constellations, Boötes has been portrayed as different characters by different cultures. One Boötes' alter egos was the Bear Watcher. As such, he and his hunting dogs followed or chased the two celestial bears around the sky. Although the hunting dogs were not recognized as a constellation by the Greeks, cartographers showed them on some celestial globes as early as 1493. Both the number and position of the dogs shown varied until Johannes Hevelius published his version in 1690. Hevelius put both α and β Canum Venaticorum in the southern dog, rather than having each in a separate dog as previous cartographers had done. He also named the two dogs, the northern one becoming Asterion and the southern one Chara. They have retained their shape and position since that time.

TO FIND: Starting with Alkaid (the end of the great bear's tail), draw an imaginary line (42°) to Denebola (the end of the Lion's tail). α Cvn lies one-third of the way (17°) from Alkaid to Denebola. β CVn is located 5° ENE of α. The remaining stars of Canes Venatici are relatively faint, and using a chart is the best way to locate them (Figs. 7.1, 7.4 and 7.5).

KEY STAR NAMES:

Name	Pronunciation	Source
α Cor Caroli	kor **kair**-oh-lee	Latin for "Heart of Charles," named by Edmund Halley in honor of his patron, King Charles II of England
β Chara	**kuh**-rah or **kay**-ruh	Χαρα (Greek), "joy" used by Hevelius to designate the southern dog
Asterion	ahs-**tee**-ree-ohn	asterion (Greek), "little star," named by Hevelius
Y La Superba	**lah** su-**per**-bah	la superb (Italian), "the superb one" by Fr. Angelo Secchi for its deep red color

Magnitudes and spectral types of principal stars:

Bayer desig.	Mag. (m_v)	Spec. type	Bayer desig.	Mag. (m_v)	Spec. type	Bayer desig.	Mag. (m_v)	Spec. type
α^1	4.61	F0 V	α^2	2.80	A0 spec	β	4.63	G0 V

Stars of magnitude 5.5 or brighter which have Flamsteed, but no Bayer designations:

Flamsteed	Mag. (m_v)	Spec. type	Flamsteed	Mag. (m_v)	Spec. type	Flamsteed	Mag. (m_v)	Spec. type
3	5.28	M0 III	5	4.76	G7 III	6	5.01	G8 III/IV
14	5.20	B9 V	20	4.72	F3 III	21	5.14	A0 V
24	4.68	A5 V	25	4.82	A7 III			

Stars of magnitude 5.5 or brighter which have no Bayer or Flamsteed designations:

Designation	Mag. (m_v)	Spec. type	Designation	Mag. (m_v)	Spec. type
HD 109317	5.42	K0 III	HD 115004	4.94	K0 III
HD 118216	4.91	F2 IV	HD 120933	4.76	K5 III

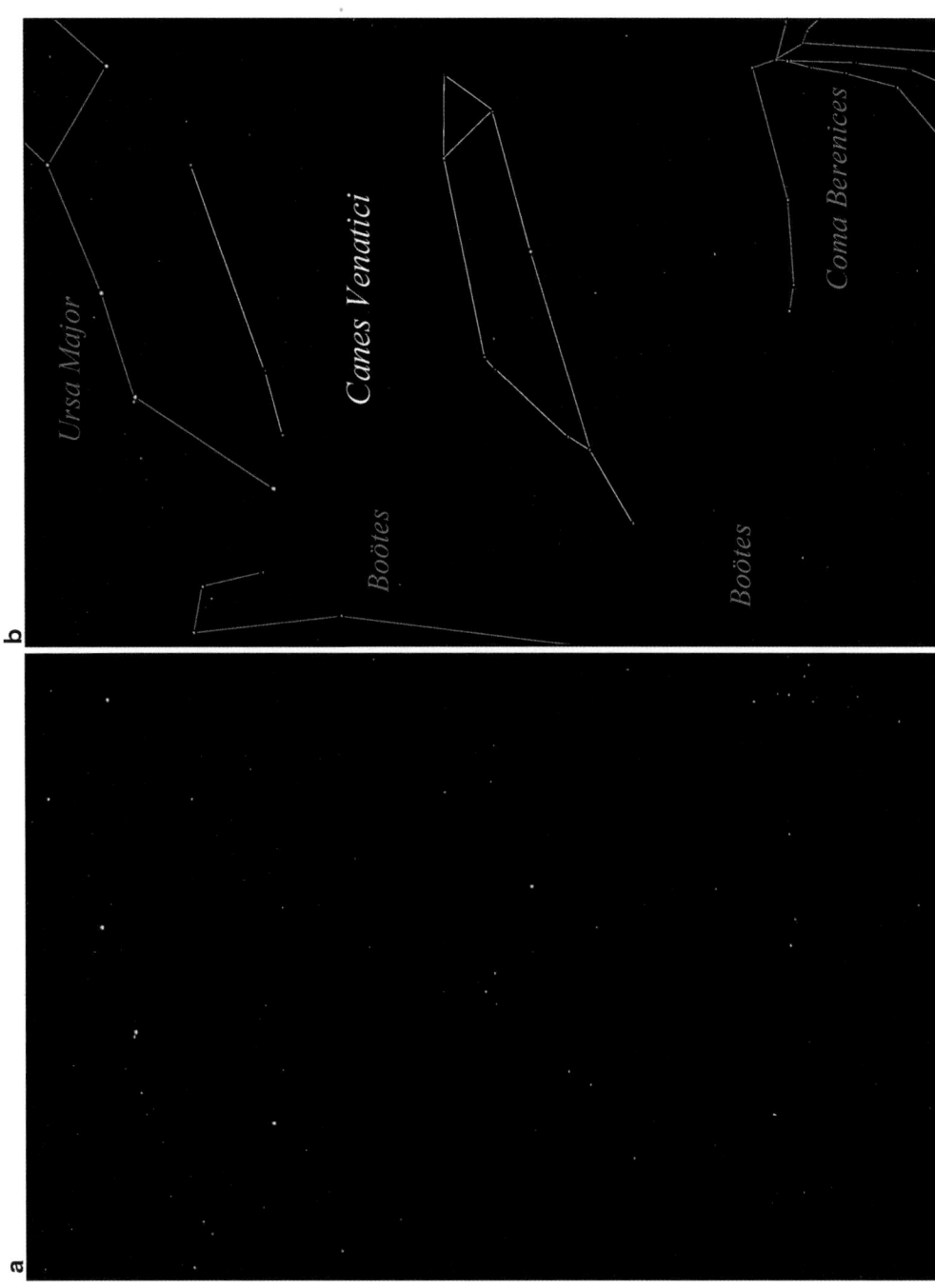

Fig. 7.4. (a) The stars of Canes Venatici and adjacent constellations. **(b)** The figures of Canes Venatici and adjacent constellations.

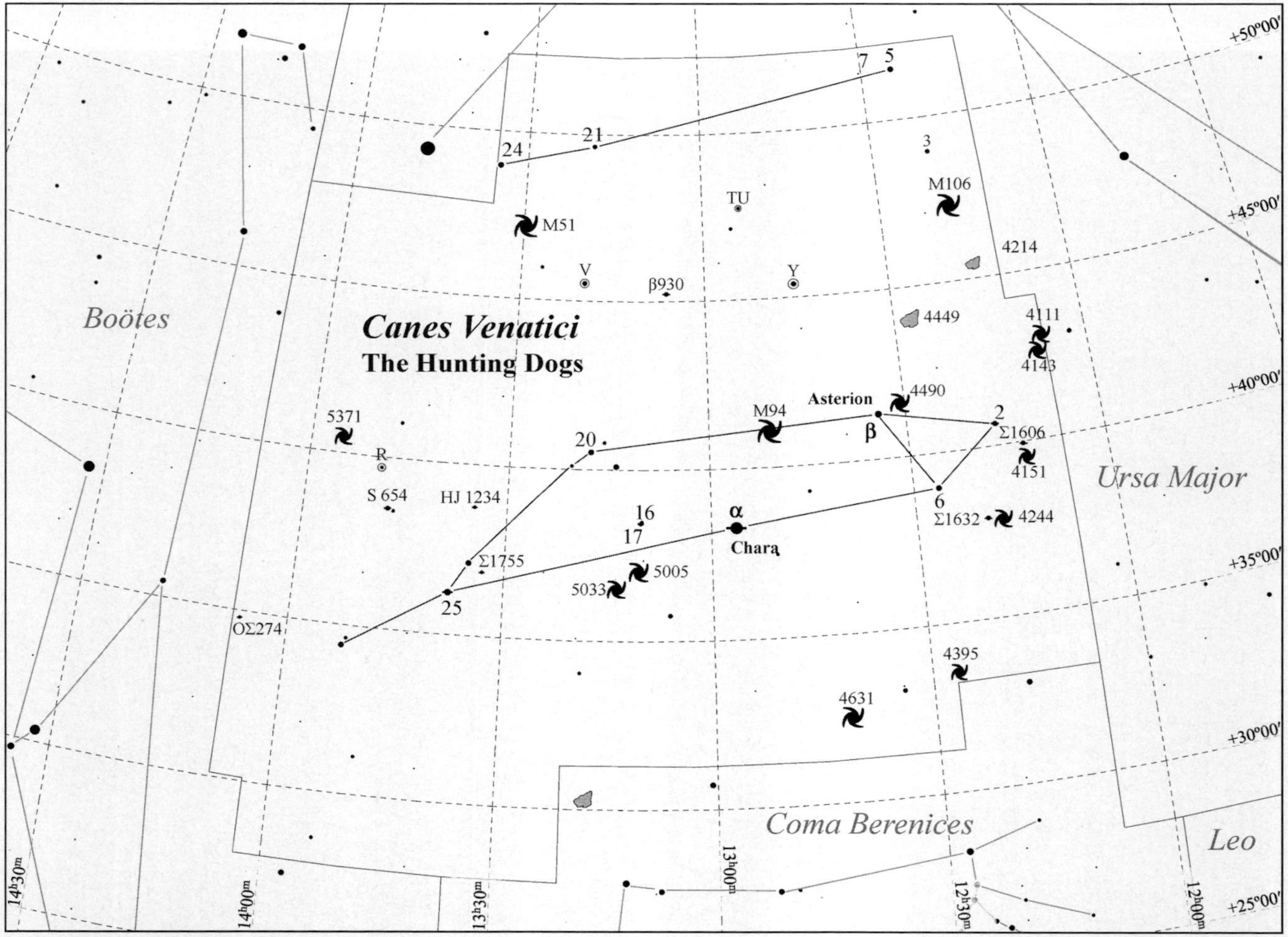

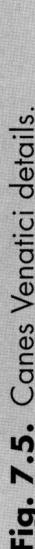

Fig. 7.5. Canes Venatici details.

Table 7.2. Selected telescopic objects in Canes Venatici.

Multiple stars:

Designation	R.A.	Decl.	Type	m_1	m_2	Sep. (")	PA (°)	Colors	Date/ Period	Aper. Recm.	Rating	Comments
Σ1606	12ʰ 10.8ᵐ	+39° 53'	A–B	6.9	7.4	**0.4**	**164**	W, ?	**1,431**	>14"	****	Near theoretical limit for 12". Slowly opening to 2.4" in 2571 A is binary, but max. sep of only 0.11" is too close for amateur scopes
2 Canum Venaticorum	12ʰ 16.1ᵐ	+40° 40'	A–B	5.9	8.7	11.5	260	gY, pB	2005	2/3"	****	Lovely color contrast, a.k.a. Σ1622
Σ1632	12ʰ 20.2ᵐ	+37° 54'	A–B	6.8	10.0	9.8	193	Y, ?	2006	2/3"	***	Binocular triple. a.k.a. βpm 143AB
7 Canum Venaticorum	12ʰ 30.0ᵐ	+51° 32'	A–B A–C	6.2 6.3	10.4 8.5	109.2 215.3	172 333	Y, Y Y, BW	1908 2002	2/3" 6×30	**** ***	a.k.a. βpm 143AC
12, α Canum Venaticorum	12ʰ 56.0ᵐ	+38° 19'	A–B	2.9	5.5	19.3	229	W, bG	2004	15×80	*****	Cor Caroli, lovely in small telescope, bright, easy to find. a.k.a. Σ1692
S930	13ʰ 05.9ᵐ	+45° 15'	A–B	5.6	10.0	2.7	123	Y, ?	1995	3/4"	****	
16 + 17 Canum Venaticorum	13ʰ 09.7ᵐ	+38° 32'	A–B	6.3	6.0	279.0	117	bW, bW	–	6×30	****	Wide binocular double and triple with 17A–B
17 Canum Venaticorum	13ʰ 10.1ᵐ	+38° 30'	A–B B–C	6.0 6.3	6.3 9.2	84.4 1.2	297 271	bW, Y Y, ?	2004 2003	6×30 6/8"	**** ***	a.k.a. ΣI24AB a.k.a. β608BC
25 Canum Venaticorum	13ʰ 37.6ᵐ	+36° 17'	A–B AB–C	4.8 5.0	6.9 11.7	**1.80** 210.2	**104°** 140	pY, pY pY, ?	**228** 2001	5/6" 3/4"	**** ***	a.k.a. Σ1768AB. Max. sep. of 1.8" in 1982, closing to min. of 0.2" in 2091
Σ1755	13ʰ 32.4ᵐ	+36° 49'	A–B	7.3	8.1	4.2	131	oY, ?	2002	2/3"	****	
h1234	13ʰ 34.4ᵐ	+38° 47'	A–B	6.4	10.51	29.1	008	W, ?	1998	15×80	****	
S654	13ʰ 47.0ᵐ	+38° 33'	A–B	5.6	8.9	70.1	236	oW, ?	2004	7×50	****	
OΣ274	14ʰ 06.7ᵐ	+34° 46'	A–B	7.1	10.47	12.8	054	oY, ?	1999	2/3"	***	

(continued)

Table 7.2. (continued)

Variable stars:

Designation	R.A.	Decl.	Type	Range (m$_v$)	Period (days)	F (f$_r$/f$_l$)	Spectral Range	Aper. Recm.	Rating	Comments
Y Canum Venaticorum	12^h 45.1^m	+45° 26'	SRb	5.2 6.6	158	–	N3	7×50	*****	La Superba. Carbon star ˙7
TU Canum Venaticorum	12^h 54.9^m	+47° 12'	SRb	5.55 6.6	50	–	M5III	7×50	****	
V Canum Venaticorum	13^h 19.5^m	+45° 32'	SRa	6.5 8.6	191.89	0.50	MI4e–M6e IIIa	10×80	****	
R Canum Venaticorum	13^h 49.0^m	+39° 33'	M	6.5 12.9	328.53	0.47	M5.5e–M9e	4/6"	****	

Star clusters:

Designation	R.A.	Decl.	Type	m$_v$	Size (')	Brst. Star	Dist. (ly)	dia. (ly)	Number of Stars	Aper. Recm.	Rating	Comments
M3, NGC 5272	13^h 42.2^m	+28° 23'	Globular	5.9	16.2		27K	130	260K	4/6"	*****	3rd brightest globular cluster in N hemisphere

Nebulae:

Designation	R.A.	Decl.	Type	m$_v$	Size (')	Brst./ Cent. star	Dist. (ly)	dia. (ly)	Aper. Recm.	Rating	Comments
None											

Galaxies:

Designation	R.A.	Decl.	Type	m$_v$	Size (')	Dist. (ly)	dia. (ly)	Lum. (suns)	Aper. Recm.	Rating	Comments
NGC 4111	12^h 07.0^m	+43° 04'	SA(r)0	10.8	4.7×1.0	15M	65K	11G	8/10"	****	Bright stellar nucleus, large bright core, faint halo, elongated NNW–SSE
NGC 4143	12^h 09.6^m	+42° 32'	SAB(s)0	10.7	2.9×1.8	36M	30K	5.4G	8/10"	****	Bright, round. Small, bright, diffuse nucleus
NGC 4151	12^h 10.5^m	+39° 24'	SAB(rs)ab	10.8	5.9×4.4	44M	76K	7.4G	8/10"	***	Seyfert galaxy, variable nucleus
NGC 4214	12^h 15.7^m	+36° 20'	IAB(s)m	9.8	7.9×6.3	13M	30K	1.6G	6/8"	***	Bright, large, slightly elongated

Designation	R.A.	Decl.	Type	m_v	Size (')	Dist. (ly)	dia. (ly)	Lum. (suns)	Aper. Recm.	Rating	Comments
NGC 4242	12h 17.5m	+45° 37'	SAB(s)dm	10.8	4.8×3.8	33M	46K	4.2G	8/10"	***	Large, faint, very faint arms, irregular outline
NGC 4244	12h 17.5m	+37° 49'	SA(s)cd	10.4	17.0×2.2	10M	33K	1.7G	8/10"	*****	Unusually long and thin, mottled appearance, elongated NE–SW
M106, NGC 4258	12h 19.0m	+47° 18'	SAB(s)bc	8.4	20.0×8.4	22M	92K	27G	4/6"	*****	Tightly wound spiral, along NNW–SSE, bright core, prominent nucleus
NGC 4395	12h 25.8m	+33° 33'	SA(s)m	10.9	3.4×3.2	13M	13K	0.5G	8/10"	***	Faint, three fainter companions, NGC 4399, 4400 and 4401 nearby
NGC 4449	12h 28.2m	+44° 06'	IBm IV	9.6	5.5×4.1	10M	15K	1.8G	6/8"	****	Stellar nucleus, concentrated core, elongated NE–SW
NGC 4490	12h 30.6m	+41° 38'	SB(s)d	9.8	6.0×3.2	27M	47K	7.0G	6/8"	****	Very large, bright, very elongated
NGC 4631	12h 42.1m	+32° 32'	SB(s)d	9.2	15.5×3.3	22M	71K	17G	4/6"	*****	"Whale Galaxy." One of the best edge-on galaxies in the N. skies *8
M94, NGC 4736	12h 50.9m	+41° 07'	(R)SA(r)	8.2	13.0×11.0	14M	131K	46G	4/6"	****	Bright core, faint halo (arms), core resembles elliptical galaxy
NGC 5005	13h 09.6m	+37° 03'	SAB(rs)bc	9.8	5.4×2.7	47M	74K	21G	6/8"	****	Bright halo, bright core, both oval
NGC 5033	13h 13.5m	+36° 36'	SA(s)c	10.2	10.5×5.6	42M	128K	12G	8/10"	****	Large, bright, elongated. Multiple thin arms. Very bright, small nucleus
M51, NCG 5194	13h 29.9m	+47° 12'	SA(s)m	8.4	8.2×6.9	25M	86K	31G	4/6"	*****	"Whirlpool Galaxy." Impressive in larger (12–16") scopes
NGC 5371	13h 55.7m	+40° 28'	SAB(rs)bc	10.6	4.4×3.5	120M	154K	66G	8/10"	***	Large, diffuse, slightly elongated

Meteor showers:

Designation	R.A.	Decl.	Period (yr)	Duration	Max Date	ZHR (max)	Comet/Asteroid	First Obs.	Vel. (mi/km /sec)	Aper. Recm.	Rating	Comments
None												

Other interesting objects:

Designation	R.A.	Decl.	Type	m_v	Size (')	Dist. (ly)	dia. (ly)	Lum. (suns)	Aper. Recm.	Rating	Comments
None												

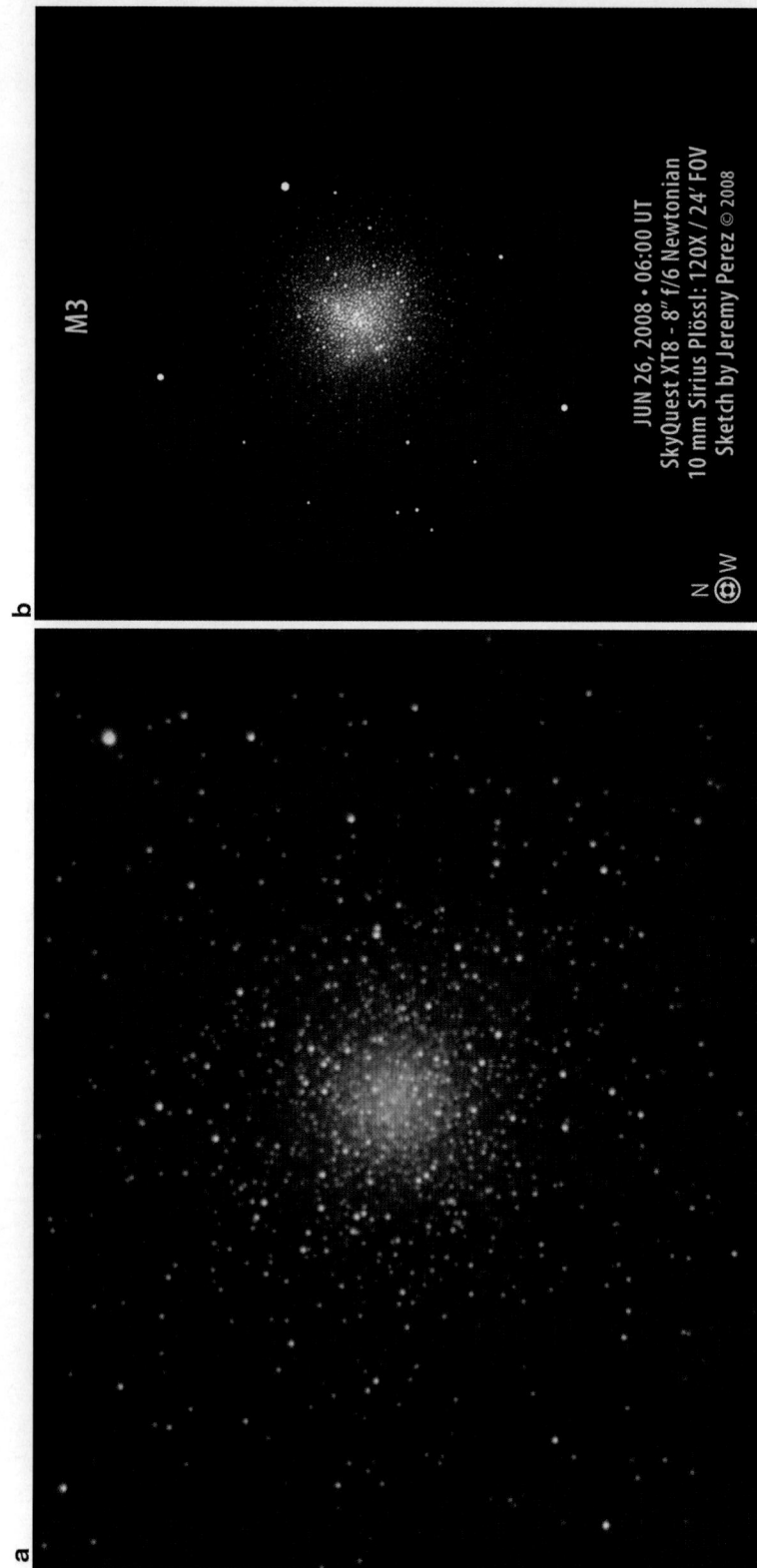

M3

JUN 26, 2008 • 06:00 UT
SkyQuest XT8 – 8″ f/6 Newtonian
10 mm Sirius Plössl: 120X / 24′ FOV
Sketch by Jeremy Perez © 2008

N
⊕W

Fig. 7.6 **(a)** M3 (NGC 5272), globular cluster in Canes Venatici. Astrophoto by Gian Michelle Ratto. **(b)** M3 drawing.

Fig. 7.7. NGC 4244, edge-on spiral galaxy in Canes Venatici. Astrophoto by Mark Komsa.

Fig. 7.8. NGC 4631, spiral galaxy, "Whale Galaxy" in Canes Venatici. Astrophoto by Mark Komsa.

Fig. 7.9. NGC 5033, spiral galaxy in Canes Venatici. Astrophoto by Mark Komsa.

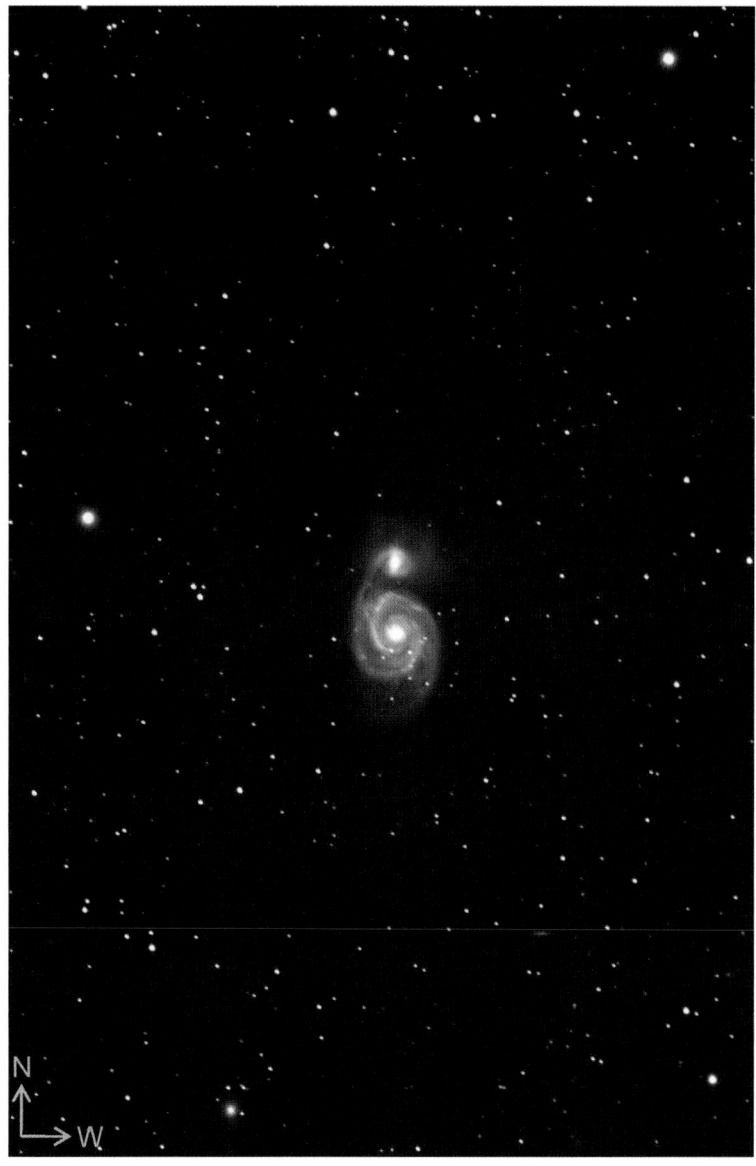

Fig. 7.10a. M51 (NGC 5194), spiral galaxy in Canes Venatici. Astrophoto by Steve Pastor.

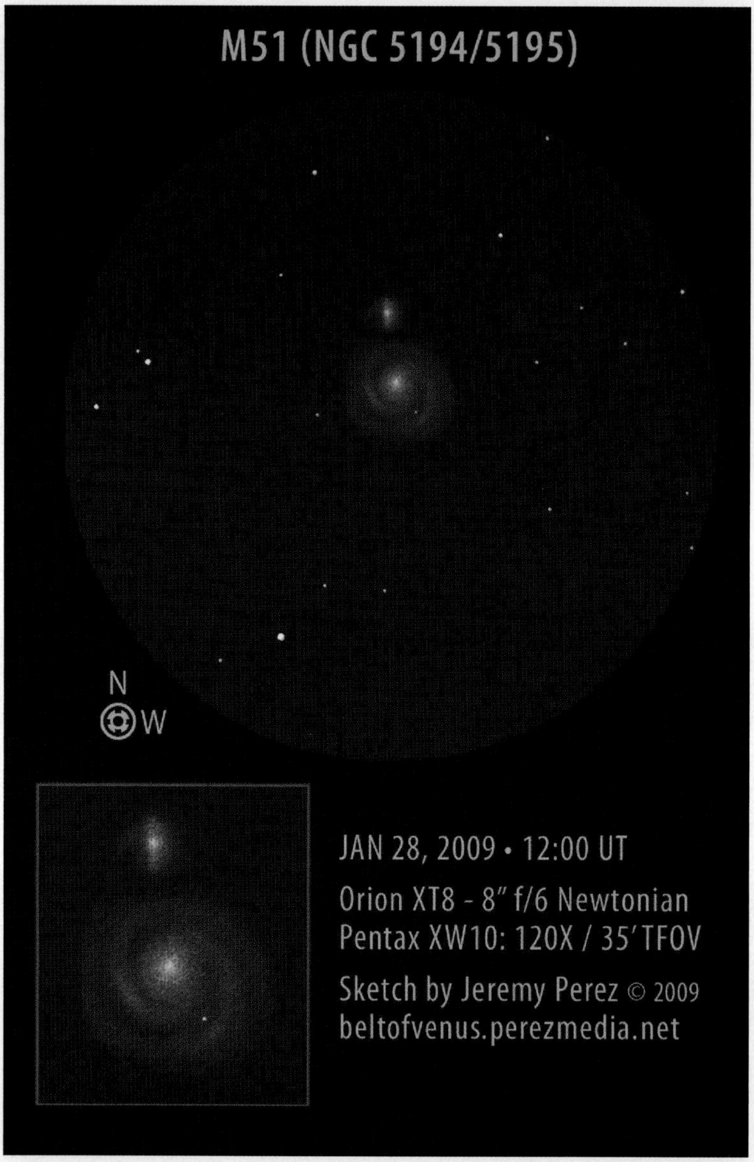

Fig. 7.10b. M51 drawing.

Canis Minor (kay-nis **my**-nor)	The Smaller Dog
Canis Minoris (**kay**-nis my-**nor**-iss)	9:00 P.M. CULMINATION: Feb. 27
CMi	AREA: 183 sq. deg., 71st in size

TO FIND: Procyon, the brightest star in Canis Minor, is easy to find. It forms a nearly perfect inverted equilateral triangle with Sirius and Betelgeuse, with Sirius at the bottom, and Procyon and Betelgeuse, respectively, to its NE and NW. Third magnitude Gomeisa lies just 4.3° NW of Procyon and 4th magnitude γ is less than 1° NNE of Gomeisa. All the other stars in Canis Minor are 5th magnitude or fainter. Procyon is also part of the "Winter Hexagon" of six 1st magnitude stars surrounding Betelgeuse. It lies between Sirius and Pollux (Figs. 7.11 and 7.12).

KEY STAR NAMES:

Name	Pronunciation	Source
α Procyon	pro-**sigh**-ohn	Προκ⌡ον (Greek), "before the dog" because it rises shortly before Sirius, the Dog Star
β Gomeisa	go-**my**-zuh or go-**may**-sah	al-ghumaisā (Ind-Arabic), "the little bleary-eyed one." An Egyptian story attributes this to the weeping of a sister who was left alone. The Greek story attributes it to the weeping of Maera over the loss of her master

Magnitudes and spectral types of principal stars:

Bayer desig.	Mag. (m_v)	Spec. type	Bayer desig.	Mag. (m_v)	Spec. type	Bayer desig.	Mag. (m_v)	Spec. type
α	0.40	F5 IV/V	β	2.89	B8 Vvar	γ	4.33	K3 III SB
δ^1	5.24	F0 III	δ^2	5.59	F2 V	δ^3	5.83	A0 Vnn
ε	4.99	G8 III	ζ	5.12	B8 II	η	5.22	F0 III
G	4.39	K2 III						

Stars of magnitude 5.5 or brighter which have Flamsteed, but no Bayer designations:

Flam-steed	Mag. (m_v)	Spec. type	Flam-steed	Mag. (m_v)	Spec. type	Flam-steed	Mag. (m_v)	Spec. type
1	5.37	A5 IV	6	4.55	K2 III	11	5.25	A1 Vnn
14	5.30	K0 III						

Fig. 7.11a. The stars of Canis Minor and adjacent constellations.

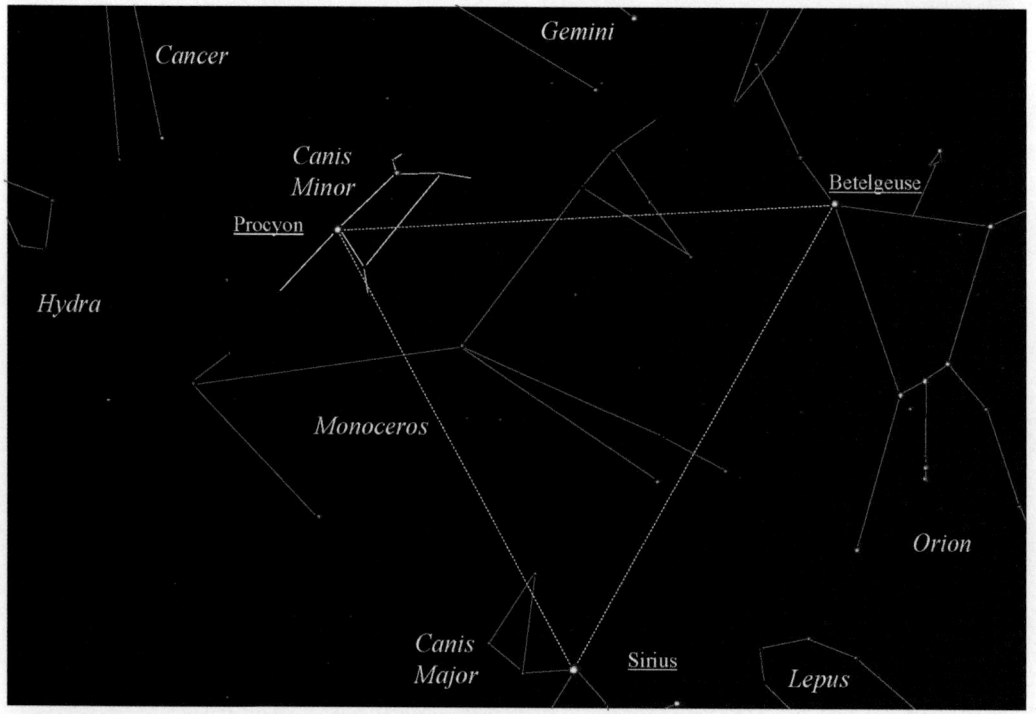

Fig. 7.11b. The figures of Canis Minor and adjacent constellations.

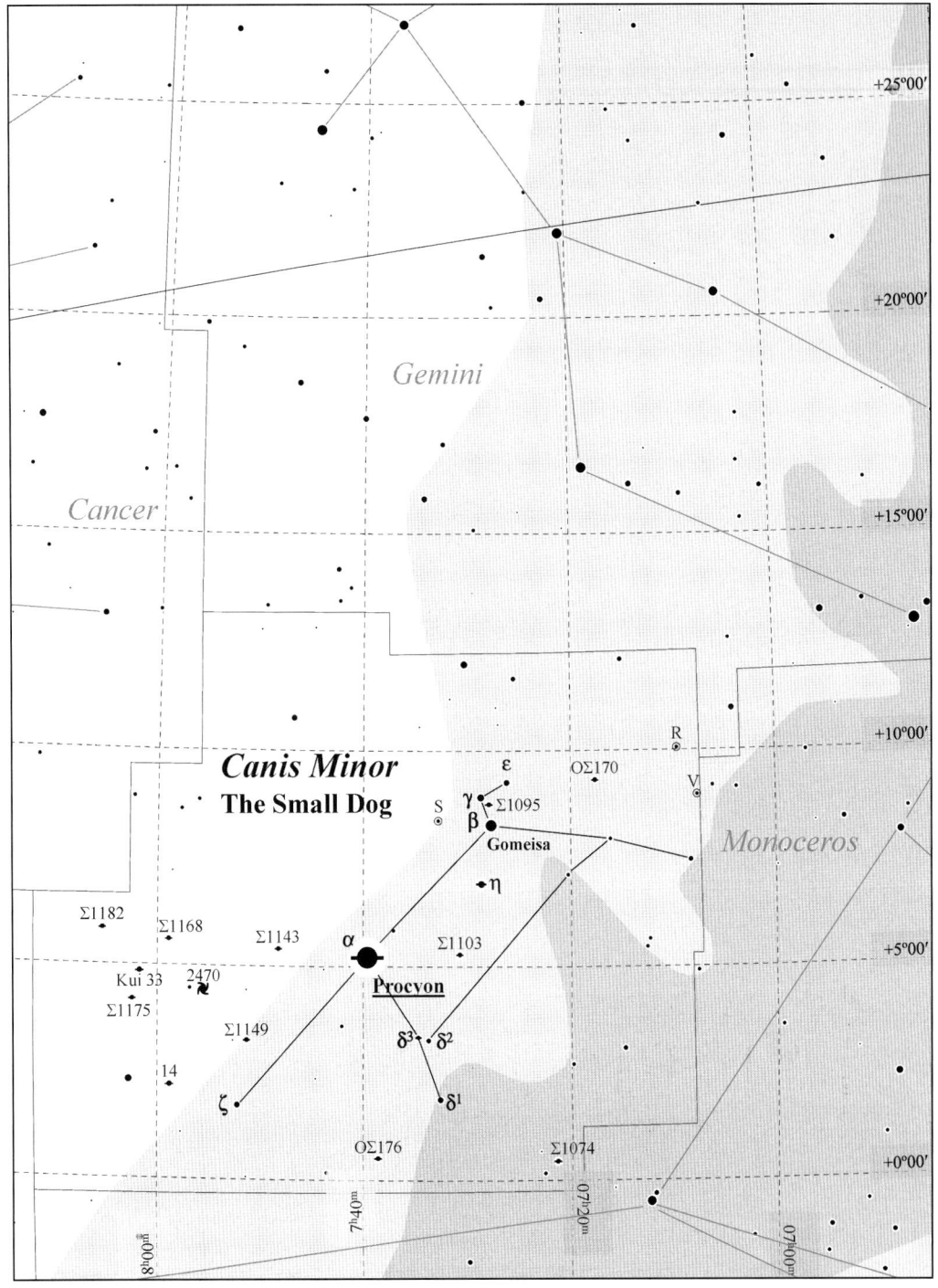

Fig. 7.12. Canis Minor details.

Table 7.3. Selected telescopic objects in Canis Minor.

Multiple stars:

Designation	R.A.	Decl.	Type	m₁	m₂	Sep. (″)	PA (°)	Colors	Date/ Period	Aper. Recm.	Rating	Comments
OΣ170	07ʰ 17.6ᵐ	+09° 18′	A–B	7.4	7.7	0.3	320	Y, –	355	36″	**	1° 15′ W (PA 270°) of ε CMi (m4.99). Min. 0.265″ in 2008, will reach 1.30″ in 2066
Σ1074	07ʰ 20.5ᵐ	+00° 24′	A–B	7.4	7.8	0.7	173	bW, –	1997	12/14″	***	1° 26′ ENE (PA 066°) of 24 Mon (m6.42). 27′ WNW of unnamed star on Norton M7
			AB–C	6.9	13.0	12.5	101	bW, –	1999	6/8″	***	a.k.a. BU 577 AB–C
			AB–D	6.9	13.1	15.3	015	bW, –	1999	6/8″	***	a.k.a. BU 577 AB–D
			AB–E	6.9	11.5	52.8	275	bW, –	1999	3/4″	***	a.k.a. BU 577 AB–E
Σ1095	07ʰ 27.4ᵐ	+08° 45′	A–B	8.6	9.4	10.4	077	W, pB	2004	2/3″	***	15′ SW (PA 228°) of γ CMi (m4.33). Very pretty
5, η Canis Minoris	07ʰ 28.0ᵐ	+06° 57′	A–B	5.3	11.1	4.1	028	W, –	1934	3/4″	***	Challenging due to closeness and faintness of secondary
Σ1103	07ʰ 30.6ᵐ	+05° 15′	A–B	7.1	8.6	3.8	246	pY, pB	1999	2/3″	***	2°11′ W (PA 271°) of Procyon
9, δ³ Canis Minoris	07ʰ 34.3ᵐ	+03° 22′	A–B	6.6	7.0	0.9	171	Y, Y	2003	10/12″	***	a.k.a. Σ1126AB
			AB–C	5.8	10.6	43.2	249	W, –	2000	3/4″	***	a.k.a. Σ1126AC
OΣ176	07ʰ 38.5ᵐ	+00° 30′	A–B	7.2	9.2	1.6	221	W, –	1995	6/8″	***	
			A–C	6.9	13.7	3.1	337	W, –	1913	8/10″	***	a.k.a. A 2529AC
10, α Canis Minoris	07ʰ 39.3ᵐ	+05° 14′	A–B	0.4	10.8	2.3	221	W, W	40.82	3/4″	*****	A–B similar to, but more difficult than Sirius, due to faintness of secondary. [9]
			A–C	0.4	11.7	148.4	017	yW, pO	1984	3/4″	***	Extremely difficult, but worth a try. A–C is much easier
Σ1143	07ʰ 48.1ᵐ	+05° 25′	A–B	6.6	11.0	9.3	152	yO, –	1825	3/4″	***	1° 33′ WNW (PA339°) of ζ CMi
Σ1149	07ʰ 49.5ᵐ	+03° 13′	A–B	7.8	9.2	21.7	041	pY, pB	2003	2/3″	***	(m5.12). Pretty colors
14 Canis Minoris	07ʰ 58.3ᵐ	+02° 13′	A–B	5.4	9.89	98.8	084	O, –	2004	2/3″	****	A–B can be seen with 7×50 binoculars, if held steady
			A–C	5.4	9.36	133.3	153	O, –	2004	2/3″	***	A–C can be seen with 10×70 binoculars, if held steady
Σ1168	07ʰ 58.8ᵐ	+05° 37′	A–B	6.8	10.6	6.0	218	W, –	1999	3/4″	***	2° 35′ NNE (PA 016°) of 14 CMi
Kui 33	08ʰ 01.2ᵐ	+04° 52′	A–B	5.7	12.4	31.1	076	W, –	2000	4/5″	***	(m5.40)
Σ1175	08ʰ 02.4ᵐ	+04° 09′	A–B	7.9	9.1	1.2	298	Y, –	1,590	8/10″	****	0° 47′ SSE (PA 158°) of KUI 33
Σ1182	08ʰ 05.4ᵐ	+05° 50′	A–B	7.5	8.8	4.4	073	W, brR	2002	7×50	***	1° 25′ NW (PA 048°) of KUI 33

Variable stars:

Designation	R.A.	Decl.	Type	Range (m$_v$)	Period (days)	F (f₂/f₁)	Spectral Range	Number of Stars	Aper. Recm.	Rating	Comments
V Canis Minoris	07^h 07.0^m	+08° 53′	M	7.4	15.1	366.1	0.39	M6e–M10		18/20″	***	Extremely red at minimum, but visible only in large amateur scopes ′10
R Canis Minoris	07^h 08.7^m	+10° 01′	M	7.3	11.6	337.78	0.48	C7, 1Je(CSep)		4/6″	***	Carbon type star, deep red
S Canis Minoris	07^h 32.7^m	+08° 19′	M	6.6	13.2	332.94	0.49	M6e–M8e		8/10″	***	

Star clusters:

Designation	R.A.	Decl.	Type	m$_v$	Size (′)	Brst. Star	Dist. (ly)	dia. (ly)	Number of Stars	Aper. Recm.	Rating	Comments
None												

Nebulae:

Designation	R.A.	Decl.	Type	m$_v$	Size (′)	Cent. star	Dist. (ly)	dia. (ly)		Aper. Recm.	Rating	Comments
None												

Galaxies:

Designation	R.A.	Decl.	Type	m$_v$	Size (′)		Dist. (ly)	dia. (ly)	Lum. (suns)	Aper. Recm.	Rating	Comments
NGC 2470	07^h 54.3^m	+04° 27′	Sab	12.7	2.0×0.6		–	–	–	12/14″	***	Bright core, elongated NW–SE

Meteor showers:

Designation	R.A.	Decl.	Period (yr)	Duration	Max Date	ZHR (max)	Comet/Asteroid	First Obs.	Vel. (mi/km/sec)	Rating	Comments
None											

Other interesting objects:

Designation	R.A.	Decl.	Type	m$_v$	Size (′)		Dist. (ly)	dia. (ly)	Lum. (suns)	Aper. Recm.	Rating	Comments
None												

Stars of magnitude 5.5 or brighter which have no Bayer or Flamsteed designations:

Desig-nation	Mag. (m$_v$)	Spec. type	Desig-nation	Mag. (m$_v$)	Spec. type
HD 055751	5.36	K0 III			

CORONA BOREALIS (kuh-**roh**-nuh bore-ee-**al**-iss)	**The Northern Crown**	
CORONAE BOREALIS (kuh-**roh**-nye bore-ee-**al**-iss)	**9:00 P.M. Culmination:** July 4	
CrB	**AREA:** 179 sq. deg., 73rd in size	

Another story about the adventures of Dionysus tells how he acquired the beautiful princess Ariadne as his wife.

After Zeus used the form of a tame white bull to carry Europa, the beautiful princess of Phoenicia, to Crete and seduce her, she gave birth to three sons. When they grew up, all three fought to become king of the island nation. Minos, one of the sons, publicly prayed for one of the gods to send a bull from the sea as a sign that they wanted him to become king. Poseidon, god of the seas, obliged and sent a beautiful white bull without any imperfections. This public display of approval convinced the other two brothers to cede any claim to the crown.

Although Minos had promised in his prayer to sacrifice the animal to the god which provided it, this bull was so beautiful that he convinced himself that Poseidon would not mind if he sacrificed another bull from his herd instead. Poseidon did not cause any immediate problem and Minos believed that his substitution was acceptable. Unfortunately, Poseidon was not forgiving. He did nothing to Minos directly, but caused Minos' wife Pasiphae to lust after the magnificent bull.

Unable to resist this urge, Pasiphae turned to the famed inventor and craftsman Daedalus who had fled to Crete and now worked for Minos. Daedalus devised a hollow wooden frame covered with a cowhide in which Pasiphae could hide. The ruse worked, and the white bull mated with her. The issue of this unnatural mating was a terrible monster with the head of a bull on the body of a man. As the apparent son of Minos, this monster became known as the Minotaur. The Minotaur turned out to be an uncontrollable man-eater which no one was able to kill.

Minos was furious with Daedalus for helping Pasipae seduce the bull, and would have had him killed except that Daedalus was too accomplished an inventor and craftsman to forfeit, and Minos was forced to turned to Daedalus for help. To hide this family embarrassment from the king's subjects, Daedalus devised the famous maze called the labyrinth. This maze was so complicated that no human venturing into it could find the way out without help.

Minos had earlier allowed Androgeus, one of his normal sons, to participate in a sports contest in Athens where he had been murdered. To avenge his son, Minos attacked and defeated Athens. As a condition of peace, he forced Athens to send seven young men and seven women selected by lot to Crete each year to be slaves. After the

Minotaur was safely concealed in the labyrinth, Minos had the young people thrown into the labyrinth as food for the Minotaur. There they wandered helplessly until the beast found and killed them.

One of the young Greek heroes who had already made a reputation for dispatching the bandits that had terrorized travelers to and from Athens decided to further his reputation by volunteering to replace one of the young men being sent to Crete. He vowed that he would bring the latest sacrificial group back safely. When the group arrived in Crete and debarked from their ship, Minos and all his family were present to verify the annual tribute. One of Minos' daughters, Ariadne, fell in love with the handsome hero Theseus. During the night she slipped into the prison and told Theseus that she would help him escape the labyrinth if he would marry her and take her away from her cruel father.

Theseus agreed to this proposal, and when he was thrown into the labyrinth, Ariadne secretly provided him with a sword and a ball of silken thread, and with instructions to unwind it as he hunted the Minotaur, and to rewind it to find his way out. Theseus was successful in finding and killing the Minotaur. Waiting until late in the night, Theseus made his way back to the entrance and freedom. He then released the other prisoners and stole a ship on which they sailed back toward Athens, taking Ariadne with him.

When they reached the island of Naxos, they decided to go ashore for rest and fresh water. All the crew and passengers had returned to the ship except Ariadne when a sudden storm caused the ship to drag its anchor into deeper water and blow away in the wind. The heartbroken Ariadne was stranded on the island while her hope of a happy life in Athens drifted out of sight.

However, Ariadne was not stranded for long. That night, Dionysus, who had been watching his favorite island, came ashore with a band of revelers. Dionysus had been attracted to Ariadne and began to court her. He succeeded in winning her heart and marrying her. In spite of his association with wine and wild parties, Dionysus loved Ariadne and was faithful to her the rest of her life. When she inevitably died, Dionysus took the jewel incrusted crown he had given her as a wedding present and threw it high into the heavens where the jewels glitter to this day as the stars of Corona Borealis, the Northern Crown.

TO FIND: Corona Borealis is fairly easy to find in spite of having only one star brighter than 4th magnitude. First, Corona Borealis has one of the most easily recognized shapes of all constellations, somewhat more than half a circle of fairly regularly spaced stars. Second, It is easily found by locating Arcturus (α Boo) and Vega (α Lyr), the two brightest stars in the summer sky. Imagine a line between Arcturus and Vega, and divide it into three equal parts. The point one-third of the way from Arcturus to Vega lies inside the arc of Corona Borealis. The point one-third of the way from Vega to Arcturus lies in the middle of an asterism of four stars called "the Keystone" from its shape. The Keystone is the "key" to recognizing Hercules. See Chap. 6 for more details (Figs. 7.1–7.3, 7.13 and 7.14).

KEY STAR NAMES:

Name	Pronunciation	Source
α Alphecca or Gemma	uhl-**fek**-kuh **jim**-muh	al-fakka (Ind-Arabic), "the broken (circle)" gemma (Latin), "jewel"
β Nusakan	**new**-suh-kan)	al-nusaq n (Ind-Arabic), two lines (of stars which marked the boundaries of the Arabic constellation of The Pasture)

Magnitudes and spectral types of principal stars:

Bayer desig.	Mag. (m_v)	Spec. type	Bayer desig.	Mag. (m_v)	Spec. type	Bayer desig.	Mag. (m_v)	Spec. type
α	2.22	A0 V	β	3.66	F0p0	γ	3.81	A1 Vs
δ	4.59	G5 III/IV	ε	4.14	K3 III	ζ^2	4.64	B7 V
η^A	4.99	F8 V	η^B	6.08	G0 V	θ	4.14	B6 Vnn
ι	4.98	A0p	κ	4.79	K0 III/IV	λ	5.43	F0 IV
μ	5.14	M2 III	ν^1	5.20	M2 III	ν^2	5.40	K5 III
ξ	4.86	K0 III	o	5.51	K0 III	π	5.57	G9 III
ρ	5.39	G2 V	σ	5.23	F8 V	τ	4.73	K0 III/IV
υ	5.89	A3 V						

Fig. 7.13a. The stars of Corona Borealis and adjacent constellations.

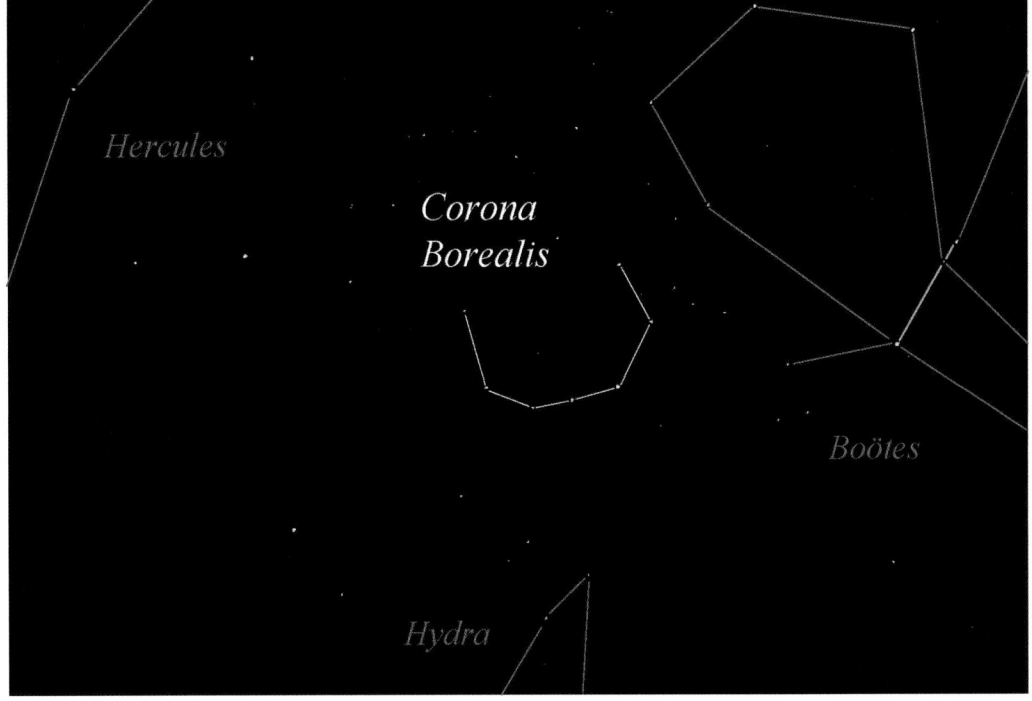

Fig. 7.13b. The figures of Corona Borealis and adjacent constellations.

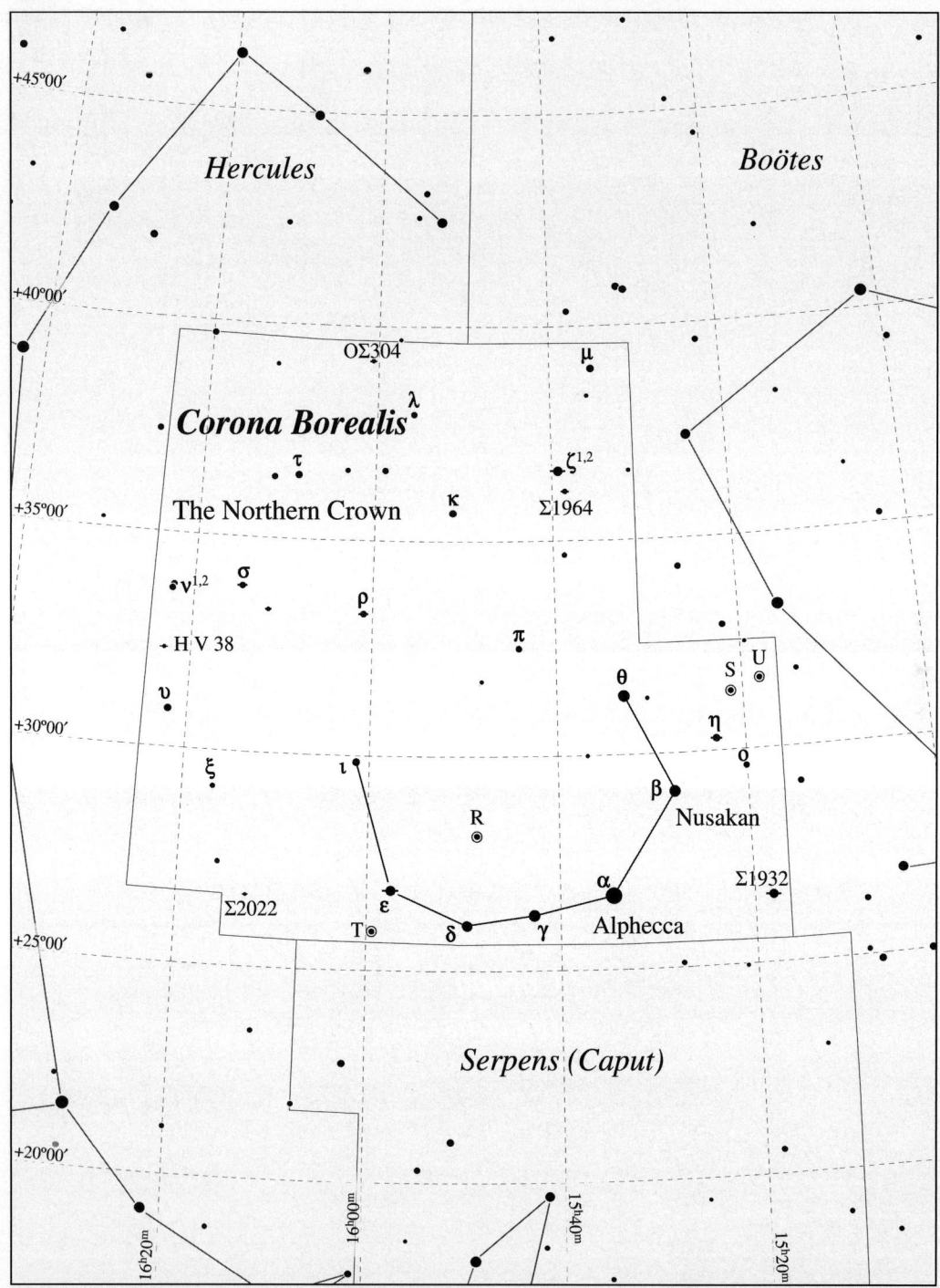

Fig. 7.14. Corona Borealis details.

Table 7.4. Selected telescopic objects in Corona Borealis.

Multiple stars:

Designation	R.A.	Decl.	Type	m_1	m_2	Sep. (")	PA (°)	Colors	Date/ Period	Aper. Recm.	Rating	Comments
Σ1932AaB	15h 18.3m	+26° 50'	Aa–B	7.3	7.4	1.6	263	Y, Y	2006	6/7"	****	3° 40' W (PA 273°) of α CrB. A nearly perfectly matched pair *11
2 = η Coronae Borealis	15h 23.2m	+30° 17'	A–B	5.6	5.9	**0.6**	**171**	pY, Y	**41.585**	16/18"	***	a.k.a. Σ1937AB. A fairly well matched pair *12
			AB–C	5.0	12.5	69.2	000	Y, ?	1984	4/5"	***	
			AB–D	5.0	10.9	217.5	049	Y, ?	2000	15×80	***	
Σ1964	15h 38.2m	+36° 14'	A–B	8.0	9.9	1.3	080	yW, ?	1995	7/8"	***	0° 28' SSW (PA 211°) of ζ CrB. Nearly perfectly matched pair
	15h 38.2m	+36° 14'	A–C	7.7	7.6	15.3	085	yW, yW	2004	2/3"	****	
7, ζ CrB	15h 39.4m	+36° 38'	A–B	5.1	6.0	6.3	306	B, Gr	2006	2/3"	*****	Colors are subtle, but definitely different. Some see bW and B
Coronae Borealis	15h 57.6m	+26° 52'	A–B	4.2	12.6	2.1	010	pO, ?	1965	5/6"	***	1° 35' NE (PA 039°) of λ CrB. Jupiter sized exoplanet. See below
OΣ304AB	16h 00.9m	+39° 10'	A–B	6.7	10.8	10.4	173	W, ?	2003	3/4"	***	
ρ Coronae Borealis	16h 01.0m	+33° 18'	A–B	5.5	8.7	135.3	071	Y, ?	2002	2/3"	***	
OΣ305AB	16h 11.6m	+33° 20'	A–B	7.3	7.4	5.6	263	yW, yW	1991	2/3"	****	Nearly perfect match
Σ2022	16h 12.7m	+26° 40'	A–B	6.4	10	2.3	151	yW, ?	1998	4/5"	***	3° 23' E (PA 093°) of ξ CrB
17, σ Coronae Borealis	16h 14.7m	+33° 51'	A–B	5.8	6.7	**7.2**	**238**	yW, oY	**888.989**	2/3"	****	Slowly opening to max. of 9.04" in 2246
			A–C	5.6	13.1	18.0	103	yW, ?	1984	6/7"	***	
			A–D	5.8	10.8	88.4	082	Y, ?	1996	15×80	***	
ν¹, ν² CrB	16h 22.4m	+33° 48'	A–B	5.4	5.6	360.8	164	Y, Y	2001	6×30	****	Very closely matched pair
H V 38	16h 22.9m	+32° 20'	A–B	6.4	9.8	31.6	017	pY, pB	2005	2/3"	****	1° 27' N (PA 007°) of ξ CrB

Variable stars:

Designation	R.A.	Decl.	Type	Range (m_J)		Period (days)	F (f_i/f_t)	Spectral Range	Aper. Recm.	Rating	Comments
U Coronae Borealis	15h 18.2m	+31° 39'	EA/SD	7.0	8.4	2.452	0.14	B6V + F8III–IV	6×30	****	##
S Coronae Borealis	15h 21.4m	+31° 22'	M	5.8	14.1	360.26	0.35	M6e–M8e	12/14"	****	##
R Coronae Borealis	15h 48.6m	+28° 09'	R CrB	5.71	14.8	–	–	C0O(F8pep)	16/18"	*****	Suddenly dims, slowly and irregularly recovers – prototype for its class. ##$ *13

(continued)

Table 7.4. (continued)

Variable stars:

Designation	R.A.	Decl.	Type	Range (m_v)	Period (days)	F (f_r/f_f)	Spectral Range	Aper. Recm.	Rating	Comments
T Coronae Borealis	15^h 59.5^m	+25^d 55'	NRa	2.0	10.8	–	M3III + pec	3/4"	****	Recurrent nova, 1866 and 1946. The "Blaze Star."##$

Star clusters:

Designation	R.A.	Decl.	Type	m_v	Size (')	Brtst. Star	Dist. (ly)	dia. (ly)	Number of Stars	Lum. (suns)	Aper. Recm.	Rating	Comments
None													

Nebulae:

Designation	R.A.	Decl.	Type	m_v	Size (')	Brtst./cent. star	Dist. (ly)	dia. (ly)	Aper. Recm.	Rating	Comments
None											

Galaxies:

Designation	R.A.	Decl.	Type	m_v	Size (')	Dist. (ly)	dia. (ly)	Lum. (suns)	Aper. Recm.	Rating	Comments
None											

Meteor showers:

Designation	R.A.	Decl.	Period (yr)	Duration	Max Date	ZHR (max)	Comet/Asteroid	First Obs.	Vel. (mi/km/sec)	Rating	Comments
No of significant showers											

Other interesting objects:

Designation	R.A.	Decl.	Type	m_v	Mass (suns)	dia. (suns)	Dist. (ly)	Planet Mass (J)	Dist. (AU)	Eccentricity	Discl.	Comments
ρ Coronae Borealis	16^h 01.0^m	+33° 18'	Exo-planet	5.4	6.8	17.43	10.4	39.845	0.22	0.04	1997	13th exoplanet discovered, 8th around a star visible to the unaided eye

VIRGO (vur-go)	The Young Woman
VIRGINIS (vur-gin-is)	9:00 P.M. Culmination: May 27
Vir	AREA: 1,294 sq. deg., 2nd in size

TO FIND: Continue the mnemonic "arc to Arcturus, spike to Spica" in a straight line 33° past <u>Arcturus</u> to <u>Spica</u>, the brightest star (mag. 0.98) in Virgo, and the only one brighter than 3rd magnitude. Virgo can be visualized as a woman lying on her back, but slightly turned toward us, with her head to the west and feet to the east. <u>Spica</u> marks her left hand which is holding a spike, or ear, of wheat. About 15° NW of Spica is γ Virginis (Porrima, mag 2.74), the second brightest star, which marks Virgo's left shoulder. Auva (δ Vir, m3.39) lies 6° NE of Porrima and forms Virgo's right shoulder. Another 8° to the NNE is Vindemiatrix (ε Vir, m2.85) which represents Virgo's right hand which holds an olive branch. Returning to γ Vir, we can go 5.5° slightly N of due West to find Zaniah (η Vir, m3.89) which lies in the back of Virgo's neck, and another 8° WNW to Zavijava (β Vir, m3.59) which designates the back of Virgo's neck. Three stars, 16, o, and ν Virginis lie above Zavijava and with it form a kite shape lying on its side. This "kite" forms Virgo's head. Returning to <u>Spica</u>, we find θ Vir (m 4.38) about 6° NW of it and 72 Vir (m6.10) about 5° NNE. These two stars, respectively, indicate the back and left hip of Virgo. From 72 Vir, Syrma (ι Vir, m4.07) about 11° due E and μ Vir (m3.87) another 7° due E form Virgo's left leg and foot. Heze (ζ Vir, m3.38) is about 6° N of 72 Vir and lies at the right hip of Virgo. Slightly N and 7° E of ζ Vir is τ Vir (m4.23) and 11° due E of τ is 109 Vir (m3.73). These two stars form Virgo's right knee and foot (Figs. 7.1, 7.15 and 7.16).

KEY STAR NAMES:

Name	Pronunciation	Source
α Spica	**spee**-ku or **spy**-kah	spica (Latin), "ear of corn or wheat"
β Zavijava	zah-veh-**jaw**-vah	zāwiyat al-ʿawwāʾ, "the angle of al-ʿawwāʾ" (ind-Arabic), originally applied to γ Vir, erroneously applied to β. Al-ʿawwāʾ was one of the ind-Arabic lunar mansions, but the meaning of the name is unknown
γ Porrima	**por**-rih-muh	Name of one of the Roman Camenae, the minor goddesses who ruled the arts and sciences, similar to the Greek Muses
δ Auva	**ow**-vah [?]	"the barker" (ind-Arabic) from an Arabic constellation consisting of a kennel of dogs barking at the lion, another Arabic constellation (not Leo)
ζ Heze	**heh**-zeh [?]	Origin uncertain
ε Vindemiatrix	ven-**day**-me-**AY**-triks	precedens vindemiatorem (Latin), "the one preceding the Grape gatherer," ultimately from the Greek word for a grape gatherer

Name	Pronunciation	Source
η Zaniah	**zah**-nee-yuh	Al Zawiah "the Angle," like β Vir, erroneously applied
ι Syrma	**sir**-muh	συρμα (Greek), "hem" (of maiden's robe)
μ Rigl al Awwa	**rye**-jel al	rijl al-ʿawwāʾ (Ind-Arabic), "the foot of al-ʿawwāʾ." See awe-**wah** (β Virginis)

Magnitudes and spectral types of principal stars:

Bayer desig.	Mag. (m_v)	Spec. type	Bayer desig.	Mag. (m_v)	Spec. type	Bayer desig.	Mag. (m_v)	Spec. type
α	0.98	B1 V	β	3.59	F8 V	γ	2.74	F0 V
δ	3.39	M3 III	ε	2.85	G8 IIIvar	ζ	3.38	A3 V
η	3.89	A2 IV	θ	4.38	A1 V	ι	4.07	F7 V
κ	4.18	K3 IV	λ	4.52	A1 V	μ	3.87	F2 III
ν	4.04	M0 III	ξ	4.84	A4 V	ο	4.92	A1p (Sr, Cr, Eu)
π	4.65	A5 V	ρ	4.88	A0 V	σ	4.78	M2 III
τ	4.23	A3 V	υ	5.14	G9 III	φ	4.81	G2 III
χ	4.66	K2 III	ψ	4.77	M3 IIIvar	ω	5.24	M4 III

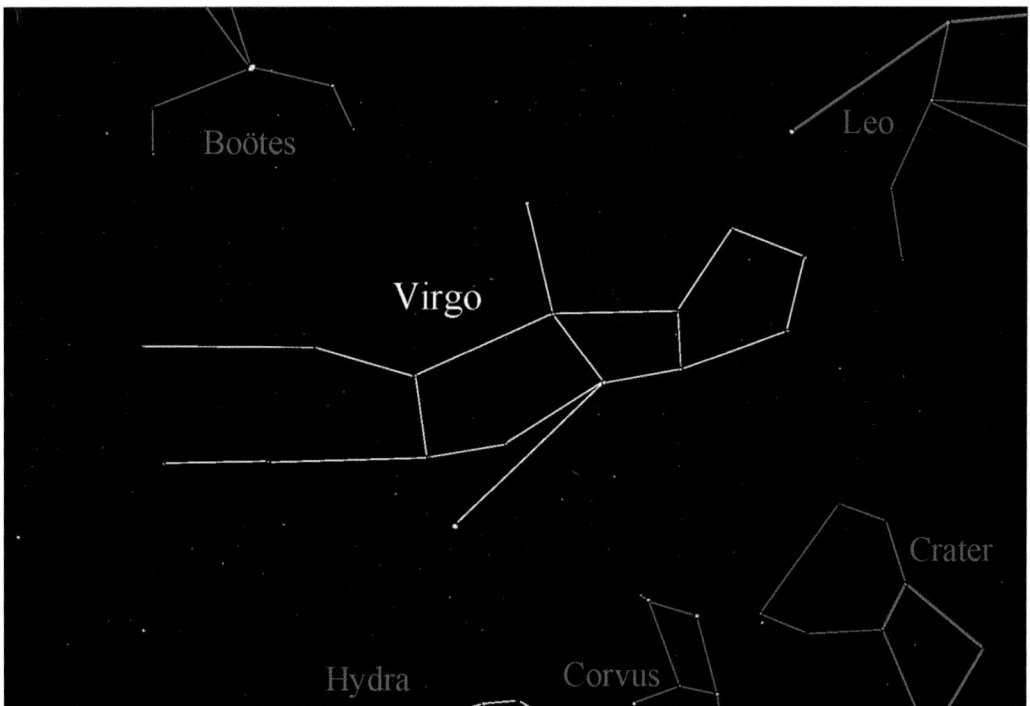

Fig. 7.15. (a, b) The stars and constellation of Virgo.

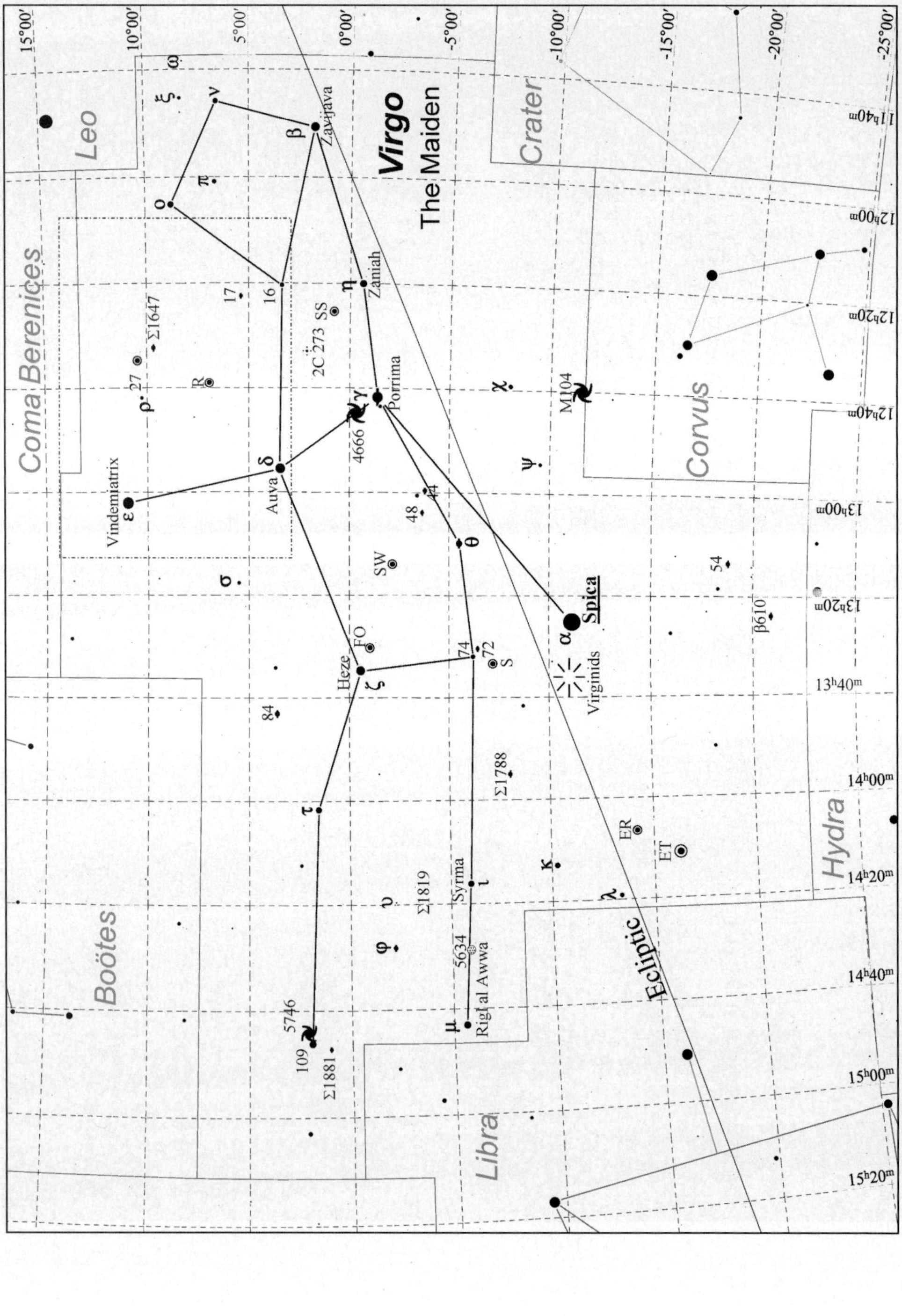

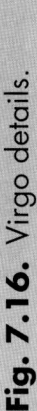

Fig. 7.16. Virgo details.

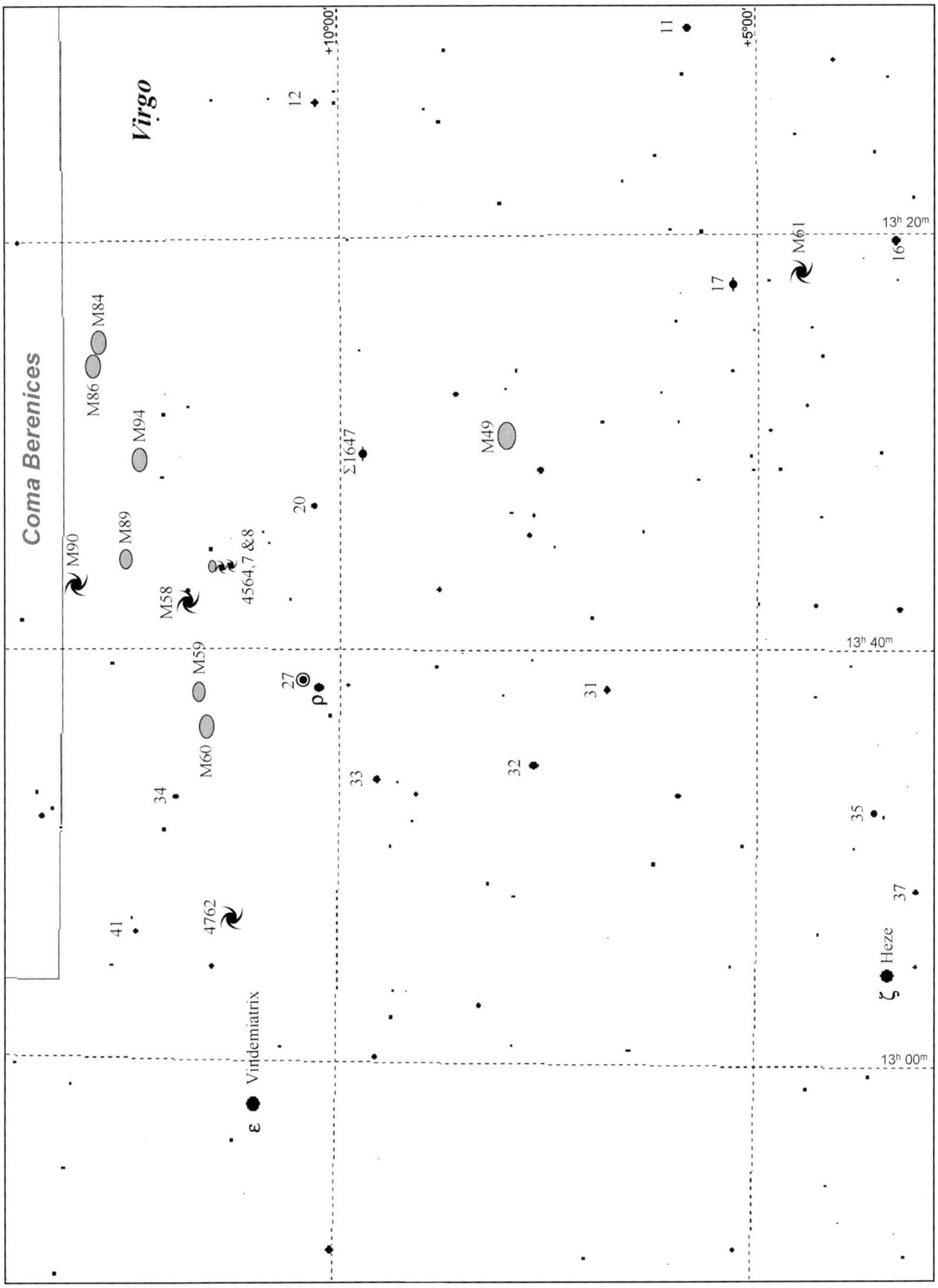

Fig. 7.17. Coma-Virgo Galaxy Cluster Details.

Table 7.5. Selected telescopic objects in Virgo.

Multiple stars:

Designation	R.A.	Decl.	Type	m₁	m₂	Sep. (")	PA (°)	Colors	Date/ Period	Aper. Recm.	Rating	Comments
17 Virginis	12ʰ 22.5ᵐ	+05° 18'	A–B	6.6	10.5	20.2	337	Y, pO	2004	2/3"	****	Nice, easy pair for small telescope. a.k.a. STF1636
Σ1647	12ʰ 30.6ᵐ	+09° 42'	A–B	7.5	7.7	**1.26**	**250**	–	**4,273**	7/8"	****	Nearly matched pair
29, γ Virginis	12ʰ 41.7ᵐ	-01° 27'	A–B	3.48	3.53	**1.46**	**020**	yW, yW	**168.9**	6/8"	*****	*Porrima.* Passed min. sep. in early 2005, rapidly separating; will reach 0.6" in 2088
44 Virginis	12ʰ 59.7ᵐ	-03° 49'	Aa–B	5.8	10.2	20.8	053	W, ?	1999	15×80	****	a.k.a. CHR 39Aa
48 Virginis	13ʰ 03.9ᵐ	-03° 40'	A–B	7.1	7.7	0.6	195	yW, yW	2005	16/18"	****	Good test object for 6" telescope. a.k.a. Bu 929
51, θ Virginis	13ʰ 10.0ᵐ	-05° 32'	Aa–B	4.4	9.4	6.9	342	W, grB	1999	2/3"	****	a.k.a. STF 1724Aa–B
H III 50			Aa–C	4.4	10.4	71.1	298	W, ?	2002	15×80		
54 Virginis	13ʰ 13.4ᵐ	-18° 50'	A–B	6.8	7.2	5.4	034	W, W	2006	2/3"	******	Easy pair. a.k.a. SHJ 151
BU 610	13ʰ 24.0ᵐ	-20° 55'	A–B	6.6	10.1	4.2	017	W, pB	1991	3/4"	****	
72 Virginis	13ʰ 30.4ᵐ	-06° 28'	A–B	6.1	10.7	29.9	017	yW, pV	1999	15×80	****	STF1750
84 Virginis	13ʰ 43.1ᵐ	+03° 32'	A–B	5.6	8.3	2.8	228	O, Y	2006	3/4"	*****	Pretty colors. a.k.a. STF1777
Σ1788 (Virgo)	13ʰ 55.0ᵐ	-08° 04'	A–B	6.7	7.3	**3.56**	**099**	pY, Y	**2,613**	2/3"	*****	Great multiple for small telescopes. a.k.a. STF1788
			A–C	6.2	10.3	117.3	297	pY, ?	1999	2/3"		
			A–D	6.2	10.9	155.0	215	pY, ?	1924	3/4"		
Σ1819 (Virgo)	14ʰ 15.3ᵐ	+03° 07'	A–B	7.0	7.1	**0.88**	**179**	Y, ?	**220**	10/12"	*****	
105, Virginis	14ʰ 28.2ᵐ	-02° 14'	A–B	4.9	9.3	5.3	112	dY, O	1998	2/3"	******	Nice system of three differently colored stars
			A–C	4.8	12.6	91.1	202	dY, B	1999	5/6"		
STF1881	14ʰ 47.1ᵐ	+00° 58'	A–B	6.7	8.8	3.5	359	bW, W	2006	2/3"	****	

Variable stars:

Designation	R.A.	Decl.	Type	Range (mᵥ)		Period (days)	F (f_r/f_f)	Spectral Range	Aper. Recm.	Rating	Comments
SS Virginis	12ʰ 25.2ᵐ	+00° 46'	M	6.0	9.6	364.14	0.48	C6,3e(Ne)	10×80	*****	One of the 10 best carbon stars. 1° 57' NE (PA 043°) of η Vir (m3.89).##
R Virginis	12ʰ 38.5ᵐ	+06° 59'	M	6.1	12.1	145.63	0.50	M3.5IIIe–M8.5e	6/8"	****	Deep red at minimum. 0° 52' WNW (PA 282°) of 31 Virginis (m5.57).## `14
27, GG Virginis	12ʰ 41.6ᵐ	+10° 26'	δ Sct	6.2	6.3	–	–	A7V	6×30	***	0° 12' NNW (PA 338°) of ρ Virginis. 12th mag. NGC 4608 is in same FOV
SW Virginis	13ʰ 14.1ᵐ	-02° 48'	SRb	6.4	7.9	150	–	SM7III	6×30	***	2° 55' NNE (PA 021°) of θ Virginis (m4.38)
FO Virginis	13ʰ 29.8ᵐ	-01° 06'	EB/KE	6.5	6.8	0.776	–	A7V	6×30	***	2° 5' NW (PA 324°) of ζ Virginis (m3.38)

	13h 33.0m	-07° 12'	M	6.3	13.2	377	—	M6IIIe–M9.5e	8/10"	****	0° 58' SSE (PA 164°) of 74 Virginis (m5.57). ##
S Virginis											
ER Virginis	14h 06.7m	-14° 12'	SRb	6.5	6.6	55	—	M4III	6×30	***	3° 7' ENE (PA 075°) of λ Virginis (m4.51).
ET Virginis	14h 10.8m	-16° 18'	SRb	4.8	5.0	80	—	M2IIIa	Eye	***	1° 32' ENE (PA 072°) of λ Virginis (m4.51).

Star clusters:

Designation	R.A.	Decl.	Type	m_v	Size (')	Brtst Star	Dist. (ly)	dia. (ly)	Number of Star	Aper. Recm.	Rating	Comments
NGC 5634	14h 29.6m	-05° 59'	Globular	9.4	4.9	—	82K	116		8/10"	***	Faint, unresolved in 14/16" scope

Nebulae:

Designation	R.A.	Decl.	Type	m_v	Size (')	Brtst./ Cent. star	Dist. (ly)	dia. (ly)	Aper. Recm.	Rating	Comments
None											

Galaxies:

Designation	R.A.	Decl.	Type	m_v	Size (')	Dist. (ly)	dia. (ly)	Lum. (suns)	Aper. Recm.	Rating	Comments
M61, NGC 4303	12h 21.9m	+04° 28'	SAB	9.7	6.0×5.9	65M	110K	44G	4/6"	*****	Bright, face-on, N–S bar. Same FOV: NGC 4294 m12.2 and 4301 m13.6
M84, NGC 4374	12h 25.1m	+12° 53'	E1	9.1	5.1×4.1	65M	100K	76G	4/6"	*****	Giant elliptical. Same FOV: M86 and 4 fainter galaxies (m10.0–11.7)
M86, NGC 4406	12h 26.2m	+12° 57'	E3	8.9	12.0×9.3	50M	175K	57G	4/6"	******	Broad central brightening
M49, NGC 4472	12h 29.8m	+08° 00'	E2	8.4	8.1×7.1	65M	153K	145G	4/6"	*****	One of the longest and brightest of Coma-Virgo cluster. Same FOV: NGC 4470, m12.1
M87, NGC 4486	12h 30.8m	+12° 24'	E0	8.6	7.1×7.1	65M	135K	135G	4/6"	*****	a.k.a. 3C 247. Giant elliptical. Same FOV: NGC 4450, m 11.6 and 4451, m12.0
M89, NGC 4552	12h 35.7m	+12° 33'	E0	9.8	3.4×3.4	65M	64K	40G	4/6"	*****	Giant elliptical. Same FOV: M90, or NGC 4550 and 4551, m 11.6 and 12.0
M90, NGC 4569	12h 36.8m	+13° 10'	SAB	9.5	10.5×4.4	50M	150K	30G	4/6"	******	Giant spiral, mottled appearance. Same FOV: M89
M58, NGC 4579	12h 37.7m	+11° 49'	SAB	9.7	6.0×4.8	55M	85K	2.9G	4/6"	****	Bright stellar nucleus, moderately bright core, faint halo

(continued)

Table 7.5. (continued)

Galaxies:

Designation	R.A.	Decl.	Type	m_v	Size (')	Dist. (ly)	dia. (ly)	Lum. (suns)	Aper. Recm.	Rating	Comments
M104, NGC 4594	12h 40.0m	−11° 37'	SA	9.6	8.6×4.2	65M	160K	2.4G	4/6"	*****	"Sombrero Galaxy." Shape and dust lane visible even in small scopes
M59, NGC 4621	12h 42.0m	+11° 39'	E3	9.6	5.3×4.0	41M	60K	2.3G	4/6"	*****	Same FOV: M60 + NGC 4606, 4607, 4638, and 4647, m11.2–12.8
M60, NGC 4649	12h 43.7m	+11° 33'	E1	8.8	7.6×6.2	41M	80K	5.5G	4/6"	*****	Same FOV: M59 + NGC 4606, 4607, 4638, and 4647, m11.2–12.8
NGC 4666	12h 45.1m	−00° 28'	SB	10.7	4.1×1.3	45M	44K	1.3G	8/10"	*****	Nice edge-on galaxy. Larger scopes show two dust lanes. Mag. 13.1 NGC 4668 in FOV
NGC 4762	12h 52.9m	+11° 14'	SB	10.3	9.1×2.2	55M	95K	1.6G	8/10"	*****	Lovely NNE-SSW needle with bright surrounding star field
NGC 5746	14h 44.4m	+01° 41'	SA?	11.9	6.8×1.0	95M	162K	7.6G	8/10"	*****	Beautiful edge-on galaxy with central bulge. Dust lane glimpsed in larger scopes

Meteor showers:

Designation	R.A.	Decl.	Period (yr)	Dura-tion	Max Date	ZHR (max)	Comet/Asteroid	First Obs.	Vel. (mi/km/sec)	Rating	Comments
α Virginids	13h 36m	−11°	Annual	Mar. 10 – May 6	Nov. 4	4–6	Unknown	1895	18	***	Discovered by A. S. Herschel. See Note A for additional details

Other interesting objects:

Designation	R.A.	Decl.	Type	m_v	Size (')	Dist. (ly)	dia. (ly)	Lum. (suns)	Aper. Recm.	Rating	Comments
3C 273	12h 29.1m	+02d 03m	Quasar	12.8		2.44G	~100AU	4T	8/10"	***	First quasar to be identified as very distant and extremely luminous

Second Other interesting objects (Galaxy Clusters):

Designation	R.A.	Decl.	Type	m_v	Size (')	Dist. (ly)	dia. (ly)	Lum. (suns)	Aper. Recm.	Rating	Comments
Virgo Galaxy Cluster	12ʰ 31ᵐ	+12° 24'	Gal. Clust.		480 (8°)	55M	15M	–	–	*****	This cluster has ~300 large galaxies, and ~2,000 smaller ones *16
NGC 4564	12ʰ 36.4ᵐ	+11° 26'	Close	11.1	3.2 × 1.8	55M	43K	8.1G	10/12"	***	NGC 4564, 67, and 68 are often overlooked because they are somewhat fainter than the surrounding galaxies, but 4567 and 4568 (sometimes called the "Siamese Twins") appear to overlap, and NGC 4564 is only 10' to the north and visible in the same FOV. Group is ~1° 48' NW (PA 312°) of ρ Virginis
NGC 4567	12ʰ 36.5ᵐ	+11° 15'	Trio of	11.3	3.0 × 2.1	55M	45K	7.4G	10/12"	***	
NGC 4568	12ʰ 36.6ᵐ	+11° 14'	galaxies	10.8	4.5 × 2.0	55M	70K	13G	10/12"	***	

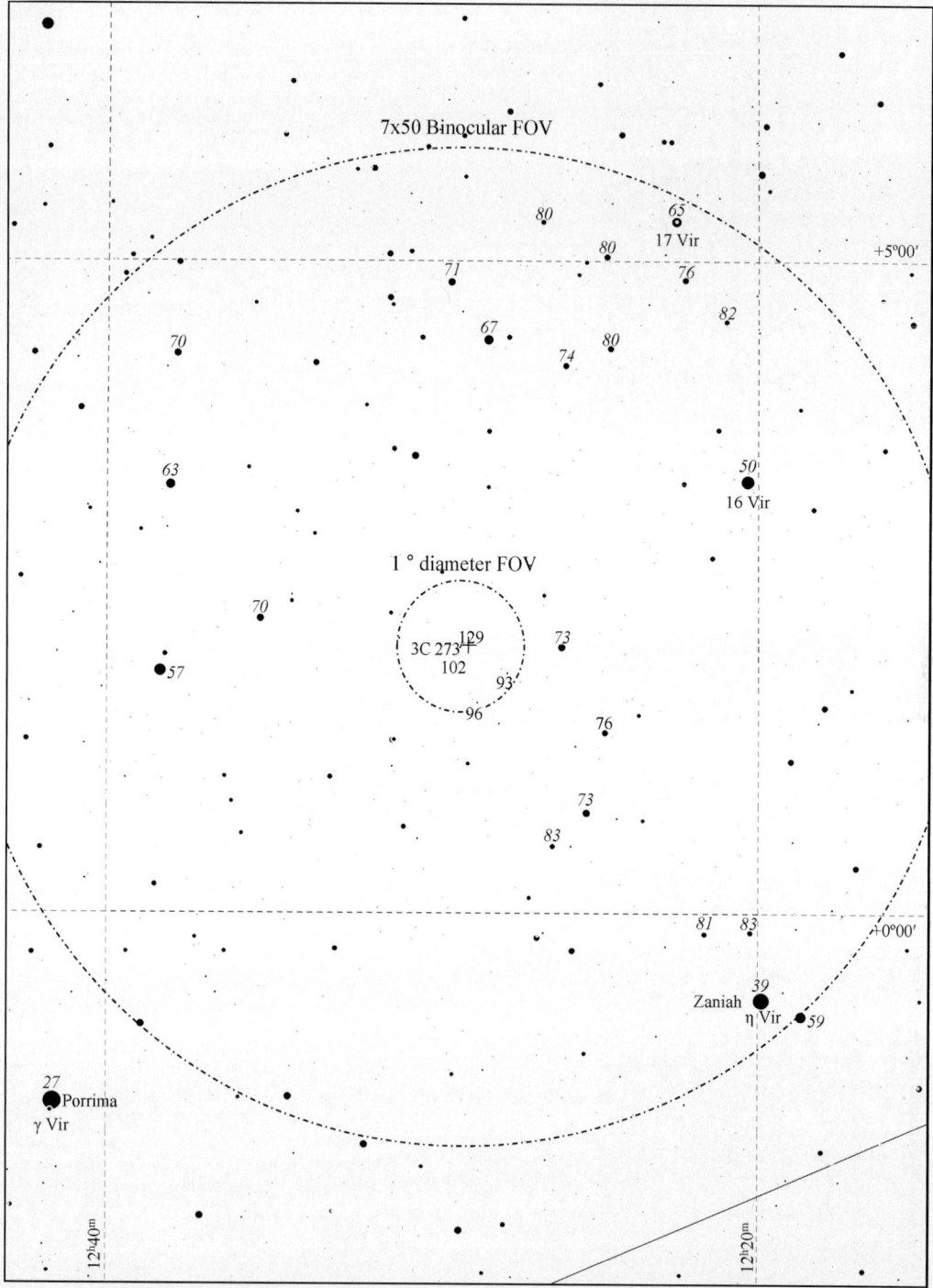

Fig. 7.18. 3C 273 Finder chart.

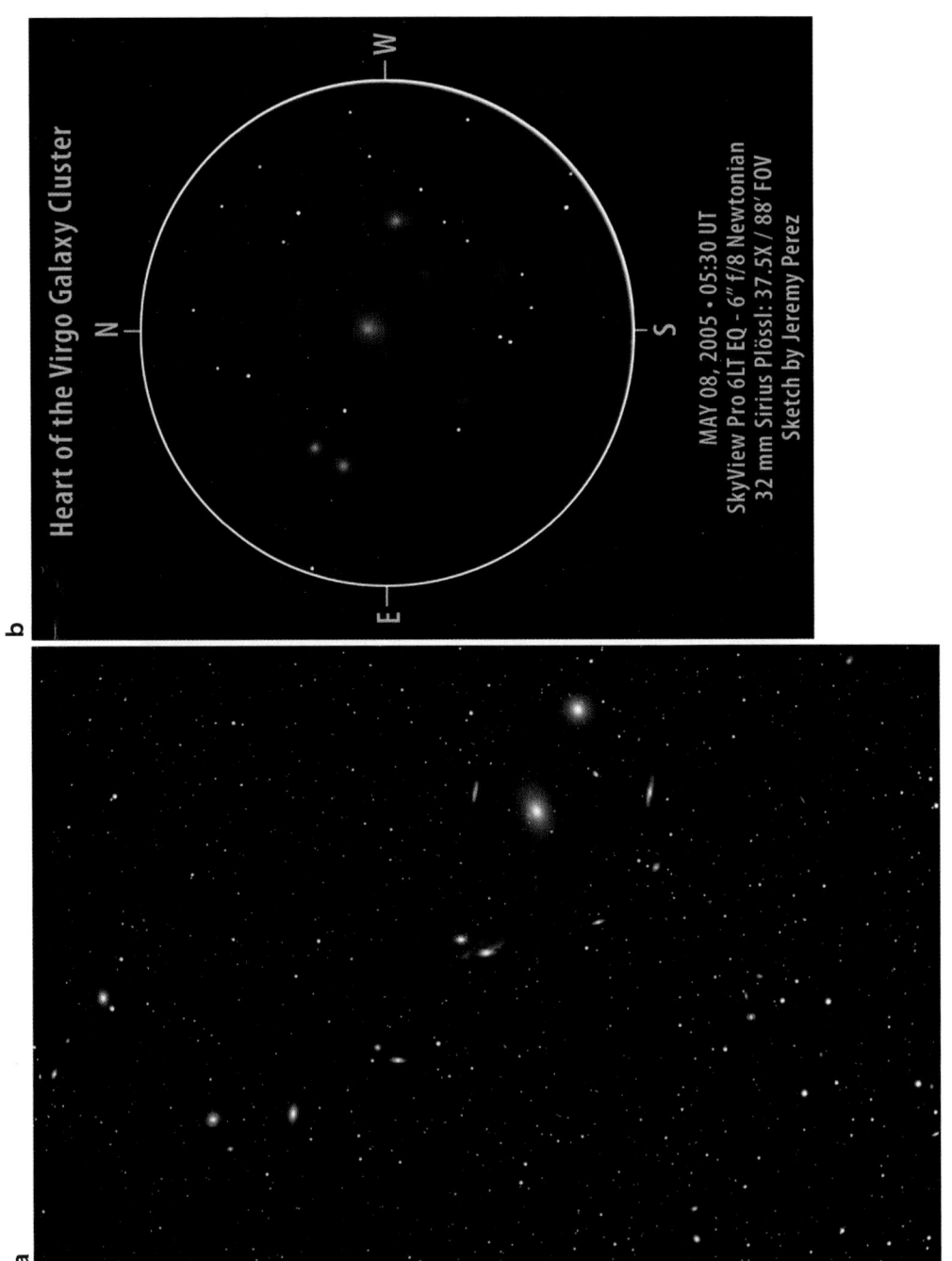

Fig. 7.19. (a) Central portion of Virgo/Coma Cluster, a.k.a. Makarian's Chain. M84 (NGC 4374) is the brightest galaxy, at center right. Astrophoto by Stuart Goosen. (b) Central portion of Virgo/Coma Cluster. M86 (NGC 4406) is in the center, M84 to right, NGC 4435 (upper) and NGC 4438 (lower) are on the left. Drawing by Jeremy Perez.

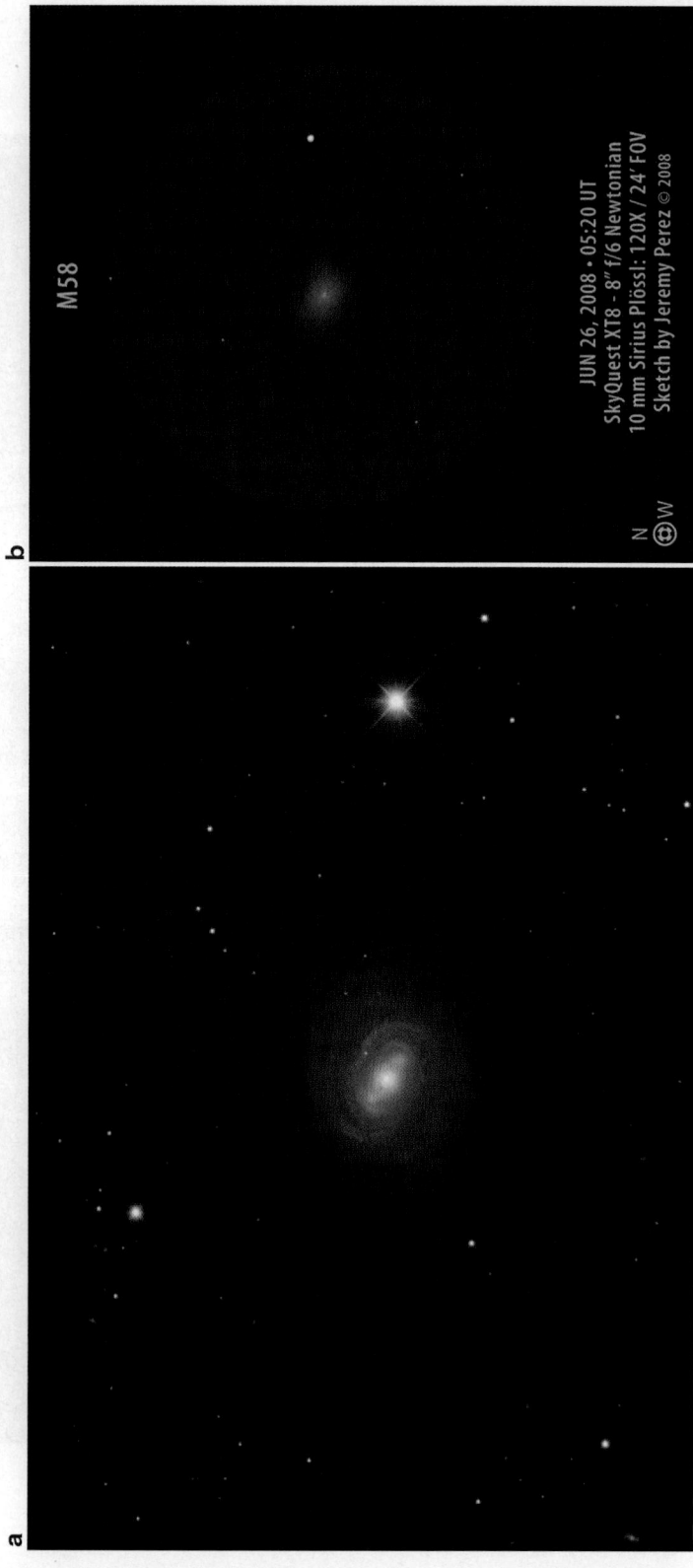

M58

JUN 26, 2008 • 05:20 UT
SkyQuest XT8 - 8" f/6 Newtonian
10 mm Sirius Plössl: 120X / 24' FOV
Sketch by Jeremy Perez © 2008

N
⊕W

Fig. 7.20. (a) M58 (NGC 4579), spiral galaxy in Virgo. Astrophoto by Rob Gendler. (b) M58. Drawing by Jeremy Perez.

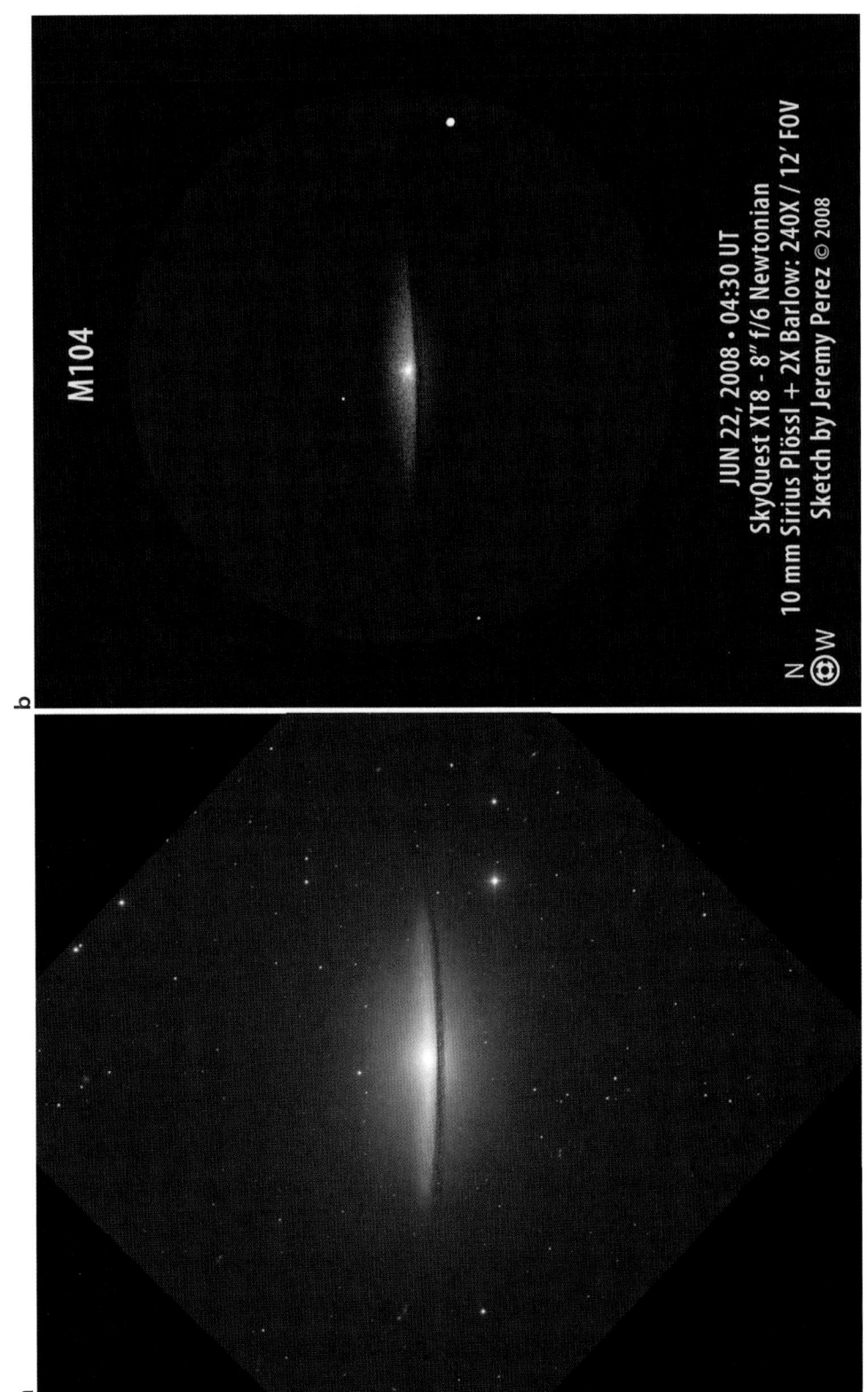

M104

JUN 22, 2008 • 04:30 UT
SkyQuest XT8 - 8" f/6 Newtonian
10 mm Sirius Plössl + 2X Barlow: 240X / 12′ FOV
Sketch by Jeremy Perez © 2008

N
⊕W

Fig. 7.21. **(a)** M104 (NGC 4594), the "Sombrero Galaxy" in Virgo. Astrophoto by Rob Gendler. **(b)** M104. Sketch by Jeremy Perez.

Fig. 7.22a. M59 (NGC 4621), elliptical galaxy in Virgo. Astrophoto by Mark Komsa. The scale of this photograph is approximately 5 times that of the drawing.

Fig. 7.22b. M59, elliptical galaxy in Virgo. Drawing by Jeremy Perez.

Fig. 7.23a. Elliptical galaxy M60 (NGC 4649, on left) and spiral galaxy NGC 4647 on the right. Astrophoto by Mark Komsa. The scale of this astrophoto is approximately 2.3 times that of the drawing.

Fig. 7.23b. M60 and NGC 4647. Drawing by Jeremy Perez.

LIBRA (lie-brah)	**The Balance (Scales)**
LIBRAE (lie-bree)	**9:00 P.M. Culmination:** May 1
LIB	**AREA:** 538 sq. deg. 29th in size

Libra is a fairly faint constellation. Even its two brightest stars, β at magnitude 2.61, and α at magnitude 2.75, fall slightly below 2nd magnitude status. Libra is well known only because it is one of the 12 constellations of the zodiac. It is unique among the zodiacal constellations because it is the only one which does not represent a living creature.

Although many believe that Libra is a relatively recent constellation which originated when Julius Caesar promulgated his "new" calendar and a zodiac in which Libra replaced the Greek constellation Chaelai (The Claws), both his calendar and a constellation including a balance or scales are older and were copied from earlier civilizations. The ancient Sumerians called this group of stars "Zi-Ba An-No," the Balance of Heaven, and at least as early as 2084 BC, the Babylonians recognized the scales and associated them with judgment (balancing of good vs. bad) of the living and dead.

The concept of a balance almost certainly came from the fact that from about 2400 BC until around 700 BC the autumnal equinox occurred during their month of *Tisri* (comprising parts of our September and October) when the sun was in this constellation, and the length of the day and night were equal or "balanced." Interestingly, this same period nearly matches the period when the Babylonian empire was the Middle East's most prominent. This section of the sky was also depicted as a balance by the Chinese, Indians, and Egyptians before the Romans copied it. The Chinese had a law requiring an annual regulating of the weights used on balances when Libra appeared in the sky, and the Indians saw Libra as a kneeling man holding a pair of scales. Later, the Romans depicted their constellation as a man (usually one of the Caesars) holding the scales of justice.

Although the Babylonian constellation was passed on to the early Greeks by the Phoenicians and Hittites, for much of their history, the Greeks saw these stars as the "Chaelai" or claws of a much enlarged constellation of Scorpius, or as a separate constellation representing only the claws. Even when they saw claws in this region, the claws often encircled a pair of scales, indicating that at earlier times they too had recognized a balance here.

The following Greek myth explaining the origin of the scales is much older than the revived Libra of the Romans.

Astraea was the daughter of Zeus and Themis, and was the goddess of justice. In the beginning, like the other gods of that time, Astraea lived among the early humans on the earth. She counseled the humans and helped them to write their laws. This was a golden age, during which men did not wage war, fight among themselves, or commit other crimes. However, the descendants of these early humans, while still basically good, were not quite so noble. The golden age had given the way to the silver age. The humans began to quarrel among themselves and to wage war on their neighbors. This displeased the gods, so they left the earth and established their new home on Mt. Olympus. Only Astrea continued to visit the humans to help write new laws and customs, and to settle disputes. Humans continued their downward path, bringing on a bronze age with further degradation. These people were so corrupt that even Astraea abandoned them and watched from the heavens. She continued to use a balance or pair of scales to compare the good parts of human souls with the bad. These scales were represented in the heavens as Libra.

The earliest Greeks recognized Libra as a god or goddess holding a pair of scales, just as many of their neighbors and predecessors had done. Later in their history, the Greeks began to recognize a giant scorpion occupying the space of both the present day constellations of Scorpius and Libra. They knew that it took the sun about twice as long to pass through this section of the zodiac as it did for the other constellations, and therefore they sometimes broke this super-constellation into two, with one consisting of the body and tail, and the other, the claws of the scorpion. They called the latter constellation "Chelai," or "The Claws." They continued to recognize the Claws until the time of Julius Caesar.

In 46 B.C.E., Julius Caesar used the help of the Alexandrian astronomer Sosigenes to revise the calendar. The Julian calender, often believed to be a new invention by Sosigenes, was actually almost identical to an earlier one described in the "Decree of Canopus" written in 238 BC under the rule of Polemy III (Euergertes). Sosigenes also borrowed one of the Eqyptian Zodical constellations depicting Libra as the figure of a man holding a pair of scales. This man was often portrayed as the current Caesar or leader. Both the constellation and a Roman standard of weight shared the same name. The Roman pound, called a libra, was made up of 12 uncias, each equal to about 1.1 aviordupois ounces. Our abbreviation for the pound (lb) comes from the libra.

The reason that Libra no longer represents a living creature is an accident of history. During the Dark Ages of Europe, much of the astronomical knowledge known by Europe was preserved and extended by Arab astronomers. Since many Muslims believed that the Koran forbade the making of any image or representation of a god or human figure, they changed the figure of Libra to just a pair of scales. During the renaissance, knowledge flowed from the middle east back to Europe in many forms, including star charts showing the Arabic constellation figures and star names. The Arabic figure of the Scales, without the man holding it, has survived to the present day. In addition, the names of the stars of Libra show that a large fraction of our star names are the result of translations of the Greek descriptions of star positions into Arabic, and then brought back to Europe without being retranslated. If you look at an alphabetical list of star names, you will notice that a large number begin with "Al" which is Arabic for "the."

TO FIND: Locate the first magnitude stars <u>Antares</u> (in Scorpius) and <u>Spica</u> (in Virgo). Zubeneshamali, α Librae, lies 4.5° north of the midpoint of the line joining <u>Antares</u> and <u>Spica</u>. Zubenelgenubi, β Librae, is 9° NE of Zubeneshamali. γ Librae, sometimes called Zubenelakrab, lies 11° east, and very slightly north of Zubeneshamali. υ and τ Librae, almost equally bright at magnitudes 3.66 and 3.60 and separated by only 1.7°, are easily recognized 14° south of β. The slightly brighter (m3.25) σ Librae lies 8° WNW of υ and τ.

KEY STAR NAMES:

Name	Pronunciation	Source
α Zubenelgenubi	zoo-**ben**-el-jeh-**NEW**-bee	Al Zub n al Jan biyyah (Arabic), the southern claw
β Zubeneschamali	zoo-**ben**-eh-shoe-**MAH**-lee	Al Zub n al Sh maliyyah (Arabic), the northern claw
γ Zubenelakrab	zoo-**ben**-el-**AH**-crab	Al Zub n al Akrabi (Arabic), the scorpion's claw
δ Zubenelakribi	zoo-**ben**-el-ah-**CRIB**-ih	A variation of Zubenelakrab (unofficial)
η Zubenhakrabi	zoo-**ben**-hah-**CRAB**-ih	A variation of Zubenelakrab (unofficial)
ν Zubenhakrabim	zoo-**ben**-hah-**CRAB**-em	A variation of Zubenelakrab (unofficial)
σ Brachium	brah-**key**-umm	Bracchium (Latin), upper arm of a person, or a limb of any living creature. Ultimately from βραχ ων [brachion] (Greek), "arm"

While all the stars in Libra are relatively faint today, this may not have been the case in the past. In "The Constellations," psuedo-Eratosthenes stated that the brightest star in the Scorpius super-constellation was the westernmost star in the northern claw (β Librae or Zubeneschamali). Modern measurements show that α, δ, ε, θ, κ, and λ Scorpii are all brighter than β Librae, so the only logical conclusion is that β Librae has dimmed since that time. About 350 years later, Ptolemy called it the same magnitude as Antares. Meanwhile, Antares has increased its brightness, since in his list of stars, Ptolemy placed it between σ and τ Scorpii without mentioning any difference in brightness. Until the end of the fifteenth or the beginning of the sixteenth century, Antares was listed as being of 2nd magnitude, then until the late nineteenth century, it was listed as 1st magnitude. In 1880, it was given a magnitude of 1.7, and today it is variable, ranging between magnitudes 0.88 and 1.8. Although it is rare to see permanent changes in star brightnesses (except for nova and supernova), here is case where two stars relatively close together may have significantly changed (Figs. 7.1, 7.24 and 7.25).

Magnitudes and spectral types of principal stars:

Bayer desig.	Mag. (m_v)	Spec. type	Bayer desig.	Mag. (m_v)	Spec. type	Bayer desig.	Mag. (m_v)	Spec. type
α^1	5.15	F3 V	α^2	2.75	A3 IV	β	2.61	B8 V
γ	3.91	K0 III	δ	4.91	B9.5 V	ε	4.92	F5 IV
ζ	5.53	B3 V	η	5.41	A6 IV	θ	4.13	K0 III
ι	4.54	A0 pSi	κ	4.75	K5 III	λ	5.04	B3 V

Bayer desig.	Mag. (m_v)	Spec. type	Bayer desig.	Mag. (m_v)	Spec. type	Bayer desig.	Mag. (m_v)	Spec. type
μ	5.32	Ap	ν	5.19	K5 III	ξ^1	5.74	G7 III
ξ^2	5.48	K4 III	o	6.14	F2 V	π	–	–
ρ	3.66	B2.5 V	σ	3.25	M3.5 III	τ	3.66	B2.5 V
υ	3.60	K3 III	φ	–	–	χ	–	–

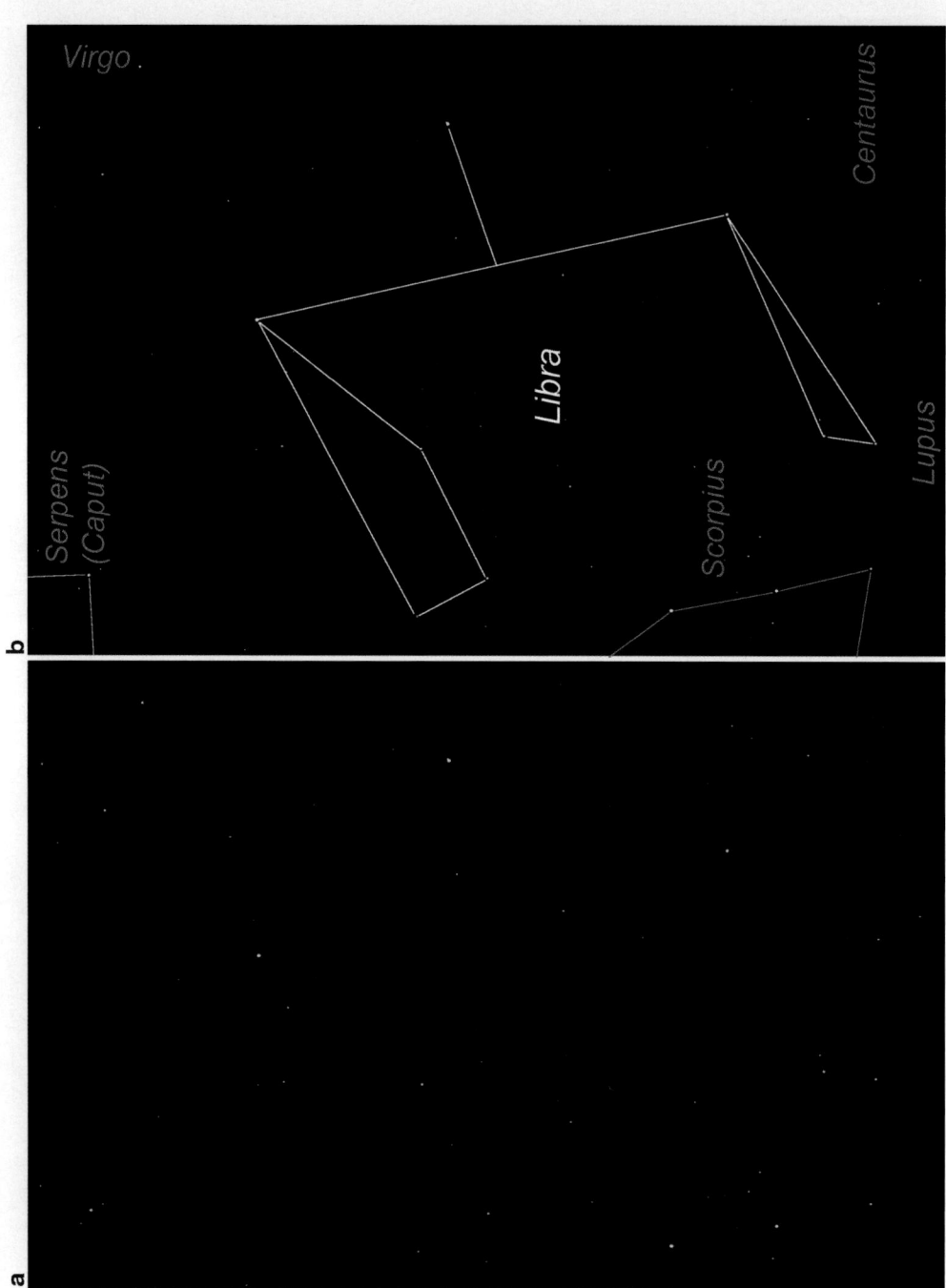

Fig. 7.24. (**a**) The stars of Libra and adjacent constellations. (**b**) The figures of Libra and adjacent constellations.

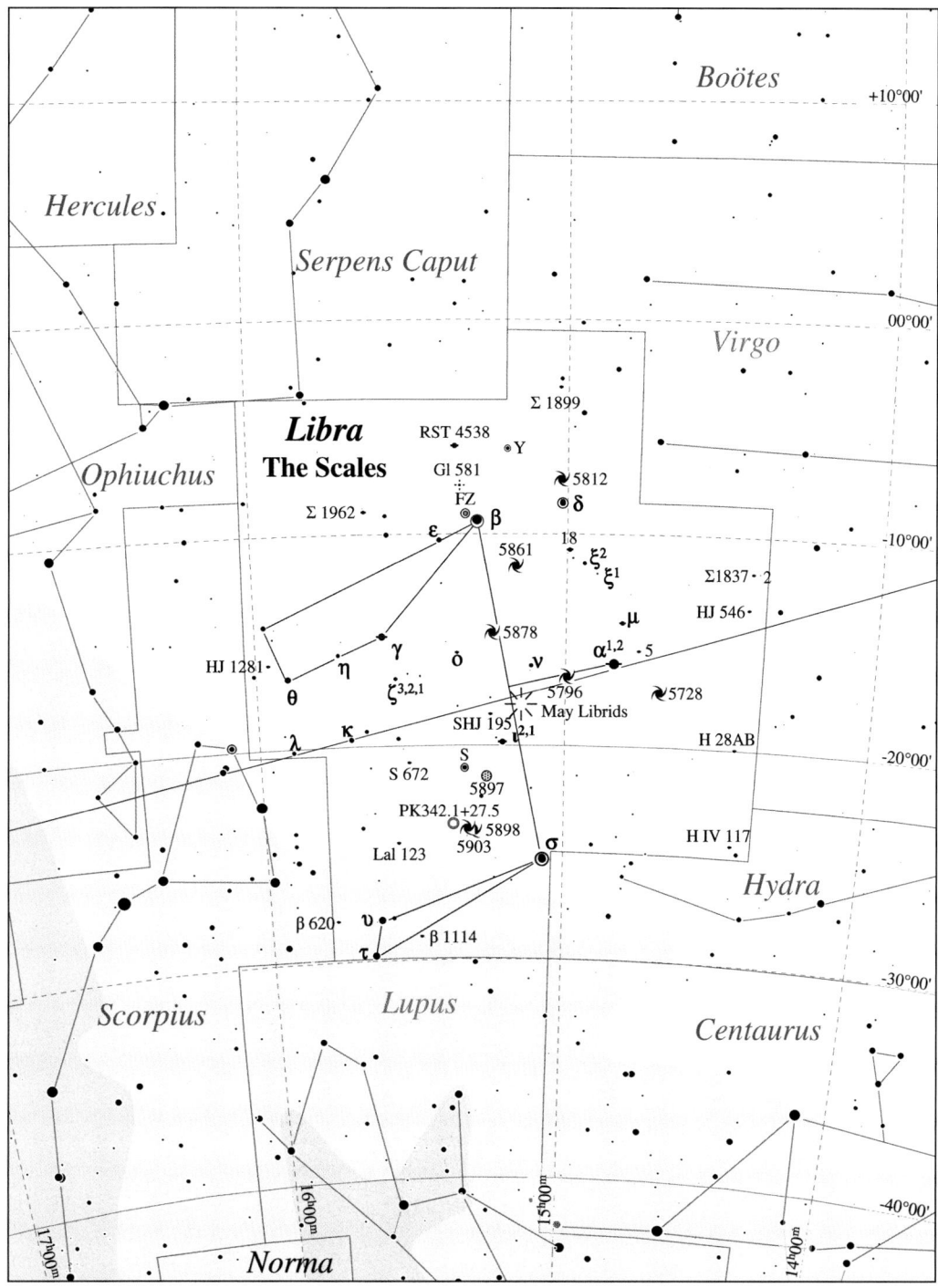

Fig. 7.25. Libra details.

Table 7.6. Selected telescopic objects in Libra.

Multiple stars:

Designation	R.A.	Decl.	Type	m_1	m_2	Sep. (″)	PA (°)	Colors	Date/Period	Aper. Recm.	Rating	Comments
Σ1837	14ʰ24.7ᵐ	−11°40′	A–B	6.9	7.94	1.1	274	Y, –	2006	8/10″	****	
HJ 546	14ʰ25.3ᵐ	−13°21′	A–B	6.6	10.0	41.6	046	–, –	2003	2/3″	***	Slightly unequal pair
SHJ 179	14ʰ25.5ᵐ	−19°58′	A–B	6.6	7.2	34.2	295	rW, rW	2003	7×50	****	
			B–C	7.2	8.4	1.2	088	–, –	1999	8/10″	***	a.k.a. HID 20
5 librae	14ʰ46.0ᵐ	−15°28′	A–B	6.4	10.1	3.1	252	O, W	1962	3/4″	***	Two additional stars, m 14.4 and 13.9, lie at 15 and 25″. A large scope is needed for these
7, μ Librae	14ʰ49.3ᵐ	−14°09′	A–B	5.6	6.6	1.8	340	pY, bGr	2006	5/6″	******	
H IV 117	14ʰ49.3ᵐ	−24°15′	A–B	5.8	8.6	63.9	221	O, B	2998	7×50	*****	Nice colors
				8.7	11.1	2.7	338	yW, –	1944	4/6″	***	
9 and 8, α², α¹ Librae	14ʰ50.9ᵐ	−16°02′	A–B	2.7	5.2	231.1	315	pY, sW	2002	Eye	****	Zubenelgenubi. a.k.a. SHJ 186AB. B is a close (~0.4″) double
H 28-AB	14ʰ57.5ᵐ	−21°25′	A–BC	5.9	8.2	**25.4**	**306.5**	O, R	**2,130**	10×50	*****	Beautiful! Unusual color combination
			B–C	8.2	9.5	**0.08**	**188.9**	R, R	**0.8457**	–	–	With a maximum separation of 0.195″, B–C is too close for amateur instruments
18 librae	14ʰ58.9ᵐ	−11°09′	A–B	5.9	9.9	19.7	039	O, W	1999	15×80	****	a.k.a. Σ1594AB
Σ1899	15ʰ01.6ᵐ	−03°10′	A–B	6.7	10.2	28.6	067	O, ?	2002	15×80	***	
24, ι Librae	15ʰ12.2ᵐ	−19°47′	Aa	5.1	5.6	**0.15**	**005**	prR, prR	**23.469**	–	***	Aa is extremely difficult to split, position angle varies rapidly, maximum separation
			Aa–B	4.5	10.4	57.8	109	pY, rPr	2002	15×80	****	Aa–B a.k.a. H. of 0.161″ in late 2006
			B–C	10.4	10.9	2.0	014	–, –	2000	5/6″	****	a.k.a. β 618BC
SHJ 195	15ʰ14.5ᵐ	−18°26′	A–B	6.8	8.3	47.4	140	pGld, –	2002	7×50	****	
29, o librae	15ʰ21.0ᵐ	−15°33′	A–B	6.2	10.0	44.4	349	yO, Pr	2003	15×80	****	Good color contrast. a.k.a. β 1447
RST 4538	15ʰ21.1ᵐ	−05°49′	A–B	5.5	12.7	11.9	004	–, –	1958	5/6″	***	
β 1114	15ʰ29.0ᵐ	−28°52′	A–B	7.0	7.8	0.8	317	Y, –	1998	12/14″	****	
			AB–C	7.0	9.6	9.5	009	–, –	1834	2/3″	***	
S 672	15ʰ31.7ᵐ	−20°10′	A–B	6.3	8.9	11.4	280	pY, R	2002	2/3″	***	
Lal 123	15ʰ33.2ᵐ	−24°29′	A–B	6.9	7.0	9.0	300	–, –	1998	2/3″	******	Matched pair
38, γ Librae	15ʰ35.5ᵐ	−14°47′	Aa–B	4.0	11.2	41.8	155	pO, –	1960	4/6″	****	Aa is too close for amateur telescopes
Σ1962	15ʰ38.7ᵐ	−08°47′	A–B	6.4	6.5	11.5	190	yW, yW	2006	2/3″	******	Matched pair. Some see Y, Y in small telescopes
β 620	15ʰ46.2ᵐ	−28°04′	Aa–B	6.5	7.0	0.6	170	yW, yW	1998	16/18″	****	Slightly unequal pair
h 4803			A–Ca	6.6	9.0	50.6	213	–, –	1999	7×50	***	
HJ 1281	15ʰ57.1ᵐ	−16°02′	A–B	6.6	13.7	35.2	237	–, –	2005	8/10″	***	

'17

Variable stars:

Designation	R.A.	Decl.	Type	Range (m.)		Period (days)	F (f./f.)	Spectral Range	Aper. Recm.	Rating	Comments
19, δ Librae	15h 01.0m	−08° 31'	EA/SD	4.92	5.90	2.327	0.23	A0IV-V	6×30	****	Algol type, partial eclipse, decline takes only 6 h
20, σ Librae	15h 04.4m	−25° 18'	SRb	3.46	3.20	20	–	M3.5IIIa	Eye	****	Variation subtle, but easily discernable to unaided eye
Y Librae	15h 11.7m	−06° 02'	M	14.7	7.6	275.7	0.41	M5e	16/18"	***	3.6° NNW (PA 338°) of β Lib. AAVSO charts are available
FZ Librae	15h 19.7m	−09° 10'	Lb	7.24	6.73	–	–	M4III	6×30	****	Easily located 38' ENE (PA 68°) of β Lib
S Librae	15h 21.4m	−20° 23'	M	13.0	7.5	192.7	0.49	M1.0e-M6.0e	8/10"	***	2.2° ESE (PA 105°) of Lib (m4.5)

Star clusters:

Designation	R.A.	Decl.	Type	m_v	Size (')	Brst. Star	Dist. (ly)	dia. (ly)	Number of Stars	Aper. Recm.	Rating	Comments
NGC 5897	15h 17.4m	−21° 01'	Globular	8.6	11.0	13.6	41K	131		8/10"	***	Low-surface brightness, several stars can be seen with averted vision

Nebulae:

Designation	R.A.	Decl.	Type	m_v	Size (')	Brst./Cent. star	Dist. (ly)	dia. (ly)	Aper. Recm.	Rating	Comments
PK342.1+27.5	15h 22.3m	−23° 38'	Planetary	11.6	7"	Cent. **15**	16.03	14K	14/16"	***	300× needed to see the tiny disk with star-like center

Galaxies:

Designation	R.A.	Decl.	Type	m_v	Size (')	Dist. (ly)	dia. (ly)	Lum. (suns)	Aper. Recm.	Rating	Comments
NGC 5728	14h 42.4m	−17° 15'	SA	11.3	3.7×2.6	138M	84K	72G	8/10"	***	
NGC 5796	14h 59.4m	−16° 37'	SB	11.6	2.7×1.7	137M	84K	36G	8/10"	***	
NGC 5812	15h 01.0m	−07° 27'	SB	11.2	2.3×1.9	103M	75K	26G	8/10"	***	
NGC 5861	15h 09.3m	−11° 19'	SB	11.6	2.8×1.8	94M	82K	32G	10/12"	***	
NGC 5878	15h 13.8m	−14° 16'	SAB	11.5	3.0×1.4	103M	90K	38G	8/10"	***	
NGC 5898	15h 18.2m	−24° 06'	SAB	11.4	2.6×2.3	106M	87K	25G	8/10"	***	
NGC 5903	15h 18.6m	−24° 04'	SA	11.1	3.4×2.7	117M	102K	35G	8/10"	***	

(continued)

Table 7.6. (continued)

Meteor showers:

Designation	R.A.	Decl.	Period (year)	Duration	Max Date	ZHR (max)	Comet/ **Asteroid**	First Obs.	Vel. (mi/ **km**/sec)	Rating	Comments
May Librids	15ʰ 08ᵐ	−18°	Annual	May 1	May 9	May 6/7	6	–	1929	–	***	A reliable minor meteor shower, cometary association unknown

Other interesting objects:

Designation	R.A.	Decl.	Type	m_v	Mass suns	dia. suns	Dist. (ly)	Planet Mass (J)	Dist. (AU)	Period (days)	Eccentricity	Comments
Gliese 581	15ʰ 19.5ᵐ	−07° 43′	Exo-planetary system	10.55	0.31	0.38	20.44	b>0.0521	0.0406	5.366	0.02	Fifth extra-solar planet discovered around a red dwarf, Nov. 30, 2005, one of lightest. T_{ave} = 420K, 330°F
								c>0.016	0.073	12.9	0.16	
								d>0.024	0.025	83.6	0.20	

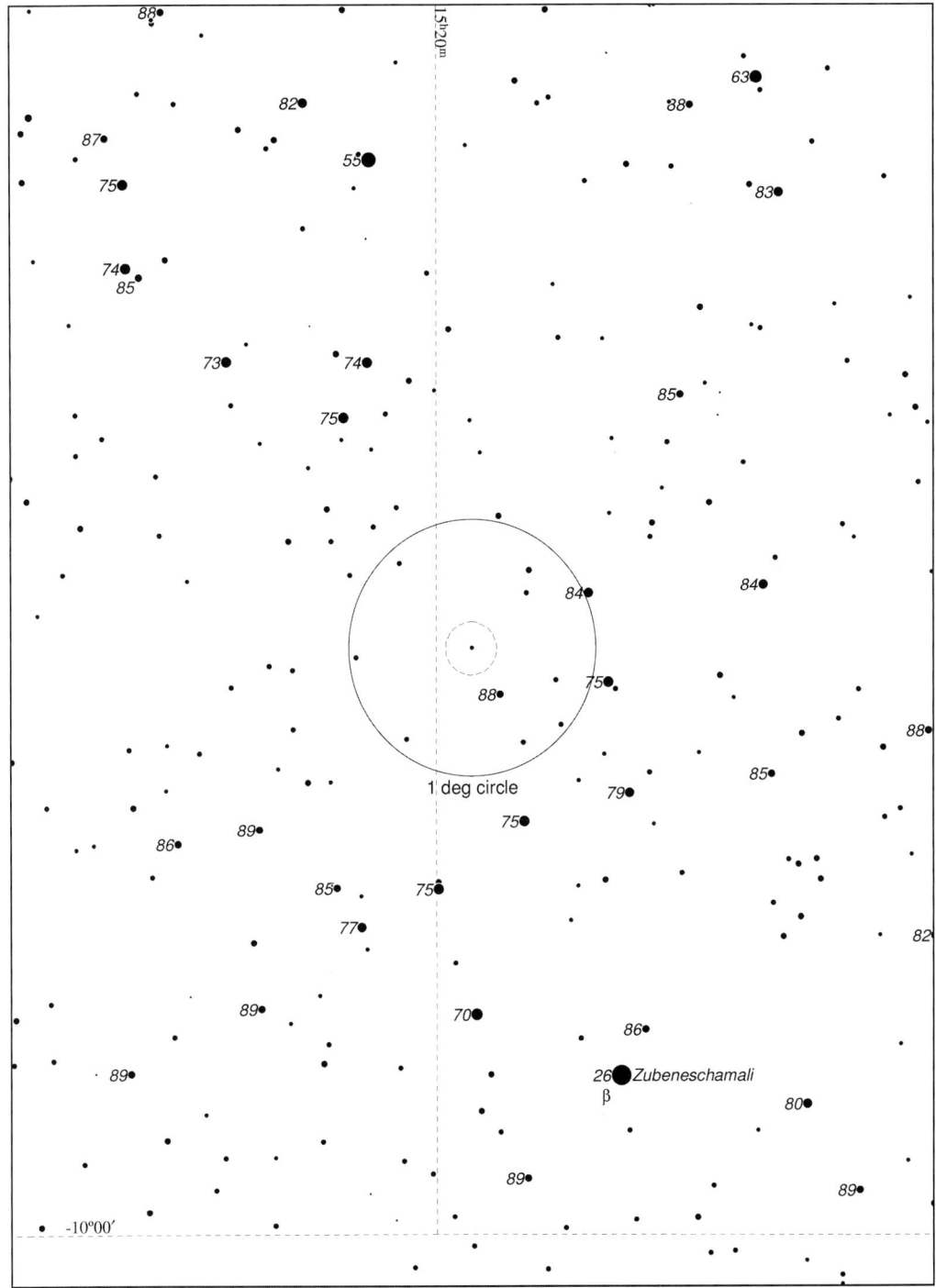

Fig. 7.26. Gliese 581 finder chart.

Bayer desig.	Mag. (m$_v$)	Spec. type	Bayer desig.	Mag. (m$_v$)	Spec. type	Bayer desig.	Mag. (m$_v$)	Spec. type
ψ	4.95	B8 Ia/Iab	ω	–	–			

NOTES:

1. 36, ε Boötis is suspected of being a binary. If it is, the orbital period is probably on the order of several thousand years.
2. 44, ι Boötis, is also known as Σ1909. The secondary is an eclipsing binary with a visual magnitude range of 5.
3. There is also evidence of a secondary period of ~137 days for V Boötis.
4. Interferometric measurements indicate that R Boötis has a diameter of 217 ± 40 times that of the sun.
5. Infrared measurements of RV Boötis indicate a possible proto-planetary disk with a diameter of about 60 AU.
6. The Quadrantids are named after the obsolete constellation of "Quadrans Muralis," the Wall Quadrant. This constellation was subsumed into Boötis by the IAU in 1922. The Quadrantids have a very sharp peak of meteors of only about 16 h during which the rate is more than half the peak. Rates also vary from year to year by a factor of nearly 3 times.
7. U Canis Venaticorum is a deep red carbon star. In the nineteenth century, Father Angelo gave it the name of "La Superba" because of its lovely color.
8. NGC 4631 is a mottled edge-on galaxy. A 12th magnitude companion, NGC 4627 (often called "the Pup"), lies approximately 2N to the northwest and a 10th magnitude companion, NGC 4656/7, lies about 31N to the southeast.
9. It is an interesting coincidence that in 1844, Friderich Bessel used precise measurements of the positions of α Canis Majoris (Sirius) and α Canis Minoris to demonstrate that both are binaries with similar periods, (49.98 and 40.84 years, respectively). Although neither companion had ever been seen at that time, they both were later found to be rather dim stars with unusually high temperatures for their sizes, and were shown to be a new type of star, both being white dwarfs with extremely high densities.
10. V Canis Minoris' period of 366.1 days insures that it will be in the same portion of its cycle about the same date each year for several consecutive years. This creates a problem for variable star observers because its repeated nearness to the sun at one part of the cycle leaves an observation gap which is hard to fill.
11. Σ1932Aa has a separation of only 0.3" and is thus too close for amateur instruments.
12. Eta (η) CrB is opening to a separation of 0.7" in 2013, closing to 0.5" in 2020, and finally opening to a maximum of 1.0" in 2032.
13. R Coronae Borealis is the prototype of one of the most unusual classes of variable stars. All variables in this class are old hydrogen-deficient and carbon-rich stars of spectral class F or G. They remain near maximum brightness for months or even years and then suddenly fade up to 8 magnitudes in brightness (a factor of >600 times) in a few weeks. Then they may recover fairly quickly or stay near minimum for up to a year or more. They recover their brightness more slowly than they fade, and may have one or more pauses or regressions before reaching maximum brightness again. The infrared luminosity remains nearly the same. The exact cause of variation is poorly understood, but the most popular theory is that the stars are ejecting carbon-rich material which forms a sooty condensate that obscures the visible light of the star. The stellar wind eventually disperses the obscuring dust and the star becomes bright again.
14. R Virginis was found by K. L. Harding of Germany in 1809, the 19th variable star to be discovered. It is in the same binocular field of view as M49. While the periods of the Mira-type variables range from 80 to 1,000 days, most have periods around 255 days, so the 145.63-day period of R Virginis is unusually short. R Virginis is also unusually red at minimum brightness.
15. The perihelion of the αVirginids lies near the orbit of Jupiter, and calculations indicate repeated close encounters over the period of 1860–2060. Orbital perturbations at such encounters are likely to spread a meteor stream and even break it up into several separate streams. Thus, the αVirginid stream is probably the parent of a large group of weak meteor showers, including the π, March, η, or Southern, μ and λ Virginids. These are relatively weak showers with maxima ranging from Jan. 10th to May 6th. The αVirginid complex has one of the longest durations of activity known. The αVirginid meteoroids have a very high

density of ~1.8 gm/cm³ and penetrate relatively deeply into the atmosphere before burning out or breaking up. No cometary parent is known for these showers, but several asteroidal or extinct cometary nuclei have been proposed as sources.

16. The Virgo Galaxy Cluster is nearly centered around the giant elliptical galaxy M57 and contains about 300 large galaxies and approximately 2,000 smaller ones. This cluster covers approximately 100 sq. deg. of the sky and is so massive that the Local Group of galaxies is being gravitationally pulled toward it.

17. In "Double Stars for Small Telescopes," Sissy Haas lists HN 28 as 33 Librae. This appears to be in error, since the R.A. of HN 28 is intermediate between those of 16 and 17 Librae. In "Lost Stars," Morton Wagman identifies Flamsteed number 33 as ζ^2 Librae (HD 137949), a magnitude 6.69 star which does not appear on the Libra chart in "Norton's 2000.0 Star Atlas," 18th edition. This error might have resulted from confusion between ξ^2 and ζ^2.

18. Gliese 581 is a red dwarf with a mass approximately one-third of our sun. It is ~20.4 ly from the earth, and has the fifth planet discovered orbiting a red dwarf. Two more planets were discovered later, the smallest one a terrestrial (rocky) planet only 1.5 times the diameter and 5.1 times the mass of the Earth. This planet orbits Gl 581 in ~13 days and lies in the habitable zone. If it has a very cloudy atmosphere similar to Venus, the calculated temperature would be about 25°F (−3°C) and, if it has an atmosphere and surface similar to the earth, a temperature of about 104°F (40°C). The surface gravity would be about 2.27 times the earth's gravity.

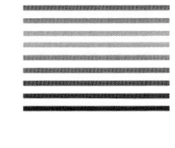

CHAPTER EIGHT

The Eternal Thirst

April–May

CORVUS	CRATER	HYDRA
The Crow	The Cup	The Water Snake

Each of the Greek gods and goddesses had a set of animals, birds, and other aspects of nature that were associated with them and over which they exercised special powers. The bird associated with Apollo was the crow (or raven). Various myths tell us that the crow had originally been a white bird with a beautiful singing voice, but a series of punishments for repeating gossip and telling false tales resulted in his present condition.

Apollo became the lover of a mortal princess named Coronis and left the crow to guard her while he went to his temple at Delphi. Coronis had long loved a mortal named Ischys and began seeing him, even though she was already pregnant with Apollo's son Asclepius. When the crow took the news to Apollo, he expected to be praised for his vigilance, but Apollo was irate that the crow had not driven Ischys away by pecking out his eyes and cursed the crow. The fury of Apollo's words immediately turned the crow's feathers black. The feathers of all his descendants have been black ever since. A similar myth explains how the crow's habit of gossiping and carrying bad tales resulted in his voice being changed to match the unpleasant news he bore.

Nevertheless, the crow remained Apollo's charge and was frequently sent on errands. One such errand occurred while Apollo was preparing a sacrifice. He became hot and thirsty from the work, so he sent the crow to his favorite spring for a cup of cool, pure water.

P. Simpson, *Guidebook to the Constellations: Telescopic Sights, Tales, and Myths*,
Patrick Moore's Practical Astronomy Series, DOI 10.1007/978-1-4419-6941-5_8,
© Springer Science+Business Media, LLC 2012

On the way, the crow spotted a grove of fig trees with nearly ripe fruit. The crow perched on a branch of one of the trees and waited for the figs to reach the peak of ripeness. This took several days, and when the crow finally reached the spring and filled the cup, he realized that Apollo would be angered by the delay. When he happened to see a harmless water snake in the spring, he thought of the stratagem of seizing the snake in his claws and pretending that the snake had delayed him in getting the water.

Of course, Apollo (who was the god of truth) was not fooled by this ploy and again cursed the crow. This time, Apollo made the crow suffer a thirst similar to his own. Later, when the crow died, Apollo wished to remind mortals of the folly of lying to cover up their own failings, so he placed the crow, the cup, and the water snake in the heavens as Corvus, Crater, and Hydra. Corvus is shown as holding the snake in his claws, and Crater is shown behind Corvus. By order of Apollo, the cup is tilted toward the crow, and water continually flows out, running down the crow's tail, while the crow permanently faces away from the cup, so that he can never slake his thirst with the cool water.

Hydra is sometimes called "the female water snake," while the southern constellation of Hydrus is called the "male water snake." This error arises from misunderstanding the names. In Latin, all nouns have a gender, but this gender applies only to the word, not necessarily to the object represented. Hydrus is the Latin word for any water snake. Hydra is the Greek name of the multiheaded serpentine monster killed by Heracles (Latin Hercules). The confusion comes from the fact that the word Hydra looks as if it could be a first declension Latin word for a female snake, since Latin nouns ending in -a are usually feminine in form but not necessarily in actuality, e.g., the feminine word "agricola" for farmer. A similar argument is true for masculine words like hydrus and amicus (friend) which are second declension words and can represent animals or people of either gender (Figs. 8.1–8.3).

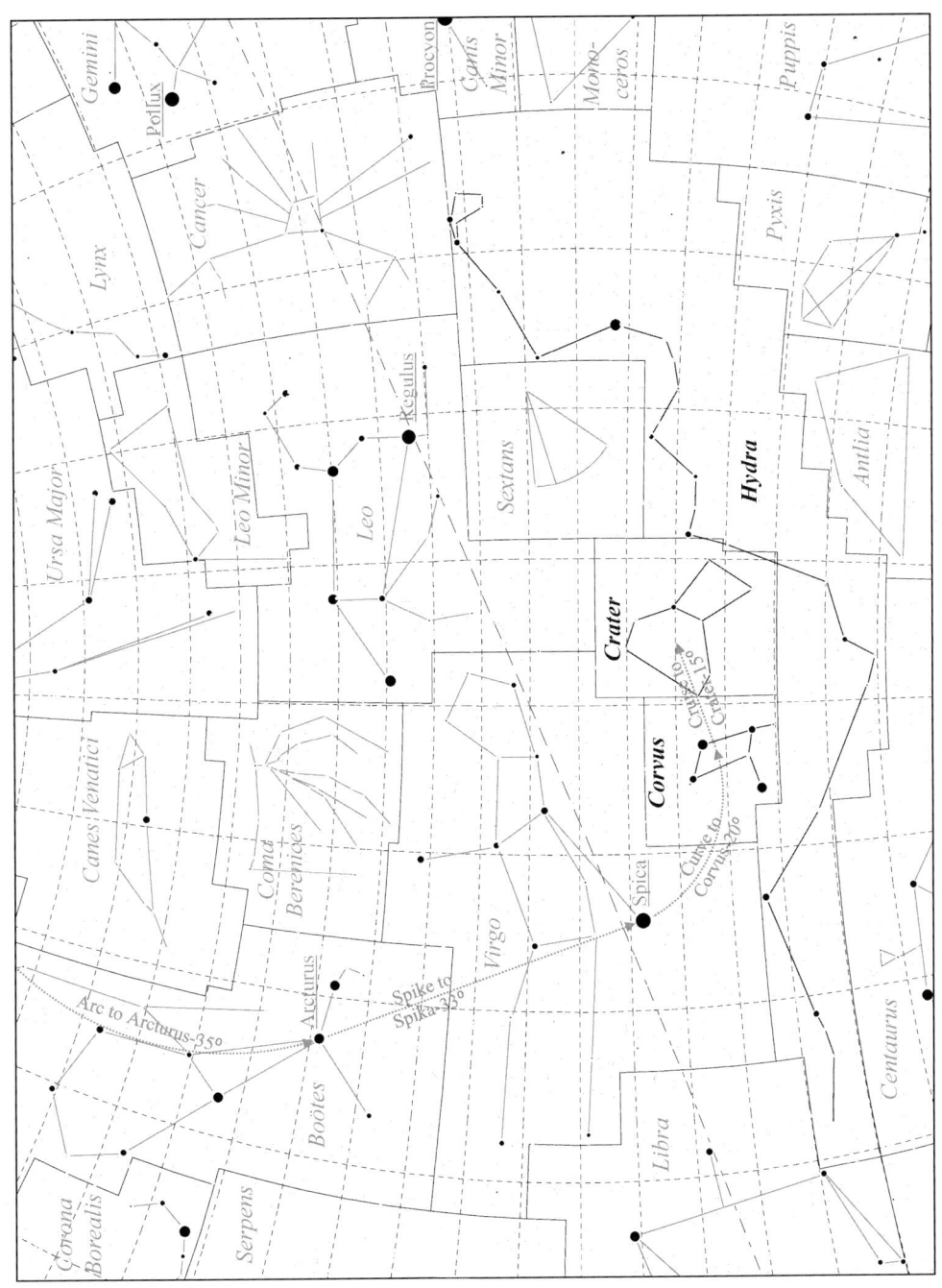

Fig. 8.1. Corvus group finder.

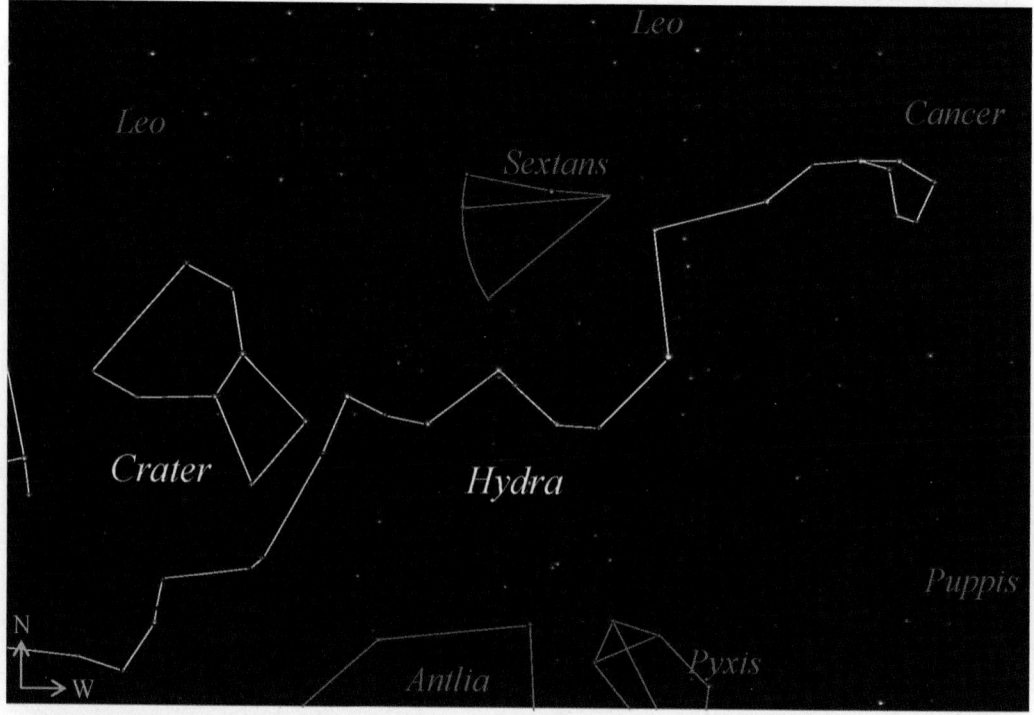

Fig. 8.2. The stars of Crater and western Hydra.

Fig. 8.3. The stars of Corvus, Crater, and eastern Hydra.

CORVUS (kor-vus)CORVI	The Crow
(**kor**-vee)	**BEST SEEN** (at 9:00 **P.M.**): May 28
Crv	**AREA**: 184 sq. deg., 70th in size

TO FIND: Look for a kite-shaped quadrilateral formed by Corvus' four brightest stars about 17° southwest of <u>Spica</u>. Also apply the mnemonic "Arc to <u>Arcturus</u>, spike to <u>Spica</u>, curve to Corvus, and cruise to Crater" by starting at the great bear's tail (big dipper's handle) and following the curves and straight lines with the arc and the curve being in the same direction as the curve formed by the great bear's tail (the big dipper's handle) (Figs. 8.1, 8.4 and 8.5).

KEY STAR NAMES:

Name	Pronunciation	Source
α Alchiba	al-**key**-buh,or uhl-kih-bah	(ind-Arabic), "the Tent," and Arabic asterism consisting of β, γ, δ, and ε Corvi. Later transferred to α
β Kraz	kraz?	Origin and pronunciation uncertain*[1]
γ Giena	**jee**-nuhor jeh-**nah**	al Janah al Ghurab al Aiman (sci-Arabic) the right wing of the raven (as the crow was known to the Arabs)
δ Algorab	al-**goe**-rabor uhl-go-**rahb**	al Janah al Ghurab al Aiman (sci-Arabic) the right wing of the raven (as the crow was known to the Arabs)
ε Minkar	**men**-caror **men**-cur	Al-minkhar (Arabic), the nostrils. A variation of the spelling of Menkar (see α Ceti)

Magnitudes and spectral types of principal stars:

Bayer desig.	Mag. (m_v)	Spec. type	Bayer desig.	Mag. (m_v)	Spec. type	Bayer desig.	Mag. (m_v)	Spec. type
α	4.02	F0 IV/V	β	2.65	G5 II	γ	2.58	B8 III
δ	2.94	B9.5 V	ε	3.02	K2 III	ζ	5.20	B8 V
η	4.32	F2 V*[2]						

Stars of magnitude 5.5 or brighter which have Flamsteewd but no Bayer designations:

Flam-steed	Mag. (m_v)	Spec. type	Flam-steed	Mag. (m_v)	Spec. type
3	5.45	A1 V	31	5.28	B2 IV

Stars of magnitude 5.5 or brighter which have no Bayer or Flamsteed designations:

Desig-nation	Mag. (m_v)	Spec. type
HD 107418	5.14	K0 III

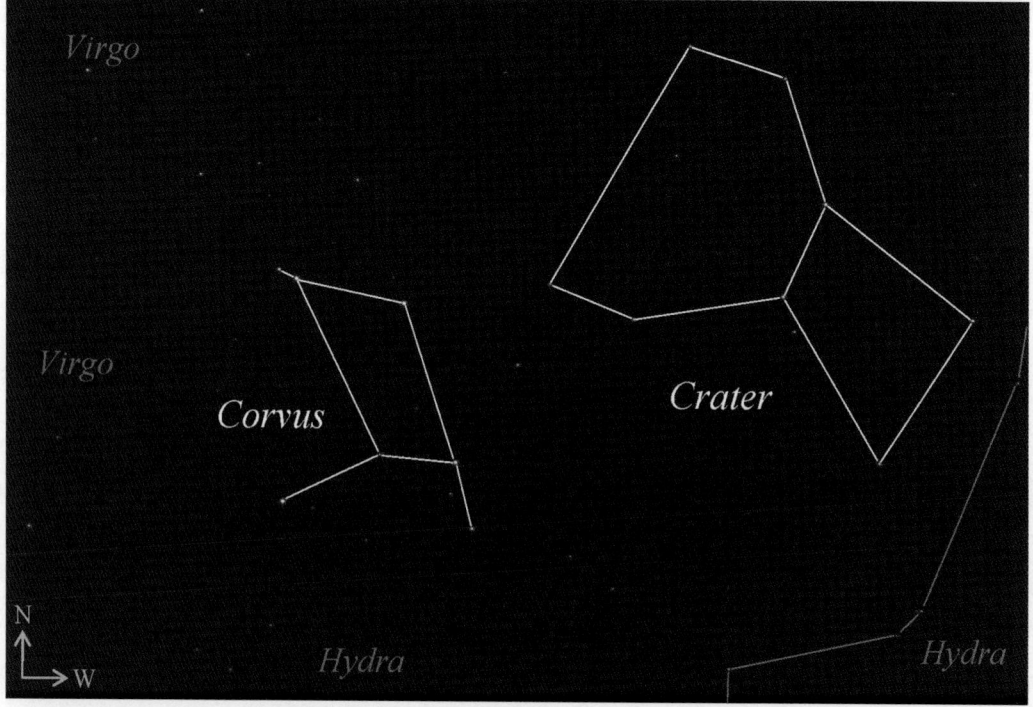

Fig. 8.4. The stars of Corvus and Crater.

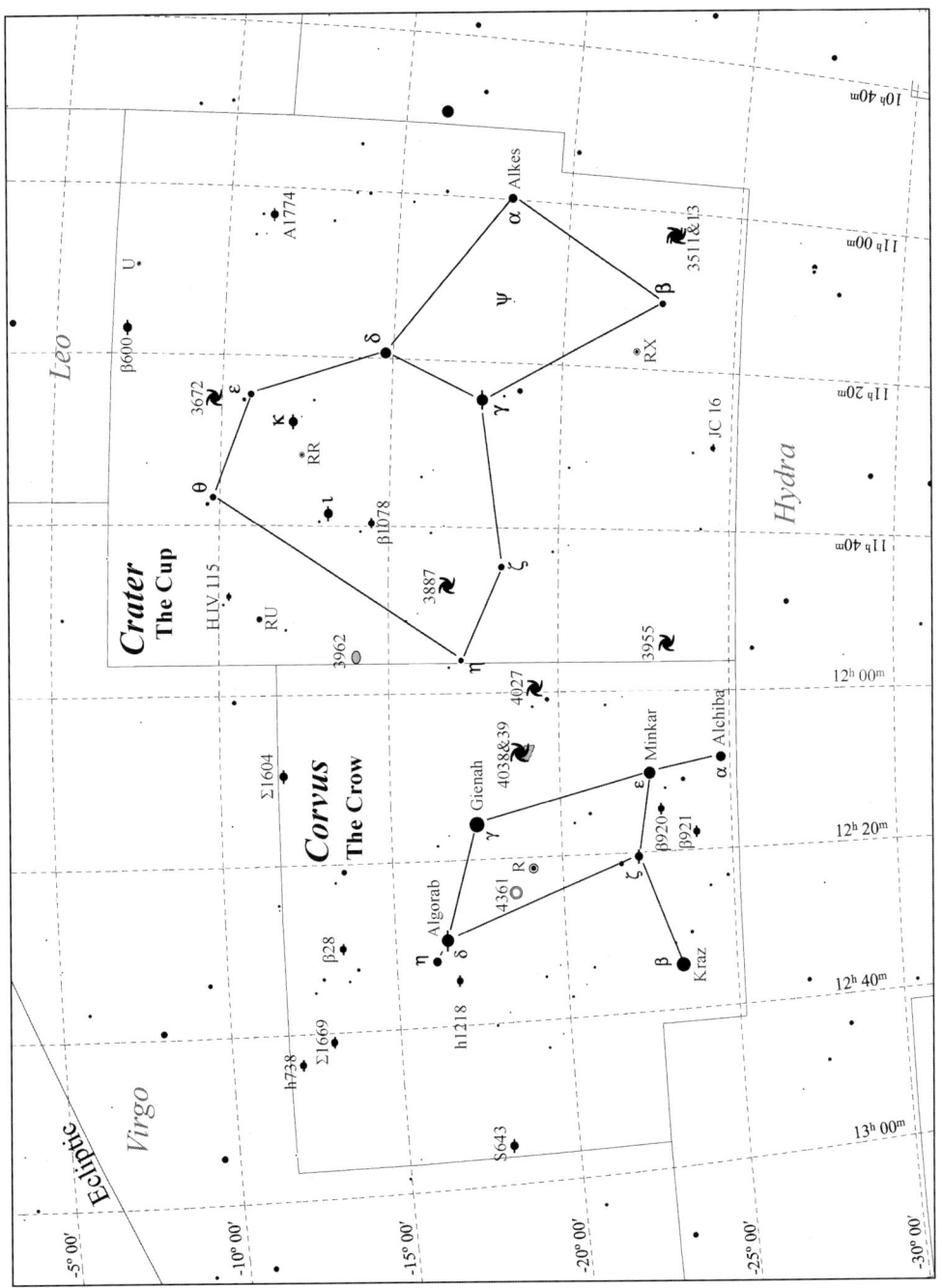

Fig. 8.5. Corvus and Crater details.

Table 8.1. Selected telescopic objects in Corvus.

Multiple stars:

Designation	R.A.	Decl.	Type	m₁	m₂	Sep. (")	PA (°)	Colors	Date/ Period	Aper. Recm.	Rating	Comments
Σ1604	12ʰ 09.5ᵐ	−11° 51'	A–B	6.6	9.4	9.2	089	oY, B	2006	2/3"	*****	Wonderful! Three-colored triplet
			A–C	6.5	8.1	10.2	023	oY, W	2006	2/3"	****	
β920	12ʰ 15.8ᵐ	−23° 21'	A–BC	6.9	8.2	**0.76**	**090**	Y, W	**175**	12/14"	****	
β921	12ʰ 17.9ᵐ	−24° 01'	A–BC	7.0	10.7	3.4	221	W, ?	1991	3/4"	****	
ζ Corvi	12ʰ 20.7ᵐ	−22° 13'	A–B	5.2	13.7	11.2	066	bW, ?	1946	8/10"	***	a.k.a. β1245
7, δ Corvi	12ʰ 29.9ᵐ	−16° 31'	A–B	3.0	8.5	24.3	216	W, prR	2004	15×80	*****	Showcase pair
β28	12ʰ 30.1ᵐ	−13° 24'	A–B	6.5	9.6	**2.16**	**329**	Y, W	**151**	2/3"	****	2005 data calculated from orbital parameters
h1218	12ʰ 35.7ᵐ	−16° 50'	A–B	6.6	11.0	11.60	260	yO, ?	1998	3/4"	***	Very pretty matched pair.
Σ1669	12ʰ 41.3ᵐ	−13° 01'	A–B	5.9	5.9	5.2	313	Y, Y	2006	2/3"	******	Secondary is slightly fainter and slightly more orange than primary
			A–C	5.1	10.5	58.5	235	Y, ?	1988	3/4"	***	
Hu 738	12ʰ 43.8ᵐ	−12° 01'	A–B	6.8	11.8	9.1	259	yO, ?	1959	3/4"	***	
S643	12ʰ 54.0ᵐ	−18° 02'	A–B	7.1	8.2	23.7	295	yW, W	2002	10×50	****	

Variable stars:

Designation	R.A.	Decl.	Type	m_v	Range (m_v)	Period (days)	F (f_r/f_f)	Spectral Range	Aper. Recm.	Rating	Comments
R Corvi	12ʰ 19.6ᵐ	−19° 15'	Mira	6.7	14.4	317.0	0.41	M4.5e–M9e	14/16"	****	

Star clusters:

Designation	R.A.	Decl.	Type	m_v	Size (')	Brtst. Star	Dist. (ly)	dia. (ly)	Number of Stars	Aper. Recm.	Rating	Comments
None												

Nebulae:

Designation	R.A.	Decl.	Type	m_v	Size (')	Brtst./ Cent. star	Dist. (ly)	dia. (ly)	Aper. Recm.	Rating	Comments
NGC 4361	12ʰ 24.5ᵐ	−18° 48'	Planetary	10.9	~45"	13.2	–	–	8/10"	****	Disk is fairly bright at center and dimming toward the outer edge

Galaxies:

Designation	R.A.	Decl.	Type	m_v	Size (')	Dist. (ly)	dia. (ly)	Lum. (suns)	Aper. Recm.	Rating	Comments
NGC 4027	11ʰ 59.5ᵐ	–19° 16'	SB	11.2	3.8×2.3	26M	78K	26G	8/10"	****	NNW–SSE elongation. 40' SW of NGC 4038–4039
NGC 4038	12ʰ 01.9ᵐ	–18° 52'	SB	10.5	5.4×3.9	83M	238K	42G	8/10"	****	Together, NGC 4038 and 4039 form the famous "Ringtail Galaxy" *3
NGC 4039	12ʰ 01.9ᵐ	–18° 53'	Irregular	10.3	5.4×2.5	25M	200K	–	8/10"	****	Together, NGC 4038 and 4039 form the famous "Ringtail Galaxy" *3

Meteor showers:

Designation	R.A.	Decl.	Period (yr)	Duration	Max Date	ZHR (max)	Comet/ Asteroid	First Obs.	Vel. (mi/ km/sec)	Rating	Comments
None											

Other interesting objects:

Designation	R.A.	Decl.	Type	m_v	Size (')	Dist. (ly)	dia. (ly)	Lum. (suns)	Aper. Recm.	Rating	Comments
None											

Fig. 8.6. NGC 4038–4039, the Ringtail or Antennae galaxies. Astrophoto by Rob Gendler.

CRATER (**kray**-ter)	**The Chalice** or **Cup**
CRATERIS (kray-**ter**-iss)	**BEST SEEN** (at 9:00 **P.M.**): May 13
Crt	**AREA:** 282 sq. deg., 53rd in size

TO FIND: Alkes, the brightest star in Crater lies about 20° west of the top stars in Corvus. Alkes and beta Crateris, the second brightest star, form the bottom of the cup or chalice. Alternatively, follow the mnemonic given under Corvus with the cruise (straight line) continuing the same direction as the end of the preceding curve. The "cruise" section ends near the middle of the large arc of faint stars forming the bowl of the chalice (Figs. 8.1 and 8.3).

KEY STAR NAMES:

Name	Pronunciation	Source
α Alkes	ul-**kes** or **al**-kehz	al-Ka's (Arabic), "the cup"

STAR LIST:

	Mag.	Spec.		Mag.	Spec.		Mag.	Spec.
α	4.08	K1 III	β	4.46	A1 V	γ	4.06	A9 V
δ	3.56	Ko III	ε	4.81	K5 III	ζ	4.71	G8 III
η	5.17	A0 V	θ	4.70	B9.5 Vn	ι	5.48	F7 V*4
κ	5.93	F4 III/IV	λ	5.08	F3 IV	ψ	6.11	A0 V*5

Table 8.2. Selected telescopic objects in Crater.

Multiple stars:

Designation	R.A.	Decl.	Type	m_1	m_2	Sep. (")	PA (°)	Colors	Date/Period	Aper. Recm.	Rating	Comments
A 1774	$11^h\,03.2^m$	−11°18'	A–B	5.5	10.8	3.8	274	Y,gB	1952	3/4"	***	
β 600	$11^h\,17.0^m$	+07°08'	A–B	6.1	10.5	1.0	210	yW,–	1959	8/10"	****	
			A–C	6.1	8.2	54.3	099	–,–	2002	7×50	****	
γ Crateristeris	$11^h\,24.9^m$	−17°41'	A–B	4.1	7.9	5.3	093	W,–	1968	2/3"	****	a.k.a. h 840
κ Crateristeris	$11^h\,27.2^m$	−12°21'	A–B	5.9	13.0	254.0	343	yO,–	1998	6/8"	*****	a.k.a. Kui 57
JC 16	$11^h\,29.6^m$	−24°28'	A–B	5.8	8.6	8.2	082	W,–	2000	2/3"	****	
			A–C	5.8	8.9	169.5	116	W,yW	2000	2/3"	***	
ι Crateristeris	$11^h\,38.7^m$	−13°12'	A–B	5.6	11.0	1.7	228	pY,Y	1958	5/6"	****	
β 1078	$11^h\,39.9^m$	−14°28'	A–B	6.2	12.2	8.1	051	W,–	1998	4/6"	****	
H 6 115	$11^h\,48.4^m$	−10°19'	A–B	6.3	9.2	89.9	066	pY,–	2002	10×50	****	
			B–C	9.2	12.8	101.6	357	Y,–	1999	5/6"	***	

Variable stars:

Designation	R.A.	Decl.	Type	Range (m_v)	Period (days)	F (f_r/f_t)	Spectral Range	Aper. Recm.	Rating	Comments
U Crateristeris	$11^h\,12.8^m$	−07°18'	M	9.9	>13.5	169	0.45	M0e	8/10"	****	1.2° WSW of 61 Crateris
RX Crateristeris	$11^h\,17.8^m$	−22°09'	SRb	7.3	7.7	300	–	M3	6×30	***	1.6° NE of β Crateris
RR Crateristeris	$11^h\,31.7^m$	−12°23'	SRb	9.0	10.5	–	–	M5	2/3"	***	1.1° E of 16, κ Crateris
RU Crateristeris	$11^h\,51.1^m$	−11°12'	Lb	8.5	9.5	–	–	M3	10×50	***	

Star clusters:

Designation	R.A.	Decl.	Type	m_v	Size (')	Brtst. Star	Dist. (ly)	dia. (ly)	Number of Stars	Aper. Recm.	Rating	Comments
None												

Nebulae:

Designation	R.A.	Decl.	Type	m_v	Size (')	Brtst./Cent. star	Dist. (ly)	dia. (ly)	Aper. Recm.	Rating	Comments
None											

Galaxies:

Designation	R.A.	Decl.	Type	m_v	Size (')	Dist. (ly)	dia. (ly)	Lum. (suns)	Aper. Recm.	Rating	Comments
NGC 3511	11^h 03.4^m	−23° 05'	SAB(s)c	11.1	5.6 × 1.9	51M	82K	18G	8/10"	****	Elongated E–W, same FOV as NGC 3513
NGC 3513	11^h 03.8^m	−23° 15'	SB(rs)c	11.5	2.9 × 2.2	47M	52K	10G	8/10"	***	Almost circular, same FOV as NGC 3511
NGC 3672	11^h 25.0^m	−09° 48'	SA(rs)c	11.4	3.7 × 1.7	85M	86K	46G	8/10"	****	Elongated WNW–ESE
NGC 3887	11^h 47.1^m	−16° 51'	SAB(rs)bc	10.6	3.5 × 2.4	63M	61K	14G	6/8"	****	Little concentration toward center, stellar nucleus visible in large scopes
NGC 3955	11^h 54.0^m	−23° 10'	S0/a pec	11.5	3.3 × 1.0	58M	51K	7G	8/10"	****	Elongated ENE–WSW, little concentration toward center, faint nucleus
NGC 3962	11^h 54.7^m	−13° 58'	E2/SA (r?)	10.7	2.6 × 2.2	28M	88K	30G	6/8"	****	Fairly bright halo, bright core in 8/10", stellar nucleus in >12"

Meteor showers:

Designation	R.A.	Decl.	Period (yr)	Duration	Max Date	ZHR (max)	Comet/ Asteroid	First Obs.	Vel. (mi km/sec)	Rating	Comments
η Crateristerids	11^h 44^m	−17°	?	Jan 11 – Jan 22	Jan 16/17	<4	–	1892	(Fast)	Weak	First noticed by Henry Corder (U.K.). Best in S. hemisphere. Most meteors are faint

Other interesting objects:

Designation	R.A.	Decl.	Type	m_v	Size (')	Dist. (ly)	dia. (ly)	Lum. (suns)	Aper. Recm.	Rating	Comments
None											

HYDRA (**high**-druh)	**The Water Snake**
HYDRAE (**high**-dry)	**BEST SEEN** (at 9:00 **P.M.**): Apr 30
Hya	**AREA:** 1,302 sq. deg. 1st in size*²

TO FIND: One way to find Hydra is to start at <u>Procyon</u> (the brightest star in Canis Minor). A 5° wide group of one 3rd and 6 4th magnitude stars lies between 14° and 19° east of Procyon. This group represents the head of the water snake (or serpentine monster described in Chapter 6). The snake's body can be traced ESE through three fainter stars (ω, θ, and ι), and then SSW to Alphard ~17° SSE of the Hydra's head(s). From Alphard, the snake's body goes SSE again, almost exactly parallel with the line from the snake's head to Alphard about 6° to κ Hya. From κ, the form goes just a little south of due east for ~17° to ν Hya which is only 3° NW of Alkes (α Cra). Then, the form dips south and east again to pass underneath Crater and Corvus, slightly north to pass under Virgo, and finally east to end just south of the western part of Libra. Although the line of stars is long and somewhat jagged, the stars form a line which is not difficult to see and recognize in moderately dark skies. It may also help to start with the four brightest stars in Hydra, α ("the solitary one" 30° ESE of <u>Procyon</u>), γ (19° nearly due E of <u>Procyon</u>), ζ (12° S of <u>Spica</u>) and ν (3° NW of Alkes). The fainter stars connecting these four can then be traced out to fill in the figure (Figs. 8.1–8.3 and 8.7–8.9).*⁶

KEY STAR NAMES:

Name	Pronunciation	Source
α Alphard	uhl-**furd** or **al**-fard	al-fard al Shuj (Arabic), "the Solitary One in the Serpent" because of the faintness of the surrounding stars
γ Deneb al Shuja	de-neb al **shoe**-jah	Dhanab al Shuj (Arabic), "tail of the Serpent" from its position near the end of the tail
ν Pleura	**plue**-rah	pleura (Latin), "the sides" from its position between the head and the middle of the serpent

Magnitudes and spectral types of principal stars:

Bayer desig.	Mag. (m_v)	Spec. type	Bayer desig.	Mag. (m_v)	Spec. type	Bayer desig.	Mag. (m_v)	Spec. type
α	1.99	K3 III	β	4.29	Ap Si	γ	2.99	G8 III
δ	4.14	A1 Vnn	ε	3.38	G0 III/IV	ζ	3.11	G8 III/IV
η	4.30	B3 V	θ	3.89	B9.5 V	ι	3.90	K3 IIIvar
κ	5.07	B4 IV/V	λ	3.61	K0 III	μ	3.83	K4 III
ν	3.11	K0/1 III	ξ	3.54	G8 III	ο	4.70	B9 V
π	3.25	K2 III	ρ	4.35	A0 Vn	σ	4.45	K2 III
$τ^1$	4.59	F6 V	$τ^2$	4.54	A3 V	$υ^1$	4.11	G6/G8 III
$υ^2$	4.60	B8 V	$φ^1$	4.91	G8 III	$φ^2$	6.01	M1 III
$χ^1$	4.92	F3 IV/V	$χ^2$	5.69	B8 V	ψ	4.94	K0 III
ω	4.99	K2 II/III						

Stars of magnitude 5.5 or brighter which have Flamsteed but no Bayer designations:

Flamsteed	Mag. (m_v)	Spec. type	Flamsteed	Mag. (m_v)	Spec. type	Flamsteed	Mag. (m_v)	Spec. type
2	4.68	K3 III	6	4.98	K3 III	9	4.87	K0 IIICN
12	4.32	G8 III	14	5.30	B9 MNpec	17	4.93	F8 V
20	5.47	G8 II	23	5.24	K2 III	24	5.49	B9 III
27	4.80	F5 V	44	5.08	K5 III	47	5.20	B8 V
50	5.07	K2 III	51	4.78	K3 III	54	5.15	F0 V
56	5.23	G8/K0 III	58	4.42	K3 III			

Stars of magnitude 5.5 or brighter which have no Bayer or Flamsteed designations:

Designation	Mag. (m_v)	Spec. type	Designation	Mag. (m_v)	Spec. type
HD 068312	5.36	G8 III	HD 071155	3.91	A0V
HD 074395	4.63	G2 Ib	HD 074988	5.28	A3 V
HD 079931	5.30	F2/3 IV/V	HD 081799	4.72	K1 III
HD 081809	5.38	G2 V	HD 085859	4.87	K2 III
HD 085951	4.94	K5 III	HD 089953	4.76	B5 V
HD 092036	4.87	M1 III	HD 093397	5.44	A3 V
HD 094388	5.23	F6 V	HD 096819	5.43	A1 V
HD 100393	5.13	M2/3 III	HD 101666	5.20	K5 III
HD 102620	5.10	M4 III	HD 103462	5.26	G8 III
HD 109799	5.41	F0V	HD 110666	5.46	K3 III
HD 122430	5.47	K2/3 III			

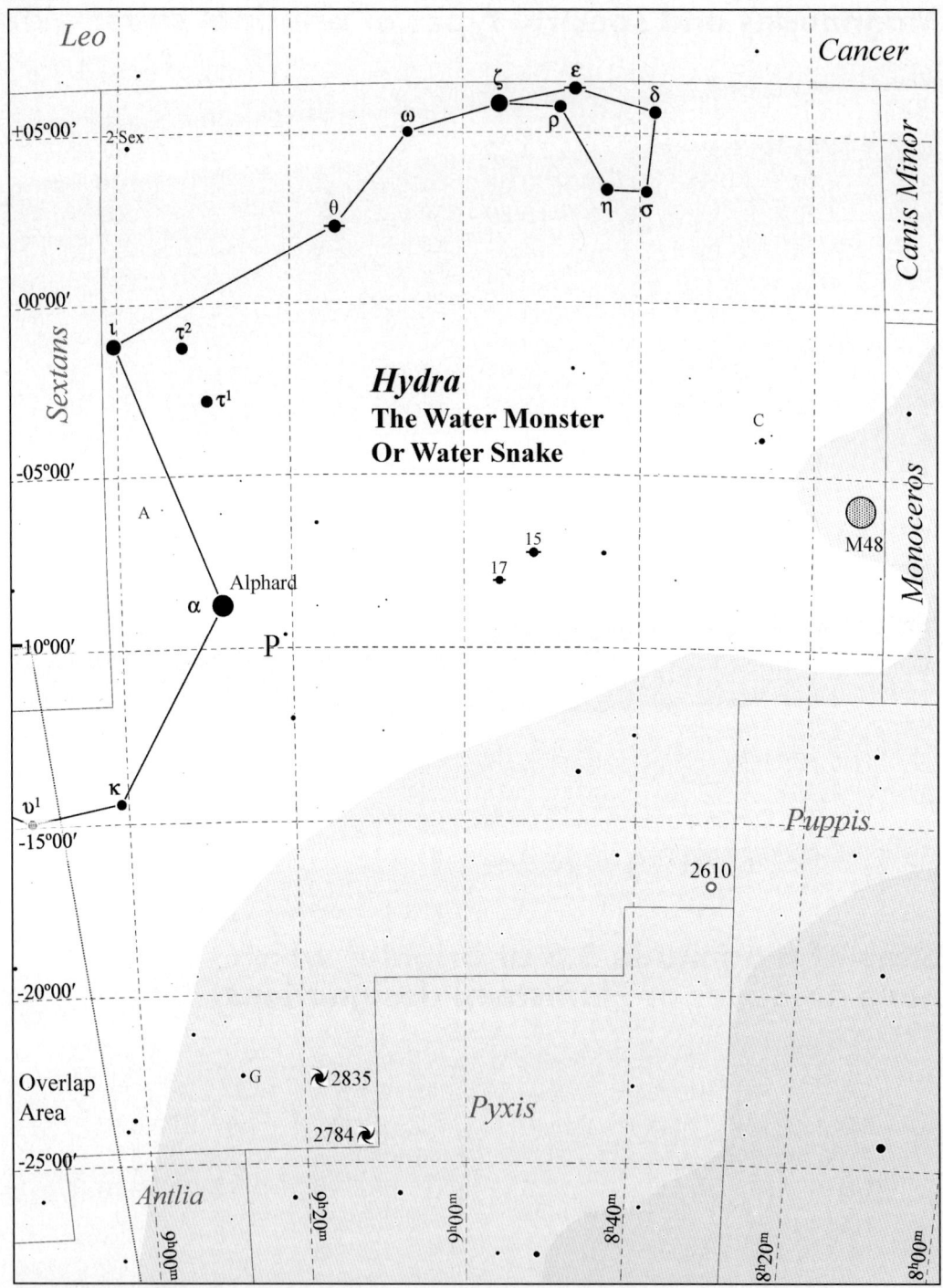

Fig. 8.7. Hydra west details.

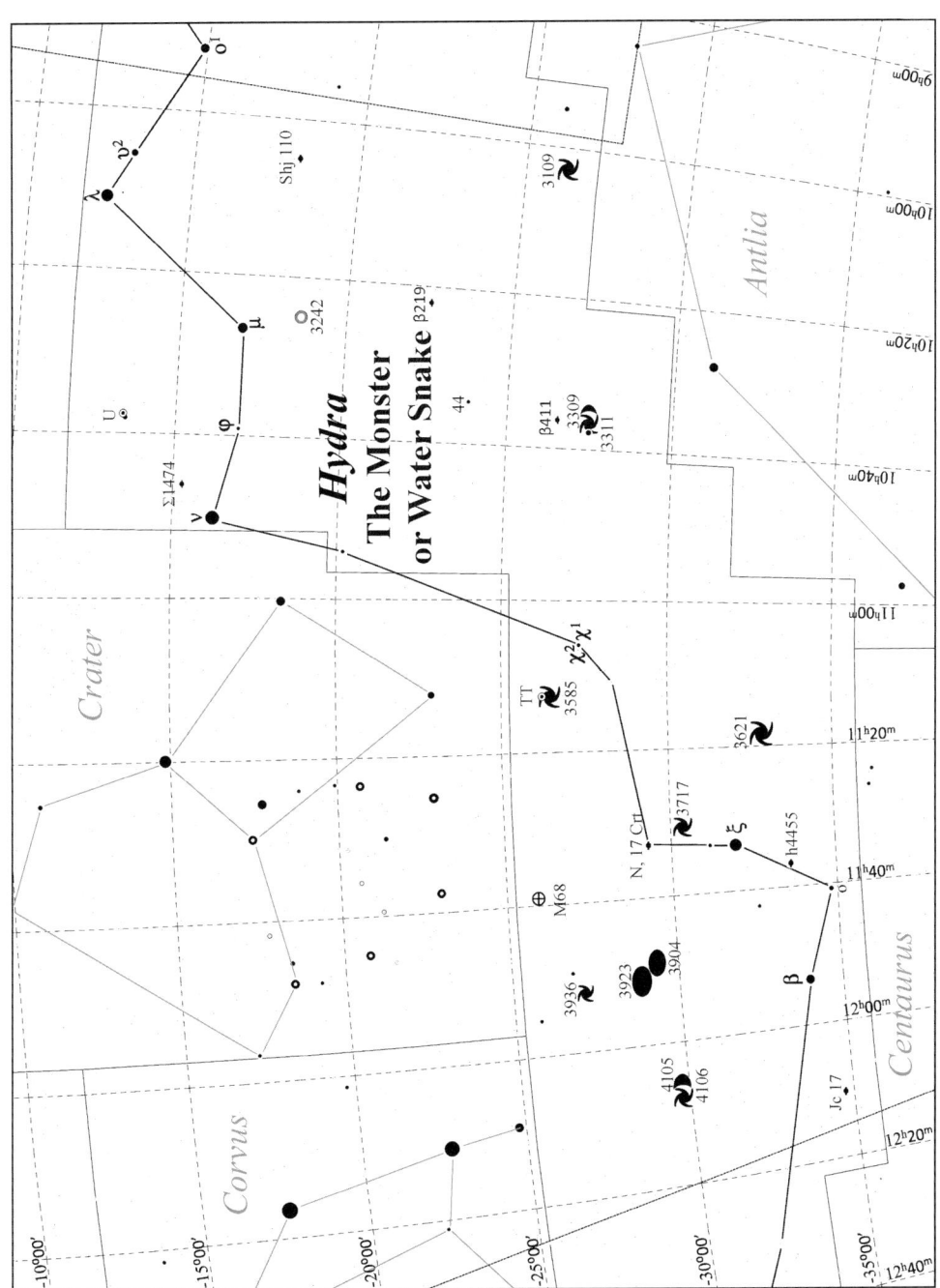

Fig. 8.8. Hydra center details.

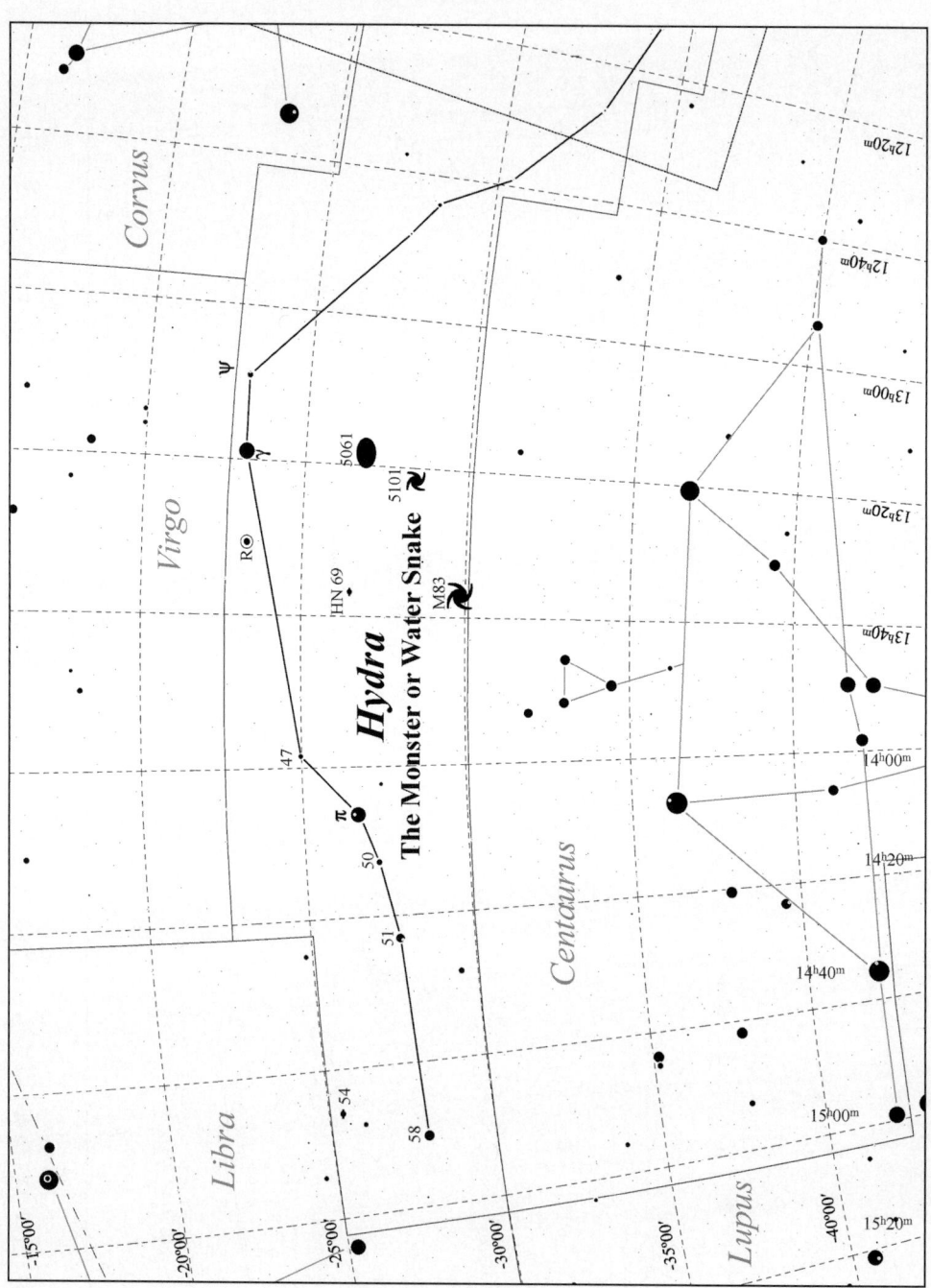

Fig. 8.9. Hydra east details.

Table 8.3. Selected telescopic objects in Hydra.

Multiple stars:

Designation	R.A.	Decl.	Type	m₁	m₂	Sep (")	PA (°)	Colors	Date/Period	Aper. Recm.	Rating	Comments
F Hydrae	08ʰ 43.7ᵐ	−07° 14′	A–B	4.7	8.2	79.0	311	Y, grB	2002	7×50	*****	Easy, colorful binocular object
			A–C	4.7	12.7	57.1	340	Y, ?	1903	5/6″	***	
Σ1270	08ʰ 45.3ᵐ	−02° 36′	A–B	6.9	7.5	4.6	265	rW, rW	2005	2/3″	****	Close, not too dissimilar pair
11, ε Hydrae	08ʰ 46.8ᵐ	+06° 25′	AB–C	3.8	5.3	**2.89**	**305**	yW, Y	**990**	3/4″	*****	a.k.a. STF1273. A–B is too close for amateur instruments
15 Hydrae	08ʰ 51.6ᵐ	−07° 11′	A–B	5.8	7.4	1.1	121	W, Y	1998	8/10″	****	Nice triplet. Each increase of aperture brings new sight
			AB–C	5.5	9.7	45.7	005	W, ?	1998	15×80	****	
			AB–D	5.5	10.8	54.7	056	W, ?	1998	3/4″	****	
S 585	08ʰ 55.2ᵐ	−18° 14′	A–B	5.9	7.2	64.2	151	oY, ?	2002	7×50	****	Easy binocular object
17 Hydrae	08ʰ 55.5ᵐ	−07° 58′	A–B	6.7	6.9	4.0	004	W, yW	2003	2/3″	*****	Almost matched pair. Tight cat's eyes
22, θ Hydrae	09ʰ 14.4ᵐ	+02° 19′	A–B	3.9	9.8	20.7	234	bW, ?	2000	15×80	****	Quadruple system, with 1 bright star and 3 faint companions
			A–C	3.9	11.9	59.6	292	bW, ?	1062	3/4″	****	
			A–D	3.9	12.4	84.3	147	bW, ?	2000	4/6″	****	
26 Hydrae	09ʰ 19.8ᵐ	−11° 58′	A–B	4.8	12.4	3.4	010	Y, ?	1960	4/6″	***	a.k.a. B 2529. Large brightness difference
Σ 1356	09ʰ 28.5ᵐ	+09° 03′	A–B	5.4	6.1	**0.74**	**103**	dY, ?	**118.227**	12/14″	****	Opening to 1.07″ in 2034. Watch to determine resolution limit of your scope
31, τ¹ Hydrae	09ʰ 29.1ᵐ	−02° 46′	A–B	4.6	7.3	66.2	005	yW, ?	2002	7×50	****	Easy binocular pair
S 604	09ʰ 35.7ᵐ	−19° 35′	A–B	6.3	9.4	51.4	092	W, ?	2002	15×80	****	300:1 brightness ratio
h20	09ʰ 41.3ᵐ	−23° 35′	A–B	4.8	11.0	51.9	293	bW, ?	1998	3/4″	****	Nice, similar pair for large binoculars
Shj 110	10ʰ 04.0ᵐ	−18° 06′	A–C	6.2	6.97	20.9	274	?, ?	2003	10×50	****	
β 219	10ʰ 21.6ᵐ	−22° 32′	A–B	6.7	8.5	1.8	186	W, ?	1996	5/6″	****	Near maximum separation of 1.40″, but will close to 0.15″ in 2120
β 411	10ʰ 36.1ᵐ	−26° 41′	A–B	6.7	7.8	**1.38**	**307**	yW, ?	**170.14**	6/8″	****	
Σ1474	10ʰ 47.6ᵐ	−15° 16′	A–B	6.7	7.1	66.2	028	bW, yW	2002	6×30	*****	Attractive trio, almost straight line. Binocular pair becomes triplet in small scope
			A–C	6.7	7.6	73.1	028	bW, ?	2002	7×50	****	
			B–C	7.1	7.5	7.0	022	yW, ?	2002	2/3″	****	
17 Crateris	11ʰ 32.3ᵐ	−29° 16′	A–B	5.6	5.7	9.4	210	yW, yW	2003	2/3″	*****	Nearly perfect matched pair. Ideal critical test for 2″ scope resolution. a.k.a. H3 96
h 4455	11ʰ 36.6ᵐ	−33° 34′	A–B	6.0	7.8	3.4	241	yO, ?	1991	2/3″	****	m13.5 companion 45.7″ away
Jc 17	12ʰ 10.1ᵐ	−34° 42′	A–B	6.4	8.0	3.2	017	W, ?	2000	3/4″	****	

(continued)

Table 8.3. (continued)

Multiple stars:

Designation	R.A.	Decl.	Type	m_1	m_2	Sep (")	PA (°)	Colors	Date/**Period**	Aper. Recm.	Rating	Comments
H N 69	$13^h\ 36.8^m$	−26° 30'	A–B	5.7	6.6	10.2	191	yW, ?	1999	2/3"	***	3 faint companions ~200" away
54 Hydrae	$14^h\ 46.0^m$	−25° 27'	A–B	5.1	7.3	5.1	124	yW, ?	1998	2/3"	****	

Variable stars:

Designation	R.A.	Decl.	Type	Range (m_v)	Period (days)	F (f_r/f_f)	Spectral range	Aper. Recm.	Rating	Comments
U Hydrae	$10^h\ 37.6^m$	−13° 23'	SRb	7	450	–	C6,5,3(N5) (Tc)	7×50	****	##$
TT Hydrae	$11^h\ 13.2^m$	−26° 28'	EA/SD	7.3	6.953	0.11	A5IIIe + G5IV	6×30	****	#
R Hydrae	$13^h\ 29.7^m$	−23° 17'	M	3.5	388.87	0.49	M6e–M9eS(Tc)	3/4"	****	#$

Star clusters:

Designation	R.A.	Decl.	Type	m_v	Size (')	Brtst. Star	Dist. (ly)	dia. (ly)	Number of Stars	Aper. Recm.	Rating	Comments
M48, NGC 2548	$08^h\ 13.8^m$	−05° 48'	Open	5.8	55	8.1	2K	22	150	7×50	****	Best in giant binoculars
M68, NGC 4590	$12^h\ 39.5^m$	−26° 45'	Globular	7.7	11	12.6	31K	100	–	7×50	****	

Nebulae:

Designation	R.A.	Decl.	Type	m_v	Size (')	Brtst./**Cent. star**	Dist. (ly)	dia. (ly)	Aper. Recm.	Rating	Comments
NGC 2610	$08^h\ 33.4^m$	−16° 09'	Planetary	12.7	37"	15.9	5600	63K AU	8/10"	***	"Ghost of Jupiter", Brightest N.Hemis. spring PN. Long exp. photos show dia. 40"
NGC 3242	$10^h\ 24.8^m$	−18° 38'	Planetary	7.7	16"	12.3	2600	13K AU	7×50	******	

Galaxies:

Designation	R.A.	Decl.	Type	m_v	Size (')	Dist. (ly)	dia. (ly)	Lum. (suns)	Aper. Recm.	Rating	Comments
NGC 2835	$09^h\ 17.9^m$	−22° 21'	SB(rs)c	10.5	6.3×4.4	27M	50K	31G	6/8"	****	
NGC 3109	$10^h\ 03.1^m$	−26° 10'	SB(s)m sp	9.8	17.7×3.7	6M	25K	1.2G	4/6"	****	Bright, edge-on galaxy

Designation	R.A.	Decl.	Type	m_v	Size (')	Dist. (ly)	dia. (ly)	Lum. (suns)	Aper. Recm.	Rating	Comments
NGC 3309	10h 36.6m	−27° 31'	E2	11.4	3.3 × 2.1	160	116K	81G	8/10"	***	
NGC 3311	10h 36.7m	−27° 32'	S0	11.2	3.2 × 1.2	140M	130K	68G	8/10"	***	
NGC 3585	11h 13.3m	−26° 45'	SB(s)a pec	10.4	4.1 × 2.3	54M	46K	31G	6/8"	****	
NGC 3621	11h 18.3m	11.5	SA(s)d III	9.3	11.9 × 6.2	20M	74K	8.8G	4/6"	****	
NGC 3717	11h 31.5m	−30° 19'	SAb: sp III	11.2	7.0 × 2.1	72M	108K	13G	8/10"	****	Companion galaxy IC 2913 is nearby
NGC 3904	11h 49.2m	−29° 17'	E2	10.9	2.7 × 1.9	68M	65K	15G	6/8"	****	NGC 3923 is 37' to NE (PA 040°)
NGC 3923	11h 51.0m	−28° 48'	E2	10.2	5.9 × 4.0	67M	117K	7.4G	6/8"	****	Moderately large, bright SW–NW elongation
NGC 3936	11h 52.3m	26° 54'.	SB(s)bc	11.9	4.2 × 0.7	81M	82K	52G	8/10"	***	
NGC 4105	12h 06.7m	−29° 46'	E2	10.7	3.0 × 1.7	83M	67K	29G	6/8"	****	
NGC 4106	12h 06.8m	−29° 46'	SB(s)0+	10.6	2.3 × 1.7	92M	54K	36G	6/8"	****	
NGC 5061	13h 18.1m	−26° 50'	E2	10.4	3.8 × 3.1	66M	82K	43G	6/8"	****	
NGC 5101	13h 21.8m	−27° 26'	SB(rs)a	10.8	5.4 × 4.6	81M	128K	33G	6/8"	****	
M83, NGC 5236	13h 37.0m	−29° 52'	Sc	7.6	12.9 × 11.5	15M	59K	21G	7 × 50	*****	Large, bright, core and arms obvious in 8"

Meteor showers:

Designation	R.A.	Decl.	Period (yr)	Duration	Max Date	ZHR (max)	Comet/Asteroid	First Obs.	Vel. (mi/km/sec)	Rating	Comments
None											

Other interesting objects:

Designation	R.A.	Decl.	Type	m_v	Size (')	Dist. (ly)	dia. (ly)	Lum. (suns)	Aper. Recm.	Rating	Comments
None											

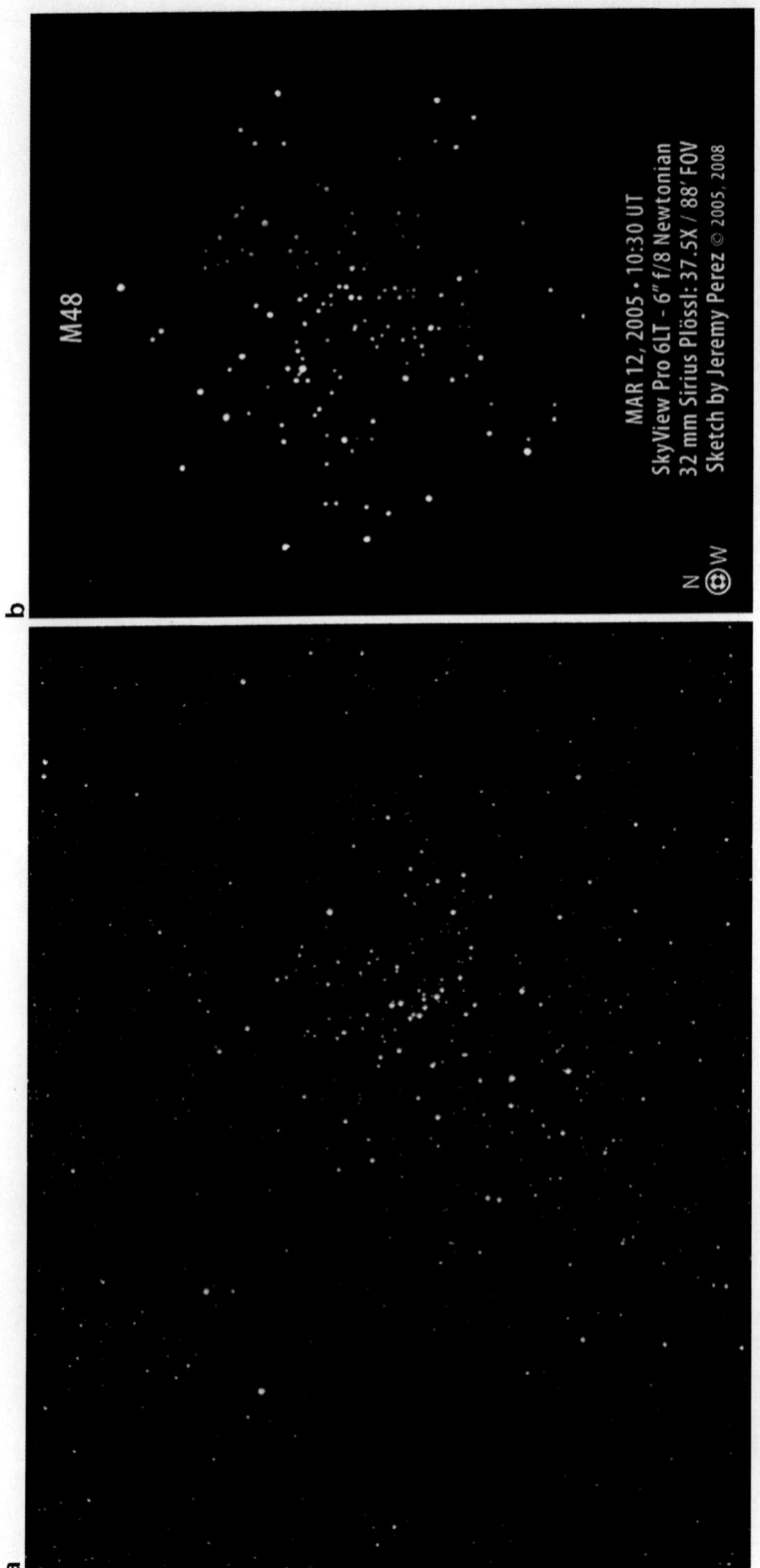

Fig. 8.10. **(a)** M48 (NGC 2548), open cluster in Hydra. Astrophoto by Mark Komsa. **(b)** M48 (NGC 2548), open cluster in Hydra. Drawing by Jeremy Perez.

MAR 12, 2005 · 10:30 UT
SkyView Pro 6LT – 6" f/8 Newtonian
32 mm Sirius Plössl: 37.5X / 88' FOV
Sketch by Jeremy Perez © 2005, 2008

M48

Fig. 8.11. M68 (NGC 4590), globular cluster in Hydra. Astrophoto by Mark Komsa.

Fig. 8.12. M83 (NGC 5236), spiral galaxy in Hydra. Astrophoto by Mark Komsa.

Fig. 8.13. NGC 3109, irregular galaxy in Hydra. Astrophoto by Mark Komsa.

NOTES:

1. Kraz is located in the crow's foot. The origin of the name is obscure, but there are references to it possibly meaning the "claw."
2. Bayer did not use any letters beyond η in Corvus.
3. NGC 4038 and 4039 appear to be a pair of interacting galaxies. In short exposure photographs, the dominant features are a ring of bright spots and a curved trail of nebulosity which looks like a "tail." In very long exposures, the ring of spots is burned out, and two faint curved extensions are seen. The name "Ring Tail Galaxy" describes the first case and "The Antennae Galaxy" describes the second.
4. Bayer did not use the letter ι in Crater, it was added by Flamsteed.
5. Bayer did not use the letter ψ in Crater, it was added by Bode. Bode also added several other letters, none of which is still used today.
6. Hydra is a very long constellation, stretching from $11^h 12^m$ to $15^h 03^m$ R.A., a distance of over 67°. If the center of Hydra were directly overhead, the constellation would stretch over 1/3 of the distance from horizon to horizon.

CHAPTER NINE

The Eagle and its Errands

August, October

AQUARIUS	AQUILA	SAGITTA
The Water Bearer	The Eagle	The Arrow

As leader of the Olympian gods and goddesses, Zeus was served by the eagle, one of the most majestic birds, and a symbol of his power. The eagle became Zeus' servant shortly after his birth. When his mother hid the young Zeus in a cave to prevent having him swallowed by Chronus, he was first suckled by the goat Amalthea, and later brought food by the eagle. Although Amalthea played no role later in Zeus' life, the eagle continued to play a significant role in many of the stories told about him.

When Zeus had reached adulthood, the eagle carried his thunderbolts, and ran errands for him. During the battle to overthrow Chronus, and the war against the Titans, the eagle frequently flew back to the smithy of the Cyclopes to fetch new supplies of thunderbolts for Zeus to hurl at his opponents, and held them until they were needed. After the war was won, the eagle tortured the chained Prometheus for his rebellion against Zeus. Throughout Zeus' reign, the eagle used his keen eyes to keep watch on people and events on earth, and performed many other important tasks for his master.

One such assignment was the abduction of Ganymede. The youngest son of the eponymous King Tros of Troy, Ganymede was supposed to be the handsomest youth alive. Zeus saw the young man and decided to make Ganymede his personal companion. To accomplish this, he sent the eagle to Mount Ida, where Ganymede

P. Simpson, *Guidebook to the Constellations: Telescopic Sights, Tales, and Myths*, Patrick Moore's Practical Astronomy Series, DOI 10.1007/978-1-4419-6941-5_9, © Springer Science+Business Media, LLC 2012

was watching his father's flocks. The eagle swooped down over the unsuspecting lad, and catching the boy's clothing with his claws, lifted and carried him back to Olympus. Zeus then compensated Tros by sending him a magnificent golden-leaved vine fashioned by Haephestus, the smithy and jeweler of the gods, and a pair of divine snow-white horses.

Later, Zeus found a pretext to justify promoting Ganymede to be the cup-bearer for the gods. Hebe was the daughter of Zeus and Hera, and as such, was one of the 12 Olympians. She was the goddess of youth but had few major responsibilities. She assisted Hera in her daily activities, helped Ares don his armor in preparation for battle, and acted as Aphrodite's herald. Her most important activity seems to have been in association with the serving of food and drink, especially at ceremonial occasions. She was the official cup-bearer of the gods, dispensing their water and Nectar. Nectar was the drink which was the source of their strength and vitality. It also gave immortal life to anyone who drank it. During one sacred ceremony, Hebe stumbled and fell, spilling the Nectar. Zeus used this incident as the excuse to relieve her of her duties and replace her with Ganymede. Eventually, Zeus placed the figure of Ganymede in the heavens as the Constellation of Aquarius, the Water Bearer. Aquarius is depicted as holding a long cylinder filled with water, which he is pouring out, as if into a chalice held by a god. However, in the skies, Piscis Austrinus, the Southern Fish is waiting for the stream with an open mouth.

Heracles' final labor (see Chap. 6) provides the explanation for the placement of Aquila (the eagle) and Sagitta (the arrow) in the skies. For this task, Heracles had to bring back the golden apples of the Hesperides. No mortal knew the location of the Garden of the Hesperides, where the tree bearing the golden apples grew. Heracles traveled over most of the known world trying to find out where the Garden of the Hesperides was located, without success. He was finally advised to ask Prometheus, the wisest of the Titans. After much searching, Heracles found Prometheus on Mount Caucasus, where he was chained to a rock on the orders of Zeus.

Prometheus (foresight) and his brother Epimetheus (aftersight) had created the human race and all the animals. Epimetheus bestowed various gifts on the animals to help them live, e.g., claws, strength, heavy fur or speed. Humans were created last, and Prometheus discovered that his brother had used up all the gifts and left none for man except for intelligence. Left without fur for warmth, claws or fangs to use for catching and killing food animals, or any effective protection from their enemies, Prometheus feared that men would not survive. He appealed to the Olympians for more things to help humans but was refused because the gods were already afraid that humans would use their intelligence to become too powerful and would then threaten the gods themselves.

Since the gods would not help, Prometheus took things into his own hands. He taught men how to farm, build wagons and houses, to tell time from the stars, and how to read, write, count and calculate. He taught them navigation, metallurgy and navigation, plus many other arts. At last, he realized that even with all these gifts men would have a difficult time without fire to keep him warm and to cook his food. He appealed to Athena, who helped him sneak into Olympus, where he stole a glowing coal, concealed it in a reed, and took it back to earth where he gave it to humans.

Zeus was furious at Prometheus' presumption in helping the lowly humans. Zeus was already angry at Prometheus because the Titan knew but would not reveal the secret of a prediction that affected Zeus. It had been foretold that Zeus would marry someone who would bring about his downfall. Zeus was fearful that he would unknowingly fulfill the prophesy and suffer the same fate he had inflicted on his father Chronos, so he decided to punish Prometheus until he revealed the secret. Zeus ordered that Prometheus be chained with an eagle tearing at his liver each day. Prometheus' torment was unending since, as an immortal his liver regenerated each night only to be subjected to another attack when morning came. Zeus would have gladly released Prometheus from his bondage if Prometheus would reveal which marriage would result in Zeus' downfall, but Prometheus would not and thus became a symbol of strength in resisting unjust punishment.

When Heracles approached Prometheus and asked his help in finding the golden apples, Prometheus willingly told him that the father of the Hesperides, Atlas knew the location of their garden. Prometheus asked for no favor, but Heracles was grateful for his help and attempted to kill the eagle which tormented Prometheus. He usually did not miss his target, but this time Zeus caused his arrow to go astray and pass harmlessly by the eagle. After Zeus and Prometheus were reconciled, Zeus placed his favorite son, Heracles, in the sky along with the eagle and the errant arrow, which we see passing harmlessly just above Aquila (Fig. 9.1).

376

Fig. 9.1. Aquila group finder.

AQUILA (ak-will-uh)	**The Eagle**
AQUILAE (ak-will-eye)	**9:00 P.M. Culmination:** Aug. 31
Aql	**AREA:** 653 sq. deg., 22nd in size

TO FIND: Locate the "Summer Triangle" by finding the summer milky way. Almost straight overhead in mid-northern latitudes, you should see a bright blue–white star in the middle of the milky way. This is <u>Deneb</u>. About 25° to the WSW, you should see an even brighter white star, <u>Vega</u>. Turning 90° southward from the Deneb–Vega line, about 35° away will be bright yellowish white <u>Altair</u> on the eastern edge of the milky way. <u>Deneb</u>, <u>Vega</u>, and <u>Altair</u> form an asterism called the Summer Triangle. <u>Altair</u> lies between two fainter stars, each about 2½ degrees away. This small group, or asterism, is sometimes called "the balance beam." You can trace out the rest of Aquila by comparing the stars in the sky to the sky chart (Figs. 9.2 and 9.3).

KEY STAR NAMES:

Name	Pronunciation	Source
α Altair	al-**tair**	Al Nasr al Tair, "The Flying Eagle"
β Alshain	al-**shane**	Al Shahin tara zed, "Star-Striking Vulture"
γ Tarazad	**tar**-ah-zed	Al Shahin tara zed, same as above

Magnitudes and spectral types of principal stars:

Bayer desig.	Mag. (m_v)	Spec. type	Bayer desig.	Mag. (m_v)	Spec. type	Bayer desig.	Mag. (m_v)	Spec. type
α	0.76	A7 IV/V	β	3.71	G8 IV	γ	2.72	K3 II
δ	3.36	F0 IV	ε	4.02	K2 III	ζ	2.99	A0 Vn
η	3.87	–*[1]	θ	3.24	B9.5 III	ι	4.36	B5 III
κ	4.93	B0.5 III	λ	3.43	B9 Vn	μ	4.45	K3 III
ν	4.64	F2 Ib	ξ	4.71	K0 III	o	5.12	F8 V
π	4.97	F2 V	ρ	4.94	A2 V	σ	5.18	B3 V
τ	5.51	K0 III	υ	5.89	A3 IV	φ	5.28	A1 IV
χ	5.28	F3 V	ψ	6.25	B9 III/IV	ω[1]	5.28	F0 IV
ω[2]	6.03	A2 V	4	5.02	B9 V	11	5.27	F8 V

Stars of magnitude 5.5 or brighter which have Flamsteed, but no Bayer designations:

Flamsteed	Mag. (m_v)	Spec. type	Flamsteed	Mag. (m_v)	Spec. type	Flamsteed	Mag. (m_v)	Spec. type
14	5.40	A1 V	15	5.40	K1 III	18	5.07	B8 III
19	5.23	F0 III/IV	20	5.35	B3 V	21	5.14	B8 II/IIIp
23	5.10	K2 II/III	26	4.98	G8 III/IV	27	5.46	B9 III
31	5.17	G8 IVvar	42	5.45	F3 IV	36	5.03	M1 IIIvar
37	5.12	G8 III	51	5.38	F0 V	66	5.44	K5 III
69	4.91	K2 III	70	4.91	K5 II	71	4.31	G8 III

Stars of magnitude 5.5 or brighter which have other catalog designations:

Draper Number	Mag. (m_v)	Spec. type
HD 194013	5.30	G8 III/IV

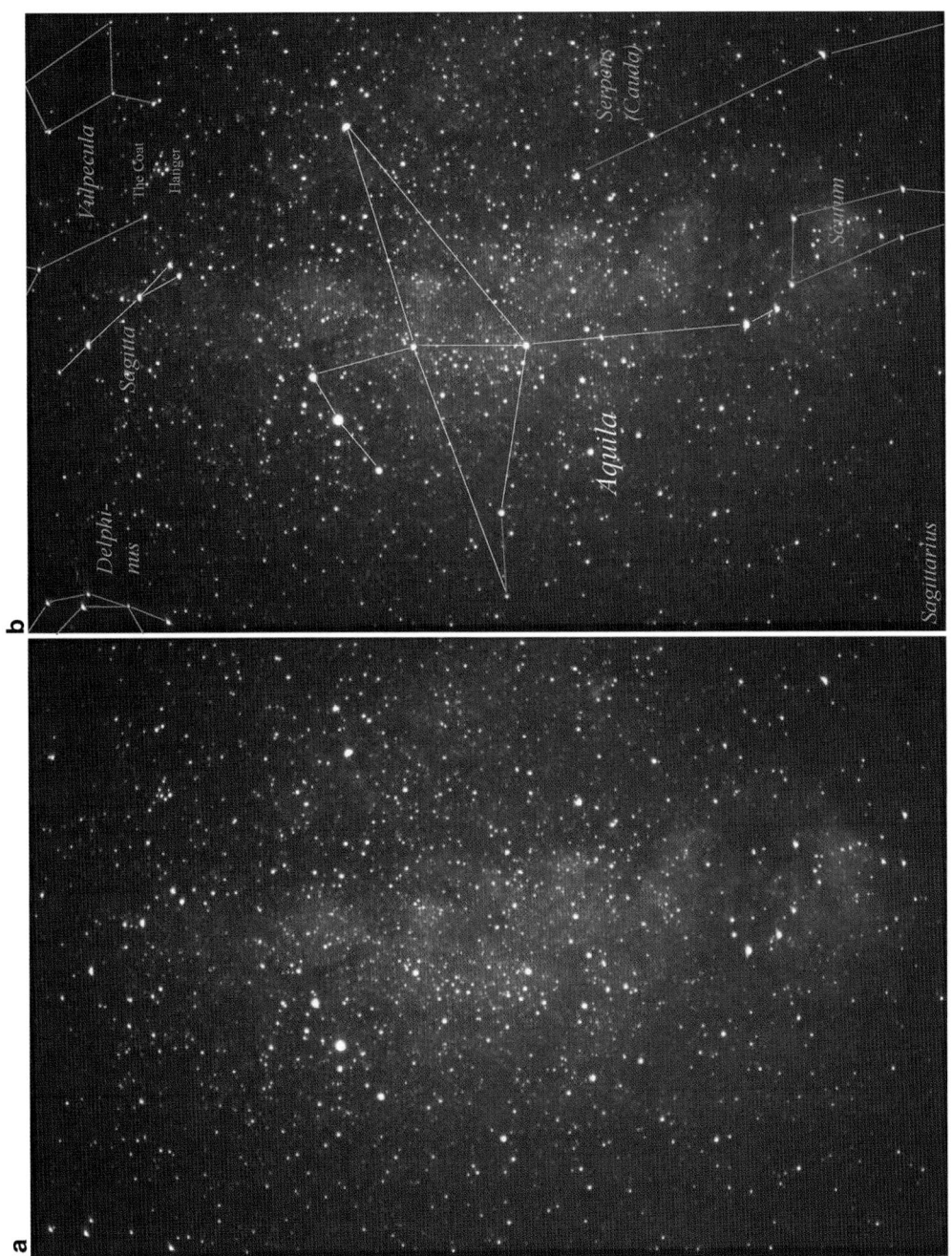

Fig. 9.2. (a) The stars of Aquila and Sagitta. (b) The constellations of Aquila and Sagitta.

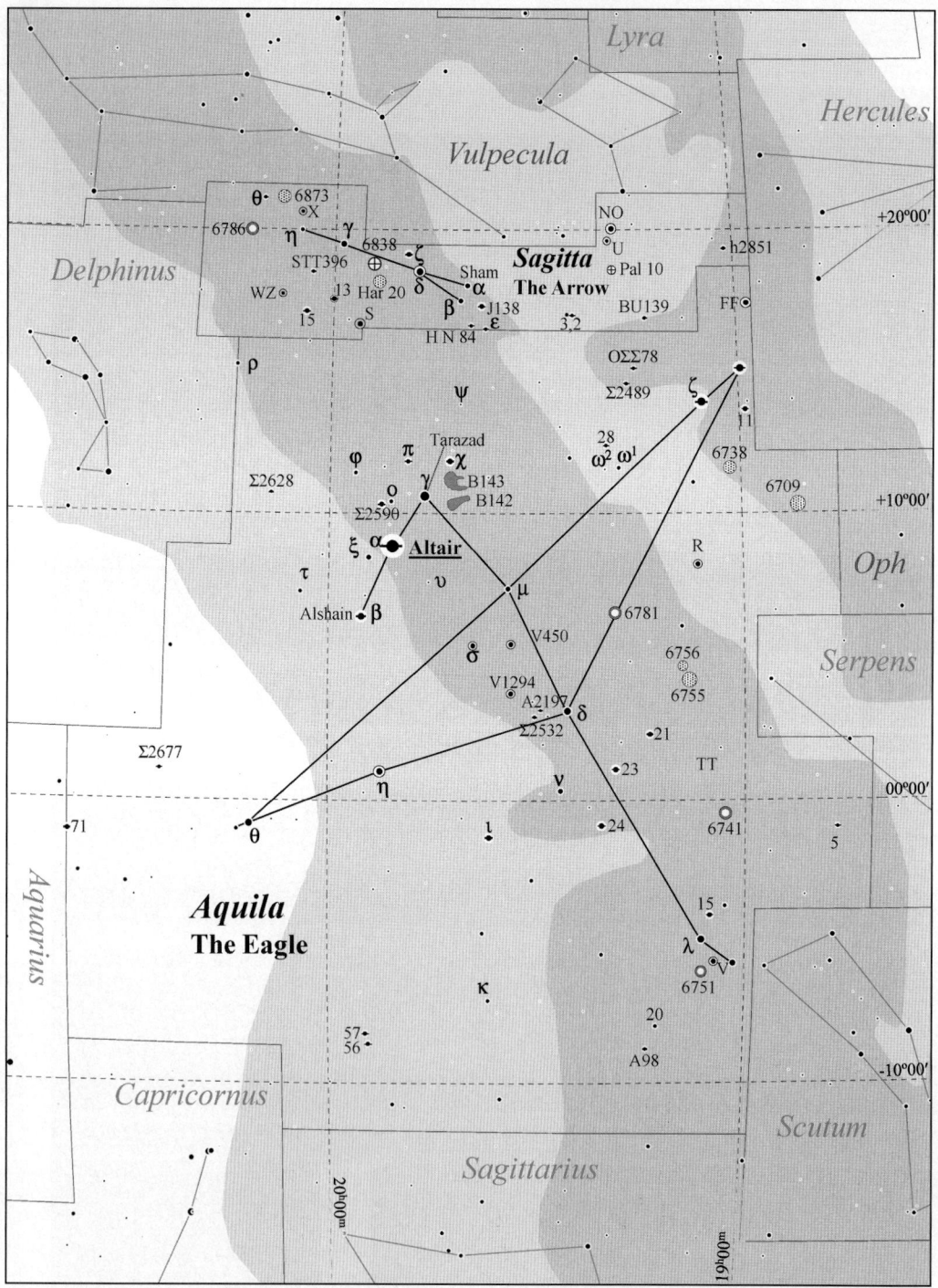

Fig. 9.3. Aquila details.

Table 9.1. Selected telescopic objects in Aquila.

Multiple stars:

Designation	R.A.	Decl.	Type	m₁	m₂	Sep. (")	PA (°)	Colors	Date/Period	Aper. Recm.	Rating	Comments
5 Aquilae	18h 46.6m	−00° 58'	Aa-B	5.9	7.0	11.2	121	yW, rP	2006	2/3"	*****	a.k.a. Σ2379
			Aa-C	5.9	11.8	24.1	147	yW, B	1999	3/4"	**	
11 Aquilae	18h 59.1m	+13° 37'	A-B	5.3	9.3	21.5	300	Y, wB	2006	2/3"	****	Nice color contrast. a.k.a. Σ2424
			A-C	5.3	12.4	78.7	270	Y, –	1997	4/5"	**	
15 Aquilae	19h 05.0m	−04° 02'	A-B	5.5	7.0	40.2	210	pY, dY	2006	7×50	*****	Easy pair, very subtle color difference. a.k.a. Shj286
17, ζ Aquilae	19h 05.4m	+13° 52'	A-B	3.0	12.0	6.5	053	W, –	1934	3/4"	***	a.k.a. β287
21 Aquilae	19h 13.7m	+02° 17'	A-B	5.2	12.3	36.8	287	W, –	1999	4/5"	***	a.k.a. h879
A 98	19h 14.2m	−08° 42'	A-BC	6.5	11.0	22.8	129	Y, –	2003	3/4"	***	Close, nearly matched pair for mid-sized scopes
			B-C	11.0	11.1	1.1	060	–	1997	8/9"	***	
OΣΣ178	19h 15.3m	+15° 05'	Aa-B	5.7	7.6	89.9	268	Y, W	2001	6×30	****	Easy binocular pair. A-a is a close binary, sep. 0.03" to 0.10"
Σ2489	19h 16.4m	+14° 33'	A-B	5.7	9.3	8.2	347	W, pB	2002	2/3"	****	Nice color contrast. a.k.a. Σ2492. Webb called these "yellow and blue"
			A-C	5.7	10.9	59.1	283	W, –	2000	3/4"	***	
23 Aquilae	19h 18.5m	+01° 05'	A-B	5.3	8.3	3.0	004	dY, pGr	1997	3/4"	****	
			A-C	5.3	13.5	10.1	070	dY, –	1999	7/8"	**	B is too faint to show color except in large scopes
24 Aquilae	19h 18.8m	+00° 20'	A-a	6.4	12.6	49.5	160	W, –	1911	5/6"	***	Wide binocular double. a.k.a. Σ40-AB
			Aa-B	6.4	6.8	426.9	317	oW, oW	2001	6×30	*****	a.k.a. Σ40-AB
			B-C	6.8	11.5	59.5	298	O, –	2002	15×80	***	a.k.a. S 717
28 Aquilae	19h 19.7m	+12° 22'	A-B	5.6	9.0	59.7	176	yW, B	2002	7×50	***	In same FOV with double. a.k.a. Σ40-AP
A 2197AB	19h 29.5m	+03° 05'	A-B	8.0	13.0	2.7	248	W, –	1984	6/7"	**	
Σ2531AC	19h 30.2m	+02° 54'	A-C	8.0	10.2	31.4	029	W, –	1999	10×50	***	Σ2532
Σ2532	19h 36.1m	+11° 09'	A-B	6.1	10.6	32.7	003	O, B	1983	3/4"	****	Lovely pair
β1257	19h 36.7m	−01° 17'	A-B	6.7	12.8	3.6	174	W, –	1933	5/6"	**	
41, ι Aquilae	19h 36.9m	+11° 16'	A-B	4.4	13.0	45.6	161	bW, –	1998	6/7"	**	a.k.a. J 118
J 133			A-B	6.1	14.0	17.1	057	Y, –	1959	9/10"	***	
47, χ Aquilae	19h 42.6m	+11° 50'	A-B	5.8	6.6	0.4	078	W, pY	2003	–	***	OΣ380
			AB-C	5.4	12.3	81.1	047	W, –	2000	4/5"	**	
			AB-D	5.4	11.5	5.0	315	W, –	1941	3/4"	***	

(continued)

Table 9.1. (continued)

Multiple stars:

Designation	R.A.	Decl.	Type	m_1	m_2	Sep. (″)	PA (°)	Colors	Date/Period	Aper. Recm.	Rating	Comments
52, π Aquilae	19ʰ 48.7ᵐ	+11° 49′	A–B	6.3	6.8	1.3	110	Y, Y	2006	7/8″	****	Close double, good test of atmospheric stability. a.k.a. Σ 2583
53, α Aquilae	19ʰ 50.8ᵐ	+08° 12′	AB–C	5.8	12.2	35.5	298	Y, –	1999	4/5″	***	a.k.a. ΣB 10
			A–B	0.8	9.8	191.6	287	yW, pV	2003	2/3″	***	
			A–C	0.8	10.1	247.2	100	yW, –	1959	15×80	***	
			A–D	0.8	11.7	32.1	096	yW, –	2005	3/4″	***	
Σ2590	19ʰ 52.3ᵐ	+10° 21′	A–B	6.5	10.3	13.4	308	W, B	2003	2/3″	***	
			C–D	11.6	12.2	6.6	272	–, –	1909	4/5″	***	
56 Aquilae	19ʰ 54.1ᵐ	–08° 34′	A–B	5.8	12.3	46.5	078	dY, –	1906	4/5″	***	a.k.a. h900
57 Aquilae	19ʰ 54.6ᵐ	+08° 52′	A–B	5.7	6.4	36.3	170	bW, bW	2006	7×50	****	a.k.a. β2594
60, β Aquilae	19ʰ 55.3ᵐ	+06° 24′	A–B	3.7	11.8	13.0	006	pO, –	2002	3/4″	***	a.k.a. OΣ532
Σ2628	20ʰ 07.8ᵐ	+09° 24′	A–B	6.6	8.7	3.1	339	yW, –	2006	3/4″	****	
Σ2677	20ʰ 24.6ᵐ	+01° 04′	A–a	6.2	13.1	16.4	013	W, –	2000	6/7″	***	
			A–B	6.2	10.6	31.4	030	W, –	2000	15×80	***	
71 Aquilae	20ʰ 38.3ᵐ	–01° 06′	A–B	4.3	11.0	35.5	284	pO, –	1998	3/4″	***	a.k.a. β672

Variable stars:

Designation	R.A.	Decl.	Type	Range (m$_v$)		Period (days)	F (f$_r$/f$_f$)	Spectral Range	Aper. Recm.	Rating	Comments
FF Aquilae	18ʰ 58.2ᵐ	+17° 22′	Cδ	5.2	6.7	4.471	0.48	F5Ia–F8Ia	6×30	*****	0.7° ESE of 44 Aq. Double, 5.4+11.6, 6.8″, primary is variable. #$
V Aquilae	19ʰ 04.4ᵐ	–05° 41′	SRb	6.6	8.4	353	–	C5,4–C6,4(N6)	6×30	*****	#$
R Aquilae	19ʰ 06.4ᵐ	+08° 14′	M	5.5	12.0	284	0.42	M5e–M9e	4/6″	***	330 luminosity of sun at max., 1,100 ly dist. 0.7° S of 63 Aql. #$
TT Aquilae	19ʰ 08.2ᵐ	+01° 18′	δ Cep	6.5	7.7	13.7546	0.34	F6–G5	6×30	****	#
U Aquilae	19ʰ 29.4ᵐ	–07° 03′	δ Cep	6.1	6.9	7.024	0.30	F5I/II–G1	6×30	****	#
V450 Aquilae	19ʰ 33.8ᵐ	+05° 28′	SRb	6.3	6.7	64.2	0.52	M5III–M8III	6×30	****	$
V1294 Aquilae	19ʰ 36.6ᵐ	+03° 46′	γ Cas	6.8	7.2	–	–	B0.5IV	6×30	*****	$
44, σ Aquilae	19ʰ 39.2ᵐ	+05° 24′	EB/DM	5.0	5.1	1.95	–	B3V+B3V	Eye	***	β Lyra type, small variation is hard to detect visually
55, η Aquilae	19ʰ 52.5ᵐ	+01° 00′	Cδ	3.5	4.4	7.18	0.32	F6Ib–G4Ib	Eye	****	3rd brightest Cepheid, compare with β Aql, m=3.71. #

Star clusters:

Designation	R.A.	Decl.	Type	m_v	Size (')	Brst. Star	Dist. (ly)	dia. (ly)	Number of Stars	Aper. Recm.	Rating	Comments
NGC 6709	18ʰ 51.5ᵐ	+10° 21'	Open	6.7	13	9.1	2,900	11	111	7×50	****	Rich, easily resolved. 5° SW of ζ Aql (m2.99)
NGC 6738	18ʰ 52.6ᵐ	+11° 36'	Open	8.3	15	9.5	49K	215	75	15×80	***	2.5° SSW of ζ Aql (m2.99). Halfway between ζ and 6709, slightly south of joining line
NGC 6755	19ʰ 07.8ᵐ	+04° 16'	Open	7.5	14	10.2	5,000	20	157	11×80	****	Moderately rich, easily resolved. 3° E of Ayla (θ1.2 Ser Cda)
NGC 6756	19ʰ 08.7ᵐ	+04° 41'	Open	10.6	4	11.5	–	–	40	6/8"	**	30' NNE of 6755. In same LP FOV

Nebulae:

Designation	R.A.	Decl.	Type	m_v	Size (')	Brst./Cent. star	Dist. (ly)	dia. (ly)	Aper. Recm.	Rating	Comments
NGC 6741	19ʰ 02.6ᵐ	-00° 27'	Planetary	11.4	6"	17.6	5,200	0.2	8/10"	****	Tiny disk. Greenish-blue
NGC 6751	19ʰ 05.9ᵐ	-06° 00'	Planetary	11.9	20"	15.4	6,500	0.6	6/8"	****	Moderate sized planetary, disk brightens toward center
NGC 6781	19ʰ 18.4ᵐ	+06° 33'	Planetary	11.4	109"	16.2	2,600	1.37	6/8"	****	Large, but faint planetary, spectacular in photographs
Barnard 142	19ʰ 41.0ᵐ	+10° 31'	Dark	–	40	–	–	–	10/12"	****	Extremely dark patch in Milky Way. 1.5° W of γ Aql (m2.72)
Barnard 143	19ʰ 41.4ᵐ	+11° 01'	Dark	–	60×40	–	–	–	10/12"	****	Extremely dark patch in Milky Way. 1.5° W of γ Aql (m2.72)

Galaxies:

Designation	R.A.	Decl.	Type	m_v	Size (')	Brst.	Dist. (ly)	dia. (ly)	Lum. (suns)	Aper. Recm.	Rating	Comments
None												

(continued)

Table 9.1. (continued)

Meteor showers:

Designation	R.A.	Decl.	Period (yr)	Duration	Max Date	ZHR (max)	Comet/ **Asteroid**	First Obs.	Vel. (mi/ **km**/sec)	Rating	Comments
None											

Other interesting objects:

Designation	R.A.	Decl.	Type	m$_v$	Size (')	Dist. (ly)	dia. (ly)	Lum. (suns)	Aper. Recm.	Rating	Comments
Milky Way Rift – west side	19^h 05.4^m	+13° 52'	Dark neb						eye	*****	Coordinates are for 17, ζ Aql
Milky Way Rift – east side	19^h 46.3^m	+10° 37'	Dark neb						eye	*****	Coordinates are for 31 Aql. Rift lies between 31 and ζ Aql

Fig. 9.4. NGC 6781, planetary nebula in Aquila. Astrophoto by Mark Komsa.

Fig. 9.5. M71 (NGC 6838), open cluster in Sagitta. Astrophoto by Mark Komsa.

SAGITTA (suh-**jee**-tuh)	**The Arrow**
SAGITTAE (suh-**jeet**-eye)	**9:00 P.M. Culmination:** Aug. 31
Sge	**AREA:** 80 sq. deg., 86th in size

The small and relatively dim constellation of Sagitta is quite ancient. It was known as an arrow to the ancient Persians, Hebrews, Arabians, Armenians, Greeks, and Romans. Because of this history, there are several different stories of how it came to be in the sky, including being shot by Apollo, Cupid, or Sagittarius. However, probably the most common story is that it was one of Heracles' poisoned arrows as described in Chap. 6.

TO FIND: Locate Vega, Deneb, and Altair, the three 1st magnitude stars of the Summer Triangle. Altair is the southernmost of the three stars, and Sagitta lies about 10° due north of it, and in the center of the smallest angle of the Summer Triangle. Although Sagitta is small and rather faint, it is easily recognized because of its distinctive shape. The tail of the arrow lies almost half (actually 42%) of the distance from Altair to Albireo (β Cyg) and the head of the arrow lies almost directly between Altair and Deneb (α Cyg) about 12° NNE of Altair (Figs. 9.1–9.3).

KEY STAR NAMES:

Name	Pronunciation	Source
α Sham or Alsahm	shomm al-**sahm**	Al Sahm (Arabic) "the arrow." Arabic name for the whole constellation, now applied to α

Magnitudes and spectral types of principal stars:

Bayer desig.	Mag. (m$_v$)	Spec. type	Bayer desig.	Mag. (m$_v$)	Spec. type	Bayer desig.	Mag. (m$_v$)	Spec. type
α	4.37	G0 II	β	4.37	G8 II	γ	3.47	K5 III
δ	3.82	M2 II	ε	5.66	G8 III	ζ	5.00	A3 V
η	5.10	K2 III	θ	6.48	F5 IV	13	5.33	M4 IIIa

Stars of magnitude 5.5 or brighter which have Flamsteed, but no Bayer designations:

Flamsteed	Mag. (m$_v$)	Spec. type
13	5.33	M4 IIIa

Table 9.2. Selected telescopic objects in Sagitta.

Multiple stars:

Designation	R.A.	Decl.	Type	m_1	m_2	Sep. (")	PA (°)	Colors	Date/Period	Aper. Recm.	Rating	Comments
h2851	19ʰ 01.9ᵐ	+19° 07'	A–B	7.1	11.8	15.3	131	dY, ?	1926	3/4"	****	
			A–C	7.1	11.4	47.8	292	dY, ?	2003	3/4"	****	
Wal 102			A–D	7.1	12.0	45.0	140	dY, ?	1944	9/10"	****	
			A–E	7.1	8.0	173.8	340	dY, ?	2002	15×80	***	
β139	19ʰ 12.6ᵐ	+16° 51'	A–B	7.1	8.0	0.6	136	W, ?	2005	3/4"	****	Wide, not too unequal, easy binocular object
OΣΣA 177AC			A–C	7.1	8.0	99.7	278	W, Y	2002	3/4"		
β 139CD			C–D	8.0	11.2	136.5	267	Y, ?	1997	3/4"	***	Easy for small binoculars. a.k.a. HDS 2753
2–3 Sagittae	19ʰ 24.4ᵐ	+16° 56'	Aa–B	6.3	6.9	341.7	078	W, W	2002	15×80	****	
4, ε Sagittae	19ʰ 37.3ᵐ	+16° 28'	A–B	5.8	8.4	87.4	082	dY, ?	2002	2/3"	*****	Lovely multiple system. a.k.a. H 6 26AB
			A–C	5.8	12.5	101.3	279	dY, ?	1998	2/3"	****	a.k.a. H 6 26AC
			A–D	5.8	9.5	160.8	343	dY, ?	1899	7×50	****	a.k.a. H 6 26AD
J 138AB	19ʰ 38.2ᵐ	+17° 14'	A–B	6.7	13.4	23.7	142	dY, ?	1998	3/4"	****	
J 138AC			A–C	6.7	11.4	37.9	130	W, ?	1998	4/5"	****	
H N 84	19ʰ 39.4ᵐ	+16° 34'	A–B	6.4	9.5	28.2	302	Or, B	2003	>20"	******	Showcase object. Beautiful color contrast
8, ζ Sagittae	19ʰ 49.0ᵐ	+19° 09'	A–B	5.6	6.0	**0.21**	**175**	W, ?	**23.22**	2/3"	–	a.k.a. AGC 11. Too close for amateur telescopes. Max sep 0.243" in 2014
			AB–C	5.0	9.0	8.4	311	pY, B	2005	4/5"	*****	Showcase object. STF2585AB–C
13 Sagittae	20ʰ 00.2ᵐ	+17° 30'	A–B	10.0	10.1	24.1	254	oY, ?	2002	–	****	a.k.a. H 4 100AB
			A–C	9.5	5.6	113.7	296	oY, oR	2000	4/5"	*****	a.k.a. H 4 100AC
			C–D	5.4	11.8	28.5	207	rO, ?	1966	7×50	****	a.k.a. β 1477CD
			C–E	5.4	11.5	4.3	116	rO, ?	2001	A5×80	****	a.k.a. β 1477CE
Σ 396	20ʰ 03.3ᵐ	+18° 30'	A–B	6.1	10.4	46.4	207	oY, ?	2002	3/4"	****	
15 Sagittae	20ʰ 04.1ᵐ	+17° 04'	A–a	5.9	11.3	53.7	360	Y, ?	2002	2/3"	****	a.k.a. βpm 202Aa
			A–B	5.9	9.5	166.9	289	Y, oY	2002	2/3"	****	a.k.a. OΣ 592AB
			A–C	5.9	6.9	213.9	333	Y, ?	2000	10/11"	****	a.k.a. OΣ592AC
17, θ Sagittae	20ʰ 09.9ᵐ	+20° 55'	A–B	6.6	8.9	12.0	329	pY, pY	2006	2/3"	*****	a.k.a. Σ2637AB
			A–C	6.6	7.5	89.8	222	yW, dY	2004	2/3"	****	a.k.a. Σ2637AC

(continued)

Table 9.2. (continued)

Variable stars:

Designation	R.A.	Decl.	Type	Range (m$_v$)		Period (days)	F (f$_i$/f$_t$)	Spectral Range	Aper. Recm.	Rating	Comments
NO Sagittae	19^h 18.0^m	+20° 01'	M	5.3	7.7	345	0.50	M5	6×30	****	1° 26' SSE (PA 163°) of 1 Vulpeculae (m4.76)
U Sagittae	19^h 18.8^m	+19° 37'	EA/SD	6.6	10.1	3.381	0.14	B8V + G2IV–III	7×50	*****	1° 35' W (PA 263°) of 1 Vulpeculae (m4.76). ##$
7, δ Sagittae	19^h 47.4^m	+18° 32'	Ib	3.8	3.8	–	–	M2.5II–III + B9V	Eye	****	55' NE (PA 37°) of 11 Sagittae (m5.54)
10, δ Sagittae	19^h 56.0^m	+16° 38'	δ Cep	5.2	6.0	8.382	0.31	F6Ib–G5Ib	6×30	****	39' N (PA358°) of η Sagittae (m5.09). #
13, VZ Sagittae	20^h 00.1^m	+17° 31'	Ib	5.3	5.6	–	–	M4IIIa	Eye	*****	
X Sagittae	20^h 05.1^m	+20° 39'	SR	7.0	9.7	196	0.50	C6,5(N3)	6×30	*****	
WZ Sagittae A	20^h 07.6^m	+17° 42'	UGSU+ E+ZZ	7.0	15.5	~11900	–	DAep(UG)	6×30– >30"	*****	Recurrent nova, large range. Outbursts in 1913, 1946, 1978, 2001. See text. 1° 3' NE (PA 53°) of 15 Sagittae (m5.8). ##$

Star clusters:

Designation	R.A.	Decl.	Type	m$_v$	Size (')	Brst. Star	Dist. (ly)	dia. (ly)	Aper. Recm.	Rating	Comments
Palomar 10	19^h 18.2^m	+18° 34'	Globular	13.2	3.5	~18	35K	40	12/14"	*	Faint patch, unresolved
Harvard 20	19^h 53.1^m	+18° 20'	Open	7.7	9	8.90	2,700	21	8/10"	***	
M71, NGC 6838	19^h 53.8^m	+18° 47'	Globular	8.4	7.2	12.1	13K	27	24×80	****	Rather loose, easily resolved in small scopes
NGC 6873	20^h 08.3^m	+21° 06'	Open	6.4	14	8.4	–	–	7×50+	***	Moderately rich, easily resolved

Nebulae:

Designation	R.A.	Decl.	Type	m$_v$	Size (')	Brst./ Cent. star	Dist. (ly)	dia. (ly)	Aper. Recm.	Rating	Comments
NGC 6886	20^h 12.7^m	+19° 59'	Planetary	11.4	4"	18	–	–	10/12"	**	Extremely small, high power needed to show disk

Galaxies:

Designation	R.A.	Decl.	Type	m_v	Size (')	Dist. (ly)	dia. (ly)	Lum. (suns)	Aper. Recm.	Rating	Comments
None											

Meteor showers:

Designation	R.A.	Decl.	Period (yr)	Duration	Max Date	ZHR (max)	Comet/ Asteroid	First Obs.	Vel. (mi/ km/sec)	Rating	Comments
None											

Other interesting objects:

Designation	R.A.	Decl.	Type	m_v	Size (')	Dist. (ly)	dia. (ly)	Lum. (suns)	Aper. Recm.	Rating	Comments
None											

AQUARIUS (uh-**qwayr**-ee-us)	**The Water Bearer**
AQUARII (uh-**qwayr**-ee-ee)	**9:00 p.m. Culmination:** Oct. 10
Aqr	**AREA:** 980 sq. deg., 10th in size

TO FIND: One convenient way to find Aquarius is to start with the Great Square of Pegasus. Starting from Alpheratz (α And) at the northeast corner, Markab (α Peg) is 20° to the southwest. Continuing along the same line, Sadalmelik (α Aqr) is about the same distance (21°) to the southwest. Again, along the same line Sadalsuud (β Aqr) is about half the length of the Great Square diagonal or 10° farther, and Albali (ϵ Aqr) is another 12° in about the same direction. α and β Aquarii represent Ganymede's shoulders, and ϵ Aquarius represents the tip of his outstretched left hand. Returning to Sadalmelik, just 6° due east is ζ Aquarii, the center of an easily seen asterism called the "Y." The four stars (ζ, γ, η, and π Aqr) of the "Y" also mark one end of the long cylindrical vessel containing the water. 14° to the east–southeast θ and χ Aquarii represent the open mouth from which the water is pouring. A pair of lines of faint stars continue to the southeast for about 10° before bending south and then southwest to end just above Fomalhaut (α PsA). Returning to Ganymede's shoulders (α and β Aqr), a pair of faint stars (25 and 26 Aqr) about 3.5° north of their midpoint represent Ganymede's head, while a pair of lines extend southeast from α and β Aquarius define the right and left sides of his torso, as well as both his legs. The line extending from Sadalmelik is slightly brighter overall, but both are relatively faint, and both end near the point where the stream of water ends just above Fomalhaut (Figs. 9.6 and 9.7).

KEY STAR NAMES:

Name	Pronunciation	Source
α Sadalmelik	**sad**-ul-**MEH**-lik	sa'd al-malik (ind-Arabic), "the Lucky (Stars) of the King," applied to α and o Aqaurii as a pair
β Sadalsuud	**sad**-ul-suh-**YOUD**	sa'd al-su'ud (ind-Arabic), "the Luckiest of the Lucky (Stars)"*3
γ Sadachbia	sad-**uhk**-bih-uh	sa'd al-akhbiya (ind-Arabic), "the Lucky (stars) of the Tent," originally applied to γ, π, ζ and η Aquarii along with 46 Capricorni
δ Skat	skate	Al-saq (Arabic), "the shin" shown in the Arabic translation of the *Almagest*
ϵ Albali	uhl-**bah**-lih or uhl-**bay**-lee	b li' (Arabic), "swallower" from the group "sa'd bula'" for ϵ, μ, and ν Aquarii[1]
θ Ancha	uhng-kuh	ancha (medieval Latin), "hip"
χ Situla	**sih**-tooh-luh	situla (medieval Latin), "pot" or "bucket," used to translate al-dalw (sci-Arabic), "the Well Bucket," which was the Arabic name for one of their constellations located where we now see Aquarius

Magnitudes and spectral types of principal stars:

Bayer desig.	Mag. (m_v)	Spec. type	Bayer desig.	Mag. (m_v)	Spec. type	Bayer desig.	Mag. (m_v)	Spec. type
α	2.95	G2 Ib	β	2.90	G0 Ib	γ	3.86	A0 V
δ	3.27	A3 V	ε	3.78	A1 V	ζ	3.22	F3 III/IV
η	4.04	B9 IV/Vn	θ	4.17	G8 III/IV	ι	4.29	B8 V
κ	5.04	K2 III	λ	3.73	M2 III	μ	4.73	A3m
ν	4.50	G8 V	ξ	4.68	A7 V	o	4.74	B7 IVe
π	4.80	B1 Ve	ρ	5.35	M8 IIIMNp	σ	4.82	A0 IVs
τ¹	5.68	B9 V	τ²	4.05	K5 III	υ	5.21	F7 V
φ	4.22	M2 III	χ	4.93	M3 II	ψ¹	4.24	K0 III
ψ²	4.41	B5 Vn	ψ³	4.99	A0 V	ω¹	4.97	A7 IV
ω²	4.49	B9 V						

Stars of magnitude 5.5 or brighter which have Flamsteed, but no Bayer designations:

Flamsteed	Mag. (m_v)	Spec. type	Flamsteed	Mag. (m_v)	Spec. type	Flamsteed	Mag. (m_v)	Spec. type
7	5.49	K5 III	1	5.15	K1 III	3	4.43	M3 IIIvar
25	5.10	K0 III	18	5.48	F0 V	21	5.48	K4 III
41	5.33	K1 III	32	5.29	A5m	38	5.43	B5 III
66	4.68	K3 III	42	5.34	K1 III	47	5.12	K0 III
86	4.48	G8 III	68	5.24	G8 III	83	5.44	F2 V
94	3.62	G6/8 IV	88	3.68	K1 III	89	4.71	A3 IV
99	4.38	K4 III	97	5.19	A3 V	98	3.96	K0 III
104	4.82	G2 Ib/II	101	4.70	A0 V	103	5.36	K4/5 III
108	5.17	Ap Si	106	5.24	B9 V	107	5.28	F0 III

Stars of magnitude 5.5 or brighter which have other catalog designations:

Draper Number	Mag. (m_v)	Spec. type
HD 222643	5.27	K3 III

Fig. 9.6a. The stars of Aquarius.

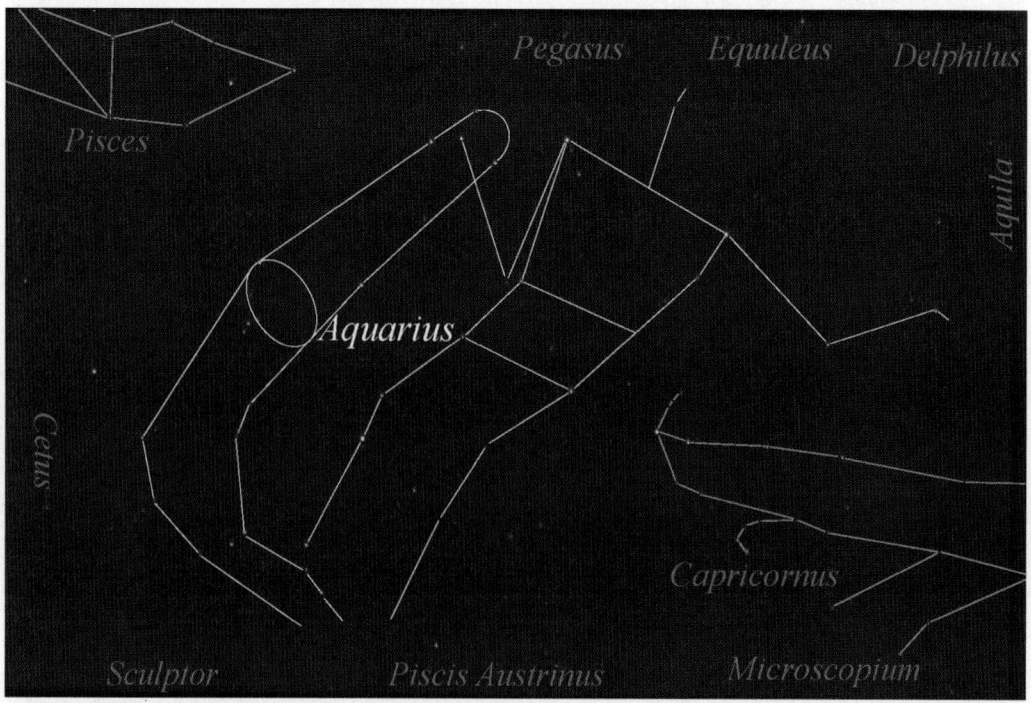

Fig. 9.6b. The constellation of Aquarius.

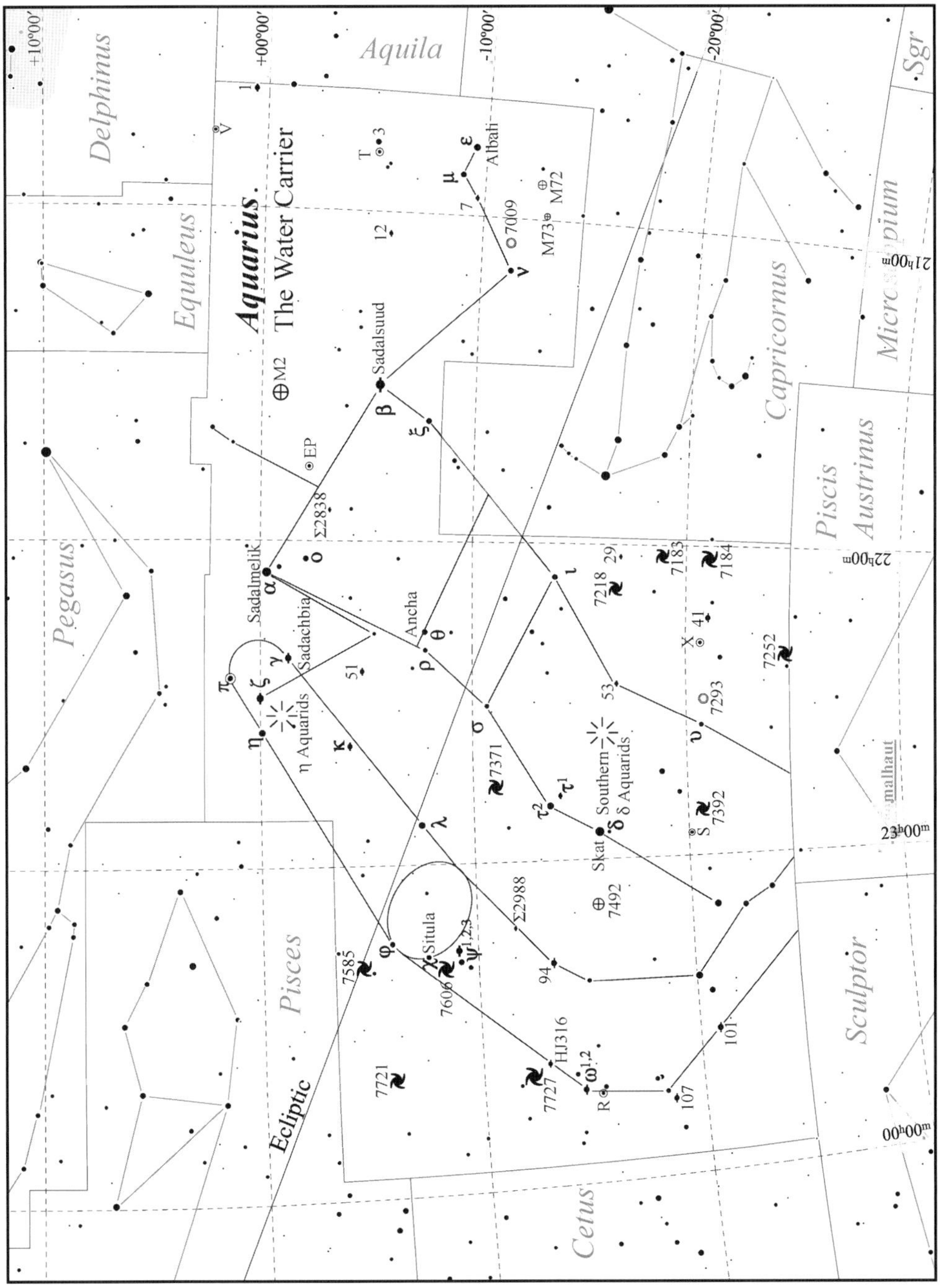

Fig. 9.7. Aquarius details.

Table 9.3. Selected telescopic objects in Aquarius.

Multiple stars:

Designation	R.A.	Decl.	Type	m_1	m_2	Sep. (")	PA (°)	Colors	Date/Period	Aper. Recm.	Rating	Comments
1 Aquarii	20ʰ 39.4ᵐ	−00° 29'	A–B	5.1	11.5	64.5	226	oY, pB	2000	3/4"	****	a.k.a. h3984. Three stars equally spaced in almost a straight line
7 Aquarii	20ʰ 56.9ᵐ	−09° 42'	A–C	5.1	11.3	66.6	030	oY, R	2000	3/4"	****	
			A–B	5.7	11.4	1.0	296	yO, –	1995	9/10"	****	a.k.a. β1034
			A–C	5.7	10.0	177.8	066	yO, –	2002	15×80	****	
12 Aquarii	21ʰ 04.1ᵐ	−05° 49'	Aa–B	5.8	7.5	2.5	194	pY, pB	2006	3/4"	****	a.k.a. Σ2745. Close, pastel colors
H 576	21ʰ 31.6ᵐ	−05° 34'	A–B	2.9	11.0	36.9	319	Y, –	1998	3/4"	****	
		−05° 34'	A–B	2.9	11.6	60.7	187	Y, –	1998	3/4"	****	
Σ2838	21ʰ 54.6ᵐ	−03° 18'	A–B	6.3	9.1	17.6	184	Y, pB	2003	15×80	****	Nice colored pair in a pretty field
29 Aquarii	22ʰ 02.4ᵐ	−16° 58'	A–B	7.2	7.2	3.5	248	W, Y	1998	2/3"	****	a.k.a. S802. Nice pair of matched stars
41 Aquarii	22ʰ 14.3ᵐ	−21° 04'	A–B	5.6	6.7	5.2	112	yO, B	2006	2/3"	****	Lovely colored pair of equal brightness stars. a.k.a. H N 143 AB
		−21° 04'	A–B	5.6	8.9	209.3	043	yW, W	1998	7×50	****	a.k.a. H N 143 AC
		−21° 04'	A–B	9.1	11.8	12.5	255	W, –	1998	3/4"	***	a.k.a. β171 CD
h3 106	22ʰ 21.7ᵐ	−01° 23'	A–B	3.8	12.2	34.2	147	W, –	1998	4/5"	****	
51 Aquarii	22ʰ 24.1ᵐ	−04° 50'	A–B	6.5	6.6	**0.42**	**324**	W, W	**190**	14/16"	****	Requires at least 10" and excellent seeing, three 9–10 mag. stars in FOV. a.k.a. β172 AB
53 Aquarii	22ʰ 26.6ᵐ	−16° 45'	AB–C	5.8	10.1	53.1	343	W, –	2005	2/3"	****	
			A–B	6.3	6.4	**2.4**	**174**	Y, Y	**3,500**	4/5"	****	a.k.a. h345AB. Matched pair. A close pair (1.8", 12.9, 13.9) is 46.7" away at PA 339°
55, ζ¹,² Aquarii	22ʰ 28.8ᵐ	−00° 01'	A–a	4.3	–	**0.04**	**140**	yW, –	**25.7**	–	****	a.k.a. Σ2909
HJ5529	22ʰ 37.8ᵐ	−04° 57'	Aa–B	4.3	4.5	**2.07**	**174**	yW, yW	**760**	4/5"	****	Another nearly matched pair, aka Σ2909
69, τ¹ Aquarii	22ʰ 47.7ᵐ	−14° 03'	A–B	5.0	8.8	93.0	243	Y, –	1980	7×50	****	
			A–B	5.7	9.6	21.5	126	B, yO	2006	15×80	****	a.k.a. Σ2943. Easy pair, reversal of the more common Y and B pairs, colors are subtle
Σ2988	23ʰ 12.0ᵐ	−11° 56'	A–B	7.93	7.95	2.6	097	Y, Y	1998	3/4"	****	Nearly perfectly matched pair
ψ¹ Aquarii	23ʰ 15.9ᵐ	−09° 05'	A–B	4.4	9.9	49.6	313	O, O	2002	2/3"	****	Nearby greenish star provides nice color contrast. a.k.a. ΣB 12A–BC
94 Aquarii	23ʰ 19.1ᵐ	−13° 28'	Aa–B	5.3	7.0	12.6	349	Y, B	1998	2/3"	****	a.k.a. Σ2998. Albireo-like in larger scopes. A–a is too close for amateur scopes

101 Aquarii	23h 33.3m	-20° 55'	A-B	4.8	7.1	0.9	126	W, –	2001	10/11"	***	Requires excellent seeing
h316	23h 37.7m	-13° 04'	A-B	5.7	9.6	30.1	098	dY, –	2003	2/3"	****	##
β279	23h 42.7m	-14° 33'	A-B	4.5	9.9	5.4	093	W, –	1998	2/3"	****	
107 Aquarii	23h 46.0m	-18° 41'	A-B	6.7	6.5	7.0	135	yW, bW	2006	2/3"	****	Tight pair for small telescopes. a.k.a. H 2 24

Variable stars:

Designation	R.A.	Decl.	Type	Range (m$_v$)		Period (days)	F (f$_r$/f$_i$)	Spectral Range	Aper. Recm.	Rating	Comments
V Aquarii	20h 46.8m	+02° 26'	SRa	7.6	9.4	244	–	M6e	7×50	****	##
T Aquarii	20h 49.9m	-05° 09'	M	7.2	14.2	202.1	0.48	M2e–M5e	12/14"	****	0° 34' ESE (PA 102 °) of 3 Aquarii (m4.43). ##
EP Aquarii	21h 46.5m	-02° 13'	SRb	6.4	6.8	55	–	M8III	6×30	****	4° 12' E (PA 089°) of o Aquarii (4.74).
X Aquarii	22h 18.7m	-20° 54'	M	7.5	14.8	311.65	0.42	S6, 3e(M4e–M6.53)	16/18"	***	##
π Aquarii	22h 25.3m	+01° 23'	γ Cas	4.4	4.7	–	–	B1Ve	Eye	****	Also a spectroscopic binary with a period of 84.1d. #
S Aquarii	22h 57.1m	-20° 21'	M	7.6	15.0	279.27	0.39	M4e–M6e	18/20"	****	##
R Aquarii	23h 43.8m	-15° 17'	M	5.8	12.4	386.96	0.42	M5e–M8.5e + pec	6/8"	****	0° 47' SSE (PA 160°) of ω² Aquarii (m4. 49). ##$ *4

Star clusters:

Designation	R.A.	Decl.	Type	m$_v$	Size (')	Brtst. Star	Dist. (ly)	dia. (ly)	Number of Stars	Aper. Recm.	Rating	Comments
M72, NGC 6981	20h 53.5m	-12° 32'	Globular	9.2	6.6	14.2	59K	113	58K	4/6"	****	Fairly bright, wide uneven core, brightest stars ~mag. 15, may be glimpsed in 7×50
M2, NGC7089	21h 33.5m	-00° 49'	Globular	6.6	16	13.1	39K	181	275K	7×50	*****	Large, bright, small core, irregular edge, req 4/6" to resolve
NGC 7492	20h 08.4m	-15° 37'	Globular	11.2	4.2	15.5	82K	100	18K	8/10"	**	May require averted vision

Nebulae:

Designation	R.A.	Decl.	Type	m$_v$	Size (')	Brtst./Cent. star	Dist. (ly)	dia. (ly)	Aper. Recm.	Rating	Comments
NGC 7009	21h 04.2m	-11° 22'	Planetary	12.8	25	–	–	–	8/10"	****	Saturn, Nebula, Bright, bluish-green, elongated, fairly even brightness
NGC 7293	22h 29.6m	-20° 48'	Planetary	13.6	12.8	–	–	–	4/6"	**	Helix, Nebula, faint visually, great photographic object! May be seen in 7×500 in dark skies

(continued)

Table 9.3. (continued)

Galaxies:

Designation	R.A.	Decl.	Type	m_v	Size (')	Dist. (ly)	dia. (ly)	Lum. (suns)	Aper. Recm.	Rating	Comments
NGC 7183	22h 02.4m	−18° 56'	SA: sp	11.9	4.2×1.1	120M	146K	9.3G	8/10"	***	Elongated NW–SE, faint, bright core
NGC 7184	22h 02.7m	−20° 49'	SB(r)b	11.2	6.1×1.4	111M	140K	54G	8/10"	***	Bright, edge-on, small bright core, southern most of 7180, 7185, and 7188 group
NGC 7218	22h 10.2m	−16° 40'	SB(r)b	12.0	2.5×0.9	72M	44K	11G	12/14"	***	Faint halo, stellar nucleus
NGC 7252	22h 20.7m	−24° 41'	SAB0° p	11.4	2.0×1.6	210M	134K	8.5G	8/10"	***	Faint round halo, no brightening toward center
NGC 7371	22h 46.1m	−11° 00'	SAB(rs)b	11.5	1.9×1.8	102M	45K	13G	8/10"	***	
NGC 7392	22h 51.8m	−20° 36'	SB?(rs)b	11.8	2.1×1.1	130M	79K	5.0G	8/10"	***	Moderately faint, slight brightening toward center
NGC 7585	23h 18.0m	−04° 39'	SA?0° pec	11.4	3.0×2.6	150M	131K	7.7G	8/10"	***	Small, round. Faint stellar nucleus
NGC 7606	23h 18.4m	−08° 29'	SB:(rs)b	10.8	5.2×4.4	94M	142K	50G	6/8	***	Elongated NW–SE, slightly brighter toward center
NGC 7721	23h 38.8m	−06° 31'	SAB(rs:)c	11.6	3.0×1.2	95M	61K	3.8G	8/10"	***	Faint, elongated N–S, slightly brighter toward center
NGC 7727	23h 39.9m	−12° 18'	SAB:(S)0/a	10.6	5.6×4.0	84M	137K	7.5G	8/10"	***	Faint, circular, stellar nucleus

Meteor showers:

Designation	R.A.	Decl.	Period (yr)	Duration	Max Date	ZHR (max)	Comet/Asteroid	First Obs.	Vel. (mi/ km/sec)	Rating	Comments
η Aquarids	22h 32m	−01°	Annual	04/21	May 5/6	60	Halley	1870	66, 41	Major	Discovered by Lt. Col. G. K. Tupman
Southern δ Aquarids	22h 36m	−16°	Annual	07/14	Jul 28/29	20	Macholz	1870	41, 25	Major	Discovered by Lt. Col. G. L. Tupman

Other interesting objects:

Designation	R.A.	Decl.	Type	m_v	Size (')	Brst. Star	Number of Stars	Dist. (ly)	dia. (ly)	Aper. Recm.	Rating	Comments
M73, NGC 6994	20h 59.0m	−12° 38'	Asterism	8.9	2.8	10	4	–	–	3/4"	***	Lazy "Y" of 4 stars mistaken for a nebula by Messier, prob. due to poor optics or seeing

Fig. 9.8. M2, globular cluster in Aquarius.

Fig. 9.9. M73, asterism in Aquarius.

Fig. 9.10a. NGC 7923 in Aquarius. Photo by Rob Gendler.

Fig. 9.10b. NGC 7923 in Aquarius. Photo by Mark Komsa, processed to simulate visual appearance.

NOTES:

1. η Aquilae is a Cepheid variable whose spectral class varies between F6 Ib and G4 Ib. See Table 9.1.

2. The Great Milky Way Rift is caused by dust near the plane of the Milky Way Galaxy, and stretches northeast to southwest ~60° from α Cygni (Deneb) to η Serpentis . It varies in width from only ~3° in northern Aquila to as much as ~10° in Vulpecula and Sagitta.

3. There are ten small groups of stars in Aquarius, Capricornus, and Pegasus which have names that start with "sa'd". Sa'd now means "luck" in Arabic, but these star names may all be associated with an early Arabic god named Sa'd, and the origin and exact meaning of these names is uncertain.

4. R Aquarii is a Mira-type variable star with a white dwarf companion. Nova-like eruptions of R Aquarii were recorded by Korean observers in 1073 and 1074.

CHAPTER TEN

The Argonaut's Adventure

February, May, December

ARIES	CARINA	PUPPIS	VELA	TRIANGULUM
The Ram	The Keel	The Stern	The Sails	The Triangle

Athamas, the ruler of Boetia, was forced by Hera to marry Nephele, a phantom which had been created by Zeus to fool a human suitor of Hera. Although Nephele was neither god nor human, she was just as beautiful as Hera and able to be a normal wife to Athamas, and she bore him a son and a daughter, Phrixus and Helle. Because of her supernatural origin, Nephele resented being married to a mortal and treated Athamas with disdain.

Seeking comfort away from home, Athamas met and fell in love with Ino, the daughter of Cadmus and Harmonia. Athamas secretly brought Ino to live near him, and she also bore him two children.

Nephele learned about her husband's mistress from the palace servants, and complained of her treatment to Hera, who proclaimed a curse on Athamas and his descendents. Nephele returned to Boetia and demanded that the people overthrow and kill Athamas, but his subjects feared their king and would not listen to her.

Ino saw an opportunity in this situation, and persuaded the women of the area to join a plot to get rid of Nephele's son Phrixus and leave the succession to the throne to her own son. At her bidding, the women secretly roasted all the seed stock so the seeds would

P. Simpson, *Guidebook to the Constellations: Telescopic Sights, Tales, and Myths*, Patrick Moore's Practical Astronomy Series, DOI 10.1007/978-1-4419-6941-5_10, © Springer Science+Business Media, LLC 2012

not grow when planted. When the proper time came, and no sprouts came through the ground, Athamas sent messengers to Delphi to learn the cause and its remedy.

Ino had anticipated this action and secretly bribed the messengers to falsely tell Athamas that he had been cursed by the gods. She also had them tell her husband that sacrifice of Phrixus was the only way to lift the curse. The people were ready to believe this story because Phrixus' aunt Demodice, who had fallen in love with the handsome young man and been rebuffed by him, had falsely accused him of attempted rape.

At the appointed time, Athamas sadly led Phrixus to a mountain top where the sacrifice was to take place. A number of people, including his sister Helle, accompanied the sacrificial party. However, just before the actual sacrifice, Zeus ordered Hermes to prevent the death of an innocent youth by sending a golden-fleeced ram to rescue Phrixus. The ram flew down to the altar and spoke to Phrixus, telling him to mount his back. As Phrixus mounted the ram, Helle ran up and demanded that the ram also take her away from their cruel stepmother. Phrixus grabbed her hand and pulled her up onto the ram's back, and the ram flew away.

The ram flew eastward toward Colchis, near the eastern shore of the Black Sea, where Helius had his palace. There, he kept his sun chariot and horses ready to start each day's journey of the sun across the sky. As they passed over the narrow strait separating Europe from Asia, Helle became dizzy and lost her grip on the ram's fleece, falling into the water below. From that time on, the strait has been known as the Hellespont in her memory.

After the ram reached Colchis with Phrixus safely on his back, Phrixus was cordially received by Helius' son, King Aeëtes. In gratitude for his escape and safe haven, Phrixus sacrificed the ram and gave the golden fleece to Aeëtes, who hung it in a tree in a garden sacred to Ares. There, the fleece was guarded by a dragon which curled around the tree's base and never slept.

Phrixus later married Chalciope, Aeëtes' oldest daughter, and lived the rest of his life in Colchis, where he and Chalciope had four sons: Argus, Melas, Phrontis, and Cytisorus. Their sons later played minor roles in the story of the Argonauts.

The Golden Fleece became the goal of one of the greatest Greek fictional or mythological adventures, the voyage of the Argonauts, which involved a huge number of the Greek heroes. This epic saga begins with the death of Cretheus, Athamas' brother and the founder of Iolcus (near modern day Volos). Cretheus had married his niece, Tyro, and adopted her two sons by Poseidon, Pelias and Neleus. They soon had a son whom they named Aeson.

After Cretheus' death, Aeson inherited the throne but was soon overthrown by Pelias, who killed most of his rivals. He spared his half-brother only because of the pleas of their mother. Instead, he imprisoned Aeson. However, Pelias was soon warned by an oracle that he would be killed by one of Aeson's descendents. At first, Pelias felt safe because Aeson had no known descendents. Unfortunately for Pelias, before Aeson was overthrown, he had married Alcimede, who was now pregnant, a fact that soon became evident.

When the time came for Alcimede to give birth, Pelias sent guards to her home to seize and kill the infant as soon as it was born. Alcimede anticipated this treachery and bribed the midwife to help her fool the guards and spirit the child away. The guards heard only the yells and groans of a woman in labor, followed by a great wailing and weeping.

Then the midwife emerged and announced that the child was still-born. While the guards returned to inform Pelias, the baby Diomedes was smuggled out by a servant and delivered to Chiron, the great centaur who mentored Asclepius, Achilles, and a host of other Greek heroes. To hide the identity of the child, Chiron called him Jason and did not reveal his real name or identity until Jason was old enough to set out on his own.

Shortly after the birth of Jason, a second oracle revealed to Pelias that he should beware of a stranger wearing only one sandal. This prophecy was not fulfilled until Jason had grown into a man and left the care of Chiron to return to Iolcus. On his way, he carried an elderly woman across a rain-swollen river and lost his left sandal in the mud. This good deed was soon rewarded handsomely, because the woman was actually Hera in disguise, and she helped him many times in his later adventures.

When Jason arrived in Iolcus, a ceremony was underway, and Pelias was startled to see the half-barefooted stranger. He would have had Jason killed immediately but was deterred by the presence of Aeson's brother Athamas. Instead, he questioned Jason, but Jason told him only the name Chiron had used and nothing about his relationship to Aeson. Pelias became even more suspicious when Jason stayed at Aeson's former home.

Jason did nothing until the sixth day, when he went to the palace and, in front of Athamas, announced his true identity and demanded that Pelias relinquish the throne and restore Jason's legacy. Again, Pelias was afraid to take action against Jason while Athamas was present, so he pretended to be concerned about a fake spell cast on Iolcus by one of the gods. Pelias told Jason that he was willing to turn over the kingdom, but that it was too dangerous to do so until the spell had been lifted by the return of the Golden Fleece from Colchis. Jason agreed to this request and began to plan his quest.

As soon as the word of the impending adventure spread, Argus, the son of Niobe and Zeus, volunteered to design, build, and outfit a ship in which to sail to Colchis. At the request of Hera, other gods and goddesses gave gifts and assistance for the Argo (as the ship was called in honor of its builder). They also instilled in various heroes of the time the desire to join the adventure. The list of volunteers quickly swelled to over 50, and it read like a roll call of the greatest men of the time. Among others, they included Heracles and his servant Hylas, Achilles, Castor and Polydeuces, Idas and Lynceus, plus many others with special talents which would prove useful on the journey.

First, they sailed northeast to the island of Lemnos in the northern Aegean Sea, which they found populated only by women. These women had neglected the worship of Aphrodite who made their husbands lose interest in them. When their husbands preferred concubines captured in war, the wives plotted to kill them and all their sons during a single night. The love-starved women of Lemnos welcomed the Argonauts and entertained them so lavishly that they remained on the island for a full year, during which several children were born who would play large roles in later wars and adventures.

From Lemnos, the Argonauts continued northeast until their ship arrived at the island of Samothrace, where they were initiated into the mysteries of "the great gods." After a brief stay, they continued on to Cyzicus (modern Marmara in NW Turkey), where they were warmly welcomed by King Cyzicus. In return, Heracles killed the hostile giants who lived nearby, and the Argonauts resumed their journey. The fortunes betrayed them, and a strong wind blew them back to Cyzicus. Their ship grounded in the darkness, and Cyzicus and his soldiers thought they were being invaded by strangers.

King Cyzicus marshalled his forces and attacked the unsuspecting Argonauts. In the battle, King Cyzicus was killed, and when the dawn allowed recognition of the forces, a truce was declared. The remaining Argonauts and Cyzicus' men jointly buried all the dead. After a period of mourning and reconciliation, the Argo resumed its journey.

To make up for the lost time, Heracles challenged all the oarsmen to see who could row the longest without stopping. When the men all took their places and the order to start was given, the ship jumped ahead as if propelled by gods. As the hours passed, the men dropped out one by one until only Jason and Heracles remained at their posts. Finally, Jason collapsed just as Heracles' oar snapped in two and he fell on the deck holding the broken handgrip. A tie was declared, and the other men resumed their places to row ashore.

They landed in Cios on the northern shore of the Sea of Marmara, where Hylas went to find fresh water and Heracles to find a tree suitable for making a new oar. He soon found a slender poplar, which he pulled from the ground and carried on his shoulder back to the ship and began shaping it into an oar. Heracles completed the oar before dark, and the Argo was ready to depart, but Hylas had still not returned. At dawn, there was a favorable wind and Tiphys urged Jason to leave because the wind would die soon and there would not be another opportunity for many days. Finally, they could wait no longer and the ship departed, but Heracles remained behind to look for his servant. A fight almost broke out over whether to return for their comrades but the sea god Glaucus surfaced beside the ship and told them that Hylas had been enchanted by a nymph and was safe, while it was Zeus' will that Heracles stay in his own country.

They soon landed in an area inhabited by a tribe of barbarians ruled by Amycos, a huge brute and son of Poseidon. He quickly challenged the Argonauts to pick a champion to box him, or they would not be allowed to leave. They regretted the absence of Heracles, but Polydeuces stepped forward to face the man who had never before been beaten. Polydeuces hurled a foul insult at Amycos, and the fight was on. The bear of a man charged forward, fists flying through the air in futility as Polydeuces wove, ducked, and sidestepped each blow. Awaiting his opening, Polydeuces caught an opening and landed a crushing blow on Amycos' temple, fracturing his skull and killing him. Amycos' men then attacked Polydeuces, but the Argonauts rallied around him, quickly drew blood with their swords and drove the barbarians away.

Their next stop was on the opposite coast of the Sea of Marmara, where an aged seer called Phineus lived. Phineus had once lived near the temple at Delphi where he learned the powers of discerning the future and was even able to learn of the councils of the gods. For this impeity, Zeus had blinded him and sent a hoard of Harpies to prevent him from eating. The Harpies were huge bat-like creatures which swooped down each time Phineus tried to eat, snatched the food from his plate, and left their odorous droppings everywhere. When the Harpies next appeared, Zetes and Calais, the winged sons of Boreas, drew their swords and pursued them far into the distance. Finally, the goddess Iris intervened, sending Zetes and Calais back to the Argo, and making the Harpies promise to never again plague Phineus.

Gratefully, Phineus offered to help the Argonauts on their quest. He warned them of the Symplegades (Clashing Rocks) near the entrance to the Black Sea. These rocks rushed together each time anything tried to pass, and had, until that time, never missed crushing the intruder. He advised them to release a dove, and if it was able to make it through the passage, to quickly row through while the rocks were recoiling to their original position. Their successful passage would then remove the rocks' magic, and they would forevermore remain fixed. All this they would do, barely making it through before the rocks again clashed, just missing the ship's stern.

Phineus also gave them detailed descriptions about the land which lay ahead, and how to avoid its pitfalls. The Argonauts carefully memorized every word he uttered and kept them in mind for the rest of their journey. Following Phineus's instructions, the Argonauts sailed on past the land of the Amazons, and the Chalybes, to the Island of Ares where they encountered the Stymphalian Birds that had been driven away from Greece by Heracles, and rescued the shipwrecked sons of Phrixus. The four men helped the Argonauts to sail up the Phasis river to Colchis.

At Colchis, King Aeëtes agreed to let them have the Golden Fleece if Jason would perform several seemingly impossible tasks. First Jason had to harness a pair of fire-breathing bulls given to Aeëtes by Hephaëstus, use them to plow a large field, sow it with dragon's teeth, and then slay the armed men who would grow from the teeth. Here, Aphrodite and Hera conspired to cause Medea, a daughter of Aeëtes who was skilled in magic, to fall in love with Jason, and to betray her father to give him help. Medea gave Jason a magic ointment which would protect him from iron and fire for a day, and advised him to throw a large stone into the middle of the armed men when they arose. When he followed her instructions, the men began fighting among themselves over possession of the stone until there were only a few injured survivors, whom Jason killed with ease.

Jason was then able to put the dragon guarding the fleece to sleep using a potion provided by Medea, take the Golden Fleece, and flee from Colchis. Furious that Jason had accomplished the seemingly impossible task, Aeëtes ordered his men to pursue the Argo to the ends of the earth. Jason successfully evaded the pursuit, using the advice of Phineus and thwarting a planned ambush by killing Aeëtes' son Apsyrtus, and demoralizing the rest of the pursuers. On the return trip, Jason and the Argonauts received protection from King Alcinoüs and Queen Arete of the Phaeacians, who were allies of Aeëtes, on the condition that Jason and Medea were married. They quickly performed the ceremony and escaped.

Before reaching Iolcus again, the Argonauts took a longer route home to avoid further pursuit. On this route, they had to pass through Libya by transporting the Argo overland for 12 days, and battling a bronze giant on the island of Crete. They defeated Talus, the bronze giant, by dislodging a bronze nail in its ankle, causing the giant to bleed to death.

When they arrived in Iolcus, Pelius refused to honor his bargain and had to be tricked by Medea. She used magic potions and spells to rejuvenate Jason's father Aeson, and an aged ram, to convince Pelius that she could restore his lost youth. After watching her cut up the ram, boil the parts in a cauldron with magic herbs and seeing a lamb emerge, then do the same with Aeson, Pelius agreed to give up the throne in return for his youth. Medea performed the ritual again but omitted a critical part, and Pelius was no more.

Although the throne was now open, Jason did not become king, since he and Medea were driven out of Iolcus by Pelias' son Acastus. The pair took refuge in Corinth, where Jason did become king through his relationship with Medea, but all did not end well. Their children died, in spite of a promise by Hera to make them immortal. Eventually, Jason and Medea divorced and went their separate ways.

ARIES (**air**-eeze)	**The Ram**
ARIETIS (air-**ee**-ay tiss)	**9:00 P.M. Culmination:** Dec. 15
Ari	**AREA:** 441 sq. deg., 39th in size

TO FIND: Although Aries (the Ram) is a relatively faint constellation with few bright stars to act as guides, the brighter stars are not difficult to find. Perhaps the easiest way is to use the two best known autumn asterisms, the Great Square of Pegasus and the Pleiades. Hamal (α Arietis), lies approximately halfway between Algenib (α Peg), the star at the southeast corner of the Great Square and the Pleiades. At magnitude 2.01, Hamal is the 47th brightest star, almost exactly as bright as Polaris (magnitude 2.02), and is easily the brightest star in its fairly dim region of the sky. Sheratan (β Arietis) lies just 4° southwest of Hamal, and Mesarthim just 1.5° south of Sheratan. Hamal and Sheratan mark the back of the ram's head, and the tip of its nose, while Mesarthim lies in the ram's extended foreleg. The star 41 Arietis lies just over 10° east–northeast of Hamal and marks the back of the ram's body and the point where the tail begins. Botein (δ Arietis) lies 9° south east of 41 and represents one of the ram's hind hooves. The remainder of the stars in Aries are relatively faint and must be identified on a dark night by using a detailed chart (Figs. 10.1 and 10.2).

KEY STAR NAMES:

Name	Pronunciation	Source
α Hamal	**hah-mahl**	al-hamal (Sci-Arabic), "the Lamb," the Arabic name for the whole constellation
β Sheratan	**sheh**-ruh-tan	al-sharatain (Ind-Arabic), the name of one of the Arabic lunar mansions which was symbolized by β and γ Arietis
γ Mesarthim	meh-zahr-**tim**	meˢsharᵉthīm (Hebrew), "Servants," a misidentification and translation by Johannes Bayer of the lunar mansion name for "The Lamb"
δ Botein	bow-**tane**	al-butain (ind-Arabic), "the Little Belly," the name of a lunar mansion

Magnitudes and spectral types of principal stars:

Bayer desig.	Mag. (m_v)	Spec. type	Bayer desig.	Mag. (m_v)	Spec. type	Bayer desig.	Mag. (m_v)	Spec. type
α	2.01	K2 III	β	2.64	A5 V	γ	3.88	A1p Si
δ	4.35	K2 III var	ε	4.63	A2 V	ζ	4.87	A1 V
η	2.84	F5 V	θ	5.58	A1 Vn	ι	5.09	K1p
κ	5.03	A2m	λ	4.79	F0 V	μ	5.74	A0 V
ν	5.45	A7 V	ξ	5.48	B7 IV	o	5.78	B9 Vn
π	5.26	B6 V	ρ	5.58	F6 V	σ	5.52	B7 IV
τ	5.27	B5 IV	41	3.61	A5 V*[1]			

Stars of magnitude 5.5 or brighter which have Flamsteed but no Bayer designations:

Flam-steed	Mag. (m$_v$)	Spec. type	Flam-steed	Mag. (m$_v$)	Spec. type	Flam-steed	Mag. (m$_v$)	Spec. type
14	4.98	F2 III	33	5.30	A3 V	35	4.65	B3 V
38	5.17	A7 III/IV	39	4.52	K1 III	41	3.61	B8 Vn
52	5.45	B7 Vn	63	5.1	K3 III	64	5.50	K4 III

Stars of magnitude 5.5 or brighter which have no Bayer or Flamsteed designations:

Designation	Mag. (m$_v$)	Spec. type
HD20644	4.47	K2 II/III[*2]

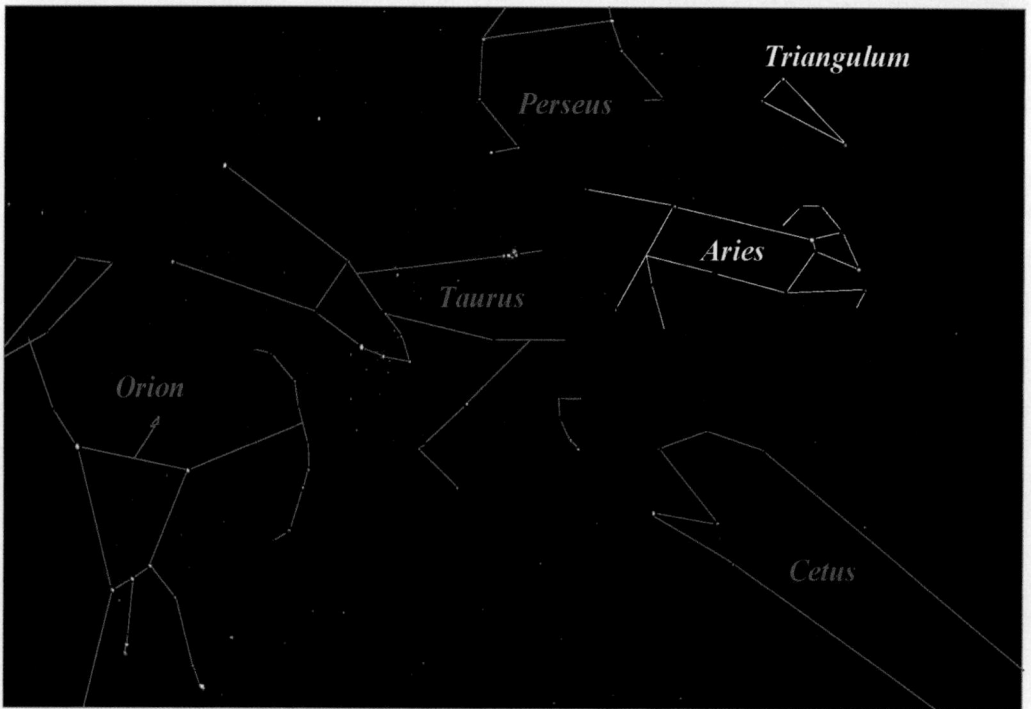

Fig. 10.1. **(a, b)** The stars and constellations of Aries and Triangulum.

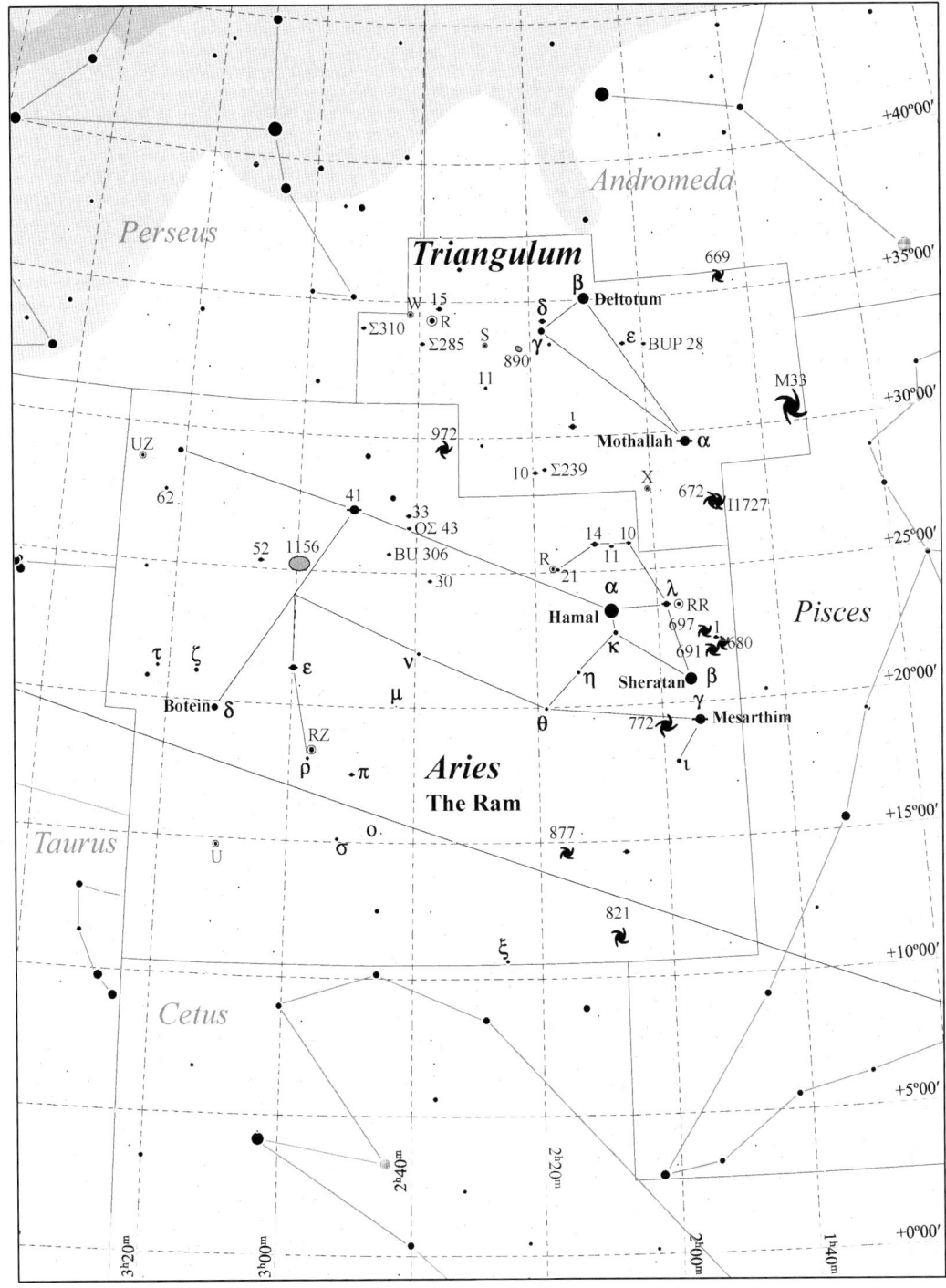

Fig. 10.2. Aries and Triangulum details.

Table 10.1. Selected telescopic objects in Aries.

Multiple stars:

Designation	R.A.	Decl.	Type	m_1	m_2	Sep. (")	PA (°)	Colors	Date/ Period	Aper. Recm.	Rating	Comments
1 Arietis	01ʰ 50.2ᵐ	+22° 16'	A–B	6.3	7.21	2.9	164	pY, B	2006	3/4"	****	a.k.a. Σ 174
γ Arietis	01ʰ 53.6ᵐ	+19° 18'	A–B	4.5	4.6	7.6	356	W, W	2006	2/3"	*****	Nearly identical pair; a.k.a. Σ 180
λ Arietis	01ʰ 57.9ᵐ	+23° 36'	A–B	4.8	6.7	38.2	047	W, yW	2006	7×50	****	Can be glimpsed in 7×50 in good conditions; a.k.a. H 5 12
10 Arietis	02ʰ 03.7ᵐ	+25° 56'	A–B	5.8	7.9	**1.42**	**344**	Y, W	**325**	7/8"	****	A miniature Boötis; a.k.a. Σ 208
			AB–C	5.7	13.5	95.0	150	Y, ?	1998	7/8"	**	
11 Arietis	02ʰ 06.8ᵐ	+25° 42'	A–B	6.0	11.5	1.4	347	W, ?	1986	7/8"	***	a.k.a. HO 312
14 Arietis	02ʰ 09.4ᵐ	+25° 56'	A–B	5.0	8.0	93.2	034	Y, B	1983	7×50	****	Great sight; a.k.a. H 69
			A–C	5.0	8.0	106.7	278	Y, ?	2003	15×80	****	
S 271	02ʰ 20.5ᵐ	+25° 13'	A–B	5.9	10.4	12.7	183	yW, ?	1999	2/3"	***	
			A–C	5.9	11.8	114.8	031	Y, bW	1999	3/4"	***	
30 Arietis	02ʰ 37.0ᵐ	+24° 39'	A–B	6.5	7.0	39.0	275	W, pB	2006	14/15"	****	a.k.a. Σ 289
33 Arietis	02ʰ 40.7ᵐ	+27° 04'	A–B	5.3	9.6	28.6	002	yW, yW	2002	15×80	****	a.k.a. STFA 5
Λ 43	02ʰ 40.7ᵐ	+26° 37'	A–B	7.9	9.0	0.7	350	Y, ?	2006	14/15"	***	Faint, but easily found 27' south of 33 Ari
BU 306	02ʰ 43.9ᵐ	+25° 38'	A–B	6.4	10.4	3.0	019	W, ?	1991	3/4"	***	
			A–C	6.4	12.1	51.0	091	W, ?	1997	3/4"	**	
ρ Arietis	02ʰ 49.3ᵐ	+17° 28'	A–B	5.3	8.0	3.5	118	yW, ?	2002	2/3"	*****	a.k.a.Σ 311AB
			A–C	5.3	10.7	25.1	109	yW, ?	2000	15×80	***	a.k.a.Σ 311AC
41 Arietis	02ʰ 50.0ᵐ	+27° 16'	Aa–B	3.6	11.0	32.1	292	W, dB	1997	3/4"	***	a.k.a. OΣ 47
			Aa–C	3.6	10.7	27.9	231	W, R	1997	15×80	***	a.k.a. H 5 116 Aa–C
			Aa–D	3.6	8.8	122.5	236	W, W	2002	7×50	**	a.k.a. H 6 5 Aa–D
Arietis	02ʰ 59.2ᵐ	+21° 20'	A–B	5.2	5.6	1.4	209	W, pB	2006	6/7"	*****	Great; Nearly equal brightness; a.k.a. Σ 333
52 Arietis	03ʰ 05.4ᵐ	+25° 15'	A–B	6.2	6.2	**0.46**	257	W, ?	227	20/21"	****	a.k.a.Σ 346AB; Pair is widening to 0.70" in 2080
			AB–C	5.5	10.8	4.7	358	W, ?	1997	2/3"	***	

Variable stars:

Designation	R.A.	Decl.	Type	Range (m_v)	Period (days)	F (f_i/f_j)	Spectral Range	Aper. Recm.	Rating	Comments
7, RR Arietis	01ʰ 55.9ᵐ	+23° 35'	EA?	6.4 6.8	47.9	0.08	KOIII	7×50	***	29' W (PA 268°) of λ Arietis (m4.79)
R Arietis	02ʰ 16.1ᵐ	+25° 03'	M	7.4 13.7	186.78	0.45	M3e–M6e	12/14"	****	2° 35' NE (PA 052°) of Hamal (α Arietis, m2.00); ##

Designation	R.A.	Decl.	Type					Spectrum		Aper. Recm.	Rating	Comments
45, RZ Arietis	02^h 55.8^m	+18° 20'	SRb	5.6	6.0	30	–	M6III		7×50	*****	21' NNW (PA 334°) of ρ Arietis (m); ##
U Arietis	03^h 11.0^m	+14° 48'	M	7.2	15.2	371.13	0.4	M4e–M9.5e		Max.	****	##
UX Arietis	03^h 26.6^m	+28° 43'	RS CVn	6.3	6.6	6.438	–	G5V – K0IVa		6×30	*****	1° 24' ESE of HD 20644 (m4.47) *3

Star clusters:

Designation	R.A.	Decl.	Type	m$_v$	Size (')	Brst. Star	Dist. (ly)	dia. (ly)	Lum. (suns)	Aper. Recm.	Rating	Comments
None												

Nebulae:

Designation	R.A.	Decl.	Type	m$_v$	Size (')	Brst./Cent. star	Dist. (ly)	dia. (ly)	Aper. Recm.	Rating	Comments
None											

Galaxies:

Designation	R.A.	Decl.	Type	m$_v$	Size (')	Dist. (ly)	dia. (ly)	Lum. (suns)	Aper. Recm.	Rating	Comments
NGC 678	01^h 49.4^m	+22° 00'	SB(s)b	12.2	4.5×0.8	130M	170K	19G	6/8"	****	Lovely edge-on spiral with dust lane
NGC 680	01^h 49.8^m	+21° 58'	E+ pec	11.9	2.2×2.0	130M	83K	24G	8/10"	***	Pretty bright, small, irregularly round, much brighter toward center
NGC 691	01^h 50.7^m	+21° 46'	SA(rs)bc	11.4	3.5×2.6	120M	117M	32G	8/10"	***	Faint, quite elongated; very gradually brighter toward middle
NGC 697	01^h 51.3^m	+22° 22'	SAB(r)c	12.0	4.4×1.5	270M	346K	92G	6/8"	***	Faint, quite large, elongated; two main arms, very bright core; paired with IC 1738
NGC 772	01^h 59.3^m	+19° 00'	S(s)b	10.3	7.1×4.5	110M	208K	81G	6/8"	***	Bright, quite large, round; small, very bright nucleus; arms almost wind around NGC 770
NGC 821	02^h 08.4^m	+11° 00'	E6?	10.8	3.0×2.0	80M	75K	19G	8/10"	***	Pretty bright, quite large, slightly elongated

(continued)

Table 10.1. (continued)

Galaxies:

Designation	R.A.	Decl.	Type	Size (')	m_v	Dist. (ly)	dia. (ly)	Lum. (suns)	Aper. Recm.	Rating	Comments
NGC 877	02ʰ 18.0ᵐ	+14° 33'	SAB(r)bc	2.3 × 1.0	11.9	180M	120K	45G	8/10"	***	Pretty faint, pretty large, slightly elongated; two arms; very small, very bright nucleus
NGC 972	02ʰ 34.2ᵐ	+29° 19'	Sab	3.4 × 1.7	11.4	70M	73K	19G	8/10"	****	Pretty bright, quite large, slightly elongated; small, faint nucleus
NGC 1156	02ʰ 59.7ᵐ	+24° 14'	IB(s)m	3.0 × 2.3	11.7	30M	–	–	6/8"	***	Pretty bright, quite large, quite elongated

Meteor showers:

Designation	R.A.	Decl.	Period (yr)	Duration	Max Date	ZHR (max)	Comet/ Asteroid	First Obs.	Vel. (mi/ km/sec)	Rating	Comments
October Arietids	02ʰ 08ᵐ	+54°	3.3?	10/06 10/10	10/08	5	Encke	1928	17/28	II	William F. Denning published details *4

Other interesting objects:

Designation	R.A.	Decl.	Type	Size (')	m_v	Dist. (ly)	dia. (ly)	Lum. (suns)	Aper. Recm.	Rating	Comments
None											

Table 10.2. Selected telescopic objects in Triangulum.

Multiple stars:

Designation	R.A.	Decl.	Type	m_1	m_2	Sep. (″)	PA (°)	Colors	Date/ Period	Aper. Recm.	Rating	Comments
2, α Trianguli	01ʰ 53.1ᵐ	+29° 35′	A–B	3.4	12.9	85.0	110	yW, –	1909	5/6″	****	Components C and D are mag. 12.2 and 10.5 at 222 and 277″, respectively
BUP 28	01ʰ 59.1ᵐ	+33° 13′	A–a	7.1	12.2 9.8	91.2	026	bW, bW	1934	4/5″	*****	Nice group in 4/6″ scopes
	01ʰ 59.1ᵐ	+33° 13′	A–B	7.2		92.8	137	bW, –	2002	15×80	****	
3, Trianguli	02ʰ 03.0ᵐ	+33° 17′	A–B	5.4	11.4	4.2	110	bW, W	1990	6/8″	****	a.k.a. Σ201
6, ι Trianguli	02ʰ 12.4ᵐ	+30° 18′	A–B	5.3	6.7	4.0	069	Y, B	2006	4/6″	****	a.k.a. Σ227; Nice color contrast; 4/6″ scope; 150x
8, δ Trianguli	02ʰ 17.1ᵐ	+34° 13′	Aa–B	4.9	13.6	65.4	337	yW, –	1902	7/8″	***	a.k.a. DOR 66AB; A–a is a very close binary, sep. <0.02″
Σ 239	02ʰ 17.4ᵐ	+28° 45′	A–B	7.1	7.8	13.8	212	yW, yW	2006	2/3″	****	Nearly same color; slightly unmatched brightness
10 Trianguli	02ʰ 19.0ᵐ	+28° 39′	A–B	5.0	11.3	57.1	205	W, –	2003	4/6″	****	Easy with the smallest optical aid
15 Trianguli	02ʰ 35.8ᵐ	+34° 41′	A–B	5.6	6.8	138.4	016	rO, W	2003	7×50	****	
Σ 285	02ʰ 38.8ᵐ	+33° 25′	A–B	7.5	8.1	1.6	162	oY, –	2005	6/7″	****	
Σ 310	02ʰ 49.4ᵐ	+33° 56′	A–B	7.4	10.4	2.2	092	W, –	2001	4/5″	****	

Variable stars:

Designation	R.A.	Decl.	Type	Range (m_v)		Period (days)	F (f_f/f_t)	Spectral Range	Aper. Recm.	Rating	Comments
X Trianguli	02ʰ 00.6ᵐ	+27° 53′	EA/SD	8.55	11.27	0.918	0.18	A5V + G0V	3/4″	***	2°21′ SE (PA 135°) of α Tri (m3.42). ##$
S Trianguli	02ʰ 27.3ᵐ	+32° 44′	M	8.90	12.40	241.6	0.48	M2e	6/8″	***	56′ N (PA 358°) of 11 Tri (m5.55). ##
R Trianguli	02ʰ 37.0ᵐ	+34° 16′	M	5.40	12.60	266.9	0.44	M4IIIe–M8e	6/8″	*****	15′ SSE (PA 152°) of 15 Tri (m6.73). # *6
W Trianguli	02ʰ 41.5ᵐ	+34° 31′	SRc	8.50	9.70	108	–	M5II	10×80	***	E (PA 100°) of 15 Tri (m6.23). #$

(continued)

Table 10.2. (continued)

Star clusters:

Designation	R.A.	Decl.	Type	m$_v$	Size (')	Number of stars	dia. (ly)	Dist. (ly)	Brtst. Star	Aper. Recm.	Rating	Comments
None												

Nebulae:

Designation	R.A.	Decl.	Type	m$_v$	Size (')	Brtst./Cent. star	Dist. (ly)	dia. (ly)	Aper. Recm.	Rating	Comments
None											

Galaxies:

Designation	R.A.	Decl.	Type	m$_v$	Size (')	Dist. (ly)	dia. (ly)	Lum. (suns)	Aper. Recm.	Rating	Comments
M33, NGC 598	01^h 33.9^m	+30° 39'	SA(c)	5.7	67.0×41.5	2.4M	50K	25G	4/6"	**	"Pinwheel Galaxy"; 4–6" scopes; very faint oval glow; spiral arms require 12" or larger
NGC 669	01^h 47.2^m	+35° 33'	Sab	12.3	3.0×0.7	–		–	12/14"	***	Elong. NE-SW (PA 036°)
IC 1727	01^h 47.5^m	+27° 20'	Sbm	11.5	6.0×2.6	23M	44K	1.3G	12/14"	**	Possibly interacting with NGC 672
NGC 672	01^h 47.9^m	+27° 26'	SBc	10.9	6.6×2.7	25M	48K	2.2G	3/4"	****	Small, faint, moderately elongated
NGC 890	02^h 22.0^m	+33° 16'	E4	11.2	2.9×2.4	180M	152K	85G	3/4"	***	Small, bright, round
NGC 925	02^h 27.3^m	+33° 35'	S(B)c	10.1	9.8×6.0	31M	88K	7.0G	3/4"	****	Very large, elongated, faint. Faint nucleus

Meteor showers:

Designation	R.A.	Decl.	Type	Period (yr)	Duration	Max Date	ZHR (max)	Comet/Asteroid	First Obs.	Vel. (mi/ km/sec)	Rating	Comments
None												

Other interesting objects:

Designation	R.A.	Decl.	Type	m$_v$	Size (')	Dist. (ly).	dia. (ly)	Lum. (suns)	Aper. Recm.	Rating	Comments
None											

Fig. 10.3. Spiral galaxies NGC 678 and 680 in Aries. Astrophoto by Mark Komsa.

Fig. 10.4. Spiral galaxy NGC 772 in Aries. Background galaxy NGC 780 is above 772. Astrophoto by Mark Komsa.

Fig. 10.5. Barred spiral galaxy NGC 877 in Aries. NGC 870/871 is at lower left, and background galaxy 876 is above 877. Astrophoto by Mark Komsa.

TRIANGULUM (try-**ang**-you-lum)	**The Triangle**
TRIANGULI (try-**ang**-you-lee)	**BEST SEEN** (at 9:00 **P.M.**): Dec. 7
Tri	**AREA:** 132 sq. deg., 78th in size

Although Triangulum is only loosely related to the other constellations of this group by one myth, it is placed here because of its physical location. One myth explaining the origin of Triangulum attributes its origin to Demeter. After the abduction of Persephone and her marriage to Hades, she gained influence as the goddess of the underworld. The people of Sicily (then called Sicilia) elevated her worship to nearly equal that of her mother, and Demeter and Persephone became the two principal deities of the island kingdom. Either at the request of the inhabitants of Sicily or on her own volition, Demeter requested that Zeus honor the triangular-shaped island with a constellation (see the chapter on Boötes and Virgo for more on Demeter and Persephone).

TO FIND: Despite its being small and relatively faint (its brightest star is magnitude 3.0), Triangulum is easy to find and recognize. First, it is a nearly perfect right triangle, and the faintest of the three stars is only one magnitude fainter than the brightest, making it easy to recognize. Second, it is tucked into a niche between three well-known constellations with fairly bright stars. The center of Triangulum lies almost exactly halfway between Almach (γ And), the easternmost bright star in Andromeda and Hamal (α Ari), the brightest star in Aries. Both Almach and Hamal are nearly standard 2nd magnitude stars and easy to pick out from their surroundings. Perseus, another relatively bright constellation, lies to the NE of Triangulum (Figs. 10.1 and 10.2).

KEY STAR NAMES:

Name	Pronunciation	Source
α Mothalla	moh-**thul**-luh	al-muthallath (Sci-Arabic), "the triangle" used only recently
β Deltotum or	del-**toe**-tum	delta (Greek), the 4th letter of the Greek alphabet originally applied to the whole constellation because of the resemblance to the shape of the capital letter "Δ"
Deltoton	**del**-toe-**tahn**	Same as above.

Magnitudes and spectral types of principal stars:

Bayer desig.	Mag. (m$_v$)	Spec. type	Bayer desig.	Mag. (m$_v$)	Spec. type	Bayer desig.	Mag. (m$_v$)	Spec. type
α	3.42	F6 IV	β	3.00	A5 III	γ	4.03	A1 Vnn
δ	4.84	4.66	ε	5.50	A2 V*5	η	5.28	A0 V
ι	4.94	F5 V						

Stars of magnitude 5.5 or brighter which have Flamsteed but no Bayer designations:

Flamsteed	Mag. (m$_v$)	Spec. type	Flamsteed	Mag. (m$_v$)	Spec. type	Flamsteed	Mag. (m$_v$)	Spec. type
6	4.94	F5V0	7	5.25	A0 V	10	5.29	A2 V
12	5.29	F0 III	14	5.15	K5 III	15	5.38	M3 III

Fig. 10.6. M33 (NGC 598) in Triangulum. Astrophoto by Mark Komsa.

Fig. 10.7. NGC 672 in Triangulum. IC 1727 is at upper left. Astrophoto by Mark Komsa.

Fig. 10.8. Barred spiral NGC 925 in Triangulum. Astrophoto by Rob Gendler.

CARINA (kuh-**ree**-nuh)	The Keel
CARINAE (kuh-**ree**-nigh)	9:00 P.M. Culmination: Mar. 18
Car	AREA: 494 sq. deg., 34th in size

TO FIND: Northern hemisphere observers south of approximately 37° north may be able to see Canopus skim the southern horizon. Canopus is about 36° south of Sirius and culminates about 23 min earlier. Being second in brightness only to Sirius, Canopus can be seen near the southern horizon if there are no clouds or fog and is unlikely to be misidentified. For those in the southern hemisphere, Canopus will be 36° above Sirius. In either case, if you are far enough south (<15° N) to see the entire constellation, Canopus is the best place to start in locating Carina. τ Puppis is located just 4.5° ENE of Canopus and serves double duty as both part of the stern and the gunwale of Carina. From τ, χ Carinae is 10.6° to the ESE, and ι Carinae (Aspidiske) is another 12.7° to the SE. Finally, the nearly equal pair of third magnitude stars, p and q Carinae, are another 8° to the ESE of Aspidiske and complete the aft portion of the Argo's gunwale. Like the constellations of Pegasus and Taurus, only a part of the figure of Argo Navis (the aft part) is represented in the sky. Miaplacidus (β Carinae) is the second brightest star in Argo Navis, and although greatly outshone by Canopus, is only 20% fainter than the limit for a first magnitude star. Miaplacidus is located 10.5° nearly due south of Aspidiske. Miaplacidus is the western point of an ENE–WSW diamond shape about 10° long and 5.6° high. This diamond forms the keel extension which kept the ship from sliding sideways when the sails were in use. A line from Canopus to the northernmost star of the diamond (υ Carinae), extended to the eastern point (θ Carinae) completes the bottom part of Carina, the Keel. The forward part of Argo Navis is supposed to be hidden behind the Clashing Rocks (Figs. 10.9–10.11).

KEY STAR NAMES:

Name	Pronunciation	Source
α Canopus	kuh-**noh**-pus	Canobus (Κανωβυσ, Greek), the Argo's chief pilot
β Miaplacidus	mee-uh-**plah**-sih-dus	Mi'ah (Arabic) placidus (Latin), placid waters
ε Avior	**ah**-vih-oar	In 1930, the Royal Air Force of the UK insisted that all its navigational stars had to have a proper name. The Nautical Almanac Office coined the name of this star, perhaps by changing avis (bird) to evoke aviator. Also α Pavonis, q.v.
ι Aspidiske	**uhs**-pih-dis-kee	aspidiske (ασπιδισκε, Greek), derived from Ptolomy's description of its location in the gunwale
Asmidiske	**uhs**-mih-dis-ke	Erroneous name, probably applied because it rhymes with Aspidiske. This star has no Bayer or Flamsteed designation but has the catalog number HD79351

Magnitudes and spectral types of principal stars:*7

Bayer desig.	Mag. (m_v)	Spec. type	Bayer desig.	Mag. (m_v)	Spec. type	Bayer desig.	Mag. (m_v)	Spec. type
α	−0.62	F0 Ib	β	1.67	A2 IV	ε	1.86	K3 III
η	6.21	Pec*8	θ	2.74	Bo Vp	ι	2.21	A8 Ib
υ^A	2.96	A8 Ib	χ	3.46	B3 IVp	ω	3.29	B8 III
a	3.43	B2 IV	b^1	4.93	B2 IV/V	b^2	5.17	F3 V
c	3.84	B8 III	d	4.31	B1.5 III	e^1	5.27	B3 V+
e^2	4.84	K0 III	f	4.50	B3 V	g	4.34	M1 III
h	4.08	B5 II	i	3.96	B3 IV	k	4.79	G6 III
l	3.69	G5 Iab/Ib	m	4.51	B9 V	n*9	–	–
o*10	–	–	p	3.30	B4 Vne	q	3.39	K3 II
r	4.45	K3/K4 II	s	3.81	F2 II	t^1	5.08	K1 III
t^2	4.69	K4/K5 III	u	3.78	K0III/IV	w	4.58	K3 Ib
x	3.93	G0 Ia0	y	4.59	A6 Ia	z^1	4.62	G8 III
z^2	5.11	B9 Ia	A	4.41	G6 II	B	4.74	F5 V
C	5.16	A+	D	4.81	B3 V	E	4.66	B2 IVe
F*11	–	–	G	4.47	F6 II/III	H	5.46	K4 III
I	3.99	F2 IV	K	4.72	A2 III	L	4.97	B8 V
M	5.15	Am	N	4.35	B9 III	P	4.65	A6 Ia
Q	4.93	K3 III	R*12	–	–	S*12	–	–

Stars of magnitude 5.5 or brighter which have Flamsteed but no Bayer designations:

Flam-steed	Mag. (m_v)	Spec. type
23	5.14	M1 III

Stars of magnitude 5.5 or brighter which have no Bayer or Flamsteed designations:

Desig-nation	Mag. (m_v)	Spec. type	Desig-nation	Mag. (m_v)	Spec. type
HD 049689	5.39	K1 II/IIp	HD 053921	5.50	B9 IV
HD 057852	5.50	F5 V	HD 057917	5.38	B9 V
HD 059219	5.09	K0 III	HD 073389	4.84	K0 III
HD 073887	5.45	K0 III	HD 077370	5.17	F3 V
HD 079351	3.43	B2 V	HD 090671	5.38	F4 V
HD 080951	5.28	A1 V	HD 082350	5.46	K2 III
HD 084121	5.30	A3 IV	HD 088323	5.26	K0 III
HD 091056	5.27	K3 Ib	HD 092063	5.08	K1 III
HD 092207	5.47	A0 Ia	HD 092964	5.36	B2.5 Ia
HD 093194	4.80	B5 Vn	HD 093540	5.33	B7 V
HD 093549	5.23	B7 IV	HD 093607	4.87	B3 IV
HD 091496	4.94	K4/5 III			

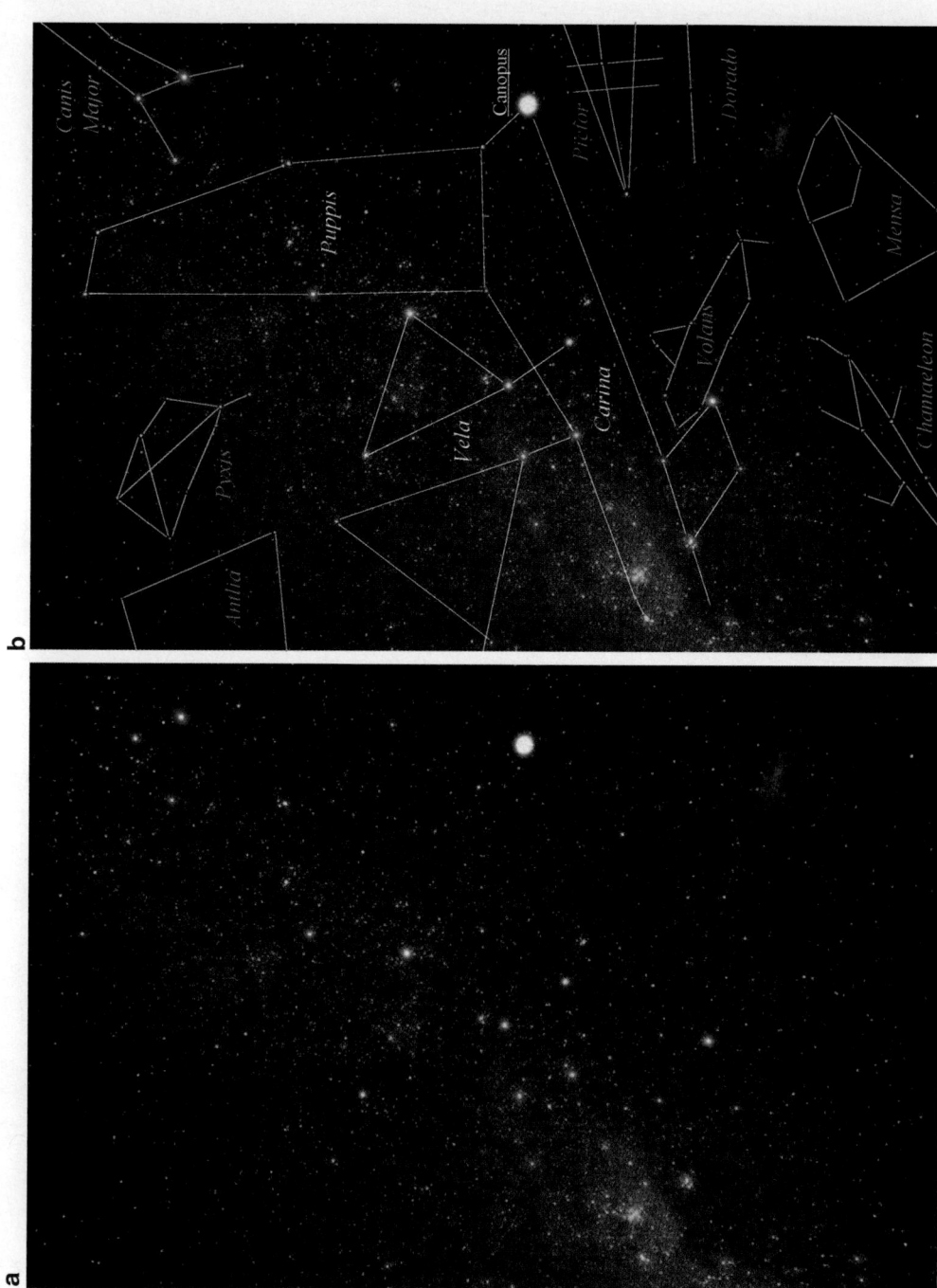

Fig. 10.9. (a, b) Stars and constellations of Argo Navis region. Astrophoto by Christopher Picking.

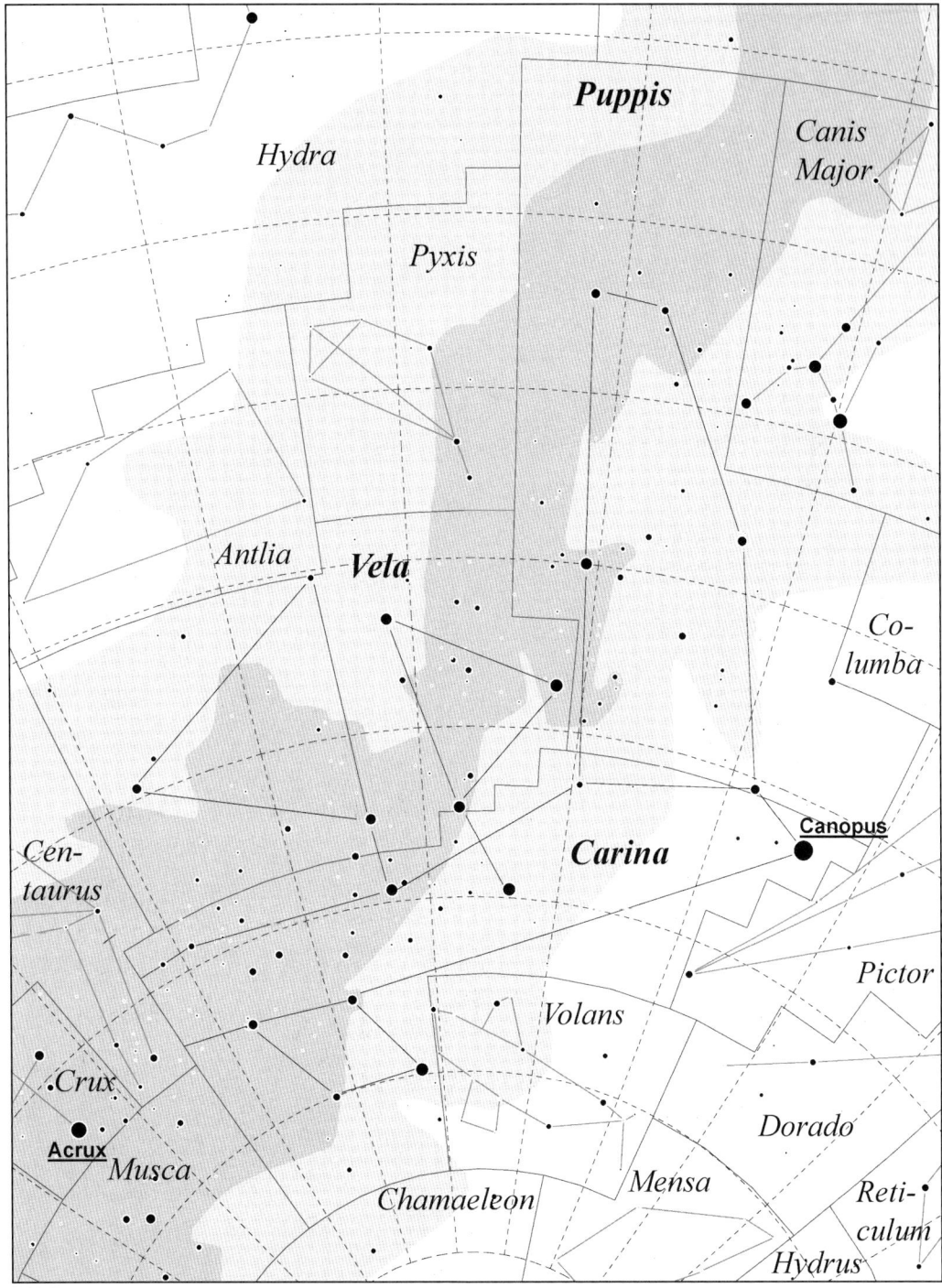

Fig. 10.10. Carina, Puppis, and Vela finder chart.

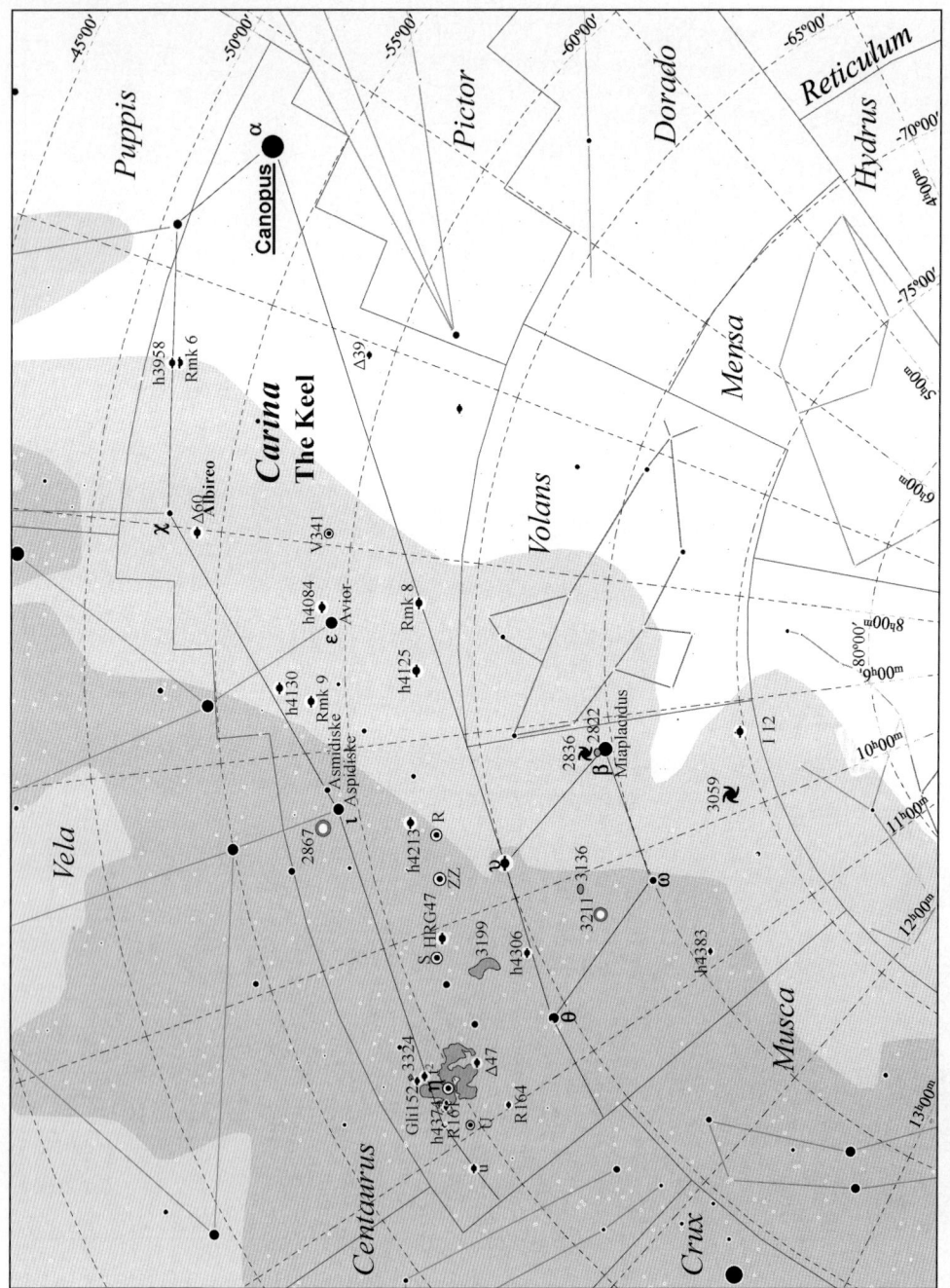

Fig. 10.11. Carina details.

Table 10.3. Selected telescopic objects in Carina.

Multiple stars:

Designation	R.A.	Decl.	Type	m_1	m_2	Sep. (")	PA (°)	Colors	Date/**Period**	Aper. Recm.	Rating	Comments
Δ 39	07ʰ 03.3ᵐ	−59° 11′	A–B	5.8	6.8	1.4	086	W, W	1997	6/7″	****	
Rmk 6	07ʰ 20.4ᵐ	−52° 19′	A–B	6.0	6.5	9.1	026	pY, pY	2000	2/3″	****	In same FOV as h 3958; southernmost of row of 3 stars
h 3958	07ʰ 20.7ᵐ	−52° 12′	A–B	6.9	9.1	28.5	279	Y, ?	2000	10×50	***	In same FOV as Rmk 6; in middle of row of three stars
Δ 60	08ʰ 01.4ᵐ	−54° 31′	A–B	6.1	7.9	40.2	162	bW, ?	1999	7×50	***	
Rmk 8	08ʰ 15.3ᵐ	−62° 55′	A–B	5.27	7.6	3.6	067	Y, ?	2000	2/3″	****	
h 4084	08ʰ 17.9ᵐ	−59° 10′	A–B	6.5	9.8	42.2	150	Y, W	2000	2/3″	***	
			B–C	9.8	9.9	3.1	087	W, ?	2000	3/4″	****	
h 4125	08ʰ 37.3ᵐ	−62° 51′	A–B	5.5	11.0	5.8	237	yO, ?	2000	3/4″	***	Rich star field
h 4130	08ʰ 40.7ᵐ	−57° 33′	A–B	6.5	8.3	3.3	245	pY, oR	1998	2/3″	****	
Rmk 9	08ʰ 45.1ᵐ	−58° 43′	A–B	6.87	6.9	3.7	293.0	bW, W	1998	2/3″	****	
			A–C	6.87	11.0	50.9	359.0	bW, ?	1913	3/4″	****	
			A–D	6.87	10.8	60.5	223.0	bW, ?	1998	3/4″	****	
I 12AB	09ʰ 17.4ᵐ	−74° 54′	A–B	5.7	6.6	0.3	265	bY, ?	1996	–	–	
h 4206AB–C	09ʰ 25.5ᵐ	−61° 57′	AB–C	5.3	9.6	7.0	345	bY, ?	2000	2/3″	***	
h 4206AB–D			AB–D	5.3	10.3	46.0	354	bY, ?	2000	2/3″	***	
h 4213			A–B	5.8	9.6	8.8	330	W, ?	2000	2/3″	****	
Avior, υ Carinae	09ʰ 47.1ᵐ	−65° 04′	A–B	3.0	6.0	5.0	129	yW, W	2000	2/3″	*****	a.k.a. Rmk 11; showcase pair; nice star field
HRG 47	10ʰ 03.6ᵐ	−61° 53′	A–B	6.3	7.9	1.2	352	W, ?	1991	8/9″	****	
H 4306	10ʰ 19.1ᵐ	−64° 41′	A–B	6.3	6.5	2.4	313	pY, ?	1997	4/5″	*****	Nice pair of close, matched "cat's eyes"
T² Carinae	10ʰ 38.7ᵐ	−59° 11′	A–B	4.9	7.5	14.5	021	yO, y	1991	2/3″	*****	a.k.a. Δ 94; showcase pair; some see deep yellow and pale yellow
Gli 152	10ʰ 39.0ᵐ	−58° 49′	A–B	6.2	8.0	26.3	080	dO, W	2000	10×50	****	Nice pair; in a rich field of background stars
Δ 97	10ʰ 43.2ᵐ	−61° 10′	A–a	5.6	7.9	19.0	205	bW, bW	1998	15×80	****	
			A–b	5.6	10.0	81.0	083	bW, ?	1998	15×80	***	
			A–b	5.6	7.9	12.0	175	bW, ?	1998	2/3″	***	
			A–C	5.6	8.1	83.6	017	bW, ?	1998	7×50	***	
h 4374	10ʰ 48.9ᵐ	−59° 27′	A–B	7.8	10.6	1.0	284	W, ?	192y	9/10″	****	On edge of NGC 3372; in same FOV as R 161, only 8′ apart
R 161	10ʰ 49.4ᵐ	−59° 19′	A–B	6.1	7.4	1.0	292	W, ?	1996	9/10″	****	

(continued)

Table 10.3. (continued)

Multiple stars:

Designation	R.A.	Decl.	Type	m₁	m₂	Sep. (')	PA (°)	Colors	Date/**Period**	Aper. Recm.	Rating	Comments
u Carinae	10ʰ 53.5ᵐ	−58° 51'	A–B	3.9	6.2	159.4	204	dY, ?	2000	6×30	***	a.k.a. Δ 102
			A–C	3.9	7.8	56.4	007	dY, ?	2000	7×50	***	
h 4383	10ʰ 53.7ᵐ	−70° 43'	A–B	6.4	7.1	1.5	288	oY, oY	1991	6/7"	****	Close pair; not too mismatched
R 164	10ʰ 59.2ᵐ	−61° 19'	A–B	6.2	9.7	3.3	078	Y, W	2000	3/4"	****	Rich star field

Variable stars:

Designation	R.A.	Decl.	Type	Range (m.)		Period (days)	F (f_f/f_t)	Spectral Range	Aper. Recm.	Rating	Comments
V341 Carinae	07ʰ 56.8ᵐ	−59° 08'	L	6.2	7.0	–	–	M1–M3, II–III	7×50	****	#
R Carinae	09ʰ 32.2ᵐ	−62° 47'	M	3.9	10.5	308.71	0.48	M4e–M8e	2/3"	***	##
ZZ, ι Carinae	09ʰ 45.2ᵐ	−62° 30'	Cδ	3.3	4.2	35.536	0.26	F6Ib–G0Ib	Eye	****	Unusual Cepheid. Takes 21 days to dim, 7 days to recover
S Carinae	10ʰ 09.4ᵐ	−61° 33'	M	4.5	9.9	149.49	0.51	K7e–M6e	10×80	***	##
η Carinae	10ʰ 45.1ᵐ	−59° 41'	Pec(e)	−0.8	7.9	–	–	Pe	7×50	******	One of the most extraordinary stars known. ##$ *13
U Carinae	10ʰ 57.8ᵐ	−59° 44'	Cδ	5.72	7.02	38.768	0.21	F6–G7Iab	7×50	****	#

Star clusters:

Designation	R.A.	Decl.	Type	m_v	Size (')	Brtst. Star	Dist. (ly)	dia. (ly)	Number of Stars	Aper. Recm.	Rating	Comments
C96, NGC 2516	07ʰ 58.0ᵐ	−60° 45'	Open	3.8	22	7.0	1,350	9	80	Eye	*****	Very bright; pretty rich; red (or orange) giant near middle
NGC 2808	09ʰ 12.0ᵐ	−64° 52'	Globular	6.2	14	13.5	30K	122	–	4/6"	****	Detached; Weak concentration; large range of brightness
NGC 3114	10ʰ 02.7ᵐ	−60° 07'	Open	4.2	34	7.3	2.9K	29	171	4/6"	****	>400 stars counted on photographic plates; many fewer will be seen in amateur scopes
IC 2581	10ʰ 27.4ᵐ	−57° 38'	Open	4.3	7	4.6	6.5K	13	25	4/6"	*****	
NGC 3293	10ʰ 35.8ᵐ	−58° 14'	Open	4.7	6	6.5	8.5K	15	90	4/6"	****	
C102, IC 2602	10ʰ 43.0ᵐ	−64° 24'	Open	1.7	50	3.0	485	7	60	Eye	*****	θ Carinae Cluster; visible to eye *1
C 91, NGC 3532	11ʰ 05.5ᵐ	−58° 44'	Open	3.0	50	7.1	1.3K	19	150	Eye	*****	Very remarkable; extremely large; little compressed; clearly visible to the eye *16

Nebulae:

Designation	R.A.	Decl.	Type	m_v	Size (')	Brst./Cent. star	Dist. (ly)	dia. (ly)	Aper. Recm.	Rating	Comments
NGC 2867	09ʰ 21.4ᵐ	−58° 19'	Planetary	9.7	14"	**15**	5.5K	0.4	4/6"	***	Very small; round; ring structure
NGC 3199	10ʰ 17.1ᵐ	−57° 55'	Diffuse	–	15×8	~8	–	–	6/8"	****	Large, round, one side brighter
NGC 3211	10ʰ 17.8ᵐ	−62° 40'	Planetary	10.7	12"	**15.5**	–	–	6/8"	***	Small, round; uniform brightness
NGC 3324	10ʰ 37.7ᵐ	−58° 14'	Diffuse	–	16×14	~11	–	–	6/8"	****	Large, bright emission nebula; with small cluster of stars
NGC 3372	10ʰ 43.8ᵐ	−59° 52'	Diffuse	–	155×128	6	–	–	15×80	*****	"Keyhole Nebula"; beautiful bright/dark nebula surrounding η Carinae

Galaxies:

Designation	R.A.	Decl.	Type	m_v	Size (')	Dist. (ly)	dia. (ly)	Lum. (suns)	Aper. Recm.	Rating	Comments
NGC 2822	09ʰ 13.8ᵐ	−69° 39'	E	10.7	2.8×1.9	55M	55K	11G	8/10"	***	Pretty faint, very small, roundish; gradually b righter toward center
NGC 2836	09ʰ 13.7ᵐ	−69° 20'	Sb	12.0	2.8×2.0	27M	55K	4.9G	10/12"	**	Faint, formless; very faint arms; similar to M81, but much smaller and fainter
NGC 3059	09ʰ 50.1ᵐ	−73° 55'	SBb	10.8	3.9	41M	63K	12G	8/10"	***	Faint, large, irregularly round
NGC 3136	10ʰ 05.8ᵐ	−67° 23'	E4	10.6	3.1×2.2	61M	58	22G	6/8"	****	Pretty bright, pretty small, round; bright nucleus; NGC 3136B and IC 2552 in same FOV

Meteor showers:

Designation	R.A.	Decl.	Period (yr)	Duration	Max Date	ZHR (max)	Comet/**Asteroid**	First Obs.	Vel. (mi/**km**/sec)	Rating	Comments
None											

Other interesting objects:

Designation	R.A.	Decl.	Type	m_v	M	Dist. (ly)	Spec. Class	Temp. (K)	Dia. (suns)	Lum. (suns)	Aper. Recm.	Rating	Comments
Canopus, α Carinae	06ʰ 24.0ᵐ	−52° 42'	F0 Ia	−0.72	−5.6	313		7,350	65	15K	Eye	****	2nd brightest star in sky; not visible north of latitude +37° *17

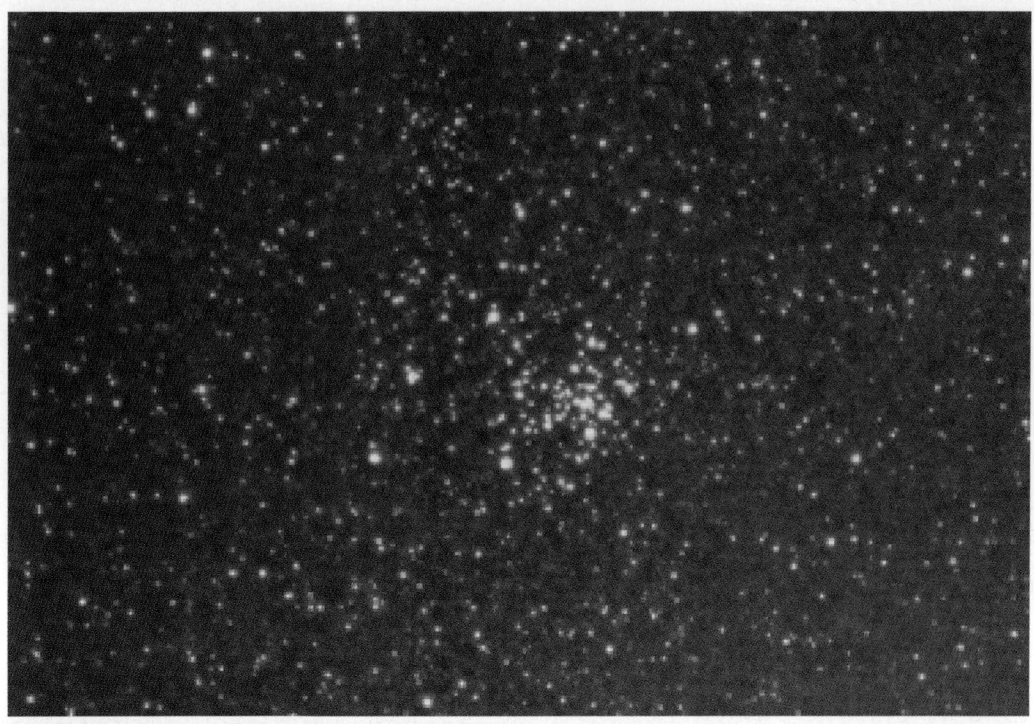

Fig. 10.12. NGC 2516, open cluster in Carina. Astrophoto by Christopher J. Picking.

Fig. 10.13. NGC 3293, open cluster in Carina. Astrophoto by Rob Gendler.

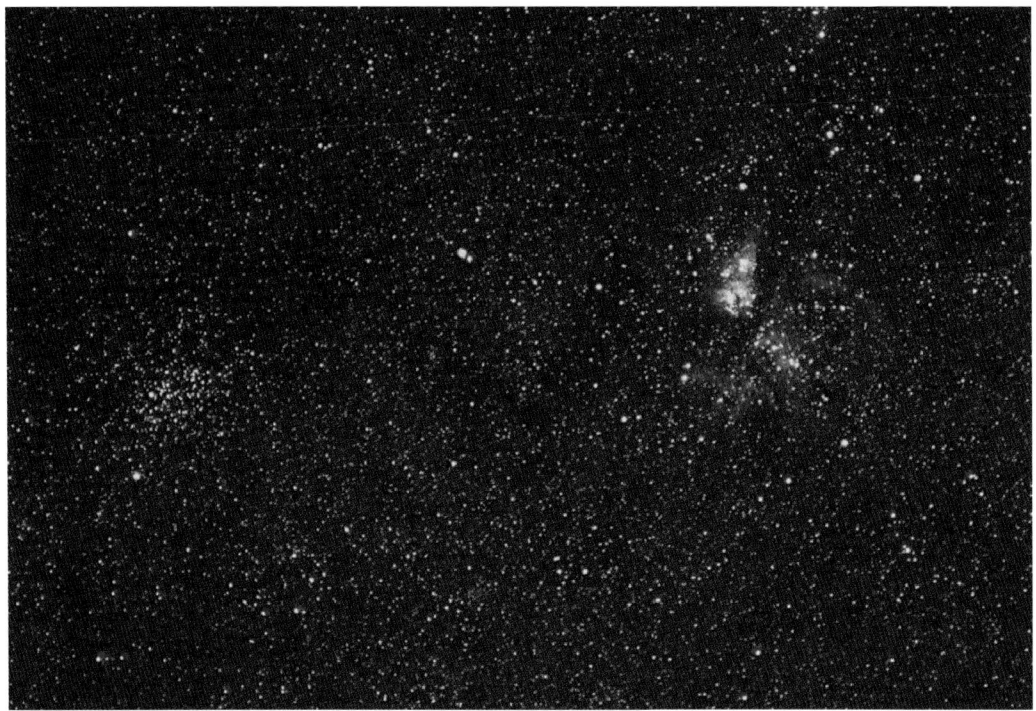

Fig. 10.14. NGC 3532 (C91), Open cluster, and 3372 (C92) eta Carina Nebula. Astrophoto by Christopher J. Picking.

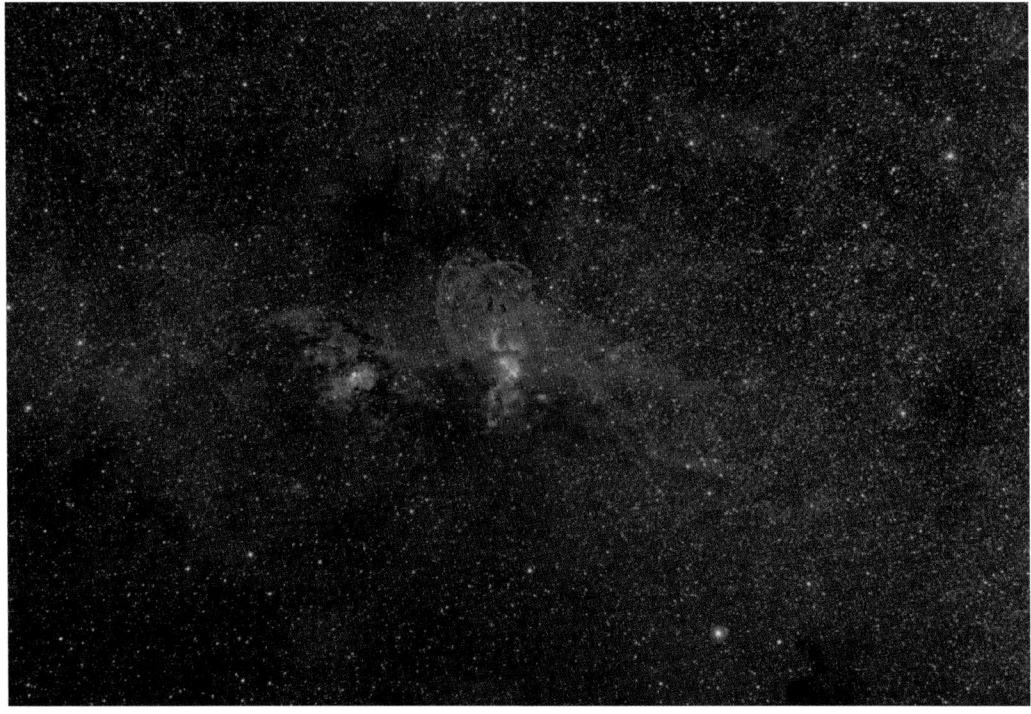

Fig. 10.15. NGC 3603, 3576, and 3579, all emission nebula in Carina. Astrophoto by Rob Gendler.

PUPPIS (pup-iss)	**The Stern**
PUPPIS (pup-iss)	**9:00 P.M. Culmination:** Feb. 23
Pup	**AREA:** 673 sq. deg., 20th in size

TO FIND: For northern observers, starting with <u>Sirius</u> is the easiest way to find Puppis. The back of Canis Major, the large dog, is formed by a 15° long line of four stars from <u>Sirius</u> to Aludra (η Canis Majoris). Extending this line 13° farther to the southeast takes you to second magnitude Naos (ζ Puppis) which marks the middle front of the Argo's stern. Naos is the brightest star in Puppis and will be the central point for finding the rest of the constellation. Third magnitude π Puppis, marking the middle back of the stern, lies 8.5° west–northwest of Naos and 8° south of Aludra. A little over 15° north of Naos lies Tureis (ρ Puppis), and just 4° to its east lies ξ Puppis. These two 3rd magnitude stars define the top of Argo's stern. The bottom of the stern is shown by another pair of third magnitude stars on an east–west line and separated by 10.5°. χ Carina is the eastern star of the pair and lies 13° south of Naos. The western star of the pair, τ Puppis is just 4.5° northeast of Canopus (Figs. 10.9–10.11).

KEY STAR NAMES:

Name	Pronunciation	Source
ζ Naos	nowz	Nauz (Greek), "the ship," from the ancient name for the whole constellation[*18]
ρ Tureis	**tuh**-race	Al-turrais (Arabic), "The Little Shield," used to describe several stars defining the stern ornament of the ship Argo

Magnitudes and spectral types of principal stars:

Bayer desig.	Mag. (m_v)	Spec. type	Bayer desig.	Mag. (m_v)	Spec. type	Bayer desig.	Mag. (m_v)	Spec. type
ζ	2.21	O5 IAf	v	3.17	B8 III	ξ	3.34	G6 Ia
π	2.71	K3 Ib	ρ	2.83	F6 II	σ	3.25	K5 III
τ	2.94	K0 III	a	3.71	G5 III	b	4.49	B2 V
c	3.62	K4 III	d¹	4.84	B3 V*18	f	4.53	B8 IV/V
h¹	4.44	K4 III	h²	4.42	K1 II/III	j	4.20	F7/F8 II
k¹	3.80	B5 IV	k²	4.62	B5 IVa	l	3.94	A2 Iab
m	4.69	B8 IV	n	5.06	F6 V	o	4.40	B1 IV:nne
p	4.65	B8 V	q	4.44	A4m	r	4.78	B2 ne
s	5.43	B5 IIIne	t	5.07	B4 IV/V	v¹	4.65	B2 V+
v²	5.11	A0 V	w	4.83	K2/3 III	x	5.27	B8/9 V
y	5.41	B3 V	z	5.42	B2 Vne	A	4.83	B3 IV/V
B	4.76	G6 II	C	5.20	Am	E	5.30	A3 p
F	5.24	A0 V	H	4.92	A4 IV	I	4.49	F0 IV
J	4.22	B0.5 Ib	L¹	4.87	Ap	L²	4.42	M5e
M	5.86	B8 II/III	N	5.08	B2.5 IV	O	5.14	K2 III
P	4.10	B0 III	Q	4.69	K0 III			

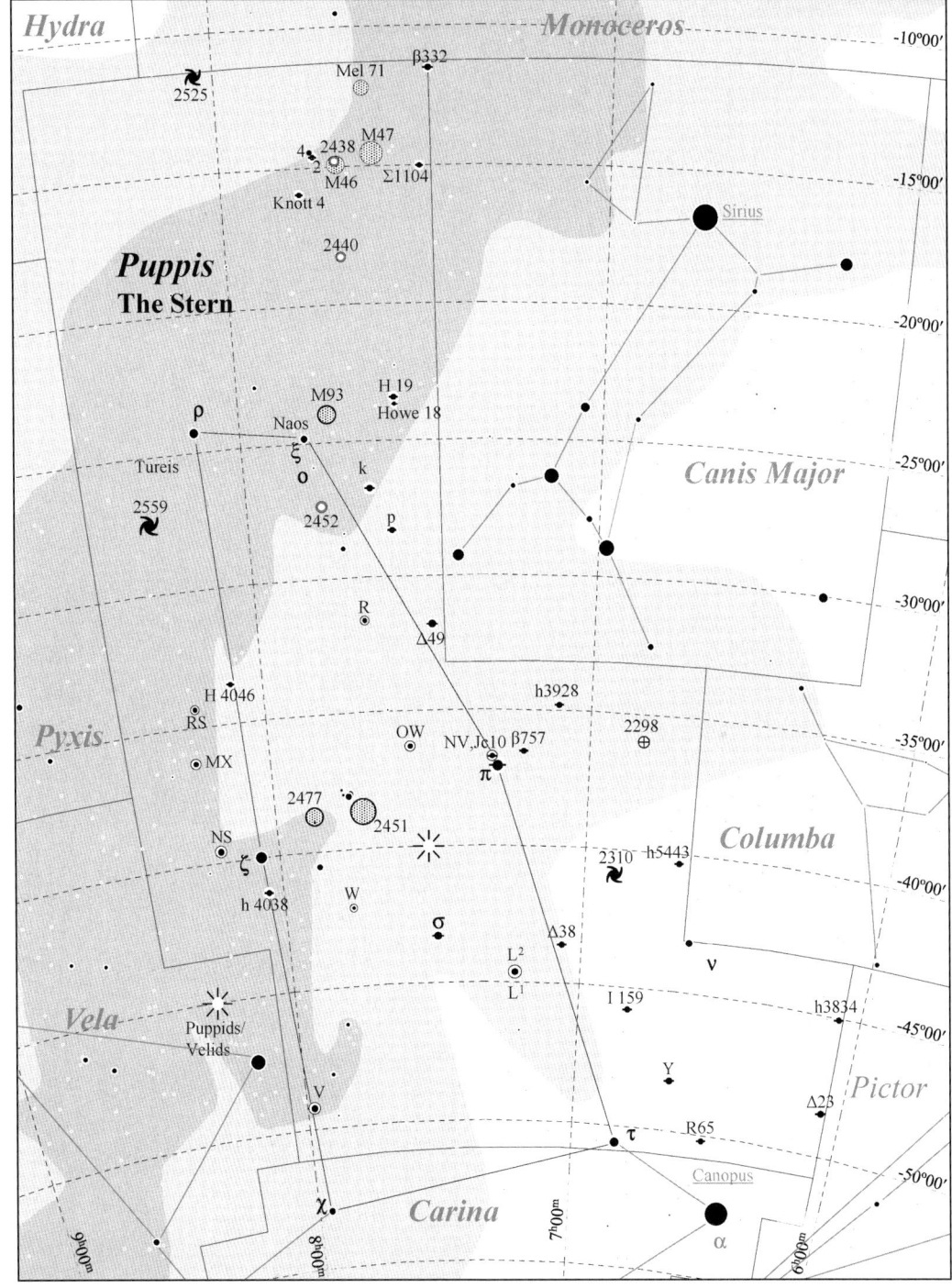

Fig. 10.16. Puppis details.

Table 10.4. Selected telescopic objects in Puppis.

Multiple stars:

Designation	R.A.	Decl.	Type	m_1	m_2	Sep. (")	PA (o)	Color	Date/Period	Aper. Recm.	Rating	Comments
h3834	06ʰ 04.7ᵐ	−45° 05′	A–B	6.0	9.0	5.7	216	Y, ?	1999	2/3″	****	
			A–C	6.0	6.4	196.2	321	Y, ?	1999	6×30	****	
Δ 23	06ʰ 04.8ᵐ	−48° 28′	A–C	6.6	7.0	**2.53**	**123**	Y. ?	**304.2**	3/4″	*****	Slowly closing to 0.390″ in 2047
R65	06ʰ 29.8ᵐ	−50° 14′	A–B	6.0	6.2	**0.61**	**260**	pY, Y	**52.9**	15/16″	****	Close, nearly equal pair; passed max. 0.81″ in 1993, closing to 0.03″ in 2021
Δ 30			AB–CD	5.3	7.5	11.9	312	pY, R	1999	2/3″	****	
HDO 195			C–D	8.0	8.7	**0.37**	**147**	R, ?	**101**	>20″	**	Opening to 0.62″ in 2052
Y Puppis, A31	06ʰ 38.6ᵐ	−48° 13′	A–B	5.1	7.4	12.9	320	Y, bW	1999	2/3″	*****	Pleasing pair; like a fainter, closer Albireo
h 5443	06ʰ 41.2ᵐ	−40° 21′	A–B	6.1	9.4	15.5	108	W, ?	1999	2/3″	****	
I 159	06ʰ 50.0ᵐ	−45° 27′	A–B	6.5	11.0	6.6	323	dY, ?	1966	3/4″	***	
Δ 38	07ʰ 04.0ᵐ	−43° 36′	A–B	5.6	6.7	20.9	126	Y, O	1999	15×80	****	Attractive triple
			A–C	5.6	8.8	184.9	335	Y, O	1999	7×50	***	
H 3928	07ʰ 05.5ᵐ	−34° 47′	A–B	6.5	7.8	2.7	145	oY, oY	2001	3/4″	****	
			A–C	6.5	9.7	37.8	291	oY, ?	1999	15×80	****	
			A–D	6.5	10.8	42.7	119	oY, ?	1999	3/4″	***	
			A–E	6.5	13.5	16.9	305	oY, ?	1999	7/8″	***	
β 757	07ʰ 12.4ᵐ	−36° 33′	A–B	6.0	8.4	2.5	069	pY, pY	1993	3/4″	****	
π Puppis	07ʰ 17.1ᵐ	−37° 06′	A–a	2.9	6.5	0.7	152	O, W	1991	13/14″	***	a.k.a. HDS 1008
			A–B	2.9	7.9	68.9	213	O, ?	1999	2/3″	****	a.k.a. Δ 43
v¹, ic 10	07ʰ 18.3ᵐ	−36° 44′	A–B	4.7	5.1	240.0	098	W, W	1991	Eye	****	Moderately unequal pair; eastern component is variable NV Puppis
β 757			B–C	5.1	8.7	117.2	216	W, ?	1999	2/3″	****	
			C–D	8.7	9.2	2.3	206	?, ?	1999	4/5″	****	26′ NNE (PA 036°) of π Puppis
Σ 1097	07ʰ 27.9ᵐ	−11° 33′	A–B	6.2	7.4	0.7	171	Y, V	2003	13/14″	***	
			A–C	5.9	8.5	19.8	313	pR, dB	2003	15×80	******	Nice colors
β 332AD			A–D	5.9	9.5	22.8	157	pR, ?	1998	2/3″	****	
β 332AE			A–E	5.9	12.2	32.2	043	pR, ?	1998	4/5″	***	
Δ 49	07ʰ 28.9ᵐ	−31° 51′	A–B	6.3	7.0	9.0	053	W, W	1999	2/3″	****	
σ Puppis	07ʰ 29.2ᵐ	−43° 18′	A–B	3.3	8.8	22.2	074	rO, Y	1981	15×80	*****	a.k.a. Δ 51
Σ 1104	07ʰ 29.4ᵐ	−15° 00′	A–B	6.4	7.6	2.4	029	oY, oY	2005	4/5″	******	A–B are a beautiful pair
			A–C	6.1	10.7	20.7	338	oY, ?	1916	3/4″	****	
			A–D	6.1	11.2	42.4	008	oY, ?	1916	3/6″	****	

Name	R.A.	Decl.	Pair	m_1	m_2	Sep (")	P.A.	Year	Colors	Aper. Recm.	Rating	Comments
Howe 18	07h 34.0m	−23° 42'	A–B	8.1	8.9	1.9	204	1991	Y, Y	5/6"	****	H 19 is in same FOV, 15' north of Howe 18; showcase pair
H 19	07h 34.3m	−23° 28'	A–B	5.8	5.9	9.8	117	1998	W, W	2/3"	*****	Nearly matched pair
p Puppis	07h 35.4m	−28° 22'	A–B	4.6	9.1	38.3	150	1999	W, ?	7×50	***	
			B–C	9.1	10.4	43.2	130	1999	W, ?	15×80	****	
k Puppis	07h 38.8m	−26° 48'	A–B	4.4	4.6	10.5	318	2002	bW, bW	2/3"	*****	Showcase pair; very similar in color and brightness; a.k.a. h 3982
2 Puppis	07h 45.5m	−14° 41'	A–B	6.0	6.7	16.6	340	2003	rY, rY	15×80	****	a.k.a. STF1138; lovely background field; B has a m13.8 companion at 7.2"
Knott 4	07h 47.8m	−16° 01'	A–B	6.5	6.6	129.4	132	2002	yR, yR	6×30	*****	Very wide, matched pair; great object for binoculars; A has a m14.3 comp at 10.5"
h4038	08h 02.7m	−41° 19'	A–B	5.5	9.0	25.2	345	1993	pY, pR	10×50	****	John Herschel 4038; very nice color
h4046	08h 05.7m	−33° 34'	A–B	6.1	11.8	27.8	050	1999	Gld, W	3/4"	****	Colorful triplet
I189			B–C	8.4	11.0	7.0	009	1999	W, dR	3/4"	****	

Variable stars:

Designation	R.A.	Decl.	Type	Range (m_v)		Period (days)	F (f_r/f_f)	Spectral Range	Aper. Recm.	Rating	Comments
L² Puppis	07h 13.5m	−44° 39'	SRb	2.6	6.2	140	0.40	M5IIIe–M6IIIe	6×30	****	26' NNE (PA 33°) of π Puppis (m2.71). See v¹, ic 10
NV Puppis	07h 18.3m	−36° 44'	γ Cas	4.6	4.8	–	–	B3V	Eye	****	
OW Puppis	07h 33.9m	−36° 20'	γ Cas	5.4	5.6	–	–	B3Vne	6×30	****	
R Puppis	07h 40.9m	−31° 40'	γ Cas	6.5	6.7	–	–	G2Ia-0	6×30	****	
W Puppis	07h 46.0m	−42° 12'	M	7.2	13.6	119.7	0.47	M1e–M6e	10/12"	****	##
V Puppis	07h 58.2m	−49° 15'	EB/SD	4.4	5.9	1.454	–	B1V + B3	6×30	****	°1 24' SE (PA 145°) of J Puppis (m4.34). #
NS Puppis	08h 11.4m	−39° 37'	Lc	4.4	4.5	–	–	K3Ib	Eye	****	1°32' ENE (PA 76°) of ζ Puppis (m4.40). Small range, photometric measurements required
RS Puppis	08h 13.1m	−34° 35'	Cepheid	6.5	7.6	41.38	0.24	F7–G7	6×30	****	#
MX Puppis	08h 13.5m	−35° 54'	γ Cas	4.6	4.9	–	–	B1.5III–IVe	Eye	****	1°16' NW (PA 306°) of q Puppis (m4.40)

(continued)

Table 10.4. (continued)

Star clusters:

Designation	R.A.	Decl.	Type	m_v	Size (')	Brtst. star	Dist. (ly)	dia. (ly)	Number of Stars	Aper. Recm.	Rating	Comments
NGC 2298	06h 49.0m	−36° 00'	Globular	9.3	5	13.4	40K	58	–	6/8"	****	Can be seen by unaided eye; partially resolved in 10×50 binoculars *19
M47, NGC 2422	07h 36.6m	−14° 30'	Open	4.4	25	5.7	1.5K	11	120	7×50	*****	
Melotte 71	07h 37.5m	−12° 03'	Open	7.1	8	10.2	9.0K	21	80	6/8"	****	Rich cluster of uniformly faint stars; 1.5° E of M47; contains PN NGC 2438
M46, NGC 2437	07h 41.8m	−14° 49'	Open	6.1	20	8.7	5.3K	31	186	4/6"	*****	
M93, NGC 2447	07h 44.6m	−23° 53'	Open	6.2	21	8.2	3.6K	22	80	4/6"	****	Wedge-shaped cluster
NGC 2451	07h 45.4m	−37° 58'	Open	2.8	45	3.5	850	11	40	7×50	****	Very bright, loose cluster
NGC 2477	07h 52.3m	−38° 33'	Open	5.8	26	9.8	4.2K	31	160	2/3"	*****	Beautiful rich cluster of faint stars; >250 vis in 12/14"; Over 1900 counted on photos
NGC 2539	08h 10.6m	−12° 49'	Open	6.5	22	9.2	4.2K	27	50	4/6"	****	Beautiful rich cluster of faint stars; >250 vis in 12/14"

Nebulae:

Designation	R.A.	Decl.	Type	m_v	Size (')	Brtst./Cent. star	Dist. (ly)	dia. (ly)	Aper. Recm.	Rating	Comments
NGC 2438	07h 41.8m	−14° 44'	Planetary	10.1	66"	**17.5**	3,300	1.1	4/6"	****	Pale green disk; foreground object of M46, not a member
NGC 2440	07h 41.9m	−18° 12'	Planetary	9.4	79×45"	**17.6**	4,000	1.5	8/10"	****	Bright green, oval disk, brighter in center; small telescopes do not show full extent
NGC 2452	07h 47.4m	−27° 20'	Planetary	12.0	19"	**16.1**	8,800	0.8	8/10"	****	Bright green, oval disk, brighter in center

Galaxies:

Designation	R.A.	Decl.	Type	m_v	Size (')	Dist. (ly)	dia. (ly)	Lum. (suns)	Aper. Recm.	Rating	Comments
NGC 2310	06h 53.9m	−40° 52'	S0 sp	11.8	4.4×	42M	34K	4.4G	12/14"	***	Pretty bright, pretty large; small; bright nucleus; flattened bulge; dark lane; 10' W of m8*
NGC 2525	08h 05.6m	−11° 26'	SB(s)c II	11.6	3.0×2.0	60M	52K	14G	12/14"	***	Pretty large, quite faint, round; strong arms; extremely small, bright nucleus
NGC 2559	08h 17.1m	−27° 28'	SB(s)bc	10.9	4.0×2.0	60M	105K	13G	12/14"	***	Very bright, knotty inner arms; very faint outer arms; m9.4 star on east edge

Meteor showers:

Designation	R.A.	Decl.	Period (yr)	Duration	Max Date	ZHR (max)	Comet/ Asteroid	First Obs.	Vel. (mi/ km/sec)	Rating	Comments
π Puppids	07h 20m	−45°	5	04/15 04/28	04/23	2–40	**	1972	11/18	III	Predicted by H. B. Ridley; observed by B. Edwards USA; ** Comet Grigg-Skjellerup
Puppid/ Velids	08h 12m	−45°	–	12/01 12/15	12/07	10	–	–	40/25	II	One of ~25 apparently related minor showers radiating from a large area around Puppis

Other interesting objects:

Designation	R.A.	Decl.	Type	m_v	M_v	Dist. (ly)	Spec. Class	Temp. (K)	Lum. (suns)	Aper. Recm.	Rating	Comments
ζ Puppis	08h 03.7m	−40° 00′	Star	2.3	−5.9	1,400	O4		20,000		*21	Hottest and brightest of the main sequence stars visible to the eye

Fig. 10.17. M47, open cluster in Puppis. Astrophoto by Mark Komsa.

Fig. 10.18. M46, open cluster in Puppis, including planetary nebula NGC 2438.

Fig. 10.19. NGC 2440, planetary nebula in Puppis.

Fig. 10.20. M93, open cluster in Puppis. Astrophoto by Mark Komsa.

VELA (**vee**-luh)	**The Sails**
VELORUM (veh-**low**-rum)	**9:00 P.M. Culmination:** Mar. 31
Vel	**AREA:** 500 sq. deg., 32nd in size

TO FIND: If you have not already found Puppis, see the directions for finding ζ Puppis and continue the line from Aludra (η Canis Majoris) to ζ Puppis for another 10°. This line will end near the center of a triangle of second magnitude stars roughly 10° on a side. These three stars are δ, γ, and λ Velorum which form the aft sail of the vessel Argo. The forward sail is a somewhat larger but also somewhat fainter, triangle centered about 12° or 13° east of the first, and facing the opposite direction. The roughly north/south lines formed by λ and δ Velorum in the aft (western) sail and by ψ and κ Velorum in the forward (eastern) sail point south to ε and ι Carinae where the masts holding the sails emerge from the deck of the ship Argo (Figs. 10.21–10.25; Table 10.5).

KEY STAR NAMES:

Name	Pronunciation	Source
γ Regor	**ree**-gore	Roger (English) spelled backwards as a prank. See γ Cassiopieae (Navi) for a full explanation
χ Markab	**mur**-keb	Markab (Arabic), "a ship or vehicle," apparently an Arabic translation of the Greek constellation name but not confirmed by scientific Arabic sources
λ Suhail	**suh**-hale	suhail al-wazn (or a similar Sci-A phrase used to designate stars in this vicinity)

Magnitudes and spectral types of principal stars:

Bayer desig.	Mag. (m_v)	Spec. type	Bayer desig.	Mag. (m_v)	Spec. type	Bayer desig.	Mag. (m_v)	Spec. type
δ	1.93	A1 V	γ^1	4.27	O9 I	γ^2	1.75	WC8*[21]
κ	2.47	B2 IV	λ	2.23	K4 Ib/II	μ	2.69	G5 IIISb
o	3.60	B3 IV	φ	3.52	B5 Ib	ψ	3.60	F2 IV
a	3.87	A1 III	b	3.77	F3 Ia	c	3.75	K2 III
d	4.05	G5 III	e	4.11	A6 II	f	5.09	B0 III
g	4.94	A2 III	h	5.47	A2/3 IV	i	4.37	A3 IV
k^2	4.63	F3/5 V*[22]	l	4.92	K1 II	m	4.58	G5 Ib
n	4.74	A5 V	p	3.84	A3m+	q	3.85	A2 V
r	4.82	K1 IIIvar	s	5.76	B9 V	t	5.02	K4 III
u	5.09	B7 III	w	4.45	Fpec	x	4.29	G2 II
y	5.25	G8 II	z	5.24	B4 V+	A	5.33	B2 IV
B	4.79	B1 V	C	5.01	K1/2 II	D	5.15	B0 IIIn

Bayer desig.	Mag. (m$_v$)	Spec. type	Bayer desig.	Mag. (m$_v$)	Spec. type	Bayer desig.	Mag. (m$_v$)	Spec. type
E	5.79	B8 Si	F	5.08	A9/F0 III/IV	H	4.68	B5 V
I	5.09	B6 V	J	4.50	B3 III	K	5.26	B7/8 III
L	5.01	B1.5 IV	M	4.34	A5 V	N	3.16	K5 III
O	5.56	A0 V	Q	4.85	B3 IV			

Stars of magnitude 5.5 or brighter which have no Bayer or Flamsteed designations:

Designation	Mag. (m$_v$)	Spec. type	Designation	Mag. (m$_v$)	Spec. type
HD 067582	5.04	K3 III	HD 068217	5.10	B2 IV/V
HD 068324	5.23	B1 I've	HD 069144	5.14	B2.5 IV
HD 071510	5.18	B2 V	HD 072127	4.99	B2 IV
HD 074067	5.20	Ap (SiCr)	HD 074071	5.45	B5 V
HD 074146	5.18	B4 IV	HD 074371	5.20	B5 Iab
HD 074455	5.48	B1.5 Vn	HD 074535	5.49	B8s
HD 074560	4.83	B3 IV	HD 075149	5.43	B3 Ia
HD 076360	5.31	Am	HD 077140	5.17	Am
HD 077653	5.23	Ap (Si)	HD 079186	5.02	B5 Ia
HD 079846	5.26	Gi II/III	HD 080108	5.12	K3 Ib
HD 080170	5.31	K2 III	HD 082419	5.45	B8 V
HD 082694	5.35	G8 III	HD 082984	5.12	B4 IV
HD 083520	5.44	A2/3 V	HD 087783	5.06	K0 IV
HD 088824	5.27	A7 V	HD 089682	4.59	K3 II
HD 093563	5.14	B8/9 III			

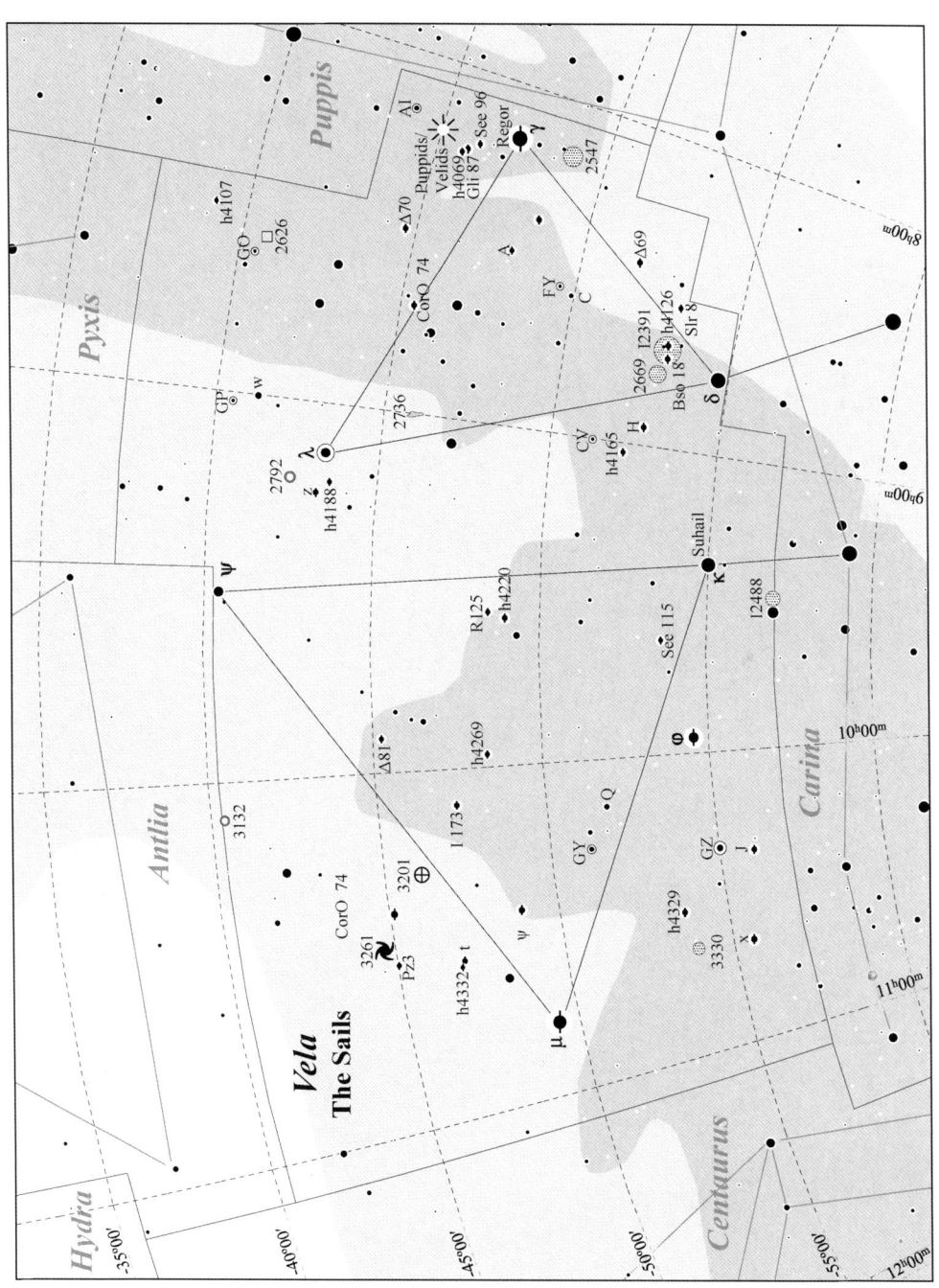

Fig. 10.21. Vela details.

Table 10.5. Selected telescopic objects in Vela.

Multiple stars:

Designation	R.A.	Decl.	Type	m_1	m_2	Sep. (")	PA (°)	Color	Date/**Period**	Aper. Recm.	Rating	Comments
γ Velorum	08h 09.5m	−47° 20'	A–B	1.8	4.1	41.0	219	bW, bW	2002	10×50	*****	Dun 65; easy; primary is a Wolf–Rayet star; T = 30,000 K; L = 15,000 suns *18
			A–C	1.8	7.3	62.6	152	bW, –	2000	7×50	****	
			A–D	1.8	9.4	93.9	142	bW, –	2000	15×80	****	
See 96	08h 12.5m	−46° 16'	Aa–B	6.2	7.7	0.6	276	bW, –	1997	16/18"	**	A–a is too close for amateur instruments
Gli 87	08h 13.6m	−45° 59'	A–B	5.1	10.1	34.9	343	Y, –	1999	15×80	****	
h4069	08h 14.4m	−45° 50'	A–B	5.8	8.7	32.1	249	W, –	1999	10×50	****	CorO 69, 8.1–8.6, 8.7", Y, Y is in same FOV
B Velorum	08h 22.5m	−48° 29'	A–B	5.1	6.1	0.7	137	bW, W	1998	13/14"	***	a.k.a. 167
Δ 69	08h 25.5m	−51° 44'	A–B	5.1	9.6	25.7	219	bW, –	1998	15×80	****	
A Velorum	08h 29.1m	−47° 56'	Aa–B	5.5	7.2	3.0	250	bW, –	1999	3/4"	***	a.k.a. h4104; A–a (Fin 315) is too close for amateur telescopes
Δ 70	08h 29.5m	−44° 43'	A–B	5.5	9.2	18.9	040	bW, W	1999	15×80	****	
h4107	08h 31.4m	−39° 04'	A–B	5.2	7.0	4.3	349	W, W	2002	2/3"	*****	Bright, easy pair
Slr 8	08h 32.1m	−53° 13'	A–B	6.5	8.2	4.3	330	bW, –	1999	2/3"	***	
			A–C	6.5	9.1	30.1	099	bW, –	1999	15×80	****	
h4126	08h 40.0m	−53° 03'	A–B	6.1	7.1	0.8	285	dY, W	1991	12/14"	***	
CorO 74	08h 40.3m	−40° 16'	A–B	5.2	8.7	16.8	015	bW, –	1998	15×80	****	
			A–B	5.2	9.1	4.1	066	W, –	1999	2/3"	***	
Bso 18AB	08h 42.4m	−53° 06'	A–B	4.8	5.5	76.1	311	bW, W	1998	7×50	*****	
β 1625			B–C	5.6	7.9	0.6	153	W, –	1991	16/18"	**	
Bso 18BD			BC–D	5.5	9.9	60.4	267	W, yW	1998	15×80	****	
δ Velorum	08h 47.7m	−54° 43'	A–B	2.0	5.6	**0.51**	**305**	W, –	**142**	18/20"	**	Closing to 0.38" in 2013, opening to 2.91" in 2067
			A–C	2.0	11.0	71.6	055	W, –	1998	3/4"	***	
			C–D	5.6	11.0	5.0	088	–, –	1998	3×4"	***	
H Velorum	08h 56.3m	−52° 43'	A–B	4.7	7.7	2.6	335	W, pY	1995	3×4"	***	
h4165	09h 01.7m	−52° 11'	A–B	5.6	6.6	0.7	136	W, –	1996	13/14"	***	In same FOV with z Vel., only 31' away
h4188	09h 12.5m	−43° 37'	Aa–B	6.0	6.8	2.8	281	W, W	1989	3/4"	***	a.k.a. h4191
z Velorum	09h 14.6m	−43° 14'	A–B	5.3	9.2	5.7	014	W, –	1999	2/3"	***	a.k.a. Cop 1; opening to max. of 1.10" in 2015
ψ Velorum	09h 30.7m	−40° 28'	A–B	3.9	5.1	**0.85**	**101**	yW, pY	**33.95**	10/12"	*****	
h4220	09h 33.7m	−49° 00'	A–B	5.5	6.2	1.6	217	bW, W	1996	6/8"	***	
R 125	09h 36.4m	−48° 45'	A–B	6.3	10.3	3.0	188	pY, –	2000	4/5"	***	
See 115	09h 37.2m	−53° 40'	A–B	6.1	6.3	0.7	008	pY, pY	1997	13/14"	***	
Δ 81	09h 54.3m	−45° 17'	A–B	5.8	8.2	5.2	240	Y, B	1999	2/3"	*****	Beautiful color contrast
φ Velorum	09h 56.9m	−54° 34'	A–B	3.6	12.4	37.9	010	bW, –	2000	4/5"	***	a.k.a. 1396

Name	R.A.	Decl.	Pair	m₁	m₂	Sep	PA	Colors	Year	Aper.	Rating	Comments	
h 4269	09ʰ 57.7ᵐ	−48° 25′	A–B	6.1	10.1	13.8	321	W, –	1999	2/3″	***		
I 173	10ʰ 06.2ᵐ	−47° 22′	A–B	5.3	7.1	**1.00**	**007**	dY, Y	**232**	9/10″	***	Opening to max. of 1.10″ in 2049	
J Velorum	10ʰ 20.9ᵐ	−56° 03′	A–B	4.5	7.2	7.1	102	W, W	2000	2/3″	***		
			A–C	4.5	9.2	36.2	191	W, –	2000	2/3″	***		
h 4329	10ʰ 31.4ᵐ	−53° 43′	A–B	5.0	8.6	73.9	103	pY, rP	2000	7×50	****		
Pz 3	10ʰ 31.9ᵐ	−45° 04′	A–B	5.6	6.0	13.6	219	W, W	1999	2/3″	****	Nice, easy, almost equal pair	
t Velorum, h 4330	10ʰ 32.9ᵐ	−47° 00′	Aa–B	5.2	8.6	40.4	097	O, –	1999	7×50	****	With h 4332, forms a long, very narrow trapezoid with a slight taper	
h 4332	10ʰ 33.5ᵐ	−46° 59′	A–B	7.1	9.8	28.4	162	W, –	1999	2/3″	***		
x Velorum	10ʰ 39.4ᵐ	−55° 36′	A–B	4.4	6.1	51.7	105	Y, –	2000	6×30	****		
			B–C	6.2	12.1	20.1	174	W, –	2000	4/5″	***		
μ Velorum	10ʰ 46.8ᵐ	−49° 25′	A–B	2.7	6.4	**2.58**	**056**	Y, Y	**138**	3/4″	****	a.k.a. R 155; Opening to max. of 2.63″ in 2020, closing to 0.20″ in 2087″	*23

Variable stars:

Designation	R.A.	Decl.	Type	Range (m_v)		Period (days)	F (f_r/f_t)	Spectral Range	Aper. Recm.	Rating	Comments	
AI Velorum	08ʰ 14.1ᵐ	−44° 35′	δ Sct	6.2	6.8	0.112	–	A2p–F2pIV/V	7×50	****	Prototype of a subclass of δ Scuti variables	*24
FY Velorum	08ʰ 32.4ᵐ	−49° 36′	EB/GS	6.8	7.1	33.72	–	B2Ibpe	7×50	****	1°25′ W (PA 270°) of λ Vel. (m2.14)	
GO Velorum	08ʰ 37.7ᵐ	−40° 26′	SRb	6.6	7.0	75	–	M4III–M5III	7×50	***		
CV Velorum	09ʰ 00.1ᵐ	−51° 33′	EA/DM	6.7	7.2	6.889	–	B2.5V + B2.5V	7×50	***		
GP Velorum	09ʰ 02.1ᵐ	−40° 33′	Mult.*	6.8	7.0	8.965	–	B0.5Iaeq	7×50	*****	a.k.a. Vel. X-1, or the first X-ray source found in Vela	*25
λ Velorum	09ʰ 08.0ᵐ	−43° 26′	Lc	2.1	2.3	–	–	K4Ib/IIa	Eye	*****	Al Suhail or Suhail	*26
GY Velorum	10ʰ 16.7ᵐ	−51° 12′	Lb	6.2	6.5	–	–	M4/5III	7×50	***		
GZ Velorum	10ʰ 19.6ᵐ	−55° 02′	Lc	3.4	3.8	–	–	K3II	Eye	****		

(continued)

Table 10.5. (continued)

Star clusters:

Designation	R.A.	Decl.	Type	m_v	Size (')	Brtst. Star	Dist. (ly)	dia. (ly)	Number of Stars	Aper. Recm.	Rating	Comments
NGC 2547	08h 10.7m	−49° 16'	Open	4.7	19	6.5	1,300	7.2	80	7×50	****	Rich; large brightness range; strong concentration toward center; some nebulosity
IC 2391	08h 40.2m	−53° 04'	Open	2.5	50	3.6	600	8.7	30	7×50	***	a.k.a. Omicron Velorum Cluster; moderately rich; large brightness range
NGC 2669	08h 46.3m	−52° 56'	Open	6.1	12	7.6	3,300	11.5	40	15×80	***	a.k.a. Harvard 3; moderately rich; large brightness range; no concentration toward center
IC 2488	09h 27.6m	−56° 57'	Open	7.4	18	10	8,100	10	70	2/3"	***	Rich; moderate brightness range; weak concentration toward center
NGC 3201	10h 17.6m	−46° 25'	Globular	6.8	20	11.7	15.6K	95	–	2/3"	***	a.k.a. C79; large, irregularly round, low concentration; 10" scope resolves brighter stars
NGC 3330	10h 38.6m	−54° 09'	Open	7.4	7	8.8	4,500	9.2	30	2/3"	***	Moderately rich; moderate brightness range; no central concentration

Nebulae:

Designation	R.A.	Decl.	Type	m_v	Size (')	Brtst./Cent. star	Dist. (ly)	dia. (ly)	Aper. Recm.	Rating	Comments
NGC 2626	08h 35.6m	−40° 40'	Em+Ref	–	5	10	–	–	4/6"	***	6" shows weak circular glow around a m10 *; 12" shows triangle with apex at S end
NGC 2736	09h 00.4m	−45° 54'	SNR	–	>20'	–	800	–	2/3"	***	Can be glimpsed in 7×50 on clear, dark night *27
NGC 2792	09h 12.4m	−42° 26'	Planetary	11.7	13"	**15.7**	8,200	0.52	8/10"	***	10" at HP shows bright, uniform spot; matched pair of m11 (36", 100°) ~3' to SE
NGC 3132	10h 07.7m	−40° 26'	Planetary	9.4	84×53	**10**	2,600	65×40	4/6"	****	a.k.a. "Eight Burst Nebula"; very bright, very large, smooth disk, slightly elongated

Galaxies:

Designation	R.A.	Decl.	Type	m$_v$	Size (')	Dist. (ly)	dia. (ly)	Lum. (suns)	Aper. Recm.	Rating	Comments
NGC 3261	10^h 29.0^m	−44° 39'	SB(rs)b	11.2	3.9 × 3.1	99M	134K	44G	8/10"	**	Small, faint; two main arms; very bright nucleus

Meteor showers:

Designation	R.A.	Decl.	Period (yr)	Duration	Max Date	ZHR (max)	Comet/ Asteroid	First Obs.	Vel. (mi/ km /sec)	Rating	Comments
Puppid/Velids	08^h 12^m	−45°	–	12/01	12/07	10	–	–	40/25	II	Several weak meteor showers have radiants in a large area around Puppis and Vela *28

Other interesting objects:

Designation	R.A.	Decl.	Type	m$_v$	Size (')	Dist. (ly)	dia. (ly)	Lum. (suns)	Aper. Recm.	Rating	Comments
None											

Fig. 10.22. Gamma Velorum and NGC 2547 (~1.25[degree symbol] south [right] of gamma). Astrophoto by Christopher J. Picking.

Fig. 10.23. NGC 2547, open cluster in Vela. North is up. Astrophoto by Rob Gendler.

Fig. 10.24. Gum Nebula, entire. The bright star on the right is Canopus. Astrophoto by Chrsitopher J. Picking.

Fig. 10.25. C85 (NGC 2391), open cluster in Vela. North is up. Astrophoto by Christopher J. Picking.

NOTES:

1. In spite of being the third brightest star in Aries, 41 has no Bayer designation. In the time of Bayer, it may have been considered either an "unformed star," or simply not part of the constellation of Aries. It is listed here because it forms the end of the tail in the figure chosen to represent the Ram of Aries.

2. HD20644 is another oddity like 41 Arietes. It is the sixth brightest star in the constellation but has no Bayer or Flamsteed designation.

3. UX Arietis may have a magnetic/sunspot cycle analogous to that of our sun. The sun exhibits a sunspot cycle of approximately 11 years, with the magnetic polarity repeating every 22 years, or two cycles. In addition, there is a poorly determined cycle of 100 ± 10 years. UX Arietis has a magnetic cycle which reverses every 25.5 days or a 51 day complete cycle. This short-term cycle is modulated by a cycle of about 158 days, or three magnetic cycles. UX Arietis thus mimics the sun but at a much faster rate. At least one extreme ultraviolet flare has been detected by a satellite.

4. Since its discovery, the October Arietid shower has been observed several times by radar, but visual observations have been infrequent. This is probably due to the fact that the Southern Taurid shower occurs about the same time and is much more active than the Arietids, perhaps masking the Arietid shower. It also appears that the two showers are related, since the two radiants are only a few degrees apart and tend to drift across the sky in a very similar fashion.

5. Bayer lettered the stars of Triangulum from α to ε. Flamsteed added several letter and number designations, but only η and ι have survived.

6. R Trianguli is a Mira type cool pulsating variable. It can usually be seen with the unaided eye when near maximum, and in binoculars for several weeks before and after maximum.

7. When LaCaille split Argo Navis into Carina, Puppis, and Vela, he kept the Bayer Greek letter designations of the individual stars, so each of the new constellations has only a part of the Greek alphabet designations. However, he assigned the entire complement of Roman letters for each of the new constellations. One change he made was to use the sequence a, b, …, z, A, B, …, Q, while Bayer had used the sequence A, b, c, …, z, a, B, C, …, Q to reduce the confusion between the Greek "α" and the Roman "a." In both the lower and upper case sequences, the letters "j" and "v" were skipped to avoid confusion with "i" and "u." Because of the small number of original Bayer designations, the extended list made by LaCaille is included here. Puppis and Vela are treated similarly.

8. Although η Carinae is fainter than the magnitude 5.5 limit, it is included here because it is the central star of the famous "Eta Carinae" nebula. Two other stars assigned letters by Lacaille, υ^B and O, are too faint to be included.

9. The star originally designated "n" by LaCaille had its designation dropped because of its faintness (magnitude 6.05).

10. This star was located on the edge of the open cluster NGC 3114, and was very faint, so the LaCaille letter "o" was later dropped.

11. F Carinae was transferred to Chameleon by Gould and left undesignated because of its faintness.

12. Lacaille assigned the letters "R" and "S" to faint stars in Carina, but they were latter dropped to avoid confusion with variable stars.

13. η Carinae is a luminous blue variable which has undergone extraordinary changes. During most of the 18th and the first four decades of the 19th century, η Carinae ranged between 2nd and 4th magnitude, but then brightened to magnitude -1 in 1848. It then slowly faded to magnitude 8 by 1880, a brightness change of $\sim 1/4,000$! It then recovered to magnitude 6, where it has remained. Spectral data indicates that η Carinae may be a binary system and that the variation and mass ejection are due to the unseen star. η Carinae's distance is estimated to be 8,000 light years, which leads to brightness estimates from 900 to 4,500,000 suns. The companion star is believed to be a very evolved star which may become a supernova within a few hundred thousand years. If this happens, it should appear as bright as or brighter than the full moon.

14. NGC 2516 (C96) was discovered by Nicolas-Louis de LaCaille during his survey of the southern skies. It is a fairly young cluster at 60 million years and is one of seven similar clusters (including the Pleiades) moving through space on similar paths.

15. NGC 2602 is almost a twin of the Pleiades, being about the same size as the original Pleiades, but at only 30 million years, only half its age. Since it is about 20% more distant, it appears about 20% smaller and 40% fainter to the eye.

16. NGC 3532 (C91) was discovered in 1752 by Nicolas-Louis de LaCaille with a ½" telescope while survey-ing the southern skies. It is now recognized as one of the best star clusters in the southern sky. NGC 3532 contains over 600 stars, with about a tenth of them visible in binoculars. Located in a rich Milky Way field, this is a spectacular object at any aperture.

17. Canopus was once thought to be as much as 1,200 light years distant, and therefore one of the most lumi-nous stars in our galaxy, measurements by the Hipparchos satellite have shown it to be only about one-third that distance. It is still an impressive star, a supergiant with a luminosity of 13,500 suns. At the distance of Sirius, Canopus would appear 15 times as bright as Venus at its brightest. It is a rare F0 Ia type star which may be evolving away from the main sequence on its way to becoming a supernova.

18. Lacaille assigned d^1 to d^4 Puppis to a close group of four stars, but only d^1 is brighter than the magnitude 5.5 limit included here. Stars previously designated e, g, i, u, E, G, and K are also too faint to be included.

19. M47 lies just 1.3° west–northwest of M46. Both the individual stars and the overall cluster M47 are brighter than for M46. It also has a greater variation in brightness, consisting of both bright and faint stars, in con-trast with M46's much more uniform distribution of faint stars.

20. At apparent magnitude 2.21, Naos (ζ Puppis) is the brightest star in Puppis. It is relatively bright in spite of its distance of ~1,400 light years because it is a very hot supergiant, with a temperature of 42,500 K (76,800 F). The combination of size and temperature gives it an absolute brightness of 20,000 suns. Also, because its temperature is so high, most of the energy it emits is in the invisible ultraviolet region of the spectrum, so its total energy output actually 790,000 times that of the sun. Naos is also a "runaway star," having formed in a nebulous region in nearby Vela. It was probably part of a binary system which encoun-tered another star or star system and was ejected at a high velocity. Even though it is a young star only four million years old, it has traveled over 400 light years from its place of origin.

21. Like typical Wolf–Rayet stars, γ Velorum's spectrum shows emission (bright) lines of hydrogen, oxygen, carbon, and silicon. See the Glossary of Terms.

22. Lacaille assigned k^1 and k^2 to two adjacent stars, k^1 is fainter than the magnitude 5.5 limit adopted in this book.

23. Although orbital motion has not yet been established, the components of μ Velorum have a common proper motion, so they probably form a binary system. The separation was 2.8″ in 1880, and is continuing to decrease by ~0.02″ per year.

24. AI Velorum is the brightest, and the prototype, of a class of variable stars which are similar to the variables, but which have a larger range of variation (>0.3 vs. ≤0.2 magnitudes), and periods of ≤6 h. Although they are often called "high amplitude δ Scuti stars," they also differ in their modes of variation in that they usually have only two or three distinct modes of variation, while normal δ Scuti stars may have several simultaneous radial modes and, as many as a dozen nonradial modes of variation.

25. GP Velorum/Vela X-1 is an eclipsing binary system consisting of a supergiant OB star with a mass of 23 suns and a pulsar (neutron star) with a mass of 1.77 suns. Each eclipse hides the pulsar for about two days.

26. A.k.a. Suhail or Al Suhail. The L class of variables designates an irregular variable with slow changes and the "c" subclass indicates a rare supergiant type star, of which only about 50 are known.

27. NGC 2736 is a small part of the Vela Supernova Remnant, the remains of a supernova explosion about 11,000–12,000 years ago. The entire remnant spreads over more than 8°, and its edges are so thin that it is difficult to tell where the remnant ends.

28. The Puppid/Velid meteor shower is one of about 25 apparently related minor showers radiating from a large area around Puppis and Vela. The large number of showers, and the small number of meteors from each make it difficult to determine the exact radiants or other characteristics of each individual shower or even the exact number of showers.

CHAPTER ELEVEN

Unbelievable Music, Unbearable Sorrow

June–August

LYRA	ERIDANUS
The Harp	The River

There are several myths explaining the origin of the lyre and its placement in the heavens as the constellation Lyra. One of the most charming (and humorous) origin stories tells about the first days of the Greek god Hermes.

Hermes, one of the more mischievous and precocious of the Greek gods, jumped out of his cradle on the day of his birth and went exploring. Encountering a herd of cattle, Hermes stole them and drove them to a cave, where he grabbed each one by the tail, and forced it to walk backward into the cave. His aim was to fool the owner of the herd into thinking that the cattle were walking away from the cave.

Unfortunately, Hermes picked the wrong herd to rustle. Apollo was the owner and, being the god of truth, he was not fooled for a minute. Apollo soon confronted Hermes. Hermes vigorously protested, saying that as a mere infant, he could not be responsible for the theft. All of Apollo's arguments and threats got no results.

With no other way to influence Hermes, Apollo petitioned Zeus for a council of all the Olympian gods and goddesses. After examining all the evidence and testimony, the council decided that Hermes was indeed responsible, and must not only return Apollo's herd but also pay a fine to Apollo. Being a newborn, Hermes had nothing

P. Simpson, *Guidebook to the Constellations: Telescopic Sights, Tales, and Myths*, Patrick Moore's Practical Astronomy Series, DOI 10.1007/978-1-4419-6941-5_11, © Springer Science+Business Media, LLC 2012

to give Apollo, but while wandering about, he spied an empty tortoise shell. The ingenious young god stretched some strings across the open face of the shell, and drawing them tight, made the first lyre, a musical instrument related to the harp, but with only seven strings. Hermes then endowed the lyre with magical powers to charm all living creatures, and even to affect inanimate objects so that they tried to move toward the magical music.

When Hermes returned to Mt. Olympus, he plucked out a lovely melody on his invention. Apollo heard the beautiful sound and offered to settle Hermes' debt in exchange for the lyre. Apollo thus became the musician of the gods and played beautiful soothing music at all their gatherings, and also for his own pleasure.

Later, Apollo had a son by Calliope, the Muse of lyric poetry. Their son was named Orpheus and Calliope taught him the arts of poetry and music. When Orpheus grew into a young man, Apollo saw his musical talent and gave him the enchanted lyre. With the talents inherited from both his parents and a magical instrument, Orpheus naturally became the most famous poet and musician who ever lived. Through his playing, he won the hand of the beautiful Eurydice. Although they were extremely happy, they were doomed to a short marriage with a tragic ending.

One day, as Eurydice walked in their garden, she was pursued by an amorous minor river god. Fleeing, Eurydice accidentally stepped on a poisonous snake and was bitten. The would-be suitor fled, leaving Eurydice to die. When Orpheus found her body hours later, her spirit had already been escorted to the underworld, and it was too late for even Apollo to help.

Orpheus was despondent and wandered about the earth playing and singing sad songs in memory of his beloved wife. At last, he arrived at the entrance to the underworld, and played such beautiful sad songs that the doorman allowed him to pass through the gates of Hades and Orpheus became one of the very few living men to enter the land of the dead. Continuing to play sad, but incomparably beautiful songs, Orpheus charmed Charon into ferrying him across the river Styx, and similarly charmed Cerebus the three-headed guard dog and the judges of the dead into allowing him to pass into the inner regions of the underworld.

When he reached the throne of Hades and Persephone, the rulers of the underworld, his ballad of loss and mourning, sung to the enchanting music, charmed them into allowing Eurydice to return to the living world with him. However, one condition was placed on the return. Orpheus was not allowed to look at the spirit of Eurydice until she had left the underworld and the sun restored her body. Orpheus played the lyre as he slowly retraced his path. Near the end, he became concerned that he could not hear her footsteps and anxiously looked back to see if she was still behind him. She was, but as he glanced back, even as she called out him longingly, she began to fade and drift back into the underworld. Orpheus was not allowed to return to the underworld a second time, and resumed his lonely wandering through the mortal world.

For years, Orpheus wandered about the land playing beautifully moving songs in Eurydice's memory. Many women were attracted to him, but he remained faithful to his lost wife and rejected them all. Orpheus' end came when he encountered a group of Maenads, or worshipers of Dionysus. They tried to induce him into joining them in their partying but to no avail. When Orpheus refused to join them in their revels, they became enraged and attacked him. After much taunting, while he ignored them and played on, they began to throw stones at him, but even the stones were charmed by his music and fell harmlessly before him. Finally, their rage increased until their screams drowned out the music and the stones began to strike Orpheus. The Maenads charged Orpheus and literally tore him to pieces. As they did, they threw the body parts and the lyre into the nearby river and watched as they floated out of sight toward the ocean. As the lyre floated away, it continued to play a mournful song, and the head of Orpheus repeatedly cried out "Eurydice, Eurydice."

After Orpheus' death, Apollo retrieved the lyre, and placed it in the sky as Lyra to honor his son. It was also said that the river which bore the lyre and Orpheus' head away was also placed in the sky as the constellation Eridanus.

Many other myths have involved rivers which were supposedly changed into a heavenly river. Several researchers have tried to identify which earthly river should be associated with Eridanus, including the Po, the Nile, and both the Tigris and Euphrates, but none could be made to match the ancient descriptions which said it bore amber and emptied into the north sea (not today's North Sea). The best explanation may be that it was a north-flowing river described by traders who brought tin and amber to ancient Greece.

Figure 11.1 introduces the summer asterism known as "The Summer Triangle" and surrounding constellations. This asterism consists of the brightest stars in the constellations of Aquila (Altair), Cygnus (Deneb), and Lyra (Vega). The Summer Triangle is easy to find because it is a large right triangle with Vega, the 5th brightest star, in the sky at the right angle. It is directly opposite the sun at 9:00 PM (Standard Time) about July 15, which is very close to midway between the summer solstice and the autumn equinox, which define summer in the USA. The Summer Triangle can be used as a guide to finding not only Lyra but also Cygnus, Aquila, Vulpecula, Sagitta, and Delphinus (Fig. 11.1).

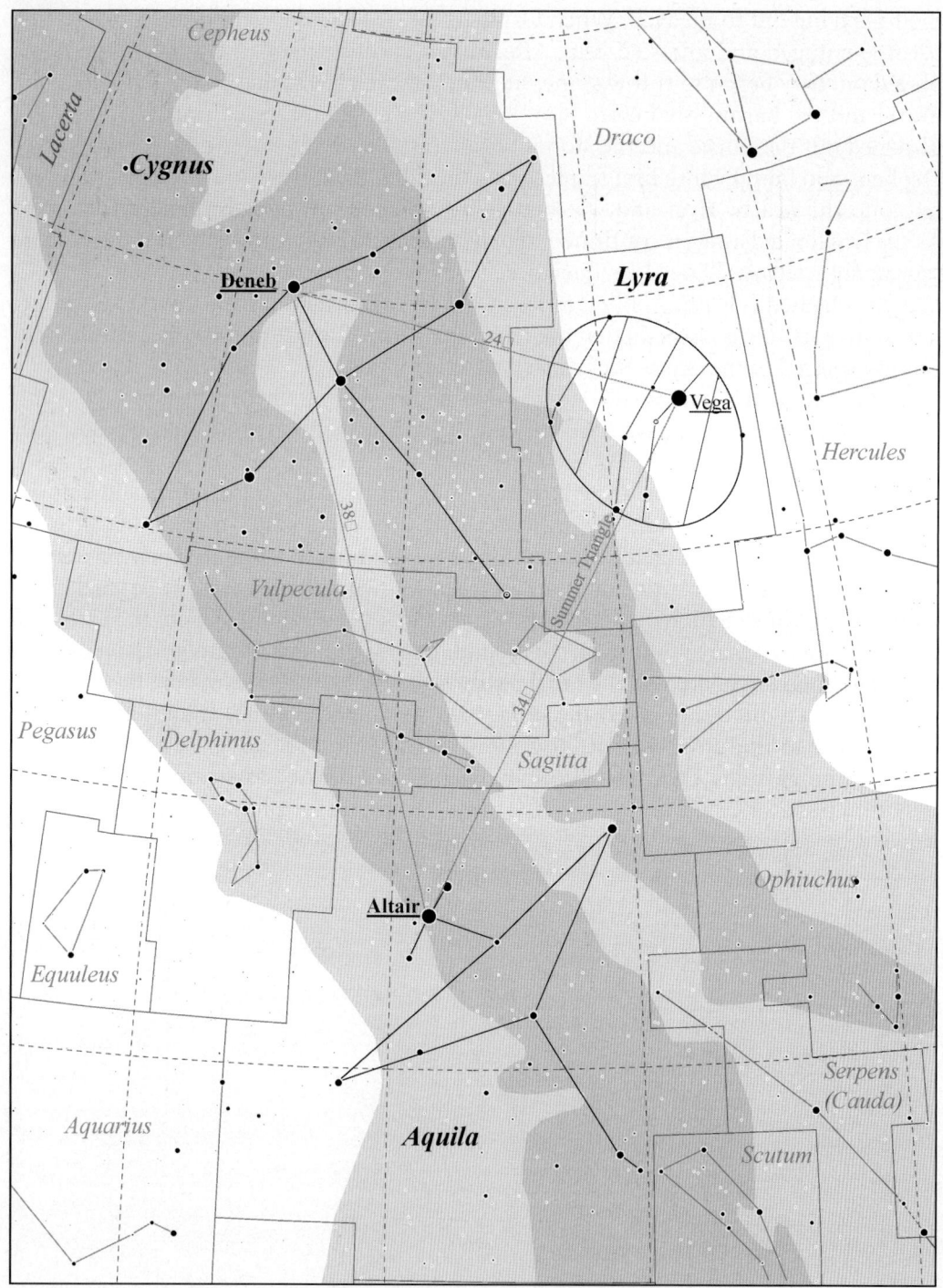

Fig. 11.1. The Summer Triangle.

LYRA (**lie**-rah)	**The Lyre**
LYRAE (**lie**-rye)	**BEST SEEN (9:00 P.M.)**: August 28
Lyr	**AREA**: 286 sq. deg., 52nd in size

TO FIND: At the time listed above, face south, then look almost overhead (85° elevation). Vega is by far the brightest star in this region of the sky. Alternately, find Altair in Aquila the eagle and follow the line formed with the two nearest (fainter) stars to the northwest. Vega is about 34° from Altair. Vega, Altair, and Deneb form the large and easily recognizable "Summer Triangle." About 2° NE and SE are $\varepsilon^{1,2}$ and ζ Lyrae. ζ is the NW corner of a parallelogram with roughly $2 \times 4°$ sides.

The small group formed by these six stars is easy to find and recognize. They form the central part of the tortoise shell and strings. θ Lyrae lies about 7.5° E of Vega, and 13 Lyrae lies about 6° NE of Vega. These two mark the eastern and northern extremities of the tortoise shell. Relatively faint κ lies about 4.5° SW of Vega and marks the western extremity of the shell (Figs. 11.1–11.4).

KEY STAR NAMES:

Name	Pronunciation	Source
α Vega	**vee**-gah	from Nasr al Waki (Arabic), "the stone eagle of the desert"
β Sheliak	**shell**-yak	from Salbāq (Persian), "the Harp"
γ Sulafat	**sue**-lah-fat	from al-Sulahfāt (Arabic), "the tortoise"

The name of Vega comes from the Arabic constellation consisting mainly of α, ζ, and ε Lyrae which represented an eagle or vulture sitting with folded wings, in contrast to Aquila which represented a flying eagle or vulture.

Magnitudes and spectral types of principal stars:

Bayer desig.	Mag. (m_v)	Spec. type	Bayer desig.	Mag. (m_v)	Spec. type	Bayer desig.	Mag. (m_v)	Spec. type
α	0.03	A0 V	β	3.45	B7 V	γ	3.24	B0 V
δ^1	5.58	B3.5 V	δ^2	4.30	M4 II	ε^1	4.69	A3 V+
ε^2	4.50	A5 V	ζ^1	4.36	Am	ζ^2	5.73	F0 V+
η	4.39	B2 IV	θ	4.36	K0 III	ι	5.28	B6 IV
κ	4.30	K2 IIIab	λ	4.93	K2.5 III	μ	5.12	A3 IVn
ν^2	5.25	F9 V						

Stars of magnitude 5.5 or brighter which have Flamsteed but no Bayer designations:

Flam-steed	Mag. (m_v)	Spec. type	Flam-steed	Mag. (m_v)	Spec. type
16	5.00	A7 V	17	5.20	F0 V

Stars of magnitude 5.5 or brighter which have no Bayer or Flamsteed designations:

Desig-nation	Mag. (m_v)	Spec. type	Desig-nation	Mag. (m_v)	Spec. type
HD 171301	5.47	B8 IVO	HD 172044	5.41	B8 II/III
HD 173780	4.83	K3 III	HD 175740	5.46	G8 III
HD 176051	5.20	G0 V			

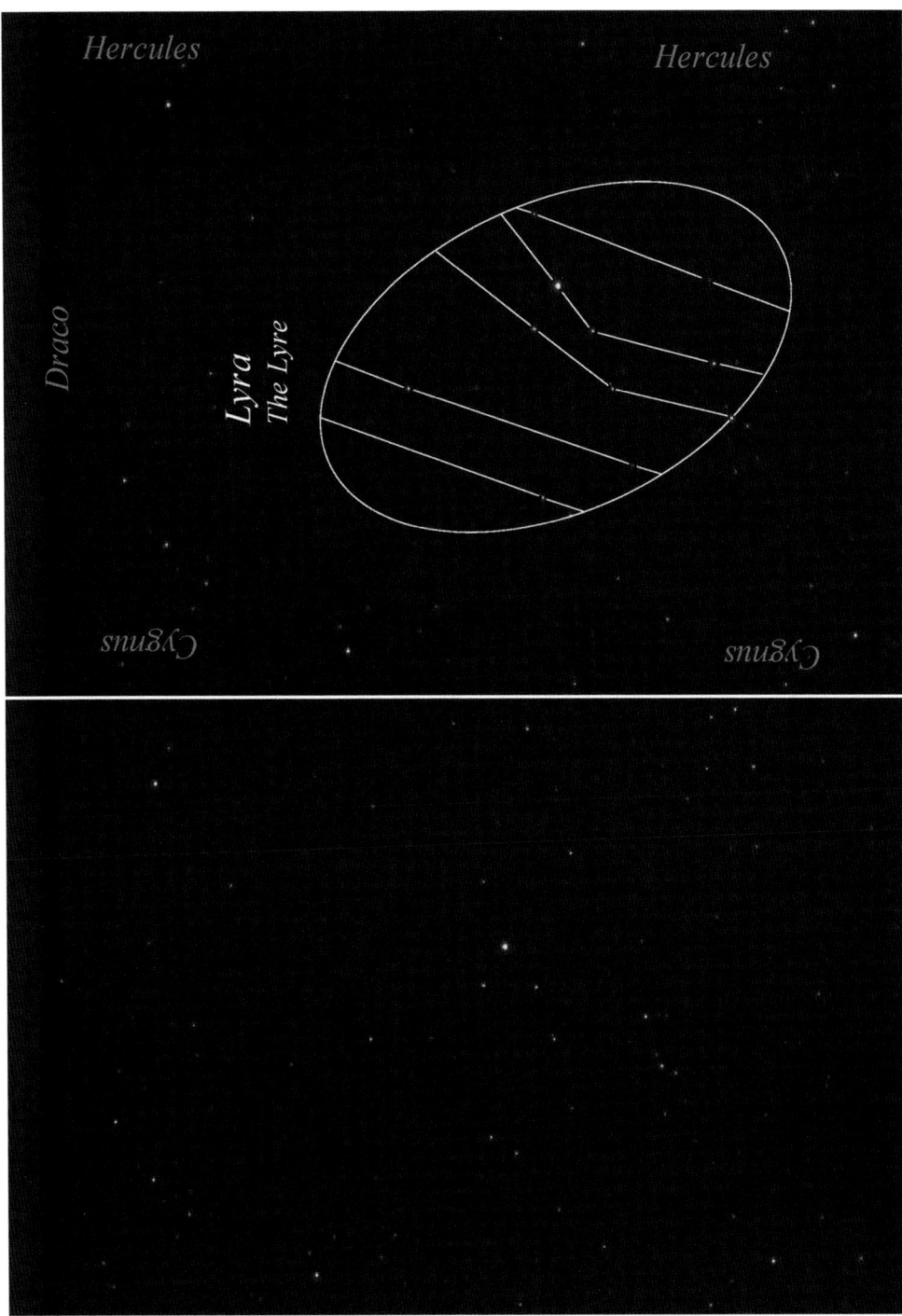

Fig. 11.2. The stars and constellation of Lyra.

Table 11.1. Selected telescopic objects in Lyra.

Multiple stars:

Designation	R.A.	Decl.	Type	m_1	m_2	Sep. (")	PA (°)	Colors	Date/Period	Aper. Recm.	Rating	Comments
Vega, α Lyrae	18ʰ 36.9ᵐ	+38° 47'	A–B	0.0	9.5	78.2	182	W, pB	2000	15×80	****	C and E components are not physically related to Vega
			A–C	0.0	11.0	51.2	293	W, ?	1881	3/4"	***	
			A–E	0.0	9.5	96.4	039	W, ?	1985	15×80	***	
ε¹ and ε² Lyrae	18ʰ 44.3ᵐ	+39° 40'	AB–CD	5.0	5.3	209.0	172	yW, yW	2000	6×30	*****	The famous "Double-Double." The most spectacular example of this type of multiple star in the sky!
4, ε¹ Lyrae			A–B	5.0	6.1	**2.51**	**094**	yW, yW	**1,165.6**	3/4"	*****	
5, ε² Lyrae			C–D	5.3	5.4	**2.37**	**173**	yW, yW	**724.3**	4/5"	*****	
7, ζ² Lyrae	18ʰ 44.8ᵐ	+37° 36'	A–C	4.3	13.3	46.1	272	yW, bGr	1923	6/7"	****	The B component, at magnitude 15, is too faint for most amateur telescopes
			A–D	4.3	5.6	49.0	150	yW, ?	2003	6×30	****	
			A–E	4.3	11.5	61.8	301	yW, ?	1960	3/4"	****	
8, ν¹ Lyrae	18ʰ 49.8ᵐ	+32° 49'	A–B	5.9	10.9	34.8	073	yW, ?	1924	3/4"	*****	Smyth called the three companions "blue"
			A–C	5.9	10.3	59.2	122	yW, ?	1934	2/3"	***	
			C–D	10.3	11.5	18.4	213	?, ?	1965	3/4"	***	
β Lyrae	18ʰ 50.1ᵐ	+33° 22'	A–B	3.6	6.7	45.8	149	W, W	2002	6×30	****	A varies 3.4–4.3
δ¹ and δ² Lyrae	18ʰ 54.1ᵐ	+36° 57'	AB–CD	5.6	4.3	619.0	115	bW, R	1995	Eye	****	A second "Double-Double" in Lyra. Not as evenly matched or as symmetrical as ε Lyrae
11, δ¹ Lyrae	18ʰ 53.7ᵐ	+36° 58'	A–B	5.6	9.93	175.8	020		2002	2/3"	****	
12, δ² Lyrae	18ʰ 54.5ᵐ	+36° 56'	C–D	4.3	11.2	88.3	349	R, ?	1995	3/4"	****	
			C–E	11.2	11.6	2.0	142	R, ?	1995	5/6"	***	
β648 (Lyra)	18ʰ 57.0ᵐ	+32° 54'	A–B	5.3	8.0	**1.01**	**255**	Y, ?	**61.15**	9/10"	*****	A rapid binary for moderate to large amateur telescopes
17 Lyrae	19ʰ 07.4ᵐ	+32° 30'	A–B	5.3	9.1	3.7	290	pY, pB	2001	2/3"	****	
Σ2470 and Σ2474 (Lyra)	19ʰ 09.0ᵐ	+34° 41'	A–B	6.4	6.6	622.0	158	bW, pY	1933	7×50	****	
Σ2470 (Lyra)	19ʰ 08.8ᵐ	+34° 46'	A–B	7.0	8.4	13.7	269	bW, bW	2001	15×80	****	
Σ2474 (Lyra)	19ʰ 09.1ᵐ	+34° 36'	A–B	6.8	7.9	15.9	262	pY, O	2002	15×80	****	
20, η Lyrae	19ʰ13.8ᵐ	+39° 09'	A–B	4.4	8.6	28.4	080	pB, pV	2000	10×50	****	Colors are subtle, and observer reports differ
21, θ Lyrae	19ʰ 16.4ᵐ	+38° 08'	A–B	4.5	10.1	98.9	070	Y, B	2001	2/3"	****	

Variable stars:

Designation	R.A.	Decl.	Type	Range (m$_v$)	Period (days)	F (f$_r$/f$_f$)	Spectral Range	Aper. Recm.	Rating	Comments
W Lyrae	18^h 14.9^m	+36° 40'	M	7.3	197.88	0.48	M2e–M8e	8/10"	****	Nearly symmetrical rise/fall. 1°10' WNW [PA 302°] from K Lyr. ##
XY Lyrae	18^h 38.3^m	+39° 41'	Lc	5.80	–	–	M4-5Ib-II	6×30	****	0°55' NNE [PA 014°] from Vega, α Lyr [m0.03]. #
10, β Lyrae	18^h 50.1^m	33° 22'	β Lyr	3.25	12.914	–	B8III–IIIep	Eye	****	Prototype of eclipsing ellipsoidal binaries. Stars are very close, and distorted by
12, δ^2 Lyrae	18^h 54.7^m	+36° 54'	SRc	4.22	–	–	M4II	Eye	****	Orange. Color shows better in binoculars
13, R Lyrae	18^h 55.3^m	+43° 57'	SRb	3.88	46	–	M5III	Eye	****	Lovely, deep red-orange. Color shows better in binoculars
RR Lyrae	19^h 25.7^m	+42° 48'	RR Lyr	7.06	0.567	0.19	A5.0 + F7.0	6×30	****	Prototype of "cluster variables" similar to Cepheids. # *1

Star clusters:

Designation	R.A.	Decl.	Type	m$_v$	Size (')	Brtst. Star	Dist. (ly)	dia. (ly)	Aper. Recm.	Rating	Comments
Stephenson 1	18^h 53.5^m	+36° 55'	Open	3.8	40	4.3	1,030	12	7×50	****	Large, loose group, ~15 stars. Between δ^1 and δ^2 Lyr, which are probably cluster
M56, NGC 6779	19^h 16.6^m	+30° 11'	Globular	8.4	8.8	13.2	38 K	90	4/6"	****	Fuzzy, unresolved, bright core, fading toward edges. In rich star field
NGC 6791	19^h 20.7^m	+37° 51'	Open	9.5	10	15.0	17.1 K	50	6/8"	**	Rich cluster of faint stars. Several dozen resolved in 10" telescope. Age ~8 billion years

Table 11.1. (continued)

Nebulae:

Designation	R.A.	Decl.	Type	m_v	Size (')	Brtst./Cent. star	Dist. (ly)	dia. (ly)	Aper. Recm.	Rating	Comments
M57, NGC 6720	18h 53.6m	+33° 02'	Planet.	8.8	52×86"	15	2.3 K	1.0	4/6"	*****	The famous "Ring Nebula" One of the best planetaries for small scopes

Galaxies:

Designation	R.A.	Decl.	Type	m_v	Size (')	Dist. (ly)	dia. (ly)	Lum. (suns)	Aper. Recm.	Rating	Comments
NGC 6703	18h 47.3m	+45° 33'	SA0+	11.3	2.5×2.3	180M	100K	78G	8/10"	***	Fairly bright halo, stellar nucleus. Visible in 6", but faint

Meteor showers:

Designation	R.A.	Decl.	Period (yr)	Duration		Max Date	ZHR (max)	Comet/ Asteroid	First Obs.	Vel. (mi/ km/sec)	Rating	Comments	
Lyrids	18h08m	+33°	Annual	04/16	04/25	04/22	10	*1	687 BC	30/	I	*1: Comet Thacher (1861 I) has very similar orbital parameters	*2
η Lyrids	19h28m	+40°	14.8?	05/03	05/8-10	05/12	4	*2	1895 *3	27	IV	*2: IRAS–Araki–Alcock. *3: H. Corder	*3
June Lyrids	19h28m	+35°	Annual	06/10	06/21	06/15	3	Unknown	1966	19	IV	Discovered by Stan Dvorak (CA)	*4

Other interesting objects:

Designation	R.A.	Decl.	Type	m_v	Size (')	Dist. (ly)	dia. (ly)	Lum. (suns)	Aper. Recm.	Rating	Comments
None											

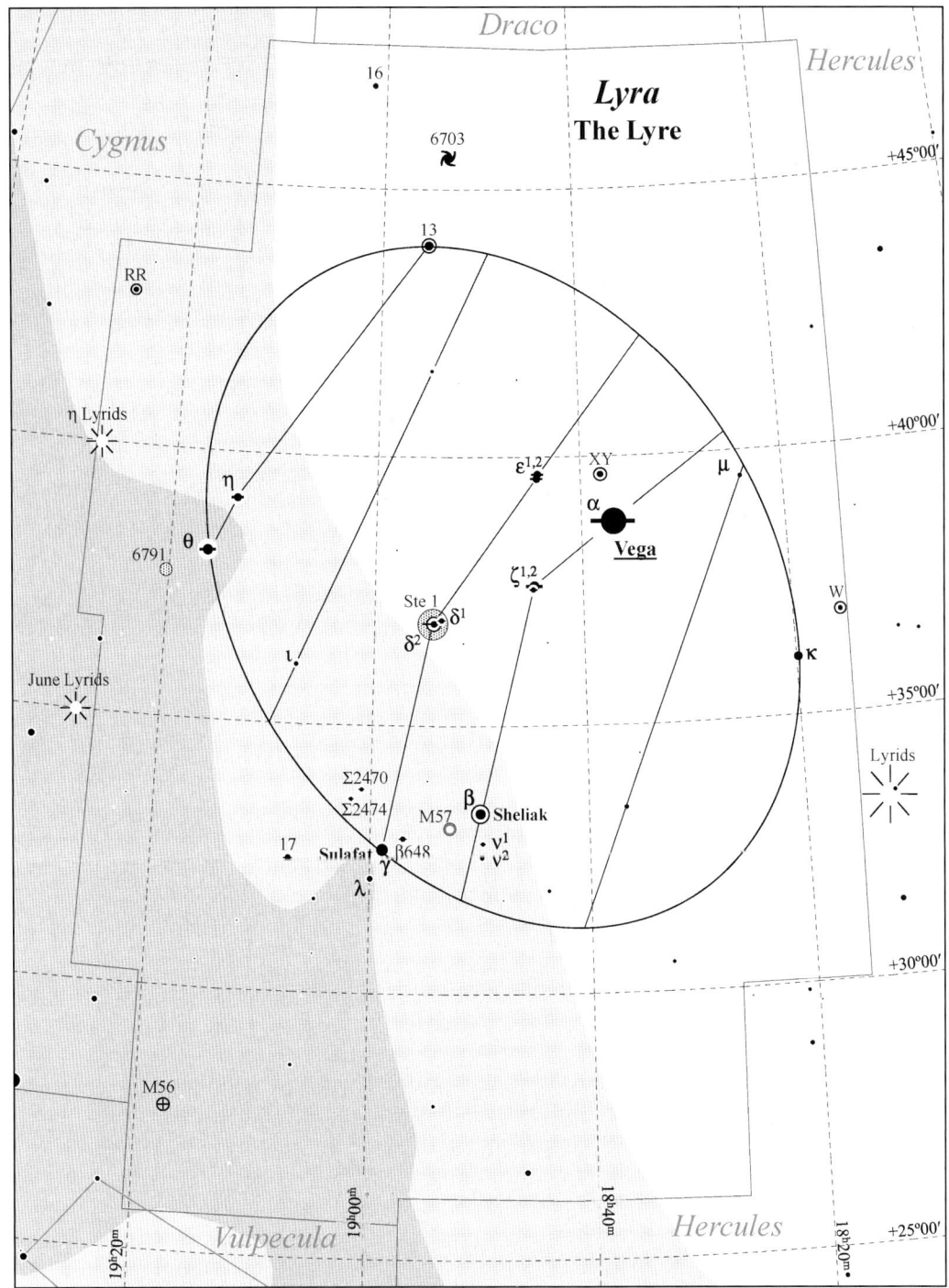

Fig. 11.3. Lyra details.

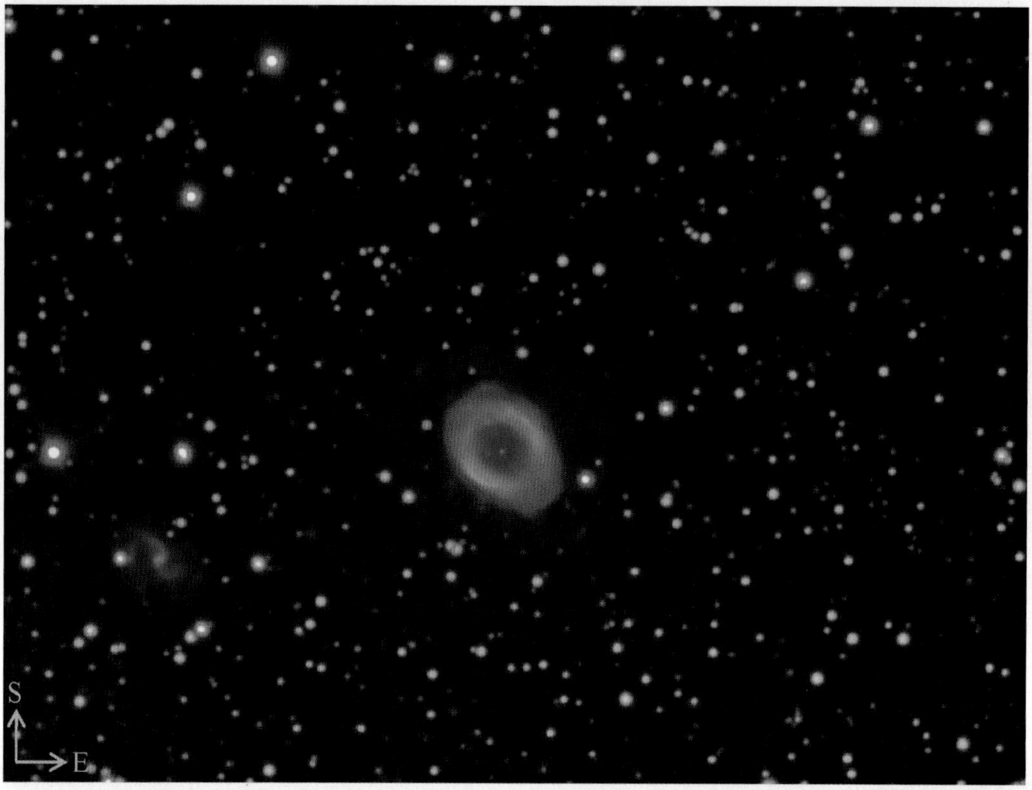

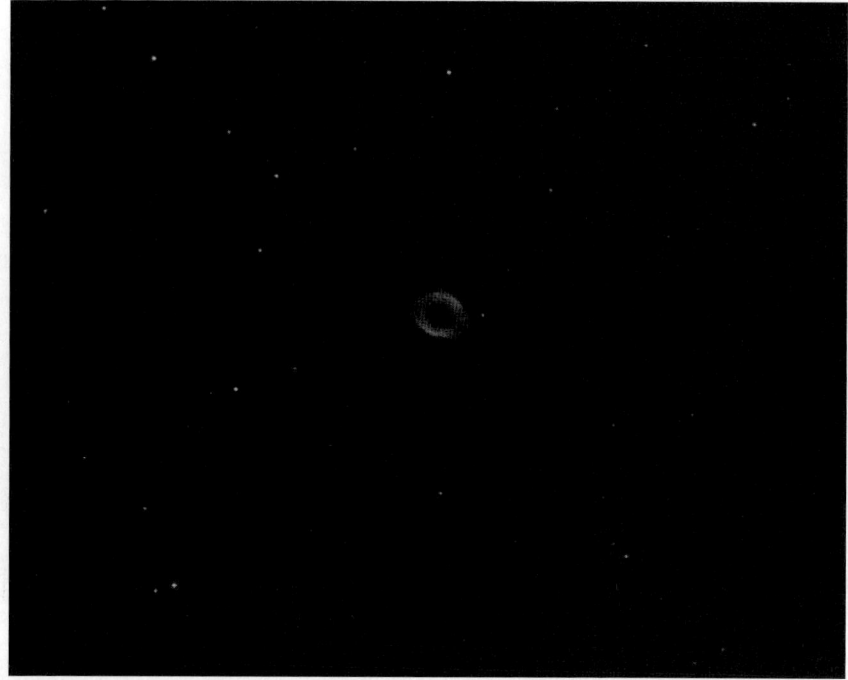

S
↑
└──→ E

Fig. 11.4a. M57, the Ring Nebula, planetary nebula in Lyra. The spiral galaxy is 14th magnitude IC 1596, too faint for most amateur telescopes. Astrophoto by Stuart Goosen.

Fig. 11.4b. M57, image processed to show approximate appearance in moderate-sized amateur telescope. Original astrophoto by Gian Michelle Ratto.

Fig. 11.5. M56, globular cluster in Lyra. Astrophoto by Mark Komsa.

ERIDANUS (eh-**rid**-dan-us)	**The River**
ERIDANI (eh-**rid**-dan-ee)	**BEST SEEN:** (9:00 P.M.): Dec. 26
Eri	**AREA:** 1,138 sq. deg., 6th in size

TO FIND: Look for the second brightest star in Eridanus, magnitude 2.8 β Eridani, just 3.5° NW of <u>Rigel</u>. From there, trace the faint line of stars – ω, μ, ν, o¹, γ, π, δ, ε, ζ, η, π Cet, τ¹–τ⁹, υ¹–υ⁴, g, f, h, y, e, θ, ι, s, κ, φ, χ, and finally to <u>Acherner</u> (α Eridani). Those who live north of about 28° latitude will have to be content to end the river short of Acherner because of the horizon. They will, however, be able to follow the river to its original ending at Acamar, and whose name is similar in origin and pronunciation to <u>Acherner</u>, the modern river's end. In tracing out Eridanus, remember that almost all its stars are relatively faint, especially those designated by English letters, i.e., f, g, h, etc. (Figs. 11.6 and 11.7).

KEY STAR NAMES:

Name	Pronunciation	Source
α <u>Achernar</u>	(**ay**-ker-nahr)	Al <u>Ahir</u> al <u>Nahr</u> (Latin) "the end of the river"
β Cursa	(**kur**-suh)	Al <u>Kursiyy</u> al Jauzah (Arabic) "the footstool of the central one"
γ Zaurak	(**zaw**-rack)	Al Nā'ir al <u>Zaurak</u> (Arabic) "the bright star of the boat"
δ Rana	(**rah**-nuh)	<u>rana</u> (Latin) "Frog"
η Azha	(**uh**-zuh)	<u>Azhā</u> (Arabic) "Nest or hatching place"
ζ Zibal	(**zih**-bahl)	Al <u>Zibāl</u> (Arabic) "the little ostriches" applied to several nearby stars
θ Acamar	(**ay**-kuh-mahr)	Same as alpha, from the time when theta was end of the constellation
o¹ Beid	(bide)	Al <u>Baid</u> (Arabic) "the egg" from the arabic constellation, the Ostrich's Nest
o² Keid	(kide)	Al <u>Kaid</u> (Arabic) "the broken eggshells"
τ² Anchat	(?)	Al <u>Hināyat</u> al Nahr (Arabic) "bend of the river"
υ³ Theemin	(**thee**-min)	Possibly distorted version of Al <u>Thalīm</u> (Arabic) "the Ostrich," from the constellation named above
53 Sceptrum	(**skep**-trum)	<u>sceptrum</u> (Latin) "dominion, kingdom" from the obsolete constellation of "Sceptrum Branden-burgicum" introduced in 1688 by Gottfried Kirch

Magnitudes and spectral types of principal stars:

Bayer desig.	Mag. (m_v)	Spec. type	Bayer desig.	Mag. (m_v)	Spec. type	Bayer desig.	Mag. (m_v)	Spec. type
α	0.45	B3 Vp	β	2.78	A3 III var	γ	2.97	M1 IIIb
δ	3.52	K0 IV	ε	3.72	K2 V	ζ	4.58	A5m
η	3.89	K1 III/IV	θ	2.88	A4 III	ι	4.11	K0 III
κ	4.24	B5 IV	λ	4.25	B2 IVn	μ	4.01	B5 IV
ν	3.93	B2 III	ξ	5.17	A2 V	o	4.04	F2 II/III
π	4.43	M1 III	ρ¹	5.75	K0 II	ρ²	5.32	K0 II/III
ρ³	5.26	A8 V	σ	–	–*5	τ¹	4.47	F5/F6 V
τ²	4.76	K0 III	τ³	4.08	A4 V	τ⁴	3.70	M3/M4 III
τ⁵	4.26	B9 V	τ⁶	4.22	F3/F5 IV	τ⁷	5.24	A1 V
τ⁸	4/64	B5 V	τ⁹	4.62	Ap Si	υ¹	4.49	K0 III
υ²	3.81	G8 III	υ³	3.97	K4 III	υ⁴	3.55	B9 V
φ	3.56	B8 IV/V	χ	3.69	G5 IV	ψ	4.80	K0 III
ω	4.36	A9 IV	b	5.50	B6 V	c	5.22	F0 V
d	3.97	K4 III	f	4.30	A1 V	f	4.86	B9 V
g	4.17	G8 III	h	4.59	K2 IIICN	i	5.11	B6/7 V
l	3.86	K1 III	q²	5.04	A1 V	s	4.74	A2 V
v	4.74	B9 Vs	w	4.46	G8 III	y	4.57	K0 III
A	4.87	K3 III						

Stars of magnitude 5.5 or brighter which have Flamsteed but no Bayer designations:

Flamseed.	Mag. (m_v)	Spec. type	Flamseed.	Mag. (m_v)	Spec. type	Flamseed.	Mag. (m_v)	Spec. type
4	5.44	A5 IV/V	15	4.86	K0 III	20	5.24	B8/9 III
24	5.24	B7 V	30	5.48	B8+	35	5.28	B5 V
37	5.44	G8 III	40	4.43	K1 V	45	4.91	K3 II/III
47	5.20	M3 III	54	4.32	M3/4 III	58	5.49	G3 V
60	5.03	K0 III	63	5.39	G4 V	64	4.78	F0 V
66	5.12	B9.5 V	68	5.11	F2 V	82	4.26	G8 V

Stars of magnitude 5.5 or brighter which have no Bayer or Flamsteed designations:

Designation	Mag. (m_v)	Spec. type	Designation	Mag. (m_v)	Spec. type
HD 018331	5.16	A3 Vn	HD 018543	5.22	A2 IV
HD 019349	5.23	M3 III	HD 020894	5.50	G8 III
HD 025457	5.38	F5 V	HD 026326	5.45	B5 IV
HD 027616	5.38	A0 V	HD 029065	5.24	K4 III
HD 029573	4.99	A0 V	HD 029613	5.46	K0 III

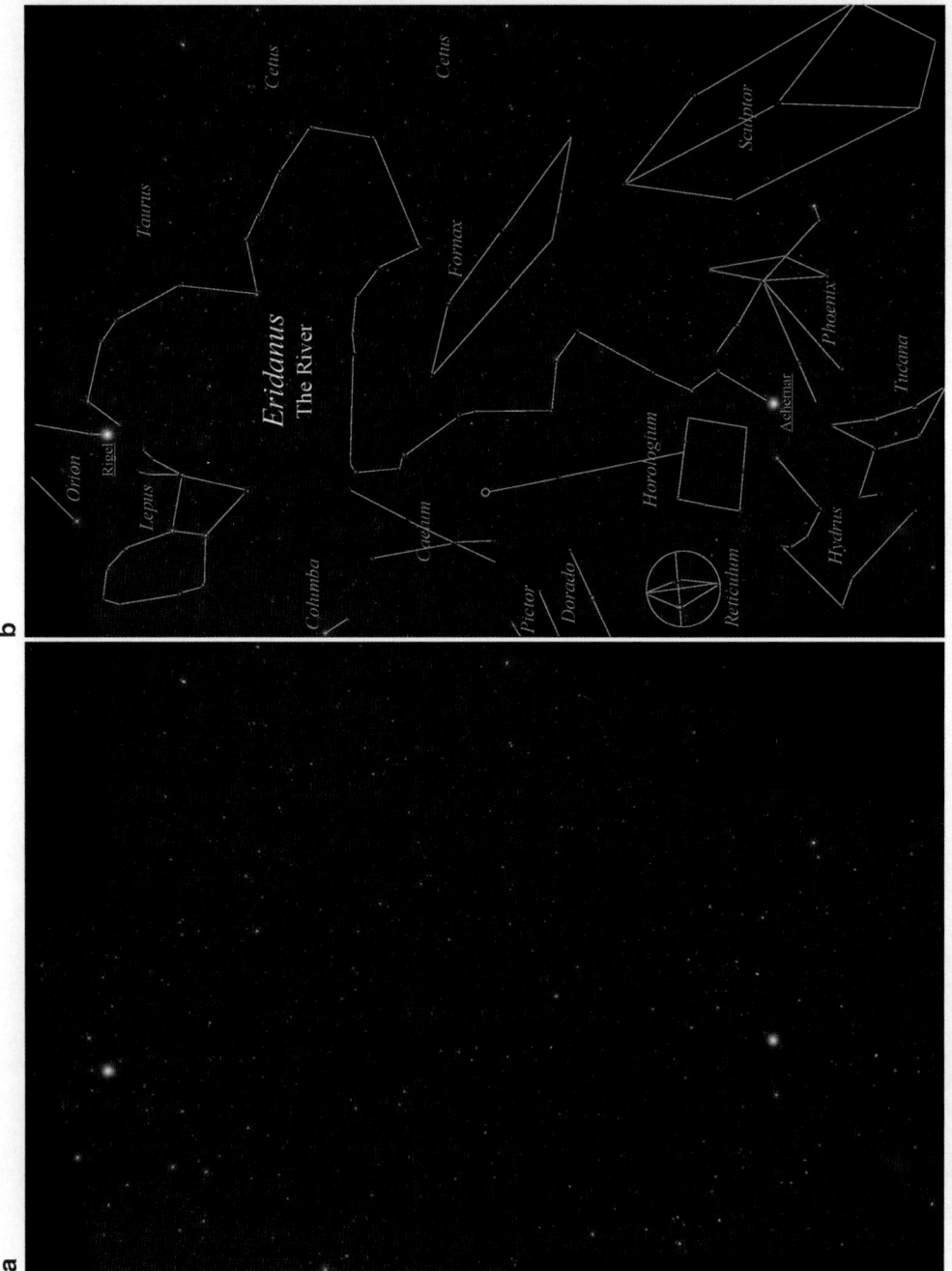

Fig. 11.6. **(a)** The stars of Eridanus and adjacent constellations. Composite astrophoto by Christopher J. Picking. **(b)** Eridanus and adjacent constellations.

Table 11.2: Selected telescopic objects in Eridanus.

Multiple stars:

Designation	R.A.	Decl.	Type	m_1	m_2	Sep. (")	PA (°)	Colors	Date/Period	Aper. Recm.	Rating	Comments
p Eridani	01h 39.8m	−56° 12'	A–B	5.8	5.9	**11.66**	**188**	Y, Y	**483.66**	2/3"	****	a.k.a. Dun 5. Nice matched pair
χ Eridani	01h 56.0m	−51° 37'	A–B	3.7	10.7	4.9	204	dY, ?	1956	2/3"	****	a.k.a. h 3473
9, θ Eridanus	02h 58.3m	−40° 18'	A–B	3.2	4.1	8.4	090	W, W	2002	2/3"	****	Acamar. Nice pair for small scopes. One of best southern doubles. a.k.a. h 3473
ρ² Eridani	03h 02.7m	−07° 41'	A–B	5.4	8.9	1.5	066	Y, Y	1990	6/7"	****	Nice pair in a nice star field. a.k.a. Pz 2
Jc 8	03h 12.4m	−44° 25'	A–B	6.4	7.4	**0.66**	**160**	yW, W	**45.2**	14/16"	****	a.k.a. Jc 8
			AB–C	6.4	8.8	3.7	190	yW, yW	2002	2/3"	****	a.k.a. h 3556
h 3565	03h 18.7m	−18° 34'	A–B	5.9	8.2	8.2	121	yW, −	2006	2/3"	****	
16, τ⁴ Eridani	03h 19.5m	−21° 45'	A–B	4.0	9.5	5.9	289	R, −	1987	2/3"	****	Four 10th mag. stars also lie at distances of 40 to 160" from A. a.k.a. Jc 1
f Eridani	03h 48.6m	−37° 37'	A–C	4.0	10.8	39.8	114	R, −	1964	3/4"	***	a.k.a. A 16.
30 Eridani	03h 52.7m	−05° 22'	A–B	4.8	5.3	7.9	212	pY, pY	2002	2/3"	****	
32 Eridani	03h 54.3m	−02° 57'	A–B	5.5	10.4	8.3	134	Y, pB	2000	2/3"	***	In larger scopes, it appears deep yellow and greenish white. a.k.a. Σ 470
			A–B	4.8	5.9	6.9	348	pO, pB	2006	2/3"	******	A–B [a.k.a. Fin 344 – 6.5, 6.5, 0.1"] is too close for amateur instruments
Hdo 188	03h 54.4m	−40° 21'	AB–C	5.8	12.7	23.5	168	dY, W	1999	5/6"	**	a.k.a. h 3608
γ Eridani	03h 58.0m	−13° 31'	A–B	3.2	12.7	56.2	253	R, −	2000	5/6"	**	
39 Eridani	04h 14.4m	−10° 15'	A–B	5.0	8.5	6.3	141	oY, W	1996	2/3"	**	Appears deep yellow and pale blue in larger scopes
40, o² Eridani	04h 15.3m	−07° 39'	A–BC	4.4	9.7	83.0	104	yO, B	2002.00	15×80	******	Very nice colored triple star. 38th nearest star system at 16.4 ± 0.1 ly *6
			B–C	9.5	11.2	**9.16**	**334**	R, W	**252.10**	3/4"	*****	a.k.a. Σ 518. Interesting pair – a red dwarf and a white dwarf
u⁴ Eridani	04h 17.9m	−33° 48'	Aa–B	3.9	9.9	5.3	163	W, −	1999	2/3"	***	a.k.a. Hub 3. A–a (a.k.a. I 270) is too close for amateur instruments
			Aa–C	3.6	11.8	49.0	008	W, −	1998	3/4"	***	a.k.a. h 3636
h 3642	04h 19.0m	−33° 54'	A–B	6.5	8.7	6.0	159	W, W	1998	2/3"	****	
h 3644	04h 42.5m	−25° 44'	AB–C	6.0	12.0	38.8	003	pY, −	1951	3/4"	***	A–B [a.k.a. β 744 – 7.26, 6.56, 0.3"] is too close for amateur instruments
			AB–D	6.1	8.2	44.6	041	pY, dY	1999	7×50	****	
51 Eridani	04h 37.6m	−02° 28'	A–B	5.2	11.8	29.1	073	pY, −	1998	3/4"	****	a.k.a. β 88
			A–C	5.2	11.2	66.5	163	pY, −	1998	3/4"	***	a.k.a. Wal 32

(continued)

Table 11.2. (continued)

Multiple stars:

Designation	R.A.	Decl.	Type	m$_1$	m$_2$	Sep. (")	PA (°)	Colors	Date/ Period	Aper. Recm.	Rating	Comments
55 Eridani	04^h 43.6^m	–08° 48'	A–B	6.7	6.8	9.4	317	dY, pY	2006	2/3"	****	Nearly equal in brightness and color. a.k.a. Σ 590
62 Eridani	04^h 56.4^m	–05° 10'	A–B	5.5	9.6	0.6	247	pY, pP	1991	16/18"	****	a.k.a. Hds 641 Aa
66 Eridani	05^h 06.8^m	–04° 39'	A–B	5.2	9.4	1.6	233	W, –	1999	6/7"	***	Fairly easy binocular double. a.k.a. Hub 5.
			A–C	5.1	10.7	53.1	010	W, –	1998	15×80	****	a.k.a. Σ 642
Σ 649	05^h 08.3^m	–08° 40'	A–B	5.8	9.0	21.6	068	W, rBr	2004	15×80	****	
			B–C	9.7	9.7	81.0	350	rBr, –	2000	2/3"	****	

Variable stars:

Designation	R.A.	Decl.	Type	Range (m$_v$)	Period (days)	F (f$_r$/f$_t$)	Spectral Range	Aper. Recm.	Rating	Comments
Z Eridani	02^h 47.9^m	–12° 28'	SRb	6.3	7.9	80	–	M4III	6×30	****	
RR Eridani	02^h 52.2^m	–08° 16'	SRb	6.8	8.2	97	–	M5III	6×30	****	
T Eridani	03^h 55.2^m	–24° 02'	M	7.4	13.2	252.29	0.45	M3e–M5e	8/10"	****	
W Eridani	04^h 11.5^m	–25° 08'	M	7.5	14.5	376.63	0.40	M7e	14/16"	****	
RZ Eridani	04^h 43.8^m	–10° 41'	EA/DS/ RS	7.8	8.7	39.282	0.05	A5–F5Vm + sgG8	7×50	****	Eclipsing binary with one variable star

Star clusters:

Designation	R.A.	Decl.	Type	m$_v$	Size (')	Brtst. Star	Dist. (ly)	dia. (ly)	Aper. Recm.	Rating	Comments
None											

Nebulae:

Designation	R.A.	Decl.	Type	m$_v$	Size (')	Cent. star	Dist. (ly)	dia. (ly)	Aper. Recm.	Rating	Comments
NGC 1535	04^h 14.2^m	–12° 44'	Planetary	9.6	21×17"	11.6	1,660	0.26	3/4"	*****	Bluish or greenish-blue disk with fainter halo *7
IC 2118	05^h 06.9^m	–07° 13'	Reflection	–	180×50	–	–	–	4/6"	****	Witch head Nebula. Illuminated by Rigel. Good photographic target

Galaxies:

Designation	R.A.	Decl.	m$_v$	Type	Size (')	Dist. (ly)	dia. (ly)	Lum. (suns)	Aper. Recm.	Rating	Comments	
NGC 1084	02^h 46.0^m	–07° 35'	10.7	SB(s)c II	2.8×1.4	51	50	16	6/8"	***	Fairly bright halo, elongated core, PA 040°	
NGC 1187	03^h 02.6^m	–22° 52'	10.8	SB(r)c I-II	4.2×3.2	58	71	13	8/10"	****	Large, faint, not concentrated. Elongated PA 135°. Very faint in 6"	
NGC 1232	03^h 09.8^m	–20° 35'	9.9	SAB(rs)c I	7.4×6.5	72	150	45	4/6"	****	Faint, diffuse in smaller scopes, bright in 12" or larger. Face-on spiral	
D487, NGC 1291	03^h 17.3^m	–41° 08'	8.5	(R)SB(s)0	11.0×9.5	29	93	27	4/6"	***	Stellar nucleus, bright core, faint halo	*9
NGC 1300	03^h 19.7^m	–19° 25'	10.4	SB(rs)bc I	6.2×4.1	52	110	21	4/6"	****	Large, faint, diffuse disk, Elongated PA 100°. One of the best face-on barred spirals	
NGC 1332	03^h 26.3^m	–21° 20'	10.3	S(s)-; sp	4.5×1.4	64	84	25	4/6"	****	Stellar nucleus, bright core, faint halo. Elongated PA 110°	
NGC 1407	03^h 40.2^m	–18° 35'	9.7	E0	4.6×4.3	75	100	59	12/14"	***	Elongated NW–SE, diffuse halo with little central brightening	
NGC 1421	03^h 42.5^m	–13° 29'	11.4	SAB(rs)bc	3.5×0.9	90	92	18	8/10"	****	Edge-on, narrow core ~1' long, moderately bright halo. Elongated PA 000°	
NGC 1531	04^h 12.0^m	–32° 51'	12.5	S0- pec	1.3×0.7	47	18	1.8	8/10"	***	Diffuse halo, slightly elongated PA 135°. 1.8' NE of NGC 1532	
D600, NGC 1532	04^h 12.1^m	–32° 52'	9.8	SB(s)b pec	11.2×3.4	44	140	19	8/10"	****	Mod bright halo, distinct core. Same LP FOV as NGC 1531. Elongated PA 035°	*10

Meteor showers:

Designation	R.A.	Decl.	Period (yr)	Type	Duration	Max Date	Max ZHR (max)	Dist. (ly)	Comet/ Asteroid	First Obs.	Vel. (mi/ km/sec)	Rating	Comments
None													

Other interesting objects:

Designation	R.A.	Decl.	m$_v$	Type	Mass (suns)	Dia. (suns)	Dist. (ly)	Planet Mass (J)	Dist. (AU)	Period (days)	Eccentricity	Comments	
ε Eridani	03^h 32.9^m	–09° 28	3.7	ExoPlanet	0.824	0.85	10.5	1.06	3.38	2,500	0.25	Third closest unaided eye star system, sixth overall	*11

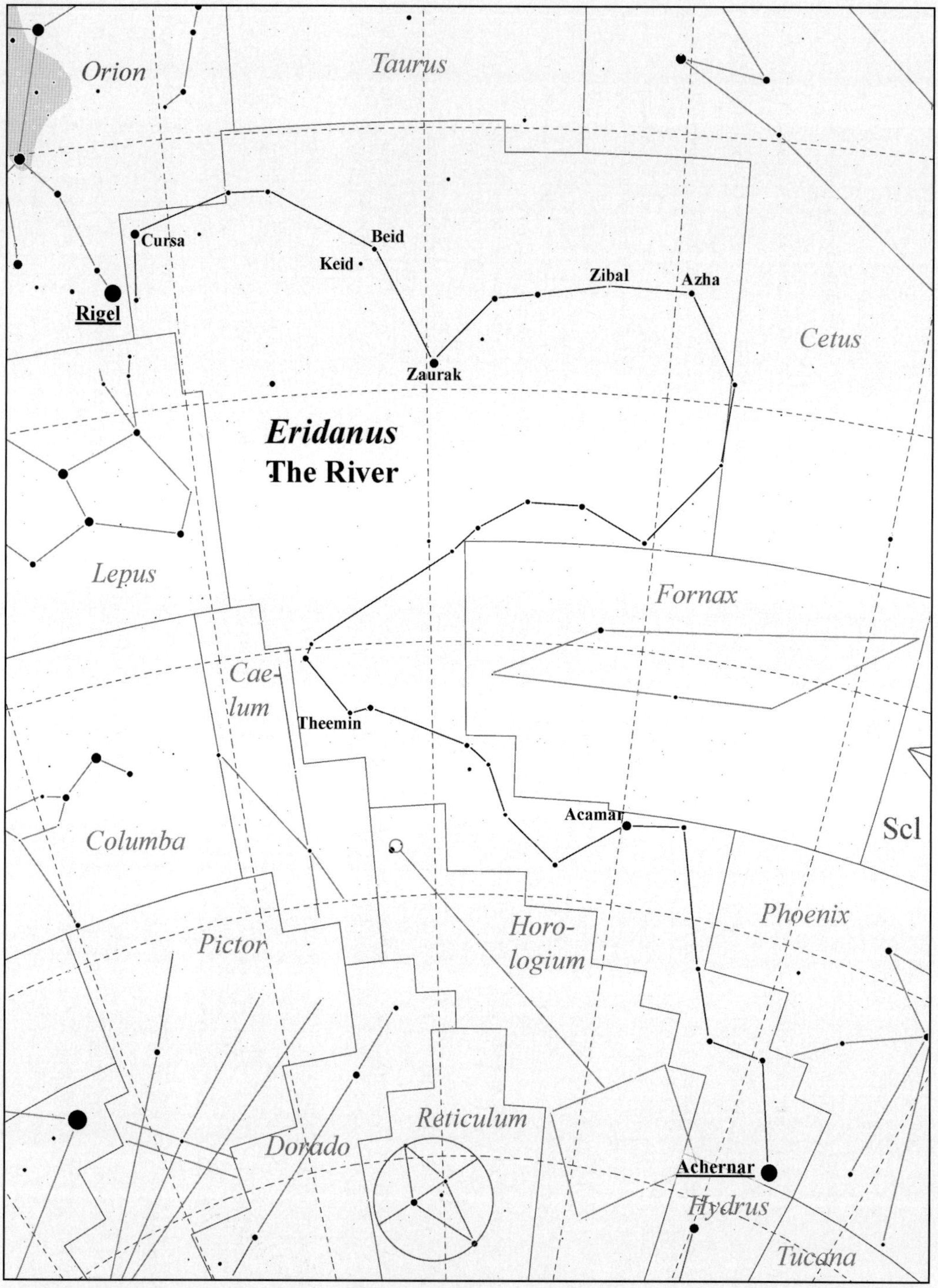

Orion   *Taurus*

Cursa   Beid
                   Keid
                                           Zibal   Azha
Rigel   *Cetus*

                              Zaurak

                      Eridanus
                      The River

Lepus   *Fornax*

               Cae-
               lum
                      Theemin

Columba   Acamar
                                                              Scl

                                                      Phoenix
                      Horo-
Pictor   *logium*

                              Reticulum
        Dorado   Achernar

                                        Hydrus
                                               Tucana

Fig. 11.7. Eridanus constellation, entire.

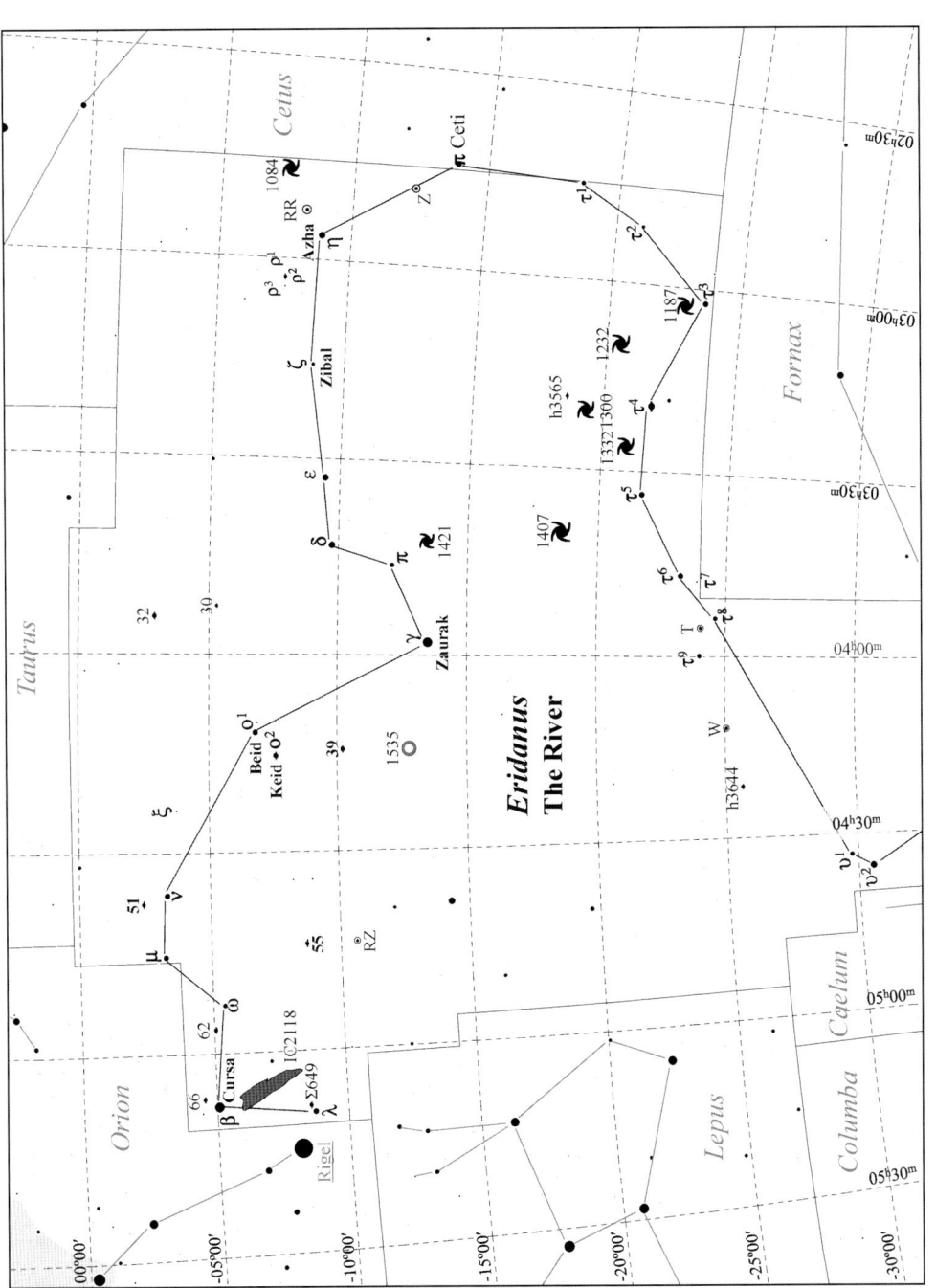

Fig. 11.8. Eridanus details, north.

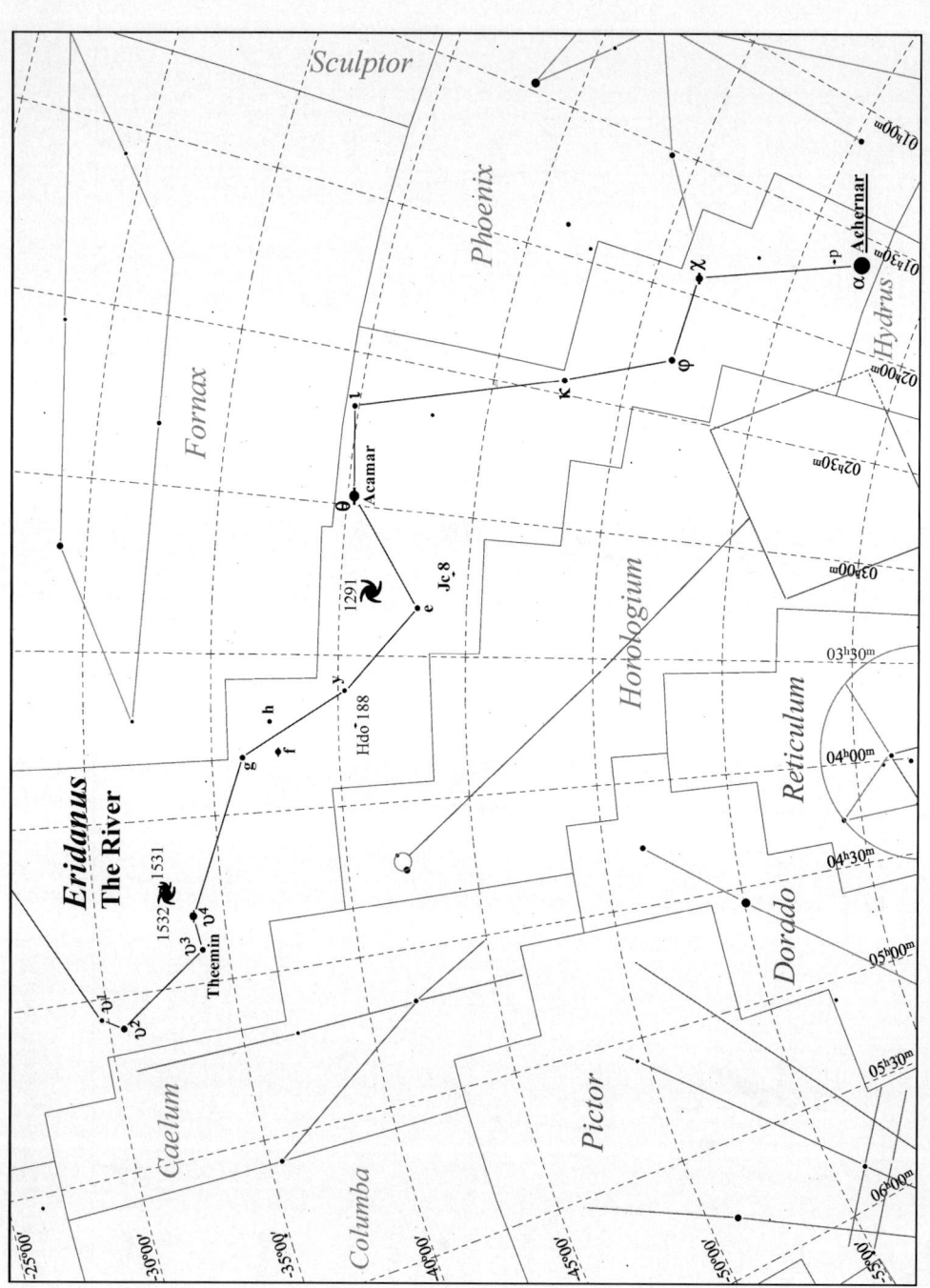

Fig. 11.9. Eridanus details, south.

Fig. 11.10. NGC 2118, Witch head Nebula in Eridanus. Astrophoto by Rob Gendler.

NOTES:

1. RR Lyrae is the prototype of its class, which is also called "cluster variables." These variables have observational characteristics similar to the "classical cepheids" and were first thought to be just classical cepheids with shorter periods and lower luminosity. They were thought to follow the same period–luminosity relationship. It was later found that the RR Lyrae stars were all about the same luminosity, regardless of the variational period.

2. The orbital period of the Lyrids appears to be different from Comet Thatcher, but the Lyrid period may be in error. Meteors are observed almost every year, but a storm of ~700/h observed on April 10–20, 1803 may be from the Lyrids, and other showers of more than 100/h have been observed since 1800.

3. The η Lyrids were predicted by Jack Drummond of Stewart Observatory on May 9, 1983 based on the discovery of Comet IRAS–Araki–Alcock and the fact that its orbital path passed close to the earth. The shower was confirmed by several observers. A review of historical records identified a possible early sighting from H. Corder on May 1–4, 1895. His records indicated a radiant at 18 h 32 m, +43°.

4. While camping in the San Bernardino mountains on June 15, 1966, Stan Dvorak plotted 13 meteors in 1.5 h from a radiant estimated to be 18 h 32 m, +30°. A few hours later, F. W. Talbot of Cheshire, England independently discovered the same shower and estimated the radiant to be 18 h 22 m, +30°. Moderate levels of activity continued until 1969, when about 9/h were observed. Since then, activity has been mostly minor. A comet association has not been established, but a similarity with the orbit of Comet Mellish (1915 II) has been noted.

5. Sigma (σ) appears to be one of the truly "lost" stars. Bayer assigned the letter, and several astronomers have tentatively identified various stars as being candidates, but all such candidates are either too faint or too far from Bayer's position to be acceptable.

6. Σ518, also called o² BC, is a rather faint but interesting binary. It has just passed its maximum separation of 9.21″ in mid-2004, and is now closing to a minimum of 1.53″ in 2085.

7. Bayer assigned the designation of σ Eridani to a star which Ptolemy had included in his catalog as magnitude 4. Other astronomers have been unable to find a 4th magnitude star at Bayer's location. The star HR 859 is at the right location but has a magnitude of only 6.32. Several other examples of Bayer's ability to see very faint stars are known.

8. NGC 1535 is an often overlooked planetary nebula. While the main part of the disk can easily be seen in a 3″–4″ telescope, a 6″–8″ telescope is needed to see the outer part of the disk and the central star. In a small telescope, it often resembles a ghostly Neptune, illustrating why early observers called this type of object "planetary" nebulae.

9. D487 is one of the entries in the catalog created by James Dunlop, who is sometimes called the "Messier of the Southern Skies" for his work in observing and cataloging both double stars and deep sky objects. Born in Scotland in 1793, Dunlop was hired by Sir Thomas Brisbane, then the Governor of New South Wales, as an observer in Brisbane's private observatory in 1821. When Brisbane returned to Scotland, he donated some of his instruments to Dunlop. Brisbane later hired Dunlop to work at his observatory there, and Dunlop was awarded the RAS Gold Medal in 1828. He returned to Australia in 1831 as superintendent of Brisbane's former observatory, now operated by the government, and remained in Australia until his death in 1848.

10. NGC 1532 and 1531 resemble a much smaller and fainter version of M31 and M32 for southern observers.

11. Because of its similarity to our sun, and its nearness, ε Eridani has figured prominently in searches for extraterrestrial life and exoplanets. It has also been used as the setting for science fiction books, movies, television shows, and computer games.

CHAPTER TWELVE

LEDA AND HER LOVER

September, February

CYGNUS	GEMINI
The Swan	The Twins

Leda was the beautiful daughter of Thestius and Eurythemis, the king and queen of Aeotolia, a region of northwestern Greece. After King Tyndareus was deposed and expelled from Lacedaemon, he took refuge at the court of his friend Thestius. Eventually, Tyndareus wooed and wed Leda.

In those days, to be a beautiful woman meant you were in constant danger of attracting the attention of the gods, especially their philandering leader, Zeus. This attention could be welcome or unwelcome, but was nearly always irresistible. It could also be beneficial or disastrous to the human involved, usually through no fault of the human.

In this case, Zeus caught sight of Leda and was immediately attracted to her. He knew that she was faithful to her husband, and that it would not be easy to approach her, so he devised one of his sneakiest plots. Knowing that Leda was fond of animals and always kind to them, he disguised himself as a beautiful white swan. He then had the eagle, which normally carried his thunderbolts in his claws, pursue him. Flying over Leda and catching her attention, Zeus then pretended to be injured and landed near her.

Leda immediately comforted the swan, and drove the eagle away. Zeus acted very tame and friendly, and stayed on to allow his supposedly broken wing to heal. During

P. Simpson, *Guidebook to the Constellations: Telescopic Sights, Tales, and Myths*, Patrick Moore's Practical Astronomy Series, DOI 10.1007/978-1-4419-6941-5_12, © Springer Science+Business Media, LLC 2012

this time, he was always near Leda, and very affectionate. Finally, Zeus saw his chance and seduced Leda, still maintaining the form of a swan. He then departed, without Leda ever knowing his true identity.

It soon became apparent that Leda was pregnant, but nothing was thought suspicious about this until the time came for her to give birth. Instead of a normal child, Leda delivered two giant eggs. These eggs soon hatched and revealed two pair of nonidentical twins. In one egg, there was Polydeuces (better known by his Roman name, Pollux), and his sister Helen. In the other egg were Castor and his sister Clytemnestra. Castor and Clytemnestra were the children of Tyndareus, while Pollux and Helen were the children of Zeus. All four played significant roles in later Greek mythology. Castor became a famous horseman and Pollux became an expert boxer. Together they participated in many adventures, including the voyage of the Argonauts. Clytemnestra married the ill-fated Tantalus and, after his death, Agamemnon. Helen grew up to be the most beautiful woman in the known world, and by abandoning her husband Menelaus to flee with Paris, became the cause of the Trojan war. Hers was "the face that launched a thousand ships," and started a 10-year war between the Greeks and the Trojans.

As they grew up, Castor and Pollux were inseparable. They became close friends with their twin cousins Idas and Lynceus. The two cousins even joined Castor and Pollux on the expedition of the Argonauts to recapture the golden fleece. Their friendship suffered a great strain when Castor and Pollux kidnapped the intended brides of Idas and Lynceus and married the girls themselves. Later they patched up their differences enough to join in a raid to steal a herd of cattle.

After this raid, Idas was chosen to divide the spoils. He set out four portions of meat and ruled that half the cattle would go the person who finished eating his portion first, and the remainder to the one who finished second. While the meat was being divided, and before the others were ready to begin, Idas bolted his portion, and then helped Lynceus finish his before either Castor or Pollux could finish.

Feeling cheated, Castor and Pollux waited until Idas and Lynceus were away, and stole the cattle for themselves. When they heard that their cousins had returned and were planning a retaliation raid, they hid and lay in wait to ambush them. The ambush was foiled when the sharp-eyed Lynceus spotted them, and a pitched battle ensued. Castor and Lynceus were killed, and Pollux severely injured. Before Idas could dispatch Pollux, Zeus stepped in to protect his son, and slew Idas with a thunderbolt.

As the son of Zeus, Pollux was immortal. However, he pleaded with Zeus to either return Castor to life, or to allow him to join Castor in the underworld. Even the mighty Zeus could not return people from the underworld realm of his brother Hades, but he did negotiate a compromise in which the two could be together again, with both spending half their time in the heavens and half in the underworld. This is why we see the twins together in the sky, holding hands so they can never be separated again, but above the earth only for half the year and below the earth and invisible for the other half.

Zeus also placed the figure of his swan disguise in the sky as the constellation Cygnus to commemorate his affair with Leda, and the outstanding children which resulted from this liaison.

CYGNUS (**sig**-nus)	**The Swan**
CYGNI (**sig**-nee)	**9:00** P.M. **Culmination:** Sep. 15
Cyg	**AREA:** 804 sq. deg., 16th in size

TO FIND: Locate the "Summer Triangle" by finding the summer Milky Way. Almost straight overhead in mid-northern latitudes, you should see a bright blue-white star in the middle of the milky way. This is <u>Deneb</u>. About 25° to the WSW, you should see and even brighter white star, <u>Vega</u>. Turning 90° southward from the <u>Deneb</u> to <u>Vega</u> line, about 35° away will be bright yellowish white <u>Altair</u> on the eastern edge of the milky way. <u>Deneb</u>, <u>Vega,</u> and <u>Altair</u> form an asterism called the Summer Triangle. Using the chart, you can trace the body and long neck of the swan SW through the middle of the Summer Triangle to Albireo. The rest of Cygnus can then be found by comparing the chart to the sky. At the same time, you can also locate the Northern Cross by finding δ Cygni to the northwest of γ Cygni, and ε Cygni to the Southeast. These three stars represent the horizontal bar and the line of stars from Deneb to Albireo represent the vertical bar of the cross (Figs. 12.1–12.3).

KEY STAR NAMES:

	Name	Pronunciation	Source
α	Deneb	**den**-eb	Al Dhanab al Dajajah, Arabic, "The Hen's Tail"
	Arided	**ah**-rih-dead	Al Ridhadh, Arabic, uncertain meaning, applied to the whole constellation by J. Scalinger, sixteenth century French scholar
	Aridif	**ah**-ri-diff	Al Ridf, Arabic, "The Hindmost"
β	Albireo	al-**bir**-ee-oh	Ab Ireo, misunderstood Latin phrase from the 1515 Almagest
γ	Sadr	**sad**-er	Al Sadr al Dajajah, Arabic, "The Hen's Breast"
ε	Gienah	**gee**-nuh	Al Janah, Arabic, "The Wing"
o¹,²	Ruchba	**ruhk**-buh	Al Rukbah al Dajajah, Arabic, "The Hen's Knee"
π¹	Azelfafage	ah-**zel**-fuh-faj	Al-sulahfat, scientific Arabic, "The Tortoise"

Magnitudes and spectral types of principal stars:

Bayer desig.	Mag. (m_v)	Spec. type	Bayer desig.	Mag. (m_v)	Spec. type	Bayer desig.	Mag. (m_v)	Spec. type
α	1.25	A2 Ia	β	3.05	K3 II	γ	2.23	F8 Ib
δ	2.86	B9.5 III	ε	2.48	K0 III	ζ	3.21	G8 II
η	3.89	K0 IIIvar	θ	4.49	F4 V	ι	3.76	A5 V
κ	3.80	K0 III	λ	4.53	B6 IV	μ¹	4.49	F6 V
μ²	6.08	G2 V	ν	3.94	A1 V	ξ	3.72	K5 Ib
o¹	3.80	K2 II	o¹	3.96	K3 Ib	π¹	4.69	B3 IV
π²	4.23	B3 III	ρ	3.98	G8 III	σ	4.22	b9 Iab
τ	3.74	F1 IV	υ	4.69	G8 III/IV	φ	4.68	G8 III/IV
χ	4.23	K0 III	ψ	4.91	A4 V	ω¹	4.94	B2.5 IV
ω²	5.44	M2 III	b¹	5.38	K0 IV	b²	4.93	B2.5 Ve
b³	4.93	A2 V	c	5.03	K3 III:var	e	5.06	K1 II/III
A	5.04	O8e	P	4.77	B2pe			

Stars of magnitude 5.5 or brighter which have Flamsteed, but no Bayer designations:

Flam-steed.	Mag. (m_v)	Spec. type	Flam-steed	Mag. (m_v)	Spec. type	Flam-steed	Mag. (m_v)	Spec. type
2	4.99	B3 IV	4	5.17	B9sp	8	4.74	B3 IV
9	5.39	A0 V	14	5.41	B9 V	15	4.98	G8 III
19	5.18	M2 IIIa	22	4.93	B2.5 V	23	5.14	B5 V
25	5.15	B3 IV	30	4.80	A5 IIIn	33	4.28	A2 IV/V:var
35	5.14	F5 Ib	39	4.43	K3 III	51	5.38	K0 IV
52	4.22	K0 III	55	4.81	B3 Ia	56	5.06	A4me
57	4.80	B5 V	60	5.38	B1 V	61	5.20	K5 V
70	5.30	B3 V	72	4.87	K1 III	74	5.04	A5 V
75	5.09	M1 III:var						

Stars of magnitude 5.5 or brighter which have no Bayer or Flamsteed designations:

Desig-nation	Mag. (m_v)	Spec. type	Desig-nation	Mag. (m_v)	Spec. type
HD 184875	5.34	A2 V	HD 185351	5.17	K0 III
HD 189178	5.46	B5 V	HD 193092	5.27	K5 II
HD 199098	5.48	K0 II	HD 199101	5.47	K5 III
HD 204411	5.29	A6pe			

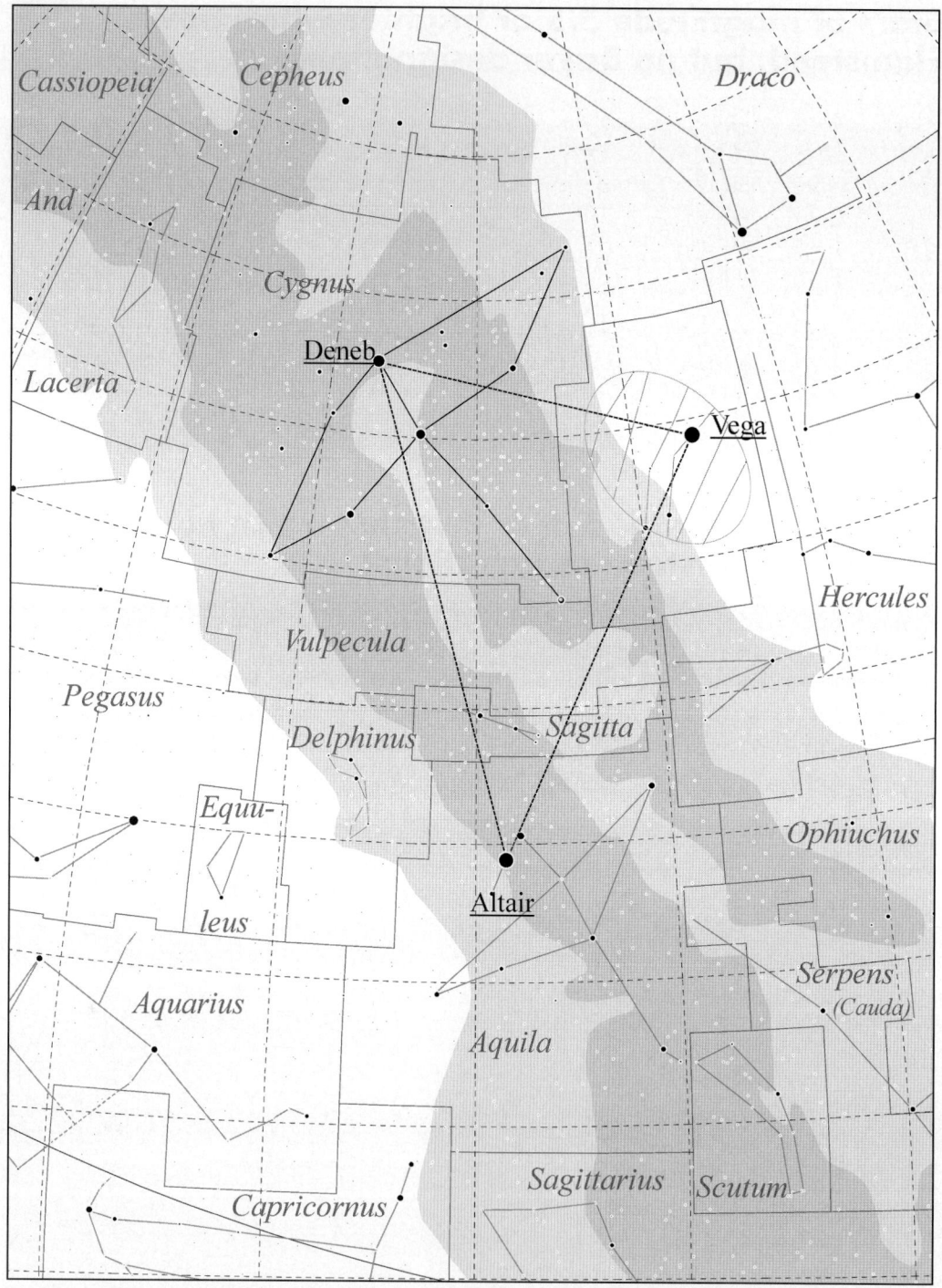

Fig. 12.1. The Summer Triangle.

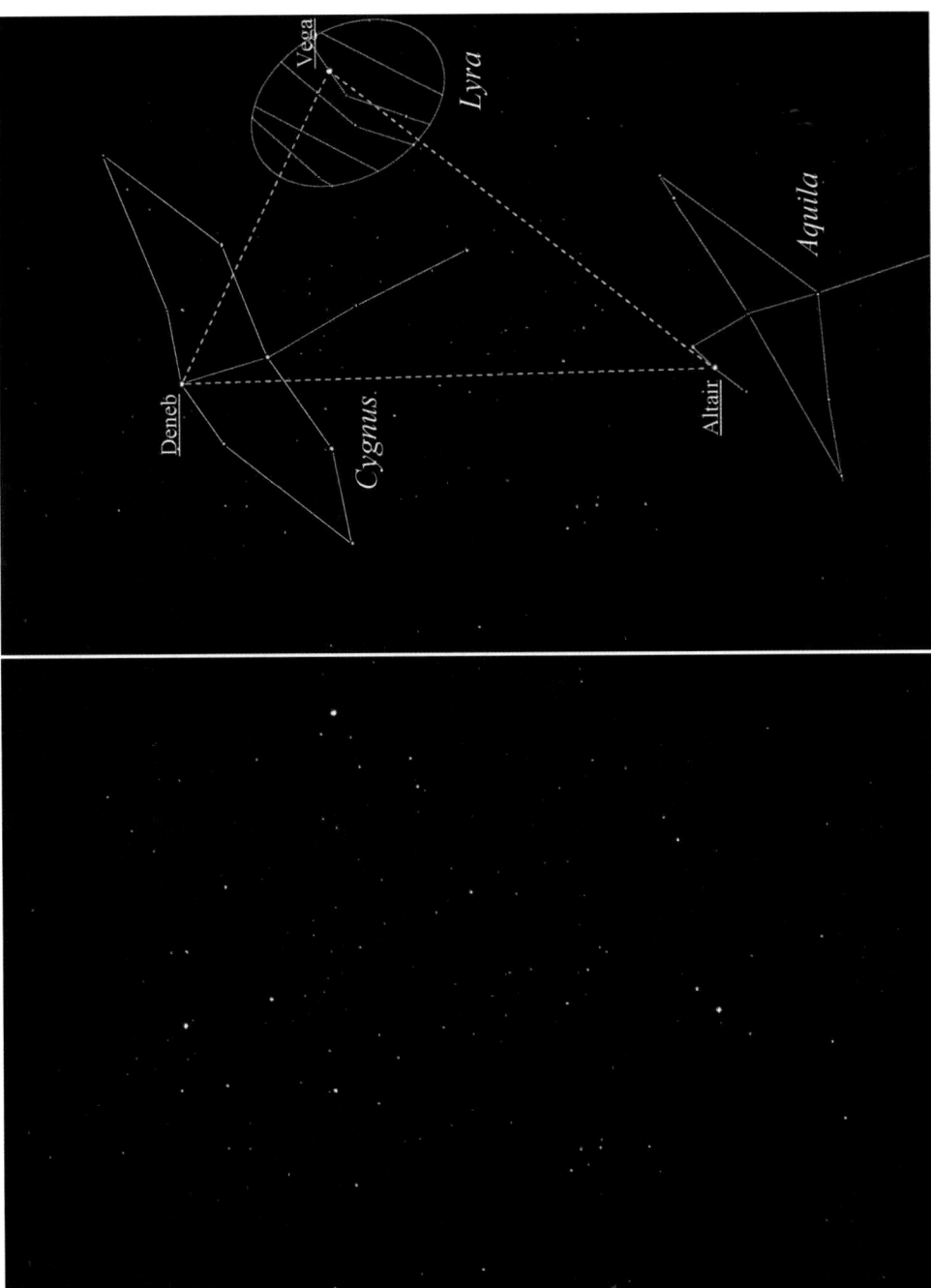

Fig. 12.2. (a, b) The stars and constellations of the Summer Triangle.

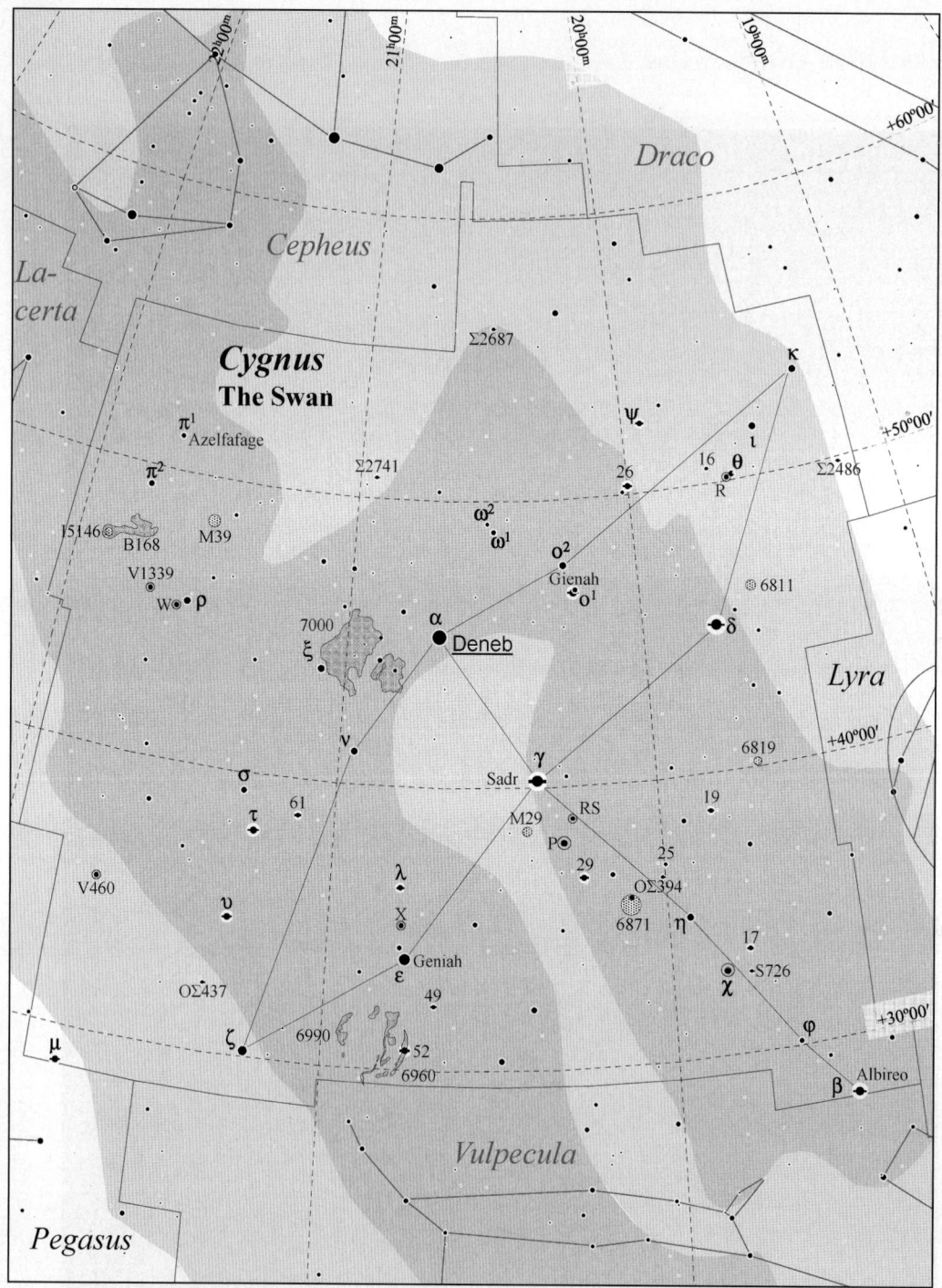

Fig. 12.3. Cygnus details.

Table 12.1. Selected telescopic objects in Cygnus.

Multiple stars:

Designation	R.A.	Decl.	Type	m₁	m₂	Sep. (")	PA (°)	Colors	Date/ Period	Aper. Recm.	Rating	Comments
Σ2486	19ʰ 12.1ᵐ	+49° 51′	A–B	6.5	6.7	**7.33**	205	Y, Y	**3,100**	2/3″	****	Showcase pair. One of the best. May be the most observed double. a.k.a. OΣ43
Albireo, β Cygni	19ʰ 30.7ᵐ	+27° 58′	A–B	3.2	4.7	35.3	054	gY, B	2006	7×50	*****	Showcase pair. Plan. Neb. NGC 6826 is in same LP FOV. a.k.a. Σ2579
16 Cygni	19ʰ 41.8ᵐ	+50° 32′	A–B	4.0	6.0	**39.68**	133	Y, dY	**13,512.7**	6×30	****	
18, 8 Cygni	19ʰ 45.0ᵐ	+45° 08′	A–B	2.9	6.3	**2.68**	220	W, bW	**780.27**	3/4″	*****	More challenging and overshadowed by Albireo, but still a great sight. a.k.a. Σ2579
17 Cygni	19ʰ 46.4ᵐ	+33° 44′	A–B	5.1	9.3	26.3	067	oY, pB	2001	2/3″	****	Nice colored pair. a.k.a. S2580
S726	19ʰ 46.6ᵐ	+32° 53′	A–B	6.2	9.2	29.2	192	oY, pB	2003	15×80	*****	Very similar to 17 Cyg, but primary is ~1 mag. fainter
19 Cygni	19ʰ 50.6ᵐ	+38° 43′	A–B	5.4	10.5	56	115	R, B	2005	15×80	*****	Wide, Mars-like red color primary
24, ψ Cygni	19ʰ 55.6ᵐ	+52° 26′	A–B	5.0	7.5	3.1	174	pY, dB	2006	3/4″	****	Lovely
OΣ394	20ʰ 00.2ᵐ	+36° 25′	A–B	7.1	10.3	10.9	295	O, B	1998	2/3″	****	An Albireo-like pair for medium to large amateur scopes
26 Cygni	20ʰ 01.4ᵐ	+50° 06′	A–B	5.2	8.9	41.5	150	rO, pV	2005	7×50	****	a.k.a. H5 47
29 Cygni	20ʰ 14.5ᵐ	+36° 48′	A–B	5.0	6.7	215.8	153	bW, yO	2002	6×30	****	Wide, colored pair for binoculars. a.k.a. ENG 72
31, o¹ Cygni	20ʰ 15.5ᵐ	+47° 43′	A–B	4.2	8.4	208.4	177	oY, B	2002	7×50	****	a.k.a. S 743
37, γ Cygni	20ʰ 22.2ᵐ	+40° 15′	A–B	2.2	9.6	140.4	196	yW, ?	1980	15×80	****	a.k.a. Sadr and BU 665
			A–D	10.0	11	1.9	302	yW, ?	1929	5/6″	***	
Σ2687	20ʰ 26.4ᵐ	+56° 38′	A–B	6.4	8.31	27.0	117	Y, pBGr	2006	15×80	****	
49 Cygni	20ʰ 41.0ᵐ	+32° 18′	Aa–B	5.6	8.1	3.0	046	yO, W	2003	3/4″	****	Nice tight pair. Closeness accentuates color contrast. A–a sep. is only 0.2″. a.k.a. Σ2716
52 Cygni	20ʰ 45.7ᵐ	+30° 43′	Aa–C	5.6	11.8	67.3	093	dY, ?	1998	3/4″	***	
			A–B	4.2	8.7	6.4	069	O, B	1997	2/3″	****	
54, λ Cygni	20ʰ 47.4ᵐ	+36° 29′	Aa–B	4.5	6.3	**0.9**	**002**	pB, pB	**391.3**	10/11″	****	a.k.a. Σ2726
Σ2741	20ʰ 58.5ᵐ	+50° 28′	A–B	5.9	6.8	2.0	026	bW, bW	2006	4/5″	***	a.k.a. OΣ413
61 Cygni	21ʰ 06.9ᵐ	+38° 45′	A–B	5.4	6.1	**31.35**	**152**	rO, rO	**659**	15×80	******	Showcase pair. Easy, 13th closest star system, high PM, only 11.2 ly distant

(continued)

Table 12.1. (continued)

Multiple stars:

Designation	R.A.	Decl.	Type	m₁	m₂	Sep. (")	PA (°)	Colors	Date/Period	Aper. Recm.	Rating	Comments
65, τ Cygni	21ʰ 14.8ᵐ	+38° 03′	A–B	3.8	6.6	0.7	274	γW, γW	2004	13/14″	***	Easy triple for small scopes, matched pair of companion stars
			AB–C	3.8	12.2	29.5	228	γW, ?	1914	4/5″	***	
			AB–D	3.8	9.9	205.4	208	γW, ?	2004	15×80	***	
66, υ Cygni	21ʰ 17.9ᵐ	+38° 45′	A–B	4.4	10.0	15.1	220	bW, bW	1998	2/3″	****	
			A–C	4.4	10.0	21.2	183	bW, bW	1998	15×80	****	
OΣ437	21ʰ 20.8ᵐ	+32° 26′	A–B	7.2	7.42	2.3	022	Y, B	2003	4/5″	****	Faint, but beautiful
78, μ Cygni	21ʰ 44.1ᵐ	+28° 45′	A–B	4.8	6.2	**1.66**	**316**	W, Y	**789**	5/6″	****	a.k.a. Σ2822
			AB–C	4.5	11.5	72.6	290	W, ?	1999	2/3″	***	
			A–D	4.5	6.9	197.5	045	W, ?	2002	6×30	***	

Variable stars:

Designation	R.A.	Decl.	Type	Range (mᵥ)		Period (days)	F (fᵣ/fₜ)	Spectral Range	Aper. Recm.	Rating	Comments
R Cygni	19ʰ 36.8ᵐ	+50° 12′	M	6.1	14.2	426.45	0.35	S3.9e–S6	14/16″	****	8.1 magnitude variation or ~1,700 times in brightness. ##$
V1339 Cygni	19ʰ 42.1ᵐ	+45° 46′	SRb	5.9	7.1	35	–	M3–M6	6×30	****	##$
χ Cygni	19ʰ 50.6ᵐ	+32° 55′	M	3.3	14.2	406.9	0.41	S6IIe–S10	14/16″	*****	Extremely large range, 10.9 magnitude variation or ~23,000 times! ##$ *1
RS Cygni	20ʰ 13.4ᵐ	+38° 44′	SRa	6.5	9.3	417.39	–	N0pe(C8 2e)	2/3″	****	Spectral class N0, strong carbon absorption gives it a very red color. ##$
P Cygni	20ʰ 17.8ᵐ	+38° 02′	S Dor	3.0	6.0	296.5	–	DB1 Iapec	6×30	****	Very luminous blue giant. ##
X Cygni	20ʰ 43.4ᵐ	+35° 35′	Cδ	5.9	6.9	16.38	0.35	F7Ib–G8Ib	6×30	****	##$
W Cygni	21ʰ 36.0ᵐ	+45° 23′	SRb	5.7	7.0	131.1	0.50	M4e–M6e	6×30	****	##$
V460 Cygni	21ʰ 42.0ᵐ	+35° 31′	SRb	5.6	7.0	–	–	N1(C6, 5)	6×30	****	#$

Star clusters:

Designation	R.A.	Decl.	Type	Size (′)	mᵥ	Brtst. Star	Dist. (ly)	dia. (ly)	Number of Stars	Aper. Recm.	Rating	Comments
NGC 6811	19ʰ 37.2ᵐ	+46° 23′	Open	15	6.8	11.0	3,500	15	250	8/10″	***	Moderately rich, small brightness range, fairly even distribution, detached *2
NGC 6819	19ʰ 41.3ᵐ	+40° 11′	Open	5	7.3	11.5	7,600	11	150	8/10″	****	Rich concentration of faint stars, m>11.5

Designation	R.A.	Decl.	Type	m_v	Size (')	Brst. star	Dist. (ly)	dia. (ly)	No.	Aper. Recm.	Rating	Comments
NGC 6871	20h 06.5m	+35° 47'	Open	5.2	30	6.8	5,400	47	66	8/10"	***	Few stars, slight central concentration, moderate brightness range *3
M29, NGC 6913	20h 23.9m	+38° 32'	Open	6.6	10	8.6	4,400	13	80	4/6"	*****	Compact group of 8–10 bright stars with many fainter ones intermingled
M39, NGC 7092	21h 32.2m	+48° 26'	Open	4.6	31	6.8	800	31	25	8/10"	*****	~20 bright stars in triangular shape, fills low-medium power eyepiece
IC 5146	21h 53.4m	+47° 14'	Open	7.2	9	9.6	3,000	20	110	4/6"	***	a.k.a Col 470. Cluster has nebulosity involved *4

Nebulae:

Designation	R.A.	Decl.	Type	m_v	Size (')	Brst./Cent. star	Dist. (ly)	dia. (ly)	Aper. Recm.	Rating	Comments
NGC 6826	19h 44.8m	+50° 31'	Planetary	8.8	25"	10.6v	5,100	0.6	8/10"	*****	"The Blinking Planetary." In same LP VOV as double star 16 Cygni *5
NGC 6888	20h 12.0m	+38° 21'	Emission	10	19×9	7.4	5,000	28	6/8"	****	The Crescent Nebula" Illuminating star is a hot Wolf-Rayet star
Veil Nebula	20h 51.5m	+31° 00'	SNR	7.0	2.75°	–	1,860	90	4/6"	******	The next 7 lines are segments of the large, complex Veil nebula *6
NGC 6960	20h 45.7m	+30° 43'	SNR	–	70×6	–	1,860	40	18×80"	****	Veil Nebula, W part of Great Cygnus Loop. Requires v. dark skies
Pickering's Triangular Wisp	20h 48.5m	+31° 09'	SNR	–	45×30	–	1,860	25	8/10"	***	Very faint Veil Nebula segment. W of NGC 6960, with long tail to SSE
NGC 6974	20h 50.8m	+31° 52'	SNR	–	6×4	–	1,860	3	8/10"	***	Extends from SE edge, and semidetached from NGC 6879
NGC 6979	20h 51.0m	+32° 09'	SNR	–	7×3	–	1,860	4	8/10"	***	Mostly north segment of Great Cygnus Loop. Semidetached from NGC 6874
IC 1340	20h 56.2m	+31° 04'	SNR	–	25×20	–	1,860	14	18×80"	***	Joined to NGC 6992 and 6995
NGC 6992	20h 56.4m	+31° 43'	SNR	–	60×8	–	1,860	33	18×80"	****	Veil Nebula, E part of Great Cygnus Loop. Requires dark skies
NGC 6995	20h 57.1m	+31° 13'	SNR	–	12	–	1,860	7	18×80"	****	Continuation of NGC 6992, Use OIII filter to improve view
IC 5067	20h 50.8m	+44° 21'	Emission	–	25×10	–	4,000	30	4/6"	**	Together, IC 5067 and 5070 form the "Pelican Nebula" *7
IC 5070	20h 58.8m	+44° 20'	Emission	–	80×70	–	4,000	90	4/6"	**	

(continued)

Table 12.1. (continued)

Nebulae:

Designation	R.A.	Decl.	Type	m_v	Size (')	Brst./ Cent. star	Dist. (ly)	dia. (ly)	Aper. Recm.	Rating	Comments
NGC 7000	20ʰ 58.8ᵐ	+44° 20'	Emission	–	120×100	–	4,000	140	7×50	****	"North America Nebula." View is best in RFT with UHC filter *8
B168	21ʰ 53.2ᵐ	+47° 12'	Dark	–	121×65	–	4,000	140	8/10"	***	IC 5146 lies in, and illuminates eastern end
IC 5146	21ʰ 53.4ᵐ	+47° 16'	Emission	–	10×10	–	4,000	12	8/10"	***	"Cocoon Nebula." Resembles a faint Triffid Nebula (M20 in Sagittarius)

Galaxies:

Designation	R.A.	Decl.	Type	m_v	Size (')	Dist. (ly)	dia. (ly)	Lum. (suns)	Aper. Recm.	Rating	Comments
None											

Meteor showers:

Designation	R.A.	Decl.	Type	Period (yr)	Duration	Max Date	ZHR (max)	Comet/ **Asteroid**	Vel. (mi/ **km**/sec)	Rating	Comments
None											

Other interesting objects:

Designation	R.A.	Decl.	Type	m_v	Size (')	Brst./ Cent. star	Dist. (ly)	dia. (ly)	Aper. Recm.	Rating	Comments
V1357, Cygnus X-1	19ʰ 58.4ᵐ	+35° 12'	X-ray	8.9			8,000			*****	~13' ESE of η, Cygni *9

Fig. 12.4. M29, open cluster in Cygnus. Photograph by Mark Komsa.

Fig. 12.5. M39, open cluster in Cygnus. Photograph by Mark Komsa.

Fig. 12.6a. Veil Nebula, entire. Photograph by Mark Komsa.

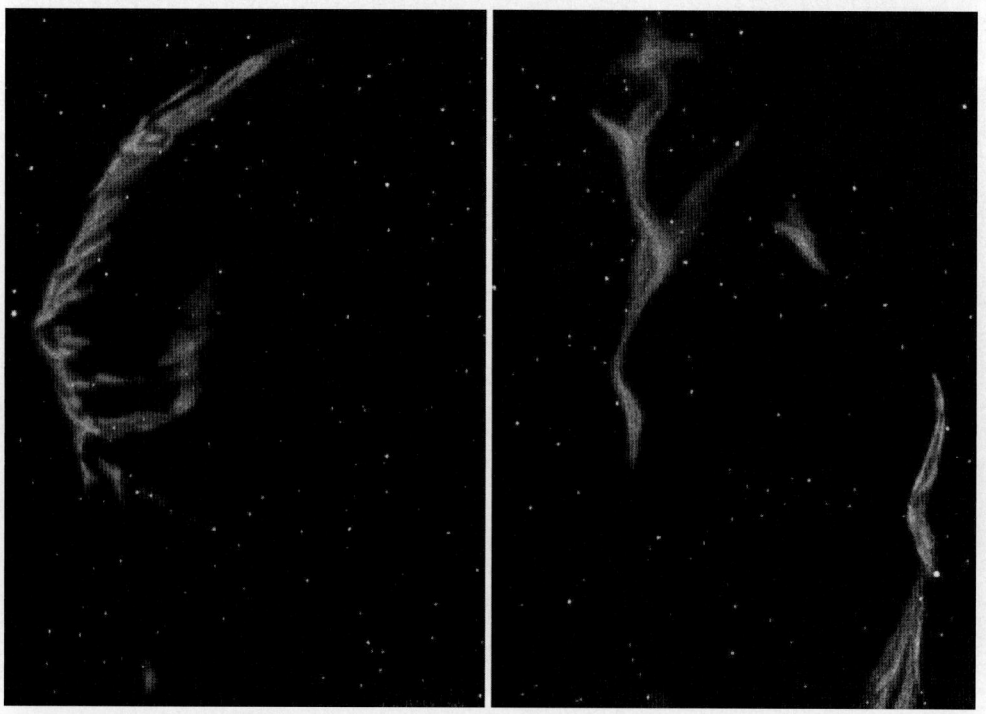

Fig. 12.6. (b) Veil Nebula West (NGC 6992) and **(c)** Veil Nebula East (NGC 6960). Sketches by Nicolas Biver, 10″ telescope at ×42.

Fig. 12.7. North America Nebula (NGC 7000) and Pelican Nebula (9C 5067 and 5070). Astrophoto by Rob Gendler.

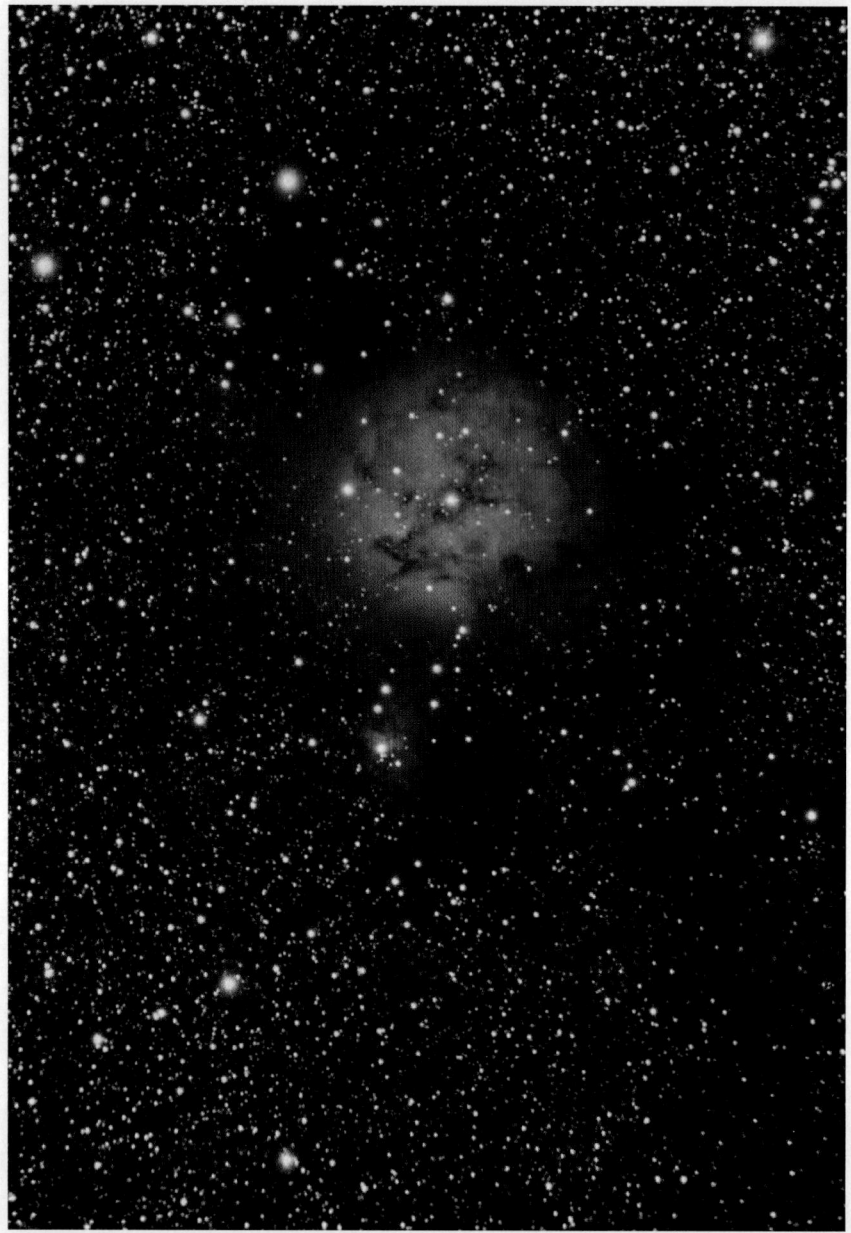

Fig. 12.8. Cocoon Nebula (IC 5146) in Cygnus. Astrophoto by Warren Keller.

Fig. 12.9. The Crescent Nebula (NGC 6888) in Cygnus. Photograph by Rob Gendler.

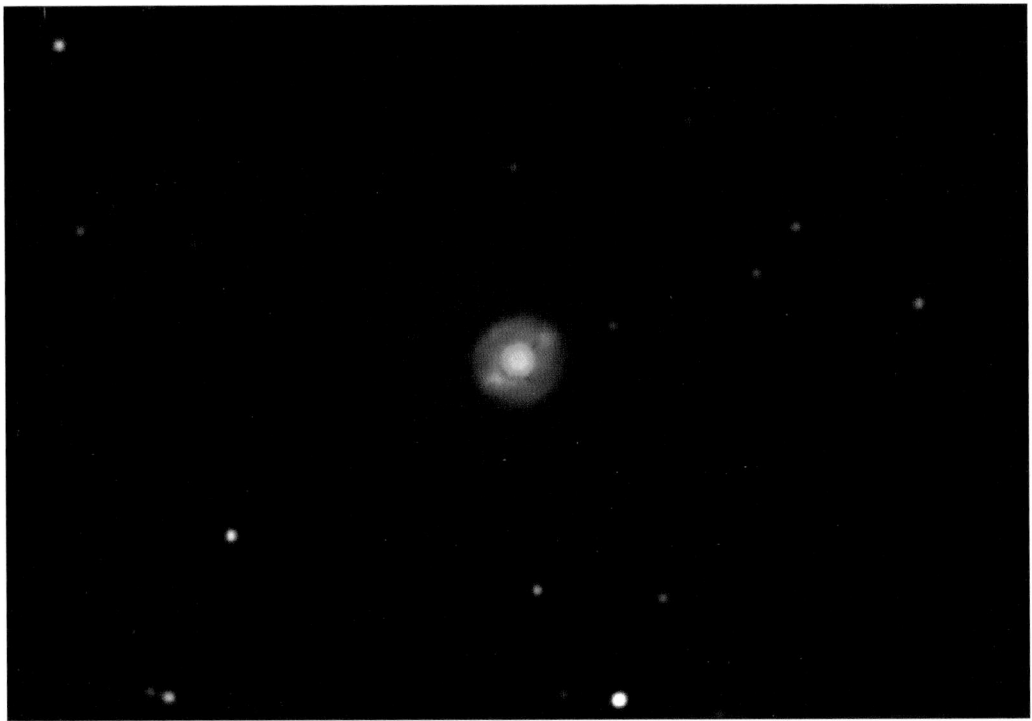

Fig. 12.10. The Blinking Planetary in Gemini. Photograph by Gian Michelle Ratto.

GEMINI (**jem**-ih-nigh),	The Twins
GEMINORUM (jem-uh-**nor**-uhm)	9:00 P.M. CULMINATION: Feb. 20
Gem	AREA: 514 sq. deg., 30th in size

TO FIND: Face south and look high overhead. Find Orion first, then follow a line from Rigel through the belt and Betelgeuse. Extend this line for about 30°. Just to the left of this line, two stars, close together, and of nearly equal brightness mark the twins' heads, while their bodies and feet extend back toward Orion's Club. See Fig. 4.1, Orion Group Finder Chart, for the location of Gemini relative to the constellations in the Orion Group (Figs. 12.11 and 12.12).

KEY STAR NAMES:

Name	Pronunciation	Source
α Castor	**kus**-tore, **kas**-ter	Κάστωρ (Kastor), ancient Greek name
β Pollux	**pol**-luks	Πολυδευχης (Polydeuces), ancient Greek name
γ Alhena	ul-**heh**-nuh	al-hanʿa (Ind-Arabic), a lunar mansion (zodiac region)
δ Wasat	**wuh**-suht, **way**-seht	wasat (Arabic), "middle" of the sky, perhaps meaning the celestial equator and applying to an old constellation including most of today's Orion
ε Mebsuta	meb-**sue**-tuh	dhiraʿal-asad al-mabsuta (Arabic), "the lion's out-stretched paw," applied to ζ Geminorum in recent times
ζ Mekbuda	mek-**bue**-duh	dhiraʿ al-asad al-maqbuda (Arabic), "the lion's folded paw," applied to ζ Geminorum in recent times
η Propus	**pro**-puhs	προπουζ (Greek), "forward foot" from Ptolemy's description of its location
μ Tejat	teh-**yawt**	tihyat (Arabic), probably singular of al-tahayi, a name of unknown meaning

Magnitudes and spectral types of principal stars:

Bayer desig.	Mag. (m_v)	Spec. type	Bayer desig.	Mag. (m_v)	Spec. type	Bayer desig.	Mag. (m_v)	Spec. type
α	1.58	A2 V	β	1.16	K0 III	γ	1.93	A0 IV
δ	3.50	F0 IV	ε	3.06	A3	ζ	4.01	G3 Ib
η	3.31	M3 III	θ	3.60	A3 III	ι	3.78	G9 III
κ	3.57	G8 III	λ	3.58	A3 V	μ	2.87	M3 IIIvar
ν	4.13	B5 III	ξ	3.35	F5 IV	ο	4.89	F3 III
π	5.14	M0 III	ρ	4.16	F0 V	σ	4.23	K1 III
τ	4.41	K2 III	υ	4.06	K5 III	φ	4.97	A3 V
χ	4.94	K2 III	ψ	–	–*10	ω	5.20	G5 II

Stars of magnitude 5.5 or brighter which have Flamsteed, but no Bayer designations:

Flam-steed.	Mag. (m_v)	Spec. type	Flam-steed.	Mag. (m_v)	Spec. type	Flam-steed.	Mag. (m_v)	Spec. type
1	4.16	G7 III	26	5.20	A2 V	28	5.42	K4 III
30	4.49	K1 III	36	5.28	A2 V	38	4.73	F0 Vp
45	5.47	G8 III	51	5.07	K3 V	56	5.09	M0 III
57	5.04	G8 III	63	5.20	F4 IV/V	64	5.07	A4 V
65	5.01	K2 III	68	5.27	A1 Vn	74	5.04	M0 III
76	5.30	K5 III	81	4.23	K1 III	85	5.38	A0 Vs

Stars of magnitude 5.5 or brighter which have no Bayer or Flamsteed designations:

Desig-nation	Mag. (m_v)	Spec. type	Desig-nation	Mag. (m_v)	Spec. type
HD 052960	5.14	K3 III	HD 059686	5.45	K2 III
HD 060318	5.34	K0 III			

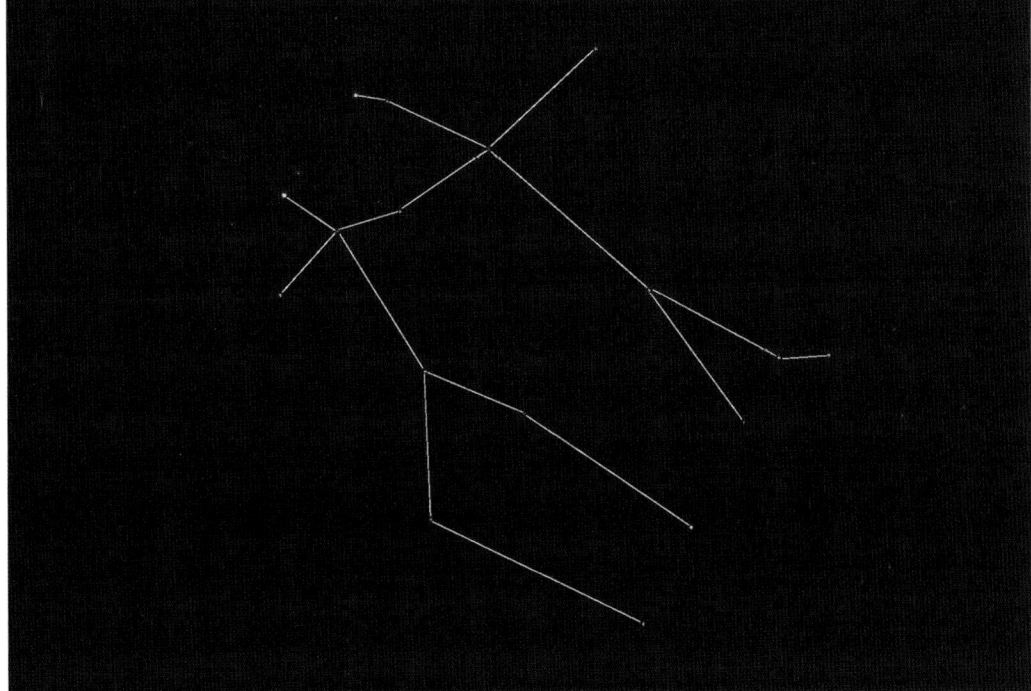

Fig. 12.11. (**a**, **b**) The stars and constellation of Gemini.

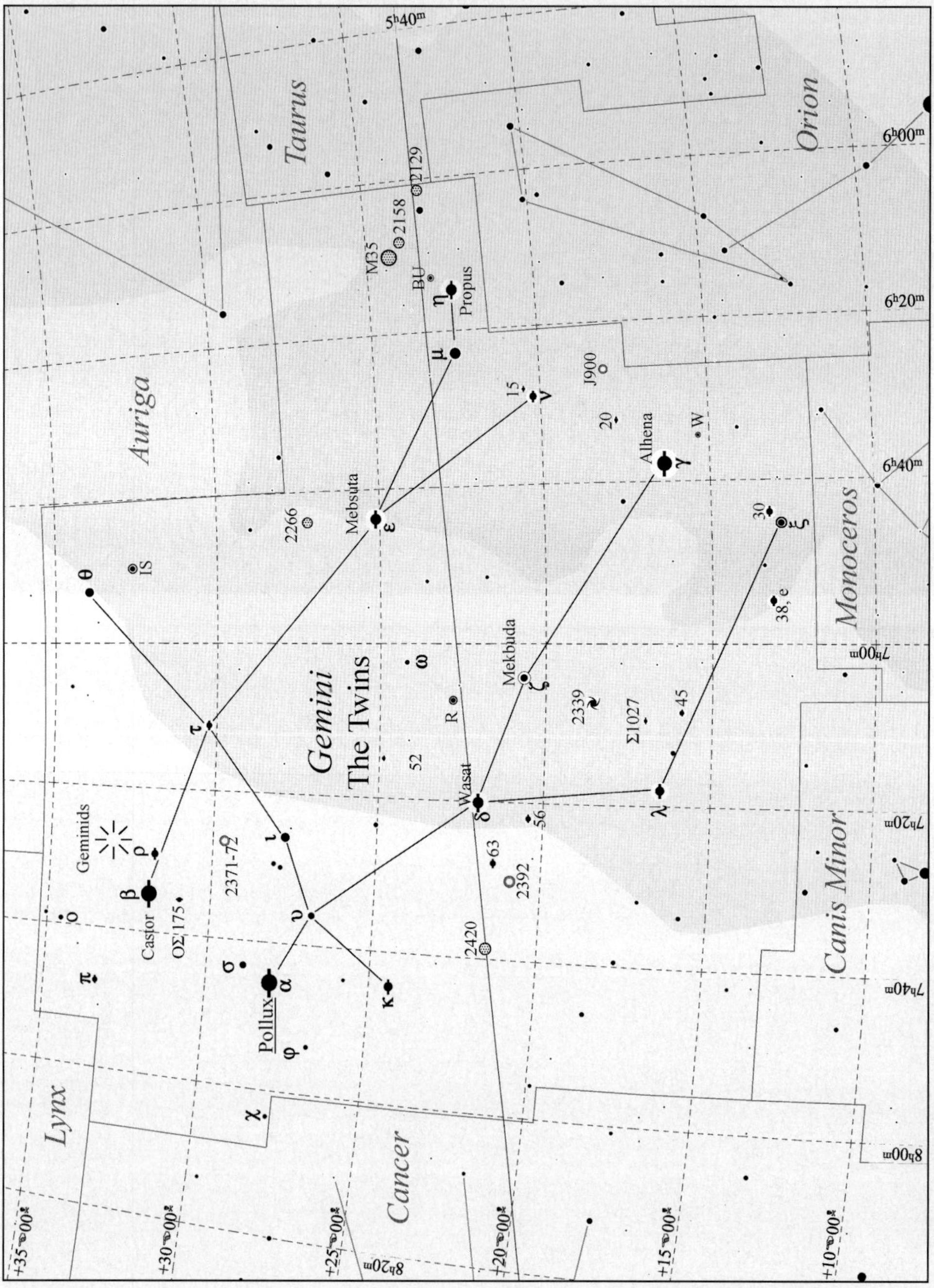

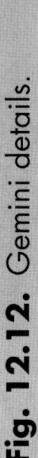

Fig. 12.12. Gemini details.

Table 12.2. Selected telescopic objects in Gemini.

Multiple stars:

Designation	R.A.	Decl.	Type	m_1	m_2	Sep. (")	PA (°)	Colors	Date/**Period**	Aper. Recm.	Rating	Comments
7, η Geminorum	06h 14.9m	+22° 30'	A–B	3.5	6.15	**1.60**	**254**	O, Y	**473.7**	5/6"	*****	a.k.a. BU 1008. Max separation, 1.62" in 2034
15 Geminorum	06h 27.8m	+20° 47'	A–B	6.7	8.2	25.2	203	Y, B	2004	10×50	*****	Lovely colors. a.k.a. SHJ 70
18, ν Geminorum	06h 29.0m	+20° 13'	Aa–BC	4.1	8.0	111.6	029	Y, pB	2002	7×50	*****	a.k.a. STTA 77. Showcase pair for binoculars. A–a and B–C are too close for amateur
20 Geminorum	06h 32.3m	+17° 47'	A–B	6.3	6.9	19.7	211	Y, W	2002	15×80	****	Nearly equal brightness, slight color difference. a.k.a. Σ924
24, γ Geminorum	06h 37.7m	+16° 24'	A–B	1.9	11.2	140.5	296	W, ?	1997	3/4"	***	a.k.a. BUP 90AC
			A–C	1.9	10.9	143.5	335	W, ?	1907	3/4"	***	a.k.a. BUP 90AC
27, Geminorum	06h 43.9m	+25° 08'	A–B	3.1	9.6	110.6	095	Y, B	2002	2/3"	*****	a.k.a. Σ533. Beautiful for small scopes
30 Geminorum	06h 44.0m	+13° 14'	A–B	4.5	11.1	27.2	184	O, ?	1094	3/4"	****	a.k.a. LAM 3
38, e Geminorum	06h 54.6m	+13° 11'	A–B	4.8	7.8	7.0	145	Y, prB	2005	2/3"	****	Showcase pair. Relatively easy double for small telescopes
45 Geminorum	07h 08.4m	+15° 56'	A–C	4.7	7.8	**6.98**	**325**	Y, ?	**1,943.8**	2/3"	****	a.k.a. Σ982. *11
Σ1027	07h 08.8m	+16° 54'	A–B	5.4	11.3	13.3	130	Gld, ?	1992	3/4"	*****	Σ165
46, τ Geminorum	07h 11.2m	+30° 15'	A–B	8.5	8.6	6.9	357	O, O	2002	2/3"	******	Nearly matched pair
52 Geminorum	07h 14.7m	+24° 53'	A–B	4.4	11.0	1.9	177	oY, ?	1925	5/6"	****	a.k.a. BU1009
			A–B	6.0	12.7	23.9	265	dR, ?	1910	5/6"	***	Primary is a Carbon Star
54, λ Geminorum	07h 18.1m	+16° 32'	A–B	3.6	10.7	9.7	033	bW, B	1997	2/3"	*****	a.k.a. Σ1061
55, δ Geminorum	07h 20.1m	+21° 59'	A–B	3.6	8.2	**5.80**	**226**	Y, prR	**2,239**	2/3"	******	Very nice contrasting colors. a.k.a. Σ1066
56 Geminorum	07h 22.0m	+20° 27'	A–B	5.1	12.2	8.2	202	dR, ?	1935	3/4"	***	a.k.a. BU1413
63 Geminorum	07h 27.7m	+21° 27'	Aa–B	5.3	10.9	43.0	324	W, –	1997	3/4"	***	a.k.a. SHJ 328 Aa–B. A–a has a separation of only 0.1"
			Aa–C	5.3	10.7	134.6	221	W, –	1997	3/4"	***	a.k.a. SHJ 328 Aa–C
			A–D	5.3	9.3	3.9	098	W, W	1991	2/3"	***	a.k.a. SHJ 328 Aa–D
62, ρ Geminorum	07h 29.1m	+31° 46'	A–B	4.2	12.5	3.4	008	yW, ?	1935	4/5"	***	a.k.a. AlC 3AC
Castor, 66, α Geminorum	07h 34.6m	+31° 53'	A–B	1.9	3.0	**4.65**	**057**	yW, yW	**444.95**	2/3"	****	a.k.a. S1110. All three components are spectroscopic doubles, C is an eclipsing binary
			A–C	1.9	9.83	72.9	164	W, pP	2001	2/3"	***	(YY Geminorum), range 9.1–9.6, period 0.814 day
OΣ175	07h 35.1m	+30° 58'	AB–C	5.5	10.4	82.0	194	dY, W	2002	15×80	****	A–B has a separation of only 0.1", too close for amateur instruments

(continued)

Table 12.2. (continued)

Multiple stars:

Designation	R.A.	Decl.	Type	m₁	m₂	Sep. (")	PA (°)	Colors	Date/Period	Aper. Recm.	Rating	Comments
77, κ Geminorum	07ʰ 44.4ᵐ	+24° 24′	A–B	3.7	8.2	7.2	241	O, pB	2002	2/3″	*****	a.k.a. Λ 179. Nice colors
78, β Geminorum	07ʰ 45.3ᵐ	+28° 02′	A–B	1.3	13.7	29.7	280	pO, W	1900	7/8″	****	a.k.a. BU580AB
80, π Geminorum	07ʰ 47.5ᵐ	+33° 24′	A–B	5.1	11.4	19.3	215	dR, pB	1998	4/5″	****	a.k.a. S1135AB
			A–C	5.1	10.4	92.0	343	dR, W	1998	15×80	****	a.k.a. S1135AC

Variable stars:

Designation	R.A.	Decl.	Type	Range (m_v)	Period (days)	F (f_r/f_f)	Spectral Range	Aper. Recm.	Rating	Comments	
BU Geminorum	06ʰ 12.3ᵐ	+22° 54′	Lc	5.7	–	–	M1–M2, Ia–Iab	6×30	****		
7, η Geminorum	06ʰ 14.9ᵐ	+22° 30′	SRa+EA	3.2	232.9	0.05	M3III	Eye	***	Propus	
W Geminorum	06ʰ 35.0ᵐ	+15° 20′	Cδ	6.5	7.914	0.30	F5–G1	6×30	****		
IS Geminorum	06ʰ 49.7ᵐ	+32° 36′	SRc	5.5	47	–	gK3	6×30	****		
43, ζ Geminorum	07ʰ 04.1ᵐ	+20° 34′	Cδ	3.7	10.151	0.15	Me	Eye	****	Mekbuda. One of brightest Cepheid variables	*12
R Geminorum	07ʰ 07.4ᵐ	+22° 42′	M	6.0	369.81	0.36	S2.9e–S8.9e	12/14″	****	Large range, ~1,500× in visible brightness	*13

Star clusters:

Designation	R.A.	Decl.	Type	m_v	Size (")	Brtst *	Dist. (ly)	dia (ly)	Aper. Recm.	Rating	Comments
NGC 2129	06ʰ 01.0ᵐ	+23° 18′	Open	6.7	6	7.4	6,500	9	4/6″	****	Can be seen in 7×50, 8/10″ shows ~25 stars, 12/14″ ~40–45
NGC 2158	06ʰ 07.5ᵐ	+24° 06′	Open	8.6	5	12.4	16K	12	8/10″	***	Seen in 2.5″, resolved in 6″, ~20–30 stars seen in 8″
M35, NGC 2168	06ʰ 08.9ᵐ	+24° 20′	Open	5.1	28	8.2	2,875	25	7×50	*****	Beautiful in 8″, 14″ shows ~150 stars
NGC 2266	06ʰ 43.2ᵐ	+26° 58′	Open	9.5	5	11	11K	16	8/10″	***	Rich, compressed, roughly triangular outline
NGC 2420	07ʰ 38.5ᵐ	+21° 34′	Open	8.3	10	11.1	8,500	25	8/10″	***	Moderately rich, small brightness range, strong central concentration. Age, ~4G years

Nebulae:

Designation	R.A.	Decl.	Type	Size (')	Brtst./Cent. star	m_v	Dist. (ly)	dia. (ly)	Aper. Recm.	Rating	Comments	
IC 443	06ʰ 16.9ᵐ	+22° 47'	SNR	50×39	–	12	5,000	70	14/16"	***	Crescent or Jellyfish Nebula. Challenging. Use nebula or red filter	*14
IC 444	06ʰ 19.4ᵐ	+23° 16'	Reflect.	7×4	–	7.0	8,580	18	16/18"	**	This small nebula is illuminated by the star 12 Geminorum	
J900	06ʰ 25.9ᵐ	+17° 47'	Planetary	12×10"	**16.3**	11.7	7,500	0.4	12/14"	***	Tiny, greenish gray, 10" smudge in 8/10" scopes, greenish oval in larger ones	*15
NGC 2371-72	07ʰ 25.6ᵐ	+29° 29'	Planetary	44"	**14.8**	11.3	4,400	0.9	6/8"	***	Small, bright disk with two lobes, NE and SW of central star	
NGC 2392	07ʰ 29.2ᵐ	+20° 55'	Planetary	54"	**10.5**	9.1	3,000	0.8	4/6"	*****	Eskimo or Clown Face Nebula. Round, bluish disk, prominent CS	*16

Galaxies:

Designation	R.A.	Decl.	Type	m_v	Size (')	Dist. (ly)	dia. (ly)	Lum. (suns)	Aper. Recm.	Rating	Comments
NGC 2339	07ʰ 08.3ᵐ	+18° 47'	SAB(rs)bc	11.8	2.7×2.0	100M	80K	32G	10/12"	**	Almost face-on galaxy

Meteor showers:

Designation	R.A.	Decl.	Period (yr)	Duration	Max Date	ZHR (max)	Comet/Asteroid	First Obs.	Vel. (mi/km/sec)	Rating	Comments
Geminids	07ʰ 28ᵐ	+33°	Annual	12/07 12/17	12/14	120	**Phaethon**	1861	22/35	I	Disc: Robert Philips Greg (England). Seen every year since, activity may be increasing

Other interesting objects:

Designation	R.A.	Decl.	Type	m_v	Mass (suns)	Dia (suns)	Dist. (ly)	Planet Mass (J)	Period (days)	Eccentricity	Comments
Pollux, β Gemini	07ʰ 45.3ᵐ	+28° 02'	ExoPlnt	1.14	1.7	8.8	34	2.41	589.64	0.02	Brightest star with a planet. Planet has a diameter ~11 times the Earth's

Fig. 12.13. M35 (NGC 2168) and NGC 2158, open clusters in Gemini Rob Gendler.

Fig. 12.14. IC 444 and 443, emission in Gemini. Astrophoto by Gian Michele Ratto.

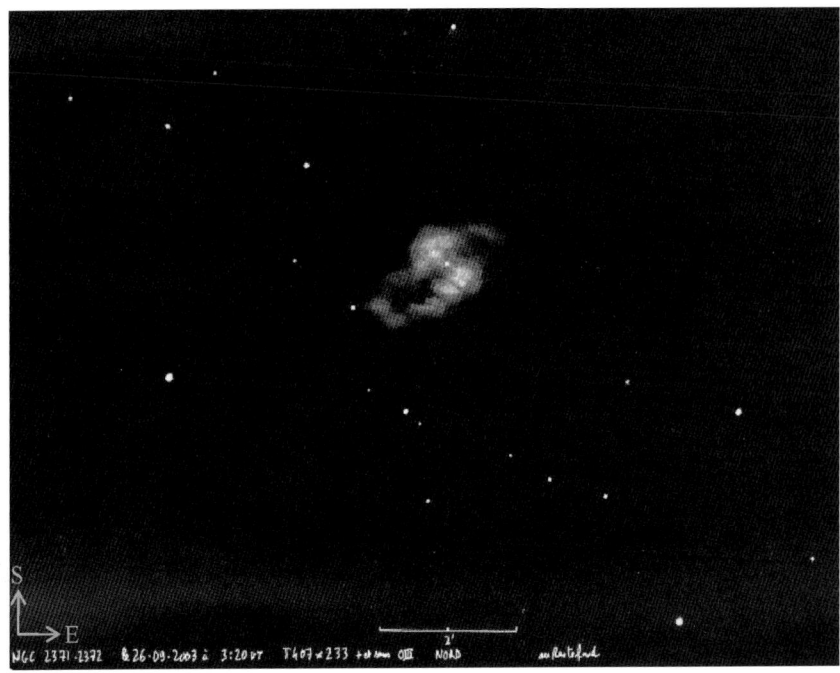

Fig. 12.15. NGC 2371-72, planetary nebula in Gemini. Drawing by Nocolas Biver.

Fig. 12.16. NGC 2420, open cluster in Gemini. Astrophoto by David Ridhard et al.

Fig. 12.17. The "Eskimo Nebula" or "Clown Face Nebula" (NGC 2392) in Gemini. Astrophoto by Gian Michele Ratto.

NOTES:

1. χ Cygni has one of the largest brightness variations, 10.9 magnitudes or ~23,000 times. It goes from easily visible to the unaided eye to barely visible in a 10" telescope.

2. NGC 6811. Walter Scott Houston called this the "Hole in a Cluster" because of reports from many observers of a hole or dark band through the cluster's center. A 6 cm (2.4") telescope will resolve about 20 stars, while a 15 cm (5.9") will show about 50 stars.

3. NGC 6871. In small telescopes (2/3"), this cluster is almost invisible against the rich Milky Way background, but it stands out well in larger (8/10") telescopes.

4. IC 5146 is a star cluster with involved nebulosity. The nebula is called the Cocoon Nebula because the stars look as if they are enveloped in a diaphanous cocoon.

5. NGC 6826 is also called the "Blinking Planetary" because of its tendency to disappear when looked at directly and reappear when the vision is slightly averted. It is easy to find since it lies in the same low-power field as the double star 16 Cygni.

6. The Veil Nebula is the remnant of a supernova explosion which occurred about 20,000 years ago. The debris from this explosion is still expanding at about 30 miles per second, and now covers a roughly circular area of ~3° in diameter. The expanding gasses have fragmented into several segments which bear the designations of NGC 6960, 6974, 6979, 6992, 6995, and IC 1340. In addition, several nebulous areas which have no NGC or IC designations lie between NGC 6960 and NGC 6996, the western and eastern edges of the nebula. All of these are relatively faint, with NGC 6960 being the brightest and easiest to find because the magnitude 4.2 star 52 Cygni lies near the middle of its thin cresent shape. Dark skies and nebular filters are recommended for all these objects. a.k.a. Bridal Veil, Sirrus, Lacework, Network or Filamentary Nebula, and the Cygnus Loop.

7. The Pelican Nebula consists of IC 5067 and IC 5070. IC 5067 forms the NW 1/3, and IC 5070 forms the SE 2/3. Both sections show up best on red-sensitive photographic plates. Some sections of IC 5067 are very bright, but others are faint to very faint. IC 5070 is mostly faint and very faint. While both of these can be seen with the unaided eye under exceptional conditions, a small telescope with a HII (deep red) filter and a LP eyepiece are generally required.

8. NGC 7000 has a striking resemblance to its namesake, the North American continent, especially the east coast, including an easily recognizable Gulf of Mexico. It can be seen with the unaided eye under extremely dark and clear skies.

9. Cygnus X-1 designates the first X-ray source found in the constellation of Cygnus. It is visible as a slightly variable star of magnitude 8.83. Observations have shown that the visible star moves toward and away from the Earth with a period of 5.6 days. The starlight passes through a large amount of dust on its way to the earth, so the brightness and spectral class of the star cannot be precisely determined. However, reasonable assumptions indicate a spectral class of O9.7 and a luminosity of about 400,000 suns. If these are correct, they imply a supergiant star with a mass of ~40 suns and a radius of ~23 suns. This in turn implies that the invisible companion must have a mass of 20 ± 5 suns. Since a normal star of this mass would be easily detectable, the companion must be a white dwarf, a neutron star, or a black hole. White dwarf and neutron stars of this mass are not stable and would collapse into black holes. Further support for the nature of the companion comes from rapid brightness variations which must come from a source 3,000 km (1,865 mi). Gas and dust falling into an object of this size and mass would be heated to millions of degrees and radiate strongly in the X-ray part of the spectrum, thus accounting for the source of the X-rays.

10. If you examine the Gemini chart, you will find every Bayer (Greek letter) designation except ψ. When the IAU formally defined the boundaries of all the constellations, they often put some of the fainter stars formerly attributed to one constellation in a different constellation to make the borders more regular. Usually these were stars which may have had a Flamsteed number, but not a Bayer letter. ψ Geminorum was one of the rare exceptions to this rule – it became 15 Cancer.

11. Struve (use capital Greek letter sigma) 982AB is a long-period binary which has not been observed over a long enough time to allow precise calculation of its orbital parameters. Two significantly different orbits match equally well the measurements for the period over which it has been observed. One has a period of of 1943.8 years and an eccentricity of 0.15, while the other has a period of 3190 years and an eccentricity of 0.485.

12. Mekbuda (ζ Geminorum) is one of the three brightest Cepheid variables we see, being approximately as bright as η Aquilae and δ Cephei itself. Intrinsically, ζ Geminorum is brighter than either of the others, but

appears about the same because it is somewhat more distant than the other two. If you want to track its variations, κ Cep (m3.57) and μ Cep (m4.06) can be used for comparison stars, even though they are not quite as close to ζ as one would like.

13. R Geminorum has an unusually large range of visible brightness, ~1,500 to 1. It declines from visible to the unaided eye to barely visible in a 10–12 telescope in ~11 months, then rises back to maximum in ~1 month.

14. IC 443 lies just west of η Gemini (Propus). A neutron star with an estimated age of 30,000 years has been detected in the nebula. It is moving away from the center of the nebula at a speed of about 800,000 km/h, and is probably the source of the supernova explosion which created the nebula.

15. NGC 2371-72 is sometimes called the "Gemini Nebula" or the "Peanut Nebula" because of its shape. It was discovered by John Herschel in the eighteenth century, and in spite of its faintness, this object is listed by the Royal Astronomical Society of Canada as one of the 110 best NGC objects in the sky.

16. NGC 2392 is a round, bluish disk with a prominent central star. A 12–14 telescope will show a separated ring ~45 in diameter. It is much easier to see with averted vision. NGC is known as the "Eskimo Nebula" because moderate resolution photographs look like a face surrounded by the hood of a parka. It is also sometimes called the "Clown Nebula."

CHAPTER THIRTEEN

Creatures of the Celestial Sea

September–November

CAPRICORNUS	DELPHINUS	PISCES	PISCIS AUSTRINUS
The (Sea) Goat	The Dolphin	The Fishes	The Southern Fish

Although the constellations in this portion of the sky are now explained by separate myths or stories, they probably were at one time closely related in the mythology of the time of their origination. All these constellations, plus Cetus, which has been included in the Chap. 3 (The Andromeda Group), represent creatures living in or associated with water. One of them, Capricornus, is the oldest known constellation and has existed almost unchanged since very early times. Some, or all, of the others may have originated about the same time as Capricornus but have evolved more and their origins cannot be traced back as far. In any case, the time at which these constellations evolved appears to be one when the summer rains came while the sun was passing through this "watery" portion of the sky.

The form of Capricornus is shown on the oldest known records, the boundary stones of the ancient Babylonians, and other evidence indicates that the origin of the constellation may go back all the way to 13000 BC. At that time, the summer solstice occurred when the sun was in Capricornus. Sumerian cuneiform inscriptions called Capricornus "the father of light." Their principal god, Ea (or Enki), was associated with Capricornus, and their priests wore goatskins as magical garments. Ea was also sometimes depicted as either half man, half fish or half goat, half fish. In addition, the cuneiform for Ea

P. Simpson, *Guidebook to the Constellations: Telescopic Sights, Tales, and Myths*,
Patrick Moore's Practical Astronomy Series, DOI 10.1007/978-1-4419-6941-5_13,
© Springer Science+Business Media, LLC 2012

indicated that his home was in the water. All these are consistent with the constellation being first conceived, or at least known, by that time.

All subsequent cultures of the Middle East recognized some form of Capricornus, sometimes the same form, sometimes purely goat, and sometimes another half mammal (e.g., ram), half fish form. The Assyro-Mesopotamians saw the same constellation as the Sumerians, but associated it with their god of wisdom, Oannes. The ancient Greeks recognized Capricornus in both the half goat, half fish, and the purely goat forms and had two different myths to explain their respective presences in the sky.

The purely goat form was explained in the story of Zeus' birth and revolt against his father, Chronus. A prophecy had foretold that Chronus would be overthrown by one of his children. Immediately after his wife, Rhea, gave birth to each of his children, Chronus would swallow them. Being immortal, they continued to live in his stomach but could not grow or escape. Furious about this, when she gave birth to Zeus, Rhea hid him and gave Chronus a stone swaddled in soft blankets. Chronus swallowed the stone, thinking that he had protected his role as ruler of the heavens.

Rhea gave Zeus to the she-goat Amalthea to suckle. Amalthea was hidden in a cave on Mt. Ida, and the Corybantes and Curetes (two groups known for holding noisy festivals and dances) were enlisted to celebrate near the cave so the noise would mask the cries of the baby Zeus. This stratagem worked, and Zeus grew up to lead a revolution which overthrew Chronus and the other Titans, freed the swallowed children, and established the Olympians as the supreme gods. After his victory, Zeus placed Amalthea in the heavens as Capricornus.

An alternate version had Pan, a demigod who had the upper body of a man and the lower body of a goat, helping Zeus and the other gods in their revolt. During the revolt, Rhea sent a gigantic monster called Typhon to attack the Olympians. When Typhon attacked, the Olympians were surprised by his size and ferocity, and except for Zeus, they all fled. Each tried different tactics to escape. Pan dove into a river and tried to change himself into a fish, but was unable to complete the transformation. Typhon managed to overcome the unaided Zeus. He cut the sinews from Zeus' arms and legs. He then hid the sinews in a cave guarded by Delphyne, a half woman, half serpent. Zeus lay helpless until Pan returned to his original form and with the help of Apollo managed to distract Delphyne, steal the sinews and replace them in Zeus' limbs. With his mobility restored, Zeus rallied his followers and defeated the Titans, chaining them or burying them under huge boulders so they could not attack again. After his victory, Zeus placed the figure of a half goat, half fish in the sky to show his appreciation for Pan's help (Fig. 13.1).

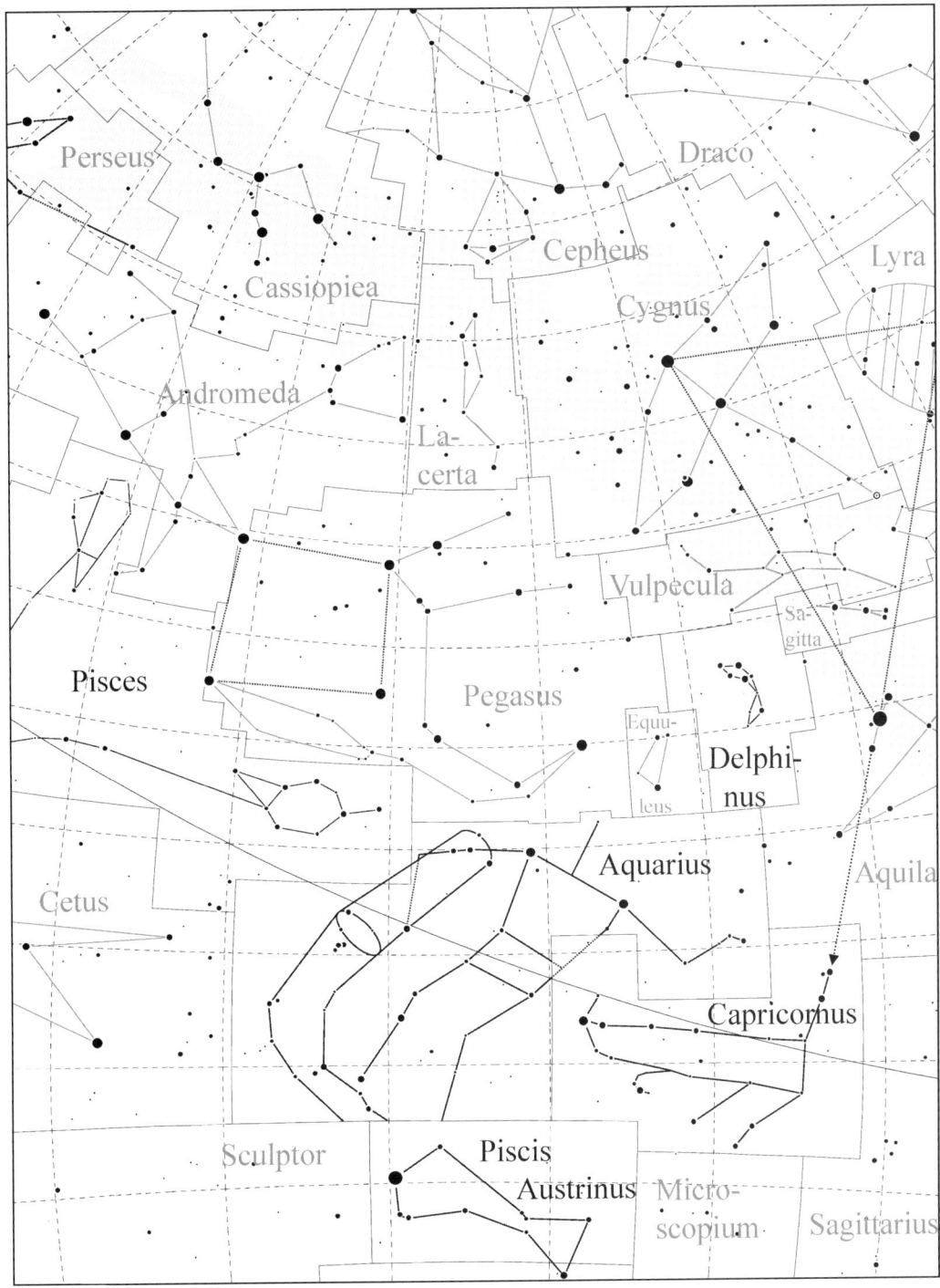

Fig. 13.1. Creatures of the sea finder chart.

Capricornus (kap-rih-**korn**-nus)	**The Goat**
Capricorni (kap-rih-**korn**-ee)	**9:00 P.M. Culmination**: Sep. 23
Cap	**AREA**: 414 sq. deg., 40th in size

TO FIND: Start with <u>Altair</u>, the brightest star in Aquila. <u>Altair</u> is flanked by two fainter stars. Follow the line defined by these three stars to the southeast (lower left) for about 20°–25° (about the distance between the tips of the thumb and little finger when your spread hand is held at arm's length). You should be able to easily find α and β Capricorni, which lie approximately along the same line and point in almost the same direction. Both these stars are wide doubles. α is fairly easily separated without optical aid, and β can be seen as a double by some people with sharp eyesight. They are not very bright, being only magnitude 3.6 and 3.1, but lie in a relatively star poor region and are fairly easy to recognize. Continue along the same line through a small triangle (ο, π, and ρ Capricorni) to ψ and ω Capricorni, about 10°–15° from the α–β pair. These stars form the goat's tail, hind quarters, and hind hooves. From the ο, π, and ρ triangle, follow a line to the E by ENE (left and slightly up) through θ, ι and γ to δ Capricorni. This row defines the goat's back from its rump to its nose. About 5° south of ι, you should see a curve formed by five to seven mostly very faint stars, including η, χ, φ, and ζ. These form the goat's front legs. Three very faint stars above δ form the goat's horns and a pair of faint stars below γ form its chin and beard (Figs. 13.1 and 13.2).

KEY STAR NAMES:

Name	Pronunciation	Source
α Algedi	uhl-**jeh**-dee	Al Giedi, or just Giedi, all from <u>Al Jady</u> (Arabic), "The Goat"
β Dabih	**dah**-bih	Al Sa'd al Dhabih (Arabic), "The Lucky One of the Slaughterers"
γ Nashura	**nah**-shi-rah	Al Sa'd al Nashirah, "The Fortunate One"
δ Deneb Algedi	**dih**-neb uhl-**jeh**-dee	Al Dhanab Al Jady (Arabic), "The Goat's Tail"

Magnitudes and spectral types of principal stars:

Bayer desig.	Mag. (m$_v$)	Spec. type	Bayer desig.	Mag. (m$_v$)	Spec. type	Bayer desig.	Mag. (m$_v$)	Spec. type
α^1	4.30	G3 Ib	α^2	3.58	G6/G8 III	β	3.05	A5:n
γ	3.69	A7 III	δ	2.85	A5 IV	ϵ	4.51	B3 V:p
ζ	3.77	G5 Ibp	η	4.82	A5 V	θ	4.08	A1 V
ι	4.28	G8 III	κ	4.72	G8 III	λ	5.57	A1 V
μ	5.08	F3 IV	ν	4.77	B9 IV	ξ	5.84	F5 V
o	5.94	A1 V	π	5.08	B4 V	ρ	4.77	F3 V
σ	5.28	K2 III	τ	5.24	B7 III	υ	5.15	M1 III
φ	5.17	K0 II/III	χ	5.30	A0 V	ψ	4.13	F5 V
ω	4.12	K4 III						

Stars of magnitude 5.5 or brighter which have Flamsteed but no Bayer designations:

Flamsteed	Mag. (m$_v$)	Spec. type	Flamsteed	Mag. (m$_v$)	Spec. type	Flamsteed	Mag. (m$_v$)	Spec. type
24	4.49	K5/M0 III	29	5.31	M2 III	30	5.40	B5 II/III
33	5.38	K0 III	36	4.50	K0 III	41	5.24	K0 III
42	5.16	G2 V	46	5.10	G8 II/III			

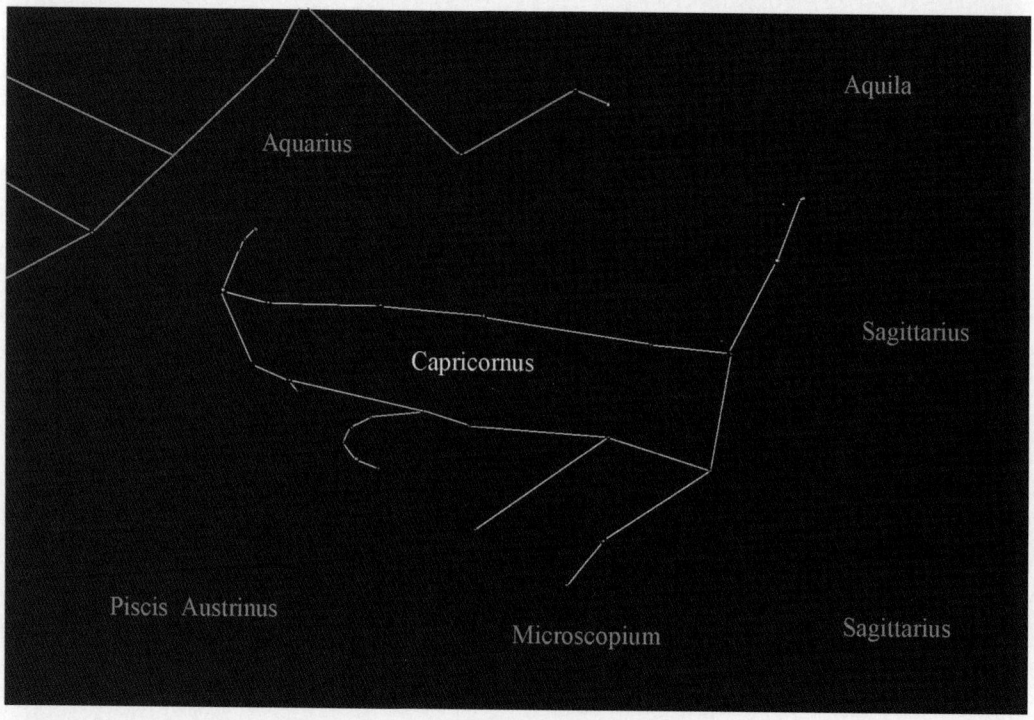

Fig. 13.2. (**a**, **b**) The stars and figure of Capricornus.

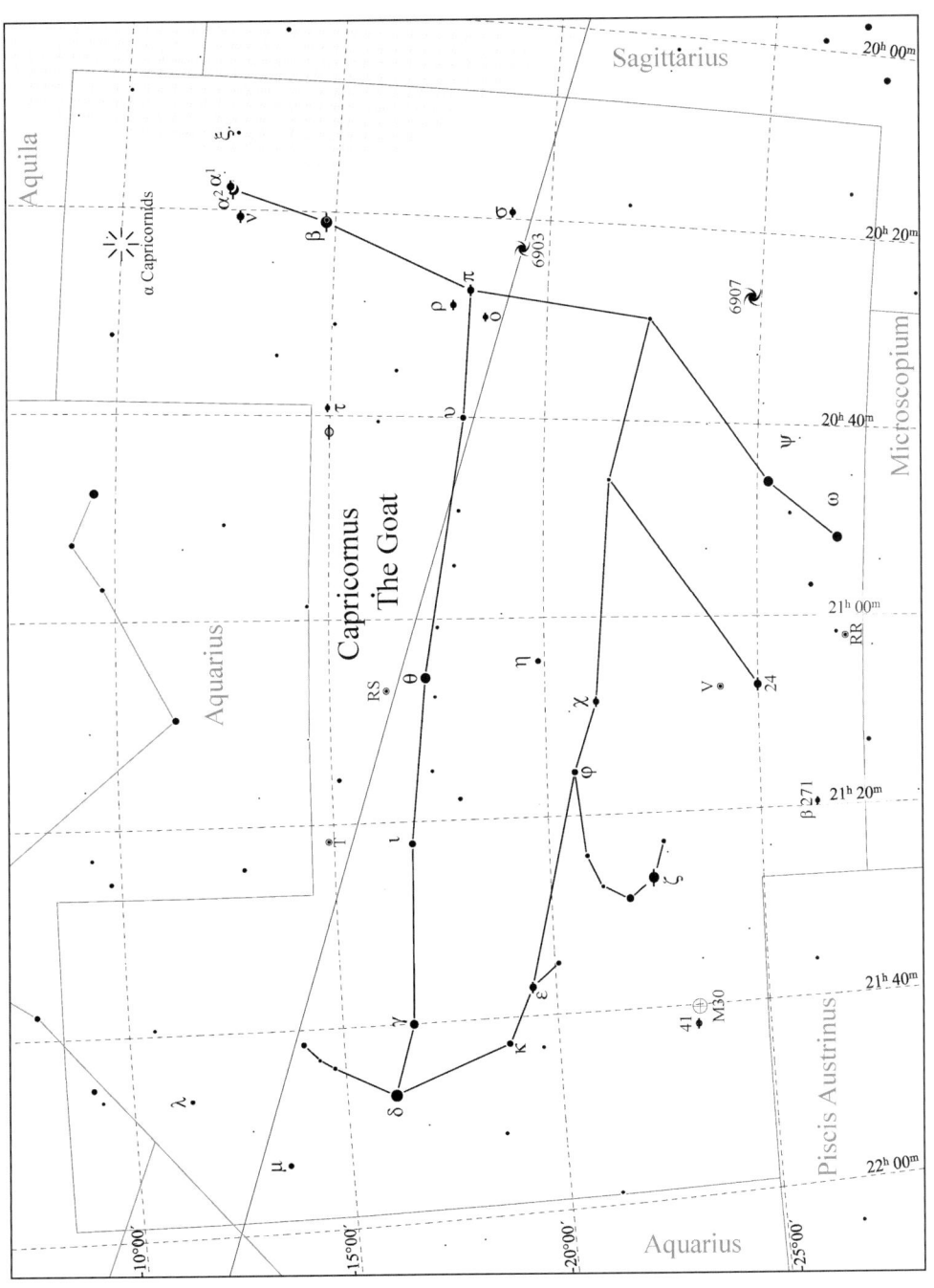

Fig. 13.3. Capricornus detail chart.

Table 13.1. Selected telescopic objects in Capricornus.

Multiple stars:

Designation	R.A.	Decl.	Type	m_1	m_2	Sep. (")	PA (°)	Colors	Date/ **Period**	Aper. Recm.	Rating	Comments
5, 6, α¹, α² Capricorni	20ʰ 18.1ᵐ	−12° 33′	A–B	3.6	4.2	377.3	291	yO, yO	2000	Eye	*****	Showcase pair. Striking in smallest optical instruments
5, α¹ Capricorni, Al Giedi	20ʰ 17.6ᵐ	−12° 30′	A–a	4.4	8.6	0.6	355	dY, R	1991	16/17″	****	a.k.a Wirtz15-Aa
			A–B	4.2	14.1	44.3	182	dY, R	1960	9/10′	****	a.k.a. β295-AB
			A–C	4.2	9.6	45.6	224	dY, R	2002	10×80	*****	a.k.a. h607-AB
			C–D	9.6	14.2	29.3	110	?, ?	1905	10/11″		a.k.a Wirtz15-CD
6, α² Capricorni	20ʰ 18.1ᵐ	−12° 33′	A–BC	3.7	11.2	6.6	172	dY, ?	2002	3/4″	******	a.k.a. h608-AB
			A–D	3.7	10.5	153.1	160	dY, R	2002	10×80		a.k.a. AGC12-AD
			B–C	11.2	11.5	1.3	243	dY, ?	1959	7/8″		a.k.a. AGC12-cd
7, σ Capricorni	20ʰ 19.4ᵐ	−19° 07′	A–B	5.4	9.4	56.3	180	yO, pB	2002	15×80	****	Showcase pair. Beautiful color contrast. a.k.a. hV87
8, ν Capricorni	20ʰ 20.7ᵐ	−12° 46′	A–B	4.8	11.8	53.7	121	W, ?	1998	3/4″	***	a.k.a. See Lam5-BC
9, β¹, β² Capricorni	20ʰ 21.0ᵐ	−14° 47′	Aa–Bb	3.87	6.08	206	267	oY, pB	2002	6×30	*****	Showcase pair. a.k.a. ΣA 52Aa-Bb
9, β¹ Capricorni	20ʰ 21.0ᵐ	−14° 47′	A–a	3.10	4.9	**0.0**	**039**	oY, pB	**3.76**	–	**	A–a is a very rapid and close binary, max separation ~0.1″. a.k.a. BLA 7
9, β² Capricorni	20ʰ 21.0ᵐ	−14° 47′	B–b	6.16	9.14	0.7	075	W, ?	1991	13/14″	***	a.k.a. BAR 12
10, π Capricorni	20ʰ 27.3ᵐ	−18° 13′	Aa–B	5.3	8.5	3.2	146	W, W	2002	3/4″	****	a.k.a. Chr 184. A–a is a very close pair, 0.1″
11, ρ Capricorni	20ʰ 27.3ᵐ	−18° 13′	Aa–C	5.3	14.1	38.1	040	W, ?	1998	9/10″	****	Fairly close double. Widening to 3.1″ in 2123
	20ʰ 28.9ᵐ	−17° 49′	A–B	5.0	6.88	**1.5**	**192**	Y, P	**149**	6/7″		
	20ʰ 28.9ᵐ	−17° 49′	A–C	5.0	13.2	9.0	54.6	Y, ?	**1998**	6/7″		a.k.a. β61-AC
	20ʰ 28.9ᵐ	−17° 49′	A–D	5.0	6.68	258.7	150	Y, R	2001	6×30	****	Nicely colored binocular pair. a.k.a. S.h323-AD
12, o Capricorni	20ʰ 28.9ᵐ	−17° 49′	D–E	6.6	?	54.3	096	W, ?	1998	6×30		a.k.a. Dob13-AD
	20ʰ 29.9ᵐ	−18° 35′	A–B	5.9	6.7	22.6	239	bW, W	2006	15×80	****	Beautiful pair, easy in small scopes
14, τ Capricorni	20ʰ 39.3ᵐ	−14° 57′	A–B	5.8	6.3	**0.3**	**119**	bW, ?	**420**	–	****	Challenging for larger amateur telescopes, requires superb seeing. Slowly closing
24 Capricorni	21ʰ 07.1ᵐ	−25° 00′	A–B	4.6	11.7	26.2	186	rO, ?	1996	3/4″	***	a.k.a. See439.
25, χ Capricorni	21ʰ 08.6ᵐ	−21° 12′	Aa–Bb	5.28	11.0	69.8	061	W, ?	1998	3/4″	****	a.k.a. h3009-Aa. Star a is 15th magnitude and beyond most amateur telescopes

(continued)

β271	21ʰ 08.6ᵐ	−21° 12′	B-b	11.4	12.2	18	088	?, ?	1998	4/5″	****	a.k.a. h3009-Bb
β271	21ʰ 08.6ᵐ	−26° 21′	A-B	6.7	9.8	1.9	269	dY, W	1991	5/6″	****	Lovely in medium to large telescopes. 12th mag companions C, D, E are at 82, 247 and 181″ respectively
34, ζ Capricorni	21ʰ 26.7ᵐ	−22° 25′	A-B	3.74	12.5	17.3	012	dY, ?	1998	4/5″	****	a.k.a. See 446
39, ε Capricorni	21ʰ 37.1ᵐ	−19° 28′	A-B	4.5	10.1	67.2	47	B, ?	2002	15×80	****	a.k.a. HVI6
41 Capricorni	21ʰ 42.0ᵐ	−23° 16′	A-B	5.2	11.2	5.5	212	dY, ?	1973	3/4″	****	a.k.a. See 454

Variable stars:

Designation	R.A.	Decl.	Type	Range (m_v)	Period (days)	F (f_i/f_f)	Spectral Range	Aper. Recm.	Rating	Comments
RR Capricorni	21ʰ 02.3ᵐ	−27° 05′	M	7.8 / 15.5	277.54	0.40	M5e–M6e	–	***	Mira type variable, visible in binoculars/small telescopes over much of its range
RS Capricorni	21ʰ 07.3ᵐ	−16° 25′	SRb	**8.3** / **10.3**	340	–	M4	2/3″	****	Because of its redness, this variable should be visible in 7×50 binoculars at maximum
V Capricorni	21ʰ 07.6ᵐ	−23° 55′	M	8.2 / 14.4	275.72	0.42	M5e–M8.2	13/14″	****	
T Capricorni	21ʰ 22.0ᵐ	−15° 10′	M	8.4 / 14.3	269.28	0.44	M2e–M8.2	12/13″	****	
49, δ Capricorni	21ʰ 47.0ᵐ	−16° 08′	EA	2.81 / 3.05	1.02	0.08	A7IIIm	Eye	***	Variations barely visible to eye, more suitable for photometric measurements

Star clusters:

Designation	R.A.	Decl.	Type	m_v	Size (′)	Brtst. Star	Dist. (ly)	dia. (ly)	Number of Stars	Aper. Recm.	Rating	Comments
M30, NGC 7099	21ʰ 40.4ᵐ	−23° 11′	Globular	7.3	11′	12.1	12.1	24K	77	7×50	*****	Lum = 55K suns, highly resolved in medium sized telescopes

Nebulae:

Designation	R.A.	Decl.	Type	m_v	Size (′)	Cent. star	Dist. (ly)	dia. (ly)	Aper. Recm.	Rating	Comments
None										*	

Table 13.1. (continued)

Galaxies:

Designation	R.A.	Decl.	Type	m$_v$	Size (')	Dist. (ly)	dia. (ly)	Lum. (suns)	Aper. Recm.	Rating	Comments
NGC 6903	20^h 23.6^m	−19° 19'	S0 pec	11.9	2.1 × 1.9	150M	113K	31G	16/18"	***	Detectable in 10" scopes, more needed to see clearly
NGC 6907	20^h 25.1^m	−24° 49'	SB	11.1	3.2 × 2.3	140M	145K	57G	8/10"	***	Can be detected in 6" as a faint, 1' circular spot

Meteor showers:

Designation	R.A.	Decl.	Period (yr)	Duration	Max Date	ZHR (max)	Comet/Asteroid	First Obs.	Vel. (mi/km/sec)	Rating	Comments
α Capricornids	20^h 25^m	−10°	Annual	07/15 09/11	8/1–2	4	2002 EX2	1871	14/**9**	II	First noted by N. de Konkoly (Hungary). High percentage of bright meteors

Other interesting objects:

Designation	R.A.	Decl.	Type	m$_v$	Size (')	Dist. (ly)	dia. (ly)	Lum. (sums)	Aper. Recm.	Rating	Comments
None											

Fig. 13.4. M30. Astrophoto by Mark Komsa.

Delphinus (dell-**fee**-nus)	**The Dolphin**
Delphini (dell-**fee**-nee)	**9:00 P.M. Culmination**: Sep. 14
Del	**AREA**: 189 sq. deg., 69th in size

Although the constellation of the dolphin was known to the Indians, and probably the Babylonians, long before Greek civilization arose, it was the Greeks who provided the most interesting story of its origin. At least one of the characters in this tale is known to have been a real person. Periander ruled Corinth from 625 to 585 BCE, and his court poet and musician Arion may also have been an actual figure.

Arion was born in Methymna on the island of Lesbos, the son of the nymph Oncaea, and by some accounts, the son of Poseidon. He was reputed to have been the best musician of all time and to have invented or perfected the form of dithyrambic poetry. As with most people of his time who showed great talent or skill, Arion's talent was attributed to the gift of a god, in this case Apollo the god of music. Periander was a well-known tyrant who ruled Corinth with harsh laws, and severely punished anyone who offended him. However, he liked Arion and his music so much that he had Arion brought to Corinth as his court poet and musician, and allowed him much greater freedom than other citizens of his kingdom.

One day, Arion heard of a musical contest to be held in Sicily. This contest offered great rewards to the winner, and Arion was anxious to make his own fortune so he would no longer be dependent on Periander. Periander tried to dissuade Arion from making the trip, but Arion was determined, and Periander eventually granted him permission to go.

At the contest, Arion did win a great prize, and, in addition, wealthy patrons who heard him gave him many other gifts of gold and jewelry. Arion paid for passage on a Corinthian ship which was returning home from the Sicilian port of Tarentum. When the sailors learned that his trunk contained gold and jewels, they plotted to throw him overboard. As Arion slept that night Apollo appeared to him in a dream and warned him that he was in danger but also promised to help him.

The next day, the crew seized Arion and prepared to throw him overboard. Arion pled for his life to no avail and finally asked to be allowed to play his lyre for one last song. The sailors thought that this request could certainly cause no harm and agreed to allow it. Arion donned his finest minstrel garb and began to sing a song honoring Apollo. As he played, a school of dolphins gathered around the ship to listen to the beautiful music. Frightened by this turn of events, the sailors seized Arion and immediately threw him overboard.

To their surprise, one of the dolphins allowed Arion to climb on its back, the other dolphins surrounded the pair and they all swam away as a group. Still the amazed sailors saw no danger because they expected Arion to soon fall off and drown. However, the dolphins swiftly carried Arion to Cape Taenarus (today Cape Malapan, the southernmost point of Europe). After he went ashore, Arion made an offering to Apollo and returned to Corinth, where he told Periander of his misadventure. At first, Periander did not believe Arion's story but agreed to check it out.

When the ship arrived, Periander had the crew brought to him immediately. He questioned them about the whereabouts of Arion. They replied that he had stayed in Sicily.

When Arion suddenly appeared and went to the ruler's side, the men were so frightened that they immediately confessed to their crime and were quickly ordered to be executed by Periander.

To celebrate his good fortune, and to thank Apollo for his help, Arion commissioned a small statue of a dolphin and placed it in Apollo's temple in Corinth. Apollo was pleased with Arion, his music, and his tribute, so he placed the dolphin in the sky as the constellation Delphinus to honor the service the dolphins had done in saving the respected musician. In this myth, Apollo also placed Arion's lyre in the heavens, giving an alternate explanation for the constellation of Lyra.

TO FIND: First find Altair (α Aquila). First magnitude Altair is relatively isolated and flanked by a 3rd and a 4th magnitude star, making it easy to recognize. About 14° ENE (PA 060°) of Altair are four 4th and 5th magnitude stars which form a small diamond (~2.7° long) with its long direction pointed toward Altair. These four stars represent the major part of the dolphin's body. Due south of the southwest point of the diamond is fainter diamond of similar size and shape, which forms the tail of the dolphin (Figs. 13.1, 13.5 and 13.6).

KEY STAR NAMES:

Name	Pronunciation	Source
α Sualocin	**swah**-low-sin	Reversed name Nicolaus*[1]
β Rotanev	**roe**-tuh-nev	Reversed name Venator*[1]
ε Deneb al Dulfim	**deh**-neb ul **dull**-fim	al Deneb al Dulfim (Arabic) "The Tail of the Dolphin"

Magnitudes and spectral types of principal stars:

Bayer desig.	Mag. (m_v)	Spec. type	Bayer desig.	Mag. (m_v)	Spec. type	Bayer desig.	Mag. (m_v)	Spec. type
α	3.77	B9 V	β	3.63	F5 IV	γ^1	5.22	F8 IV/V
γ^2	4.27	G5 IV	δ	4.43	A7 III	ε	4.03	B6 III
ζ	4.64	A3 V	η	5.39	A3 IV	θ	5.69	K3 Ib
ι	5.42	A2 V	κ	5.07	G5 IV			

Stars of magnitude 5.5 or brighter which have Flamsteed but no Bayer designations:

Flamsteed	Mag. (m_v)	Spec. type
17	5.19	K0 III

Fig. 13.5a. The stars of Delphinus, Sagitta, and adjacent constellations.

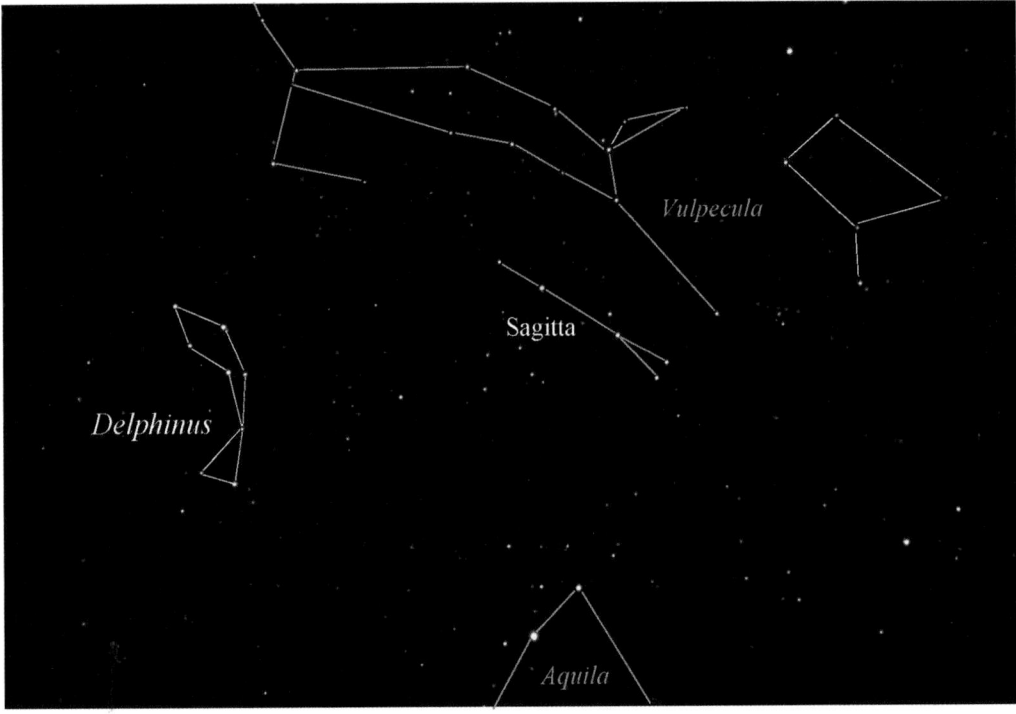

Fig. 13.5b. The figures of Delphinus, Sagitta, and adjacent constellations.

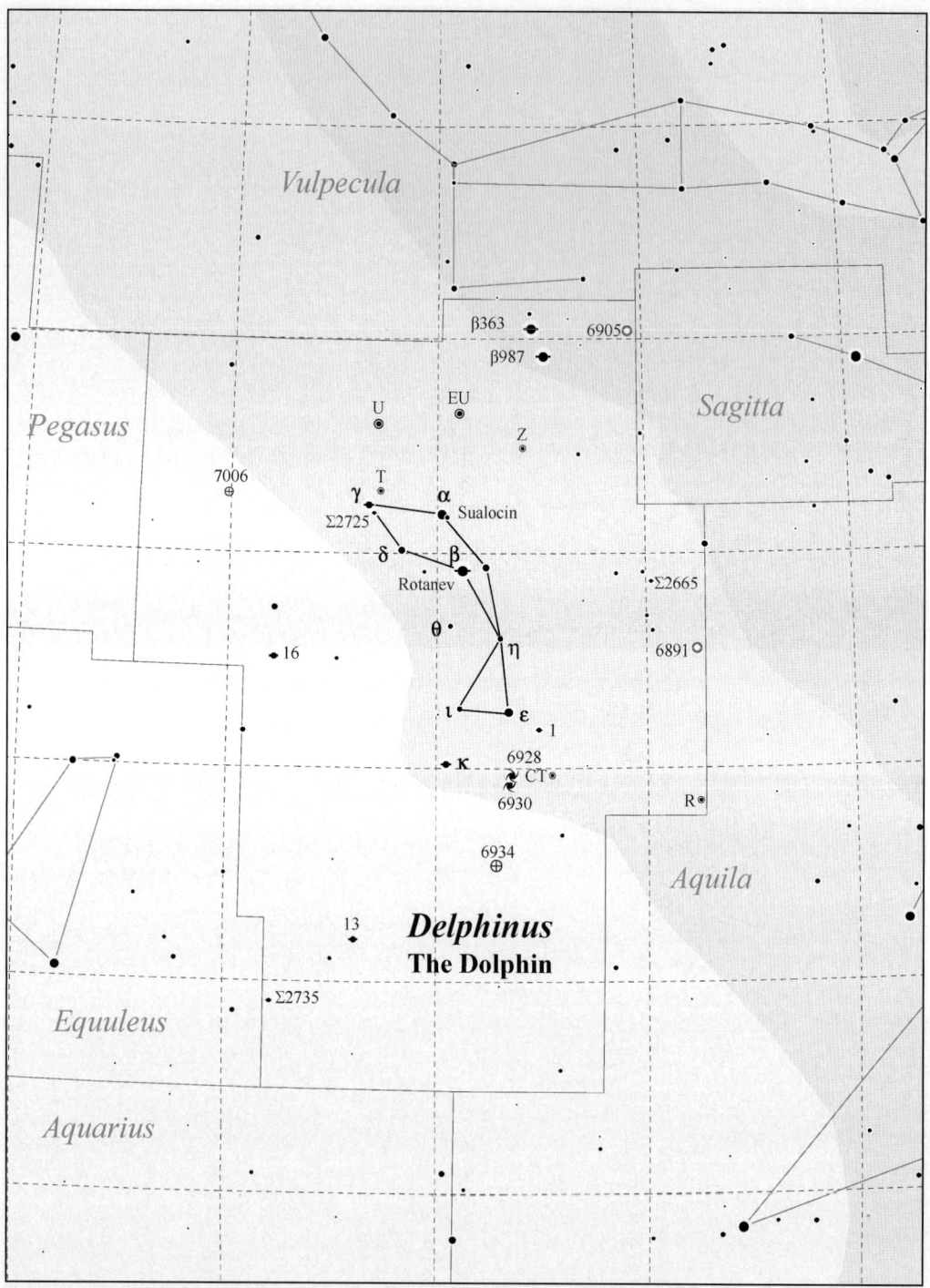

Fig. 13.6. Delphinus detail chart.

Table 13.2. Selected telescopic objects in Delphinus.

Multiple stars:

Designation	R.A.	Decl.	Type	m_1	m_2	Sep. (")	PA (°)	Colors	Date/ Period	Aper. Recm.	Rating	Comments
Σ2665	20h 19.4m	+14° 22'	A–BC	6.9	9.6	3.4	013	W, dY	2003	3/4"	****	1° 29' SE (PA 124°) of ρ Aql (m4.94). BC is too close for amateur instruments
β987AB	20h 30.2m	+19° 25'	A–B	6.6	10.9	2.6	126	W, W	1986	4/5"	*****	Very interesting quintuple star system. All 5 components visible in binoculars or small telescopes
S752AC			A–C	6.6	7.3	106.7	288	W, –	2001	7×50		
Fox 254AD			A–D	6.6	10.9	56.4	296	W, –	1913	3/4"		
Fox 254Aa			A–a	6.6	10.5	21.1	070	W, –	2003	2/3"		
1 Delphini	20h 30.3m	+10° 54'	AC–B	6.2	8.0	0.9	349	W, –	2003	10/11"	****	a.k.a. β63AB
β297AC			A–C	6.2	14.1	17.2	340	W, –	2000	9/10"		
β363	20h 31.0m	+20° 36'	A–B	6.2	10.0	11.8	070	W, –	1962	2/3"	****	
			A–C	6.2	12.8	53.3	206	W, –	2000	5/6"		
OΣΣ207, β Delphini	20h 37.5m	+14° 36'	A–B	4.1	5.0	0.37	024	dY, dB	26.65	>20"	******	A–B is a very rapid binary, too close for amateur scopes except near greatest separation, e.g., 2003/4, 2029/30, 2056, etc.
h5545AB–C			AB–C	3.7	13.1	18.7	126	B, –	1961	6/7"		
Σ2704AB–D			AB–D	3.7	11.0	46.0	319	B, –	1999	3/4"		
Lbz 1AB–D			AB–E	3.7	11.6	109.6	271	B, –	1991	3/4"		
7, κ Delphini	20h 39.1m	+10° 05'	A–B	5.0	11.8	34.7	282	Y, Y	1970	3/4"	****	a.k.a. OΣ533
			A–C	5.0	8.6	215.9	100	Y, pR	2001	7×50		
Σ2725 (Delphinus)	20h 46.2m	+15° 54'	A–B	7.5	8.2	6.15	011	Y, yW	2,851	6/7"	****	"Ghost double" (of γ Del), 0° 14.5' SSW (PA 206°) of Del (m3.77)
12, γ Delphini	20h 46.7m	+16° 07'	A–B	4.4	5.0	9.07	265	oY, gB	3,249	2/3"	*****	Lovely colors. Some observers see yellow and green. a.k.a. Σ2727
13 Delphini	20h 47.8m	+06° 00'	A–B	5.6	8.2	1.5	199	W, –	1991	6/7"	****	a.k.a. β65
16 Delphini	20h 55.6m	+12° 34'	A–B	5.6	11.8	34.9	015	W, –	2000	3/4"	****	a.k.a. h1592
Σ2735	20h 55.7m	+04° 32'	A–B	6.5	7.5	2.0	282	Y, –	2006	4/5"	****	

(continued)

Table 13.2. (continued)

Variable stars:

Designation	R.A.	Decl.	Type	Range (m_v)		Period (days)	F (f_a/f_1)	Spectral Range	Lum. (suns)	Aper. Recm.	Rating	Comments
R Delphini	20^h 14.9^m	+09° 05'	M	7.6	13.8	285.07	0.45	M5e–M6e		1:17"	***	3°13' ENE (PA 055°) of τ Aql (m5.51), 5°58' E of Altair (α Aql).##
CT Delphini	20^h 29.4^m	+09° 53'	Lb	6.8	8.5	–	–	M7		7×50	****	1°41' SW (PA 213°) of Del (m4.03).#
Z Delphini	20^h 32.7^m	+17° 27'	M	8.3	15.3	304.48	0.48	S5.2.5e–S2,7e		Large	****	3°07' W (PA 259°) of α Del (m3.77).##
EU Delphini	20^h 37.9^m	+18° 16'	SRb	5.8	6.9	59.7	–	M6.4III		6×30	****	2°23' N (PA 350°) of α Del (m3.77). A pulsating red giant similar to the Mira
T Delphini	20^h 45.3^m	+16° 24'	M	8.5	15.2	332.14	0.45	M3e–M6e		Large	****	0°25' NW (PA 311°) of γ2 Del (m3.77).##
U Delphini	20^h 45.5^m	+18° 05'	SRb	5.9	7.7	110	–	M5II–III		6×30	****	2°35' NNE (PA 032°) of α Del (m3.77).##$

*2

Star clusters:

Designation	R.A.	Decl.	Type	m_v	Size (')	Brst. Star	Dist. (ly)	dia. (ly)	Lum. (suns)	Aper. Recm.	Rating	Comments
NGC 6934	20^h 34.2^m	+07° 24'	Globular	8.7	5.9	13.8	49K	100	52K	4/6"	***	Partially resolved in 10"
NGC 7006	21^h 01.5^m	+16° 11'	Globular	10.6	3.6	15.6	127K	133	75K	4/6"	***	One of the more distant globulars known

Nebulae:

Designation	R.A.	Decl.	Type	m_v	Size (')	Cent. star	Dist. (ly)	dia. (ly)	Aper. Recm.	Rating	Comments
NGC 6891	20^h 15.2^m	+12° 42'	Planet.	10.5	14"	**12.44**	–	–	4/6"	****	May be mistaken for a star at low power. Bluish-green color
NGC 6905	20^h 22.4^m	+20° 05'	Planet.	11.1	39"	**15.5**	–	–	8/10"	****	"Blue Flash Nebula", lovely in larger scopes. Bluish color

Galaxies:

Designation	R.A.	Decl.	Type	m$_v$	Size (')	Dist. (ly)	dia. (ly)	Lum. (suns)	Aper. Recm.	Rating	Comments
NGC 6928	20^h 32.8^m	+09° 56'	SB(s)ab	12.2	2.0 × 0.6	190M	110K	40G	12/14"	**	1° 23' S (PA 184°) of Del (m4.03)
NGC 6930	20^h 33.0^m	+09° 52'	SB(s)ab?	12.8	1.1 × 0.5	190M	60K	20G	12/14"	**	~4' SSE of NGC 6928, in same mod. power FOV *3

Meteor showers:

Designation	R.A.	Decl.	Period (yr)	Duration	Max Date	ZHR (max)	Comet/ **Asteroid**	First Obs.	Vel. (mi/ **km**/sec)	Rating	Comments
None											

Other interesting objects:

Designation	R.A.	Decl.	Type	m$_v$	Size (')	Dist. (ly)	dia. (ly)	Lum. (suns)	Aper. Recm.	Rating	Comments
Harrington 9, θ Delphini	20^h 38.7^m	+13° 19'	Asterism	<5.7	30×40	–			7 × 50+	***	At least 7 stars can be seen in 7 × 50, more in larger binoculars

Fig. 13.7. NGC 6891.

PISCES (**pie**-seez) PISCIUM (**pish**-ee-um) Psc	**The Fishes** **9:00 P.M. Culmination**: Oct. 12 **AREA**: 889 sq. deg., 14th in size

After Zeus attacked and deposed his father Chronos, he and the other Olympian gods had to also fight the remaining 11 Titans who were brothers and sisters of Chronos. After they had won this fight, Zeus and his brothers and sisters imprisoned the Titans in the realm of Tartarus, one of the four primal entities in Greek mythology.

The realm of Tartarus, which also was known by its ruler's name, was believed to be as far below Hades as Olympus was above Hades. It was also the abode of terrible monsters and a place of terrible tortures, and perhaps the model for the subsequent Christian idea of hell.

Gaea (earth) was the mother of all the Titans and resented the imprisonment of her children. She fought back by mating with Tartarus (the primal being) and producing Typhon (or Typhoeus), the largest and most terrible monster ever seen. His upper limbs consisted of multiple serpents and when outspread spanned 600 miles. His lower limbs also consisted of multiple serpents. His head resembled that of an ass and touched the stars. Fire spurted from his eyes and flaming stones from his mouth.

When he attacked Olympus, all the gods save one fled in fear. Each changed into the form of an animal to hide. Aphrodite and her son Eros tied the ends of their sashes together, and changed themselves to fishes and dove into the waters of the Nile river to hide among the normal fish.

Athene was the only god or goddess to stay and fight Typhon. As she did, she taunted Zeus about his cowardice. Finally, the embarrassed Zeus resumed his form and began hurling thunderbolts at Typhon. Injured and in pain, Typhon fled to Mount Casius in the present day Syria. Zeus, sensing victory, followed him. Typhon turned on Zeus, and wrapping his many serpent limbs around Zeus' body, immobilized him. He then wrested Zeus's sickle away from him and used it to cut out the tendons which moved his arms and legs. Although Zeus was immortal, he was now helpless and Typhon dragged him into a cave. He then hid the tendons in a bearskin and had his sister monster, Delphine, guard them.

The other Olympians were dismayed and fled in disarray. Only Hermes and Pan took positive action. Together, they entered the cave, where Pan gave out one of his terrible screams which had earlier caused the Titans to flee (the origin of our word "panic"). Frightened by the sudden noise, Delphine temporarily abandoned her post, and Hermes was able to find the tendons and reinsert them into Zeus.

Zeus then returned to Olympus, mounted his chariot drawn by winged horses and set out in pursuit of Typhon again. With an enormous volley of thunderbolts, Zeus again wounded Typhon who once again fled in pain. This time, Zeus hurled Mt. Aetna onto Typhon. Trapped, Typhon continued to spew fire and rocks out of Mt. Aetna.

Afterward, the figures of the two fish were placed in the sky as the constellation Pisces.

Although faint, this Zodiacal constellation has two groups of stars representing the two fish, and a "V" shaped chain of stars which represent the knotted sashes which held Aphrodite and Eros together.

TO FIND: Look high in the south at about 9:00 PM a week or two before or after Oct. 12, and about two-thirds the distance from the horizon to the zenith, depending on your latitude. The Great Square of Pegasus should be almost directly overhead. Just south of the bottom of the square, you should be able to see a group of six faint stars forming a rough circle about 5° in diameter. This is the so-called circlet in Pisces. It represents one of the two fishes. From the northeastern star of the circlet, there is a train of seven faint stars going eastward and then bending slightly southeast. This line is about 40° long, and ends at α Pisces which marks the "knot" where the two sashes are tied together. From α, follow another train of four or five stars north–northwest for about 25°. This line ends at φ Piscium, which is the southeastern star of another group of stars forming a rough oval. This oval is about 50% larger than the "circlet." This oval represents the second fish. This oval lies about halfway between α Andromedae and α Arietis (Figs. 13.1, 13.9 and 13.10).

KEY STAR NAMES

Name	Derivation and Meaning
α Alrescha (al-ree-**shah**)	from al-rishia, the rope*[4]

Magnitudes and spectral types of principal stars:

Bayer desig.	Mag. (m_v)	Spec. type	Bayer desig.	Mag. (m_v)	Spec. type	Bayer desig.	Mag. (m_v)	Spec. type
α	3.82	A2	β	4.48	B6 Ve	γ	3.70	G7 III
δ	4.44	K5 III	ε	4.27	K0 III	ζ	5.21	A7 IV
η	3.62	G8 III	θ	4.27	K1 III	ι	4.13	F7 V
κ	4.95	A0 Ip	λ	4.49	A7 V	μ	4.84	K4 III
ν	4.45	K3 III	ξ	4.61	K0 III	o	4.26	K0 III
π	5.54	F0 V	ρ	5.35	F2 V	σ	5.50	B9.5 V
τ	4.51	K0 III/IV	υ	4.74	A3 V	φ	4.67	K0 III
χ	4.66	K0 III	ψ¹	4.68	A1 Vn	ψ²	5.56	A3 V
ψ³	5.57	G0 III	ω	4.03	F4 IV			

Stars of magnitude 5.5 or brighter which have Flamsteed but no Bayer designations:

Flam-steed.	Mag. (m_v)	Spec. type	Flam-steed.	Mag. (m_v)	Spec. type	Flam-steed.	Mag. (m_v)	Spec. type
2	5.43	K1 III	5	5.42	G8 IV	19	4.95	C5 II
20	5.49	G8 III	27	4.88	G9 III	29	5.13	B7 III/IV
30	4.37	M3 III	33	4.61	K1 III	41	5.38	K3 III
47	5.01	M3 III:var	52	5.83	K0 III	57	5.36	M4 III
64	5.07	F8 V	68	5.44	G6 III	82	5.15	F0 V
89	5.13	A3 V	91	5.23	K5 III	94	5.50	K1 III
107	5.24	K1 V						

Fig. 13.8a.   The stars of Pisces and adjacent constellations.

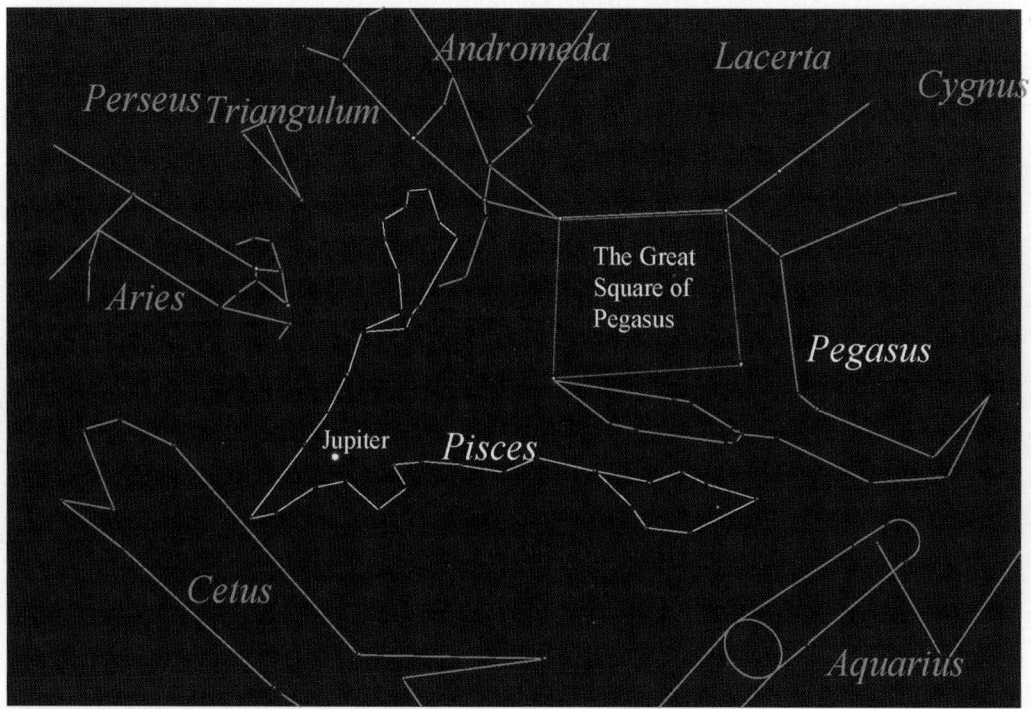

Fig. 13.8b.   The figures of Pisces and adjacent constellations.

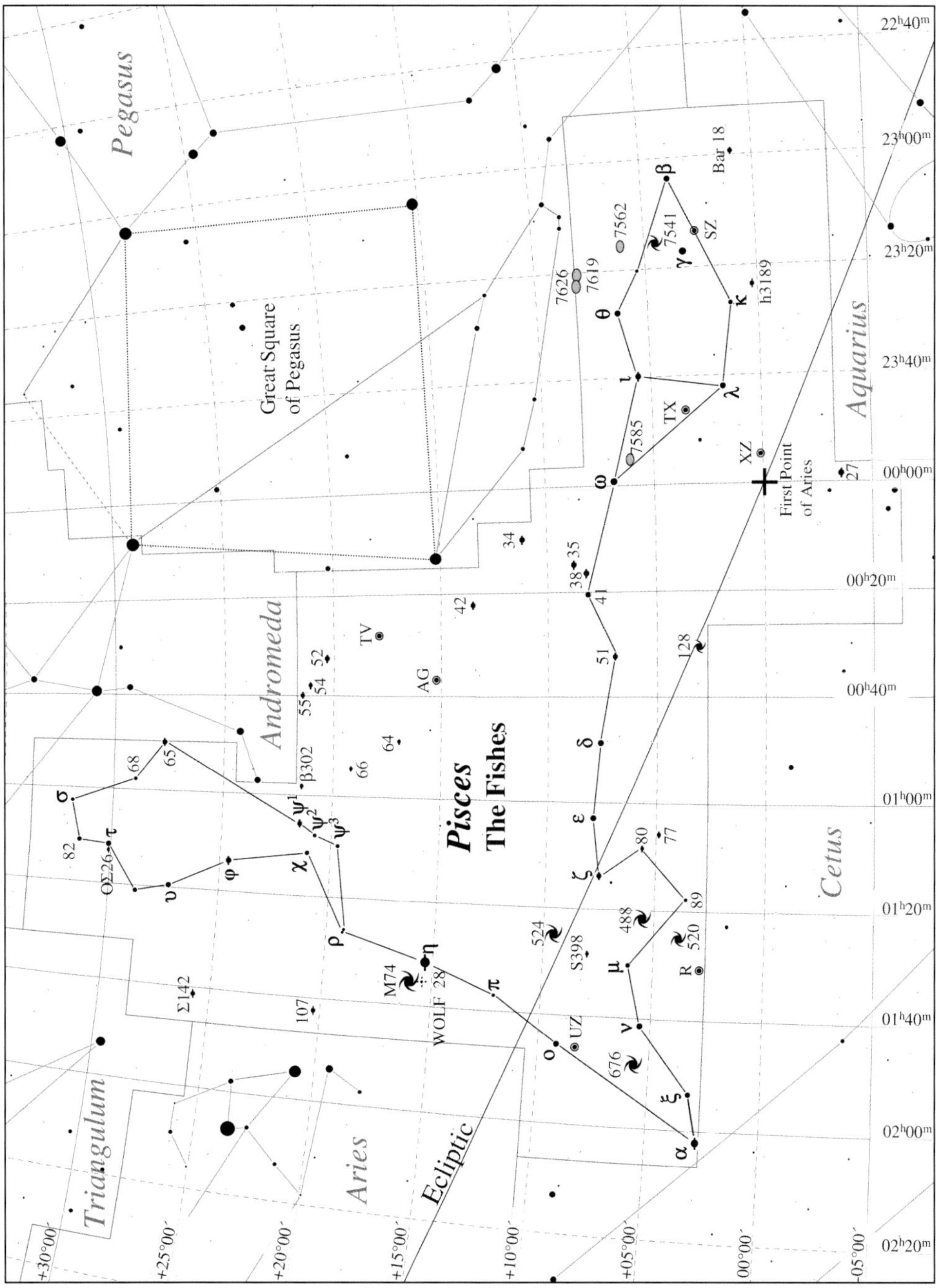

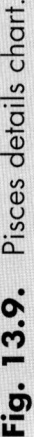

Fig. 13.9. Pisces details chart.

Table 13.3. Selected telescopic objects in Pisces.

Multiple stars:

Designation	R.A.	Decl.	Type	m_1	m_2	Sep. (")	PA (°)	Colors	Date/ Period	Aper. Recm.	Rating	Comments
2 Piscium	22h 59.5m	+00° 58'	A–B	5.4	13.3	3.9	088	oY, ?	1961	6/7"	***	a.k.a. Bar 18
h 3189	23h 23.5m	+00° 17'	A–B	6.3	10.5	39.2	148	dY, ?	2003	15×80	****	
17, ι Piscium	23h 39.9m	+05° 38'	A–B	4.1	13	119.2	305	yW, ?	2002	6/7"	***	a.k.a. Bup 240AB
27 Piscium	23h 58.7m	–03° 33'	A–C	4.9	8.9	**0.89**	**323**	dY, ?	**695**	10/11"	****	a.k.a. β 730. Slowly closing to min. of 0.84" in 2029, then opening to 3.53" in 2389
34 Piscium	00h 10.0m	+11° 09'	A–B	5.5	9.4	7.9	162	bW, pB	2003	2/3"	****	a.k.a. Σ5
35, UU Piscium	00h 15.0m	+08° 49'	A–B	6.1	7.5	11.6	148	yW, pY	2006	2/3"	****	Nice colored pair. a.k.a Σ12
38 Piscium	00h 17.4m	+08° 53'	AB–C	7.3	7.7	4.0	235	Y, Y	2004	2/3"	*****	Attractive pair in a pretty field. a.k.a. Σ22AB–C. AB has a separation of ~0.1"
			AB–D	7.3	11.5	65.7	151	yW, ?	2004	15×80	***	a.k.a. Σ22AB–D
42 Piscium	00h 32.4m	+13° 29'	A–B	6.3	10.3	30.6	315	dY, Gr	2006	15×80	****	Easy, lovely colored pair. a.k.a. Σ27
51 Piscium	00h 32.4m	+06° 57'	A–B	5.7	9.5	27.1	082	bW, bGr	2003	15×80	****	Easy, lovely colored pair. a.k.a. Σ36Aa–B A–a is a close double, sep. 0.3"
52 Piscium	00h 32.6m	+20° 18'	A–B	5.5	11.7	54.4	299	Y, Y	1982	3/4"	***	a.k.a. h 1982
54 Piscium	00h 39.4m	+21° 16'	A–B	6.0	11.5	167.6	080	oY, ?	1997	15×80	****	a.k.a. OΣ550
55 Piscium	00h 39.9m	+21° 26'	A–B	5.6	8.5	6.6	196	yO, B	2003	2/3"	*****	Beautifull Compare with Albireo at 3× aperture and power. a.k.a. Σ46
64 Piscium	00h 49.0m	+16° 56'	A–B	5.1	12.6	82.9	330	yW, ?	1998	5/6"	***	a.k.a. Bup 12AB
			A–B	5.1	13.0	64.2	162	yW, ?	1998	6/7"	***	a.k.a. Bup 12AC
65 Piscium	00h 49.9m	+27° 43'	A–B	6.3	6.3	4.4	115	oY, oY	2006	2/3"	*****	Lovely matched pair
66 Piscium	00h 54.6m	+19° 11'	A–B	6.1	7.2	**0.56**	**182**	pY, pB	**334.76**	17/18"	****	a.k.a. OΣ20AB. Passed min. of 0.46" in 1973, opening to max of 0.86" in 2099
β 302	00h 58.3m	+21° 24'	AB–C	5.8	12.8	150.6	009	W, ?	2003	5/6"	***	a.k.a. β 1352AB–C
			A–B	6.6	8.8	0.3	191	W, ?	2003	–	**	A real challenge for larger scopes

Name	R.A.	Decl.	Type					Colors		Aper. Recm.	Rating	Comments
72, Piscium	01h 05.1m	+14° 57'	A–B	5.7	12.6	56.9	256	yW, ?	1997	5/6"	***	a.k.a. h 1068
74, γ¹ Piscium	01h 05.7m	+21° 28'	A–B	5.3	5.5	30.3	159	W, W	2006	10×50	*****	Easy pair of nearly matched stars. Lovely. Very nice field. a.k.a. Σ88
77 Piscium	01h 05.8m	+04° 55'	A–B	6.4	7.3	33.8	083	Y, Y	2006	7×50	****	Wide pair, easy in any scope. a.k.a. Σ90AB
80 Piscium	01h 08.4m	+05° 40'	A–B	5.5	12.2	86.1	286	yW, ?	1907	4/5"	***	a.k.a. Bup 15
OΣ26	01h 13.0m	+30° 04'	A–B	6.3	10.5	10.6	258	dY, Y	1999	2/3"	****	
85, φ Piscium	01h 13.7m	+24° 35'	A–B	4.7	9.1	7.8	229	O, ?	2003	2/3"	***	Distinctly orange primary, secondary is faint and color hard to distinguish. a.k.a. Σ100
86, ζ Piscium	01h 13.7m	+07° 35'	A–B	5.2	6.2	23.3	069	W, Y	2006	10×50	****	Wide pair, easy in any scope, slight color difference
S 398	01h 28.4m	+07° 58'	A–B	6.3	8.0	69.1	100	pR, pB	2001	7×50	***	
99, η Piscium	01h 31.5m	+15° 21'	A–B	3.8	7.5	0.6	054	dY, ?	2000	16/17"	***	
Σ145	01h 41.3m	+25° 45'	A–B	6.3	10.9	10.5	031	pY, ?	1997	3/4"	****	
			A–C	6.3	11.2	84.1	337	pY, ?	1997	3/4"	***	
107 Piscium	01h 42.5m	+20° 16'	A–B	5.3	11.7	19.0	248	dY, W	1910	3/4"	****	a.k.a. h 2071
113, α Piscium	02h 02.0m	+02° 46'	A–C	5.3	12.1	104.4	353	dY, ?	1924	3/4"	***	
			A–B	4.1	5.2	**1.77**	**265**	W, W	**933.05**	5/6"	****	a.k.a. Σ202AB. Slowly closing to 1.0" in 2096, then min. of 0.70" in 2112

Variable stars:

Designation	R.A.	Decl.	Type	Range (m_v)		Period (days)	F (f_1/f_2)	Spectral Range	Aper. Recm.	Rating	Comments
SZ Piscium	23h 13.4m	+02° 41'	EA/DS/RS	7.2	7.7	3.966	0.11	K1IV-V + F8V	6×30	****	Very red. Brightest class "N" star in northern hemisphere
19, TX Piscium	23h 46.4m	+03° 29'	Lb	4.8	5.2	–	–	N0(C6.2)	Eye	****	
XZ Piscium	23h 46.8m	+00° 07'	Lb	5.6	6.0	–	–	M5Ib	6×30	****	1°39' SE (PA 126°) of 21 Piscium

(continued)

Table 13.3. (continued)

Variable stars:

Designation	R.A.	Decl.	Type	Range (m_v)	Period (days)	F (f_r/f_f)	Spectral Range	Aper. Recm.	Rating	Comments
47, TV Piscium	00^h 28.0^m	+17° 54'	SR	4.7	49.1	–	M3III–M4IIIb	Eye	****	
53, AG Piscium	00^h 36.8^m	+15° 14'	β Cep	5.8	0.08	–	B2.5IV	6×30	****	
R Piscium	01^h 30.6^m	+02° 53'	M	7.1	344.5	0.44	M3e–M6e	18/20"	****	3° 16' S (PA 178°) of μ Piscium (m4.84)
UZ Piscium	01^h 45.7^m	+08° 34'	α CVn	6.4	4.1327	0.50	A2p (Cr–Eu–Sr–Si)	6×30	*****	36' S (PA 173°) of O Piscium (m4.26). Spectral lines show Si, Cr, Eu, and Sr

Star clusters:

Designation	R.A.	Decl.	Type	m_v	Size (')	Brtst. Star	Dist. (ly)	dia. (ly)	Number of Stars	Aper. Recm.	Notes
None											

Nebulae:

Designation	R.A.	Decl.	Type	m_v	Size (')	Brtst./Cent. star	Dist. (ly)	dia. (ly)	Aper. Recm.	Comments
None										

Galaxies:

Designation	R.A.	Decl.	Type	m_v	Size (')	Dist. (ly)	dia. (ly)	Lum. (suns)	Aper. Recm.	Rating	Comments
NGC 7541	23^h 14.7^m	+04° 32'	Sb	11.7	3.2×1.0	109M	86K	35G	8/10"	****	Large, bright core, mod. bright elong. halo, ESE–WNW *5
NGC 7562	23^h 16.0^m	+06° 41'	E2.5	11.6	2.2×1.7	170M	109K	25G	8/10"	***	Fairly bright halo, brighter toward center
NGC 7619	23^h 20.2^m	+08° 12'	E	11.1	2.8×2.5	170M	138K	84G	8/10"	***	Fairly bright, in same FOV and near twin of NGC 7626
NGC 7626	23^h 20.7^m	+08° 13'	Epec	11.1	2.8×2.4	160M	130K	74G	8/10"	***	Fairly bright, two fainter galaxies may be visible in large scopes

Designation	R.A.	Decl.	Type	m_v	Size				Aper. Recm.	Rating	Comments	
NGC 7785	23h 55.3m	+05° 55'	E5.5	11.6	2.0×1.3	180M	105K	59G	8/10"	***	Halo elongated NW–SE, more difficult than its magnitude suggests	*6
NGC 128	00h 29.2m	+02° 51'	S0	11.8	2.8×0.9	190M	155K	55G	8/10"	****	Fairly bright, elongated N–S, slightly brighter toward center	
NGC 488	01h 21.8m	+05° 15'	Sa	10.3	5.4×4.5	96M	145K	55G	8/10"	****	Fairly bright, elongated N–S, concentrated core. Tightly wound arms	
NGC 520	01h 24.6m	+03° 48'	Pec	11.4	4.6×1.9	91M	103K	19G	12/14"	***	a.k.a. Arp 157. Actually a pair of overlapping irreg gal, elong NW–SE, sep. by dark streak	
NGC 524	01h 24.8m	+09° 32'	Sa	10.2	3.2×3.2	105M	98K	46G	8/10"	****	Northernmost of an appr. equilateral triangle with two 13 m gal. ~14' away	*7
M74, NGC 628	01h 36.7m	+15° 47'	Sc	9.4	11.0×11.0	32M	67K	21G	8/10"	****	Fine example of face on galaxy, lge brt core, faint arms	
NGC 676	01h 49.0m	+05° 54'	S0	9.6	4.6×1.7	64M	70K	28G	8/10"	***	Lenticular, surprisingly difficult to see given its total magnitude	

Meteor showers:

Designation	R.A.	Decl.	Period (yr)	Duration	Max Date	ZHR (max)	Comet/ **Asteroid**	First Obs.	Vel. (mi/ **km**/sec)	Rating	Comments
None											

Other interesting objects:

Designation	R.A.	Decl.	Type	m_v	Dist. (ly)	Planet Mass (J)	Spec. Class	Lum. (suns)	Aper. Recm.	Rating	Comments	
Wolf 28	00h 49.1m	+05° 25'	White Dwarf	12.4	13.64	2.99"/year	DZ8	0.00016	8/10"	**	One of the few white dwarf stars visible in amateur telescopes	*8
1st point of Aries	00h 00.0m	+00° 00'	Ref Pt.	–	–	–	–	–	Eye	***	Point where sun crosses celestial equator defining start of spring	*9

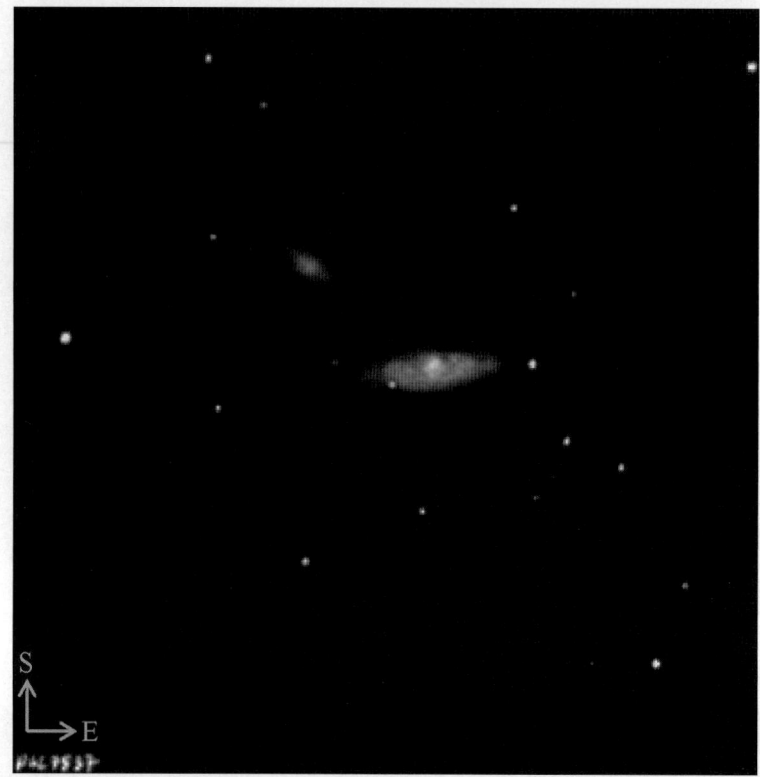

Fig. 13.10. NGC 7541 and 7537 (in background), spiral galaxies in Pisces. Sketch by Nicoas Biver.

Fig. 13.11. NGC 520, overlapping galaxies in Pisces. Astrophoto by Mark Komsa.

Fig. 13.12. M74 (NGC 628), spiral galaxy in Pisces. Astrophoto by Stuart Goosen.

PISCIS AUSTRINUS (**pis**-eez os-**try**-nus) (**pie**-siss os-**try**-nus is the more common US pronunciation)	**The Southern Fish**
PISCIS AUSTRINI (**pis**-eez **awe**-stry-nigh) (**pie**-siss **os**-tree-nee is the (more common US pronunciation)	**9:00 P.M. Culmination**: Oct. 10
PsA	**AREA**: 245 sq. deg., 60th in size

Piscis Austrinus is a very ancient constellation and was recognized by many of the ancient middle eastern civilizations. These civilizations always saw this constellation as a fish or half fish. However, each civilization seems to have had its own explanation for which god or goddess the constellation represented. The Babylonians saw the form of Ea, the first of their gods, whose home was in a fresh water sea, and the creator of mankind. The Sumerians of Eridu recognized the constellation as Ea, the god of fresh water rather than the creator. Others saw the figure of Oannes, Dagon (Dagan), or goddesses such as Decerto (Dekerto), Aphrodite, or Venus.

Probably the most familiar version is that the fish represented the Semitic god Dagon, who is mentioned in the Bible. Dagon was seen as a half man/half fish, and his female counterpart was recognized by some cultures as Derceto (or Derketo). These two may have been the inspiration for tales of mermen and mermaids supposedly seen by ancient sailors. In one story, Derceto fell in love with a mortal and became pregnant by him. When the baby was born, Derceto was so embarrassed that she abandoned it and threw herself into the water and changed into the fish or half fish that we see in Piscis Austrinus today. The baby girl was fed by doves until she was found and raised by a royal shepherd named Simmas. She grew up to become Semiramis, a legendary Assyrian queen who conquered much of Asia.

TO FIND: Although most of the stars in Piscis Austrinus are relatively faint, it is easy to find because of the 1st magnitude star <u>Fomalhaut</u>, which is in the middle of a region with very few bright stars. From <u>Fomalhaut</u>, the nearest first magnitude star is <u>Achernar</u> (α Eridani) which is about 40° to the southeast but is not visible from much of the northern hemisphere. The nearest bright star visible to northern hemisphere observers above 30° latitude is <u>Altair</u> (α Aquilae) which is nearly 60° to the northwest. An alternate way to find <u>Fomalhaut</u> is to find the two westernmost stars in the Square of Pegasus, Markab (α Pegasi) and Sheat (β Pegasi). The line from Sheat to Markab extended southward 45° ends just east of Fomalhaut (Figs. 13.1, 13.13 and 13.14).

NAMED STARS:

Name	Pronunciation	Source of name
α Fomalhaut	**foe**-mal-hought	Fum al Hut, (Arabic) "The Fish's Mouth"

Magnitudes and spectral types of principal stars:

Bayer desig.	Mag. (m$_v$)	Spec. type	Bayer desig.	Mag. (m$_v$)	Spec. type	Bayer desig.	Mag. (m$_v$)	Spec. type
α	1.17	A3 V	β	4.29	A1 V	γ	4.46	A0 III
δ	4.20	G8 III	ε	4.18	B8 V	ζ	6.43	K1 III
η	5.43	B8/B9 V	θ	5.02	A1 V	ι	4.35	B9.5 V
κ	_._	_ _*10	λ	5.45	B7 V	μ	4.50	A2 V
ν	_._	_ _*11	ξ	_._	_ _*12	ο	_._	_ _*12
π	5.12	A9 V*13	ρ	_._	_ _*14	σ	_._	_ _*14
τ	4.94	F6 V*15	υ	4.99	K4 III*15			

Stars of magnitude 5.5 or brighter which have Flamsteed but no Bayer designations:

None.

Stars of magnitude 5.5 or brighter which have no Bayer or Flamsteed designations:

Desig-nation	Mag. (m$_v$)	Spec. type
HD 2120171	5.37	A5 IV

Fig. 13.13a. The stars of Piscis Austrinus and adjacent constellations.

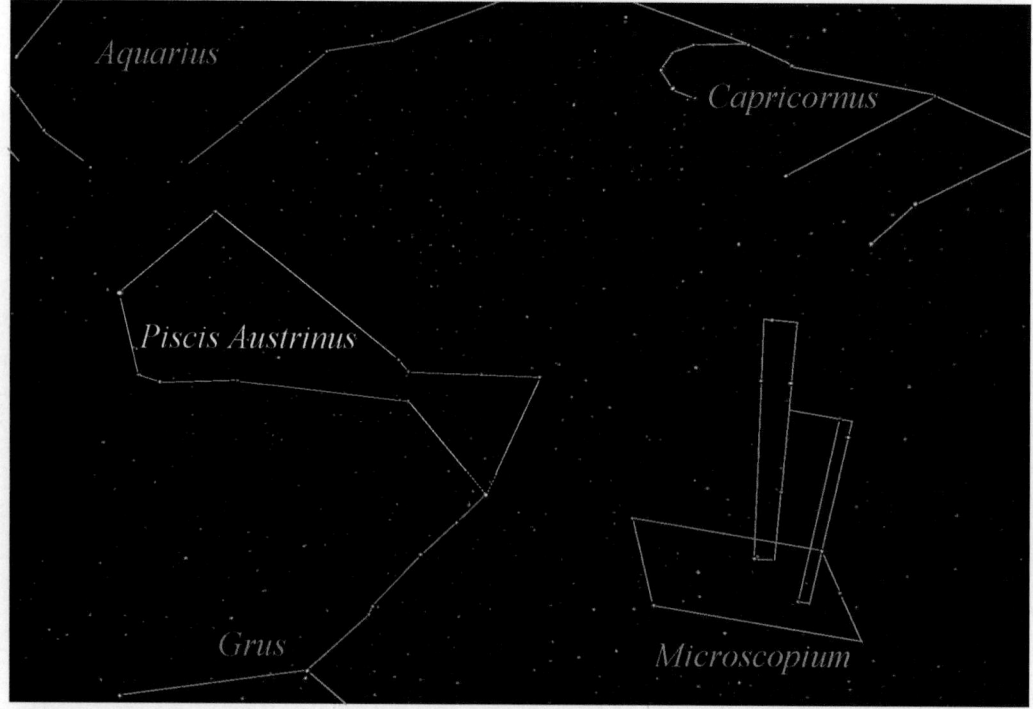

Fig. 13.13b. The figures of Piscis Austrinus and adjacent constellations.

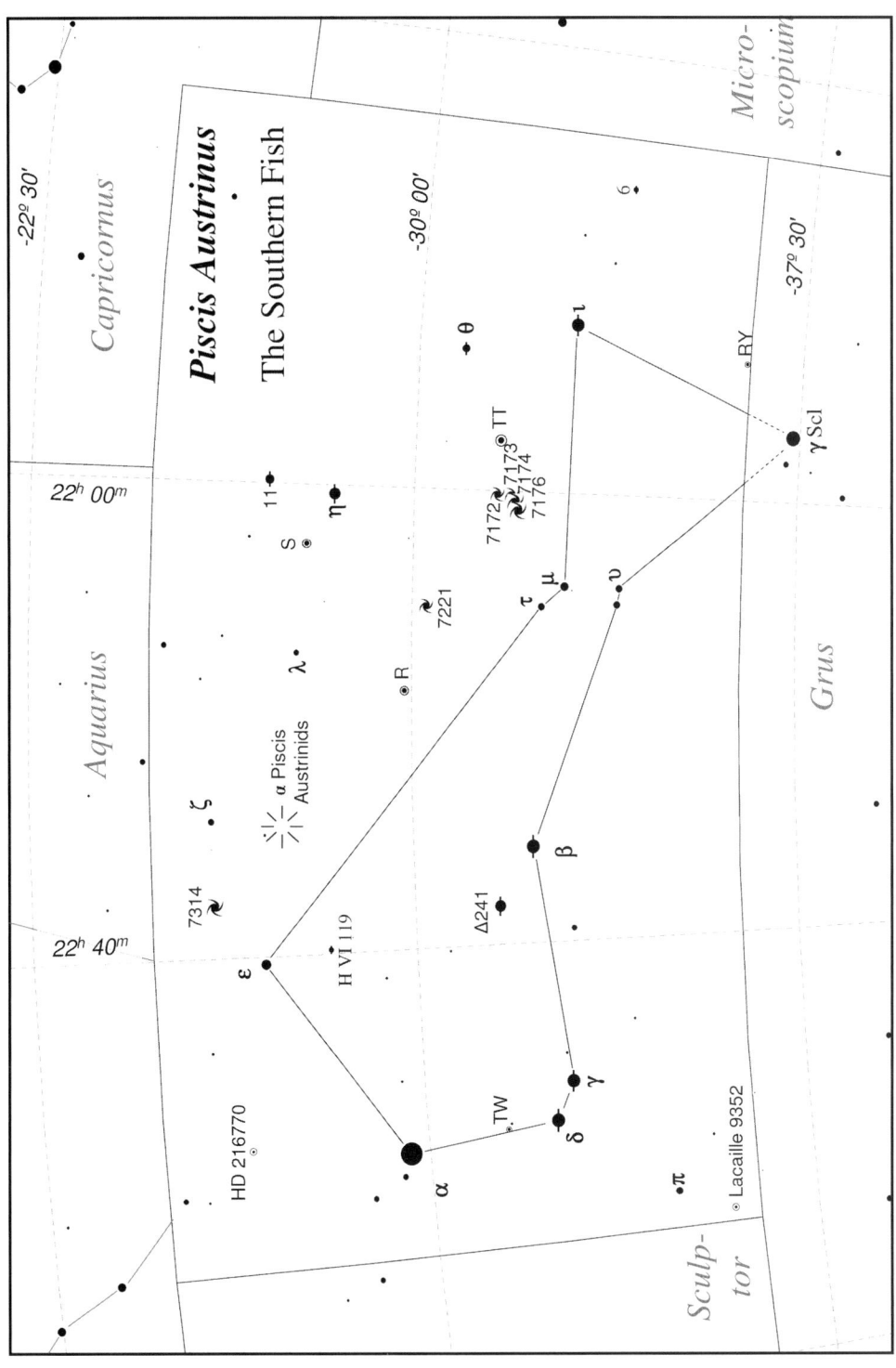

Fig. 13.14. Piscis Austrinus details.

Table 13.4. Selected objects in Piscis Austrinus.

Multiple stars:

Designation	R.A.	Decl.	Type	m_1	m_2	Sep. (")	PA (°)	Colors	Date/Period	Aper. Recm.	Rating	Comments
6 Piscis Austrini	21h 32.2m	−33° 57'	A–B	6.0	13.3	6.2	058	W, ?	1999	7/8"	***	a.k.a. I 1051
9, ι Piscis Austrini	21h 44.9m	−33° 02'	A–B	4.3	11.4	20.0	290	W, ?	1910	3/4"	****	a.k.a. h 5296. AB = Fin 330AB, separation 0.s1"
10, θ Piscis Austrini	21h 47.8m	−30° 54'	AB–C	5.1	11.3	35.6	339	W, ?	1998	3/4"	****	
11 Piscis Austrini	21h 59.6m	−27° 38'	A–B	7.3	10.5	11.4	036	Y, ?	2000	2/3"	****	a.k.a. β 276
12, η Piscis Austrini	22h 00.8m	−28° 27'	A–B	5.7	6.8	1.4	112	bW, bW	2006	6/7"	****	
17, β Piscis Austrini	22h 31.5m	−32° 21'	A–B	4.3	7.1	30.6	173	Y, pB	2006	10×80	*****	a.k.a. Pz 7. Showcase double, similar to Albireo but colors are less vivid
Δ 241	22h 36.6m	−31° 40'	A–B	5.9	7.6	93.3	031	O, dY	1999	7×50	******	Showcase double, nice color combination
H VI 119	22h 39.8m	−28° 21'	A–B	6.4	7.5	86.5	159	dY, pY	1991	7×50	****	
			B–C	7.5	8.6	3.1	069	pY, pY	2000	3/4"	****	
γ Piscis Austrini	22h 52.5m	−32° 53'	A–B	4.5	8.2	4.0	258	pY, dY	1992	2/3"	****	a.k.a. h5367. 1°17' ENE (PA 058°) of β PsA
23, δ Piscis Austrini	22h 55.9m	−32° 32'	A–B	4.2	9.2	5.2	243	Y, ?	1966	2/3"	****	a.k.a. Hwe 91

Variable stars:

Designation	R.A.	Decl.	Type	Range (m_v)		Period (days)	F (f_1/f_2)	Spectral Range	Aper. Recm.	Rating	Comments
RY Piscis Austrini	21h 47.5m	−36° 12'	M	9	<14.0	224	–	Me	2/18"	***	1°44' NW (PA 312°) of ? Gru (m3.00).##
TT Piscis Austrini	22h 03.3m	−31° 27'	Lb	7.0	7.5	–	–	M7 III	7×50	****	1°50' NW (PA 307°) of ? PsA (m3.94).#
S Piscis Austrini	22h 03.8m	−28° 03'	M	8.0	14.5	172	0.42	Me IIe–M5 IIe	02/20"	***	0°46' ENE (PA 058°) of ? PsA (m5.43).##
R Piscis Austrini	22h 18.0m	−29° 36'	M	8.3	14.7	298	0.38	Me IIe–M5 IIe	02/20"	***	2°00' SSE (PA 156°) of ? PsA (m5.45).##
TW Piscis Austrini	22h 56.4m	−31° 34'	BY Dra	6.4	6.5	10	–	K5 Ve	6×30	****	0°58' N (PA 005°) of ? PsA (m4.20) *16

Star clusters:

Designation	R.A.	Decl.	Type	m_v	Size (')	Brtst. Star	Dist. (ly)	dia. (ly)	Number of Stars	Aper. Recm.	Rating	Comments
None												

Nebulae:

Designation	R.A.	Decl.	Type	m_v	Size (')	Dist. (ly)	dia. (ly)	Brtst./Cent. star	Aper. Recm.	Rating	Comments
None											

Galaxies:

Designation	R.A.	Decl.	Type	m_v	Size (')	Dist. (ly)	dia. (ly)	Lum. (suns)	Aper. Recm.	Rating	Comments
NGC 7172	22h 02.0m	−31° 52'	SB	11.9	2.5×1.4	111M	58K	17G	12/14"	***	1°51' WNW (PA 291°) of † PsA. In same FOV as 7173, 7174 and 7176
NGC 7173	22h 02.6m	−31° 58'	SB	12.0	4.4×2.6	104M	37K	10G	12/14"	***	1°47' WNW (PA 289°) of † PsA (m4.94). Small, but fairly bright
NGC 7174	22h 02.1m	−31° 59'	S0pec	11.3	3.4×2.2	116M	44K	11G	12/14"	***	1°47' WNW (PA 288°) of † PsA. Faint, diffuse halo, no nucleus
NGC 7176	22h 02.1m	−31° 59'	S0pec	11.1	3.5×2.2	105M	40K	11G	16/18"	***	1°47' WNW (PA 289°) of † PsA
NGC 7221	22h 11.3m	−30° 37'	SB(rs)pec	12.1	2.0×1.5	190M	110K	42G	8/10'	***	1°37' ESE (PA 113°) of 13 PsA (m6.45)
NGC 7314	22h 35.8m	−26° 03'	SBbc	10.8	4.6×2.0	60M	66K	18G	8/10"	***	1°06' E (PA 089°) of z PsA (m6.43). Diffuse halo makes it appear faint

(continued)

Table 13.4. (continued)

Meteor showers:

Designation	R.A.	Decl.	Period (yr)	Duration	Max Date	ZHR (max)	Comet/ **Asteroid**	First Obs.	Vel. (mi/ **km**/sec)	Rating	Comments
Piscis Austrinids	22^h 44^m	−16°	Annual	07/15 08/10	07/28	5	Unknown	1865	22/**35**	****	High % of bright and fireballs, S. hemisphere sites more favorable *17

Other interesting objects:

Designation	R.A.	Decl.	Type	m_v	Mass (suns)	Dia. (suns)	Dist. (ly)	Planet Mass (J)	Dist. (AU)	Period (days)	Eccentricity	Comments
HD 216770	22^h 55.9^m	−26° 40'	ExoPlnt	8.1	0.9	0.7	124	>0.647	0.456	118.45	0.37	Planet: M>0.65 J, d=0.46 AU, e=0.37±0.06

Designation	R.A.	Decl.	Type	m_v	M_v	Lum. (suns)	Mass (suns)	Dia. (suns)	Dist. (ly)	Aper. Recm.	Rating	Comments
Lacaille 9352	23^h 05.7^m	−35° 51'	Red dwf	7.34	9.8	0.011	0.47	0.53	10.75	7 × 50	****	Brightest Red Dwarf. a.k.a. Gl 887, AKS Gl 887, PM of 6.9"/ year is 4th highest known

NOTES:

1. Giuseppe Piazza's Palermo Observatory assistant, Niccolo Cacciatore, translated his name into the Latin "Nicolaus Venator" (Nicolas Hunter in English) and put the reversed spellings on a star chart of Delphinus that he was preparing for Piazza. No one detected the false names until they had become well established. This appears to be the only case of someone successfully attaching his own name to a star, or in this case, to two stars.

2. U Delphini is a somewhat uncommon type of red giant. Its spectrum shows lines of technetium, an extremely rare radioactive element whose longest-lived isotope has a half-life of only 4.2 million years. On the earth, all the technetium present at its formation has decayed, and it is found only in uranium ores as a product of spontaneous fission with an abundance of about 1 part in 1 quadrillion (1/1,000,000,000,000,000). In red giant stars, it may be the product of spontaneous fission of uranium present at their formation, but some astronomers believe it is evidence of later formation of uranium (and other heavy elements) in the star itself. In addition to the main period of 110 days, an additional period of 1,100 days has been reported, and the 110-day period is sometimes replaced by a period of about 160–180 days.

3. NGC 6930 and 6928 are also accompanied by two fainter galaxies, NGC 6927 and 6927A, in the same medium power FOV.

4. Although the name Alrescha (the rope) seems appropriate, it is actually a mistake. This name originally referred to β Piscium which was located on the rope in an Arabic constellation consisting of a well and bucket. It was apparently erroneously transferred to α sometime in the middle ages.

5. NGC 7541 is accompanied by a fainter companion, NGC 7537, just 3′ away at PA 225°.

6. NGC 128 is the brightest of a group of five galaxies, only two of which are visible in 8 telescopes.

7. M74 (NGC 628) often resembles a globular cluster in small telescopes.

8. Wolf 28 is also called "Van Maanen's Star." It is the 26th closest star to our sun, and coincidentally, has the 26th highest proper motion.

9. The "First Point of Aries" is the point where the sun's apparent motion around the sky crosses the celestial equator in a south to north direction. This crossing also defines the vernal (spring) equinox, or one of the two times of the year when the daylight and nighttime are exactly equal everywhere on the earth. It is so named because, when it was first determined several thousand years ago, it was located in the constellation of Aries. Precession of the earth's axis causes this point to move along the ecliptic about 1° every 70 years. It will move into Aquarius in approximately 2600 AD.

10. In the Hipparcho–Ptolemy star list for Piscis Austrinus, a star was listed as "the one at the end of the tail." Bayer designated this star κ, but when the IAU set formal boundaries to the constellations, this star lay in Grus and was redesignated as γ Gruis. Also at that time, the stars with Flamsteed designations of 1, 2, 3, and 4 became part of Microscopium, so Piscis Austrinus now has no stars designated 1, 2, 3, or 4. The Hipparcho–Ptolemy star list did not include π PsA as either being part of the figure or as one of the "unformed stars" nearby, so it may have been added by the modern definition of the constellations, but I have not been able to confirm this.

11. Bayer did not assign the letter ν in PsA. Lacaille assigned this letter to 13 PsA, but the designation was later dropped due to the faintness of 13 PsA (magnitude 6.47).

12. ξ PsA and o PsA were not assigned by Bayer.

13. π PsA was not used by Bayer. Lacaille assigned this designation.

14. ρ ad σ are not used in PsA.

15. τ and υ were not assigned by Bayer or Lacaille. Bode added these letters, as well as several others, but only τ and υ have survived.

16. TW PsA is an orange-red flare star with a luminosity of only 1/8th that of the sun. It is of spectral class K4-5 V, hotter and brighter than the typical flare star, which is usually a red dwarf of class M. TW PsA is only about 0.9 light years from Fomalhaut and shares a common space motion, so it is believed that the two are gravitationally associated and may have formed in the same nebula or star cluster.

17. In his book "Meteor Showers" and on his corresponding Web site, Gary W. Kronk attributes the discovery of this shower to Alexander S. Herschel and lists six other radiants within Piscis Austrinus and one in Grus, all with slightly varying positions and times of maxima. This may imply that a single parent comet broke up into several discrete bodies which followed independent orbits, or that the meteor stream itself made repeated passes near a planet which deflected different parts of the stream into multiple orbits. The fact that all these streams intersect the earth's orbit at approximately the same point could also imply that the defecting body was the earth itself.

CHAPTER FOURTEEN

A Fatal Chariot Ride

February

AURIGA

The Charioteer

Auriga is one of the oldest constellations, being known in ancient Mesopotamia as a charioteer at least as early as 2000 BC and probably much earlier. Even then, Auriga was depicted as holding a goat or goats, much as it is shown today. Because of its extreme age, the stories told about Auriga have incorporated the myths and legends of several cultures, with each one modifying the earlier stories to correspond to some local person or diety.

The Greeks had two separate and distinct stories about the origin of Auriga. The first of these portrays Auriga as Myrtilus, the son of Hermes. Myrtilus became the charioteer to Oenomaus, king of Pisa. King Oenomaus had a beautiful daughter named Hippodamia, who was courted by the most eligible bachelors of the country. However, an oracle had told Oenomaus that the man who married Hippodamia would indirectly cause his death. In fear, Oenomaus set up a challenge for potential suitors. Each had to bring his own chariot and horses. The suitors were allowed to take Hippodamia in their chariot and given a head start as far as the isthmus of Corinth (about half an hour's ride) before Oenomaus began his pursuit. If the suitor escaped, he won Hippodamia as his bride. However, if Oenomaus overtook him, the suitor was executed. Although sounding fair,

P. Simpson, *Guidebook to the Constellations: Telescopic Sights, Tales, and Myths*,
Patrick Moore's Practical Astronomy Series, DOI 10.1007/978-1-4419-6941-5_14,
© Springer Science+Business Media, LLC 2012

this setup was actually a deadly trap. Oenomaus possessed horses and armor given to him by his father Ares, the god of war. These horses were as fast as the wind and would quickly overtake any normal horses. At least a dozen suitors had died while trying to win the lovely Hippodamia before Pelops, the son of Tantalus, wandered into Pisa. He and Hippodamia promptly fell in love, and Pelops decided that he had to take the dangerous challenge.

Pelops had been warned about the danger and appealed to Poseidon who was his patron. Poseidon came to Pelops' aid and provided him with a golden chariot and tireless winged horses to even out the odds, but even this did not insure success. Hippodamia, for the first time, was so much in love with Pelops that she also came to his aid. She persuaded Myrtilus, her father's charioteer, to help her. Myrtilus also loved Hippodamia and wanted her for his own wife. However, this was a chance to get into her good graces, and he provided the winning advantage. Before the race, he made wax linchpins which were indistinguishable from the bronze ones normally used. When the fake linchpins failed and the chariot overturned, Oenomaus became tangled in the horses' reins and was dragged along the ground. Before he died, Oenomaus realized what had happened and cursed Myrtilus.

Myrtilus' strategy worked, at least for a while, because after the couple wed, he became their chariot driver and chauffeured them on their honeymoon. However, being near the object of his desire was too much for him and one day while Pelops was away, Myrtilus let his passion overcome him and tried to rape Hippodamia. Pelops returned while the attempt was in progress. He promptly killed Myrtilus and threw his body into the sea.

Myrtilus was placed in the sky as the constellation Auriga as a visible warning to humans of the punishment which treacherous people would suffer. In the sky, Auriga is shown kneeling with either a whip or set of reins in one hand. He is also supporting a she-goat and two kids with his other hand. These represent the secondary duty of chariot drivers of the time, caring for the domestic animals usually kept in pens near their master's house.

The second myth portrays Auriga as the son of Hephaestus, Erichthonius, who became the King of Athens. Hephaestus was the son of Zeus and Hera but was so ugly and weak when he was born that the embarrassed Hera threw him out of Olympus. Hephaestus became a blacksmith who made such beautiful armor, jewelry, and other useful metal objects that Hera relented and brought him back to Olympus. However, he was again thrown out of Olympus by Zeus for interfering in an argument he and Hera were having. This time he fell for a full day, broke both legs in the landing and was crippled ever afterward. Hephaestus then made marvelous devices to help himself get around in his shop and in the outer world. Because of a misunderstanding, Hephaestus tried to make love to the virgin goddess Athene but wound up fathering a child by Mother Earth instead.

Their son, Erichthonius, was born as handicapped as his father. In one version of the myth, he was only crippled, but in another he had serpents' tails instead of legs. Although he was not her child, Athene adopted him and saw to his care and education. When he grew up, Erichthonius invented the four horse chariot for transportation. At his death, Zeus placed his image in the sky to honor his contribution.

AURIGA (or-EYE-guh)	**The Charioteer**
AURIGAE (or-EYE-guy)	**9:00 P.M. Culmination:** Feb. 4
Aur	**AREA:** 657 sq. deg., 21st in size

TO FIND: Auriga lies approximately 40° due north of Orion (see Fig. 4.1). Bright yellow Capella, the sixth brightest star in the sky, is easy to identify. It lies at the northern tip of the slightly elongated "Winter Hexagon" and 54° due north of Rigel. The triangle of faint stars representing the baby goats lies just a few degrees SW of Capella, and is also easy to find, in spite of its relative faintness. Menkalinan (β Aur) lies about 10° to the ESE of Capella. Capella and Rigel mark Auriga's shoulders. Bright El Nath (β Tau) represents one of Auriga's feet and lies nearly halfway from Capella to the small triangle marking Orion's head. Nearly 8° NW of El Nath is ι Aur which marks Auriga's other foot (Figs. 14.1 and 14.2).

KEY STAR NAMES:

Name	Pronunciation	Source
α <u>Capella</u>	kuh-**pell**-luh	capella (Latin), a female goat
Amalthea	am-al-**thee**-uh	Amalthea (Greek), the goat that nursed the infant Zeus
Cornu Copiae	**kor**-nuu **cope**-ih-uh	cornu copiae (Latin), horn of plenty
β Menkalinan	**men**-kul-ih-**NAN**	Al Mankib dhi'l 'Inan, "The shoulder of the Rein Holder"
ε Almaaz	al-**mah**-ahz	Al Ma'az (Arabic), "The He-Goat"
ζ Sadatoni	sahd-ah-**tone**-ih	Al Said al Thani, "The Second Arm" (of the Rein Holder)
γ Alkab*[1]	al-**kahb**	Al Ka'b dhi'l Inan (Arabic), "Heel" (of the Rein Holder)

Magnitudes and spectral types of principal stars:

Bayer desig.	Mag. (m_v)	Spec. type	Bayer desig.	Mag. (m_v)	Spec. type	Bayer desig.	Mag. (m_v)	Spec. type
α	0.08	G8 III	β	1.90	A2 V	γ	–	–*[1]
δ	3.72	K0 III	ε	3.03	F0 Ia	ζ	3.69	K4 II
η	3.18	B3 V	θ	2.65	A0p Si	ι	2.69	K3 II var
κ	4.32	G8 III	λ	4.69	G0 V	μ	4.82	A4m
ν	3.97	K0 III	ξ	4.96	A2 V	o	5.46	A0p
π	4.30	M3 II	ρ	5.22	B5 V	σ	5.02	K4 III
τ	4.51	G8 III	υ	4.72	M1 III	φ	5.08	K4 IIIp
χ	4.71	B5 Iab	ψ[1]	4.80	K3 III	ψ[2]	4.92	K5 Iab var
ψ[3]	5.34	B8 III	ψ[4]	5.04	K5 III	ψ[5]	5.24	G0 V
ψ[6]	5.22	K1 III	ψ[7]	4.99	K3 III	ψ[8]	6.46	B9.5sp
ψ[9]	5.85	B8 III	ω	4.93	A1 V			

Fig. 14.1a. The stars of Auriga and adjacent constellations.

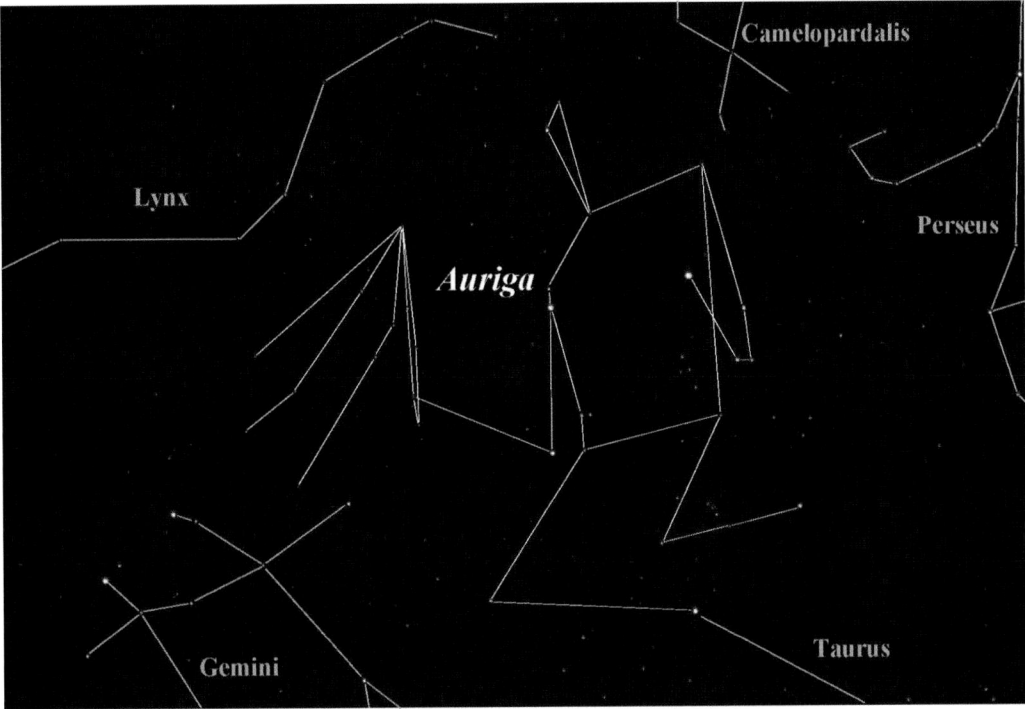

Fig. 14.1b. The figures of Auriga and adjacent constellations.

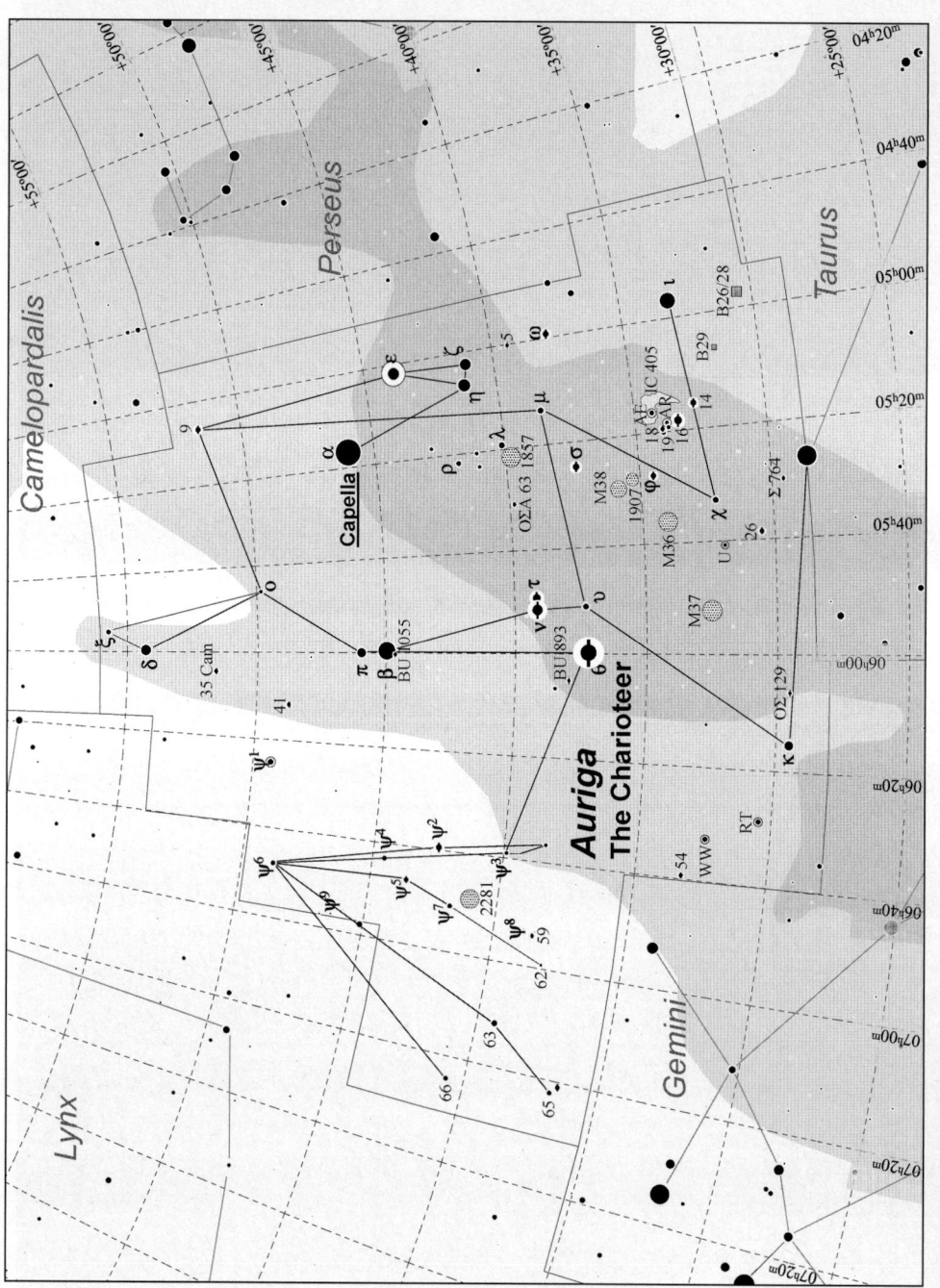

Fig. 14.2. Auriga details.

Table 14.1. Selected telescopic objects in Auriga.

Multiple stars:

Designation	R.A.	Decl.	Type	m_1	m_2	Sep. (")	PA (°)	Colors	Date/**Period**	Aper. Recm.	Rating	Comments
4, ω Aurigae	04h 59.3m	+37° 53'	A–B	5.0	8.2	4.7	003	Y, B	2004	2/3"	*****	a.k.a. Σ 616
5 Aurigae	05h 00.3m	+37° 18'	Aa–B	6.0	9.7	3.7	265	Y, Y	2002	2/3"	****	a.k.a. OΣ 92. A–a is too close for amateur instruments
9 Aurigae	05h 06.7m	+51° 36'	A–B	5.0	12.2	6.3	082	yW, ?	1958	4/5"	***	a.k.a. BU 1046-AB
			A–C	5.0	10.0	89.9	090	yW, ?	2002	11×80	***	a.k.a. H 35-AC
14 Aurigae	05h 15.4m	+32° 41'	A–B	5.0	9.0	9.8	342	pY, ?	2004	2/3"	*****	a.k.a. Σ 653 AB
			A–C	5.0	7.3	15	227	pY, dB	2006	18×80	*****	a.k.a. Σ 653 AC
16 Aurigae	05h 18.2m	+33° 22'	A–B	4.8	10.6	4.4	056	yO, ?	1936	3/4"	****	a.k.a. OΣ 103
18 Aurigae	05h 19.4m	+33° 59'	A–B	6.5	12.5	3.9	169	yW, ?	1963	4/5"	***	a.k.a. Ho 18
21, σ Aurigae	05h 24.7m	+37° 23'	A–B	5.0	12.0	8.7	167	yO, ?	1922	3/4"	***	a.k.a. β 888
			A–C	5.0	13.3	27.4	331	yO, ?	1914	6/7"	***	
24, φ Aurigae	05h 27.6m	+34° 28'	A–C	5.2	10.9	62.0	071	yO, ?	1998	15×80	**	
			A–D	5.2	8.1	210.3	016	yO, ?	2002	6×30	***	
OΣ A 63	05h 30.8m	+39° 50'	A–B	6.5	7.7	75.9	277	yO, yO	1998	6×30	***	
26 Aurigae	05h 38.6m	+30° 30'	A–B	6.3	6.2	**0.22**	**330**	brY, ?	**53.275**	–	*****	AB pair is too close for most amateur telescopes
[Σ 753]			AB–C	5.5	8.4	12.3	269	brY, dB	2005	2/3"	****	
[β 90]			AB–D	5.5	11.5	34.7	113	brY, ?	2002	15×80	****	
Σ 764	05h 41.3m	+29° 29'	A–B	6.4	7.1	26.1	015	wY, bW	2004	10×50	*****	
29, τ Aurigae	05h 49.2m	+39° 11'	A–B	4.5	11.6	39.3	356	Y, ?	1928	15×80	***	a.k.a. BU 90
			A–C	4.5	11.6	47.9	036	Y, ?	2000	15×80	***	H 21-AC
32, ν Aurigae	05h 51.5m	+39° 09'	A–B	4.0	11.4	55.9	206	dY, oR	2003	3/4"	****	a.k.a. H 90
37, θ Aurigae	05h 59.7m	+37° 13'	A–B	2.7	7.2	3.8	308	B, yW	2004	2/3"	****	Very difficult. OΣ 545AB
			AB–C	2.7	10.7	54.0	300	B, ?	1998	15×80	****	a.k.a. OΣ 545AC
BU 1055	06h 00.3m	+44° 36'	A–B	6.2	11.0	1.7	324	oY, ?	1953	5/6"	***	
(H 6 91)			A–C	6.2	10.3	35.6	345	oY, ?	2000	15×80	****	a.k.a. H 6 91
35 Camelopardalis	06h 04.5m	+51° 34'	A–BC	6.4	9.3	40.0	014	yW, V	2003	7×50	****	a.k.a. OΣ 128
			B–C	9.0	10.0	0.6	309	yW, ?	1999	16/18"	***	a.k.a. HU 559
BU 893	06h 05.0m	+37° 58'	A–B	6.3	12.7	18.1	131	dY, W	1998	5/6"	***	
			A–B	6.3	11.6	85.4	141	dY, ?	1998	15×80	***	
OΣ 129	06h 06.6m	+29° 31'	A–B	6.2	10.5	9.5	211	oR, ?	1998	2/3"	****	

(continued)

Table 14.1. (continued)

Multiple stars:

Designation	R.A.	Decl.	Type	m₁	m₂	Sep. (")	PA (°)	Colors	Date/ Period	Aper. Recm.	Rating	Comments
41 Aurigae	06ʰ 11.6ᵐ	+48° 43'	A–B	6.2	6.9	7.6	357	W, pB	2006	2/3"	*****	Nice pair for small scopes. a.k.a. Σ 845
52, ψ³ Aurigae	06ʰ 39.3ᵐ	+42° 29'	A–B	4.8	10.6	51.5	105	dY, ?	1999	15×80	****	a.k.a. BUP 92AB
			A–C	4.8	11.3	100.5	073	dY, ?	1999	3/4"	***	a.k.a. BUP 92AC
54 Aurigae	06ʰ 39.6ᵐ	+28° 16'	A–B	6.2	7.9	0.9	035	B, B	1995	10/11"	***	a.k.a. OΣ 152
56, ψ⁵ Aurigae	06ʰ 46.7ᵐ	+43° 35'	A–B	5.3	8.7	31.1	038	Y, pB	1998	10×50	****	Nice color contrast
			A–C	5.3	11.7	80.0	330	Y, ?	1944	3/4"	***	a.k.a. WAL 47
59 Aurigae	06ʰ 53.0ᵐ	+38° 52'	A–B	6.1	10.2	22.2	224	pY, B	1998	2/3"	****	a.k.a. Σ 974AB
			B–C	9.5	12.8	25.6	219	B, ?	1988	5/6"	***	a.k.a. Σ 974AC
65 Aurigae	07ʰ 22.1ᵐ	+36° 45'	A–B	5.1	11.7	11.5	010	dY, ?	2004	3/4"	***	a.k.a. BU 901-AB
			A–C	5.1	12.1	44.9	040	dY, ?	1998	3/4"	***	a.k.a. BU 901-AC

Variable stars:

Designation	R.A.	Decl.	Type	Range (m_v)	m₂	Period (days)	F (f_r/f_f)	Spectral Range	Aper. Recm.	Rating	Comments
7, ε Aurigae	05ʰ 02.0ᵐ	+43° 49'	EA/GS	2.9	3.8	~9,890	0.08	A8Ia-F2Iaep	Eye	*****	Exceptionally long period eclipsing variable (27.1 years). # *2
AE Aurigae	05ʰ 16.3ᵐ	+34° 19'	Ina	5.8	6.1	–	–	O9.5V	7×50	****	51' NW (PA 295) of 19 Aur (m5.05). A so-called runaway star. # *3
46, ψ¹ Aurigae	06ʰ 27.9ᵐ	+49° 17'	Lc	4.7	5.0	–	–	K5 – M0Iab-Ib	6×30	****	Small range, but lovely orange color vis. in binoculars and small scopes
48, RT Aurigae	06ʰ 26.8ᵐ	+30° 30'	Cδ	5.0	5.8	3.73	0.25	F4Ib-G1Ib	7×50	****	Classic Cepheid variable. ~1,600 ly distant, ave. L ~ 1,000 L☉. ##
WW Aurigae	06ʰ 32.5ᵐ	+32° 27'	EA/DM	6.0	6.5	2.5202	0.10	A3m + A3m	6×30	****	2° 08' NNE (PA 023°) of 48-RT Aur (m5.00–5.82). M = 1.9 and 1.8 M☉. ##
UV Aurigae	05ʰ 18.6ᵐ	+32° 28'	M	7.4	10.6	394.42	–	C6,2-C8,2Jep	3/4"	****	1° 09' E (PA 082°) of 16 Aur (m4.54).##$
U Aurigae	05ʰ 42.2ᵐ	+32° 02'	M	7.5	15.5	408.09	0.39	M7e–M9e	Max.	****	2° 00' E (PA 094°) of 25-χ Aur (m4.71). # *4
17, AR Aurigae	05ʰ 18.3ᵐ	+33° 46'	EA/DM	6.2	5.8	4.1347	0.07	Ap(Hg-Mn) + B9V	6×30	****	A chemically peculiar star (excess mercury and manganese). #$

Star clusters:

Designation	R.A.	Decl.	Type	m_v	Size (')	Brst. Star	Dist. (ly)	dia. (ly)	Number of Stars	Aper. Recm.	Rating	Comments
NGC 1857	05ʰ 20.2ᵐ	+39° 21'	Open	7.0	5	7.4	6,200	6	30	8/10"	****	Nice compact cluster, ~30 stars 9–12 m
NGC 1907	05ʰ 28.0ᵐ	+35° 19'	Open	8.2	5	11	4,300	6	110	4/6"	****	32' SSW of M38. In same LP FOV
M38, NGC 1912	05ʰ 28.7ᵐ	+35° 50'	Open	6.4	21	9.5	4,025	25	100	4/6"	*****	The total cluster luminosity is estimated at 4,125 suns
M36, NGC 1960	05ʰ 36.3ᵐ	+34° 08'	Open	6.0	10	8.86	4,000	12	60	4/6"	*****	Young cluster with many very hot, blue B2 and B3 type stars. Luminosity ~5,100 suns
M37, NGC 2099	05ʰ 52.4ᵐ	+32° 33'	Open	5.6	15	9.2	4,375	19	150	4/6"	*****	The total luminosity is estimated at 8,800 suns
NGC 2281	06ʰ 48.8ᵐ	+41° 05'	Open	5.4	25	8.0	1,500	11	120	3/4"	****	"Broken Heart Cluster." Often over-looked, but very pretty. Use low power

Nebulae:

Designation	R.A.	Decl.	Type	m_v	Size (')	Brst./Cent. star	Dist. (ly)	dia. (ly)	Aper. Recm.	Rating	Comments
Barnard 26–28	04ʰ 55.2ᵐ	+30° 35'	Dark	–	20'	–	–	–	16/18	****	Z shape NW in rich star field NW of AB Aur
Barnard 29	05ʰ 06.2ᵐ	+31° 44'	Dark	–	10'	–	–	–	16/18	****	Dark streak ~2° SE of Z Aur
IC 405	05ʰ 16.2ᵐ	+34° 16'	Emission	–	20×30'	AE Aur	–	–	14/18"	**	The "Flaming Star Nebula." Very faint, Good photographic object. Note Aur C

Galaxies:

Designation	R.A.	Decl.	Type	m_v	Size (')	Dist. (ly)	dia. (ly)	Lum. (suns)	Aper. Recm.	Rating	Comments
None											

Other interesting objects:

Designation	R.A.	Decl.	Type	m_1	m_2	Sep. (")	PA (°)	Colors	Date/Period	Aper. Recm.	Rating	Comments
UV Aurigae	05ʰ 21.8ᵐ	+32° 31'	Mira B9 V	7.4–10.6 1.5	11.5	3.4	004	R, bw	394.42	6/8"	****	A red carbon variable and bW companion

*5

Fig. 14.3. M36, NGC 1960, open cluster in Auriga. Astrophoto by Mark Komsa.

Fig. 14.4. M37, NGC 2209, open cluster in Auriga. Astrophoto by Mark Komsa.

Fig. 14.5. M38, NGC 1912, open cluster in Auriga. Astrophoto by Mark Komsa.

Fig. 14.6. IC 405, bright nebula in Auriga. Astrophoto by Rob Gendler. (IC 405 is on the *left*, IC 410 and 417 art in the *middle* and on the *right*, but are fainter and are not included in the object list).

NOTES:

1. Like several other stars, Alkab (γ Aurigae) was originally considered to be part of two constellations, in this case, it was designated as both γ Aurigae and β Tauri. When the IAU formalized the constellation boundaries in 1930, it was assigned to Taurus and designated β Tauri only. The proper name also became Elnath.

2. ε Aurigae is an exceptionally slow eclipsing variable with a period of 27.1 years. It is also very puzzling in that the partial phase of the eclipse lasts several months. Many different proposals have been made, including one of a large cloud of gas and dust orbiting the primary like a planet.

3. AE Aurigae is an O class variable with an average luminosity of 10,000 suns. One of the three "runaway stars," AE Aurigae is moving away from the central Orion region at 109 km/s (about 245,000 mi/h)! The other two "runaway stars" are 53 Arietes with a velocity of 26 km/s (58,000 mi/h) and μ Colombae with a velocity of 116 km/s (261,000 mi/h). All three probably were once members of massive binary pairs which escaped when the pair was disrupted by a near collision with another member or the supernova explosion of their partner about 3 million years ago. Although they appear to be about the same age, the latest measurements show that they did not originate at the same place and therefore are probably not directly connected.

4. UV Aurigae is a binary consisting of a red carbon variable star and a blue–white companion. Although it is faint, it is worth the effort of finding to watch the apparent (and real) color changes as the brighter star changes its brightness. UV Aurigae can be found by using AAVSO charts. It is shown on Chart 136 of the Millennium Star Atlas and plotted but not identified on chart 59 of the Uranometria 2000.0 Deep Sky Atlas. It is about 1.25° from 14 and 16 Aurigae and forming an approximately isosceles triangle with them.

5. The "Flaming Star Nebula" is illuminated by the hot star AE Aurigae, which is not connected to the nebula but only passing through the diffuse gas at the present time.

CHAPTER FIFTEEN

Plancius, the Minister Turned Map Maker

February, May

CAMELOPARDALIS	COLUMBA	CRUX	MONOCEROS
The Giraffe	The Dove	The Cross	The Unicorn

Petrus Plancius (1552–1622) was born Pieter Platevoet (literally "Peter Flatfoot") in Dranouter, a town in what is now the Flanders region of Belgium. From a wealthy family, he studied theology in Germany and England, and at the age of 24 became a Calvinist minister. As was the custom among scientists and prominent people of his time, Platevoet later Latinized his name and became Petrus Plancius. Fearing religious persecution from the inquisition, Plancius fled Brussels in 1585 after the Spanish put down a rebellion and King Phillip II forced all non-Catholics to join the Catholic Church or leave the country.

After traveling to Amsterdam, Plancius became interested in navigation and cartography and soon became an expert on routes to India. It is not certain when he acquired an interest in astronomy, but it is likely that his interest was associated with his work as a cartographer and navigator. Plancius was one of the founders of the Dutch East India Company and created over a hundred maps for them. Plancius was the first to use the Mercator projection on navigation charts, and is considered by many to be the second greatest map maker of the time, surpassed only by Mercator himself. On his maps, he often included elaborate decorations, including maps of the northern and southern hemisphere heavens.

Plancius used many sources for the maps he made, and in 1589 he created a revision of an earlier celestial globe by Jacob Floris van Langren on which he added two new

P. Simpson, *Guidebook to the Constellations: Telescopic Sights, Tales, and Myths*,
Patrick Moore's Practical Astronomy Series, DOI 10.1007/978-1-4419-6941-5_15,
© Springer Science+Business Media, LLC 2012

constellations, Crux and Triangulus Antarticus, and the two Magellanic Clouds to the southern sky. In 1592, Plancius published a new celestial globe which included the new constellation, Columba Nohae, Noah's Dove. As part of planning new trips, Plancius taught navigators Pieter Keyser, Frederik de Houtman, and several crewmen to measure and plot the positions of the southern stars. For a 1595–1597 southern expedition for the company, he directed these sailors to measure and to plot the portion of the sky which could not be seen from Europe. Plancius used the measurements for another celestial globe published in 1612. On this globe, he included not only the new southern constellations, but also eight new northern constellations formed from stars plotted by Tycho Brahe, himself, and others. Only the reduplicatively named Gyraffa Camelopardalis (both Dutch and Latin names for giraffe) and Monoceros Unicornis (Dutch and Latin for unicorn) survive today with their now singleton names of Camelopardalis and Monoceros.

CAMELOPARDALIS (kam-uh-low-**par**-dah-liss)	**The Giraffe**
CAMELOPARDALIS (kam-uh-low-**par**-dah-liss)	**9:00 P.M. Culmination**: Feb. 7
Cam	**AREA:** 757 sq. deg., 18th in size

Camelopardalis is a large, but faint, constellation created in the seventeenth century to fill a large gap surrounded by the constellations of Auriga, Cassiopeia, Draco, Lynx, Perseus, Ursa Major, and Ursa Minor. Its creation was long attributed to Bartschius (Jakob Bartsch) who included an outline of a giraffe on his celestial globe of 1614. However, it is now correctly attributed to Petrus Plancius.

In addition to the 12 constellations he created with the help of Keyser and de Houtman, Plancius alone created 4 more constellations which we still use today, Camelopardalis, Columba, Crux, and Monoceros.

The reduplicative name Gyraffa Camelopardalis came directly from the scientific name for the species which includes 11 subspecies of giraffes. It is actually a redundant name since gyraffa is an old form of giraffe, and camelopardalis is the Latin name adapted from the Greek name describing a "camel-like animal with leopard-like spots." The name was shortened and spelled in various ways until it was finally officially declared Camelopardalis by the IAU.

Camelopardalis is quite deficient in Bayer designation stars, with only the α, β, and γ designations assigned on Jacob Bartsch's planisphere of 1624. Flamsteed numbers 1 through 58 were apparently assigned by Flamsteed himself, or perhaps by the editors of his posthumously published catalog. In any case, all of the numbers are confined to the south-central part of the constellation, leaving a considerable part of the constellation with no stars of any designation except for variables and the stars included in the Henry Draper (HD) and Smithsonian Astrophysical Observatory (SAO) catalogs. It was later found that 13 and 27 Camelopardalis did not exist and were probably erroneous duplicates of other stars. Also, as the constellation boundaries changed over the years, 32, 33, and 35 Camopardalis wound up in Auriga, 44, 45, 46, 48, 50, 52, 54, and 55 Camelopardalis in Lynx and 55, 56, and 57 Camelopardalis in Ursa Major.

TO FIND: Camelopardalis is one of the faintest constellations, with its brightest star (β) being only magnitude 4.03, and averaging only a single fourth or fifth magnitude star per 17 sq. deg. Picking out the brightest stars is challenging, and visualizing the figure of a giraffe is mainly an exercise in imagination. For those up to the challenge, try the following directions. Just as the two-end stars of the big dipper point to Polaris, the two-end stars of the little dipper point to SAO 07522 (a.k.a. SAO 7522)[*1] (m5.14), and the faint star at the tip of the Giraffe's nose. The line from Pherkad (γ UMi, m3.00) to Kochab (β UMi, m2.06) can be extended 10° to end at SAO 07522. SAO 07522 is at the right angle of a nearly perfect right triangle with Polaris (m1.95) and Kochab, the two brightest stars of Ursa Minor. At the other end of Camelopardalis, the stars can be identified by starting with Cassiopeia. The star 50 Cassiopieae (m3.95) marks the jewel in Queen Cassiopeia's crown. Just under 9° E (PA 095) of 50 Cassiopieae, γ Camelopardalis can be found, which marks the camel's back near the hindquaters. ι Cassiopeiae (m4.46) is the top of Queen Cassiopeia's throne, and about 5.5° ESE of it is SAO 12704 (m4.84) which is where the giraffe's tail joins the body. From γ Cam to α Cam, marking the camel's belly, is only 7.5° to the SE. β Camelopardalis, the brightest star in the constellation, lies just 6° SSE of α. β Camelopardalis is not used in the figure of the camel, lying instead between the fore and hind legs. However, it can be used as a reference point to find other surrounding stars. After finding these stars and marking the key parts of the figure, the best procedure is to consult a star chart and find the other stars by comparing the chart with the sky. All these stars are fairly faint, so I wish you clear, dark skies and good luck! (Figs. 15.1–15.3).

NAMED OBJECTS:	Pronunciation	Source of name
CAMELOPARDALIS	kam-uh-low-**par**-dah-liss	camelos (Greek), "camel" + leopardos (Greek), "leopard" for "a leopard-spotted camel."
Kemble's Cascade		Named in honor of Father Lucian J. Kemble, the Canadian amateur astronomer who first described it

KEY STAR NAMES: None.

Magnitudes and spectral types of principal stars:

Bayer desig.	Mag. (m_v)	Spec. type	Bayer desig.	Mag. (m_v)	Spec. type	Bayer desig.	Mag. (m_v)	Spec. type
α	4.29	O9.5 Ia	β	4.03	G0 Ib	γ	4.63	A2 Ivn

Stars of magnitude 5.5 or brighter which have Flamsteed but no Bayer designations:

Flamsteed	Mag. (m_v)	Spec. type	Flamsteed	Mag. (m_v)	Spec. type	Flamsteed	Mag. (m_v)	Spec. type
2	5.36	A8 V	3	5.07	K0 III	4	5.29	A3m
5	5.52	B9.5 V	7	4.47	A1 V	11	5.22	B2.5 V
16	5.24	A0 V	17	5.43	M2 III	31	5.20	A2 V
36	5.36	K2 II/III	37	5.35	G8 III	40	5.37	K3 III
42	5.14	B4 IV	43	5.11	B7 III	BE	4.39v	M1 III

Other stars magnitude 5.5 or brighter which have only catalog designations:*1

Designation	Mag. (m_v)	Spec. type	Designation	Mag. (m_v)	Spec. type
HD 021389	4.55	A0 Ia	HD 049878	4.55	K4 III
HD 029336	4.74	B2.5 V	HD 042818	4.76	A0 V
HD 023089	4.78	A3 V	HD 024479	4.95	B9.5 V
HD 024480	4.99	K3 I/II	HD 025291	5.00	F0 II
HD 033564	5.08	F6 V	HD 021447	5.09	A1 V
HD 106111	5.14	A5m	HD 026764	5.20	A2 V
HD 090089	5.25	F2 V	HD 027022	5.26	G5 III
HD 064307	5.37	K3 III	HD 112028	5.38	A1 III
HD 064486	5.39	A0p	HD 023277	5.40	A2m
HD 027245	5.40	M0 III	HD 032650	5.44	B9p
HD 046588	5.44	F8 V	HD 030442	5.47	M2 III

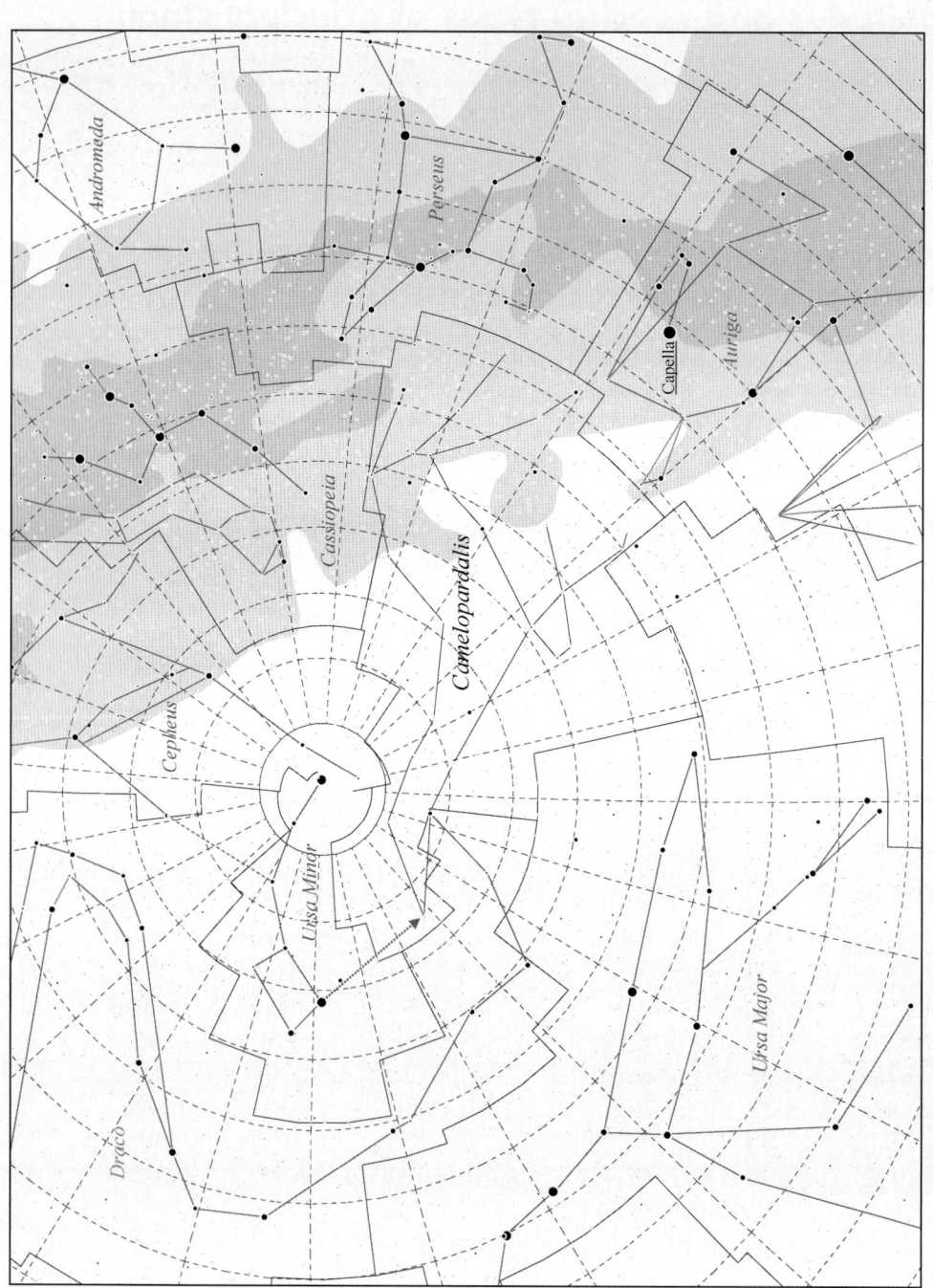

Fig. 15.1. Camelopardalis Finder Chart.

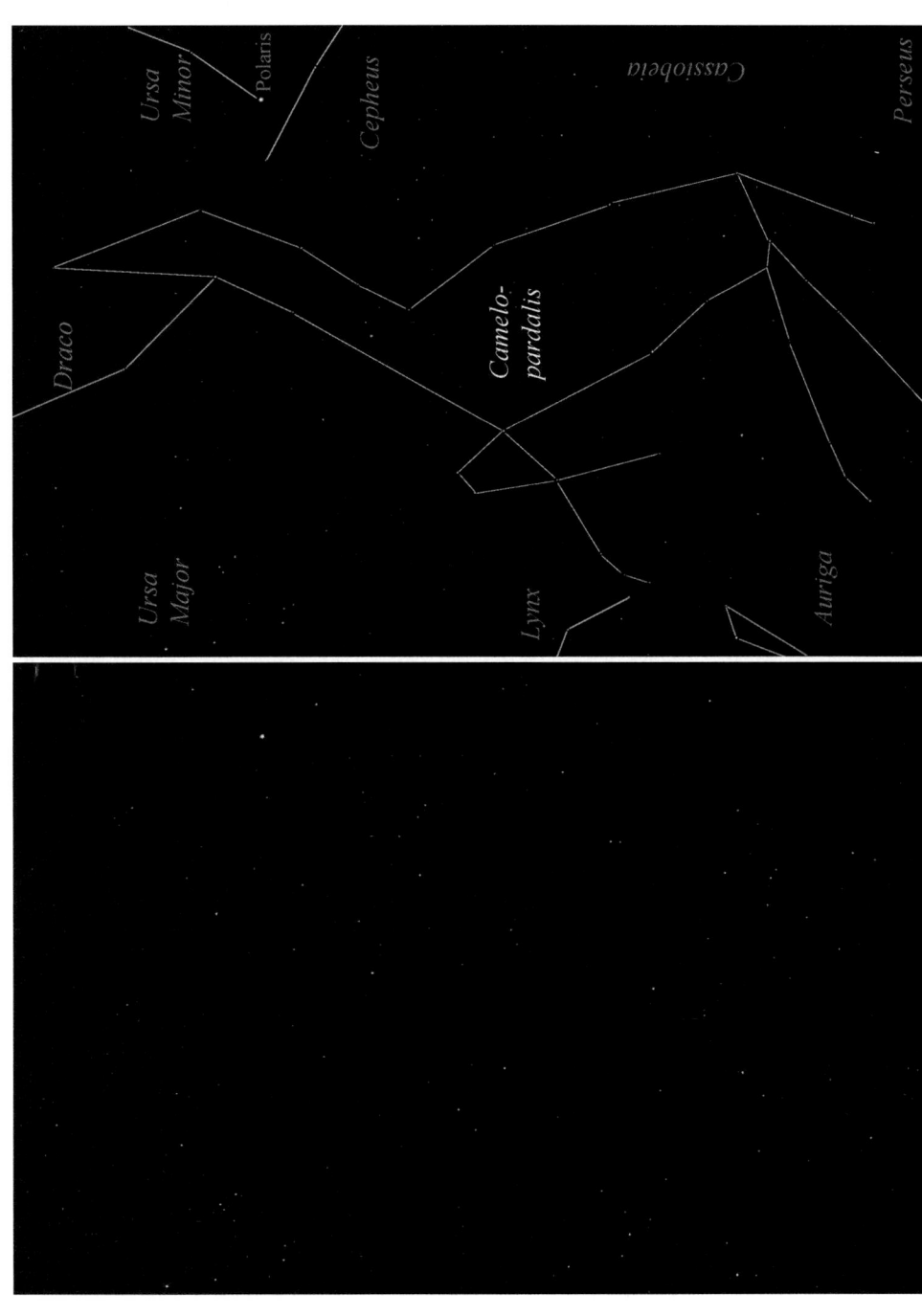

Fig. 15.2 (a) The stars of Camelopardalis and adjacent constellations. **(b)** The figures of Camelopardalis and adjacent constellations.

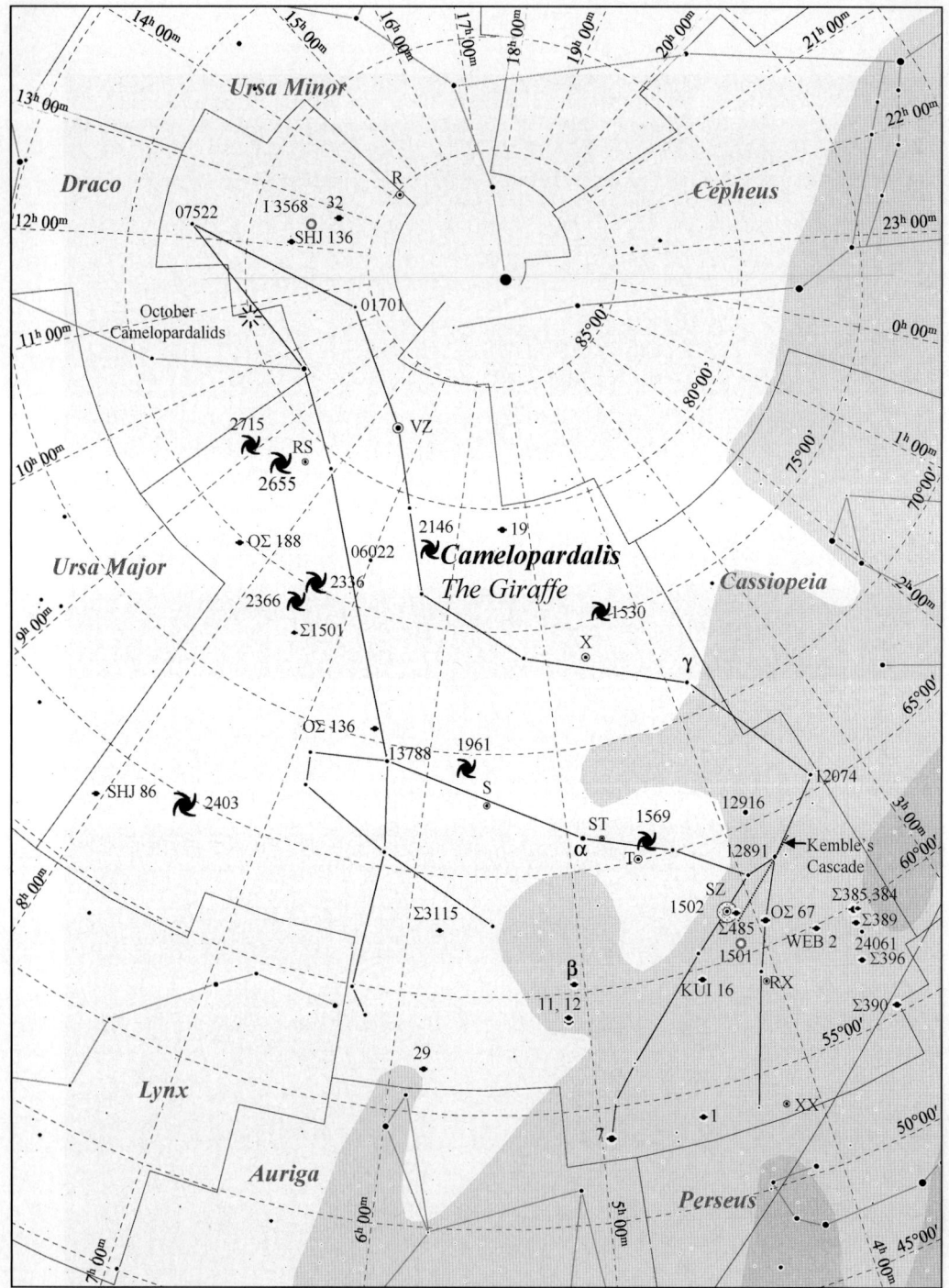

Fig. 15.3. Camelopardalis details.

Table 15.1. Selected telescopic objects in Camelopardalis.

Multiple stars:

Designation	R.A.	Decl.	Type	m_1	m_2	Sep. (")	PA (°)	Colors	Date/Period	Aper. Recm.	Rating	Comments
Σ384	03h 28.5m	+59° 54'	A–B	8.1	8.9	2.0	274	Gld, B	1998	4/6"	****	Faint but very attractive pair which forms a double-double with Σ285
Σ385	03h 29.1m	+59° 56'	A–B	4.2	7.8	2.5	159	Y, ?	1991	3/4"	****	The separations are similar to ι Lyrae, but the brightnesses are unequal
Σ390	03h 30.0m	+55° 27'	A–B	5.1	10.0	14.6	160	W, P	1999	2/3"	****	
Σ389	03h 30.2m	+59° 22'	A–B	6.4	7.9	2.6	071	W, W	2004	3/4"	****	
Σ396	03h 33.5m	+58° 46'	A–B	6.4	7.7	20.1	244	O, rO	2004	15×80	*****	Lovely orange/deep orange pair
Web 2	03h 42.7m	+59° 58'	A–B	5.9	8.5	55.3	036	brO, bGr	2002	7×50	******	Beautiful color contrast pair
OΣ67	03h 57.1m	+61° 07'	A–B	5.3	8.1	1.7	048	Y, Gr	1991	5/6"	******	Very nice color contrast
Σ485	04h 07.9m	+62° 20'	A–E	6.9	6.9	17.7	305	bW, bW	2004	2/3"	****	Nearly perfectly matched pair, stands out against background of NGC 1502 [*2]
KUI 16	04h 23.0m	+59° 37'	A–B	6.2	10.9	1.1	212	W, ?	1969	8/10"	****	In an FOV rich in stars
Arg 100			AB–C	6.2	9.3	32.1	060	W, B	2002	10×50	****	
1 Camelopardalis	04h 32.0m	+53° 55'	A–B	5.7	6.8	10.3	304	W, pB	2002	2/3"	****	Easy, showcase object in a barren field
7 Camelopardalis	04h 57.3m	+53° 45'	A–B	4.4	7.8	0.56	359	W, ?	284	16/18"	***	A–B is close for amateur instruments, but is opening ~0.018"/year
10, β Camelopardalis	05h 03.4m	+60° 27'	AB–C	4.4	11.3	25.8	240	W, O	1999	4/6"	***	Brilliant colors, in a streak of dark nebulosity. Use low power for best view
			A–B	4.1	7.4	83.0	210	yO, B	2003	2/3"	******	
11 and 12 Camelopardalis	05h 06.1m	+58° 58'	A–B	5.2	6.2	179.0	009	pY, yO	2002	6×30	*****	Great binocular double. People with excellent eyesight may resolve by eye
19 Camelopardalis	05h 22.6m	+79° 14'	A–B	5.1	9.2	25.8	135	W, L	2001	10×50	******	AKA Σ634. Pale but lovely color in secondary
Σ3115	05h 49.1m	+62° 48'	A–B	6.6	7.5	1.0	342	W, yW	2005	8/10"	******	Separation decreasing. A good test of seeing
29 Camelopardalis	05h 50.6m	+56° 55'	A–B	6.5	10.4	26.1	128	W, ?	2001	2/3"	****	Magnitude difference makes secondary color hard to distinguish
OΣ136	06h 28.2m	+70° 32'	A–B	6.0	11.0	5.0	080	W, ?	1999	3/4"	****	Attractive trio of very similar stars. In a relatively barren region of the sky
Σ1501	07h 26.6m	+73° 05'	A–B	7.6	9.08	1.1	284	oW, W	1999	8/10"	*****	
SHJ 86	08h 02.5m	+63° 05'	AB–C	7.4	7.79	31.5	82	oW, ?	1999	7×50	*****	At 0.2", A–a is too close for amateur instruments. Nice Aa–B color contrast
			Aa–B	6.2	7.5	51.1	081	Rw, pB	1999	7×50	*****	
OΣ188	08h 22.1m	+74° 49'	A–B	6.5	10.5	10.5	194	dY, ?	1999	2/3"	*****	Easy pair for small telescopes

(continued)

Table 15.1. (continued)

Multiple stars:

Designation	R.A.	Decl.	Type	m1	m2	Sep. (")	PA (°)	Colors	Date/Period	Aper. Recm.	Rating	Comments	
SHJ 136	12h 11.0m	+81° 43'	A–B	6.2	8.3	70.8	074	W, W	2000	7×50	****	May be seen by some in 6×30 binoculars	
32 Camelopardalis	12h 49.2m	+83° 25'	A–B	5.3	5.7	21.5	328	W, W	2003	15×80	****	Nearly matched pair. Easy, visible all year in northern hemisphere. a.k.a. Σ1694	

Variable stars:

Designation	R.A.	Decl.	Type	Range (m$_v$)		Period (days)	F (f$_r$/f$_l$)	Spectral Range	Aper. Recm.	Rating	Comments	
XX Camelopardalis	04h 04.8m	+53° 14'	R CrB	8.1	9.8	–	–	G1I(C0,2,0)	15×80	****	3° 02' N (PA 006°) of 47, λ Persei (m4.25).##$	*3
RX Camelopardalis	04h 05.0m	+58° 40'	δ Cep	7.3	8.1	7.91202	0.28	F6Ib–B2Ib	6×30	****	2° 38' NNW (PA 339°) of SAO 12968 (m4.99).#	*4
SZ Camelopardalis	04h 07.9m	+62° 20'	EA/DM	7.0	7.3	2.699	0.17	O9.5V+B0	7×50	*****	a.k.a. Σ485. 1° 24' ESE (PA 121°) of SAO 12969 (m4.95); (in NGC 1502)	*5
T Camelopardalis	04h 35.2m	+68° 09'	M	7.3	14.4	373.2	0.47	S4,7e–S8.5,8e	12/14"	***	1° 25' W (PA 264°) of 9, α Cam (m4.26).#	*6
X Camelopardalis	04h 45.7m	+75° 06'	M	7.4	14.2	143.56	0.49	K8–M8e	12/14"	***	##$	
ST Camelopardalis	04h 51.2m	+68° 10'	SRb	9.2	12.0	300	–	C5,4(N5)	4/6"	****	#$	
S Camelopardalis	05h 41.0m	+68° 48'	SRa	7.7	11.6	327.26	0.51	C7,3e(R8e)	3/4"	****		
VZ Camelopardalis	07h 31.0m	+82° 25'	SR	4.9	5.4	23.7	–	MIIIa	Eye	****		
RS Camelopardalis	08h 50.8m	+78° 58'	SRb	7.9	9.7	88.6	0.45	M4III	7×50	*****		
R Camelopardalis	14h 17.8m	+84° 04'	M	7.0	14.4	270.22	0.45	S2,8e–S8,7e	12/14"	***	Spectrum shows zirconium oxide.##	

Star clusters:

Designation	R.A.	Decl.	Type	m$_v$	Size (')	Brtst. Star	Dist. (ly)	dia. (ly)	Number of Stars	Aper. Recm.	Rating	Comments
NGC 1502	04h 07.7m	+62° 20'	Open	5.7	20	6.9	2,650	15	63	3/4"	****	Small, bright, moderately rich. Can be seen in 7×50. a.k.a. "Golden Harp Cluster."

Nebulae:

Designation	R.A.	Decl.	Type	m$_v$	Size (')	Brst./Cent. star	Dist. (ly)	dia. (ly)	Aper. Recm.	Rating	Comments
NGC 1501	04^h 07.0^m	+60° 55'	Planetary	11.5	52	14.5	3K	0.8	8/10"	***	Large, bright, bluish tint. CS requires larger aperture to be seen
I 3568	12^h 33.1^m	+82° 34'	Planetary	11.6	6.0	12.9	6.9K	0.2	8/10"	***	

Galaxies:

Designation	R.A.	Decl.	Type	m$_v$	Size (')	Dist. (ly)	dia. (ly)	Lum. (suns)	Aper. Recm.	Rating	Comments
NGC 1530	04^h 23.5^m	+75° 18'	SBb	11.4	4.6×2.6	120M	160K	32G	6/8"	***	Pretty bright, large. Very bright, very small nucleus
NGC 1569	04^h 30.8^m	+65° 51'	IBm	11.0	3.0×1.9	4M	54K	51M	6/8"	***	Pretty bright, small, slightly elongated. Very bright nucleus
NGC 1961	05^h 42.1^m	+69° 23'	SAB(rs)	11.0	4.3×3.0	180M	230K	100G	6/8"	***	a.k.a. Arp 184. Large, faint, irregular shape. Very small, bright nucleus
NGC 2146	06^h 18.7^m	+78° 21'	SBbp	10.6	5.0×3.2	45M	65K	9G	4/6"	***	Bright, large, slightly elongated. Dark lane in one arm
NGC 2336	07^h 27.1^m	+80° 11'	SAB(r)	10.4	6.4×3.3	100M	190K	55G	4/6"	***	Faint halo, slight brightening toward center. Weak, knoty arms, very small nucleus
NGC 2366	07^h 28.9^m	+69° 13'	IB(s)	10.8	8.2×3.3	11M	26K	460M	4/6"	***	Highly elongated, large size makes it difficult to see in small scopes
NGC 2403	07^h 32.1^m	+65° 43'	SB	8.5	25.5×13	14M	85K	11G	3/4"	****	One of the brightest non-Messier objects in the northern sky
NGC 2655	08^h 55.6^m	+78° 13'	SAB(s)	10.1	6.0×5.3	71M	120K	37G	4/6"	***	Very bright, large, slightly elongated. Very bright, very large nucleus. Dust lane. Diffuse arms
NGC 2715	09^h 08.1^m	+78° 05'	SAB(rs)	11.2	4.6×1.6	160M	280	68G	6/8"	***	Pretty bright, large, elongated. Very small, bright nucleus. Several knoty arms

*7

Meteor showers:

Designation	R.A.	Decl.	Period (yr)	Duration	Max Date	ZHR (max)	Comet/Asteroid	First Obs.	Vel. (mi/km)/sec	Rating	Comments
October Camelopardalids	10^h 48^m	+79°	–	Sep. 25 – Dec. 6	Oct. 5/6	–	Unknown	2005	47/29	Minor	First confirmed on Oct. 5, 2005 by a team operating a camera network in Finland

Other interesting objects:

Designation	R.A.	Decl.	Type	m$_v$	Size (')	Dist. (ly)	Lum. (suns)	Aper. Recm.	Rating	Comments	
Kemble's Cascade	03^h 57.0^m	+63° 00'	Asterism	5–9	Various			7×50	****	A striking ~2.5° NW–SE chain of mostly ~mag. 8 stars	*8

Fig. 15.4. NGC 1501, planetary nebula in Camelopardalis. Astrophoto by Mark Komsa.

Fig. 15.5. NGC 1502, open cluster in Camelopardalis. Astrophoto by Mark Komsa.

Fig. 15.6. NGC 1530, barred spiral galaxy in Camelopardalis. Astrophoto by Mark Komsa.

Fig. 15.7. NGC 1961, spiral galaxy in Camelopardalis. Astrophoto by Mark Komsa.

Fig. 15.8. NGC 2146, spiral galaxy in Camelopardalis. Astrophoto by Mark Komsa.

Fig. 15.9. NGC 2336, spiral galaxy in Camelopardalis. Astrophoto by Mark Komsa.

Fig. 15.10. NGC 2366, irregular galaxy in Camelopardalis. Astrophoto by Mark Komsa.

Fig. 15.11. NGC 2403, spiral galaxy in Camelopardalis. Astrophoto by Rob Gendler.

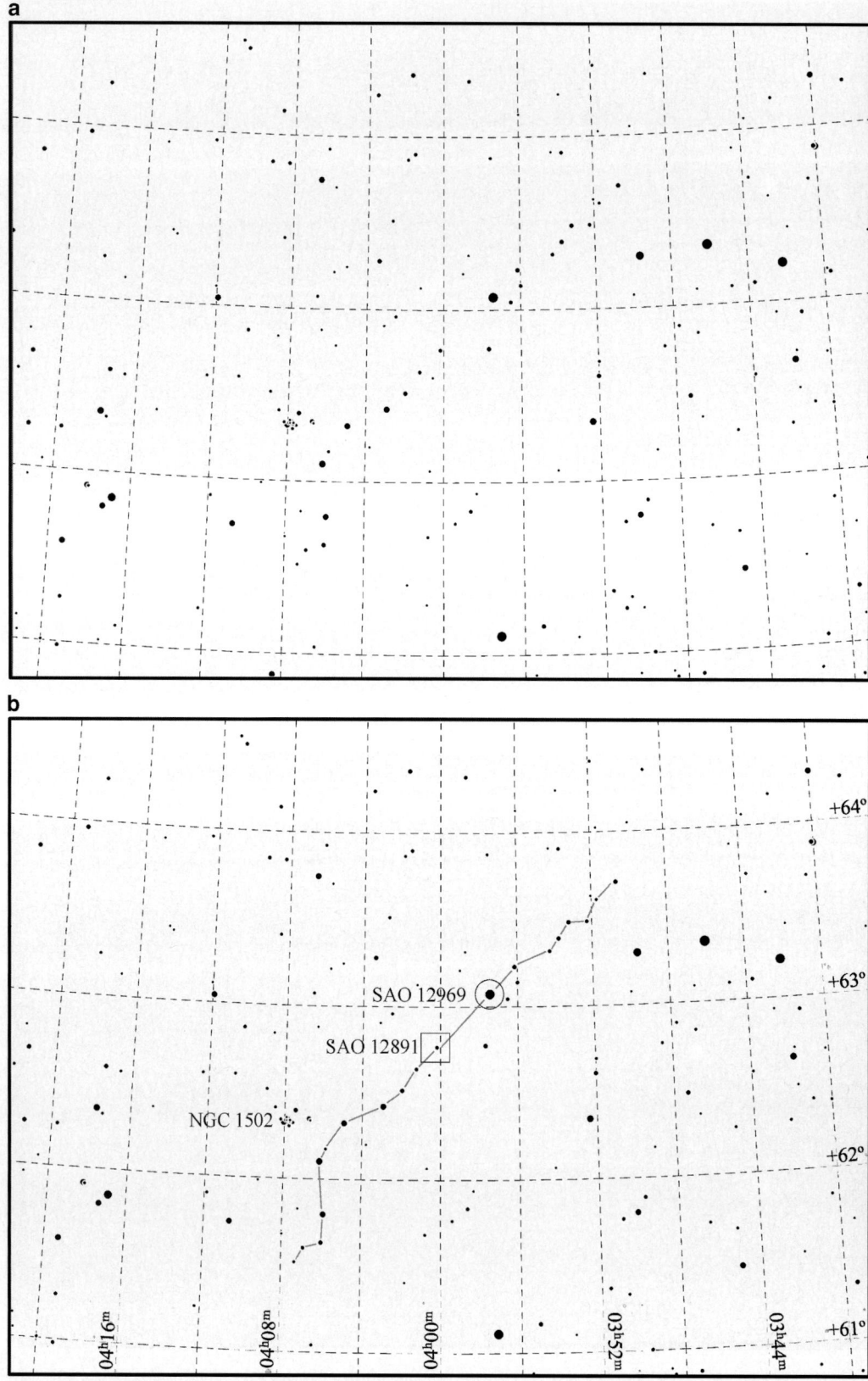

a

b

+64°

+63°

SAO 12969

SAO 12891

+62°

NGC 1502

+61°

04^{h}16^m 04^{h}08^m 04^{h}00^m 03^{h}52^m 03^{h}44^m

Fig. 15.12. (**a**, **b**) Kemble's Cascade.

COLUMBA (koh-**lum**-bah)	**The Dove**
COLUMBAE (koh-**lum**-bye)	**9:00** P.M. **Culmination:** Feb. 2
Col	**AREA:** 270 sq. deg., 54th in size

TO FIND: See Fig. 4.1, Orion Group Finder Chart, for the location of Columba just south of Canis Minor and Lepus. Columba is relatively small and has only two stars as bright as third magnitude, but is not too difficult to find because of its closeness to some moderately bright stars in Canis Major. Just as you can find <u>Arcturus</u> by the saying "arc to <u>Arcturus</u>," you can find Phact (α Columbae) by following an arc formed by η, ε (nearly first magnitude Adhara), and ζ Canis Majoris. This arc is approximately as long as Ursa Major's tail, and like it, points the way to the brightest star in another constellation, α Columbae about 10° from ζ. Continuing the arc another 2° brings you to ε Columbae. The entire arc from η CMa to ε Col is nearly the same length as the entire Big Dipper. α and ε Columbae, respectively, form the bird's head and beak. About 4° to the east–southeast of α, a small triangle, less than 2° on a side, formed by β, γ, and ξ Columbae marks the front of the dove's body. About 2° and 3° further east lie θ and κ Columbae, which complete the body. Just 2° northeast of κ Columbae and 3.4° south of ζ Canis Majoris, δ Columbae represents the dove's tail feathers (Figs. 4.1 and 15.13).

KEY STAR NAMES:

Name	Pronunciation	Source
α Phact	fact	fāhita (Arabic), "ring dove," from a Renaissance writing about Arabic names of birds
β Wazn	wuhzn	al wazn (Arabic), "the weight," significance unknown, perhaps erroneously applied (δ Canis Majoris is also called Wezen by the same source)
θ Al Kurud	al **kuh**-rud	al Kurud (Arabic), "the apes" from the Arabic constellation of the same name
κ Al Kurud	al **kuh**-rud	al Kurud (Arabic), "the apes" from the Arabic constellation of the same name

Magnitudes and spectral types of principal stars:

Bayer desig.	Mag. (m_v)	Spec. type	Bayer desig.	Mag. (m_v)	Spec. type	Bayer desig.	Mag. (m_v)	Spec. type
α	2.65	B7 IV*9	β	3.12	K1.5 III	γ	4.36	B2.5 IV
δ	3.85	G7 II	ε	3.86	K1 II/III	ζ	—.—	— —*10
η	3.96	K0 III	θ	5.00	B8 IV	ι	—.—	— —*10
κ	4.37	G8 II	λ	4.88	B5 V	μ	5.18	B1 IV/V
ν¹	6.15	F0 IV	ν²	5.28	F5 V	ξ	4.97	K1 III CN*11
o	4.81	K0 III/IV	π¹	6.15	Am*12	π²	5.50	A0 V*12
ρ	—.—	— —*13	σ	5.52	F2 III			

Stars of magnitude 5.5 or brighter which have no Bayer or Flamsteed designations:

Designation	Mag. (m_v)	Spec. type	Designation	Mag. (m_v)	Spec. type
HD 036848	5.45	K2/3 III	HD 037811	5.42	G6/8 III
HD 038170	5.29	B9/9.5 V	HD 046568	5.25	G8 III
HD 046815	5.42	K3 III			

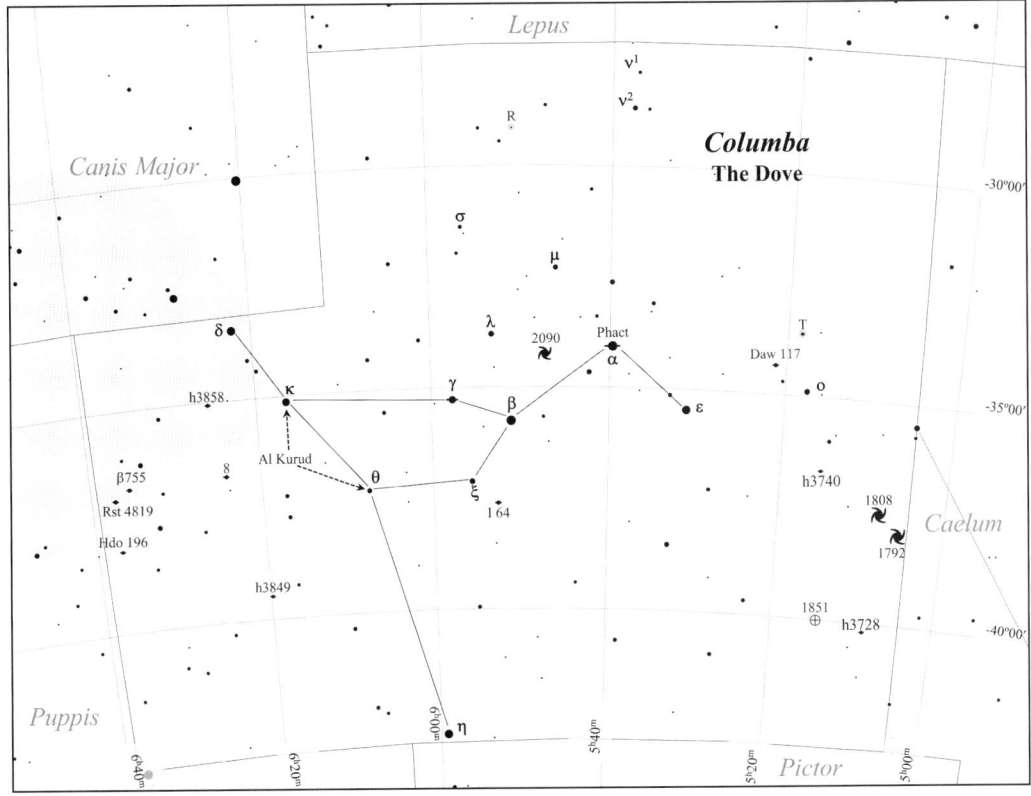

Fig. 15.13. Columba details.

Table 15.2. Selected telescopic objects in Columba.

Multiple stars:

Designation	R.A.	Decl.	Type	m₁	m₂	Sep. (″)	PA (°)	Colors	Date/Period	Aper. Recm.	Rating	Comments
h3728	05ʰ 08.5ᵐ	−40° 13′	A–B	6.8	10.3	10.1	260	Y, –	1999	2/3″	***	
h3740	05ʰ 15.1ᵐ	−36° 39′	A–B	6.9	8.3	23.7	286	dY, dY	1999	10×50	****	
Daw 117	05ʰ 21.3ᵐ	−34° 21′	A–B	6.1	10.9	2.2	007	bW, –	1969	5/6″	***	
α Columbae	05ʰ 39.6ᵐ	−34° 04′	A–B	2.6	12.5	13.5	359	bW, –	1950	4/5″	***	Challenging because of magnitude difference. a.k.a. Hdo 193
I 64	05ʰ 52.6ᵐ	−37° 38′	A–B	5.6	11.0	16.4	250	O, –	1960	3/4″	***	
			A–C		11.5	47.4	010	O, –	1999	15×80	***	
h3819	05ʰ 57.5ᵐ	−35° 17′	A–B	4.4	12.7	33.8	111	bW, –	1999	5/6″	***	a.k.a. h3819
h3849	06ʰ 19.8ᵐ	−39° 29′	A–B	6.7	8.1	39.5	054	brO, –	2000	7×50	****	
8 Mon	06ʰ 24.0ᵐ	−36° 42′	A–B	5.7	9.8	12.0	255	O, yO	1999	2/3″	*****	h3857AB
			A–C	5.7	6.9	63.6	074	O, bW	1999	6×30	*****	Δ28AC
h3858	06ʰ 25.5ᵐ	−35° 04′	A–B	6.4	7.6	131.7	047	dY, –	1991	6×30	*****	
β 755	06ʰ 35.4ᵐ	−36° 47′	B–C	7.6	8.2	3.8	309	–, –	1999	2/3″	***	
			A–B	5.9	6.9	1.5	260	W, gyW	1999	6×7″	***	Fairly rich star field
			A–C	6.0	11.5	21.0	302	W, –	1999	15×80	***	
Hds	06ʰ 37.0ᵐ	−38° 09′	A–a	6.3	8.7	0.7	213	dY, –	1991	13/14″	**	
Hdo 196	06ʰ 37.0ᵐ	−38° 09′	A–B	6.0	12.0	18.0	270	dY, –	1900	3/4″	***	
			A–C		11.2	27.7	175	dY, –	1991	3/4″	***	
			A–D		13.0	35.0	100	dY, –	1900	6/7″	***	
Rst 4819	06ʰ 37.2ᵐ	−36° 59′	A–B	5.9	7.5	0.5	004	bW, –	1999	19/20″	**	

Variable stars:

Designation	R.A.	Decl.	Type	Range (m_v)		Period (days)	F (f_r/f_f)	Spectral Range	Aper. Recm.	Rating	Comments
T Columbae	05ʰ 19.3ᵐ	−33° 42′	M	6.6	12.7	225.84	0.50	M4e–M6e	6/8″	***	Symmetrical light curve, spectral class varies from M3 to M6
R Columbae	05ʰ 50.5ᵐ	−29° 12′	M	7.8	15.0	327.62	0.39	M3e–M4e	18/20″	***	

Star clusters:

Designation	R.A.	Decl.	Type	m$_v$	Size (')	Brtst. Star	Dist. (ly)	dia. (ly)	Number of Stars	Aper. Recm.	Rating	Comments
NGC 1851	05^h 14.1^m	−40° 03'	Globular	7.2	11.0	13.2	35K	110	–	4/6"	***	Bright, dense core, loose halo, irregular periphery, some resolution in 10" *14

Nebulae:

Designation	R.A.	Decl.	Type	m$_v$	Size (')	Brtst./ Cent. star	Dist. (ly)	dia. (ly)	Aper. Recm.	Rating	Comments
None											

Galaxies:

Designation	R.A.	Decl.	Type	m$_v$	Size (')	Dist. (ly)	dia. (ly)	Lum. (suns)	Aper. Recm.	Rating	Comments
NGC 1792	05^h 05.2^m	−37° 59'	SAB	9.9	5.5×2.5	43M	69K	16G	4/6"	****	12/14" needed to show some detail
NGC 1808	05^h 07.7^m	−37° 31'	SB	9.9	5.2×2.3	33M	50K	9.5G	4/6"	****	12/14" shows tiny core, stellar nucleus, bright major axis
NGC 2090	05^h 47.0^m	−34° 15'	Sc	11.0	5.0×2.8	69M	100K	15G	10/12"	**	12/14" shows moderate core, faint halo, faint arms

Meteor showers:

Designation	R.A.	Decl.	Period (yr)	Duration	Max Date	ZHR (max)	Comet/ Asteroid	First Obs.	Vel. (mi/ km/sec)	Rating	Comments
None											

Other interesting objects:

Designation	R.A.	Decl.	Type	m$_v$	M$_v$	Dist. (ly)	Lum. (suns)	Aper. Recm.	Rating	Comments
μ Columbae	05^h 46.0^m	−32° 18'	Star	5.2	−4.4	2,700	5,000	Eye	****	"Runaway star," p.m.=73 mi/s away from Orion Nebula *15

CRUX (kruks)	**The (Southern) Cross**
CRUCIS (**krew**-siss)	**9:00** P.M. **Culmination:** May 13
Cru	**AREA:** 68 sq. deg., 88th in size

Crux, the smallest of the 88 constellations, is also the brightest in terms of brightness per unit area. It is located in one of the brightest portions of the Milky Way, has two first magnitude stars, and one only slightly fainter than the first magnitude limit. α, β, and γ Crucis are, respectively, the 13th, 19th, and 24th brightest stars in the sky. In units of total brightness per unit area, the stars of Crux outshine those of faint Sculptor by more than 30 times.

TO FIND: With its bright stars and distinctive cross shape, Crux is easy to recognize, and the presence of α and β Centauri nearby give Crux the brightest pointer stars in the sky. A line drawn from α to β Centauri is about 4.4° long, and when extended by 12.5°, points directly to γ Crucis, the northern tip of the cross. If you remember that the two bright stars in Centaurus are close to Crux, you cannot confuse Crux with the False Cross (four stars in Carina and Vela) or the Diamond Cross (four stars in a different part of Carina) which sometimes lead people astray (Figs. 6.26, 6.27, 15.14 and 15.15).

KEY STAR NAMES:

Name	Pronunciation	Source
α Acrux	ay-kruks	alpha Crucis (Greek + Latin), shortened, probably by navigators who needed a more convenient name
β Mimosa	**me**-moh-suh	mimus (Latin), "an actor," reason unknown
γ Gacrux	**guh**-kruks	gamma Crucis (Greek + Latin), Same as Acrux

Magnitudes and spectral types of principal stars:

Bayer desig.	Mag. (m_v)	Spec. type	Bayer desig.	Mag. (m_v)	Spec. type	Bayer desig.	Mag. (m_v)	Spec. type
α	0.91	B0.5 IV*[16]	β	1.25	B0.5 III	γ	1.59	M4 III
δ	2.79	B2 IV	ε	3.59	K3/4 III	ζ	4.06	B2.5 V
η	4.14	F2 III	θ¹	4.32	Am	θ²	4.72	B2 IV
ι	4.69	K1 III	κ	5.89	B5 Ia*[17]	λ	4.62	B4 Vn
μ	3.68	B2 IV/V*[18]						

Stars of magnitude 5.5 or brighter which have Flamsteed but no Bayer designations:

Flamsteed	Mag. (m_v)	Spec. type	Flamsteed	Mag. (m_v)	Spec. type
35	5.49	F5/G0 Ib	39	4.91	B6 IV

Other stars magnitude 5.5 or brighter which have only catalog designations:

Designation	Mag. (m_v)	Spec. type	Designation	Mag. (m_v)	Spec. type
HD 110956	4.62	B3 V	HD 112244	5.34	09 Ib
HD 107696	5.38	B9 V	HD 103961	5.44	B8 III

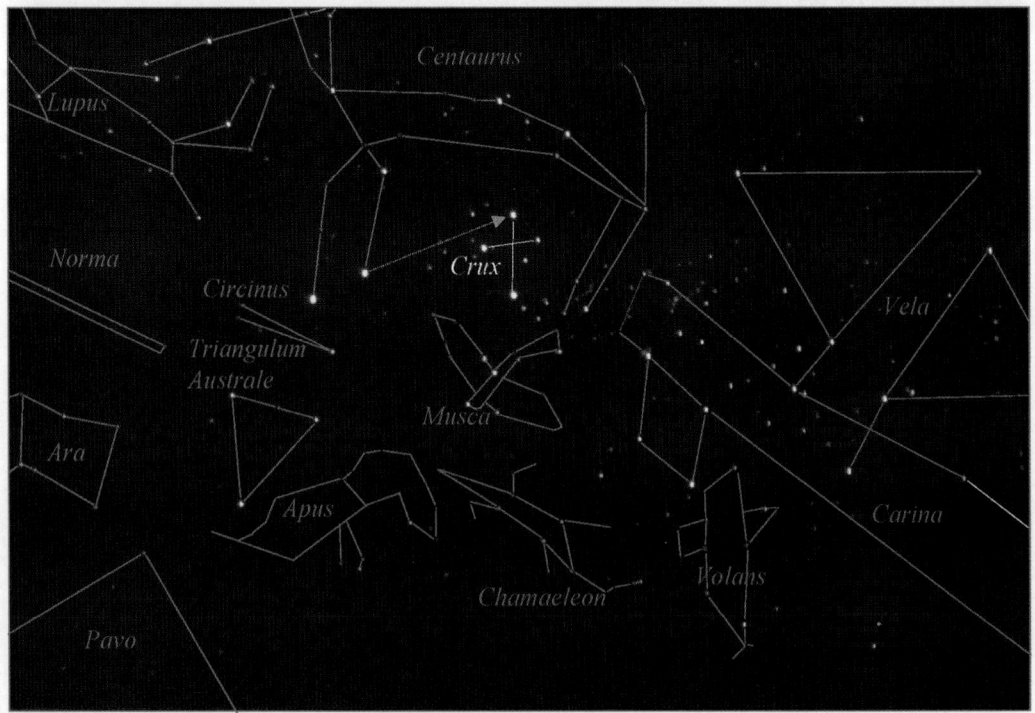

Fig. 15.14. The figures of Crux and surrounding constellations.

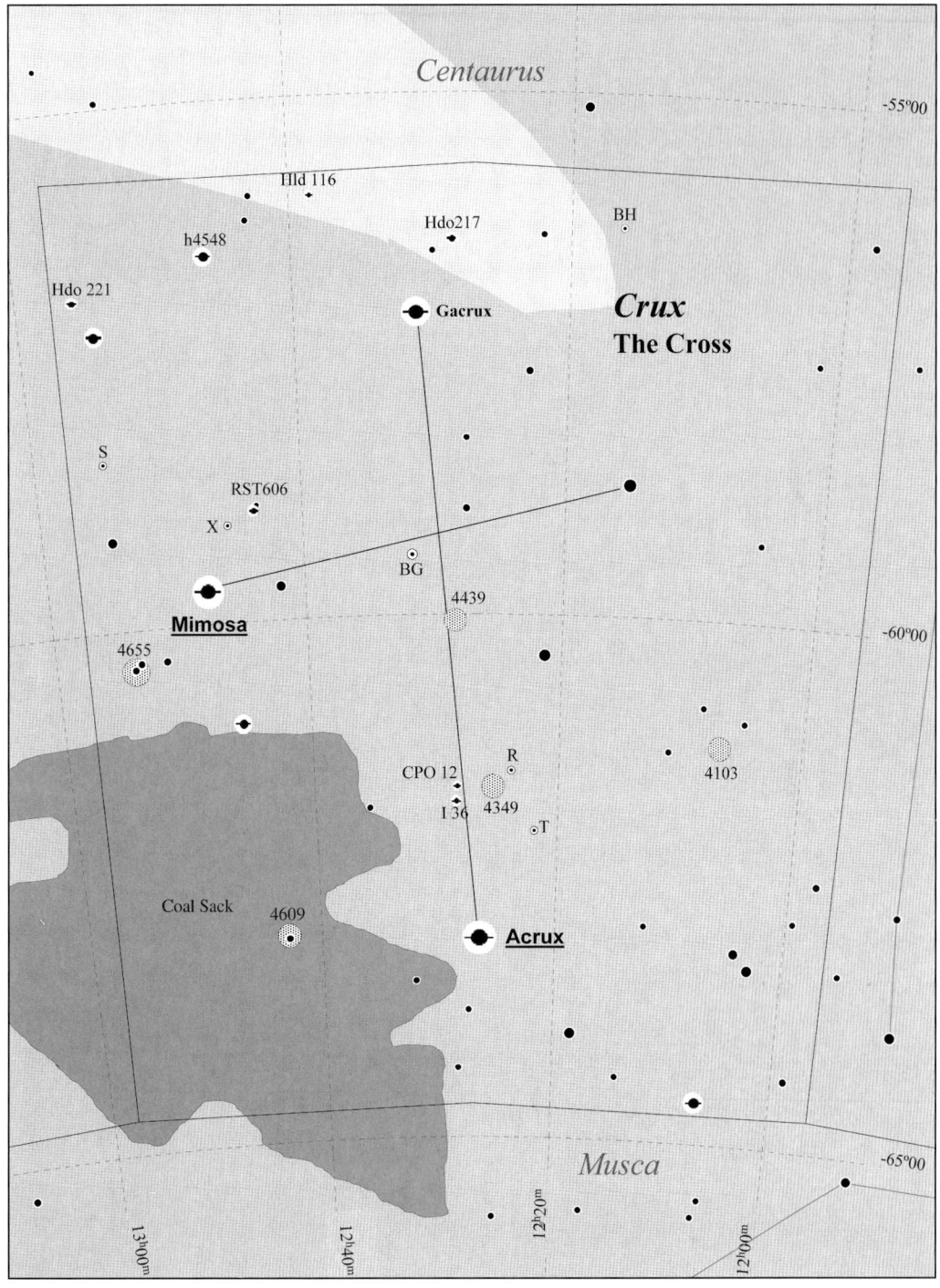

Fig. 15.15. Crux details.

Table 15.3. Selected telescopic objects in Crux.

Multiple stars:

Designation	R.A.	Decl.	Type	m₁	m₂	Sep. (")	PA (°)	Colors	Date/Period	Aper. Recm.	Rating	Comments
η Crucis	12ʰ06.9ᵐ	−64°37'	A–B	4.2	11.8	46.5	302	Y, ?	1945	3/4"	****	
Acrux, α Crucis	12ʰ26.6ᵐ	−63°06'	A–B	1.3	1.6	4.0	114	bW, bW	2002	2/3"	*****	Second in overall brightness of double only to α Centauri
		−63°06'	A–C	1.3	4.8	90.0	202	bW, R	1991	6×30	*****	With C, probably the brightest triplet in the sky
CPO 12	12ʰ28.3ᵐ	−61°45'	A–BC	7.3	8.2	**2.10**	**189**	Y, ?	**2,520**	4/5"	****	Min. sep. of 1.97" in 1920, max. of 8.84" in 3242
			B–C	8.8	8.8	**0.26**	**193**	Y, ?	**27.1**	–	**	Near secondary max. in 2010, min. of 0.19" in 2016, primary max. of 0.32" in 2025
Hdo 217	12ʰ28.7ᵐ	−56°23'	A–B	6.1	13.0	27.2	275	O, ?	1903	6/7"	***	
			A–C	6.1	12.0	49.0	290	O, ?	1902	3/4"	***	
I 36	12ʰ28.9ᵐ	−61°52'	A–B	6.9	10.9	7.0	326	bY, ?	2000	3/4"	****	
Gacrux, γ Crucis	12ʰ31.1ᵐ	−57°06'	A–B	1.6	6.5	125.2	027	O, W	2002	6×30	****	
Hld 116	12ʰ38.1ᵐ	−55°56'	A–C	1.6	9.5	167.0	071	O, ?	2002	2/3"	****	
RST 606	12ʰ43.5ᵐ	−58°53'	A–B	7.1	8.9	1.9	183	W, ?	1991	5/6"	****	
HJ 4543			A–B	6.4	10.9	2.4	157	dY, ?	1967	4/5"	****	
			A–C	6.4	9.8	36.5	095	dY, ?	2000	15×80	****	
ι Crucis	12ʰ45.6ᵐ	−60°84'	A–B	4.7	10.2	28.1	008	pY, ?	1991	15×80	****	
h4548	12ʰ46.4ᵐ	−56°29'	A–B	4.6	9.8	51.7	166	bW, ?	2000	15×80	****	
Mimosa, β Crucis	12ʰ47.8ᵐ	−60°84'	A–B	1.3	11.4	42.2	326	bW, ?	2002	3/4"	****	Spectral class O0.5, probably the hottest first magnitude star
μ Crucis	12ʰ54.6ᵐ	−57°11'	A–B	3.9	5.0	34.8	010	bW, ?	2002	7×50	****	Bright, wide, easy to view double
HdO 221	12ʰ56.0ᵐ	−56°50'	A–B	5.3	11.8	29.2	314	B, ?	2000	3/4"	***	

Variable stars:

Designation	R.A.	Decl.	Type	Range (mᵥ)		Period (days)	F (f₂/f₁)	Spectral Range	Aper. Recm.	Rating	Comments
BH Crucis	12ʰ16.3ᵐ	−56°17'	M	7.2	10.0	421	–	SC4.5e–SC7e	10×80	****	2°12' WNW (PA 291°) of γ Crucis (m1.59).#

Designation	R.A.	Decl.	Type						Aper.	Rating	Comments
T Crucis	12h 21.4m	−62° 17'	Cδ	6.3	6.8	6.733	0.34	F6-G2Ib	7×50	****	1° 01' NW (PA 323°) of α Crucis (m0.77).#
R Crucis	12h 23.6m	−61° 38'	Cδ	6.4	7.2	5.826	0.28	F6-G2, Ib-II	7×50	****	1° 31' NNW (PA 346°) of α Crucis (m0.77).#
BG Crucis	12h 31.7m	−59° 25'	Cδ	5.34	5.58	3.343	0.47	G5Ib-G0p	6×30	****	2° 03' W (PA 276°) of β Crucis (m1.21)
X Crucis	12h 46.4m	−59° 08'	Cδ	8.1	8.7	6.220	0.31	F6-G2Ib	7×50	****	0° 35' NNW (PA 343°) of β Crucis (m1.21).#
S Crucis	12h 54.4m	−58° 26'	Cδ	6.2	6.9	4.690	0.34	F6-G1, Ib-II	7×50	****	1° 31' NE (PA 035°) of β Crucis (m1.21).#

Star clusters:

Designation	R.A.	Decl.	Type	m_v	Size (')	Brtst. Star	Dist. (ly)	dia. (ly)	Number of Stars	Aper. Recm.	Rating	Comments
NGC 4103	12h 06.7m	−61° 15'	Open	7.4	7	10.0	3.9K	8	45	2/3"	****	High concentration of stars, large range of brighten, pretty large, pretty bright, irregular rounds
NGC 4349	12h 24.5m	−61° 54'	Open	7.4	15	10.9	5.5K	24	30	2/3"	****	High concentration of stars, moderate range of brighten, very large, very bright
NGC 4439	12h 28.4m	−60° 06'	Open	8.4	4	10.3	5.2K	6	20	4/6"	***	Low concentration of stars, small size
NGC 4609	12h 42.3m	−62° 58'	Open	6.9	5	9.0	4.9K	7	40	3/4"	***	Low concentration of stars, pretty large, elongated. Young cluster, ~36M years
NGC 4755	12h 53.6m	−60° 20'	Open	4.2	10	5.8	7.6K	9	100	2/3"	*****	a.k.a. Jewel Box Cluster, C94. One of the best in the southern hemisphere skies *19

(continued)

Table 15.3. (continued)

Nebulae:

Designation	R.A.	Decl.	Type	m_v	Size (')	Brtst./Cent. star	Dist. (ly)	dia. (ly)	Aper. Recm.	Rating	Comments
Coal Sack	12ʰ 53.0ᵐ	–63° 00'	Dark	–	6.7° × 5.0°	–	500	60	Eye	*****	The most famous unaided eye dark nebula. Very obvious

Galaxies:

Designation	R.A.	Decl.	Type	m_v	Size (')	Dist. (ly)	dia. (ly)	Lum. (suns)	Aper. Recm.	Rating	Comments
None											

Meteor showers:

Designation	R.A.	Decl.	Period (yr)	Duration	Max Date	ZHR (max)	Comet/Asteroid	First Obs.	Vel. (mi/ km/sec)	Rating	Comments
None											

Other interesting objects:

Designation	R.A.	Decl.	Type	m_v	Size (')	Dist. (ly)	dia. (ly)	Lum. (suns)	Aper. Recm.	Rating	Comments
None											

Fig. 15.16. α and β Centauri, Crux, and η Carinae nebula. Astrophoto by Christopher J. Picking.

Southern Cross

False Cross

Diamond Cross

Fig. 15.17. The three crosses. The false cross consists of κ and δ Velorum plus ε and ι Carinae. It is sometimes mistaken for the true cross. Astrophoto by Christopher J. Picking.

MONOCEROS (mo-**noss**-sir-us)	**The Unicorn**
MONOCEROTIS (mo-nos-er-**OH**-tis)	**9:00 P.M. Culmination:** Feb. 20
Mon	**AREA:** 482 sq. deg., 35th in size

TO FIND: See Figs. 4.1 and 7.11 for a finder chart and photograph of Monoceros. Using these charts as a guide, you can find the brightest stars in Monoceros, and then fill in the rest of the pattern. α Monocerotis lies about 15° south (PA 178°) of <u>Procyon</u>, and marks the hind paws of the unicorn. About half-way between Orion's belt and <u>Sirius</u> lies a small equilateral triangle (~3.5° on each side) with a star in the middle. The two stars on the upper left are β and γ Monocerotis, and they represent the two forepaws of the unicorn. The unicorn's nose is represented by 13 Monocerotis, which lies 9.3° due east of <u>Betelgeuse</u>. Be careful to avoid mistaking 8 Monocerotis which lies south of, and about the same distance from <u>Betelgeuse</u>, but does not form part of the unicorn's figure. About 3.5° to the east–northeast of 13 Monocerotis lies 15 and 17 Monocerotis. These two mark the rest of the unicorn's head. δ and ζ Monocerotis lie 9° southeast and 11° southwest of Procyon. These mark the unicorn's front and hindquarters, respectively. Using the above stars as starting points, you should be able to make out the large, but faint, form of the heavenly unicorn (Figs. 4.1, 7.11 and 15.18).

KEY STAR NAMES:

Name	Pronunciation	Source
α Cerastes	**keer**-ass-tez	κεραστης [kerastez] (Greek), "horned"

Magnitudes and spectral types of principal stars:

Bayer desig.	Mag. (m_v)	Spec. type	Bayer desig.	Mag. (m_v)	Spec. type	Bayer desig.	Mag. (m_v)	Spec. type
α	3.94	K0 III	β	3.76	B3 Ve	γ	3.99	K3 III
δ	4.15	A2 V	ε	4.39	A5 IV	ζ	4.36	G2 Ib

Stars of magnitude 5.5 or brighter which have Flamsteed but no Bayer designations:

Flam-steed	Mag. (m_v)	Spec. type	Flam-steed	Mag. (m_v)	Spec. type	Flam-steed	Mag. (m_v)	Spec. type
13	4.47	A0 Ib	18	4.48	K0 III	28	4.69	K4 III
17	4.77	K4 III	20	4.91	K0 III	3	4.92	B5 III
27	4.93	K2 III	19	4.99	B1 V	2	5.04	A6m
10	5.06	B2 V	25	5.14	F6 III	77	5.19	K1 II
7	5.27	B2.5 V	21	5.44	F2 V			

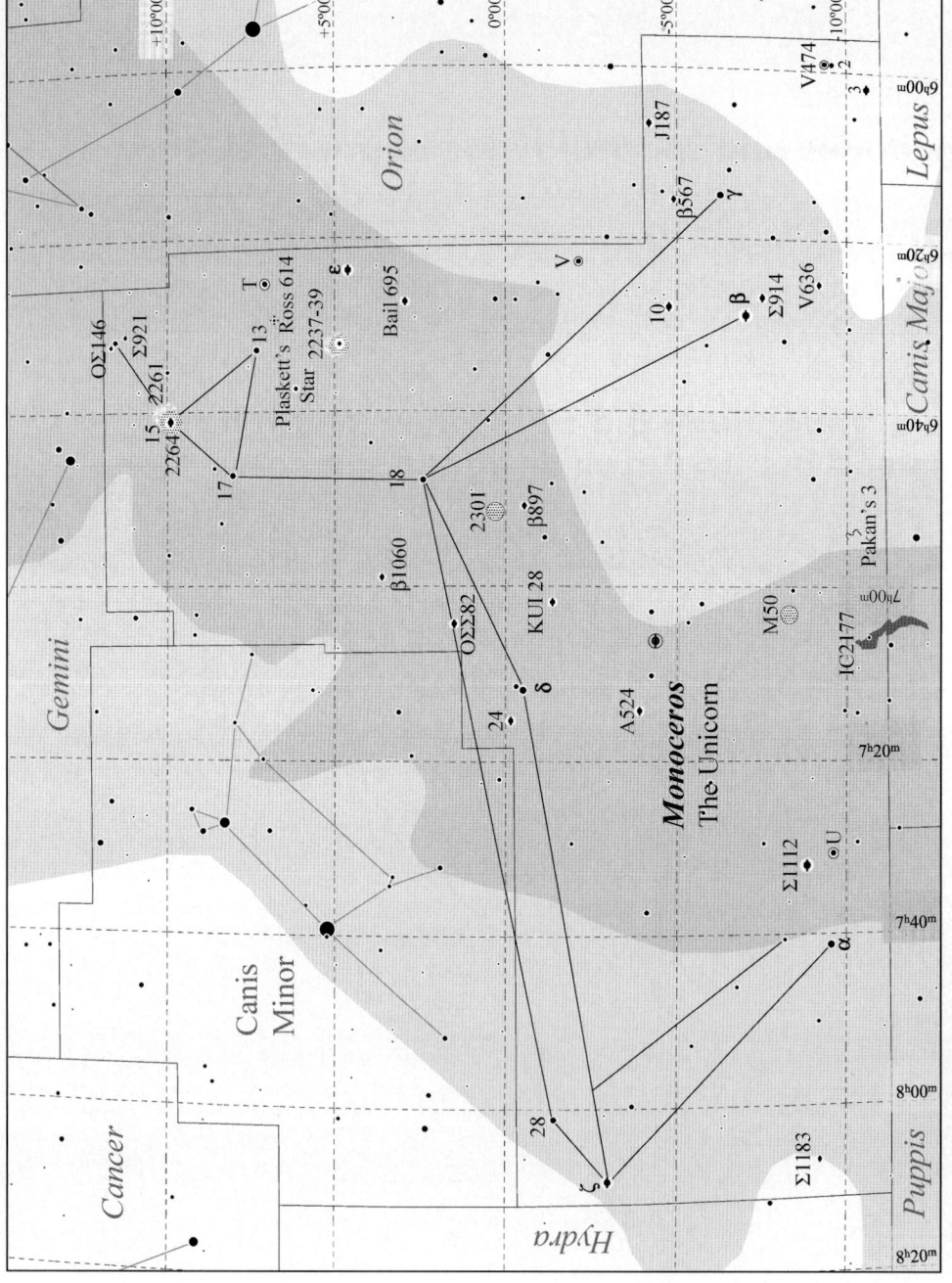

Fig 15.18. Monoceros details.

Table 15.4. Selected telescopic objects in Monoceros.

Multiple stars:

Designation	R.A.	Decl.	Type	m₁	m₂	Sep. (")	PA (°)	Colors	Date/Period	Aper. Recm.	Rating	Comments
3 Monocerotis	06ʰ 01.8ᵐ	−10° 36′	A–B	5.0	8.0	1.9	355	bW, ?	1996	5/6″	****	a.k.a. β 16
J 187	06ʰ 06.6ᵐ	−04° 12′	A–B	5.4	11.6	29.3	145	bW, ?	1998	15×80	****	a.k.a. J 187
β 567	06ʰ 15.5ᵐ	−04° 55′	A–B	6.0	10.0	3.9	241	pY, ?	1998	3/4″	****	
8, ε Monocerotis	06ʰ 23.8ᵐ	+04° 36′	A–B	4.4	6.6	12.1	029	pY, pY	2004	2/3″	*****	Showcase pair. Splitting A–B is easy. a.k.a. Σ900AB
			A–C	4.3	12.7	92.6	254	pY, bW	1999	5/6″	***	Seeing the color of C requires larger (12/14″) scope. a.k.a. 900AC
Σ914	06ʰ 26.7ᵐ	−07° 31′	A–B	6.3	9.3	21.1	298	brW, W	2003	2/3″	****	a.k.a. βpm 89AB. A nearly perfect, thin isoceles triangle with A at the point
10 Monocerotis	06ʰ 28.0ᵐ	−04° 46′	A–B	5.1	9.2	77.0	257	bW, ?	2002	7×50	****	
11, β Monocerotis	06ʰ 28.8ᵐ	−07° 02′	A–C	5.1	9.2	78.0	232	bW, ?	2004	7×50	****	a.k.a. βpm 89AC
			A–B	4.6	5.0	7.1	133	pY, bW	2006	2/3″	****	A, B, and C form an attractive thin triangle. a.k.a. Σ919AB
			A–C	4.6	5.4	9.8	125	bW, bW	2006	2/3″	****	a.k.a. Σ919AC. C has a close companion, sep. ~0.3″
Bail1695	06ʰ 29.2ᵐ	+02° 39′	A–D	4.6	12.2	25.4	047	bW, bW	1999	4/5″	***	a.k.a. β570AD
Σ921	06ʰ 31.2ᵐ	+11° 15′	A–B	6.2	10.4	19.1	300	R, ?	1999	15×80	****	
OΣ146	06ʰ 32.4ᵐ	+11° 40′	A–B	6.1	9.1	16.0	003	Y, B	2003	2/3″	******	Beautiful pair
15, S Monocerotis	06ʰ 41.0ᵐ	+09° 54′	Aa–B	4.6	7.8	2.9	214	bW, W	2002	3/3″	****	S Mon is the brightest star in NGC 2264, and is slightly variable. a.k.a. Σ950
			Aa–C	4.6	9.9	16.6	014	bW, ?	2002	15×80	****	Range 4.2–4.6. a.k.a. Σ950AC
			Aa–E	4.6	8.9	74.1	140	bW, ?	2002	2/3″	****	a.k.a. Σ950AE
			E–a	8.8	10.4	4.3	044	bW, ?	2005	3/4″	****	a.k.a. Dem 11
β 897	06ʰ 50.8ᵐ	−00° 32′	A–B	5.8	11.1	6.5	019	yW, ?	2004	3/4″	****	
β 1060	06ʰ 59.0ᵐ	+03° 36′	A–B	6.0	11.0	3.6	059	dY, ?	1953	3/4″	****	
Kui 28	07ʰ 01.9ᵐ	−01° 21′	A–B	6.3	10.1	24.3	252	dY, ?	2003	2/3″	****	
OΣΣ82	07ʰ 04.3ᵐ	+01° 29′	A–B	6.5	7.6	89.9	320	W, rY	2004	7×50	****	

(continued)

Table 15.4. (continued)

Multiple stars:

Designation	R.A.	Decl.	Type	m₁	m₂	Sep. (")	PA (°)	Colors	Date/ Period	Aper. Recm.	Rating	Comments
V569 Monocerotis	07ʰ 05.8ᵐ	−10° 40'	A–B	6.5	10.1	6.4	282	W, B	2003	2/3"	****	a.k.a. D12AB and V569 Mon. See variable star section also
			A–C	6.5	9.6	38.5	295	W, ?	2004	15×80	****	a.k.a. Σ1019AC
			C–D	9.8	11.2	0.9	302	?, ?	1963	0/11"	****	a.k.a. Rst 3491CD
A524	07ʰ 14.2ᵐ	−03° 54'	A–B	6.1	10.2	2.7	152	yO, ?	1991	3/4"	****	
24 Monocerotis	07ʰ 15.3ᵐ	−00° 09'	A–B	6.4	12.3	3.8	310	Y, ?	1946	4/5"	****	a.k.a. β1268
Σ1112	07ʰ 32.1ᵐ	−08° 53'	A–B	6.0	8.7	6.0	113	pY, ?	1998	2/3"	****	
Σ1183	08ʰ 06.5ᵐ	−09° 15'	A–B	6.2	7.8	30.6	328	Y, grW	2004	15×80	****	a.k.a. A543BC
			B–C	8.0	11.7	1.4	324	grW, ?	1991	7/8"	***	a.k.a. A543BD
			B–D	8.0	13.5	19.4	325	grW, ?	1959	7/8"	***	
29, ζ Monocerotis	08ʰ 08.6ᵐ	−02° 59'	A–B	4.3	10.1	33.0	105	oY, Y	1998	15×80	****	a.k.a. Σ1190AB
			B–C	4.3	9.7	64.7	247	oY, W	2002	15×80	****	a.k.a. Σ1190AC

Variable stars:

Designation	R.A.	Decl.	Type	Range (m_v)		Period (days)	F (f_i/f_f)	Spectral Range	Aper. Recm.	Rating	Comments
V474 Monocerotis	05ʰ 59.3ᵐ	−09° 23'	δ Sct	5.9	6.4	0.136	—	F2IV	6×30	****	11'N (PA 355°) of 1 Monocerotis (m5.04)
V Monocerotis	06ʰ 22.7ᵐ	−02° 12'	M	6.0	13.7	330.45	0.46	M5e–M8e(SE)	10/12"	***	2° 53' NNW (PA 333°) of 10 Monocerotis (m5.05)
V636 Monocerotis	06ʰ 25.0ᵐ	−09° 07'	M	6.2	9.8	—	—	C	10×80	****	Carbon star, lovely red color
T Monocerotis	06ʰ 25.2ᵐ	+07° 05'	δ Cep	5.5	6.6	27.02	0.27	F7Iab–K1Iab	6×30	***	Large spectral variation
V569 Monocerotis	07ʰ 05.8ᵐ	−04° 41'	β Cep	6.4	6.5	0.267	—	B0 V	6×30	****	See double star section also
U Monocerotis	07ʰ 30.8ᵐ	−09° 47'	RV Tauri	5.5	7.7p	91.32	0.22	F8eIb–K0p1b	6×30	****	RV Tauri variables are characterized by a double period *20

Star clusters:

Designation	R.A.	Decl.	Type	m_v	Size (')	Brst. Star	Dist. (ly)	dia. (ly)	Number of Stars	Aper. Recm.	Rating	Comments
NGC 2244	06h 32.4m	+04° 52'	Open	4.8	30	5.8	4,900	43	100	4/6"	****	A young (~500K years) cluster with several hot O-type stars. See Mon B *21
NGC 2264	06h 41.1m	+09° 53'	Open	3.9	20	4.6	2,600	30	40	7×50	****	a.k.a. "Christmas Tree" cluster because of its shape *22
NGC 2301	06h 51.8m	+00° 28'	Open	6.0	15	8.0	2,500	11	80	7×50	****	Large, bright, irregular. Several distinct star chains are visible
M50, NGC 2323	07h 03.2m	-08° 20'	Open	5.9	15	7.9	3,300	15	80	7×50	*****	Large, bright, slightly irregular

Nebulae:

Designation	R.A.	Decl.	Type	m_v	Size (')	Brst./Cent. star	Dist. (ly)	dia. (ly)	Aper. Recm.	Rating	Comments
NGC 2237-39	06h 32.3m	+05° 03'	Em.	-	80×60	-	-	-	7×50	**	Rosette Nebula. a.k.a. C49. Very large, faint. Best in binoculars *23
NGC 2261	06h 39.2m	+08° 44'	Refl.	-	2×1	-	3K	3×1	4/6"	***	a.k.a. Hubble's Variable Nebula and C46. Bright, wide cone, point at south end *24
IC 2177	07h 05.1m	-10° 42'	Em.	-	2.7°×4.1	6.2	-	-	11×80	***	a.k.a. Seagull's or Eagle's Wings. Large, bright diffuse nebula *25

Galaxies:

Designation	R.A.	Decl.	Type	m_v	Size (')	Dist. (ly)	dia. (ly)	Lum. (suns)	Aper. Recm.	Rating	Comments
None											

Other interesting objects:

Designation	R.A.	Decl.	Type	m_v	Sep. (°)	Size (')	Dist. (ly)	dia. (ly)	Lum. (suns)	Aper. Recm.	Rating	Comments
Ross 614	06h 29.4m	+06° 48'	Double	11.1-14.4	12.9							24th nearest star system *26
Plaskett's star	06h 37.4m	+06° 08'	Star	6.06								~1.5° SE of 13 Mon. A massive binary system of two giants *27
Minus 3 Group	06h 54.0m	-10° 12'	Asterism	9-10		44×23						~15 stars forming a reversed numeral 3. a.k.a. Pakan's 3. 1° 57' N (PA 351°) of θ CMa

Fig. 15.19. M50 (NGC 2323), open cluster in Monoceros. Astrophoto by Mark Komsa.

Fig. 15.20. NGC 2237, Rosette Nebula in Monoceros. Astrophoto by Mark Komsa.

Fig. 15.21. NGC 2264, Christmas Tree cluster (*upper middle*), Cone Nebula (*near top middle*), and NGC 2169, open cluster, *top left*. Astrophoto by Mark Komsa.

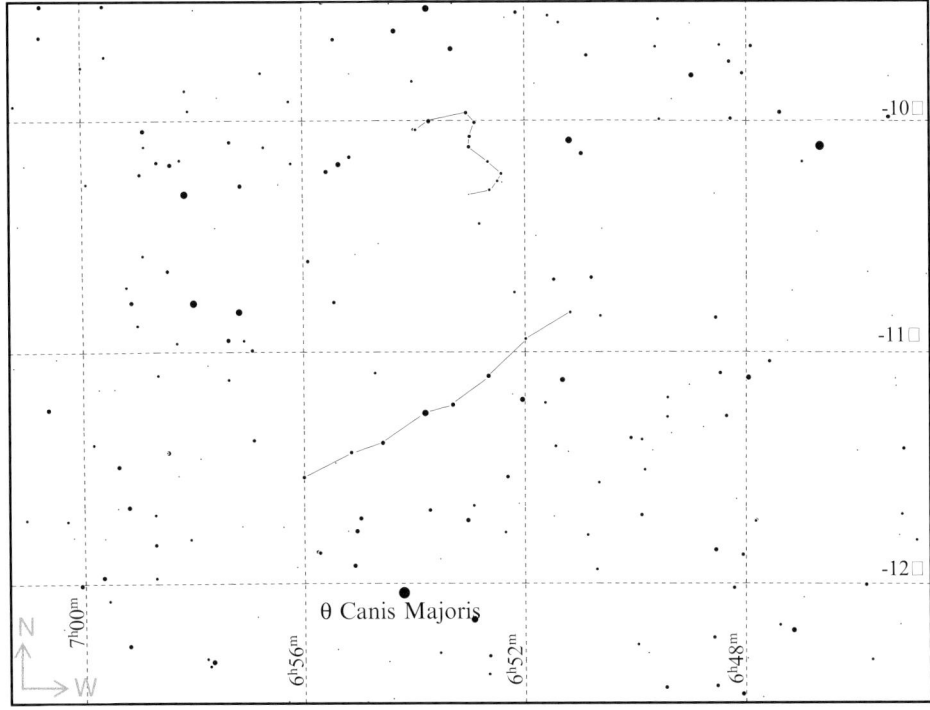

Fig. 15.22. Pakan's 3 a binocular asterism. It lies less than 2 degrees (use symbol) nearly due north of θ Canis Majoris. Also note the smaller, fainter Kemble's Cascade.

NOTES:

1. To distinguish these SAO 2264 numbers from the four-digit NGC numbers on the charts, they are all displayed as five- or six-digit numbers, even if the leading number is zero.

2. Σ485 is accompanied by at least 12 other stars ranging from 10th to 14th magnitude within 140″, including variable star SZ Cam, q.v. These may just be members of the cluster rather than actual companions of Σ485.

3. Some observers doubt the R CrB classification because at least one 94-day, symmetric minimum was observed between July 12 and Oct. 4, 1912. The spectrum reveals an excess of carbon and a deficiency of hydrogen, indicating a very old star which has nearly exhausted its hydrogen and which is nearing the end of its life.

4. RX Cam is also a spectroscopic binary with a period of 113.55 days.

5. SZ Cam is part of a multiple system. A third star orbits the eclipsing pair with a period of about 60 years.

6. T Cam is a rare type of variable which shows spectral lines or bands of zirconium oxide.

7. NGC 2403 is one of the brightest non-Messier objects in the northern sky. It rivals the famous Whirlpool Galaxy in photographs, but is visually more difficult because of its low surface brightness. It is also known as Caldwell 7. An outlying member of the M81 (Ursa Major) galaxy group, it is visible in 7 × 50 binoculars, but requires a telescope to reveal details.

8. This well-known 2.5° long chain of mostly 8th magnitude stars is named for the Canadian observer Fr. Lucien J. Kemble who first wrote about it.

9. Bayer did not consider Columba as separate constellation, but only as an asterism in Canis Major and assigned it no letters. All the Greek letter designations were assigned by Lacaille and Gould.

10. Lacaille did not use either ζ or ι in Columba.

11. Lacaille cataloged ξ Columbae, but did not assign a designation. It was later assigned by Gould.

12. Lacaille assigned these adjacent stars the single designation of π.

13. Bayer gave the designation ρ Columbae to a magnitude 6.35 star which later became part of Puppis, and which is now designated only by its catalog numbers, SAO 217702, HD 41700, or HR 2157.

14. NGC 1851 can be detected with 7 × 50 binoculars, and some resolution is possible with a 10″ telescope. It has a bright, dense core, a loose halo, and has a luminosity of about 200,000 suns.

15. μ Columba has a space velocity of about 73 mi/s (117 km/s) away from the Orion nebula, where it is thought to have originated. Two other "runaway stars" are AE Aurigae and 53 Arietis. Although it was once thought that the three stars had a common origin, the latest measurements indicate that their paths traced backward do not intersect.

16. α Crucis is a triple star with the components having individual magnitudes of 1.40, 2.09, and 4.86. The combined brightness is given in this list. The "C" component is probably a background star. It is separately visible to the unaided eye, but is sometimes designated as 256 Crucis, but the origin of the designation is not known.

17. κ Crucis is a member of the "Jewel Box" cluster, but the name is often used to mean the cluster rather than the star.

18. μ Crucis is a double star with the components having individual magnitudes of 4.03 and 5.08. The combined brightness is given in this list.

19. NGC 4755 was discovered by Lacaille and is listed as Lacaille II. It is also known as the κ Crucis cluster and as Caldwell 94. It lies only ~1° north of the Coal Sack and about the same distance south of β Crucis. κ Crucis, the star, is a red supergiant. Although it is quite young, it has already near the end of its life span, and begun to expand. The contrast between its orange color and the blue-white color of the other giant stars of the cluster is partly responsible for William Herschel's impression that it looked like a box of colored jewels. The contrast between the brilliant β, the dense cluster of stars, and the black nebula is striking in binoculars, especially large ones.

20. U Monocerotis is typical of the relatively rare RV Tauri variables in that, in addition to the 91.32-day period, it has a longer period of 2,475 days or ~6.8 years. They also often have alternating shallow and deep minima, which sometimes shift sequence by having two successive similar minima followed by alternating minima with the sequence reversed. They are usually very far advanced in their evolution. The masses of RV Tauri stars are estimated to be 25–30 times that of the sun, and their luminosities about 6,000 suns.

21. NGC 2244 is a young cluster (~500K years) with several hot-type O stars. It is surrounded by the Rosette nebula, and the radiation of the hot stars is slowly pushing the gas out of the central part of the nebula.

22. NGC 2267 is a large and bright star. It is embedded in a faint cone-shaped emission nebula which requires a 12–14 telescope to see.

23. The Rosette Nebula is very large, but very faint. It is often overlooked because of its low surface brightness, but it can be seen in 7×50 binoculars on a clear, dark night.

24. NGC 2261 is a cone-shaped nebula which is illuminated by R Monocerotis near the tip of the cone. Edwin Hubble was photographing this nebula in 1916, when he noticed changes between images taken only a few weeks or months apart. R Mon is an irregular variable star of spectral class ranging from A to F; however, the variations in the nebula were not just following the star, but were much more complicated. Some of the theories which are considered reasonable are that either streams of material emitted by the star or thick dusk clouds passing between the star and the nebula are casting irregular shadows on the nebula, causing the irregular changes in shape and brightness. Because of its shape and brightness profile, with the cone's point much brighter than the rest, the Cone Nebula is sometimes mistaken for a comet.

25. IC 2177 is also known as the "Seagull Nebula" (not to be confused with a nebula with the same name in Dorado). It is also listed by its name in various catalogs, e.g., Gum 2 and Sharpless 2-296. IC 2177 is best seen in small aperture, short focal length optical systems, e.g., a 4 f/5 telescope with a low power eyepiece or 11×80 binoculars. It is part of a large complex of dust, gas, and stars located about 2,500 light-years away. This complex includes the open clusters NGC 2335, 2343, and 2353, as well as the dark nebulae Ced 90 and vdB 93. IC 2177 straddles the border between Monoceros and Canis Major.

26. Ross 614 is the 24th nearest star system, and one of the least massive binary systems discovered. The components are estimated to be 0.14 and 0.08 solar masses.

27. In contrast to Ross 614, Plaskett's star is one of the most massive binary systems known, and may be the most massive binary in our galaxy. The two components have masses of about 60 and 40 times the sun. It is probably an outlying member of the open cluster NGC 2244. Plaskett's star is about 450 times as massive as Ross 614.

CHAPTER SIXTEEN

Keyser and de Houtman, the Map Maker's Apprentices

February–May, July, October–December

APUS	CHAMELEON	DORADO	GRUS
The Bird of Paradise	The Chameleon	The Golden Fish	The Crane

HYDRUS	INDUS	MUSCA	PAVO
The Water Snake	The Native American	The Fly	The Peacock

PHOENIX	TRIANGULUM AUSTRALE	TUCCANA	VOLANS
The Phoenix	The Southern Triangle	The Toucan	The Flying Fish

At about the same time that Petrus Plancius was preparing and publishing his new celestial sphere with significant improvement for southern hemisphere navigators (see Chap. 15), he was also planning major expeditions for the firm he helped establish, the Dutch East India Company.

Although the addition of Columba, Crux and the two Magellanic Clouds helped southern hemisphere navigators, the positions of most southern stars were still very poorly known. To remedy this lack of knowledge, Plancius trained the pilot of one voyage, Pieter Dirkszoon Keyser, to use scientific instruments to make measurements of the positions of the stars and asked him to plot the positions of all of the brighter stars too close to the south celestial pole to be seen from Europe. With the help of his colleague

P. Simpson, *Guidebook to the Constellations: Telescopic Sights, Tales, and Myths*,
Patrick Moore's Practical Astronomy Series, DOI 10.1007/978-1-4419-6941-5_16,
© Springer Science+Business Media, LLC 2012

Frederick de Houtman, Keyser catalogued 135 southern stars and arranged them into 12 new constellations. Keyser and de Houtman must have been surprised and impressed by the fauna they encountered during their voyage, for most of the constellations they created represented the exotic animals, fish and birds of the regions they explored. Only three were exceptions; Indus, Phoenix, and Triangulum Australe.

The Eerste Schipvaart (the First Voyage), as it was called, left Holland on April 2, 1595 with four ships and about 250 men. The fleet suffered several disasters and had to make several stops to replenish their supplies of water, fresh meat, and fresh fruits to ward off scurvy. Keyser and de Houtman probably made most of their observations during a 3-month stay on Madagascar forced by the very poor condition of the crew. By the time the ships returned from this voyage on February 25, 1597, they had lost over 60% of their crewmen, including Keyser. However, de Houtman survived and delivered their star catalogue to Plancius. Plancius included the new constellations on charts he prepared about 1597.

There is still some controversy about how much credit for the creation and naming of the new constellations is due to Plancius, and how much is due to Keyser and de Houtman, but the first publication of these new constellations was in the form of a globe made by Jodocus Hondius (1563–1612) and released in late 1597 or early 1598. In 1602, another globe was published by Willem Janzoon on which were copied the constellations as charted by Plancius. Neither of these globes achieved wide fame, and for many years the new constellations were attributed to Johann Bayer, a much more well-known astronomer, who had copied them from the Hondius globe and included them in his famous *Uranometria* of 1603. Unfortunately for Plancius, Bayer did not give credit for the constellations and for many years they were thought to be the creation of Bayer himself. Even more confusion arose when Jacob Bartsch published a planisphere including the new constellations in 1624. Bartsch did not know of Plancius' globe, and got the information about the new constellations from the German astronomer Isaac Habrecht II, to whom he gave credit. However, Bartsch acknowledged that this credit was often missed, and both he and Habrecht were sometimes credited with originating the constellations. All in all, a very confusing situation (Fig. 16.1).

The new constellations were:

Original name	Meaning	Modern name
Apis Indica*[1]	The Bird of Paradise	Apus
Chameleon	The Chameleon	Chameleon
Dorado	The Gided (or Golden) Fish	Dorado
Grus	The Crane	Grus
Waterslange/Hydrus	The Water Snake	Hydrus
Indus	The (American) Indian	Indus
Muia (Greek)	The fly	Musca (Latin)
Pavo	The Peacock	Pavo
Phoenix	The Phoenix	Phoenix
Triangulus Antarcticus	The Southern Triangle	Triangulum Australe
Tucana	The Toucan	Toucan
Pisces Volucris	The Flying Fish	Volans

See figure 16.1 for the locations of these constellations.

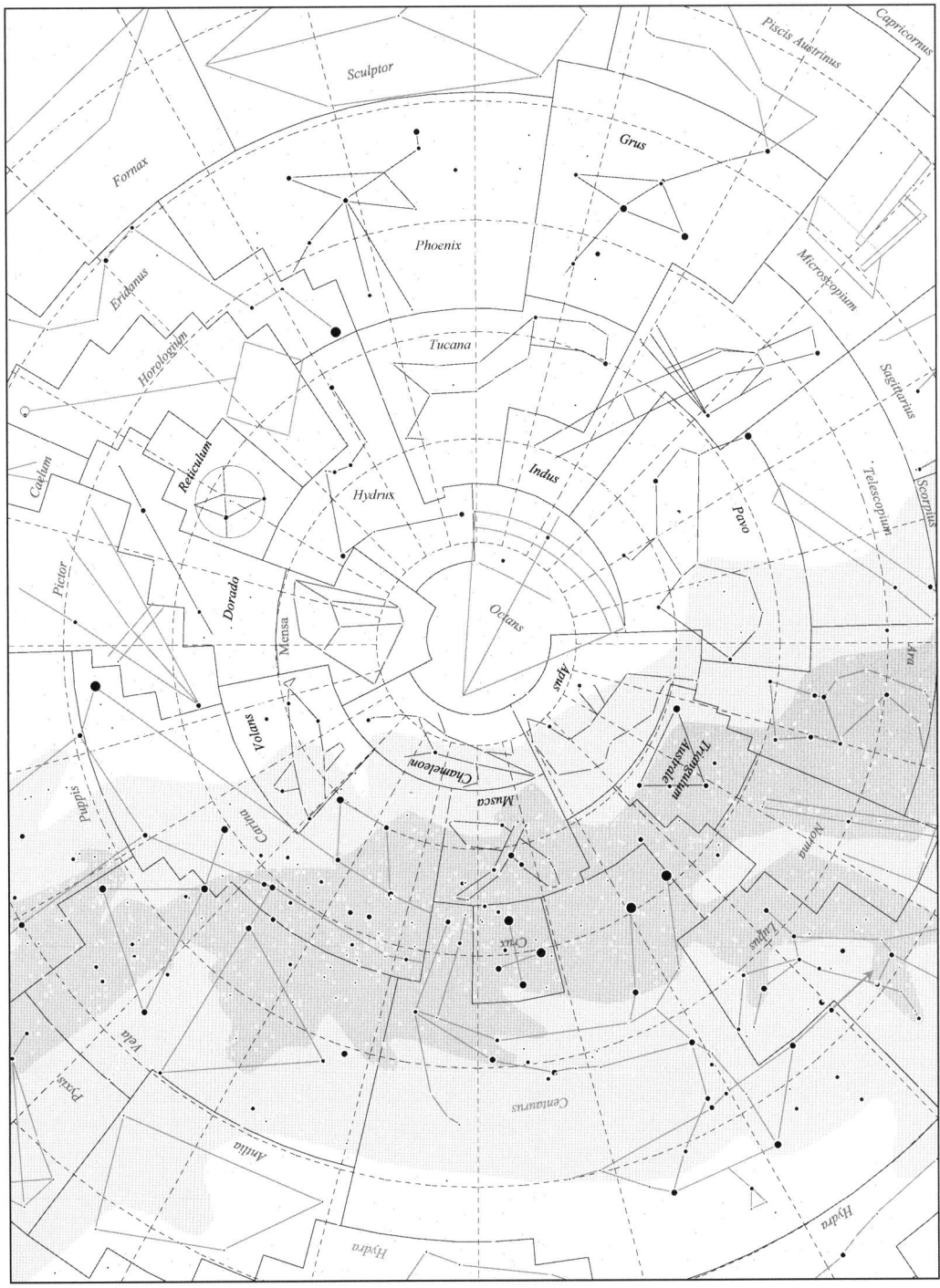

Fig. 16.1. Keyser and de Houtmann constellation finder chart.

Apus (ape-us)	**The Bird of Paradise**
Apodis (ap-oh-diss)	**9:00 P.M. Culmination:** July 6
Aps	**Area:** 206 sq. deg., 67th in size

The Great Bird of Paradise, paradisea apodea, is a very colorful creature. Its back is a reddish-brown, its head is yellow, and its neck is emerald-green. All this is matched in size, if not in mass, by a huge tail of golden-orange feathers.

Plancius gave this constellation both Dutch and Latin names; Paradys-vogel, and Apis Indica. The later was probably a translation or typographical error, since its translation is "Indian Bee," and it is likely that the name should have been Apus Indica or "Indian Bird." The Dutch name was soon dropped, and the Latin name eventually shortened to just Apus or "Bird." However, the bird represented continued to be the Bird of Paradise.

TO FIND: With its three brightest stars being only 4th magnitude, Apus is another of the very faint constellations which is difficult to find at all, much less attempt to discern any meaningful figure. For those adventurous enough to try, see Figs. 16.1 and 15.14 for the location of Apus relative to other constellations of the southern sky. Starting with α and β Centauri, you can find the center of Triangulum Australe about 10° southeast of α Centauri, with α Triangulum Australe another 5° farther along the same line. α Triangulum Australe is fairly bright, and almost a full magnitude brighter than β and γ, the other two stars forming the triangle. ζ Apodis, which marks the bird's head, lies just 3.3° east–northeast of α Triangulum Australe. The slightly brighter stars of the bird's body can be found directly south of α Centauri. From α Centauri to α Circini (magnitude 3.2) is only slightly over 4° almost due south. Continuing that line 14° farther brings you to α Apodis (magnitude 3.8) in the tail. Another 2.4° south–southwest brings you to η Apodis (magnitude 4.88) which defines the end of the large tail. Going back north–northwest to about even with α Apodis brings you to an unlettered 6th magnitude star, and from there two northward steps of about 2° each takes you to θ Apodis (magnitude 5.7) and to another unlettered 6th magnitude star. From there, a jump of 3.5° east–northeast takes you to a slightly brighter unlettered star, from which two small steps of about 1.5° each takes you to κ^1 and κ^2 (magnitudes 5.4 and 5.6). These define the juncture of the bird's tail and body. A couple of steps (marked by two faint, unlettered stars) to the northeast takes you to about 2.3° south–southwest of Atria (α Trianguli Australis, magnitude 1.9). Another step due east will take you to ι Apodis, which defines the start of the bird's neck, just 2.3° due south of ζ. Returning to κ^1 and κ^2, and going south 2.6°, you will find a magnitude 6 star which designates the lower junction of the bird's tail to its body. From there, a 3.7° step to the southeast will place you midway between β and δ Apodis. β represents one of the bird's feet, while δ, and its fainter companion γ Apodis, represents the other. The faint stars of the celestial bird of paradise pale in comparison with the colorful expanse of feathers in the real great bird of paradise (Figs. 16.1 and 16.2).

KEY STAR NAMES: None.

Magnitudes and spectral types of principal stars:

Bayer desig.	Mag. (m_v)	Spec. type	Bayer desig.	Mag. (m_v)	Spec. type	Bayer desig.	Mag. (m_v)	Spec. type
α	3.83	K5 III	β	4.23	K0 III	γ	3.86	K0 IV
δ	4.18	M4+K3[*2]	ε	5.06	B4 V	ζ	4.76	K1 III
η	4.89	A2m	θ	5.69	M6.5 III	ι	5.39	B8+B9[*3]
κ[1]	5.40	B1 npe	κ[2]	5.64	B8 IV[*4]	R	5.37	K4 III[*5]

Stars of magnitude 5.5 or brighter which have other catalog designations:

Draper Number	Mag. (m_v)	Spec. type
HD 1248488	5.50	K1 III

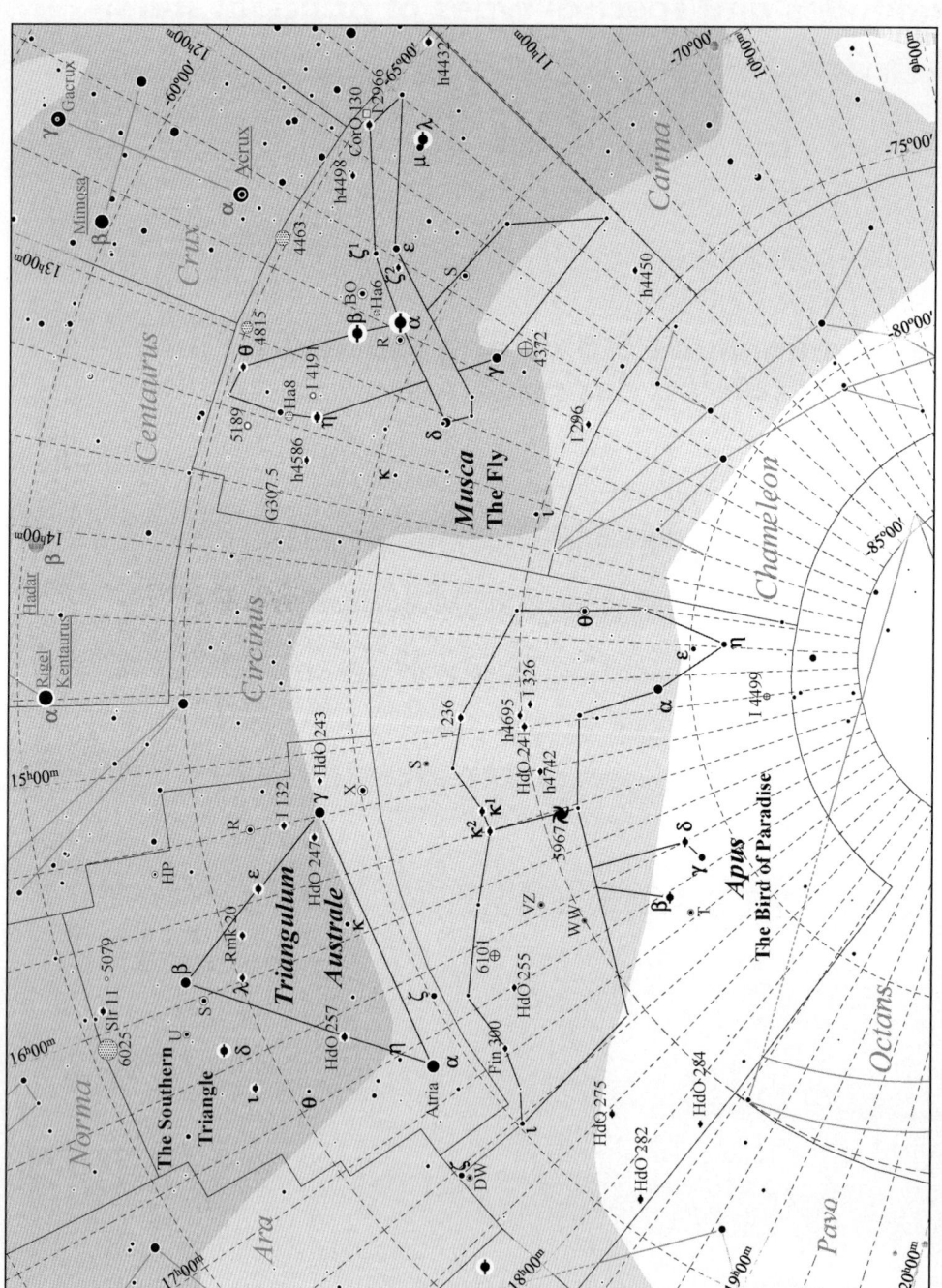

Fig. 16.2. Apus, Musca and Trianglum Australe details.

Table 16.1. Selected telescopic objects in Apus.

Multiple stars:

Designation	R.A.	Decl.	Type	m₁	m₂	Sep. ('')	PA (°)	Colors	Date/Period	Aper. Recm.	Rating	Comments
I 326 (Apus)	14ʰ 31.8ᵐ	−76° 16'	A–B	7.1	10.0	2.3	117	dY, W	1991	4/5"	****	
h4695 (Apus)	14ʰ 51.5ᵐ	−74° 56'	Aa–B	6.8	12.0	17.7	288	W, ?	2000	3/4"	***	A-a is a close double (~0.1"), a.k.a. HDS2096Aa
I 236 (Apus)	14ʰ 53.2ᵐ	−73° 11'	A–B	5.9	7.6	2.2	123	O, O	1996	4/5"	****	a.k.a. h4695AB
HdO 241 (Apus)	14ʰ 59.9ᵐ	−75° 02'	A–B	6.2	13.0	30.0	035	W, ?	1891	6/7"	***	
h4742 (Apus)	15ʰ 20.3ᵐ	−75° 34'	A–B	6.9	12.7	30.3	194	yO, ?	2000	5/6"	***	
κ¹ Apodis	15ʰ 31.5ᵐ	−73° 23'	A–B	5.5	11.3	27.3	255	bW, ?	2000	3/4"	***	
κ² Apodis	15ʰ 40.4ᵐ	−73° 27'	A–B	5.7	12.5	15.0	120	W, ?	1891	4.5"	***	a.k.a. HdO 249AB
			A–C	5.7	13.5	15.0	000	W, ?	1891	7/8"	***	a.k.a. HdO 249AC
δ¹, δ² Apodis	16ʰ 20.3ᵐ	−78° 42'	A–B	4.9	5.4	103.3	010	oY, oY	2000	6×30	****	δ¹ is a small range variable, 4.66–4.87. Common proper motion (cpm) indicates the pair may be physically associated
			A–C	4.7	12.5	91.8	070	oY, ?	2000	4/5"	***	
HdO 255 (Apus)	16ʰ 40.8ᵐ	−72° 18'	A–B	6.7	10.6	2.4	253	dY, ?	1991	4/5"	****	
P Apodis	16ʰ 43.2ᵐ	−77° 30'	A–B	4.2	12.0	91.5	072	dY, ?	2000	3/4"	***	a.k.a. h4858
Fin 300 (Apus)	16ʰ 57.0ᵐ	−71° 07'	A–B	6.6	12.1	2.8	162	W, ?	1930	3/4"	***	
HdO 275 (Apus)	17ʰ 44.3ᵐ	−72° 13'	A–B	6.9	8.1	**0.7**	**104**	yW, yW	**101.3**	14/16"	****	Opening to max of 0.762" in late 2036, then closing to min of 0.457" in 2088
HdO 282 (Apus)	18ʰ 11.3ᵐ	−75° 53'	A–B	5.9	13.0	25.9	231	dY, ?	1902	6/7"	***	
HdO 284 (Apus)	18ʰ 12.6ᵐ	−73° 40'	A–B	6.0	9.1	2.3	260	yW, ?	1991	4/5"	****	

Variable stars:

Designation	R.A.	Decl.	Type	Range [m]		Period (days)	F. (fᵣ/fₜ)	Spectral Range	Aper. Recm.	Rating	Comments
θ Apodis	14ʰ 05.3ᵐ	−76° 48'	SRb	6.4	8.6	119.0	–	M7III	7×50	*****	Reddish color stands out well against background star field.#
S Apodis	15ʰ 09.3ᵐ	−72° 04'	R CrB	9.6	15.2	–	–	C(R3)	20/22"	****	2° 07' NW (PA 306°) of κ¹ Apodis (m5.43).##
VZ Apodis	16ʰ 16.2ᵐ	−74° 02'	M	8.2	17.5	385.0	–	M5e	~57"	****	3° 12' ESE (PA 107°) of κ¹ Apodis (m5.43). Use largest aperture available.#
VW Apodis	16ʰ 31.5ᵐ	−75° 00'	M	9.0	16.8	267.0	0.5	M4(Ib)e	~41"	****	2° 37' NNW (PA 344°) of β Apodis (m4.23). Use largest aperture available.#
T Apodis	16ʰ 55.9ᵐ	−77° 48'	M	8.4	15.0	261.0	0.43	M3e	18/20"	****	0° 45' ESE (PA 114°) of β Apodis (m4.23). Use largest aperture available.##
DW Apodis	17ʰ 23.5ᵐ	−67° 56'	EA/SD	7.9	9.1	2.3	0.14	B6III	10×80	****	0° 13' SE (PA 138°) of ζ Apodis (m4.76).#

(continued)

Table 16.1. (continued)

Star clusters:

Designation	R.A.	Decl.	Type	m_v	Size (')	Brst. Star	Dist. (ly)	dia. (ly)	Number of Stars	Aper. Recm.	Rating	Comments
IC 4499	15ʰ 00.3ᵐ	−82° 13'	Globular	10.1	8.0	14.60	62K	144	–	10/12"	**	~1° N of π¹ Octantis (m = 5.65). Luminosity = 28K suns
NGC 6101	16ʰ 25.8ᵐ	−72° 12'	Globular	9.2	5.0	13.50	49K	90	–	8/10"	***	~2° S of ζ TrA (m = 4.91). Luminosity = 40K suns

Nebulae:

Designation	R.A.	Decl.	Type	m_v	Size (')	Brst./ Cent. star	Dist. (ly)	dia. (ly)	Aper. Recm.	Rating	Comments
IC 4593	16ʰ 11.7ᵐ	+12° 04'	Planetary	10.7	12×10"	**11.30**	–	–	3/4"	****	Appears stellar at low magnification, higher magnification reveals its true nature
NGC 6210	16ʰ 44.5ᵐ	+23° 49'	Planetary	8.8	25×15"	**13.70**	–	–	10×80	****	

Galaxies:

Designation	R.A.	Decl.	Type	m_v	Size (')	Dist. (ly)	dia. (ly)	Lum. (suns)	Aper. Recm.	Rating	Comments
NGC 5967	15ʰ 41.9ᵐ	−75° 31'	Sbc	12.0	2.4×1.4	~119M	>33K	33G	8/10"	***	Supergiant spiral, brightest member of Abell 2199 galaxy cluster

Meteor showers:

Designation	R.A.	Decl.	Period (yr)	Max Date	ZHR (max)	Duration	Comet/ Asteroid	Vel. (mi/ km/sec)	First Obs.	Rating	Comments
None											

Other interesting objects:

Designation	R.A.	Decl.	Type	m_v	Size (')	Dist. (ly)	dia. (ly)	Lum. (suns)	Aper. Recm.	Rating	Comments
None											

Musca (mus-kuh)	The Fly
Muscae (mus-kye)	9:00 P.M. Culmination: July 4
Mus	Area: 138 sq. deg., 77th in size

TO FIND: Although Musca is a relatively small and faint constellation, it is not difficult to find its brighter stars because of its location directly south of Crux, the southern cross. α Muscae lies only 6.2° south (PA 171°) of α Crucis. α Muscae is accompanied by slightly fainter β only 1.3° to its northeast (PA 040°). The pair look like a smaller and fainter version of α and β Crucis which are just 7° due north. A still smaller and fainter pair, ζ² and ε Muscae lie just 3.2° west (PA 281°) from the α and β pair, and look like another miniaturization of α and β Crucis. All three pairs have almost the same orientation in the sky, northeast to southwest. δ Muscae lies 3.2° southeast (PA 142°) of α, and, along with a pair of 6th magnitude unlettered stars represents the fly's head. δ, α, ζ², and ε form the forward part of the fly. A pair of 5th magnitude, unlettered stars (HD 103079 and 101379) near the boundary with Centaurus and only 2.5° southeast of λ, Centauri form the end of the fly's tail. The northern edge of the fly's western wing is formed by a line from α through β, extended 3.5° to θ Musca. Just 0.9° east (PA 080°) of θ is 6th magnitude HD 115147 which defines the tip of the western wing. That wing's southern edge is formed by a line from η Musca drawn perpendicular to the line between α and δ Musca. Like several stars in Musca, η has a near match in nearby HD 115211 (m4.86), making yet another pair of nearly matched stars in Musca. The fly's eastern wing is formed by a series of lines starting with δ Musca, going to the nearly matched pair (the seventh in Musca!), to 5th magnitude HD102839 and finally back to the fly's body opposite α Muscae. The main part of the fly's body is relatively easy to find. Tracing out the tail and wings is more difficult. You should get a feeling of satisfaction if you can make out the entire body of the fly! (Figs. 16.1–16.3).

KEY STAR NAMES: None.

Magnitudes and spectral types of principal stars:

Bayer desig.	Mag. (m_v)	Spec. type	Bayer desig.	Mag. (m_v)	Spec. type	Bayer desig.	Mag. (m_v)	Spec. type
α	2.69	B2 IV/V	β	3.04	B2 V	γ	3.84	B5 V
δ	3.61	K2 III	ε	4.06	M5 III	ζ^1	5.73	K0 III*[6]
ζ^2	5.15	Am*[6]	η	4.79	B8 V	θ	5.44	WC6 + O9.5 I
ι^1	5.04	K0 III*[7]	ι^2	6.62	B9 V*[7]	κ	6.61	K0 III*[8]
λ	3.68	A7 III*[9]	μ	4.75	K4 III*[9]			

Stars of magnitude 5.5 or brighter which have other catalog designations:

Draper Number	Mag. (m_v)	Spec. type	Draper Number	Mag. (m_v)	Spec. type
HD 114211	4.86	K2 Ib/II	HD 103079	4.89	B4 V
HD 102839	4.98	G5 Ib	HD 101379	5.01	G2 III*[10]
HD 99103	5.09	B5	HD 105340	5.17	K2 II/III*[11]
HD 104878	5.34	A0 V			

Fig. 16.3a. The stars of Musca and adjacent constellations. Astrophoto by Christopher J. Pickings.

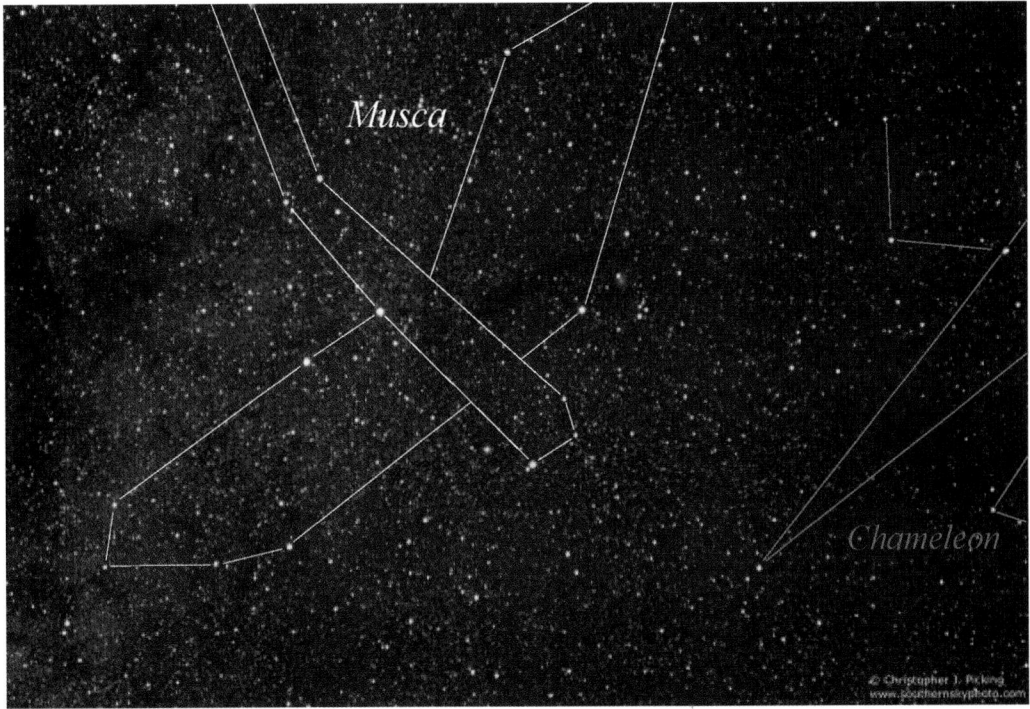

Fig. 16.3b. The figure of Musca and adjacent constellations. Astrophoto by Christopher J. Pickings.

Table 16.2. Selected telescopic objects in Musca.

Multiple stars:

Designation	R.A.	Decl.	Type	m_1	m_2	Sep. (")	PA (°)	Colors	Date/Period	Aper. Recm.	Rating	Comments
h4432	11ʰ 23.4ᵐ	−64° 57'	A–B	5.4	6.6	2.4	308	pY, pY	1991	3/4"	*****	Showcase pair. Great for small apertures
h4450	11ʰ 31.8ᵐ	−73° 54'	A–B	6.9	10.9	20.9	040	bW, −	2000	3/4"	****	
λ Muscae	11ʰ 45.6ᵐ	−66° 44'	A–B	3.6	12.8	40.6	275	W, −	1918	5/6"	***	a.k.a. h4471
CorO 130	11ʰ 51.9ᵐ	−65° 12'	A–B	5.0	7.3	1.6	159	W, −	1991	6/7"	****	
h4498	12ʰ 06.4ᵐ	−65° 43'	A–B	6.8	7.1	8.9	060	pY, yW	2000	2/3"	****	
ζ² Muscae	12ʰ 22.2ᵐ	−67° 31'	A–B	5.3	8.7	0.5	128	W, −	1991	19/20"	****	HdO 215
α Muscae	12ʰ 37.2ᵐ	−69° 08'	A–B	2.7	13.0	29.7	321	bW, −	2000	6/7"	***	a.k.a. Don 541
I 296	12ʰ 39.2ᵐ	−75° 22'	A–B	6.6	9.1	2.0	271	−, −	1991	4/5"	****	
β Muscae	12ʰ 46.3ᵐ	−68° 06'	A–B	3.5	4.0	**1.3**	**050**	bW, bW	**383.12**	7/8"	****	a.k.a. R 207
θ Muscae	13ʰ 08.1ᵐ	−65° 18'	A–B	5.7	7.6	4.9	187	Y, W	2002	2/3"	****	a.k.a. Rmk 16
η Muscae	13ʰ 15.3ᵐ	−67° 54'	A–B	5.1	9.4	2.7	125	W, −	1999	3/4"	****	
			A–C	4.8	7.2	38.4	332	W, W	2002	7×50	****	a.k.a. Δ 131
h4586	13ʰ 28.4ᵐ	−67° 52'	A–B	7.3	9.1	2.9	141	−, −	1991	3/4"	****	

Variable stars:

Designation	R.A.	Decl.	Type	Range (m_v)	Period (days)	F (f_r/f_t)	Spectral Range	Aper. Recm.	Rating	Comments
S Muscae	12ʰ 12.8ᵐ	−70° 09'	δ Cep	5.9	9.660	0.49	F6 Ib–G0	7×50	****	2° 32' NW (PA 319°) of λ Mus (m3.84).#
BO Muscae	12ʰ 34.9ᵐ	−67° 45'	SRb	5.3	132.4	−	M6 II–II	7×50	****	1° 24' N (PA 351°) of α Mus (m2.68).#
R Muscae	12ʰ 42.1ᵐ	−69° 24'	δ Cep	5.9	7.510	0.30	F7 Ib–G2	7×50	****	0° 39' SE (PA 123°) of α Mus (m2.68).#

Star clusters:

Designation	R.A.	Decl.	Type	m_v	Size (')	Brtst. Star	Dist. (ly)	dia. (ly)	Number of Stars	Aper. Recm.	Rating	Comments
NGC 4372	12ʰ 25.8ᵐ	−72° 40'	Globular	7.2	5	12.2	11K	16	−	7×50	****	Large, round, fairly faint
NGC 4463	12ʰ 30.0ᵐ	−64° 48'	Open	7.2	6	8.3	38.5K	67	30	7×50	****	Strong concentration, wide range of brightnesses
Harvard 6	12ʰ 37.9ᵐ	−68° 23'	Open	10.7	9	−	−	−	100	15×80	***	

	R.A.	Decl.	Type	m_v	Size (')		Dist. (ly)	dia. (ly)	Aper. Recm.	Rating	Comments
NGC 4815	12^h 58.0^m	−64° 57'	Open	8.6	5	9.6	–	–	100	★★★	Strong concentration, wide range of brightnesses
Harvard 8	13^h 18.2^m	−67° 05'	Open	9.5	5	11.6	5,700	7	30	★★★	

Nebulae:

Designation	R.A.	Decl.	Type	m_v	Size (')	Brst./Cent. star	Dist. (ly)	dia. (ly)	Aper. Recm.	Rating	Comments
I 2966	11^h 50.4^m	−64° 54'	Reflection	–	3.0	–	–	–	–	★★★★	Small, well defined, illuminated by m11.5 star
I 4191	13^h 08.8^m	−66° 39'	Planetary	10.6	5"	12	7,500	0.2	3/4"	★★★	Expansion velocity 9 mi/s (14 km/s)
NGC 5189	13^h 33.6^m	−65° 59'	Planetary	9.9	2.3	14	2,600	2	11×70	★★★★	Expansion velocity 19 mi/s (31 km/s)
Hourglass Nebula	13^h 39.6^m	−67° 23'	Planetary	12.9	25	13.9	–	–	12/14"	★★★	a.k.a. PK307-04.1 and He 2-95

Galaxies:

Designation	R.A.	Decl.	Type	m_v	Size (')	Dist. (ly)	dia. (ly)	Lum. (suns)	Aper. Recm.	Rating	Comments
None											

Meteor showers:

Designation	R.A.	Decl.	Period (yr)	Duration	Max Date	ZHR (max)	Comet/Asteroid	First Obs.	Vel. (mi/km/sec)	Rating	Comments
None											

Other interesting objects:

Designation	R.A.	Decl.	Type	m_v	Size (')	Dist. (ly)	dia. (ly)	Lum. (suns)	Aper. Recm.	Rating	Comments
None											

Fig. 16.4. Globular Clusters NGC 4372 and 4833 in Musca. Astrophoto by Christopher J. Picking.

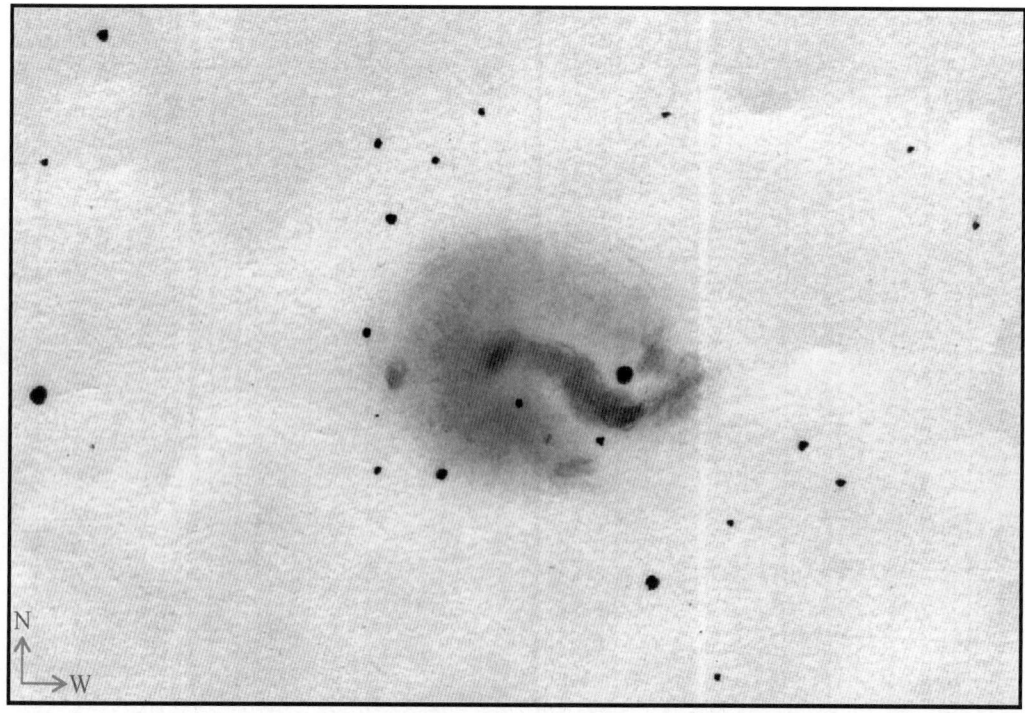

Fig. 16.5. Planetary nebula NGC 5189 in Musca. Drawing by Eiji Kato.

Triangulum Australe (try-**ang**-guu-lum os-**tray**-lis)	**The Southern Triangle**
Trianguli Australis	**9:00 P.M. Culmination:** July 8
(try-ang-guu-lye os-tray-liss)	
TrA	**Area:** 110 sq. deg., 83rd in size

TO FIND: Triangulum Australe is quite easy to find because it is relatively bright, and close to the bright stars α and β Centauri. The center of the triangle formed by the three stars α, β, and γ Triangulum Australe lies just 11° (about the distance across the knuckles of your fist at arm's length) southeast of α Centauri. The three sides of the Southern Triangle are 7.7° (α–β), 6.4° (β–γ), and 8.1° (γ–α), making it a reasonable approximation of an equilateral triangle (Figs. 16.1, 16.2 and 16.6).

KEY STAR NAMES:

Name	Pronunciation	Source
α Atria	ah-**trih**-uh or **ay**-trih-eh	A contraction of <u>a</u>lpha <u>Tri</u>anguli <u>A</u>ustralis

Magnitudes and spectral types of principal stars:

Bayer desig.	Mag. (m_v)	Spec. type	Bayer desig.	Mag. (m_v)	Spec. type	Bayer desig.	Mag. (m_v)	Spec. type
α	1.91	K2 IIb/IIIa	β	2.83	F2 III	γ	2.87	A1 V
δ	3.86	G5 II	ε	4.11	K0 III	ζ	4.90	F9 V
η¹	5.89	B7 IV[12]	η²	6.54	A1 III[12]	θ	5.50	G8/K0 III
ι	5.28	F4 IV	κ	5.11	G6 IIa	λ	5.74	B7 III[13]

Stars of magnitude 5.5 or brighter which have other catalog designations:

Draper Number	Mag. (m_v)	Spec. type	Draper Number	Mag. (m_v)	Spec. type
HD 150549	5.10	Ap	HD 148291	5.19	K0 II/III

Fig. 16.6a. The stars of Triangulum Australe and adjacent constellations. Astrophoto by Christopher J. Pickings.

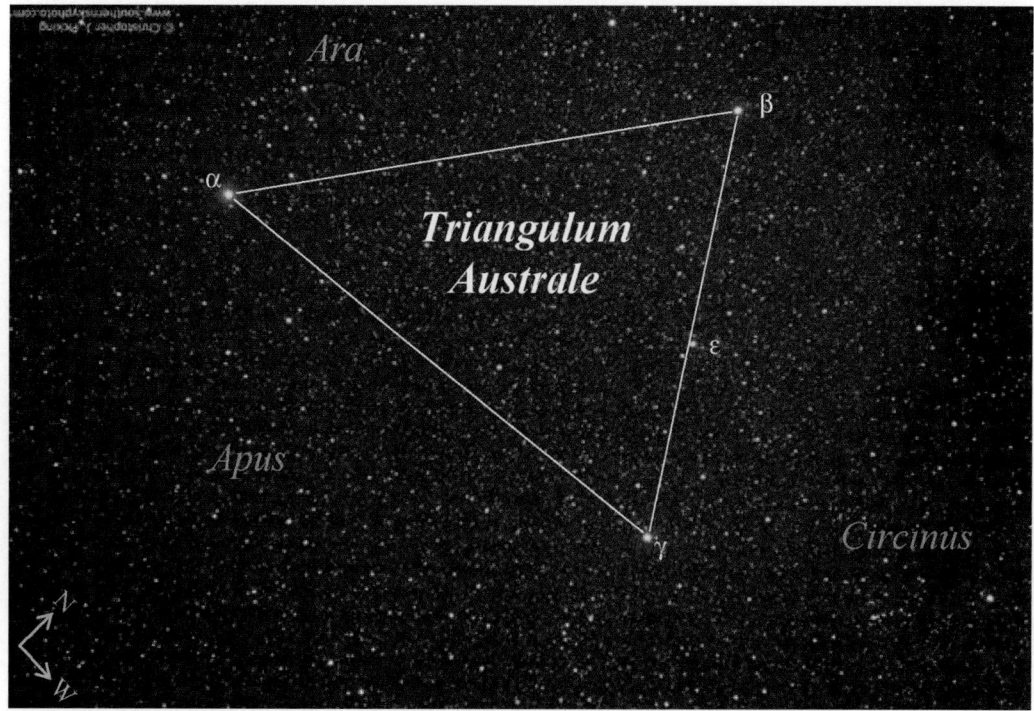

Fig. 16.6b. The figure of Triangulum Australe and adjacent constellations. Astrophoto by Christopher J. Pickings.

Table 16.3. Selected telescopic objects in Triangulum Australe.

Multiple stars:

Designation	R.A.	Decl.	Type	m₁	m₂	Sep. (″)	PA (°)	Colors	Date/ Period	Aper. Recm.	Rating	Comments
HdO 243	15ʰ 09.6ᵐ	−68° 43′	A–B	7.0	15.5	6.6	116	W, ?	1932	19/20″	****	
			A–C	7.0	14	12.7	178	W, ?	2000	9/10″	****	
			A–D	7.0	10.5	41.3	041	?, ?	2000	15×80	****	
			A–E	7.0	11.0	55.0	311	?, ?	2000	15×80	****	
I 332	15ʰ 20.7ᵐ	−67° 29′	A–B	6.4	8.2	1.1	107	pY, Y	1991	8/9″	****	
HdO 247	15ʰ 26.2ᵐ	−68° 19′	A–B	5.9	12.0	50.0	090	dY, ?	1900	2/3″	****	
ε Trianguli Australis	15ʰ 36.7ᵐ	−66° 19′	A–B	4.2	9.4	82.1	220	W, ?	2001	7×50	****	a.k.a.Δ188
Rmk 20	15ʰ 47.9ᵐ	−65° 27′	A–B	6.2	6.4	1.7	146	bW, bW	1996	5/6″	****	
SlrR 11	15ʰ 54.9ᵐ	−60° 45′	A–B	6.4	8.1	1.1	095	W, ?	1991	8/9″	****	
			A–C	6.4	10.0	44.4	048	W, ?	2000	15×80	****	a.k.a.Δ194-AC
			A–D	6.4	9.0	48.3	256	W, ?	2000	19″	****	a.k.a.Δ194-AD
λ Trianguli Australis	15ʰ 59.0ᵐ	−65° 02′	A–a	5.8	9.3	0.2	326	bW, ?	1991	–	–	a.k.a. B 584
			Aa–B	5.7	12.5	9.8	288	W, ?	2000	4/5″	****	With separation of only ~0.2″, A–a is beyond the range of amateur instruments
δ Trianguli Australis	16ʰ 15.4ᵐ	−63° 41′	A–B	4.0	12.0	20.0	185	Y, ?	1902	2/3″	****	a.k.a. R 274
ι Trianguli Australis	16ʰ 28.0ᵐ	−64° 03′	A–B	5.3	9.4	16.7	001	dY, W	2000	15×80	****	a.k.a.Δ201
HdO 257	16ʰ 46.7ᵐ	−67° 06′	A–B	5.1	11.5	25.0	123	W, ?	1900	15×80	****	
I 1621			B–C	11.5	12.0	6.0	315	?, ?	1900	3/4″	***	

Variable stars:

Designation	R.A.	Decl.	Type	Range (mᵥ)		Period (days)	F (fᵣ/fₜ)	Spectral Range	Aper. Recm.	Rating	Comments
X Trianguli Australis	15ʰ 14.3ᵐ	−70° 05′	Ib	5.0	6.4	–	–	C5,5(Nb)	6×30	****	1° 27′ SSW (PA 196°) of γ TrA (m 2.87)
R Trianguli Australis	15ʰ 19.8ᵐ	−66° 30′	δ Cep	6.3	7.0	3.389	0.30	F6Ib/I-G0	6×30	****	1° 42′ W (PA 262°) of TrA (m 4.11)
HP Trianguli Australis	15ʰ 26.1ᵐ	−63° 26′	E	8.2	8.6	2.758	–	B5III	7×50	***	0° 57′ E (PA 080°) of ε Cir (m4.84)
S Trianguli Australis	16ʰ 01.2ᵐ	−63° 47′	δ Cep	6.0	6.8	6.323	0.28	F6II-G2	6×30	****	0° 45′ ESE (PA 118°) of β TrA (m2.83)

(continued)

Table 16.3. (continued)

Variable stars:

Designation	R.A.	Decl.	Type	Range (m$_v$)	Period (days)	F (f$_r$/f$_1$)	Spectral Range	Aper. Recm.	Rating	Comments
U Trianguli Australis	16^h 07.3^m	−62° 55'	Cep(B)	7.3 8.3	2.568	0.34	F5 F8Ib/II	7×50	****	1° 12' NW (PA 309°) of δ TrA (m3.86). Also has second pulsation mode of 1.825d

Star clusters:

Designation	R.A.	Decl.	Type	m$_v$	Size (')	Brst. Star	Dist. (ly)	dia. (ly)	Number of Stars	Aper. Recm.	Rating	Comments
C95, NGC 6025	16^h 03.7^m	−60° 30'	Open	5.1	12	7.3	2,700	9	60	7×50	****	

Nebulae:

Designation	R.A.	Decl.	Type	m$_v$	Size (')	Brst./Cent. star	Dist. (ly)	dia. (ly)	Aper. Recm.	Rating	Comments
NGC 5979	15^h 47.8^m	−61° 13'	Planetary	11.5	0.3	15.3	11K	1.0		****	

Galaxies:

Designation	R.A.	Decl.	Type	m$_v$	Size (')	Dist. (ly)	dia. (ly)	Lum. (suns)	Aper. Recm.	Rating	Comments
None											

Meteor showers:

Designation	R.A.	Decl.	Period (yr)	Duration	Max Date	ZHR (max)	Comet/ Asteroid	First Obs.	Vel. (mi/ km/sec)	Rating	Comments
No significant showers											

Other interesting objects:

Designation	R.A.	Decl.	Type	m$_v$	Size (')	Dist. (ly)	dia. (ly)	Lum. (suns)	Aper. Recm.	Rating	Comments
None											

Fig. 16.7. NGC 6025 (C95), Open Cluster in Triangulum Australe. Astro-photo by Christopher J. Picking.

Chameleon (kuh-**meel**-ee-un)	**The Chameleon**
Chameleontis (kuh-meel-ee-**on**-tis)	**9:00 P.M. Culmination:** Apr. 16
Cha	**Area:** 132 sq. deg., 79th in size

The fleet's long stop in Madagascar was probably the inspiration for this constellation, since this large island is home to about half of the world's species of chameleons. Chameleons are a popular pet renowned for their ability to change their color by enlarging or contracting the differently colored cells arranged in layers below the surface of their skin.

Plancius' celestial globe of 1598 placed the chameleon just south of Musca the fly, but the orientation of the chameleon made it seem that it was uninterested in, or unaware of, one of its favorite foods just above its head. Two years later, Hondius published a new globe with the chameleon's position changed so that it was poised to catch the fly with a flick of its tongue.

TO FIND: Having Chameleon following Musca is very appropriate, since not only is the fly one of the chameleon's favorite foods but also it is placed just north of the chameleon and above it when the two are near their culmination. The four brightest stars of Musca can guide you to the brighter stars of Chameleon. A line from α to β Muscae is 3° long and can be extended by 6.8° to lead you to β Chameleontis. ε Chameleontis is just 1.4° northwest of β, and the two designate the chameleons shoulders. The similar, but slightly wider and fainter, pair γ and δ is located about 5° west of β and δ Chameleontis. They represent the lizard's hips. The tip of the chameleon's tail is a pair of 4th magnitude stars, α and θ Chameleontis which is 7.7° west of γ Chameleontis. The remaining parts of the Chameleon's body can be traced out using the detail chart (Fig. 16.9). However, terrestrial chameleons have a reputation for being well camouflaged and difficult to see. The celestial Chameleon lives up to its terrestrial cousin's reputation! (Figs. 16.1, 16.2, 16.8 and 16.9)

KEY STAR NAMES: None.

Magnitudes and spectral types of principal stars:

Bayer desig.	Mag. (m_v)	Spec. type	Bayer desig.	Mag. (m_v)	Spec. type	Bayer desig.	Mag. (m_v)	Spec. type
α	4.05	F5 III	β	4.24	B5 V	γ	4.11	M0 III
δ^1	5.46	K0 III	δ^2	4.54	B2.5 IV	δ	4.88	B9 V
ζ	5.07	B5 V	η	5.46	B9 IV	θ	4.34	K0 III/IV
ι	5.34	F3/5 IV	κ	5.04	K4 III	λ	–	–*14
μ^1	5.53	A0 Iv	μ^2	6.60	G6/8 III	ν	5.43	G8 III
ξ	–	–*15	o	–	–*15	π^1	5.64	F1 III*16
π^2	–	–*16						

Fig. 16.8a. The stars of Chameleon, Volans, and adjacent constellations. Astrophoto by Christopher J. Picking.

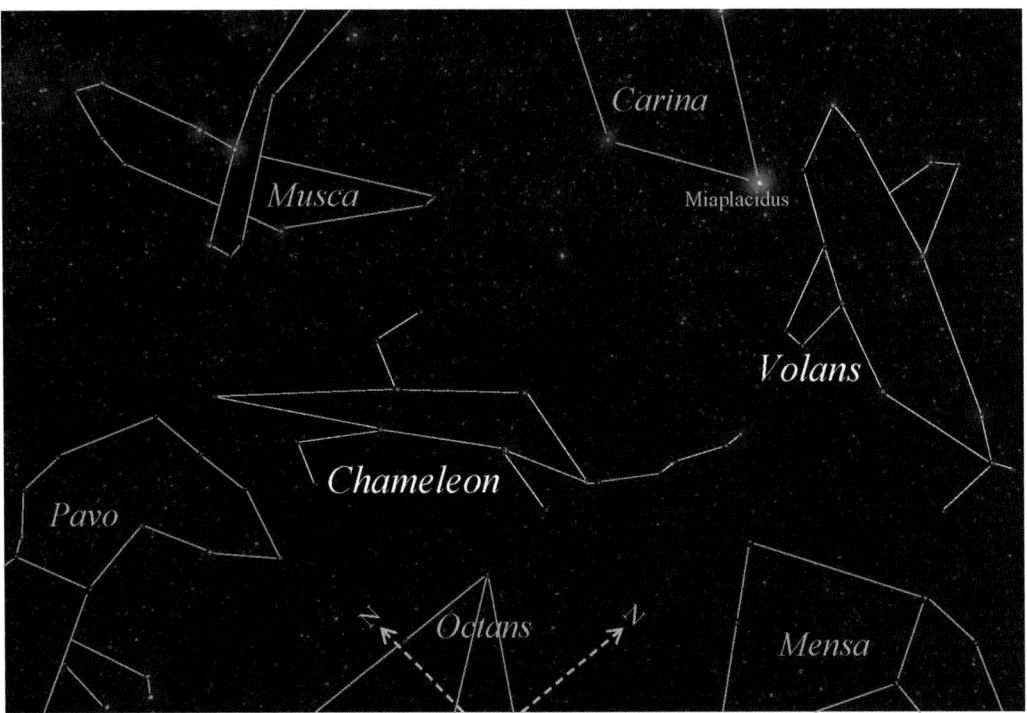

Fig. 16.8b. The figures of Chameleon, Volans, and adjacent constellations.

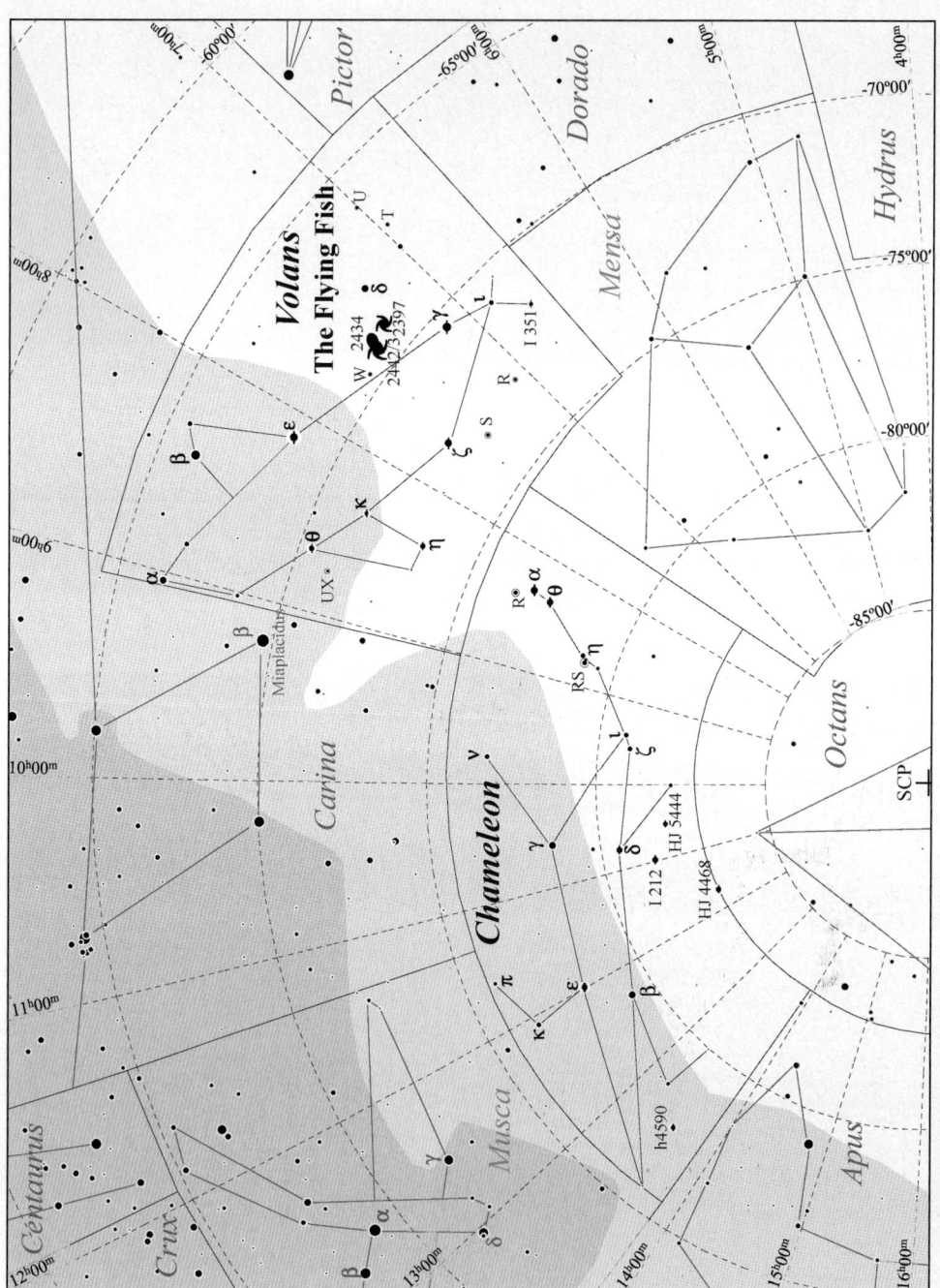

Fig. 16.9. Chameleon and Volans details.

Table 16.4. Selected telescopic objects in Chameleon.

Multiple stars:

Designation	R.A.	Decl.	Type	m_1	m_2	Sep. (")	PA (°)	Colors	Date/Period	Aper. Recm.	Rating	Comments
α, θ Chameleontis	08ʰ 18.5ᵐ	−75° 55'	A–B	4.1	4.4	34' 36"	168	yW, pO	2000	4/5"	***	Easy unaided eye double, > width of full moon separation. Binoculars help show colors
θ Chameleontis	08ʰ 20.6ᵐ	−77° 29'	A–B	4.3	12.4	331.3	250	O, –	1911	4/5"	****	
h5444	10ʰ 31.9ᵐ	−81° 55'	A–B	7.0	9.1	41.4	236	bW, –	2000	7×50	****	
δ²δ¹ Chameleontis	10ʰ 46.7ᵐ	−80° 32'	A–B	4.5	6.1	265.0	343	O, W	2000	6×30	***	Another unaided eye double, a little brighter and wider than ε Lyrae
δ¹ Chameleontis	10ʰ 45.3ᵐ	−80° 28'	A–B	6.2	6.5	33.0	85	O, –	1996	10×50	****	a.k.a. I294
I 212	10ʰ 59.2ᵐ	−81° 33'	A–B	6.7	6.71	**0.39**	**322**	yW, –	**1,210**	24"	****	Minimum separation was 0.237" in 1994, 1.000" in 2047, max 2.971" in 2422
h4468	11ʰ 41.0ᵐ	−83° 06'	A–B	6.3	11.4	25.3	140	O, –	2000	3/4"	****	
ε Chameleontis	11ʰ 59.6ᵐ	−78° 13'	A–B	5.3	6.0	0.4	211	bW, bW	1997	24"	****	Challenging even in large amateur scopes
S Chameleontis	13ʰ 33.3ᵐ	−77° 34'	A–B	6.6	9.2	22.1	133	yW, –	2000	2/3"	****	a.k.a. h4590

Variable stars:

Designation	R.A.	Decl.	Type	Range (m_j)		Period (days)	F (f_r/f_t)	Spectral Range	Aper. Recm.	Rating	Comments
R Chameleontis	08ʰ 21.8ᵐ	−76° 21'	M	7.5	14.2	334.58	0.41	M4e–M7e	12/14"	****	
RS Chameleontis	08ʰ 43.2ᵐ	−79° 04'	EA	6.02	6.68	1.6699	0.15	A5V+A7V	6×30	****	Algol type eclipsing binary

(continued)

Table 16.4. (continued)

Star clusters:

Designation	R.A.	Decl.	Type	m$_v$	Size (')	Brtst. Star	Dist. (ly)	dia. (ly)	Number of Stars	Aper. Recm.	Rating	Comments
None												

Nebulae:

Designation	R.A.	Decl.	Type	m$_v$	Size (')	Brtst./Cent. star	Dist. (ly)	dia. (ly)	Aper. Recm.	Rating	Comments
NGC 3195	10^h 09.5^m	–80° 52'	Planetary	11.6	38"	**15.3**	5.5K	1.0	8/10"	****	Disk is fairly bright at center and fades toward outer edge. ~midway between δ and ζ Cha

Galaxies:

Designation	R.A.	Decl.	Type	m$_v$	Size (')	Dist. (ly)	dia. (ly)	Lum. (suns)	Aper. Recm.	Rating	Comments
None											

Meteor showers:

Designation	R.A.	Decl.	Period (yr)	Duration	Max Date	ZHR (max)	Comet/Asteroid	First Obs.	Vel. (mi/km/sec)	Rating	Comments
No significant showers											

Other interesting objects:

Designation	R.A.	Decl.	Type	m$_v$	Size (')	Dist. (ly)	dia. (ly)	Lum. (suns)	Aper. Recm.	Rating	Comments
None											

Volans (**voe**-lanz)	**The Flying Fish**
Volantis (voe-**lan**-tis)	**9:00 P.M. Culmination:** Mar. 5
Vol	**Area:** 141 sq. deg., 76th in size

One of the creatures represented by the new constellations, the dorado is a prized game fish which is primarily found in warm tropical or subtropical waters. One of the dorado's favorite foods is the flying fish (Volans), and the dorado makes great leaps out of the water following its fleeing prey or when trying to dislodge a fishhook. The "flying" fish does not actually fly, but instead uses its speed to leap out of the water after which it extends its large pectoral fins and glides until it loses altitude and reenters the water. The dorado must often lose its prey since some species of flying fish can glide for as much as 100 m (110 yards). On Bayer's globes the dorado and the flying fish are shown during one of these leaps, with Piscis Volans (the original name) just in front of Dorado's mouth.

TO FIND: Volans is difficult to find since it has no star brighter than 4th magnitude. The five brightest stars in Volans range from m 3.77 to 4.00, only 0.23 magnitudes (24% in brightness). This difference is barely noticeable to the unaided eye. The best way to find Volans is probably to start with Miaplacidus (β Car), a magnitude 1.67 (nearly 1st magnitude) star in Carina the Keel. Miaplacidus is a navigational star, and the brightest star in its region of the sky. If you do not already know how to recognize Miaplacidus see the directions for finding Carina. The tip of the flying fish's nose is marked by α Volantis which is just 3.5° NNW (PA 342°) of Miaplacidus. β Volantis, the brightest star in Volans at only m3.77, is 3.7° due west of α and marks the front of one of Volans' outstretched fins. 6.9° WSW (PA 247°) of β is δ Volantis (m3.98) which marks the back tip of the same fin. Just 3.4° SE of δ is ζ Volantis. Some people see ε, δ, γ, and ζ forming the body of a kite with α and β forming the tail. Even though these stars are faint, they are so close to being equally bright that the "kite" asterism is fairly easy to recognize. If you are really determined, the rest of the flying fish can then be traced out using the chart of that region (Figs. 16.1, 16.8 and 16.9).

KEY STAR NAMES: None.

Magnitudes and spectral types of principal stars:

Bayer desig.	Mag. (m_v)	Spec. type	Bayer desig.	Mag. (m_v)	Spec. type	Bayer desig.	Mag. (m_v)	Spec. type
α	4.00	Am	β	3.77	K2 III	γ	3.61	G8+F2*[17]
δ	3.97	F6 II	ε	4.35	B6 IV	ζ	3.93	K0 III
η	5.28	B6 IV	θ	5.19	A0/1 IV/V	ι	5.41	F7 IV
κ	4.72	B9+A0*[18]						

Stars of magnitude 5.5 or brighter which have other catalog designations:

Draper Number	Mag. (m_v)	Spec. type	Draper Number	Mag. (m_v)	Spec. type
HD 70514	5.06	K1 III	HD 53501	5.18	K2 III
HD 76143	5.34	F5 IV			

Table 16.5. Selected telescopic objects in Volans.

Multiple stars:

Designation	R.A.	Decl.	Type	m₁	m₂	Sep. (")	PA (°)	Colors	Date/Period	Aper. Recm.	Rating	Comments
I 351	06ʰ 41.0ᵐ	−71° 47'	A–B	6.5	10.0	16.4	333	O, –	1998	2/3"	***	a.k.a. Δ42
γ², γ¹	07ʰ 08.7ᵐ	−70° 30'	A–B	3.9	5.4	14.4	296	dY, W	2002	2/3"	****	Nearly identical pair, in larger scopes like a pair of auto headlights
h3997	07ʰ 35.4ᵐ	−74° 17'	A–B	7.0	7.1	1.9	306	bW, bW	1996	4/6"	*****	
ζ Volantis	07ʰ 41.8ᵐ	−72° 36'	A–B	4.0	9.7	16.5	117	O, W	1999	2/3"	****	a.k.a. Δ57
ε Volantis	08ʰ 07.9ᵐ	−68° 37'	A–B	4.4	7.3	6.0	024	bW, yW	1999	2/3"	****	a.k.a. Rmk 7
κ¹,² Volantis	08ʰ 19.8ᵐ	−71° 31'	A–B	5.3	5.6	64.7	058	bW, bW	2002	6×30	****	a.k.a. BrsO 17AB, wide, almost equal pair, very easy. a.k.a. BrsO 17BC
κ² Volantis	08ʰ 22.1ᵐ	−73° 24'	B–C	5.6	7.7	36.7	032	bW, –	2002	7×50	****	a.k.a. h4103AB. A is a
η Volantis			A–B	5.3	11.9	27.3	285	W, –	1999	3/4"	****	spectroscopic binary, period 14.17 days
θ Volantis	08ʰ 39.1ᵐ	−70° 23'	A–C	5.3	11.7	47.2	164	W, –	1999	15×80	**	a.k.a. Δ271
			A–B	5.2	15.0	22.4	057	W, –	1999	14/16"		
			A–C	5.2	10.3	41.1	104	W, –	1999	2/3"	****	a.k.a. h4134

Variable stars:

Designation	R.A.	Decl.	Type	Range (m)	Period (days)	F (f/f₁)	Spectral Range	Aper. Recm.	Rating	Comments	
T Volantis	06ʰ 57.8ᵐ	−67° 07'	M	9.7	175	0.47	M5II/IIIe	7/8"	***	2° 0' WNW (PA 292°) of δ Volantis (m3.29).##	*19
U Volantis	07ʰ 00.3ᵐ	−66° 13'	EA	10.2	11.5	–	–	3/4"	***	2° 22' NW (PA 315°) of δ Volantis (m3.29).#	
R Volantis	07ʰ 05.6ᵐ	−73° 01'	M	8.7	453.6	0.48	C(N)e	10/12	***	2° 42' WSW (PA 257°) of ζ Volantis (m3.93).##	
S Volantis	07ʰ 29.8ᵐ	−73° 23'	M	7.7	394.8	0.54	M4e	10/12"	****	1° 10' SW (PA 227°) of ζ Volantis (m3.93).##	
W Volantis	07ʰ 37.8ᵐ	−69° 33'	EA/AR	10.9	2.758	0.17	K/M	4/6"	***	2° 28' SE (PA 132°) of ζ Volantis (m3.93)	
UX Volantis	08ʰ 46.3ᵐ	−71° 02'	Lb	9.2	–	–	SC	2/3"	****	0° 54' SE (PA 140°) of θ Volantis (m5.19)	

(continued)

Table 16.5. (continued)

Star clusters:

Designation	R.A.	Decl.	Type	m$_v$	Size (')	Brst. Star	Dist. (ly)	dia. (ly)	Number of Stars	Aper. Recm.	Rating	Comments
None												

Nebulae:

Designation	R.A.	Decl.	Type	m$_v$	Size (')	Brst./ **Cent.** star	Dist. (ly)	dia. (ly)	Aper. Recm.	Rating	Comments
None											

Galaxies:

Designation	R.A.	Decl.	Type	m$_v$	Size (')	Dist. (ly)	dia. (ly)	Lum. (suns)	Aper. Recm.	Rating	Comments
NGC 2397	07^h 21.3^m	−69° 00'	S	12.0	2.5 × 1.2	50M	36K	3.2G	12/14"	***	1° 26' WNW (PA 290°) from NGC 2442
NGC 2434	07^h 34.9^m	−69° 17'	E	11.5	2.4 × 2.3	50M	35K	5.0G	12/14"	***	17' NNW (PA 331°) from NGC 2442
NGC 2442/3	07^h 36.4^m	−69° 32'	SAB(s)bc	10.4	5.5 × 4.9	50M	80K	14G	8/10"	****	Sir John Herschel discovered *20 and named it "Meathook" galaxy

Meteor showers:

Designation	R.A.	Decl.	Period (yr)	Duration	Max Date	ZHR (max)	Comet/ **Asteroid**	First Obs.	Vel. (mi/ **km**/sec)	Rating	Comments
None											

Other interesting objects:

Name/ Designation	R.A.	Decl.	Type	m$_v$	Mass (suns)	Dia. (suns)	Dist. (ly)	Planet Mass (J)	Dist. (AU)	Period (days)	Eccentricity	Comments
HD 76700	08^h 53.9^m	−66° 48'	ExoPlnt	8.16	1.13	1.57	195	>0.233	0.051	3.971	0.00	0° 20' E (PA 92°) of HD 76143 (m5.34)

Fig. 16.10a. Spiral galaxy NGC 2442 in Volans. Astrophoto by Rob Gendler.

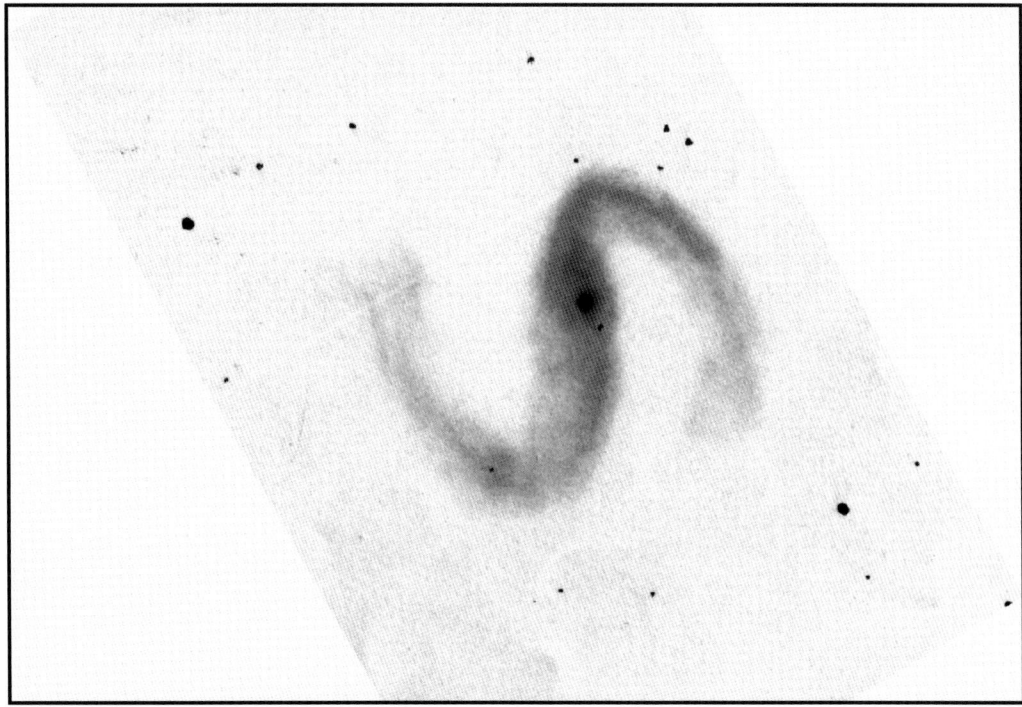

Fig. 16.10b. Spiral Galaxy NGC 2442 in Volans. Drawing by Eiji Kato.

Dorado (dor-**ah**-doe)	**The Golden Fish**
Doradus (dor-**ah**-dus)	**9:00 P.M. Culmination:**
Dor	**Area:**

Most of the stars in Dorado are fairly faint, and the figure of a fish is difficult to see. The only easily recognizable asterism is a straight line of four stars formed by α and β Doradi and two fainter stars. This straight line inspired Kepler to rename the constellation Xiphias (the swordfish).[21] The constellation is still occasionally called the swordfish. In the US, the dorado is more often known by the Hawaiian name of mahi-mahi.

TO FIND: The first magnitude stars Canopus (α Carinae) and Achernar (α Eridani) are about 40° apart. Both α Doradus and α Reticuli are nearly equal in brightness and near the center of the line joining these stars. α Doradus (m3.30) is approximately 16° from Canopus and a few degrees north of the line, while α Reticuli (m3.33) is approximately halfway between Canopus and Achernar and a few degrees south of the line. γ, β, ρ, and η² lie in a straight line stretching about 20° from NW to SE. Almost all the other visible stars in Dorado lie less than 5 or 6° from this line. On a very dark night, these stars can be visualized as forming the fairly thick body surrounding the backbone of the dorado (Fig. 16.1).

NAMED STARS: None.

Magnitudes and spectral types of principal stars:

Bayer desig.	Mag. (m_v)	Spec. type	Bayer desig.	Mag. (m_v)	Spec. type	Bayer desig.	Mag. (m_v)	Spec. type
α	3.27	A0 III	β	3.76	F8/G0 Ib	γ	4.25	F0 V
δ	4.35	A7 V	ε	5.11	B6 V	ζ	4.72	F7 V
η¹	5.71	A0 V	η²	5.01	M2 III	θ	4.83	K2.5 IIIa
ι	–	–[22]	κ	5.27	A8 III/IV	λ	5.14	G6 III
μ	–	–[22]	ν	5.06	B8 V	ξ	–	–[22]
ο	–	–[22]	π¹	5.56	K5 III	π²	5.38	G8 III

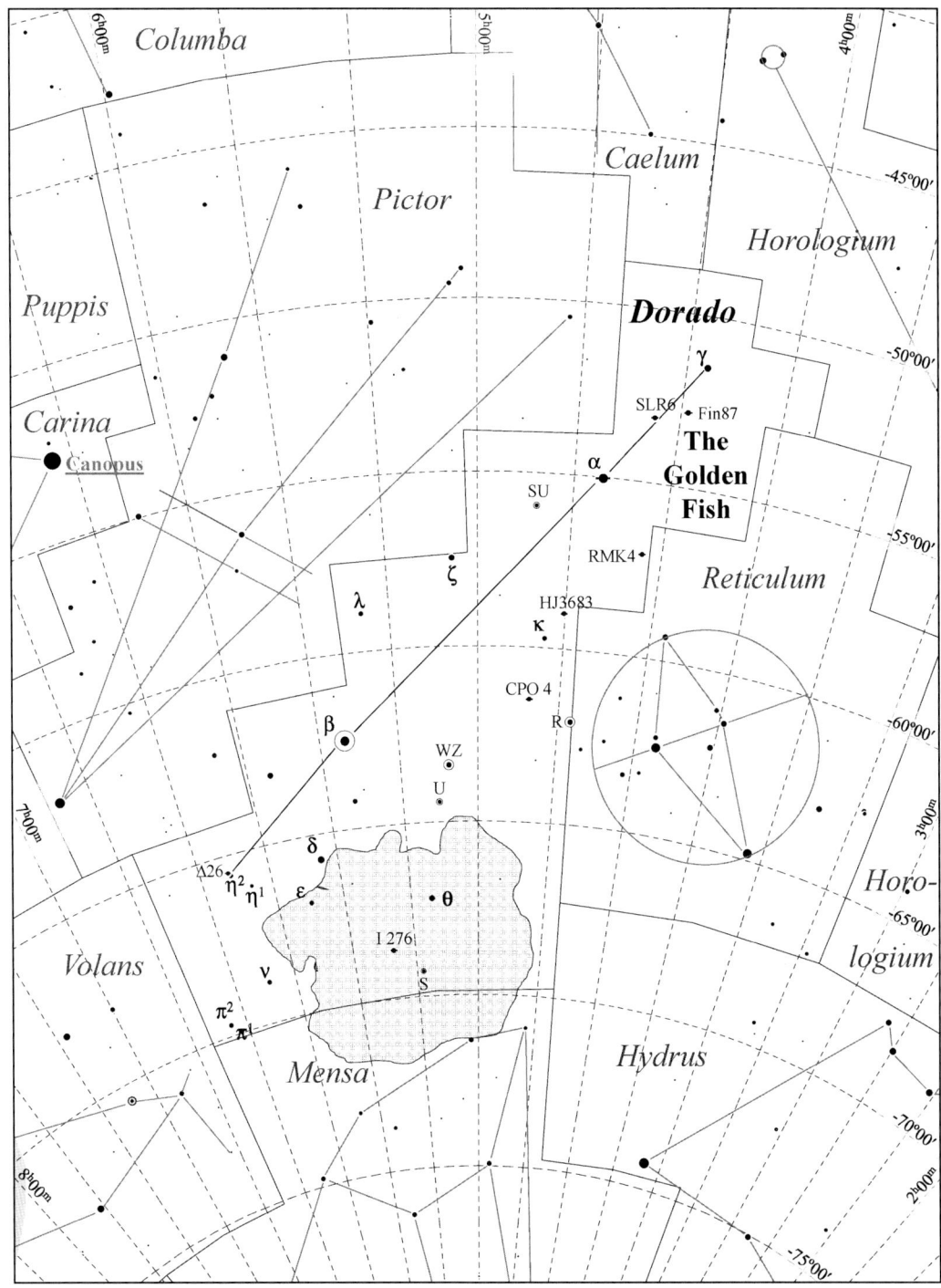

Fig. 16.11. Dorado details.

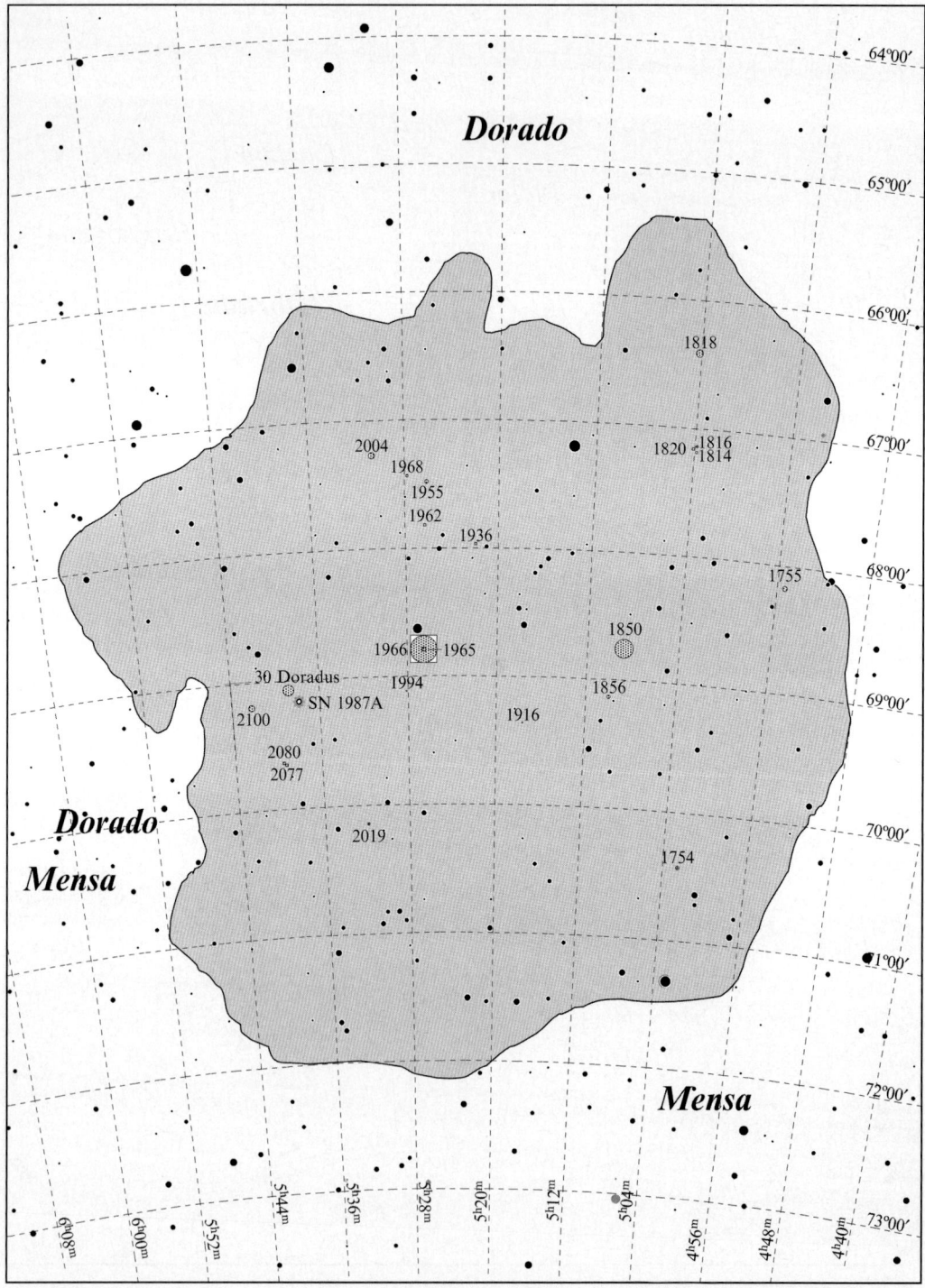

Fig. 16.12. Large Magellanic Cloud details.

Table 16.6. Selected telescopic objects in Dorado.

Multiple stars:

Designation	R.A.	Decl.	Type	m₁	m₂	Sep. (")	PA (°)	Colors	Date/Period	Aper. Recm.	Rating	Comments
Fin 87	04^h 18.7^m	−52° 52′	A–B	6.2	8.8	0.8	106	yW, ?	2001	10/12″	***	
Rmk 4	04^h 24.2^m	−57° 04′	A–B	6.9	7.2	5.5	246	Y, ?	2001	7×50	****	
Sir 6	04^h 25.5^m	−53° 07′	A–B	7.1	9.1	0.9	105	yW, ?	1991	10/12″	***	
Doradus	04^h 34.0^m	−55° 03′	AB–C	3.2	10.2	77.2	094	W, ?	1999	2/3″	***	A–B is a binary, max sep. 0.32″
h3683AB	04^h 40.3^m	−58° 57′	A–B	7.3	7.5	**3.23**	**088**	Y, ?	**240**	2/3″	*****	Separation and position angle are calculated for 2010.0 from orbital parameters
CapO 4	04^h 47.4^m	−61° 29′	A–B	7.4	10.3	2.6	046	yW, ?	2001	4/6″	***	Good color contrast, separation increasing, PA decreasing
I 276	05^h 27.0^m	−68° 37′	A–B	6.7	6.9	1.2	172	Y, B	2001	6/8″	******	
□ 26	06^h 12.2^m	−65° 32′	A–B	7.9	8.1	20.6	120	yW, W	1998	2/3″	*****	Easy, nearly matched pair

Variable stars:

Designation	R.A.	Decl.	Type	Range (m$_v$)	m₂	Period (days)	F (f₂/f₁)	Spectral Range	Aper. Recm.	Rating	Comments
R Doradus	04^h 36.8^m	−62° 03′	SRb	4.8	6.6	338.0	0.30	M8IIIe	6×30	****	2° 31′ SSW (PA 201°) of κ Dor (m5.28).#
SU Doradus	04^h 47.5^m	−55° 41′	M	8.5	14.0	235.9	–	–	10/21″	****	2° 01′ ESE (PA 110°) of α Dor (m3.26).##
WZ Doradus	05^h 07.6^m	−63° 24′	SRb	5.2 ± 0.06	–	40.0	–	M3 III	6×30	***	3° 06′ WSW (PA 250°) of β Dor (m3.46). Variation barely detectable to the unaided eye
U Doradus	05^h 10.1^m	−64° 19′	M	8.1	14.6	394.4	–	M8IIIe	14/15″	****	0° 58′ SSE (PA 163°) of WZ Dor (m5.2).##
S Doradus	05^h 18.2^m	−69° 15′	S Dor	8.6	11.5	–	–	A0.5eq	4/6″	*****	2° 06′ S (PA 169°) of θ Dor (m4.80).# *23
β Doradus	05^h 33.6^m	−62° 29′	Cδ	3.46	4.08	9.843	0.48	F4–G4 Ia/II	Eye	****	One of brightest Cepheid variables. Compare to α (m3.27) and δ (m4.35) Dor *24
SN 1987A	05^h 35.5^m	−62° 16′	SN						Eye	****	The brightest extragalactic supernovae ever observed

(continued)

Table 16.6. (continued)

Star clusters:

Designation	R.A.	Decl.	Type	m_v	Size (')	Brst. Star	Dist. (ly)	dia. (ly)	Number of Stars	Aper. Recm.	Rating	Comments
NGC 1652	04h 38.4m	−68° 40'	Open	9.3	1.2	–	180K	65	–	4/6"	***	All the listed clusters are in the LMC, and hence are at nearly the same distance
NGC 1755	04h 55.2m	−68° 12'	Open	9.9	2.6	–	180K	136	–	4/6"	***	Moderately brighter toward center
NGC 1814	05h 03.8m	−67° 18'	Open	9.0	1.0	–	180K	50	–	4/6"	***	Compact, sparse, fairly bright stars, merges into background, some nebulosity
NGC 1816	05h 03.9m	−67° 16'	OC+Neb	9.0	1.0	–	180K	50	–	4/6"	***	Sparse, fairly bright stars, merges into background, some nebulosity *25
NGC 1818	05h 04.2m	−66° 26'	Open	9.7	3.4	13.3	180K	175	–	4/6"	***	Fairly large and bright, distinct from background, partly resolved
NGC 1820	05h 04.1m	−67° 16'	Open	9.0	–	–	180K	–	–	4/6"	***	Sparse, fairly bright stars, merges into background
NGC 1850	05h 08.8m	−68° 46'	OC+Neb	9.1	3.4	14.3	180K	175	–	4/6"	***	Rich, may be up to 3 overlapping clusters
NGC 1856	05h 09.5m	−69° 08'	Open	9.7	1.8	15.0	180K	95	–	4/6"	***	Rich, strong concentration toward center, distinct, but large telescope needed to resolve
NGC 1916	05h 17.5m	−69° 23'	Globular	10.4	0.5	–	180K	25	–	6/8"	***	Small, rich, moderate concentration toward center. Use large telescopes
NGC 1955	05h 26.1m	−67° 30'	Open	9.0	1.8	11.3	180K	95	–	4/6"	***	Sparse, large range of brightnesses, E-W elongation
NGC 1968	05h 27.7m	−67° 27'	OC+Neb	9.0	1.1	12.6	180K	60	–	4/6"	***	Sparse, large range of brightnesses, E-W elongation
NGC 1994	05h 28.6m	−69° 08'	Open	9.8	0.6	–	180K	30	–	4/6"	***	Fairly bright, very compact
NGC 2004	05h 30.7m	−67° 17'	Open	9.6	2.7	13.0	180K	140	–	4/6"	***	Distinct, rich, moderately brighter toward center
30 Doradus, NGC 2070	05h 38.7m	−69° 06'	OC+Em	8.3	5	–	180K	1300	–	4/6"	*****	NGC 2070 includes both the 30 Doradus cluster and the Tarantula Nebula *26
NGC 2100	05h 42.1m	−69° 13'	Open	9.6	2.8	11.8	180K	145	–	4/6"	***	Stands out from background, resolved in larger scopes

Nebulae:

Designation	R.A.	Decl.	Type	m_v	Size (')	Brtst./Cent. star	Dist. (ly)	dia. (ly)	Aper. Recm.	Rating	Comments	
NGC 1936	05h 22.0m	–67° 59'	Em+OB	–	20"×15"	–	180K	17	12/14"	***	Brightest of a compact group (NGC 1929, 34, 35, 36) of nebulae in LMC	*27
NGC 1962	05h 26.3m	–68° 50'	Em+OC	–	20×3	–	180K	1,000	6/8"	****	NGC 1962, 65, 66, and 70 are essentially one object with OC (1962) imbedded	
NGC 1965	05h 26.6m	–68° 49'	Em+OC	–	1.3	–	180K	–	6/8"	****		
NGC 1966	05h 26.8m	–68° 49'	Em+OC	–	13×13	–	180K	700	6/8"	****		
NGC 2077	05h 39.6m	–69° 40'	Em+OC	–	15×11	–	180K	800	6/8"	****	SW of NGC 2080, together they form a dumbbell shape. Both have embedded clusters	
NGC 2080	05h 39.7m	–69° 39'	Em+OC	–	0.7	–	180K	–	6/8"	****	NE of NGC 2077	

Galaxies:

Designation	R.A.	Decl.	Type	m_v	Size (')	Dist. (ly)	dia. (ly)	Lum. (suns)	Aper. Recm.	Rating	Comments	
NGC 1515	04h 04.0m	–54° 06'	SBb	11.2	5.4×1.3	38M	51K	9.5G	8/10"	***		
NGC 1549	04h 15.7m	–55° 36'	E0	9.6	3.7×3.2	41M	33K	13G	4/6"	****	NGC 1553 is in same LP FOV	
NGC 1553	04h 16.2m	–55° 47'	S0	9	4.1×2.8	45M	50K	18G	4/6"	****	NGC 1549 is in same LP FOV	
NGC 1566	04h 20.0m	–54° 56'	SBb	9.4	7.6×6.2	45M	102K	23G	4/6"	****		
NGC 1617	04h 31.7m	–54° 36'	SBa	10.5	4.8×2.3	44M	43K	12G	6/8"	***		
NGC 1672	04h 45.7m	–59° 15'	SBb	10.1	4.8×3.9	47M	79K	13G	6/8"	******		
Large Magellanic Cloud	05h 00m	–69°	SBb/Irr	0.1	~11°×6°	180K	35K	2.4G	eye	***		*28
NGC 1947	05h 26.8m	–63° 46'	S0p	10.8	2.4×2.0	28M	39K	3.8G	6/8"	***		

Meteor showers:

Designation	R.A.	Decl.	Type	Period (yr)	Duration	Max Date	ZHR (max)	M	Comet/Asteroid	First Obs.	Vel. (mi/km/sec)	Rating	Comments
None													

Other interesting objects:

Designation	R.A.	Decl.	Type	m_v	Dist. (ly)	Aper. Recm.	Rating	Comments
Supernova remnant	05h 35.5m	–69° 16'	II P		180K			SN 1987A

Fig. 16.13a. The Tarantuila Nebula (NGC 2070) in Dorado. Astrophoto by Christopher J. Picking.

Fig. 16.13b. Tarantula Nebula (NGC 2070) in Dorado. Drawing by Eiji Kato.

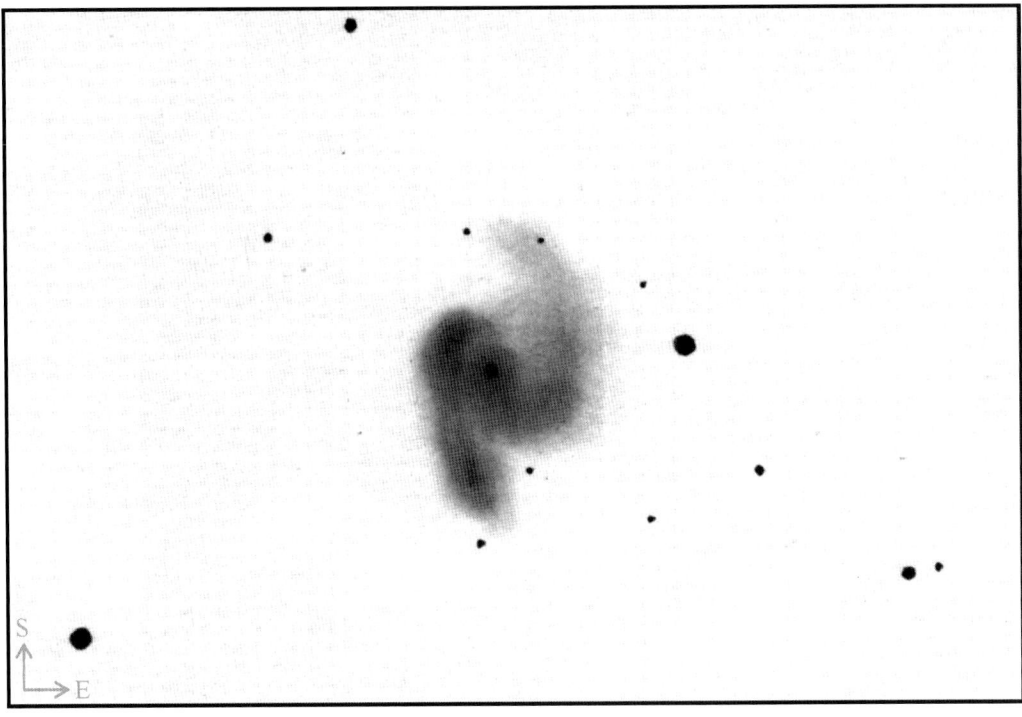

Fig. 16.14. NGC 1556, spiral galaxy in Dorado. Drawing by Eiji Kato.

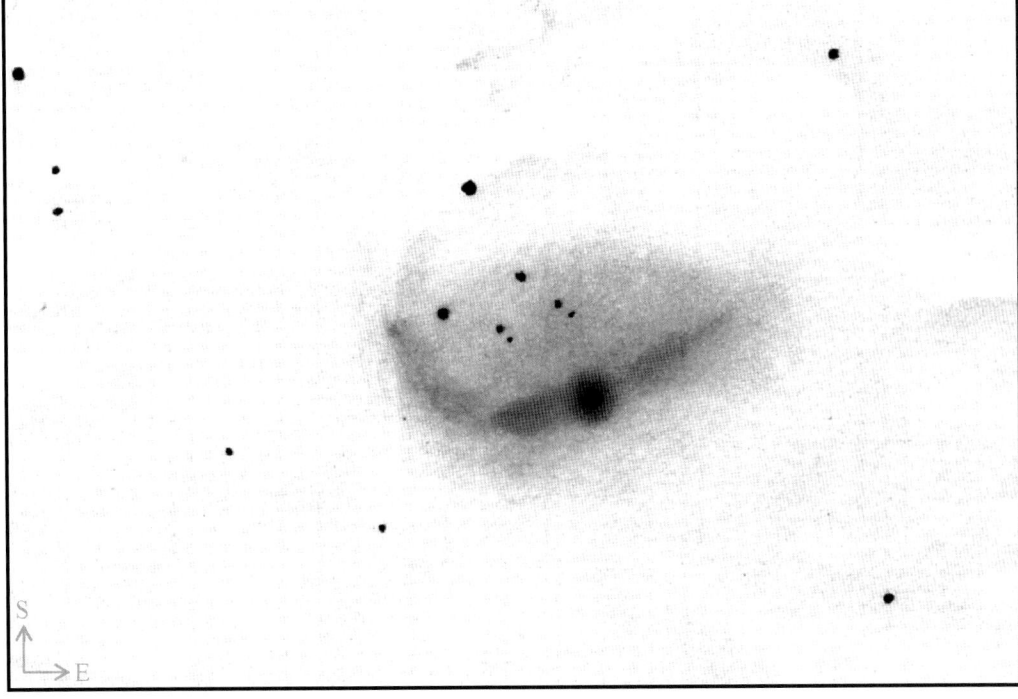

Fig. 16.15. NGC 1672, spiral galaxy in Dorado. Drawing by Eiji Kato.

Grus (grewse)	The Crane
Gruis (grew-eese)	9:00 P.M. Culmination: Oct. 13
Gru	Area: 366 sq. deg., 45th in size

Although the configuration of this constellation has remained relatively constant since its inception, it has carried different names at different times. Plancius' first globe labeled the constellation "Krane Grus," which actually means Crane, since it used both the Dutch and Latin words for Crane. A posthumous Plancius globe issued in 1625 labeled it Phoenicopterus or flamingo, and de Houtman's star catalog issued the same year as the *Uranometria* called it Den Reygher, the Heron.

Part of Grus came at the expense of Piscis Austrinus, since the star which Ptolemy called "the end of the fish's tail" (later known as κ PsA) was taken to indicate the Crane's head and became γ Grucis. Also, the Arabs had extended the fish's tail downward to include even more of the stars which later were taken away to form Grus.

TO FIND: Grus can be seen in its entirety only from latitudes south of 33° N., and not at all from latitudes north of 53° N. the first magnitude star Fomalhaut (α PsA) is a good starting point for finding Grus. For those at about 33° N., when Grus is at its highest, Fomalhaut lies about 8° east of the meridian and about 27° above the southern horizon. Of course, Fomalhaut will be lower for those farther north and higher for those farther south. β Grucis (Gruid) is 17½° south (PA 188°) of Fomalhaut. Al Nair (α Grucis) is 6° W (PA 266°) of β. γ Grucis (Al Dhanab) is 10° NNW (PA 343°) of Al Nair. Al Dhanab and Gruid form the NW end and the middle of a 20° long, nearly straight, line of 2nd, 3rd, and 4th magnitude stars which form the head, body, and tail of the crane. Al Nair lies near the tip of one wing, and θ Grucis (5.3° NE of β Grus) lies near the end of the other. Although most of Grus consists of fairly faint stars, Al Nair is 30th in brightness in the sky and Gruid is 57th, respectively, being noticeably brighter and very slightly fainter than Polaris. The combination of these two relatively bright stars and the long chain of faint stars, makes Grus fairly easy to recognize. In spite of this, Grus is one of the most poorly known constellations, especially for northern observers (Figs. 16.1, 16.16 and 16.17).

KEY STAR NAMES:

Name	Pronunciation	Source
α Al Nair	al **nah**-ir	al-nayyir min dhanab al-hut (Arabic), "the bright one from the fish's tail"
β Al Dhanab	al **dah**-nab	al-dhanab (Arabic), "the tail." Apparently from previously being in the tail of Piscis Austrinus
γ Gruid	**grew**-id	Apparently from Gruidae (Latin), the name of the family of cranes

Magnitudes and spectral types of principal stars:

Bayer desig.	Mag. (m_v)	Spec. type	Bayer desig.	Mag. (m_v)	Spec. type	Bayer desig.	Mag. (m_v)	Spec. type
α	1.73	B7 IV	β	2.07	M5 III	γ	3.00	B8 III
δ^1	3.97	G6/8 III	δ^2	4.12	M4.5 III	ε	3.49	A3 V
ζ	4.11	G8 III	η	4.84	K2 III	θ	4.28	F5m
ι	3.88	K0 III	κ	5.37	K5 III	λ	4.47	M0 III
μ^1	4.79	G8 III	μ^2	5.11	G8 III	ν	5.47	G8 III
ξ	5.29	K0 III*29	o	5.53	F4 V*29	π^1	6.42	S5,7
π^2	5.62	F3 III/IV	ρ	4.84	K0 III	σ^1	6.28	A3 V
σ^2	5.85	A1 V	τ^1	6.03	G3 IV	τ^2	6.67	F7 V
τ^3	5.72	Am	υ	5.62	A0 V	φ	5.54	F5 V*30

Stars of magnitude 5.5 or brighter which have only catalog designations:

Desig-nation	Mag. (m_v)	Spec. type	Desig-nation	Mag. (m_v)	Spec. type
HD 211415	5.36	G1 V	HD 216149	5.43	K3 III
HD 208737	5.50	K1 II/III			

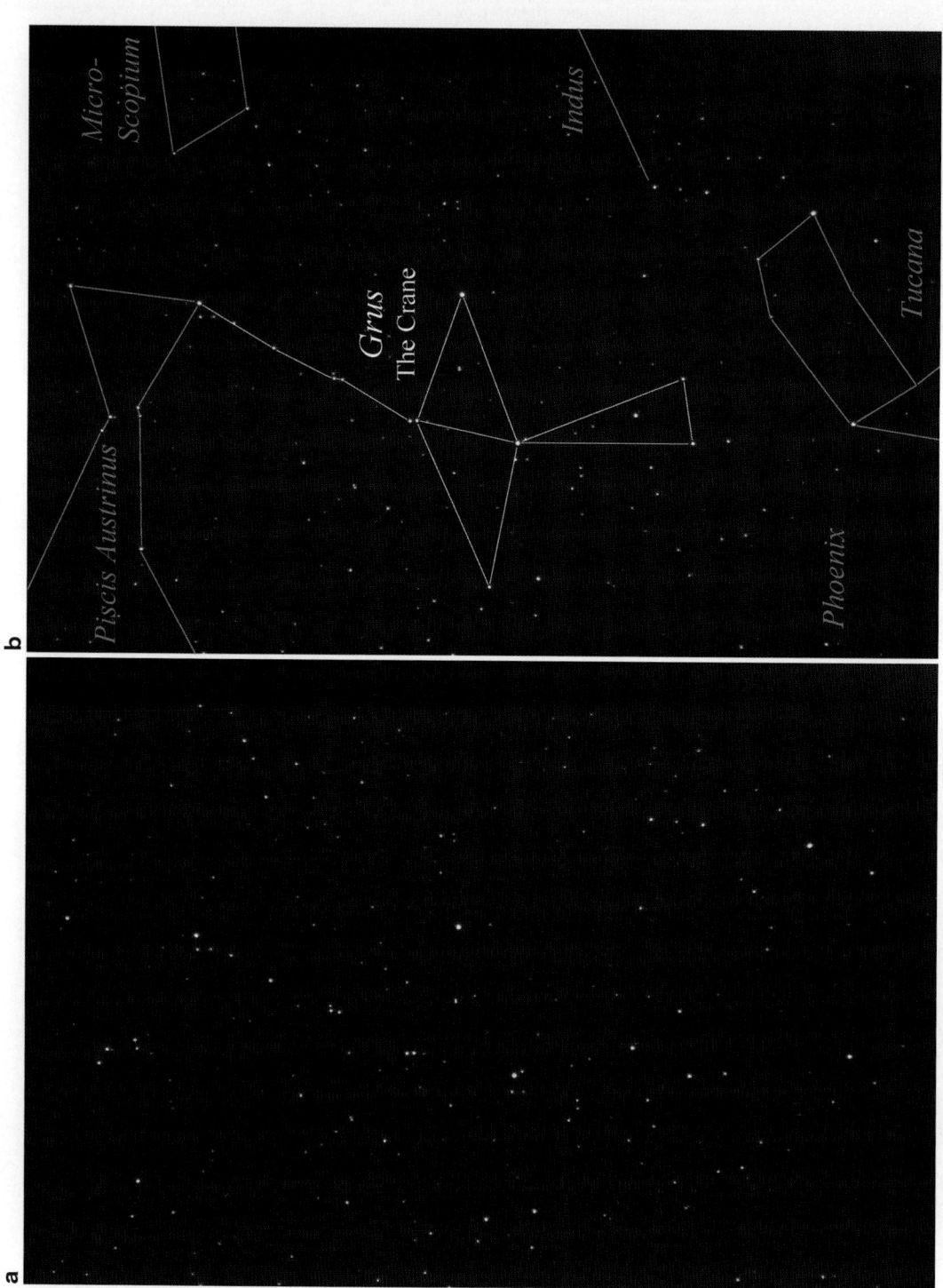

Fig. 16.16. (**a**) The stars of Grus and adjacent constellations. (**b**) The figures of Grus and adjacent constellations.

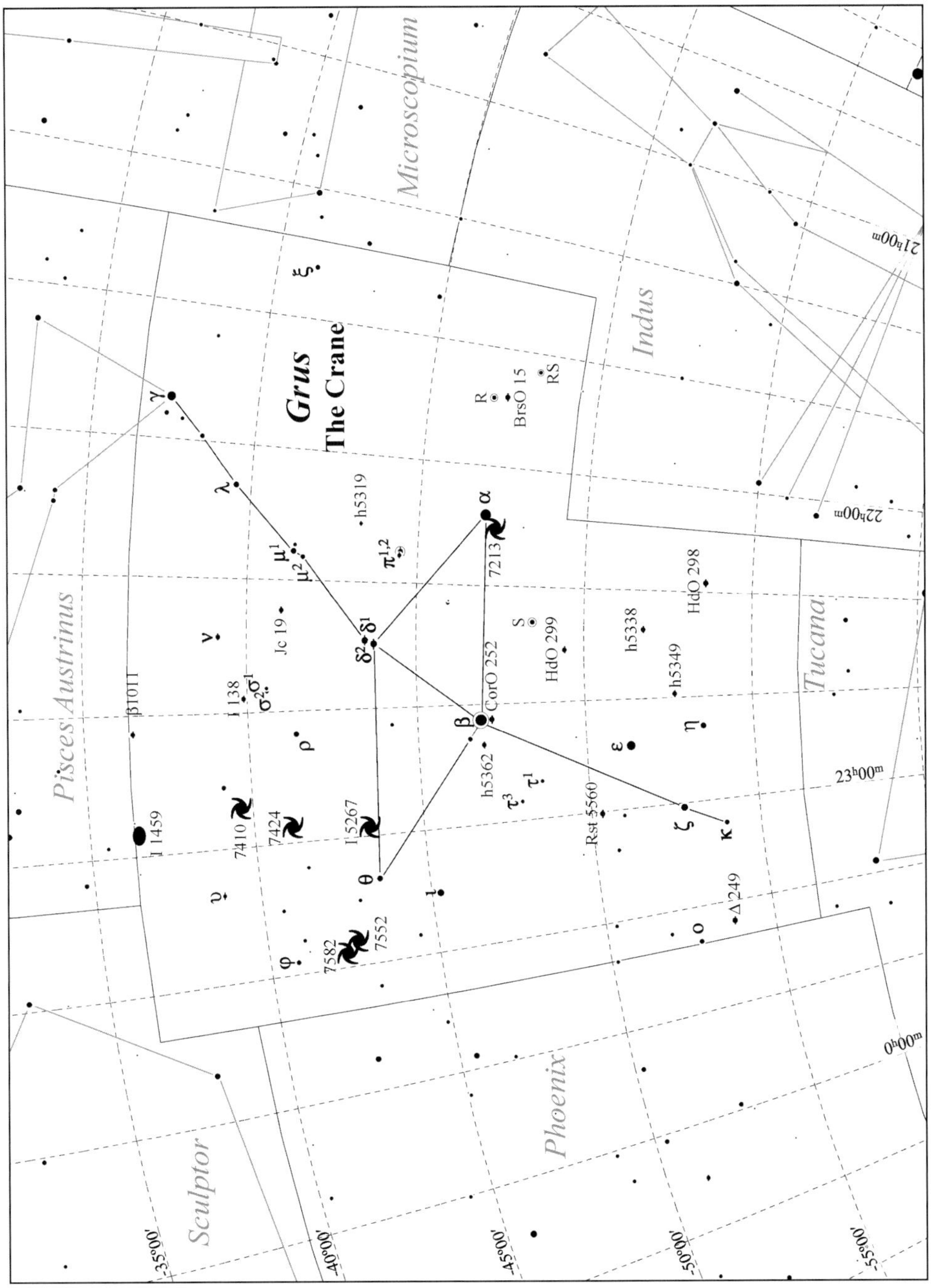

Fig. 16.17. Grus details.

Table 16.7. Selected telescopic objects in Grus.

Multiple stars:

Designation	R.A.	Decl.	Type	m₁	m₂	Sep. (″)	PA (°)	Colors	Date/Period	Rating	Aper. Recm.	Comments
BrsO 15	2ʰ 48.3ᵐ	−47° 18'	A–B	5.6	8.8	75.2	351	Y, ?	1999	****	7×50	
A Gruis	22ʰ 08.2ᵐ	−46° 58'	A–B	1.7	12.3	28.5	149	bW, ?	1947	******	4/5″	Rst5483
h 5319	22ʰ 12.0ᵐ	−38° 18'	A–B	7.7	7.7	2.1	312	yW, ?	2001	******	4/5″	Nearly matched pair
HdO 298	22ʰ 18.3ᵐ	−53° 38'	A–B	5.4	9.7	5.0	039	O, ?	1988	****	2/3″	
π¹ Gruis	22ʰ 22.7ᵐ	−45° 57'	A–B	6.5	10.7	2.8	201	O, ?	1975	****	4/5″	
π² Gruis	22ʰ 23.1ᵐ	−45° 56'	A–B	5.6	11.4	5.1	220	Y, ?	1985	***	3/4″	
jc19	22ʰ 24.7ᵐ	−41° 26'	A–B	6.7	8.2	17.8	065	Y,Y	1998	****	18×80	
h5338	22ʰ 28.5ᵐ	−51° 47'	A–B	7.0	10.8	30.1	182	pO, ?	1919	****	18×80	
ν Gruis	22ʰ 28.7ᵐ	−39° 07'	A–B	5.5	12.5	26.9	116	Y, ?	1904	***	4/5″	a.k.a. See 473
δ¹ Gruis	22ʰ 29.3ᵐ	−43° 30'	A–B	4.0	12.8	5.6	228	Y, ?	1935	***	5/6″	a.k.a. I 1054
δ² Gruis	22ʰ 29.8ᵐ	−43° 45'	A–B	4.3	9.7	60.8	211	R, ?	2000	****	7×50	a.k.a. Δ239. A is variable
HdO 299	22ʰ 31.0ᵐ	−49° 26'	A–B	6.8	12.0	21.9	359	Y, ?	1999	****	3/4″	
σ² Grucis	22ʰ 37.0ᵐ	−40° 35'	A–B	5.9	10.0	2.7	265	W, ?	1991	****	3/4″	
I 138	22ʰ 37.8ᵐ	−39° 51'	A–B	6.7	10.7	3.4	277	yW, ?	1991	***	3/4″	
h5349	22ʰ 39.1ᵐ	−52° 42'	A–B	6.7	11.5	32.6	118	yW, ?	1999	****	18×80	
CorO 252	22ʰ 42.6ᵐ	−47° 13'	A–B	6.0	11.1	7.5	125	Y, ?	1998	****	3/4″	
h5362	22ʰ 46.7ᵐ	−46° 56'	A–B	6.6	9.9	10.4	141	W, ?	1998	****	2/3″	
β1011	22ʰ 02.6ᵐ	−36° 25'	A–B	6.6	9.3	2.1	293	oY, ?	1999	****	4/5″	
υ Gruis	22ʰ 06.9ᵐ	−38° 54'	A–B	5.7	8.2	1.0	206	pY?	2001	****	9/10″	
θ Gruis	22ʰ 06.9ᵐ	−43° 31'	A–B	4.5	6.6	1.5	108	pY?	2001	****	6/7″	a.k.a. β773
			A–C	4.5	7.8	158.9	292	Y, ?	2002	****	7×50	
Δ246	22ʰ 07.2ᵐ	−50° 41'	A–B	6.3	7.1	8.8	254	pY, pY	1999	****	2/3″	
Rst5560	22ʰ 20.8ᵐ	−50° 18'	A–B	6.2	8.9	1.3	233	pY, ?	1999	****	7/8″	
			AB–C	6.1	6.6	16.8	212	pY, ?	2000	****	18×80	a.k.a. Δ248
Δ 249	23ʰ 23.9ᵐ	−53° 49'	A–B	6.1	6.6	26.5	212	yW, yW	2000	****	10×50	

Variable stars:

Designation	R.A.	Decl.	Type	Range (m.)		Period (days)	F (f./f₁)	Spectral Range	Rating	Aper. Recm.	Comments
RS Grucis	21ʰ 43.1ᵐ	−48° 11'	δSct	7.9	8.5	0.147	0.26	A6–A9IV–F0	****	7×50	1° 15'SW (PA 224°) of HD207129 (m5.57), unmarked on most charts.#
R Grucis	21ʰ 48.5ᵐ	−46° 55'	M	7.4	14.9	331.96	0.42	M5e–M7II–IIIae	***	18/20″	0° 24'N (PA 007°) of HD207129 (m5.57) unmarked on most charts.##
π¹ Grucis	22ʰ 22.7ᵐ	−45° 57'	SRb	5.4	6.7	150	–	S5.7e	****	6×30	2° 33' SSE (PA 167°) of π¹ (m5.41). A #
S Grucis	22ʰ 26.1ᵐ	−48° 26'	M	6.0	15.0	401.51	0.43	M5e–M8IIIe	****	18/20″	4,000:1 brightness range!##
β Grucis	22ʰ 42.7ᵐ	−46° 53'	Lc	2.0	2.3	–	–	M3–M5, II–III	****	Eye	Small, but readily perceptible change

Star clusters:

Designation	R.A.	Decl.	Type	m_v	Size (')	Brtst. Star	Dist. (ly)	dia. (ly)	Number of Stars	Aper. Recm.	Rating	Comments
None												

Nebulae:

Designation	R.A.	Decl.	Type	m_v	Size (')	Brtst./ **Cent.** star	Dist. (ly)	dia. (ly)	Aper. Recm.	Rating	Comments
None											

Galaxies:

Designation	R.A.	Decl.	Type	m_v	Size (')	Dist. (ly)	dia. (ly)	Lum. (suns)	Aper. Recm.	Rating	Comments
NGC 7213	$22^h\,09.3^m$	$-47°\,10'$	Sa	10.1	3.3×3.0	76M	73K	46G	4/6"	★★★★	Bright, round, dust lane
NGC 7410	$22^h\,55.0^m$	$-39°\,40'$	SBm	10.5	5.0×1.6	71M	103K	25G	4/6"	★★★★	
I 1459	$22^h\,57.2^m$	$-36°\,28'$	E3	10.0	4.9×3.5	71M	101K	40G	4/6"	★★★★	
I5267	$22^h\,57.2^m$	$-43°\,24'$	S0	10.3	5.0×3.7	74M	103K	31G	4/6"	★★★★	
NGC 7424	$21^h\,57.3^m$	$-41°\,04'$	S[B]c	10.2	9×9	37M	97K	9.1G	4/6"	★★★★	
NGC 7552	$23^h\,16.2^m$	$-42°\,35'$	SBb	10.4	3.5×3.1	71M	72K	28G	4/6"	★★★★	NGC 7552 and 7582 are on the southwest end of a string of four galaxies known as the "Grus Quartet." The other two are 11th magnitude NGC 7590 and 7599
NGC 7582	$23^h\,18.4^m$	$-42°\,22'$	SBb	10.1	5.2×2.4	62M	94K	28G	4/6"	★★★★	

Meteor showers:

Designation	R.A.	Decl.	Period (yr)	Duration	Max Date	ZHR (max)	Comet/ **Asteroid**	First Obs.	Vel. (mi/ **km**/sec)	Rating	Comments
None											

Other interesting objects:

Designation	R.A.	Decl.	Type	m_v	Mass (suns)	Dia. (suns)	Dist. (ly)	Planet Mass (J)	Dist. (AU)	Period (days)	Eccentricity	Comments
τ¹ Gruis	$22^h\,53.6^m$	$-48°\,36'$	Exo-planet	3.42	1.3	~2	109	~1.49	2.7	1443	0.34	9/17/2002, 100th announced exo-planet discovery. a.k.a. HD 216435, SAO113044

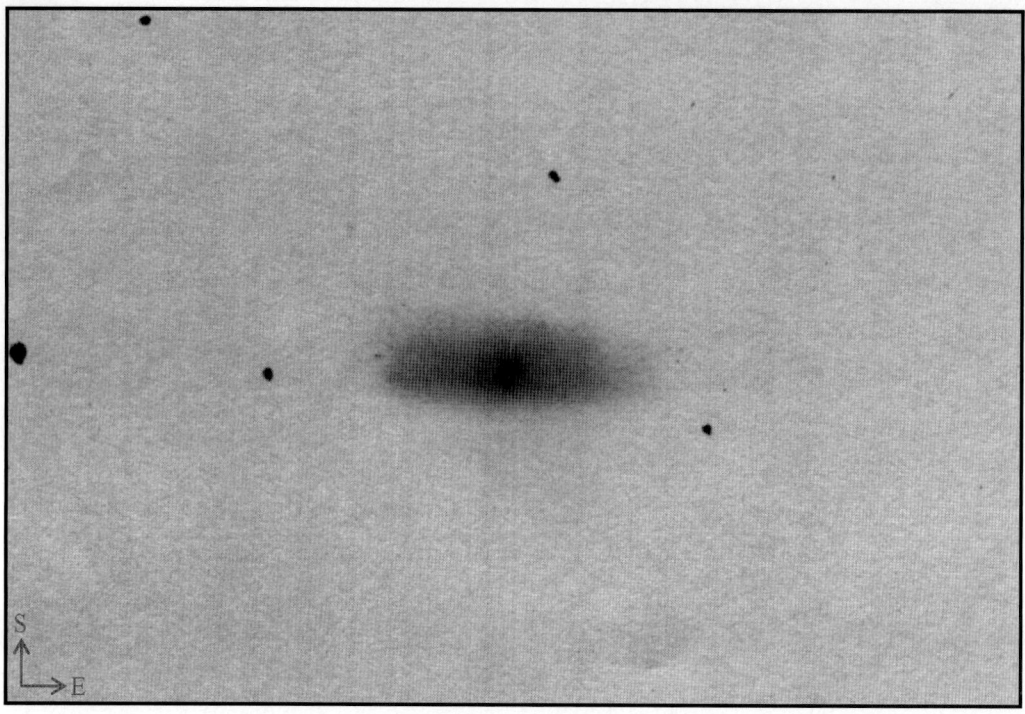

Fig. 16.18. NGC 7552, spiral galaxy in Grus. Drawing by Eiji Kato.

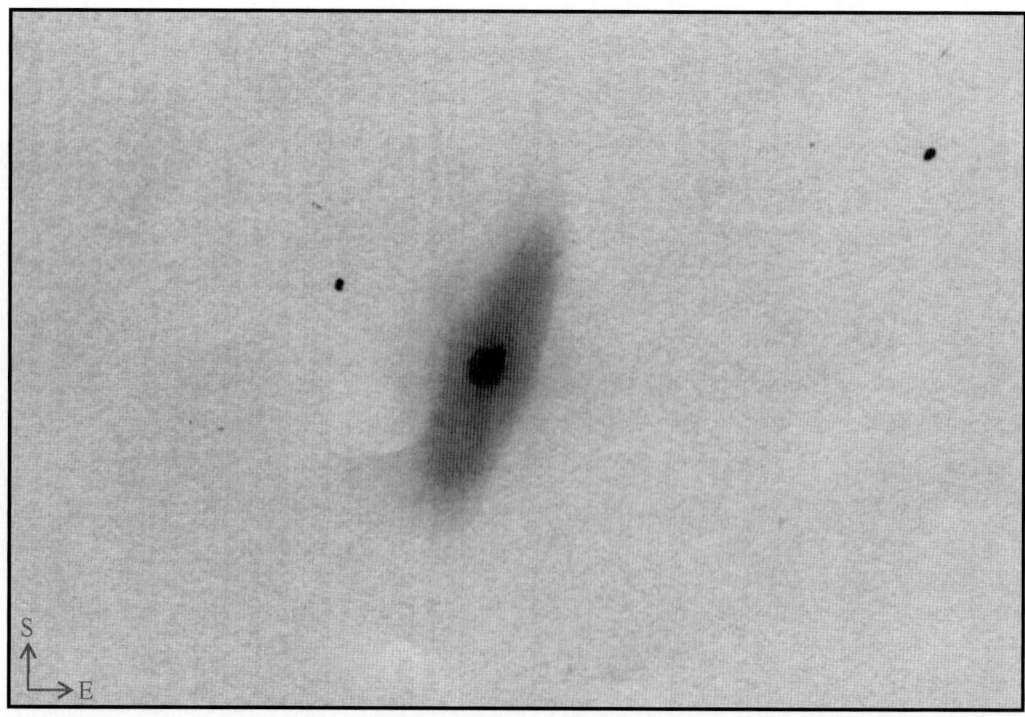

Fig. 16.19. NGC 7582, spiral galaxy in Grus. Drawing by Eiji Kato.

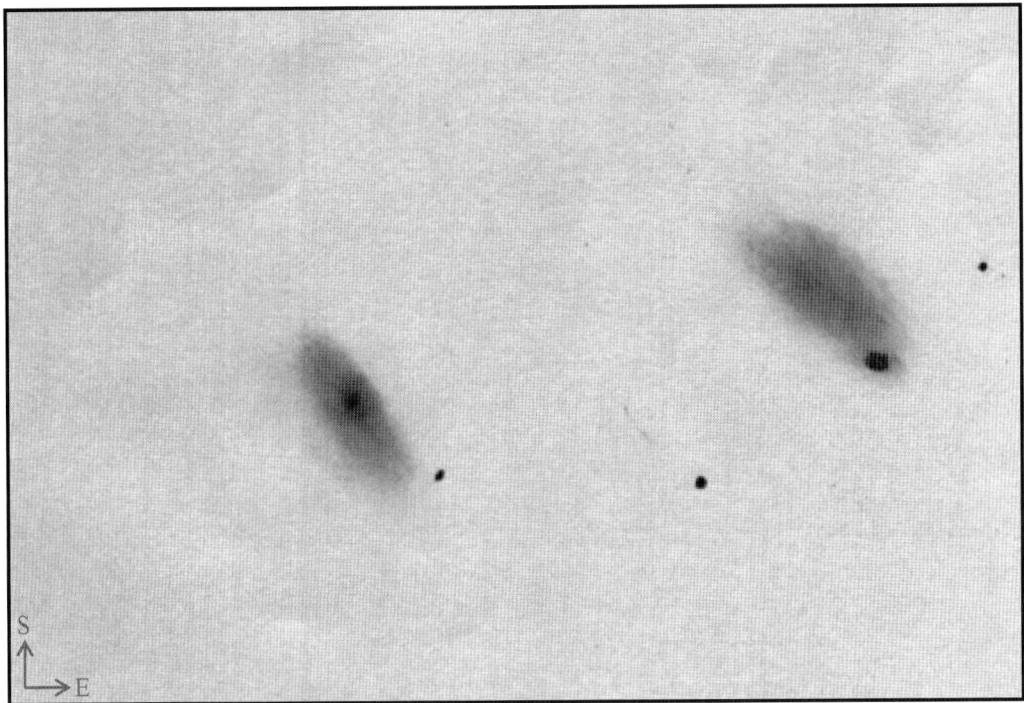

Fig. 16.20. NGC 7599 and 7590, spiral Galaxies in Grus. Drawing by Eiji Kato.

Hydrus (high-druss)	The Water Snake
Hydri (hide-rye)	9:00 P.M. Culmination: Apr. 30
Hyi	Area: 243 sq. deg., 61st in size

Hydrus is sometimes called "The Male Water Snake," and Hydra "The Female Water Snake," but this comparison is based on a misunderstanding of the names. See Chapter 8 for an explanation of the difference between the two names.

Plancius' original constellation extended well beyond its present southern limit, with the tail coming within 5½ degrees of the south celestial pole.

TO FIND: Hydrus is relatively easy to find, in spite of its lack of any bright stars, because α Hydri is only 9° south–southeast (PA 150°) of <u>Achernar</u> (α Eridani). γ Hydri is located about 12° farther along nearly the same line. β Hydri is about the same distance (21°) south (PA 191°) of <u>Achernar</u>. All three of these stars are brighter than average 3rd magnitude stars, and stand out well in rather dim surroundings. They also define the limits of the water snake's body. α represents the snake's head, β denotes the end of its tail, and γ represents the easternmost extent of the body. Once you have located the triangle, it is easy to trace the body's slight westward excursion from the α to γ line, its turn back across the line and direct path to γ. From there to β is almost a direct line, with only a slight detour to include 5th magnitude ν (Figs. 16.1, 16.21 and 16.23).

KEY STAR NAMES: None.

Magnitudes and spectral types of principal stars:

Bayer desig.	Mag. (m_v)	Spec. type	Bayer desig.	Mag. (m_v)	Spec. type	Bayer desig.	Mag. (m_v)	Spec. type
α	2.86	F0 V	β	2.82	G2 IV	γ	3.26	M2 III
δ	4.08	A3 V	ε	4.12	B9 III	ζ	4.83	A2 IV/V
η^1	6.77	G8.5 V	η^2	4.68	G5 III	θ	5.51	B8 III/IV
ι	5.51	F4 III	κ	5.99	K0 III	λ	5.09	K5 III
μ	5.27	G4 III	ν	4.76	K3 III	ξ	–	–*31
o	–	–*31	π^1	5.57	K2 III	π^2	5.67	F2 II/III
ρ	–	–*32	σ	6.15	F5/6 IV/V	τ^1	6.33	G6/8 III
τ^2	6.05	F0 III						

Indus (in-dus)	The Indian (native American)
Indi (in-dye)	9:00 P.M. Culmination: Sep. 27
Ind	Area: 294 sq. deg. 49th in size

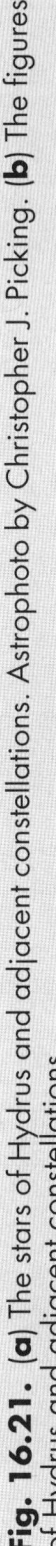

a

b

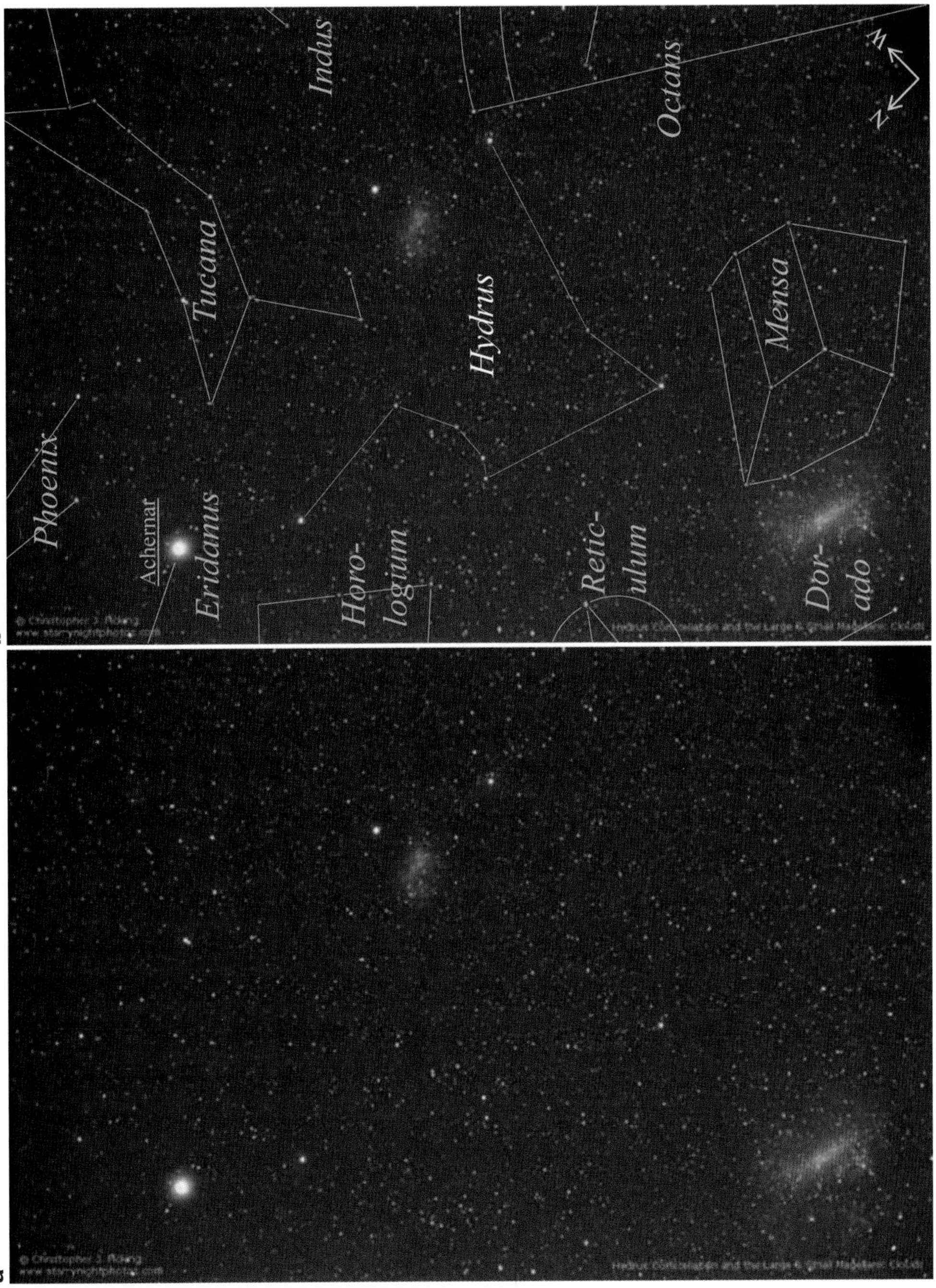

Phoenix

Achernar

Eridanus

Horo-
logium

Indus

Tucana

Hydrus

Octans

Reticulum

Mensa

Dor-ado

Fig. 16.21. **(a)** The stars of Hydrus and adjacent constellations. Astrophoto by Christopher J. Picking. **(b)** The figures of Hydrus and adjacent constellations.

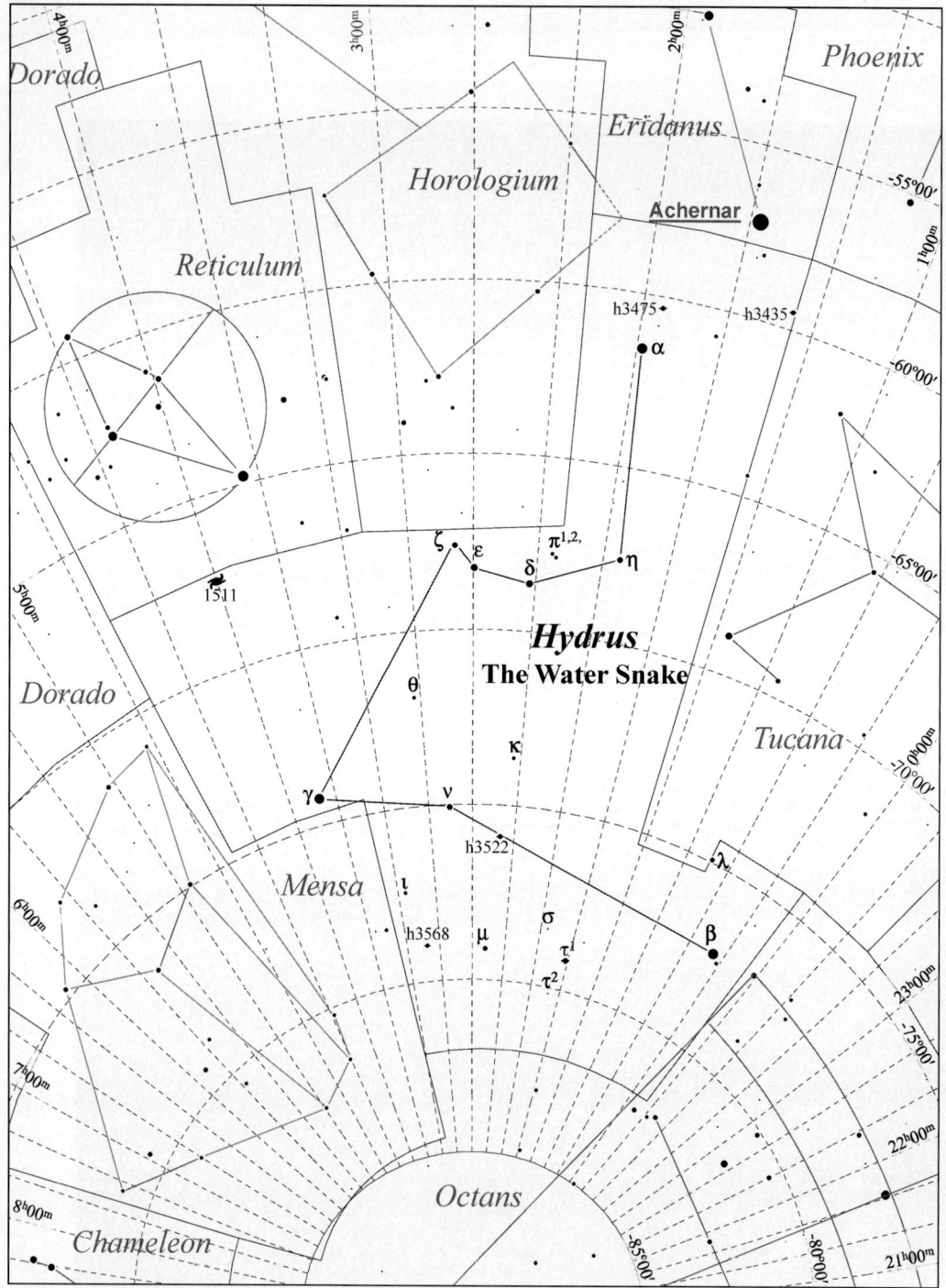

Fig. 16.22. Hydrus details.

Table 16.8. Selected telescopic objects in Hydrus.

Multiple stars:

Designation	R.A.	Decl.	Type	m_1	m_2	Sep. (″)	PA (°)	Colors	Date Period	Aper. Recm.	Rating	Comments
h3435	01ʰ 25.3ᵐ	−59° 30′	A–B	7.1	9.4	25.3	001	yW, ?	1999	15×80	****	Barren FOV except for a few faint stars
τ¹ Hydri	01ʰ 41.3ᵐ	−79° 09′	A–B	6.3	12.2	16.0	351	yO, ?	2000	4/5″	****	a.k.a. h3467
h3475	01ʰ 55.3ᵐ	−60° 19′	A–B	7.2	7.2	2.5	075	Y, Y	2002	3/4″	****	Nicely matched pair
h3522	02ʰ 33.2ᵐ	−75° 54′	A–B	6.9	11.0	39.7	286	yW, ?	1999	3/4″	***	
Dawes 185			B–C	10.7	12.6	9.7	352	?, ?	1999	5/6″	*****	
h3568	03ʰ 07.5ᵐ	−78° 59′	A–B	5.7	7.7	15.3	224	Y, bW	2000	2/3″	******	Showcase pair. Very sparse field

Variable stars:

Designation	R.A.	Decl.	Type	Range (m_v)	Period (days)	F (f_r/f_t)	Spectral Range	Aper. Recm.	Rating	Comments
None									****	

Star clusters:

Designation	R.A.	Decl.	Type	m_v	Size (′)	Brtst. Star	Dist. (ly)	dia. (ly)	Number of Stars	Aper. Recm.	Rating	Comments
None											****	

Nebulae:

Designation	R.A.	Decl.	Type	m_v	Size (′)	Brtst./ **Cent.** star	Dist. (ly)	dia. (ly)	Aper. Recm.	Rating	Comments
None										****	

Galaxies:

Designation	R.A.	Decl.	Type	m_v	Size (′)	Dist. (ly)	dia. (ly)	Lum. (suns)	Aper. Recm.	Rating	Comments
NGC 1511	03ʰ 69.6ᵐ	−67° 38′	SAa (pec)	11.5	3.9×1.8	49M	35K	3.3G	8/10″	****	

(continued)

Table 16.8. (continued)

Meteor showers:

Designation	R.A.	Decl.	Period (yr)	Duration	Max Date	ZHR (max)	Comet/Asteroid	First Obs.	Vel. (mi/km/sec)	Rating	Comments
None											

Other interesting objects:

Designation	R.A.	Decl.	Type	m_v	Mass (suns)	Dia. (suns)	Dist. (ly)	Planet Mass (J)	Dist. (AU)	Period (days)	Eccentricity	Comments
GJ 3021, HD 1237	00ʰ 16.2ᵐ	−79° 51′	Exo-planet	6.59	0.9	0.86	57	3.32	0.49	133.82	0.505	L* = 0.61L(.). Illumation at planet=3.5×solar illumination of the earth
η²	03ʰ 59.6ᵐ	−67° 38′	Exo-planet	4.68	1.91	10.1	217	6.54	1.93	711	0.4	L* = 55.5 L(.). Illumination at planet=15×solar illumination of the earth

TO FIND: Indus is another constellation made up entirely of relatively faint stars, with its brightest star being slightly less than the standard for 3rd magnitude. To find Indus, you should begin with α Pavonis (Peacock), the brightest star close to Indus. α Indi is 9.6° north–northeast (PA 012°) of Peacock. About 5.5° east–northeast of α Indi, you may be able to see magnitude 6.45 HD 200887, although it is near the limit of vision for most people with normal vision. This faint star and α Indi represent the top of the Indian's head. From those two stars, two roughly parallel, slightly bent lines run to the bottom of the Indian's feet, ending in HD 216437 (m6.08) and ν Indi. At a couple of strategic points, i.e., the neck and shoulders, there are pairs of faint stars oriented roughly parallel to α Indi and HD 200887. Between these rows and Peacock, there are two more stars representing a spear held in the Indian's left hand and paralleling the Indian's body. On the other side of the body, there is a short row of three stars representing the points of three arrows held in his right hand and nearly perpendicular to his body. Following the two lines of stars through several more faint stars, you may reach the Indian's feet. You will need sharp eyes, dark skies, and a bit of luck to pick out this well-camouflaged warrior (Figs. 16.1, 16.23 and 16.24).

KEY STAR NAMES: None.

Magnitudes and spectral types of principal stars:

Bayer desig.	Mag. (m_v)	Spec. type	Bayer desig.	Mag. (m_v)	Spec. type	Bayer desig.	Mag. (m_v)	Spec. type
α	3.11	K0 III	β	3.67	K0 III	γ	6.10	F3 V
δ	4.40	F0 IV	ε	4.69	K5 V	ζ	4.90	K5 III
η	4.51	A6 III/IV	θ	4.39	A5 V	ι	5.06	K1 II/III
κ¹	6.12	F3 VII	κ²	5.62	K4 III	λ¹	–	–*33
λ²	–	–*33	μ	5.17	K2 III	ν	5.28	A3 V*34
ξ	–	–*35	ο	5.52	M0 III	π	6.17	A3 0
ρ	6.04	G4 IV/V*36						

Pavo (**pay**-voh) **The Peacock**
Pavonis (pah-**voe**-nis) **9:00 P.M. Culmination:** Sep. 2
Pav **Area:** 378 sq. deg., 44th in size

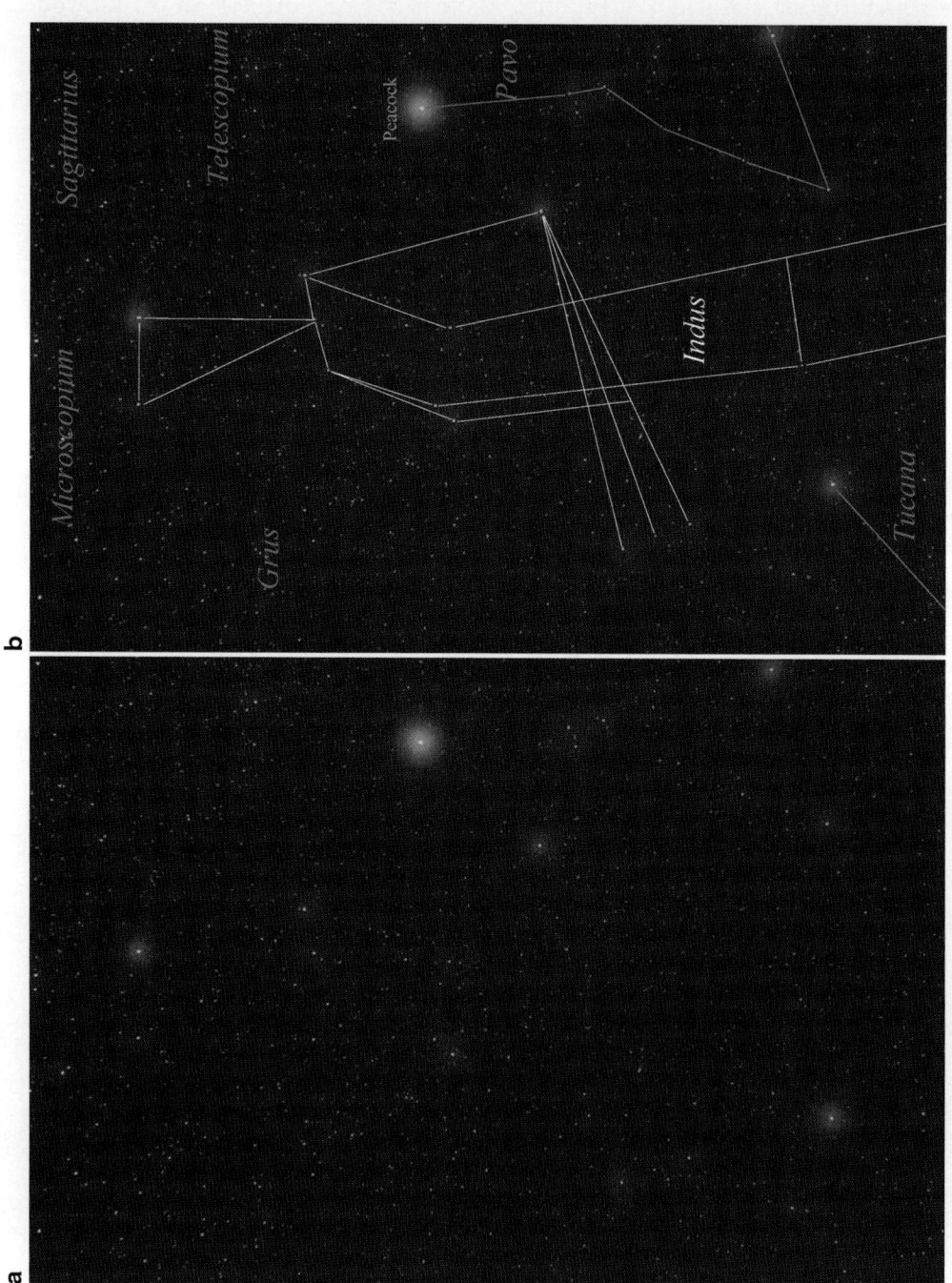

Fig. 16.23. (a) The stars of Indus and adjacent constellations. Astrophoto by Christopher J. Picking. (b) The figures of Indus and adjacent constellations.

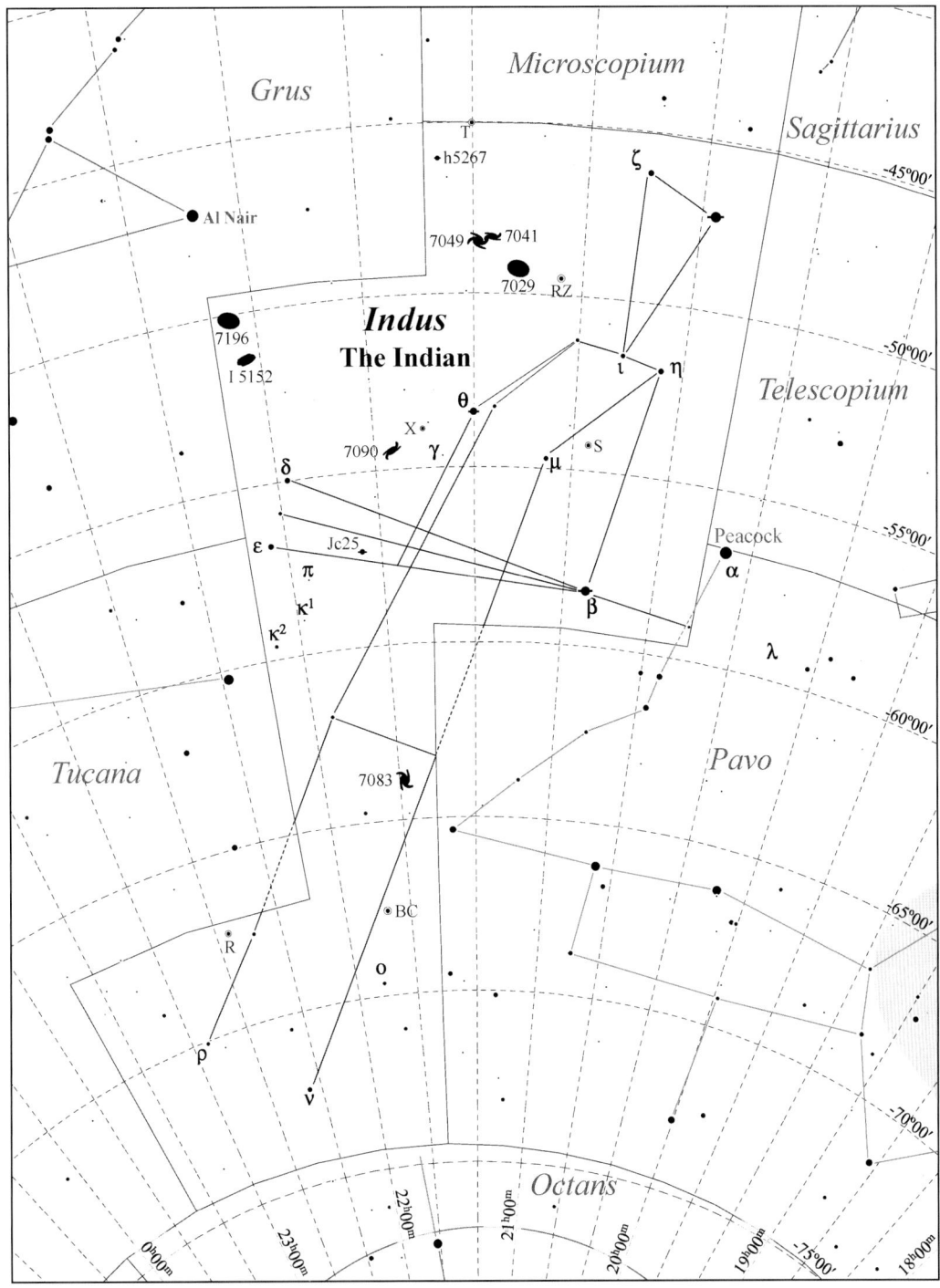

Fig. 16.24. Indus details.

Fig. 16.25. NGC 7083, spiral galaxy in Indus. Drawing by Eiji Kato.

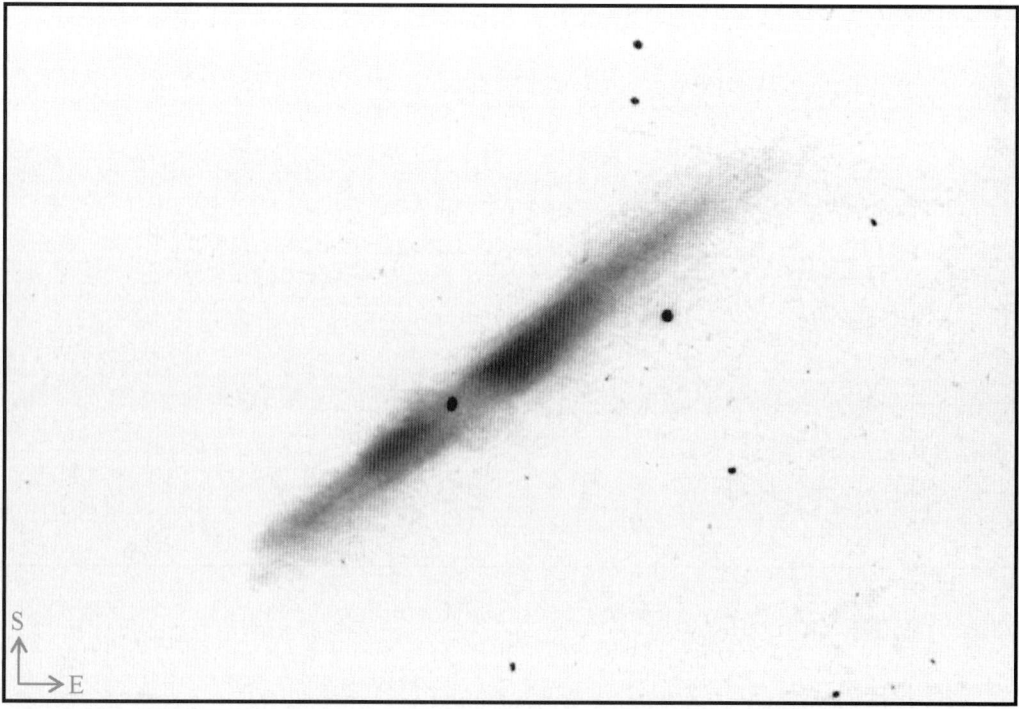

Fig. 16.26. NGC 7090, spiral galaxy in Indus. Drawing by Eiji Kato.

Table 16.9. Selected telescopic objects in Indus.

Multiple stars:

Designation	R.A.	Decl.	Type	m_1	m_2	Sep. (″)	PA (°)	Colors	Date/ Period	Aper. Recm.	Rating	Comments
α Indi	20ʰ 37.5ᵐ	−47° 17′	A–B	3.1	12.0	69.2	206	yO, ?	2002	3/4″	★★★★	a.k.a. HJ 5909-AB
			A–C	3.1	13.5	62.0	343	yO, ?	1994	7/8″	★★★	a.k.a. HJ 5909-AC
β Indi	20ʰ 54.8ᵐ	−58° 27′	A–B	3.7	12.5	17.3	104	yO, ?	1945	4/5″	★★★★	a.k.a. 2487
θ Indi	21ʰ 19.9ᵐ	−52° 27′	A–B	4.5	6.9	7.0	273	pY, dR	2002	2/3″	★★★★	a.k.a. h5258
h5267	21ʰ 26.6ᵐ	−4.6° 04′	A–B	7.3	11.0	5.0	057	yW, ?	1834	3/4″	★★★★	h5267-AB
			A–C	7.3	13.3	22.8	208	yW, ?	1914	7/7″	★★★	I 1442-A-C
			AD	7.3	10.0	44.0	182	yW, ?	1999	15×80	★★★★	h5267-AD
Jc25	21ʰ 44.0ᵐ	−57° 20′	A–B	6.5	6.9	152.3	004	Y, oY	1999	6×30	★★★★	A, B, and C have slightly different shades of yellow
			A–C	6.5	7.5	187.0	214	oY, Y	2000	7×50	★★★★	

Variable stars:

Designation	R.A.	Decl.	Type	Range (m_v)	Period (days)	F [f_s/f_i]	Spectral Range	Aper. Recm.	Rating	Comments
S Indi	20ʰ 56.8ᵐ	−54° 15′	M	7.9	399.95	0.41	M6–M8eII	Large	★★★	Very challenging, even for large telescopes. 1°21′ NW (PA|287°) of μ Indi
RZ Indi	21ʰ 03.9ᵐ	−49° 30′	M	8.4	255	–	Me	12/14″	★★★★	
T Indi	21ʰ 20.6ᵐ	−45° 00′	SRb	7.7	320	–	C 7, 2 (Na)	7×50	★★★★	Carbon star, deep red
X Indi	21ʰ 31.0ᵐ	−53° 56′	M	9.0	225.85	0.49	M4e–M5IIe	8/10″	★★★★	56′ NE (PA 41°) of λ)
BC Indi	21ʰ 47.2ᵐ	−67° 34′	αCVn	7.18	1.788	–	B9p(Si)	6×30	★★★★	2° 4′ NNW (PA 349°) of ° Indi. Suitable for photoelectric measurements only
R Indi	22ʰ 36.5ᵐ	−67° 16′	M	8.2	216.26	0.47	M2e–M4IIe	16/18″	★★★★	

Star clusters:

Designation	R.A.	Decl.	Type	Size (′)	Brtst. Star	Dist. (ly)	dia. (ly)	Number of Stars	Aper. Recm.	Rating	Comments
None											

(continued)

Table 16.9. (continued)

Nebulae:

Designation	R.A.	Decl.	m_v	Size (')	Brtst./ Cent. star	Dist. (ly)	dia. (ly)	Aper. Recm.	Rating	Comments
None										

Galaxies:

Designation	R.A.	Decl.	Type	m_v	Size (')	Dist. (ly)	dia. (ly)	Lum. (suns)	Aper. Recm.	Rating	Comments
NGC 6925	20^h 34.3^m	−31° 59'	Sb	11.3	4.4×1.2	110M	141K	29G	8/10"	****	Nearly edge-on, inclined 84° to line-of-sight
NGC 7029	21^h 11.9^m	−49° 17'	E6	11.5	2.6×1.5	120M	91K	29G	8/10"	****	Very bright nucleus
NGC 7041	21^h 16.5^m	−48° 22'	S(B)O	11.2	3.5×1.4	80M	81K	17G	8/10"	****	In same group as 7029 and 7049
NGC 7049	21^h 19.0^m	−48° 34'	SO	10.3	4.5×2.8	92M	120K	51G	6/8"	****	Very bright nucleus
NGC 7083	21^h 35.8^m	−63° 54'	SBc	11.2	3.5×2.2	130M	132K	45G	8/10"	****	
NGC 7090	21^h 36.5^m	−54° 33'	SBm	10.7	7.1×1.4	31M	64K	4G	6/8"	****	
IC 5152	23^h 02.7^m	−51° 18'	Ir	10.6	5.0×3.7	5M	7.3K	115M	6/8"	****	Nearby dwarf
NGC 7196	22^h 05.9^m	−50° 07'	E3	11.4	2.5×1.9	130M	95K	37G	8/10"	****	

Meteor showers:

Designation	R.A.	Decl.	Period (days)	Duration	Max Date	ZHR (max)	Comet/ Aasteroid	First Obs.	Vel. (mi/ km/sec)	Rating	Comments
None											

Other interesting objects:

Designation	R.A.	Decl.	Type	m_v	Mass (suns)	Spec. class	Dist. (ly)	dia. (ly)	Lum. (suns)	Aper. Recm.	Rating	Comments
ε Indi	22^h 03.4^m	56° 47'	Nearby star	4.69	0.77	K5 V	11.83	0.76	0.15	Eye	****	14th closest star, Binary brown dwarf companion, ~28 and 47 mass Jupiter

TO FIND: Peacock (α Pavonis) is a relatively bright star in a fairly barren area of the sky, so it is easy to identify once you know you are in the correct area. One convenient way to find it is to start with the scorpion's tail. Three stars in the tail, λ (m3.95), κ (m4.13), and ι¹ (m3.03), form a straight line 4° long pointing directly toward Peacock, which is 30° from ι¹. The remaining stars in Pavo are relatively faint, with the next three, β, δ, and η, being about 1.5 magnitudes fainter than α Pavonis. These can be found by going 9.7° south–southeast (PA 168°) to β, then 3.7° west (PA 274°) to δ and finally, 14.7° along roughly the same line (PA 280°) to η. These four stars define the head, breastbone, body center, and the end of the peacock's tail. ε, the 5th brightest star marks the bird's feet, and lies 6.8° south (PA 185°) of δ. The remaining features of the figure can be traced out by comparing the chart to the sky. Good hunting! (Figs. 16.1 and 16.27).

KEY STAR NAMES:

Name	Pronunciation	Source
α Peacock	**pee**-kok	English translation of the constellation name, applied to the star.[*37]

Magnitudes and spectral types of principal stars:

Bayer desig.	Mag. (m$_v$)	Spec. type	Bayer desig.	Mag. (m$_v$)	Spec. type	Bayer desig.	Mag. (m$_v$)	Spec. type
α	1.94	B2 IV	β	3.42	A4 IV	γ	4.21	F6 V
δ	3.55	G5 IV/V	ε	3.97	A0 V	ζ	4.01	K2 III
η	3.61	A0 V	θ	5.71	A9 V	ι	5.47	G0 V
κ	4.40	F5 Ib/II	λ	4.22	B2 II/III	μ¹	5.75	K0 IV
μ²	5.32	K2 IV	ν	4.63	B8 III	ξ	4.35	M1 III[*38]
o	5.06	M2 III	π	4.33	Am	ρ	4.86	Fm
σ	5.41	K0 III	τ	6.25	A6 IV/V	υ	5.14	B8 V
φ¹	4.75	F1 III	φ²	5.11	F8 V[*39]	χ	6.95	K2/3 III[*40]
ψ	–	–[*41]	ω	5.14	K1 III/IV			

Stars of magnitude 5.5 or brighter which have only other catalog designations:

Draper Number	Mag. (m$_v$)	Spec. type	Draper Number	Mag. (m$_v$)	Spec. type
HD 172555	4.78	A7 V	HD 189124	4.95	M6 III
HD 188162	5.224	B9.5 IV	HD 177389	5.31	G8/K0 III/IV
HD 186219	5.39	A4 III	HD 186957	5.41	A0 IV

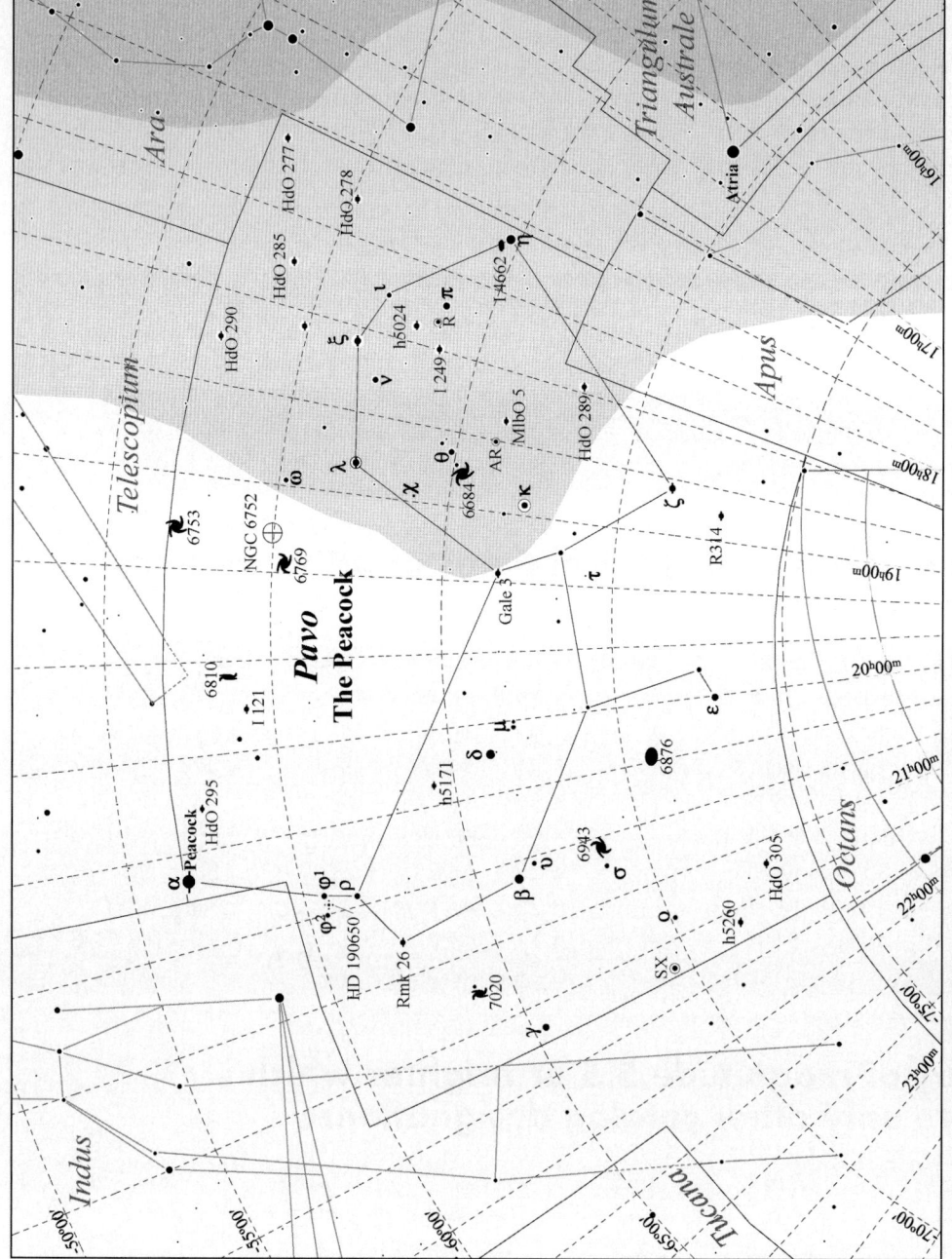

Fig. 16.27. Pavo details.

Table 16.10. Selected telescopic objects in Pavo.

Multiple stars:

Designation	R.A.	Decl.	Type	m₁	m₂	Sep. (″)	PA (°)	Colors	Date/ Period	Aper. Recm.	Rating	Comments
HdO 227	17ʰ 44.9ᵐ	−57° 33′	Aa–B	6.0	12.0	30.0	252	dY,?	1999	3/4″	***	A–a is a close (~0.2″) double
HdO 278	17ʰ 51.6ᵐ	−60° 10′	A–B	5.9	13.0	30.0	350	oY,?	1891	6/8″	***	
HdO 285	18ʰ 10.0ᵐ	−59° 02′	A–B	6.4	12.0	40.0	225	O, O	1900	3/4″	***	
H5024	18ʰ 15.7ᵐ	−63° 03′	A–B	5.7	10.8	45.3	011	oY,?	2000	3/4″	****	
HdO 289	18ʰ 18.0ᵐ	−68° 14′	A–B	6.4	9.9	2.0	305	W,?	1968	4/5″	****	
1249	18ʰ 19.7ᵐ	−63° 52′	A–B	6.2	10.8	7.8	352	Y,?	2000	3/4″	****	
ξ Pavonis	18ʰ 23.2ᵐ	−61° 30′	A–B	4.4	8.1	3.4	156	O, O	1988	2/3″	****	a.k.a GLE 2
HdO 290	18ʰ 29.9ᵐ	−57° 31′	A–B	5.8	12.5	1.1	340	oY,?	1963	9/10″	***	
			A–C	5.8	10.5	30.2	123	oY,?	1999	10×80	****	
MlbO 5	18ʰ 34.2ᵐ	−66° 17′	A–B	7.0	9.1	4.7	292	Y,?	1991	2/3″	****	
ζ Pavonis	18ʰ 43.0ᵐ	−71° 26′	A–B	4.0	12.0	55.6	356	oY,?	1917	3/4″	***	a.k.a h5048
R314	18ʰ 49.7ᵐ	−73° 00′	A–B	6.2	8.1	1.9	271	W,?	1991	5/6″	****	
λ Pavonis	18ʰ 52.3ᵐ	−62° 11′	A–B	4.2	12.4	60.9	205	bW,?	2000	4/5″	****	a.k.a. h5062
Gale 3	19ʰ 17.2ᵐ	−66° 40′	A–B	6.1	6.4	**0.52**	**343**	W, W	**156.76**	18/19″	****	Widening toward a max of 0.572″ in 2057
I 121	19ʰ 50.7ᵐ	−59° 12′	A–B	5.6	7.2	0.8	151	W,?	1991	12/13″	****	
HdO 295	20ʰ 11.1ᵐ	−57° 31′	A–B	6.8	7.8	0.7	269	W,?	1991	13/14″	****	
h 5171	20ʰ 14.6ᵐ	−64° 26′	A–B	6.9	9.8	17.5	306	W,?	2000	2/3″	****	
			A–C	7.0	10.0	33.8	336	W,?	2000	10×80	****	
α Pavonis	20ʰ 25.6ᵐ	−56° 44′	A–BC	1.9	9.1	249.0	081	W,?	1991	7×50	****	
Rmk 26	20ʰ 51.6ᵐ	−62° 26′	A–B	6.2	6.6	2.4	082	pY, pY	2002	3/4″	****	
HdO 305	21ʰ 09.4ᵐ	−73° 10′	A–B	5.8	13.5	7.2	123	Py, Y	1982	7/8″	***	
H5260	21ʰ 25.3ᵐ	−71° 48′	B–C	9.1	10.5	17.5	332	?, ?	1999	15×80	****	
			B–C	6.1	12.0	44.6	271	oY,?	1999	3/4″	***	

Variable stars:

Designation	R.A.	Decl.	Type	Range (m)		Period (days)	F (f₁/f₂)	Spectral Range	Aper. Recm.	Rating	Comments	
R Pavonis	18ʰ 12.9ᵐ	−63° 37′	M	7.5	13.8	229.46	0.48	M3e–M5(II)e	10/12″	***	0°29′ E (PA 084°) from π Pavonis (m4.33). ##	
AR Pavonis	18ʰ 20.5ᵐ	−66° 05′	EA+	7.4	13.6	604.6	–	M3III+Shell	10/12″	****	2° 43′ SSE (PA 154°) of π Pav (m4.33). #	*42
λ Pavonis	18ʰ 52.2ᵐ	−62° 11′	γ Cas	4.0	4.3	–	–	B2II–IIIe	Eye	****	D = 180 ly; Lum = 2,000 suns	
κ Pavonis	18ʰ 57.0ᵐ	−67° 14′	δ Cep	3.9	4.8	9.094	0.46	F5–G5, I–II	Eye	****	D = 250 ly; Lum = 2,000 suns	
SX Pavonis	21ʰ 28.7ᵐ	−69° 30′	SRb	5.3	6.0	50	–	M5–M7III	6×30	****	1°028 ENE (PA 067°) of Pav (m5.05)	

(continued)

Table 16.10. (continued)

Star clusters:

Designation	R.A.	Decl.	Type	m$_v$	Size (')	Brtst. Star	Dist. (ly)	dia. (ly)	Number of Stars	Aper. Recm.	Rating	Comments
NGC 6752	19^h 10.9^m	–59° 59'	Globular	5.3	28	10.5	13.7K	112	–	4/6"	****	Bright, very large, moderate concentration

Nebulae:

Designation	R.A.	Decl.	Type	m$_v$	Size (')	Brtst./ **Cent.** star	Dist. (ly)	dia. (ly)	Aper. Recm.	Rating	Comments
None											

Galaxies:

Designation	R.A.	Decl.	Type	m$_v$	Size (')	Dist. (ly)	dia. (ly)	Lum. (suns)	Aper. Recm.	Rating	Comments
IC 4662	17^h 47.1^m	–64° 38'	Irr	11.3	2.8×1.5	7M	5.7K	120M	8/10"	***	Faint, pretty small. Slightly elongated. 0° 10' ENE (PA 061°) of η Pav (m3.61)
NGC 6684	18^h 49.0^m	–65° 10'	SB0	10.4	4.1×2.6	30M	43K	7G	6/8"	****	Very, bright, large, round. 0°6' SSE (PA 160°) of θ Pav (m5.71)
NGC 6753	19^h 11.4^m	–57° 03'	Sb	11.1	2.5×2.1	130M	96K	50G	8/10"	****	Pr. brt, pr lge, rnd, sm brt nucl. 3°34' NNE (PA029°) of ω Pav (m5.14)
NGC 6769	19^h 18.4^m	–60° 30'	S(B)bp	11.5	2.2×1.5	160M	103K	51G	8/10"	****	Very bright, small, round. Dark lane. 2°28' E (PA 099°) of ω Pav (5.14) *43
NGC 6810	19^h 43.6^m	–58° 39'	Sb	11.4	3.2×0.9	75M	70K	12G	8/10"	***	Small round, bright nucleus. 5°56' WSW (PA 246°) of α Pav
NGC 6876	20^h 18.3^m	–70° 51'	E3	10.8	2.8×2.3	130M	106K	65G	6/8"	****	Pretty bright, small, round. Bright Center. 0°26' WSW (PA 246°) of α Pav (n1.94) *44
NGC 6943	20^h 44.6^m	–68° 45'	S(B)c	11.2	3.9×2.2	130M	150K	45G	8/10"	***	Pretty faint, large, elongated. Small, bright nucleus. 0°26' (PA273°) of α Pav (m5.41)
NGC 7020	21^h 11.3^m	–64° 02'	S0	11.8	3.3×1.6	130M	125K	26G	8/10"	****	Pretty. brt, sm, elong. Very bright center. 2°06' NW (PA 308°) of σ Pav (m5.41)

Meteor showers:

Designation	R.A.	Decl.	Period (yr)	Duration	Max Date	ZHR (max)	Comet/ **Asteroid**	First Obs.	Vel. (mi/ **km**/sec)	Rating	Comments
None											

Other interesting objects:

Name/ Designation	R.A.	Decl.	Type	m_v	Mass (suns)	Dia. (suns)	Dist. (ly)	Planet Mass (J)	Dist. (AU)	Period (days)	Eccentricity	Comments
HD 196050	20^h 37.0^m	–60° 38′	Exoplanet	7.5	1.15	1.0	46.9	>3.27	2.54	1,378	0.228	~Midway between ϕ^1 and ϕ^2 Pav (m4.76 and 5.12)

Fig. 16.28. NGC 6752, Globular Cluster in Pavo. Astrophoto by Christopher J. Picking.

Phoenix (fee-nicks)	**The Phoenix**
Phoenicis (feh-**nye**-siss)	**9:00 P.M. Culmination:** Nov. 19
Phe	**Area:** 469 sq. deg., 37th in size

Keyser and de Houtman created this constellation to represent the mythological bird which supposedly lived in the Arabian desert and had a life span of 500 years. The phoenix resembled an eagle but had a rich plumage of red and gold feathers. At the end of its life, it built its own funeral pyre, which was ignited by the sun. The bird was then consumed by the fire but re-emerged from the ashes rejuvenated and ready to begin another 500 year cycle.

TO FIND: Phoenix has only one 2nd magnitude star, and it is only a little brighter than the 2nd magnitude limit, but it is not difficult to find because of its very bright neighbor <u>Achernar</u> (α Eridani), the 9th brightest star in the entire sky. <u>Achernar</u> is at the mouth of the river Eridanus. χ Eridani (m3.69), the next to last star in the river lies 8.0° north–northeast (PA 027°) of <u>Achernar</u>. About 4.5° west–northwest of this pair, and parallel to it lies another pair of fainter stars, ζ and δ Phoenicis. ζ represents one of the bird's feet, and δ lies in its tail. Extending the χ Eridani to δ Phoenicis line by about 4.8° takes you to β Phoenicis, and continuing the line another 8.4° takes you to α Phoenicis. α, β, and δ Phoenicis basically define the bird's entire body. All that is left is to define the wings by adding γ and λ¹, respectively 5.2° northeast (PA 051°) and 6.2° west–southwest (PA 247°) of β, and then add ζ and η, 8.5° south (PA 178°) and 11.3° south–southwest (PA 196°) of β to mark the bird's feet (Figs. 16.1, 16.29 and 16.30).

KEY STAR NAMES:

Name	Pronunciation	Source
α Ankaa	uhn-**kah**	al-ʿanqâ (Arabic), "a fabulous bird," a modern name

Magnitudes and spectral types of principal stars:

Bayer desig.	Mag. (m$_v$)	Spec. type	Bayer desig.	Mag. (m$_v$)	Spec. type	Bayer desig.	Mag. (m$_v$)	Spec. type
α	2.40	K0 III	β	3.32	G8 III	γ	3.41	K5 II/III
δ	3.93	K0 III/IV	ε	3.88	K0 III	ζ	3.94	B6 V+B0 V
η	4.36	A0 IV	θ	6.09	A+*45	ι	4.69	A2 V
κ	3.93	A7 V	λ¹	4.76	A0 V	λ²	5.51	F6 V
μ	4.59	G8 III	ν	4.97	F8 V	ξ	5.72	Ap
o	6.97	K2 III*46	π	5.13	K1 III	ρ	5.24	F2 III
σ	5.18	B3 V	τ	5.71	G8 III	υ	5.21	A3 V
φ	5.12	A3 V	χ	5.15	K5 III	ψ	4.39	M4 III
ω	6.12	G1 III						

Stars of magnitude 5.5 or brighter which have other catalog designations:

Draper Number	Mag. (m$_v$)	Spec. type	Draper Number	Mag. (m$_v$)	Spec. type
HD 12055	4.82	A2 V	HD 6245	5.39	G8 III
HD 2490	5.42	M0 III	HD 8651	5.42	K0 III

Fig. 16.29a. The stars of Phoenix and adjacent constellations. Astrophoto by Christopher J. Picking.

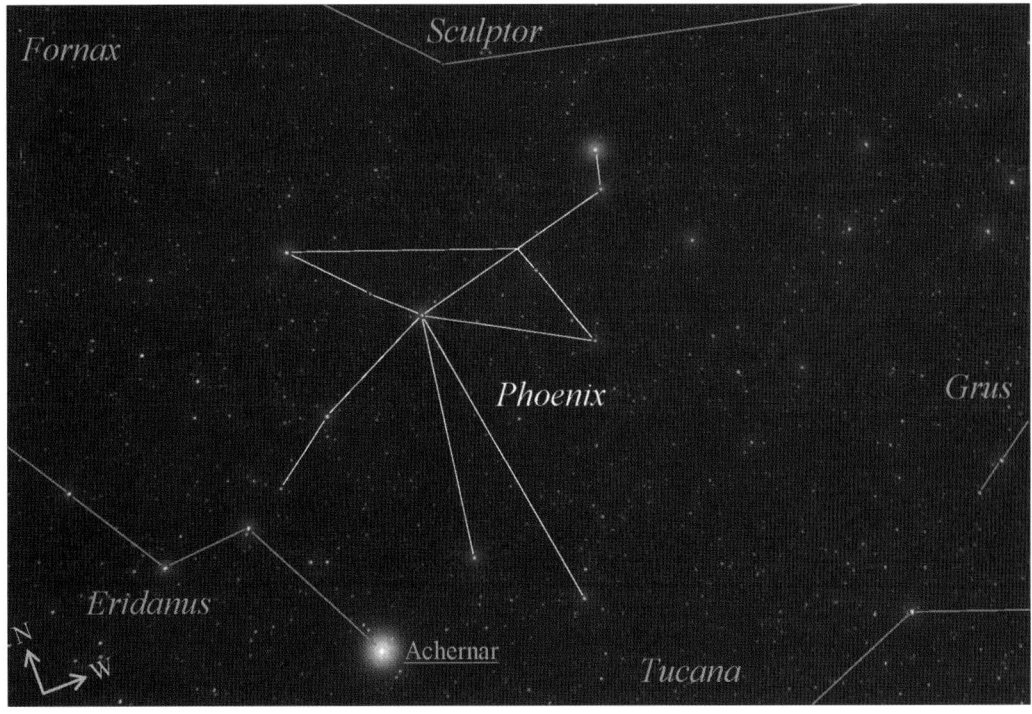

Fig. 16.29b. The figures of Phoenix and adjacent constellations. Astrophoto by Christopher J. Picking.

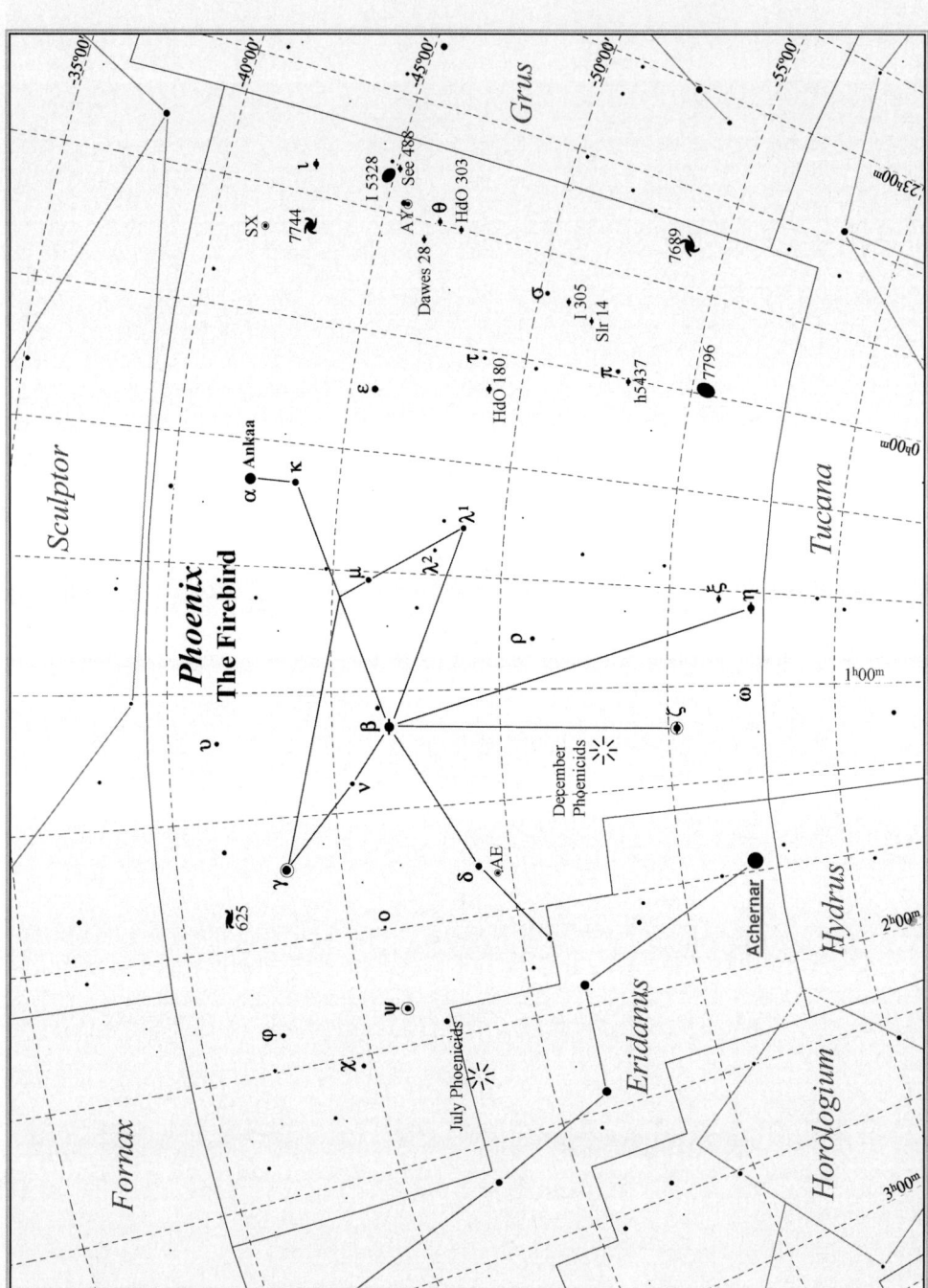

Fig. 16.30. Phoenix details.

Table 16.11. Selected telescopic objects in Phoenix.

Multiple stars:

Designation	R.A.	Decl.	Type	m₁	m₂	Sep. (")	PA (°)	Colors	Date/Period	Aper. Recm.	Rating	Comments
See 488	23ʰ 32.7ᵐ	−45° 06′	A–B	7.0	12.0	18.9	337	W, ?	1897	3/4″	***	a.k.a. B 603
ι Phoenicis	23ʰ 35.1ᵐ	−42° 37′	A–B	4.7	12.8	6.1	273	W, ?	1999	5/6″	***	a.k.a. Δ 251
θ Phoenicis	23ʰ 39.5ᵐ	−46° 38′	A–B	6.5	7.3	3.9	276	oY	2002	2/3″	***	
HdO 303	23ʰ 40.0ᵐ	−47° 20′	A–B	7.2	9.3	2.0	066	oY, ?	1991	4/5″	****	
Dawes 28	23ʰ 43.1ᵐ	−46° 19′	A–B	6.6	11.1	3.8	074	dY, ?	1933	3/4″	****	a.k.a. h5416-AC
			A–C	6.8	10.4	45.4	215	dY, ?	1999	10×70	****	
I 305	23ʰ 48.1ᵐ	−50° 53′	A–B	7.0	11.6	3.1	123	dY, ?	1991	3/4″	****	Slowly widening to reach a max of 1.03″ in 2033. See Note A. A good resolution test
Slr 14	23ʰ 50.6ᵐ	−51° 42′	A–B	7.6	7.9	0.9	072	Y, ?	118.8	10/11″	****	
h5437	00ʰ 00.5ᵐ	−53° 05′	A–B	6.8	10.3	1.9	319	y, ?	2001	5/6″	****	
Hdo 180	00ʰ 06.3ᵐ	−49° 05′	A–B	5.7	11.5	4.1	183	y, ?	2000	3/4″	****	
ξ Phoenicis	00ʰ 41.8ᵐ	−56° 30′	A–B	5.7	10.0	13.1	259	W, ?	1999	18×80	****	a.k.a. h3387
η Phoenicis	00ʰ 43.4ᵐ	−57° 28′	A–B	4.4	11.5	20.1	218	W, ?	1999	4/5″	****	
β Phoenicis	01ʰ 06.1ᵐ	−46° 43′	A–B	4.1	4.2	0.3	258	dY, yW	1999	32″	****	a.k.a. SLR-1-AB
	01ʰ 06.1ᵐ	−46° 42′	AB–C	3.4	11.5	57.9	052	dY, ?	1927	3/4″	****	a.k.a. HJ 3417-AC
ζ Phoenicis	01ʰ 08.4ᵐ	−55° 15′	A–B	4.0	6.8	0.6	095	bW, bW	1999	14/16″	****	a.k.a. Rst 1205 AB
			AB–C	4.0	8.2	6.8	242	bW, bW	2000	2/3″	****	a.k.a. Rmk 2AB-C

Variable stars:

Designation	R.A.	Decl.	Type	Range (mᵥ)		Period (days)	F (f₂/f₁)	Spectral Range	Aper. Recm.	Rating	Comments
SX Phoenicis	23ʰ 46.5ᵐ	−41° 34′	SX Phe	6.8	7.5	0.05496	–	A2V	7×50	*****	Prototype of sub-class of δ Sctui variables with very low metal content
ζ Phoenicis	01ʰ 08.4ᵐ	−41° 34′	EA/DM	3.9	4.4	1.670	0.12	B6V–B9V	Eye	****	
γ Phoenicis	01ʰ 28.4ᵐ	−43° 19′	Lb	3.4	3.5	–	–	M0IIIa	Eye	****	Variations too small for unaided eye, photoelectric magnitudes needed
AE Phoenicis	01ʰ 32.5ᵐ	−49° 32′	EW/KW	7.6	8.3	0.362	–	G0V+G0V	6×30	****	Binary pair of nearly identical stars
ψ Phoenicis	01ʰ 53.7ᵐ	−46° 18′	SR	4.3	4.4	50	–	M4III	Eye	****	
A Y Phoenicis	23ʰ 37.5ᵐ	−45° 34′	Lb	7.8	8.0	–	–	M4III	7×50	****	1° 6′ NNW (PA 355°) of θ Phe (m6.05) and 4.6″ S (PA 182°) of SAO 231707 (m4.74)

(continued)

Table 16.11. (continued)

Star clusters:

Designation	R.A.	Decl.	Size (')	Type	m_v	Brst. Star	Dist. (ly)	dia. (ly)	Number of Stars	Aper. Recm.	Rating	Comments
None												

Nebulae:

Designation	R.A.	Decl.	Size (')	Type	m_v	Brst./Cent. star	Dist. (ly)	dia. (ly)	Aper. Recm.	Rating	Comments
None											

Galaxies:

Designation	R.A.	Decl.	Size (')	Type	m_v	Dist. (ly)	dia. (ly)	Lum. (suns)	Aper. Recm.	Rating	Comments
I5328	23h 33.3m	−45° 01'	2.6×1.6	E4	11.2	130M	98K	46G	6/8"	****	Small, faint, bright nucleus
NGC 7689	23h 33.3m	−54° 06'	2.9×1.9	S(B)c	11.4	82M	69K	15G	8/10"	****	Large, pretty faint, round, patchy arms, small bright nucleus
NGC 7744	23h 45.0m	−42° 55'	2.4×1.9	S(B)0	11.5	130M	91K	34G	8/10"	****	Small, bright, elongated, small, bright nucleus
NGC 7796	23h 59.0m	−55° 27'	2.2×1.9	E2	11.5	150M	96K	45G	8/10"	****	Very small, bright, round, very bright nucleus
NGC 625	01h 35.1m	−41° 26'	3.4×1.3	SBm	11.0	83M	82K	22G	6/8"	****	Large, very faint, very elongated

Meteor showers:

Designation	R.A.	Decl.	Period (yr)	Duration	Max Date	ZHR (max)	Comet/Asteroid	First Obs.	Vel. (mi/km/sec)	Rating	Comments
July Phoenicids	02h 08m	−48°	–	06/24 07/18	07/12	3–4	–	1957	29/47	III	A.A. Weiss (Adelaide Obs.), radio echo observations
Dec. Phoenicids	01h 02m	−53°	–	12/04 12/06	12/05	Var	Blanpain, 2003WY25	Nov. 5, 1956 11/18	Var	III	Comet Blanpain (D 1819W$_1$) was discovered on 11/28/1819 but was observed for only 2 months before being lost with only an approximate orbit calculated

Other interesting objects:

Designation	R.A.	Decl.	Type	m_v	Mass (suns)	dia. (suns)	Dist. (ly)	Planet Mass (J)	Dist. (AU)	Period (days)	Eccentricity	Comments
HD 142	00h 06.3m	−49° 05'	Exoplanet	5.7	1.24	1.7	25.64	>1.28	1.04	349.6	****	

Tucana (too-**kan**-uh)	The Toucan
Tucanae (too-**kan**-ee)	9:00 P.M. **Culmination:** Nov. 2
Tuc	**Area:** 295 sq. deg., 48th in size

The toucan is a tropical bird of South America, found mostly in the forested areas of Argentina, Bolivia, and Brazil. Its most distinguishing characteristic is it extremely large beak, which can be over half the length of its body. Although the huge beak makes the bird appear unbalanced, the beak is actually very light, consisting of a thin outer shell, with a porous inside. This adaptation allows it to safely perch on larger branches and still reach soft fruits on the thinner branches. The bills of some species are very colorful. The larger species can measure up to 29 inches from the tip of the beak to the end of the tail. It was certainly spectacular enough to catch the attention of Keyser and de Houtman.

TO FIND: As with Phoenix, <u>Achernar</u> (α Eridani) can be used to find the brighter stars of Tucana. However, Tucana is even fainter than Phoenix, and thus harder to recognize. The same pair of stars, χ and α Eridani, are the starting point. A line from χ to α Eridani extended by 6° ends near ι Tucanae, and the same line extended another 4.6° ends near ρ Tucanae. The unaided eye double β^{12} and β^3 lies 4.3°, almost the same distance from ι as ρ, but 2.6° farther north. These three stars define the tail of the toucan. A pair of stars, ζ and ε, lie 2.3° south–southwest (PA 212°) and 4.3° (PA 228°), respectively. These three stars define the rest of the toucan's body. κ and $\lambda^{1,2}$ are located 4.7° southeast (PA 140°) and south–southeast (PA 165°) of ρ Tucanae and denote the bird's feet. At the other end of the constellation, α Tucanae lies at the point of the bird's huge beak. If you have previously found Indus, a line from δ Indi through ε Indi (the arrowheads) extended 4° ends at α Tucanae. γ Tucanae, 7.8° east (PA 081°) of α, and a 6th magnitude star just 2° south of γ form the other end of the beak. If you have not previously found Indus, you can find α Tucanae by looking 13.4° south (PA 174°) of α Gruis (m1.74). The remaining stars connecting the head and beak to the body are faint and can be found by comparing the chart to the sky (Figs. 16.1 and 16.31).

KEY STAR NAMES: None.

Magnitudes and spectral types of principal stars:

Bayer desig.	Mag. (m_v)	Spec. type	Bayer desig.	Mag. (m_v)	Spec. type	Bayer desig.	Mag. (m_v)	Spec. type
α	2.87	K3 III	$\beta^{1,2}$	3.68	B9/A2 V*48	β^3	5.07	A0 V
γ	3.99	F1 III	δ	4.51	B8 V	ϵ	4.49	B9 IV
ζ	4.223	F9 V	η	5.00	A1 V	θ	5.35	G5 III
ι	5.36	G5 III	κ	4.25	F6 IV	λ^1	6.22	F7 IV/V
λ^2	5.45	G7 III	μ	6.42	K3.5 III	ν	4.91	M4 III
ξ	–	–*49	o	–	–*49	π	5.50	B9 V
ρ	5.38	F6 V						

Stars of magnitude 5.5 or brighter which have only catalog designations:

Draper Number	Mag. (m_v)	Spec. type
HD 212330	5.30	F9 V

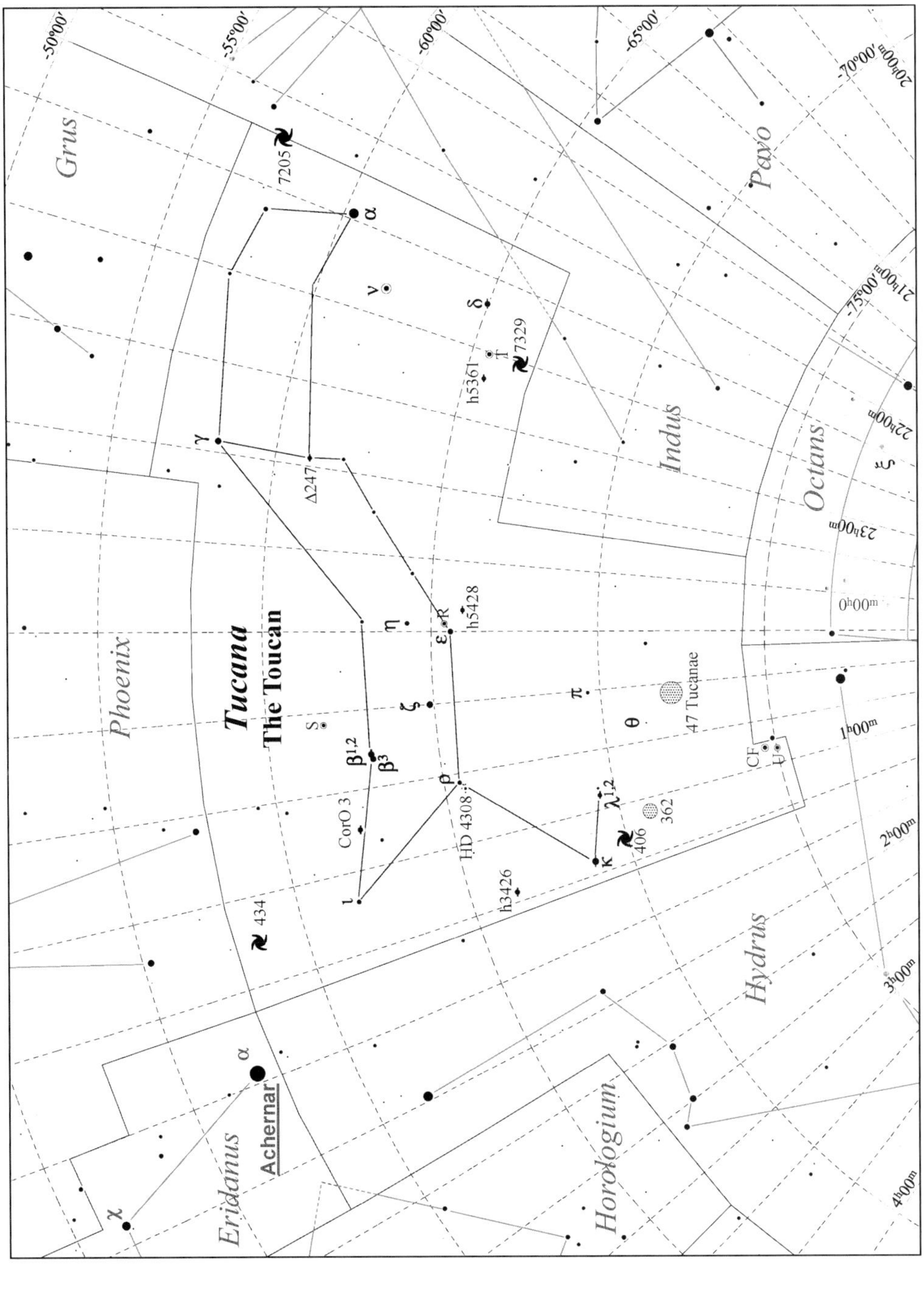

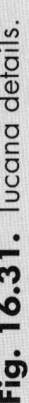

Fig. 16.31. Tucana details.

Table 16.12. Selected telescopic objects in Tucana.

Multiple stars:

Designation	R.A.	Decl.	Type	m₁	m₂	Sep. (')	PA (°)	Colors	Date/Period	Aper. Recm.	Rating	Comments
δ Tucanae	22ʰ 27.3ᵐ	−64° 58'	Aa–B	4.5	8.7	6.9	280	W, pR	2000	2/3"	*****	Showcase pair. Aa is too close for amateur instruments. a.k.a. Chr 188-Aa
h5361 (Tucana)	22ʰ 47.6ᵐ	−65° 34'	A–B	6.6	11.2	84.0	044	oY, ?	1999	3/4"	***	
Δ247 (Tucana)	23ʰ 18.0ᵐ	−61° 00'	A–B	6.9	8.2	50.0	293	Y, yW	2002	6×30	***	a.k.a. Δ247
h5428 (Tucana)	23ʰ 53.7ᵐ	−65° 56'	A–B	6.7	12.5	12.2	112	Y, ?	1999	4/5"	***	
β¹,² Tucanae	00ʰ 31.5ᵐ	−62° 57'	A–B	4.4	13.5	2.6	155	pY, ?	1999	7/8"	***	a.k.a. B 7-AB
			A–C	4.3	4.5	26.6	168	pY, yW	2002	0.2	*****	Showcase pair, a.k.a. LCL 119-AC
			C–D	4.6	6.5	0.5	270	W, W	2001	19/20"	****	a.k.a. I 260-CD
CorO 3 (Tucana)	00ʰ 44.5ᵐ	−69° 30'	A–B	6.3	8.0	2.3	066	Y, Y	2001	4/5"	****	
λ¹ Tucanae	00ʰ 52.6ᵐ	−69° 30'	A–B	6.7	7.4	20.4	081	y, W	1998	15×80	****	a.k.a. or Δ2
κ Tucanae	01ʰ 15.8ᵐ	−68° 53'	A–B	5.0	7.7	**5.35**	**233**	Y, O	**1,222.40**	2/3"	*****	Showcase pair, a.k.a. h2423-AB
			A–CD	5.0	7.9	319.5	310	Y, ?	1991	6×30	****	a.k.a. h2423-A-CD.
			C–D	7.8	8.4	1.0	280	Y, ?	2001	9/10"	****	a.k.a. h2423-CD
h3426 (Tucana)	01ʰ 17.1ᵐ	−66° 24'	A–B	6.4	8.3	2.4	330	W, W	1999	4/5"	****	

Variable stars:

Designation	R.A.	Decl.	Type	Range (m.)		Period (days)	F (fᵣ/fₜ)	Spectral Range	Aper. Recm.	Rating	Comments
ν Tucanae	22ʰ 33.0ᵐ	−61° 59'	Ib	4.8	4.9	–	–	M4III	Eye	***	Variation is at the limit of detection by the unaided eye
T Tucanae	22ʰ 40.6ᵐ	−63° 33'	M	7.5	13.8	250.3	0.46	M3IIe–M6IIe	10/12"	****	2° 59' ESE (PA 118°) of α Tuc (m2. 87).##
R Tucanae	23ʰ 57.5ᵐ	−65° 23'	M	8.5	15.2	286.06	0.41	M5e	20/22"	****	0° 19' NW (PA 307°) of ε Tuc (m4.49).##
S Tucanae	00ʰ 23.1ᵐ	−61° 40'	M	8.2	15.0	240.71	0.44	M3e–M5II–Ib	18/20"	****	1° 45' NW (PA 320°) of β Tuc (m5.07).##
CF Tucanae	00ʰ 53.5ᵐ	−74° 39'	EA/RS	7.4	7.8	2.798	0.10	F8V+K1IV	7×50	****	0° 24' E (PA 048°) of λ Hya (m5.09)
U Tucanae	00ʰ 57.7ᵐ	−75° 00'	M	8.0	14.8	264.8	0.46	M3e–M7e	16/18"	****	0° 34' E (PA 099°) of λ Hya (m.5.09).##

*50

Star clusters:

Designation	R.A.	Decl.	Type	mᵥ	Size (')	Brtst. Star	Dist. (ly)	dia. (ly)	Number of Stars	Aper. Recm.	Rating	Comments
47 Tucanae, NGC 104	00ʰ 24.1ᵐ	−72° 05'	Globular	4.0	50'	11.7	15K	220	1M	Eye	****	a.k.a. Caldwell 106
NGC 362	01ʰ 03.2ᵐ	−70° 51'	Globular	6.8	12'	12.7	29K	101	29K	7×30	****	a.k.a. Caldwell 104

Nebulae:

Designation	R.A.	Decl.	Type	m_v	Size (')	Brst./ Cent. star	Dist. (ly)	dia. (ly)	Aper. Recm.	Rating	Comments
None										****	

Galaxies:

Designation	R.A.	Decl.	Type	m_v	Size (')	Dist. (ly)	dia. (ly)	Lum. (suns)	Aper. Recm.	Rating	Comments
NGC 7205	$22^h 08.6^m$	$-57° 27'$	Sb	11.1	3.8×2.2	60M	66K	10G	6/8"	****	Exactly on the border with Indus
NGC 7329	$22^h 40.6^m$	$-66° 29'$	SBbc	11.6	3.7×2.7	130M	140K	30G	8/10"	****	Very elongated
Small Magellanic Cloud	$00^h 52.7^m$	$-72° 50'$	SBmp	2.3	4.6°×2.7°	190K	15K	350M	Eye	*****	Dozens of NGC objects in the SMC can be seen with binoculars or small/medium scopes
NGC 406	$01^h 07.4^m$	$-69° 53'$	Sc	12.3	3.0×1.3	56M	49K	3G	12/14"	****	Large but faint
NGC 434	$01^h 12.2^m$	$-58° 15'$	S(B)b	12.1	2.2×1.2	200M	128K	46G	10/12	****	

Meteor showers:

Designation	R.A.	Decl.	Period (yr)	Duration	Max Date	ZHR (max)	Comet/ Asteroid	First Obs.	Vel. (mi/ km/sec)	Rating	Comments
None											

Other interesting objects:

Designation	R.A.	Decl.	Type	m_v	Mass (suns)	Dia. (suns)	Dist. (ly)	Planet Mass (J)	Dist. (AU)	Period (days)	Eccentricity	Comments
HD 4308, SAO 248244	$00^h 44.6^m$	$-65° 39'$	Exoplanet	6.55	0.904	0.92	71.2	0.0467	0.118	15.56	0	Companion believed to be a brown dwarf. 0° 17.4' southeast (PA 129°) of ρ Tucanae

Fig. 16.32. Globular cluster 47 Tucanae (NGC 104). Astrophoto by Christopher J. Picking.

Fig. 16.33. Small Magelanic Cloud and 47 Tucanae. Astrophoto by Christopher J. Picking.

NOTES:

1. The original name meant "bee of paradise," and was probably a misspelling avis Indica, Latin for Indian bird. An alternate explanation says that the name came from apodi (Greek) Indica, meaning a footless bird of paradise. This name may have been given because first specimens brought back to Europe were only the skins of the birds with the feet removed.
2. Lacaille designated each component of δ Apodis individually, but at a separation of 103″, most people can see only one star. Both are giants of luminosity class III, so both spectra contribute to the visual appearance.
3. ι Apodis is a very close double (~0.1″) which is beyond the resolution of most amateur astronomers and difficult for professionals to take clean spectra of the individual components.
4. κ was the last letter assigned by Lacaille in Apus, and he assigned it to two stars close together. The "1" and "2" superscripts were added later.
5. R Apodis was given this designation because it was thought to be variable, but it was later found to be not variable.
6. Lacaille designated only one star as ζ Muscae. Baily thought Lacaille had erred in which star was designated, and added another (brighter) star as ζ¹ Muscae. He called Lacaille's ζ Muscae "ζ² Muscae."
7. Lacaille assigned the designation ι Muscae to two adjacent stars. The "1" and "2" superscripts were added later.
8. Lacaille thought this star was 6th magnitude and assigned it the letter κ, the last designation in Musca. Gould thought it was fainter than 7th magnitude and removed the letter. The designation has now almost disappeared from astronomical literature.
9. Lacaille observed these two stars, but placed them between Argo Navis and Centaurus without letters. Baily placed them in Musca, and assigned the letters λ and μ.
10. Lacaille assigned the single letter π to two adjacent stars in Chameleon. Gould moved π² (HD 105822) to Musca and left it unlettered because of its faintness. Lacaille did not use any subsequent letters in Musca.
11. HD 105340 was first labeled λ Chameleontis but was reassigned to Musca by Gould and left unlettered.
12. Lacaille assigned the single designation η Trianguli Australis to both stars. Baily was inconsistent, labeling only the brighter star in one catalog, and giving a single letter to both in another. Gould kept both designations because he felt that stars this close to the south celestial pole should be highlighted.
13. Lacaille designated this star λ Trianguli Australis, but Gould thought it was too faint and dropped the designation.
14. The star, originally designated λ Chameleontis, was moved by Gould to Musca and left unlettered because of its faintness.
15. The letters ξ and o were not used in Chameleon by Lacaille.
16. Lacaille designated two stars in Chameleon as π. One of them, π², was moved to Musca but left unlettered.
17. γ Volantis is a double star with the brighter component only ~1.5 magnitudes (~4 times) brighter than the fainter, so spectral lines of both show up in a combined spectra.
18. κ Volantis is a triple star in which the two brightest components differ by only 0.3 magnitudes (~1.3 times), so the spectral lines are about equally prominent in the combined spectra.
19. U Volantis is a known eclipsing variable of the Algol type. Since there is no established period, it would be a nice project for some southern hemisphere amateur variable star observer to determine its period.
20. Astronomers believe that the distorted shape of NGC 2442's arms is due to a collision with a nearby galaxy. Both NGC 2434 and 2397 appear to be part of the NGC 2442 group. However, neither one shows evidence of a collision. Long exposures show a very faint galaxy directly behind NGC 2442 which is believed to be the likely cause of the distorted arms. There are also three other galaxies in this area which show similar red-shifts and thus are at about the same distance and part of the NGC 2442 group. Some red-shift measurements have indicated a distance of 62Mly for NGC 2397. If this is correct, it would not be part of the NGC 2442 group. A Type II supernova was discovered in NGC 2397 on March 3, 2006. Measurements of it may clarify the actual distance of NGC 2397.
21. Kepler called this constellation "Xiphias" or the Sword. This caused some confusion and Dorado has sometimes been erroneously called the Swordfish. The Sword was an appropriate name for the four stars (γ, α, β, and η²) which lie in almost a straight line. As mentioned in the finding directions, this line of stars can help you recognize Dorado.

22. Lacaille did not designate any stars in Dorado as ι, ξ, or o. He did designate one as μ, but it was later dropped as being too faint (magnitude 8.92!) to warrant a letter. The remaining letters, ρ through ω, were also not used.

23. Although S Doradus is most often about magnitude 8.8 to 9.2, it has occasionally reached magnitude 8.4, at which the calculated luminosity is over 1,000,000 suns, making it one of the most luminous stars known. The main variability is intrinsic, but there is also some indication that S Doradi may be an eclipsing binary with period of about 40 years.

24. SN1987A exploded on February 23, 1987 and was discovered independently by two observers at the Las Campanas Observatory in Chile, and by an amateur astronomer in New Zealand. It was about magnitude 5.1 at its discovery, and continued to brighten for 80 days, when it reached a maximum of magnitude 2.9, the brightest extragalactic supernovae ever observed. Because of its location in the Large Magellanic Cloud, 166,000 light-years distant, and consequent brightness, it became the most studied supernovae in history. The biggest surprise from this scrutiny was that the precursor star was a fairly ordinary looking blue giant instead of the red supergiant which theory predicted. This finding resulted in significant revisions to the theory of supernovae.

25. NGC 1816, illustrates the fact that most of the clusters in the LMC appear to have some nebulosity because the angle at which we view the galaxy makes nebulae in front or behind them seem to be associated with them. Because of the distance, all the clusters appear small and faint, and use of the largest telescopes available is recommended.

26. NGC 2070 is a combination of an open cluster and a surrounding emission nebula. This complex is visible to the unaided eye and was assigned the Flamsteed number of 30 Doradus. The cluster is now called the 30 Doradus Cluster. Because the surrounding nebula has complex looped ribbon-like filaments it is called the "Tarantula Nebula." The Tarantula nebula is the largest of its type known. Even at the distance of the Large Magellanic Cloud, it spans 40×25′. If this nebula were at the distance of the Orion Nebula, it would span several degrees and be bright enough to cast distinct shadows on moonless nights. Imagine what an impressive sight that would be!

27. NGC 1936 is the brightest section in a compact group (NGC 1929, 1934, 1935, and 1936) in the Large Magellanic cloud. This complex is illuminated by an association of hot O and B stars.

28. The Large Magellanic Cloud is a satellite of the Milky Way galaxy and is the nearest of the medium sized galaxies to our own. It has a mass of about 1/10th of the Milky Way galaxy and is about 180,000 light-years distant. There are dozens of clusters and nebulae here which are visible in medium to large amateur telescopes.

29. Lacaille observed these two stars, but put ξ in Microscopium and left both unlettered. Gould assigned one ξ to one and moved it to Microscopium. He named the other o, but it later lost that designation to Grus and designated it ξ.

30. Lacaille did not use any letters beyond φ in Grus.

31. Lacaille did not use ξ or o in Hydrus.

32. Lacaille assigned the designation ρ Hydri to a star he observed but apparently made an error in its coordinates. No subsequent observer has been able to positively identify the star he designated as ρ.

33. Lacaille assigned the designation λ to two adjacent stars. Both stars are now in Pavo.

34. Lacaille inadvertently assigned ν Indi to two stars at opposite ends of the constellation. One of these was latter included in Pavo as an unlettered star.

35. Lacaille did not use the letter ξ in Indus.

36. Lacaille did not assign letters beyond ρ in Indus.

37. A standard set of 57 (or 58 if Polaris is included) navigational stars has been widely used since they appeared in the 1802 edition of Nathiel Bowditch's American Practical Navigator. The stars in this list were chosen to be easily recognizable, reasonably bright, and to be evenly distributed over the entire celestial sphere. In 1930, the Royal Air Force of the UK insisted that all its navigational stars had to have a proper name. The Nautical Almanac Office coined the name of this star and that of ε Carinae, q.v.

38. ξ Pavonis was observed by Lacaille but left unlettered. Gould added the designation.

39. Lacaille gave a single designation to these two stars.

40. χ Pavonis was included by Lacaille, but Gould dropped it from his catalog, and the designation is seldom used today.

41. Lacaille did not use the letter ψ in Pavo.

42. AR Pavonis is a symbiotic system consisting of a red giant with a mass about 2.5 times our sun, and a hot companion about the same mass as our sun. The pair is so close and the red giant so extended that its outer atmosphere is escaping and being captured by the more dense companion. The red giant eclipses

the smaller star once each orbit. In addition, there is a disk of gas surrounding the large star which adds complexity to the system, and the spectrum that astronomers see.

43. NGC 6769 is accompanied by two fainter companions, NGC 6970 and 6971. The latter may be seen in larger telescopes. All three are at about the same distance, and NGC 6769 and 6970 are interacting.

44. NGC 6876 is a large elliptical galaxy which is in a group including nearby NGC 6877 and IC 4972. Also in a group within about ½ degree of NGC 6876 are NGC 6872 and 6880 as well as the fainter galaxies IC 4945, 4960, 4967, 4970, 4981, and 4985.

45. Lacaille catalogued and lettered θ Phoenicis. Gould thought that Lacaille had actually meant the designation for a somewhat brighter star about 1° farther north. Because of the confusion, Gould, Brisbane and Baily all left both stars unlettered, and the confusion over this designation still exists.

46. Lacaille made a time error in measuring the position of o Phoenicis. Baily corrected the error, but both he and Gould dropped the letter in their catalogs.

47. Comet Blanpain's calculated orbit indicated a period of ~5 years and a close encounter with Jupiter. On Dec. 5, 1956, an unexpected meteor shower of ~100/h from the constellation of Phoenix was observed in New Zealand, Australia, the Indian Ocean, and Africa. Radar observations by A. A. Weiss of Adelaide Observatory enabled computation of an approximate orbit for the meteoroid stream. This stream was later associated with the newly discovered minor planet 2003 WY_{25} and orbit backtracking in turn connected it to Comet Blanpain.

48. $\beta^{1,2}$ and β^3 Tucanae form a naked-eye double. $\beta^{1,2}$ is actually a quadruple star, and the magnitude given represents the sum of the brightness of its four components. The 1 and 2 designations refer to the wider pair (A–C), which are almost matched in brightness and temperature, and both spectra are given.

49. Lacaille did not use either ξ or o in Tucana. Gould assigned ξ to globular cluster NGC 104, which Lacaille had observed but left unlettered because of its nebulosity.

50. CF Tucanae is a binary consisting of a G0 V star and a K4 IV star. The latter has an unusually low iron abundance, only 1/10 to 1/20th that of the sun.

CHAPTER SEVENTEEN

Hevelius, The Merchant Astronomer

March–May, August–October

CANES VENATICI	LACERTA	LEO MINOR	LYNX
the Hunting Dogs	the Lizard	the Little Lion	the Lynx

SCUTUM	SEXTANS	VULPECULA
the Shield	the Sextant	the Fox

Born on January 28, 1611, only 3 or 4 months after Galileo made his first astronomical observations, Johannes Hevelius was the first major astronomer to live his entire life in the telescopic era. In spite of that, many of his observations, especially those of the star positions were made with non-telescopic instruments.

The son of a prosperous business and brewing family of the autonomous city of Danzig, Johannes received 9 years of education in Poland to prepare him for the family business. However, when he returned home, he convinced his parents to let him study mathematics, astronomy, engraving, and instrument making under Peter Krüger. While he carried out his studies, he assisted Professor Krüger in making astronomical observations. He was so fascinated with this work that he wanted to pursue a career in astronomy but, to please his parents, he went to London and studied law instead.

When he returned to Danzig, he entered the family business, and after a couple of years, was accepted by the Brewer's guild. For the next few years, he continued to make

P. Simpson, *Guidebook to the Constellations: Telescopic Sights, Tales, and Myths*,
Patrick Moore's Practical Astronomy Series, DOI 10.1007/978-1-4419-6941-5_17,
© Springer Science+Business Media, LLC 2012

some astronomical observations, but they were severely limited because of his business duties. It was not until 1639, in response to one of Professor Krüger's deathbed wishes that Hevelius turned back to astronomy. His interest in astronomy was sealed when he observed an eclipse of the sun on June 1, 1639.

Although his wife took over much of the day-to-day administration of the brewery so he could observe more, Johannes stayed active in the Brewer's Guild and became its leader in 1643 and remained in that position for many years. The family must have been great brewers for Gdansk still has a Hevelius Brewing Co. Ltd. Hevelius Classic Lager and Hevelius Kaper are still available there as well as in the USA through importers.

After his return to astronomy, Hevelius decided to embark on the ambitious project of recharting the heavens to improve upon Tycho Brahe's 1602 catalog of 777 stars. He proceeded to build an observatory with wooden quadrants and sextants, the standard of the time, but found that they could not achieve the desired accuracy, so he built larger metal instruments, and even constructed a type of timer which counted pendulum swings for precise time measurements. He was also unable to buy a telescope which suited him, so he made his own.

While the construction of a new observatory and new instruments were in progress, Hevelius undertook a diversion in the form of a series of observations of the moon. His studies of instrument making and engraving became extremely important because he published an atlas of the moon, including plates for each day of the lunar cycle, doing all the observing, drawing, and engraving himself. His 1647 Selenographia was considered a beautiful masterpiece.

Finally, in 1657, his new observatory was ready, and he resumed his measurements of stellar positions. These were interrupted in 1663 by the death of his first wife, when Hevelius had to take over the business and household tasks that she had been doing. He remarried the following year, and his new wife, although 36 years his junior proved to be an excellent choice. A merchant's daughter like his first wife, she was able not only to care for their children, but also to assist him in both his business and astronomical affairs. She became an excellent observer herself, and proved invaluable to Hevelius.

Hevelius published several important works during the 1660s and 1670s. Unfortunately, his observatory and several other buildings burned down in 1679 along with over 40 houses in the adjoining town. Most of his records were destroyed, but the manuscript of his star catalog was saved, apparently by his oldest daughter. Already famous for his astronomical observations, Hevelius soon began constructing a new observatory with the financial help of several wealthy people. Among these were King Louis XIV of France and King Jan III Sobieski of Poland, who also awarded Hevelius a lifetime stipend to support the continuation of his astronomical work.

A constellation, like beauty, is in the eye of the beholder. Also like beauty, what form the eye picks out in a group of more-or-less randomly distributed stars is influenced by his culture and circumstances. The human eye is very good at picking out patterns from randomness, especially straight lines, arcs, and geometric forms, such as triangles, squares, or circles. Hevelius lived in a period when classical mythology was still very familiar to educated people, but new developments in science and politics were becoming increasingly important. The new constellations visualized by Hevelius reflect these disparate influences on his thinking.

The first constellation devised by Hevelius was Scutum Sobiescianum in honor of the Polish King Jan III Sobieski. This honor was partly because of King Jan's leadership and heroism in routing the Ottoman army on September 12, 1683. This surprising defeat may have been the turning point in reversing the gains of the Ottoman Empire and saving the Holy Roman Empire and the rest of Europe from being overrun by the Muslims. Another important reason for the honor of having a constellation named for him was King Jan's patronage in helping to rebuild and re-equip Hevelius' observatory. Today, this constellation is known simply as Scutum.

Hevelius either created, or popularized twelve new constellations which were included in a massive volume published by his wife in 1690. Hevelius had died on January 28, 1687, having lived exactly 76 years, but he had finished the manuscript of his catalog of over 1,500 stars, the engravings for 54 constellations, and a manuscript on spherical trigonometry, a constellation list, a list of differences between his measurements and those of Tycho Brahe, and corrections to Ptolemy's catalog. All these were included in the three part volume published by his wife.

The new constellations invented by Hevelius were:

Asterion	One of the Hunting Dogs
Cerberus	The Three-Headed Hound of Hades
Chara	The Second Hunting Dog
Lacerta sive Stellio	The Lizard or Spotted Lizard
Leo Minor	The Small Lion
Lynx	The Lynx
Mons Maenalus	Mount Maenalus
Scutum Sobieski	Sobieski's Shield
Sextans Uraniae	The Astronomical Sextant
Triangulum Minus	The Small Triangle, and
Vulpecula cum Anser	The Fox with the Goose.

Of these constellations, some have been merged into larger ones, or disappeared altogether. Asterion and Chara were joined to become Canes Venatici. Anser, the goose was originally shown being held in the mouth of Vulpecula, the fox, but now is recognized only as a part of Vulpecula. King Sobieski's name was dropped so that only Scutum, his shield is remembered today, Lacerta sive Stellio became just Lacerta and Sextans Uraniae became Sextans. Cerebus, Triangulum Minus, and Mons Maenalus (named for the son of King Arcas of Arcadia, who is represented by Ursa Minor) were dropped entirely.

CANES VENATICI (for details, see Chap. 7. Hospitality and Homicide)

When Hevelius created the constellations of Canes Venatici and Leo Minor, he did his best to connect them to traditional classical mythology. Although celestial cartographers had long shown Boötes, the bear driver, holding the leash of one or more dogs, the number of dogs varied, and they were not considered a separate constellation until Hevelius published his celestial charts.

LACERTA (lah-**sir**-tah or luh-**sir**-tah),   **The Lizard**
LACERTAE (lah-**sir**-tie or luh-**sir**-tie)   **9:00 P.M. CULMINATION:** Oct. 12
Lac   **AREA:** 201 sq. deg., 68th in size

In Demeter's search for her abducted daughter (see Chap. 7), she neglected not only her duties to make the crops grow, but also her own appearance and health. In one episode during this search, a kind lady offered Demeter a bowl of soup. When Demeter tasted the soup, it was so delicious that she noisily gulped it down. Stellio, a young boy standing nearby was so amused by this that he laughed out loud. Angered by this insult, Demeter turned on the boy, hurled the bowl with its remaining contents at him, and immediately changed him into a lizard. The drops of soup which had splattered the boy's clothing became small spots on the lizard's skin. The descendants of Stellio continue to carry the star-like spots, and now inhabit the Eastern Mediterranean Coast from Greece to Northeast Africa. Perhaps Hevelius tried, but failed, to find a more suitable mythological source, because the only explanation he gave for the choice of a lizard was that there was so little space between the classical constellations that it was the only thing which would fit.

Lacerta was formed in an area of unformed stars (stars not officially belonging to any previously existing constellation) between Andromeda and Cassiopeia on the east, Cygnus on the west, Cepheus on the north, and Pegasus on the south. Hevelius also had an alternate name for this group, *Stellio* (the Stellion) a Mediterranean newt with star-like spots on its back.

TO FIND: Lacerta consists of a few very faint stars filling a stellar void surrounded on the north by Cepheus, and going clockwise, Cygnus, Pegasus, Andromeda, and Cassiopeia. Before trying to find Lacerta, you should study Fig. 3.1, the Andromeda Group Finder, Fig. 12.2a, the Stars of Cygnus and surrounding constellations and Fig. 12.2b, Cygnus and surrounding constellations. The brightest star, α Lacertae, is only magnitude 3.77. It lies on a line from Deneb to 3 Andromedae, a faint (mag 4.65) star that marks the northernmost point in the rock to which Andromeda is chained. It is about 19° from <u>Deneb</u> and 5° from 3 Andromedae. α Lacertae and the three nearby stars form a diamond shape somewhat larger than, but fairly similar in shape and brightness to, the four main stars in Delphinus, the Dolphin. The rest of constellation figure shown on the chart is fairly difficult to see, and certainly requires very dark skies to trace out. Hevelius must have had very sharp eyes (or a very good imagination). An alternate way to find Lacerta is to imagine a line (13° 16′) between the two stars at the ends of the "W" (ε and β Cas) of Cassiopeia and extend it 1¼ times (16° 33′) its own length. This line ends near the southern part of the above diamond shape (Fig. 3.1, 12.1, 12.2 and 17.1).

KEY STAR NAMES: None

Magnitudes and spectral types of principal stars:

Bayer desig.	Mag. (m_v)	Spec. type	Bayer desig.	Mag. (m_v)	Spec. type
α	3.76	A1 V	β, 3	4.42	G9 III

Stars of magnitude 5.5 or brighter which have Flamsteed, but no Bayer designations:

Desig-nation	Mag. (m_v)	Spec. type	Desig-nation	Mag. (m_v)	Spec. type	Desig-nation	Mag. (m_v)	Spec. type
1	4.14	K3 III	2	4.55	B6 V	4	4.55	B9 Iab
5	4.34	M0 III	6	4.53	B2 IV	10	4.89	O9 V
11	4.50	K3 III	12	5.25	B2 III	13	5.11	M4 III
15	4.95	M0 III	16	5.60	B2 IV*[1]			

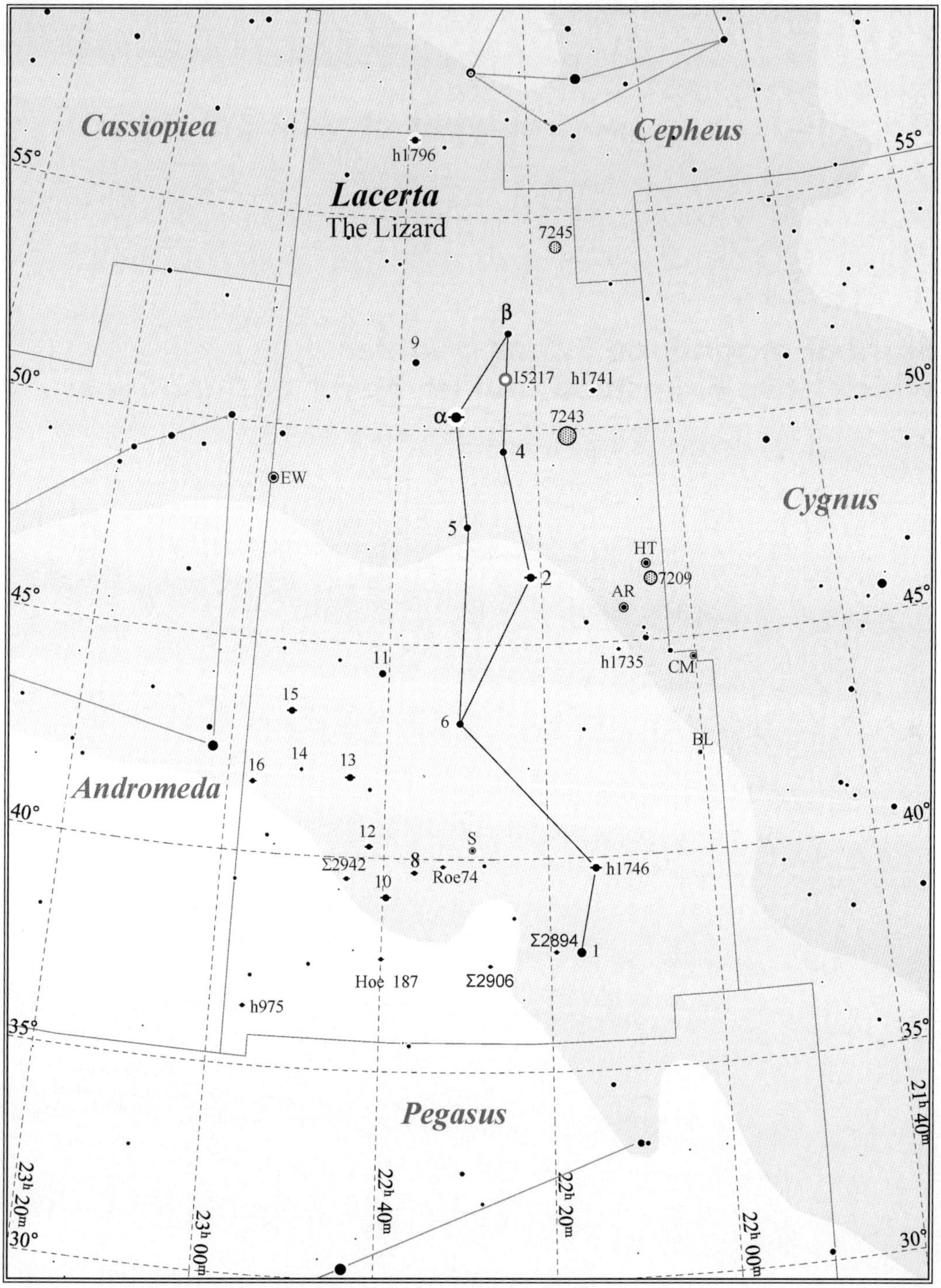

Fig. 17.1. Lacerta details.

Table 17.1. Selected telescopic objects in Lacerta.

Multiple stars:

Designation	R.A.	Decl.	Type	m_1	m_2	Sep. (")	PA (°)	Colors	Year/ Period	Aper. Recm.	Rating	Comments
h 1735	$22^h\ 09.3^m$	+44° 51'	A–B	6.7	9.7	27.6	110	W, bW	2006	15×80	∗∗∗∗∗∗	Nice triplet of stars in a nearly perfect line
	$22^h\ 09.0^m$	+44° 51'	A–D	6.7	6.8	109.6	286	W, yW	2001	6×30	∗∗∗∗	C is 15th mag.
h 1741	$22^h\ 11.2^m$	+50° 49'	A–B	5.4	10.4	36.0	288	W, –	2000	2/3"	∗∗∗∗	
	$22^h\ 11.2^m$	+50° 49'	A–D	5.4	10.0	73.7	271	W, –	2000	2/3"	∗∗∗∗	
h 1746	$22^h\ 13.9^m$	+39° 43'	A–B	4.7	10.6	30.4	188	–, –	1999	15×80	∗∗∗∗	
	$22^h\ 13.9^m$	+39° 43'	A–C	4.7	12.1	72.0	191	–, –	1998	3/4"	∗∗∗	
	$22^h\ 13.9^m$	+39° 43'	A–B	4.7	13.1	50.0	090	–, –	1944	6/7"	∗∗∗	
Σ 2894	$22^h\ 18.9^m$	+37° 46'	A–B	6.2	8.9	16.1	196	W, W	2006	15×80	∗∗∗∗	In same FOV as 1 Lacertae (m4.14), a bright orange star
2 Lacertae	$22^h\ 18.9^m$	+37° 46'	A–C	6.2	13.1	47.5	248	W, –	1998	6/7"	∗∗∗	
	$22^h\ 21.0^m$	+46° 32'	A–B	4.6	11.6	47.6	006	Y, pO	1998	15×80	∗∗∗∗∗∗	Great contrast in both brightness and color. a.k.a. h 1755
Σ 2906	$22^h\ 26.8^m$	+37° 27'	A–B	6.5	9.6	3.7	001	B, pR	2000	2/3"	∗∗∗∗	
α Lacertae	$22^h\ 31.3^m$	+50°17'	A–B	3.8	11.8	36.3	294	oY, –	1925	3/4"	∗∗∗∗	a.k.a. β 703
Roe 74	$22^h\ 32.4^m$	+39°47'	A–B	5.8	9.8	42.3	156	W, –	203	15×80	∗∗∗∗	
	$22^h\ 32.4^m$	+39° 47'	A–C	5.8	10.1	34.0	342	W, –	2000	2/3"	∗∗∗∗	
	$22^h\ 32.4^m$	+39° 47'	A–D	5.8	9.4	103.3	218	W, –	2000	15×80	∗∗∗∗	
	$22^h\ 32.4^m$	+39° 47'	D–E	9.4	9.8	6.7	180	–, –	2000	2/3"	∗∗∗∗∗	Reasonably well matched pair
8 Lacertae	$22^h\ 35.8^m$	+39° 38'	Aα–B	5.7	63	22.7	185	bW, bW	2006	15×80	∗∗∗∗∗∗	Showcase pair. a.k.a. Σ 29922-Aα-B. Several distant companions, may be a small cluster
h 1796	$22^h\ 35.8^m$	+39° 36'	Aα–C	5.7	10.5	49.4	168	bW, –	2005	15×80	∗∗∗∗	a.k.a. A 1469–Aα–C
10 Lacertae	$22^h\ 38.6^m$	+56° 48'	A–B	5.4	10.5	31.8	006	dO, –	2004	15×80	∗∗∗∗	
Ho 187	$22^h\ 39.3^m$	+39° 03'	A–B	4.8	10.3	62.2	049	W, V	2002	2/3"	∗∗∗∗	a.k.a. S 813
12 Lacertae	$22^h\ 39.6^m$	+37° 36'	A–B	6.0	13.0	20.4	283	Y, –	1998	6/7"	∗∗∗	
13 Lacertae	$22^h\ 41.5^m$	+40° 14'	A–B	5.2	10.8	69.1	015	W, B	2002	3/4"	∗∗∗∗	a.k.a. DD Lacertae
Σ 2942	$22^h\ 44.1^m$	+41° 49'	A–B	5.2	10.9	14.6	129	yW, pB	1998	3/4"	∗∗∗	a.k.a. OΣ 479
15 Lacertae	$22^h\ 44.1^m$	+39° 27'	A–B	6.2	8.9	2.7	277	O, grW	2003	3/4"	∗∗∗∗	Nice color
	$22^h\ 44.1^m$	+39° 27'	A–C	6.2	11.6	9.0	247	O, –	1998	3/4"	∗∗∗∗	a.k.a. β 450-AC
	$22^h\ 52.0^m$	+43° 19'	A–B	4.9	11.9	23.8	154	O, –	1998	3/4"	∗∗∗∗	a.k.a. β 451-AB. Two 12th mag. companions are >100" distance

(continued)

Table 17.1. (continued)

Multiple stars:

Designation	R.A.	Decl.	Type	m_1	m_2	Sep. (")	PA (°)	Colors	Aper. Recm.	Rating	Comments
h 975	22ʰ 55.7ᵐ	+36° 21'	A–B	5.7	9.2	52.3	244	W, pB	7×50	****	Wide, but unequal pair
16 Lacertae	22ʰ 56.4ᵐ	+41° 36'	A–B	5.6	11.3	27.6	345	bW, –	3/4"	****	a.k.a. EN Lacertae and Σ2960
	22ʰ 56.4ᵐ	+41° 36'	A–C	5.6	9.4	62.2	048	bW, Y	15×80	****	Easy object for small scopes or larges or large binoculars

Variable stars:

Designation	R.A.	Decl.	Type	Range (m_v)		Period (days)	F (f_r/f_f)	Spectral Range	Aper. Recm.	Rating	Comments
CM Lacertae	22ʰ 00.1ᵐ	+44° 33'	EA	8.2	9.2	1.605	0.11	A2V+A8V	10×80	****	4° 01' WSW (PA 243°) of Lac (m4.55).##. Eclipse duration = 4.0 h
HT Lacertae	22ʰ 05.5ᵐ	+46° 46'	SRb	6.1	6.4	82	–	M4IIIab	6×30	****	2° 43' W (PA 275°) of 2 Lac (m4.55)
AR Lacertae	22ʰ 08.7ᵐ	+45° 45'	EA	6.1	6.8	1.983	0.15	G2IV–K0IV	6×30	****	2° 17' WSW (PA 251°) of 2 Lac (m4.55).#$
S Lacertae	22ʰ 29.0ᵐ	+40° 19'	M	7.6	13.9	241.50	0.46	M5e–M8e	8/10"	****	2° 49' S (PA 186°) of 6 Lac (m4.51).##
EW Lacertae	22ʰ 57.4ᵐ	+48° 43'	γ Cas	5.2	5.5	–	–	B3III–IVpev	6×30	****	1° 47' SW (PA 221°) of 3 and (m4.63). ##$
BL Lac	22ʰ 00.7ᵐ	+42° 17'	BL Lac	12.4	17.2	–	–	–	>20"	******	3° 34' NW (PA 317°) of HD 211075 (m4500). ##$. *2

Star clusters:

Designation	R.A.	Decl.	Type	Size (')	m_v	Brtst. Star	Dist. (ly)	dia. (ly)	Number of Stars	Aper. Recm.	Rating	Comments
NGC 7209	22ʰ 05.2ᵐ	+46° 30'	Open	15	7.7	9.0	9.02	2,950	~100	6/8"	***	2.7° WNW of 2 Lacertae. Binocs show ~15 *s. Cluster is less distinct in lge. apertures
NGC 7243	22ʰ 15.2ᵐ	+49° 54'	Open	30	6.4	8.0	8.0	2,500	40	7×50	***	1.3° NW of 4 Lacertae. Binocs show ~10–12 *s. Cluster is less distinct in lge. apertures
NGC 7245	22ʰ 15.3ᵐ	+54° 20'	Open	6.5	9.2	12.8	12.75	6,000	~170	8/10"	***	Rich concentration of ~ 40 faint stars ~13+ mag

Nebulae:

Designation	R.A.	Decl.	Type	m$_v$	Size (')	Brtst./ **Cent.** star	Dist. (ly)	dia. (ly)	Aper. Recm.	Rating	Comments
IC 5217	22^h 23.9^m	+50° 58'	Planetary	11.3	6"	**15.5**	15.4	0.4	12/14	***	Small, bright, bluish disc. An OIII filter helps

Galaxies:

Designation	R.A.	Decl.	Type	m$_v$	Size (')	dia. (ly)	Dist. (ly)	Lum. (suns)	Aper. Recm.	Rating	Comments
None											

Meteor showers:

Designation	R.A.	Decl.	Period (yr)	Duration	Max Date	ZHR (max)	Comet/ **Asteroid**	First Obs.	Vel. (mi/ **km**/sec)	Rating	Comments
None											

Other interesting objects:

Designation	R.A.	Decl.	Type	m$_v$	Size (')	Dist (ly)	dia. (ly)	Lum. (suns)	Aper. Recm.	Rating	Comments
None											

LEO MINOR (lee-owe **my**-nor)	The Little Lion
LEONIS MINORIS (lee-**owe**-niss my-nor-**iss**)	**BEST SEEN:** (at 9:00 **P.M.**): Apr. 10
LMi	**AREA**: 232 sq. deg., 64th in size

Leo Minor first appeared in the "*Firmamentum Sobiescianum sive Uranographia*" by Johannes Hevelius. Usually abbreviated as just the "*Uranographia*," this work was published in 1690, 3 years after Hevelius' death. The stars of Leo Minor were first given designations by Flamsteed, whose catalog was later updated by Francis Bailey. Bailey also updated the "*British Association Catalog*" in 1845, in which he modified some of Flamsteed's designations. Bailey also extended Bayer's lettering system, but oddly, used only β in Leo Minor.

TO FIND: Leo Minor is very faint, with the three brightest stars being only 4th magnitude. These stars can be found southwest of the hind paws of the Big Bear. Starting with Alula Borealis (ν UMa), look for 46 LMi just over 5° to the WNW.[*3] β LMi can be found slightly less than 5° SSE of Tania Australis (34 UMa). 21 LMi lies a little more than 4° WSW of β LMi. From there, the Little Lion's tail can be traced (on a dark night) SE for about 10°. About 7° WNW of 21 lies a group of four faint stars (8, 9, 10, and 11 LMi) forming the Little Lion's head. Try this on a clear night from a very dark location. See Charts. Good luck! (Figs. 6.6, 17.2–17.4).

KEY STAR NAMES:

Name	Pronunciation	Source
20 Cor	kore	cor (Latin), "heart" (as the seat of thought, emotion)
37 Praecipua	**PRY**-sih-**pu**-uh	praecipuum, -a (Latin), "special, distinguished, extraordinary"

Magnitudes and spectral types of principal stars:

Bayer desig.	Mag. (m_v)	Spec. type	Bayer desig.	Mag. (m_v)	Spec. type	Bayer desig.	Mag. (m_v)	Spec. type
α[*4]	–	–	β	4.20	G8 III–IV	o[*5]	3.83	K0 III–IV

Stars of magnitude 5.5 or brighter which have Flamsteed, but no Bayer designations:

Flamsteed	Mag. (m_v)	Spec. type	Flamsteed	Mag. (m_v)	Spec. type	Flamsteed	Mag. (m_v)	Spec. type
8	5.39	M1 III	10	4.54	G8 III	11	5.40	IV–V
19	5.11	F6 V	20	5.37	G1 V	21	4.49	A7 V
23	5.49	A0 V	30	4.72	F0 V	37	4.68	G0 II
41	5.08	A3 V	42	5.36	A1 V			

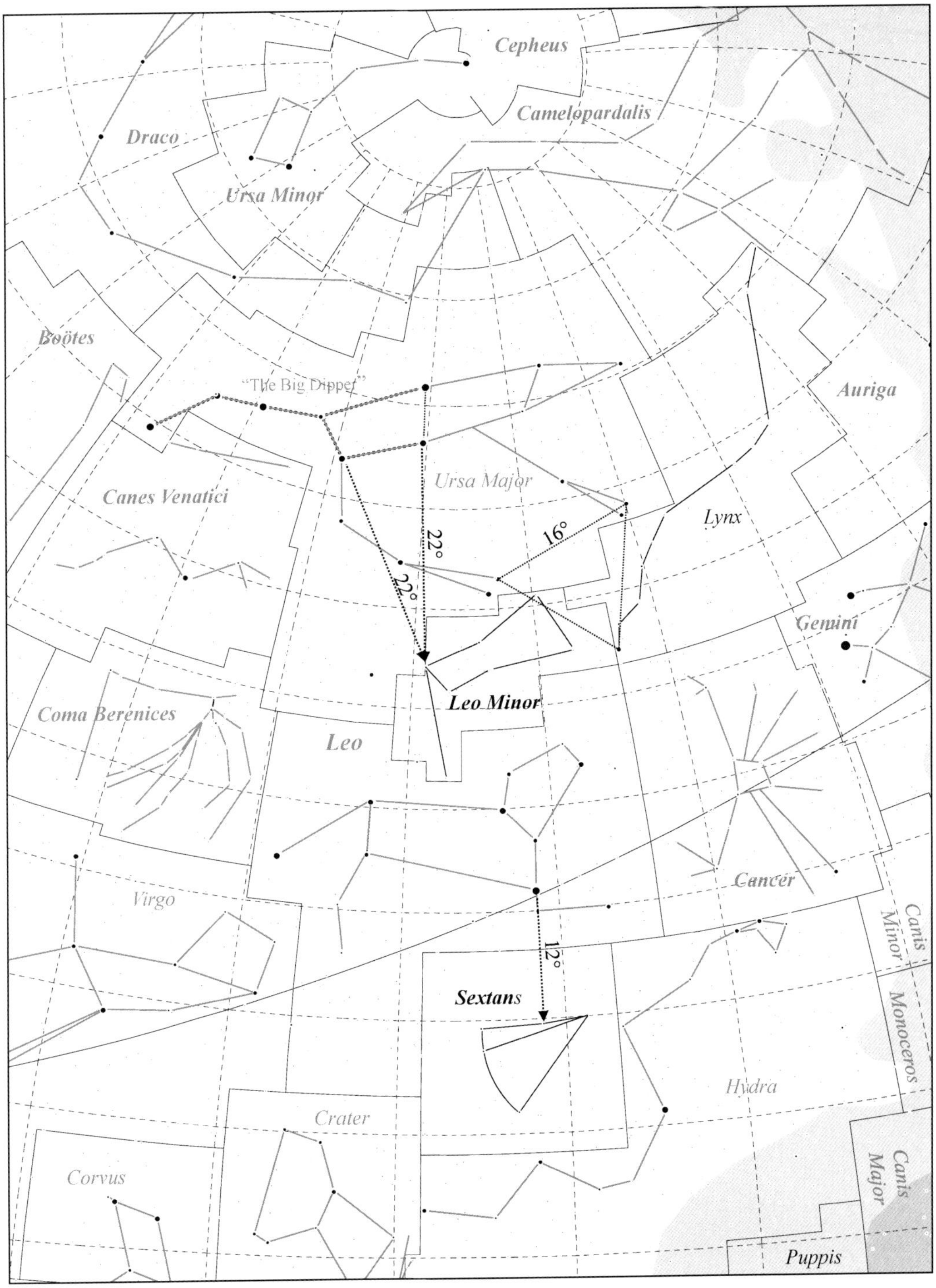

Fig. 17.2. Leo Minor, Lynx, and Sextans Finder.

Fig. 17.3a. The stars of Leo Minor and adjacent constellations.

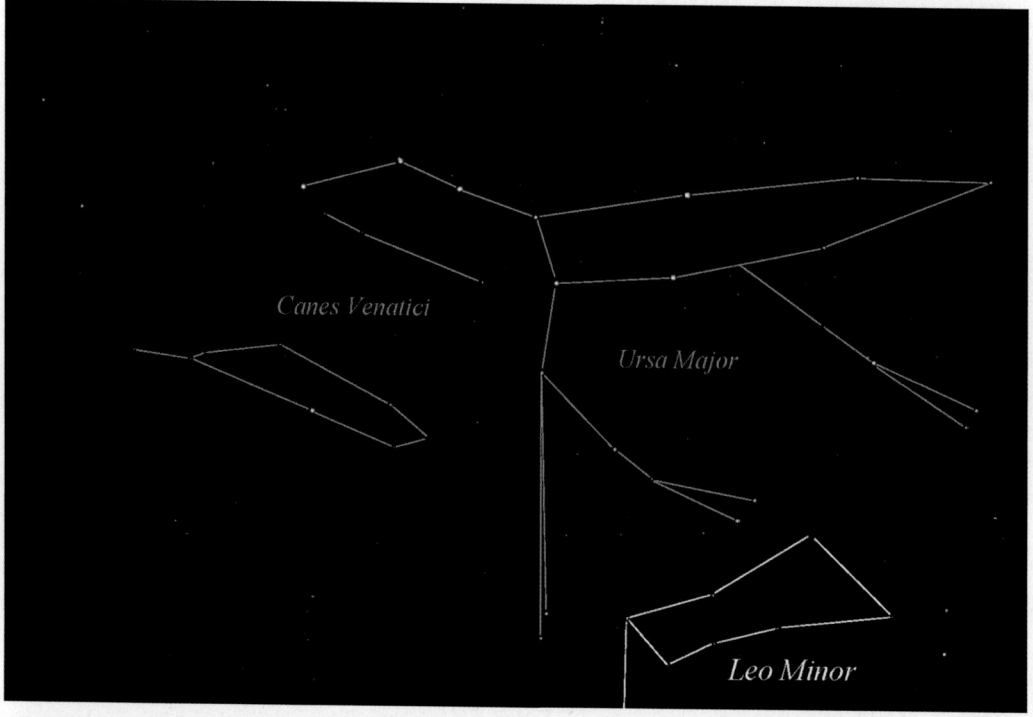

Fig. 17.3b. The figures of Leo Minor and adjacent constellations.

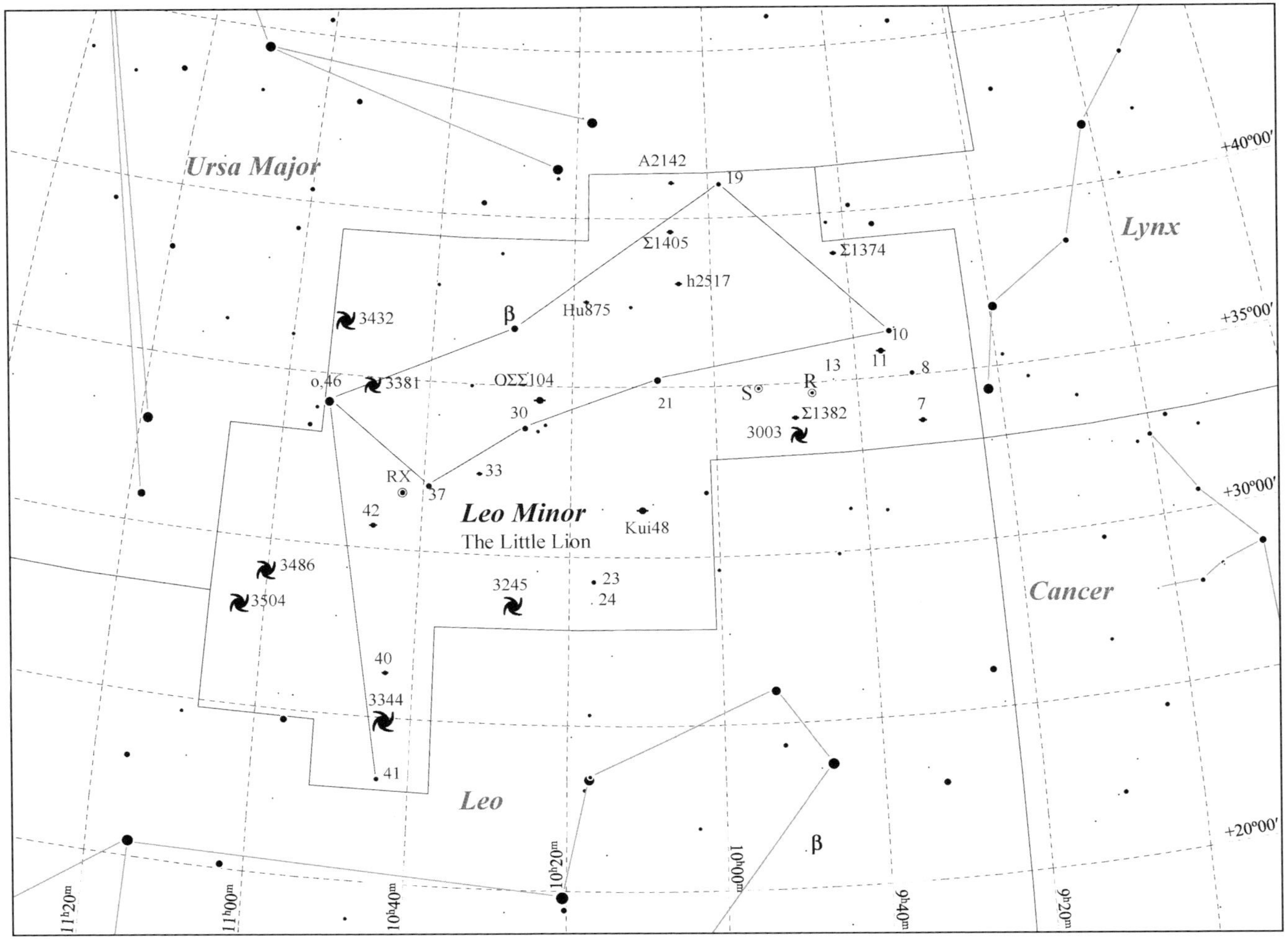

Fig. 17.4. Leo Minor details.

Table 17.2. Selected telescopic objects in Leo Minor.

Multiple stars:

Designation	R.A.	Decl.	Type	m_1	m_2	Sep. (")	PA (°)	Colors	Date/ **Period**	Aper. Recm.	Rating	Comments
7 LMi	09ʰ 30.7ᵐ	+33° 39'	A–B	6.0	9.7	63.2	128	Y, W	2004	15×80	****	a.k.a. h 1166-AB
			A–C		9.8	96.2	216	Y, –	1998	15×80	****	a.k.a. h 1166-AC
11 LMi	09ʰ 35.7ᵐ	+35° 49'	A–B	4.8	12.5	**6.2**	**051**	dY, –?	**201**	4/5"	****	Separation is increasing at ~0.05"/year. Max. will be 7.14" in 2052. a.k.a. Hu 1128
Σ 1374	09ʰ 41.4ᵐ	+38° 57'	A–B	7.3	8.7	3.0	309	brY, wB	2003	3/4"	*****	Lovely contrast
Σ 1382	09ʰ 49.1ᵐ	+34° 05'	A–B	7.1	11.1	32.6	108	W, –	1998	3/4"	****	
			A–C		11.8	29.2	263	W, –	1998	3/4"	****	
h 2517	10ʰ 03.9ᵐ	+38° 01'	A–B	6.8	13.5	52.6	153		1926	7/8"	****	
A 2142	10ʰ 05.7ᵐ	+41° 02'	A–B	8.0	8.8	0.6	297	yW, –	2005	16/17"	**	
Σ 1405	10ʰ 05.9ᵐ	+39° 35'	A–B	7.3	11.8	21.9	252	W, –	1998	3/4"	****	
Kui 48-AB	10ʰ 08.3ᵐ	+31° 06'	A–B	6.9	7.2	0.2	173	yW, –	1999	–	*	
h 475-AB-c			AB–C	6.3	13.7	27.7	172	yW, –	1999	8/9"	***	
Hu 875	10ʰ 18.4ᵐ	+37° 30'	A–B	8.3	10.3	1.1	072	W, –	1991	8/9"	****	
OΣΣ 104	10ʰ 24.4ᵐ	+34° 11'	A–B	7.2	7.3	208.4	288	oR, –	2002	7×50	*****	Very nice matched pair a.k.a. HJ482
33 LMi	10ʰ 31.9ᵐ	+32° 23'	A–B	5.9	11.8	46.9	245	W, –	1934	3/4"	****	Deep yellow and pale yellow. a.k.a. β 913
40 LMi	10ʰ 43.0ᵐ	+26° 20'	A–B	5.5	12.6	18.4	112	W, –	1958	5/6"	****	
	10ʰ 43.0ᵐ	+26° 20'	A–C		13.6	39.7	072	W, –	1998	7/8"	***	
	10ʰ 43.0ᵐ	+26° 20'	A–D		13.1	54.1	278	W, –	1919	6/7"	***	
42 LMi	10ʰ 45.9ᵐ	+30° 41'	A–B	5.3	7.8	196.5	174	W, grW	2002	7×50	******	Good color contrast. a.k.a. S 612AB
	10ʰ 45.9ᵐ	+30° 41'	A–B		8.3	424.6	094	W, –	2002	7×50	****	a.k.a. Arn 3AC

Variable stars:

Designation	R.A.	Decl.	Type	Range (m_v)		Period (days)	F (f_r/f_t)	Spectral Range	Aper. Recm.	Rating	Comments
R LMi	09ʰ 45.6ᵐ	+34° 31'	M	7.1	12.6	372.19	0.41	M6.5e-M9.0e	6/8"	****	45' SE of 6.2 m 13 LMi. At maximum nearly equal to 13 LMi; very red near minimum
S LMi	09ʰ 53.7ᵐ	+34° 56'	M	7.5	13.9	233.6	0.42	M2.0e-M8.2e	12/14"	****	2° 56' W (PA 266°) of 21 LMi
RX LMi	10ʰ 42.2ᵐ	+31° 42'	SRb	6.0	6.2	150	–	M2III	6×30	****	47' ESE (PA 110°) of 37 LMi

Star clusters:

Designation	R.A.	Decl.	Type	m_v	Size (')	Brtst. Star	Dist. (ly)	dia. (ly)	Number of Stars	Aper. Recm.	Rating	Comments
None												

Nebulae:

Designation	R.A.	Decl.	Type	m_v	Size (')	Brtst./ Cent. star	Dist. (ly)	dia. (ly)	Aper. Recm.	Rating	Comments
None											

Galaxies:

Designation	R.A.	Decl.	Type	m_v	Size (')	Dist. (ly)	dia. (ly)	Lum. (suns)	Aper. Recm.	Rating	Comments
NGC 2859	06ʰ 24.3ᵐ	+34° 31'	Sba	10.9	4.1 × 3.6	72M	86K	180G	6/8"	★★★★	Very, bright, pretty large, round; very bright nucleus, surrounding ring
NGC 3245	10ʰ 27.3ᵐ	+28° 30'	SB	10.8	3.0 × 2.0	52M	45K	10G	6/8"	★★★★	Faint, small, slightly elongated N-S. Bright core, stellar nucleus
NGC 3344	10ʰ 43.5ᵐ	+24° 55'	R(SAB)r	9.9	6.9 × 6.5	22M	44K	4.2G	3/4"	★★★	Bright, large. 3 faint arms. Very small, bright nucleus
NGC 3381	10ʰ 48.4ᵐ	+34° 43'	SBp	11.8	2.0 × 1.7	70M	41K	7.4G	6/8"	★★★	Faint, large, irregularly round. Gradually brighter toward center
NGC 3432	10ʰ 52.5ᵐ	+36° 37'	SB	11.2	6.8 × 1.7	27M	53K	1.9G	6/8"	★★★★	Bright, large, very elongated NE-SW. Nearly edge-on. Dwarf comp (U5983) at SW end
NGC 3486	11ʰ 00.4ᵐ	+28° 59'	SAB (r)c	10.5	6.9 × 5.2	29M	58K	4.2	6/8"	★★★	Bright, large, round. Multiple arms. Bright, diffuse nucleus
NGC 3504	11ʰ 03.2ᵐ	+27° 58'	(R) SAB(r)	11.0	2.6 × 2.3	64M	48K	13G	6/8"	★★★	Bright, large, elongated NNW-SSE. Extra bright diffuse nucleus with dark lines

Meteor showers:

Designation	R.A.	Decl.	Duration	Period (yr)	Max Date	ZHR (max)	Comet/ Asteroid	First Obs.	Vel. (mi/ km/sec)	Rating	Comments
None											

Other interesting objects:

Designation	R.A.	Decl.	Type	m_v	Size (')	Dist. (ly)	dia. (ly)	Lum. (suns)	Aper. Recm.	Rating	Comments
None											

Table 17.3. Selected telescopic objects in Lynx.

Multiple stars:

Designation	R.A.	Decl.	Type	m₁	m₂	Sep. (")	PA (°)	Colors	Date/Period	Aper. Recm.	Rating	Comments
4 Lyncis	06ʰ 22.1ᵐ	+59° 22'	Aa–B	6.1	7.7	0.7	143	W, W	2003	13/14"	***	a.k.a. Σ881 Aa–B. A–a separation is only 0.2", too close for amateur scopes
			Aa–C	6.1	12.9	26.4	096	W, –	1935	5/6"	***	a.k.a. Σ881 Aa–C
			Aa–D	6.1	11.9	99.9	357	W, –	1999	3/4"	***	a.k.a. Σ881 Aa–D
			Aa–E	6.1	12.2	70.0	000	W, –	1944	4/5"	***	a.k.a. Wal45AB–E
5 Lyncis	06ʰ 26.8ᵐ	+58° 25'	A–B	5.4	11.9	32.3	140	oY, B	1999	3/4"	****	a.k.a. S 514-AB
			A–C	5.4	7.9	95.8	272	oY, Pr	2002	7×50	***	a.k.a. S 514-A
			A–D	5.4	11.9	80.0	180	oY, –	1944	3/4"	***	a.k.a. Wal46AD
12 Lyncis	06ʰ 46.2ᵐ	+59° 27'	A–B	5.4	5.9	**1.7**	**065**	yW, yW	**706.1**	5/6"	***	A–B is good test for 3" scope. Binary, max 1.71" in 2091, min 1.61" in 2447. a.k.a
			A–C	5.4	7.05	8.7	309	yW, pB	2006.0	2/3"	***	Easy pair, background star
			A–D	5.4	10.5	172.2	259	yW, –	1999.0	15×80	***	D is faint for small telescopes, background star, unrelated to others
Σ958	06ʰ 48.2ᵐ	+55° 42'	A–B	6.3	6.3	4.6	257	W, yW	2006	2/3"	**	11th mag. background star is ~175" distant
15 Lyncis	06ʰ 57.3ᵐ	+58° 25'	A–B	4.7	5.7	0.6	226	Y, yW	2005	16/17"	***	a.k.a. OΣ159AB
			A–C	4.7	12.4	29.0	346	Y, –	1924	4/5"	****	a.k.a. OΣ159AC
			A–D	4.7	8.9	187.8	168	Y, –	1999	7×50	***	a.k.a. OΣ159AD
Σ1009	07ʰ 05.7ᵐ	+52° 45'	A–B	6.9	7.1	4.3	148	oY, oY	2003	2/3"	****	Nearly identical pair, 11th mag. background star at PA 181°
20 Lyncis	07ʰ 22.3ᵐ	+50° 09'	A–B	7.5	7.7	15.0	255	yW, yW	2003	2/3"	****	Nearly identical pair. Difference in brightness barely detectable by eye. A.k.a. Σ1065
19 Lyncis	07ʰ 22.9ᵐ	+55° 17'	A–B	5.8	6.7	15.1	315	W, W	2006	2/3"	****	a.k.a Σ1062
			A–D	5.8	7.6	215.3	006	W, B	2002	7×50	***	Similar to Mizar, but fainter
			B–C	6.5	10.9	74.2	288	W, –	1999	3/4"	***	
OΣ174	07ʰ 35.9ᵐ	+43° 02'	A–B	6.6	8.3	2.1	089	yW, B	2003	4/5"	****	
27 Lyncis	08ʰ 08.5ᵐ	+51° 30'	A–B	4.8	12.9	42.4	277	W, –	1998	5/6"	***	a.k.a. ES 70AB
				12.9	13.4	7.3	348	–, –	1926	7/8"	***	Easy to find faint double
S 565	08ʰ 24.7ᵐ	+42° 00'	A–B	6.2	8.6	83.9	176	oR, B	2004	7×5	**	Lovely colors
			B–C	8.7	11.2	0.4	077	B, –	1991	24"	***	

Designation	R.A.	Decl.	Pair	m₁	m₂	Sep.	PA	Colors	Date	Aper. Recm.	Rating	Comments
Σ1282	08h 50.7m	+35° 04'	A–B	7.6	7.8	3.5	279	rY, rY	2001	2/3"	★★★★	Lovely nearly identical pair. At larger apertures, both show a rich yellow color
Σ1333	09h 18.4m	–35° 22'	A–B	6.6	6.7	1.9	050	pY, pY	2004	5/6" 3/4"	★★★	Excellent test object for 3" scope in good seeing
38 Lyncis	09h 18.8m	+36° 48'	A–Bb	3.9	6.1	2.6	226	yW, grW	2004	3/4" 15×80		Nice pair. a.k.a. Σ1334. B has a close (0.2") companion of unknown magnitude
			Bb–C	6.1	10.8	80.8	216	grW, –	1998	15×80 15×80		
			Bb–D	6.1	10.7	177.9	261	grW, –	1998	15×80 9/10"		
Σ1338 [Lynx]	09h 21.0m	+38° 11'	A–B	6.1	6.4	**1.01**	304	Y, Y	**303.27**	9/10" 15×80		Min sep. 0.96" in 2044, max 1.67" in 2168

Variable stars:

Designation	R.A.	Decl.	Type	Range (m_v)		Period (days)	F (f_1/f_2)	Spectral Range	Aper. Recm.	Rating	Comments
1, UW Lyncis	06h 17.9m	+61° 31'	Lb?	4.9	5.0	–	–	M3IIIab	Eye	★★★	
2, UZ Lyncis	06h 20.1m	+59° 00'	E + δ Sct	4.4	4.7	–	–	A2V	Eye	★★★★	A δ Sct variable eclipsing and being eclipsed by a normal star. Very complex light curve
RR Lyncis	06h 26.4m	+56° 17'	EA	5.6	6.0	9.945	0.04	A3Vm	6×30	★★★	
R Lync	07h 01.3m	+55° 20'	M	7.2	>14.5	378.66	0.44	S2.5, 5e–S6,8e	14/16"	★★★	Variations are almost seasonal. 830:1 variation
UX Lyncis	09h 04.1m	+38° 43'	SRb	6.6	6.8	–	–	M6III	6×30	★★★★	37' WNW (PA 299°) of m4.55 SAO 61254

Star clusters:

Designation	R.A.	Decl.	Type	m_v	Size (')	Brst. Star	Dist. (ly)	dia. (ly)	Number of Stars	Aper. Recm.	Rating	Comments
NGC 2419	07h 38.1m	+38° 53'	Globular	10.3	4.1	17.3	272K	325	450K		★★★	a.k.a. C25 and "Intergalactic wanderer" because of distance. High concentration of Stars

(continued)

Table 17.3. (continued)

Nebulae:

Designation	R.A.	Decl.	Type	Size (')	Brtst./ **Cent.** star	Dist. (ly)	dia. (ly)	Aper. Recm.	Rating	Comments
None										

Galaxies:

Designation	R.A.	Decl.	Type	m_v	Size (')	Dist. (ly)	dia. (ly)	Lum. (suns)	Aper. Recm.	Rating	Comments
NGC 2500	08^h 01.9^m	+50° 44'	SB(rs)	11.6	2.9 × 2.6	24M	20K	1.1G	10/12"	***	Slightly elongated, N-S
NGC 2537	08^h 13.3^m	+45° 59'	SB(s)	11.7	1.7 × 1.5	19M	9.4K	600M	10/12"	***	The "Bear Claw Galaxy" elongated N W-S E
NGC 2541	08^h 14.7^m	+49° 04'	S+	11.8	6.6 × 3.5	26M	50K	1.0G	12/14"	**	Low surface brightness, elongated nearly N-S
NGC 2549	08^h 19.0^m	+57° 48'	E6/Sa	11.2	3.8 × 1.2	51M	56K	0.9G	10/12"	***	Stellar nucleus, elongated nearly N-S
	08^h 52.7^m	+33° 25'	SA(rs)	9.8	9.3 × 2.3	11M	30K	1.2G	3/4"	****	Can be seen in 7×50
	09^h 14.1^m	+40° 07'	SA B(rs)	11.6	3.8 × 2.9	110M	122K	22G	8/10"	***	Stellar nucleus, off center. Lenticular halo, SSE-NNW

Meteor showers:

Designation	R.A.	Decl.	Period (yr)	Duration	Max Date	ZHR (max)	Comet/ **Asteroid**	First Obs.	Vel. (mi/ **km**/sec)	Rating	Comments
None											

Other interesting objects:

Designation	R.A.	Decl.	Type	m_v	Size (')	Dist. (ly)	dia. (ly)	Lum. (suns)	Rating	Comments
41 Lyn, 10 UMa, Kui 37	09^h 00.6^m	+41° 47'	Stars	3.97		50				This star is a study on how the constellations have evolved. See the text for details

Fig. 17.5. NGC 2859, a ringed spiral galaxy in Leo Minor. Astrophoto by Gian Michelle Ratto.

LYNX (links)	**The Lynx**
LYNCIS (lin-sis)	**9:00 P.M. CULMINATION:** Mar. 6
Lyn	**AREA:** 545 sq. deg., 28th in size

Lynx is the name of both an ancient and a modern constellation, and the two are almost certainly not the same. Accordingly, there are two different stories about the origin of the constellation.

The old Lynx represented a character in an episode resulting from the kidnapping of Persephone, daughter of the goddess of agriculture. During her travels over the earth in search of her daughter, the distraught Demeter (Virgo) neglected her appearance until she was mistaken for an ordinary poor woman. Nevertheless, she was very hospitably entertained by Eleusis, the founder of a kingdom of the same name. In gratitude, Demeter wanted to reward him by making his young son Triptolemus immortal. Late at night, when she expected everyone would be asleep, Demeter placed Triptolemus in the fireplace to burn away his mortal parts. Unfortunately, Eleusis was still awake, and unexpectedly came into the room. Naturally, he misunderstood her intent and attacked her.

In the attack, Eleusis was killed, and the disrupted attempt to make Triptolemus immortal failed. As a consolation prize, Demeter later gave Triptolemus a chariot pulled by winged dragons, and gave him the job of sowing the seed of wheat, her favorite grain, over the entire world.

At one point, Triptolemus visited Scythia, a country ruled by King Lyncus. When Lyncus learned that Demeter's magical seed would produce a bountiful crop entirely free of weeds, he feared his people might replace him with Triptolemus. The jealous king wanted to be seen as the source of this wonderful grain himself, so he decided to entertain Triptolemus royally, but to kill him after he fell asleep and distribute the seed himself. Demeter saw what Lyncus was intending to do, and changed him into a lynx. She later put the form of a lynx in the sky as a warning to mortals, but made the constellation out of very faint stars, so no one could recognize or honor Lynceus. Today, no one knows exactly what stars made up the ancient constellation of Lynx.

The modern constellation of Lynx first appeared in the Johnannes Hevelius stellar atlas published posthumously in 1690. Hevelius formed this constellation from faint stars in the relatively dim area between Auriga, Ursa Major, Gemini, and Camelopardalis, and took several stars from these constellations. There is no real figure to be seen, only a rough line of faint stars stretching about 40° NW from α Lyn. In the entire constellation, only α Lyncis is brighter than 4th magnitude. Even Hevelius admitted that the constellations were so faint that you would have to have the eyes of a lynx to see it.

The star designated 10 UMa was originally part of Ursa Major. It was included in Hevelius' atlas as part of the new constellation, but in 1725, the English Royal Astronomer Royal John Flamsteed still included it in his catalog as part of UMa. The star was officially made a part of Lynx when the International Astronomical Union formally defined the boundaries of all 88 constellations, but still carries the designation assigned by Flamsteed. 16 Lyncis was formerly known as ψ^{10} Aurigae. After Hevelius included it in his new constellation, Flamsteed recognized the constellation and numbered it as 16th among the 44 stars he recognized in Lynx. The stars originally listed by Flamsteed as 44, 45, 46, 48, 50, 52, and 54 Camelopardalis are also now counted among those in Lynx, but were not given new letters or numbers.

Although faint, Lynx has an abundance of double and multiple stars, especially those which are very attractive because of the nearly equal brightness of their components. Lynx also has NGC 2683, one of the most overlooked bright galaxies for amateur observers, and is the home to NGC 2419, one of the most distant globular clusters.

TO FIND: Find Talitha (ι Uma) at about 9:00 PM a week or two before or after Oct. 10. Talitha is the northernmost of the close pair of stars marking the Great Bear's forepaws. Also find λ UMa (unnamed), the northernmost of the pair of stars marking the foremost of the bear's hind paws. These two stars are approximately 14° apart. α Lyncis lies south of these two and forms a very nearly equilateral triangle with them. Follow the line of 4th and 5th magnitude stars wandering northwestward from α Lyn for nearly 40° to 2 Lyn (m4.5). 2 Lyncis is about 15° NE of Capella. Good luck! You truly need dark skies and the eyes of a lynx for this task (Figs. 6.6, 17.6 and 17.7).

KEY STAR NAMES: None

Magnitudes and spectral types of principal stars:

Bayer desig.	Mag. (m_v)	Spec. type	Bayer desig.	Mag. (m_v)	Spec. type	Bayer desig.	Mag. (m_v)	Spec. type
α	3.14	K7 IIIv	β	4.20	G8 III/IV	o*5	3.83	K0 III/IV

Stars of magnitude 5.5 or brighter which have Flamsteed, but no Bayer designations:

Flamsteed	Mag. (m_v)	Spec. type	Flamsteed	Mag. (m_v)	Spec. type	Flamsteed	Mag. (m_v)	Spec. type
1	5.01	M3 III	2	4.44	A2 V			
5	5.21	K4 III	12	4.86	A3 V	13	5.34	K0 III
14	5.34	G4 III	15	4.35	G5 III/IV	16	4.90	A2 Vn
18	5.20	K2 III	21	4.61	A1 V	22	5.35	F6 V
24	4.93	A3 IVn	26	5.46	M3 III	27	4.78	A2 V
31	4.25	K4.5 III	34	5/35	G0 IV	35	5.15	K0 III
36	5.30	B8 III	38	3.82	A1 V	42	5e.28	F0 V

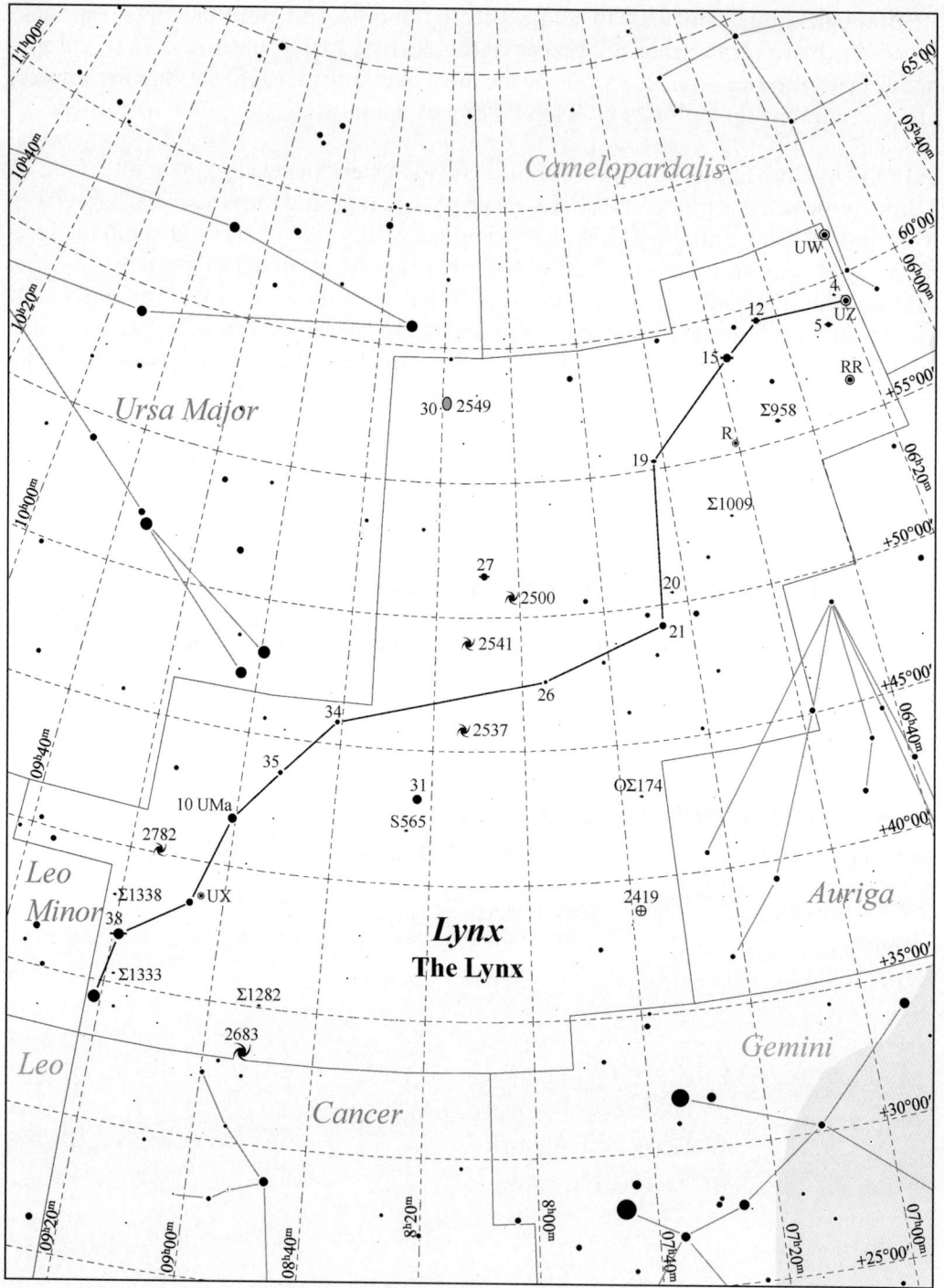

Fig. 17.6. Lynx details.

Table 17.4. Selected telescopic objects in Scutum.

Multiple stars:

Designation	R.A.	Decl.	Type	m₁	m₂	Sep. (")	PA (°)	Colors	Date/Period	Aper. Recm.	Rating	Comments
Σ 2303	18ʰ 20.1ᵐ	−07° 59'	A–B	6.6	9.3	1.6	240	Gld, B	1991	6/7"	****	Nice color contrast
Σ 2306	18ʰ 22.2ᵐ	15° 05'	A–B	8.1	8.6	9.5	222	Gld, B	2006	2/3"	****	Nice color contrast
			B–C	8.6	9.0	0.9	069	Gld, ?	2005	10/11"		
Σ 2313	18ʰ 24.7ᵐ	−06° 36'	B–C	7.5	9.7	5.9	196	Y, grW	2002	2/3"	****	In IC 1287
Σ 2325	18ʰ 31.4ᵐ	−10° 48'	A–B	5.8	9.3	12.3	256	Y, grW	2003	2/3"	****	
β 135-AB	18ʰ 38.1ᵐ	−13° 59'	A–B	6.5	10.5	2.0	173	W, ?	1953	4/5"	****	
Σ 2350	18ʰ 40.0ᵐ	−07° 47'	A–B	5.8	11.1	17.6	192	O, –	1999	3/4"	****	
δ Scuti	18ʰ 42.3ᵐ	−09° 03'	A–B	4.7	12.2	14.4	045	pY, pB	1999	4/5"	****	a.k.a. Rst 4594-AB
			A–C	4.7	10.6	52.3	131	pY, pB	1999	15×80	****	a.k.a. H 36-AC
			A–D	4.7	10.6	110.9	264	pY, ?	1999	15×80	****	a.k.a. LE 233-AD
Σ 2373	18ʰ 45.9ᵐ	−10° 30'	A–B	7.4	8.4	4.1	337	pY, dY	1999	2/3"	****	
Rst 4596	18ʰ 46.8ᵐ	−14° 28'	A–B	7.4	11.5	5.8	154	yW, ?	1998	3/4"	****	
Σ 2391	48ʰ 48.7ᵐ	−06° 03'	A–B	6.5	9.6	39.3	331	yW,	2006	15×80	****	Subtle, but pretty colors
Σ 2391	18ʰ 48.7ᵐ	−06° 00'	B–C	9.6	14.3	12.6	121	grB	1999	10/11"	****	
H VI 50	18ʰ 49.7ᵐ	−05° 55'	A–B	6.0	12.3	23.5	359	yO, ?	1934	4/5"	***	Lovely orange primary
				6.1	8.6	113.7	171	yO, W	2001	7×50	****	
h 5503	18ʰ 56.8ᵐ	−14° 51'	A–B	7.2	10.8	37.7	079	W, ?	1998	15×80	****	

Variable stars:

Designation	R.A.	Decl.	Type	Range (mᵥ)		Period (days)	F (fᵣ/fᵢ)	Spectral Range	Aper. Recm.	Rating	Comments
V430 Scuti	18ʰ 25.2ᵐ	−13° 59'	α Cyg	6.5	6.6	–	–	B1Ia-0a	6×30	****	1° 07' WNW (PA 302°) of λ Scuti (m4.67). Urano Metria Cht. 126.)
RZ Scuti	18ʰ 26.6ᵐ	−09° 13'	EA/GS	7.3	8.8	15.190	0.17	B3Ib	7×50	****	46' ESE (PA 110)° of ζ Scuti (m4.66).#
EW Scuti	18ʰ 37.8ᵐ	−06° 48'	β Cep	7.8	8.2	–	–	G5	6×30	***	1° 35' NNE (PA 24°) of α Scuti (m3.85)
RU Scuti	18ʰ 41.9ᵐ	−04° 07'	δ Cep	8.8	10.0	19.701	0.36	F4-G5	10×80	****	1° 27' WNW (PA 296°) of β Scuti (m4.22).##
δ Scuti	18ʰ 42.3ᵐ	−09° 03'	415	4.6	4.8	0.194	–	F2IIIp	Eye	***	##
R Scuti	18ʰ 47.5ᵐ	−05° 42'	RVa	4.5	8.2	146.5	–	G0eIa-K0pIb	7×50	****	57' S (PA 175°) of β Scuti (m4.22).##$

(continued)

Table 17.4. (continued)

Star clusters:

Designation	R.A.	Decl.	Type	m_v	Size (')	Brst. Star	Dist. (ly)	dia. (ly)	Number of Stars	Aper. Recm.	Rating	Comments	
NGC 6664	18h 36.7m	−08° 13'	Open	7.8	12.0	10.2	4,500	16	50	8/10"	***	0° 22' E (PA 086°) of α Scuti (m3.85). Can be seen in 7×50	
M26, NGC 6694	18h 45.2m	−09° 24'	Open	8.0	10.0	11.0	4,950	13	120	4/6"	****	0° 48' ESE (PA) 116° of δ Scuti (m4.70)	*6
M11, NGC 6705	18h 51.1m	−06° 16'	Open	5.8	11.0	11.0	5,460	20	682	7×50	*****	1° 48' SSE (PA 147°) of β Scuti (m4.22)	*7
NGC 6712	18h 53.1m	−08° 42'	Globular	8.1	9.8	13.3	22K	63	–	8/10"	**	Very low luminosity for a globular cluster	
NGC 6683	18h 42.2m	−06° 17'	Open	9.4	3.0	11.7	3,900	15	20	12/14"	***		
Trumpler 35	18h 42.9m	−04° 08'	Open	9.2	6.0	11.4	5,900	11	87	12/14"	***		

Nebulae:

Designation	R.A.	Decl.	Type	m_v	Size (')	Brst./Cent. star	Dist. (ly)	dia. (ly)	Aper. Recm.	Rating	Comments
IC 1287	18h 30.4m	−10° 48'	Refl.	–	43×33	6	5.8	694	12/14"	**	1° 45' SSW (PA 200°) of α Scuti (m3.85). Around Σ2325
Barnard 100	18h 32.6m	−09° 12'	Dark	–	16	–	–	–	4/6"	**	0° 52' SW (PA 221°) of α Scuti (m3.85). Lies on galactic equator
Barnard 101	18h 32.6m	−06° 57'	Dark	–	13×4	–	–	–	4/6"	**	1° 05' SW (PA 215°) of α Scuti (m3.85)
Barnard 110	18h 50.1m	−08° 48'	Dark	–	11	–	–	–	4/6"	**	B110 is a much darker spot near the center of B111
Barnard 111	18h 50.1m	−08° 48'	Dark	–	5×15	–	–	–	4/6"	**	Long exposure photos show extent of 1° E/W ×2° N/S
Barnard 113	18h 51.4m	−04° 19'	Dark	–	11	–	–	–	4/6"	**	B113 is a much darker spot near the north end of B111

Galaxies:

Designation	R.A.	Decl.	Type	m_v	Size (')	Dist. (ly)	dia. (ly)	Lum. (suns)	Aper. Recm.	Rating	Comments
None											

Meteor showers:

Designation	R.A.	Decl.	Period (yr)	Duration	Max Date	ZHR (max)	Comet/ **Asteroid**	First Obs.	Vel. (mi/ **km**/sec)	Rating	Comments
June Scuids	18^h 32^m	–04°	Annual	6/2 7/29	6/27	2–4	–	1935	–	II	Published in 1935 list of 320 radiants by Ronald A. McIntosh, poss. seen in 1869–1872

Other interesting objects:

Designation	R.A.	Decl.	Type	m_v	Size (')	Dist. (ly)	dia. (ly)	Lum. (suns)	Aper. Recm.	Rating	Comments
None											

Fig. 17.7. NGC 2683, spiral galaxy in Lynx. Astrophoto by Rob Gendler.

Stars of magnitude 5.5 or brighter which have other catalog designations:

Draper Number	Mag. (m_v)	Spec. type	Draper Number	Mag. (m_v)	Spec. type
10 UMa	3.96	F5 V	HD 77912	4.56	G8 Ib-II
HD 82741	4.81	K0 III	HD 56169	5.00	A4 IIIn
HD 62647	5.15	M3 III	HD 61931	5.31	A0 IIIn
HD 54895	5.46	M3 III			

SCUTUM (skoo-tum)	**the Shield**
SCUTI (skoo-tee)	**9:00 P.M. CULMINATION:** Aug. 28
Sct	**AREA**: 286 sq. deg., 52nd in size

One of the greatest Polish heroes was John (Jan) III Sobieski (1629–1696), the oldest son of the Castellan (Governor) of Crakow. As many of the elite of that time did, Sobieski joined the army as an officer. He was a brilliant military commander, and by 1665 had been promoted to Field Commander of the Polish army.

In 1656, the Ottoman Empire established by Muslims reversed its decline and began to expand again, taking Crete from Vienna in 1669. In November 1673, Sobieski was leading the battle against the Ottoman army which was advancing into Eastern Europe when he received word that the king had died. Sobieski left the battlefield and returned to Warsaw to compete for the elective office of King. He was elected in May of 1674, and soon returned to the front to lead the Polish army once again. The war continued indecisively until the battle of Kalenberg occurred on September 12, 1683. The Holy Roman Empire had a total of 74,000 troops to face the 140,000 man Ottoman army. Sobieski commanded 20,000 troops, and personally led a cavalry charge against the Turkish army. This charge proved to be the tipping point, preventing a Turkish siege of Vienna. Hungary was soon freed, and the Turkish and other Muslim armies were forced to retreat. Sobieski was hailed as "the hero of all Christendom."

In 1679, Hevelius' observatory (Stellaburgum) was destroyed by fire while he was absent, and all his instruments and records were lost except for a few books which his daughter managed to save. Hevelius had many benefactors who recognized the importance of his work, but the most important one was King Jan III. One of the manuscripts which was saved was Hevelius' star catalog, including all of his new constellations, with the exception of Scutum, Sextans, and Cerberus. In honor of King Jan, Hevelius added the constellation of *Scutum Sobiescianum* to his atlas. The name has evolved and been shortened until today it is only *Scutum*. Although it is a small and fairly dim constellation, it is the only surviving constellation honoring a modern historical person. Many of the stars of Scutum came from the now extinct constellation of Antonius. The name comes from scutum (Latin), "a large quadrangular shield."

TO FIND: The stars of Scutum are not very bright, but the constellation of Aquila can be used as a guide. Starting with <u>Altair</u> (m0.77), look for δ Aquilae (m3.36) ~8.5° SW, and then λ Aquilae (m3.43) another 9° to the SW. From there, β Scuti is slightly less than 5° due W. α Scuti is about the same distance to the SW of 03β2 (Figs. 9.1–9.3, 17.8 and 17.9).

KEY STAR NAMES: None.

Magnitudes and spectral types of principal stars:

Bayer desig.	Mag. (m$_v$)	Spec. type	Bayer desig.	Mag. (m$_v$)	Spec. type	Bayer desig.	Mag. (m$_v$)	Spec. type
α	3.85	K3III	β	4.22	G5II	γ	3.70	A3Vn
δ	4.72	F3III	ε	4.90	G8II	ζ	4.68	G9IIIb
η	4.83	K2V						

Stars of magnitude 5.5 or brighter which have other catalog designations:

Draper Number	Mag. (m$_v$)	Spec. type	Draper Number	Mag. (m$_v$)	Spec. type
HD 175156	5.08	B3 III	HD 171391	5.12	G8 III
HD 170975	5.47	K3 Iab			

a

b

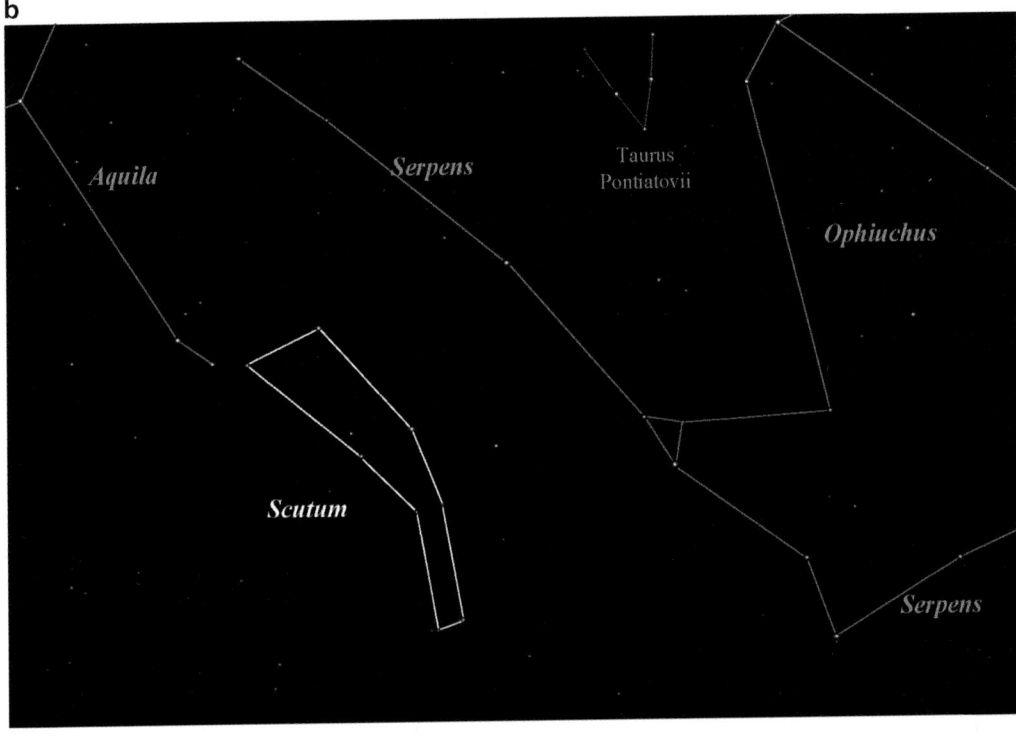

Aquila

Serpens

Taurus
Pontiatovii

Ophiuchus

Scutum

Serpens

Fig. 17.8. (**a**) Scutum stars. (**b**) Scutum constellation.

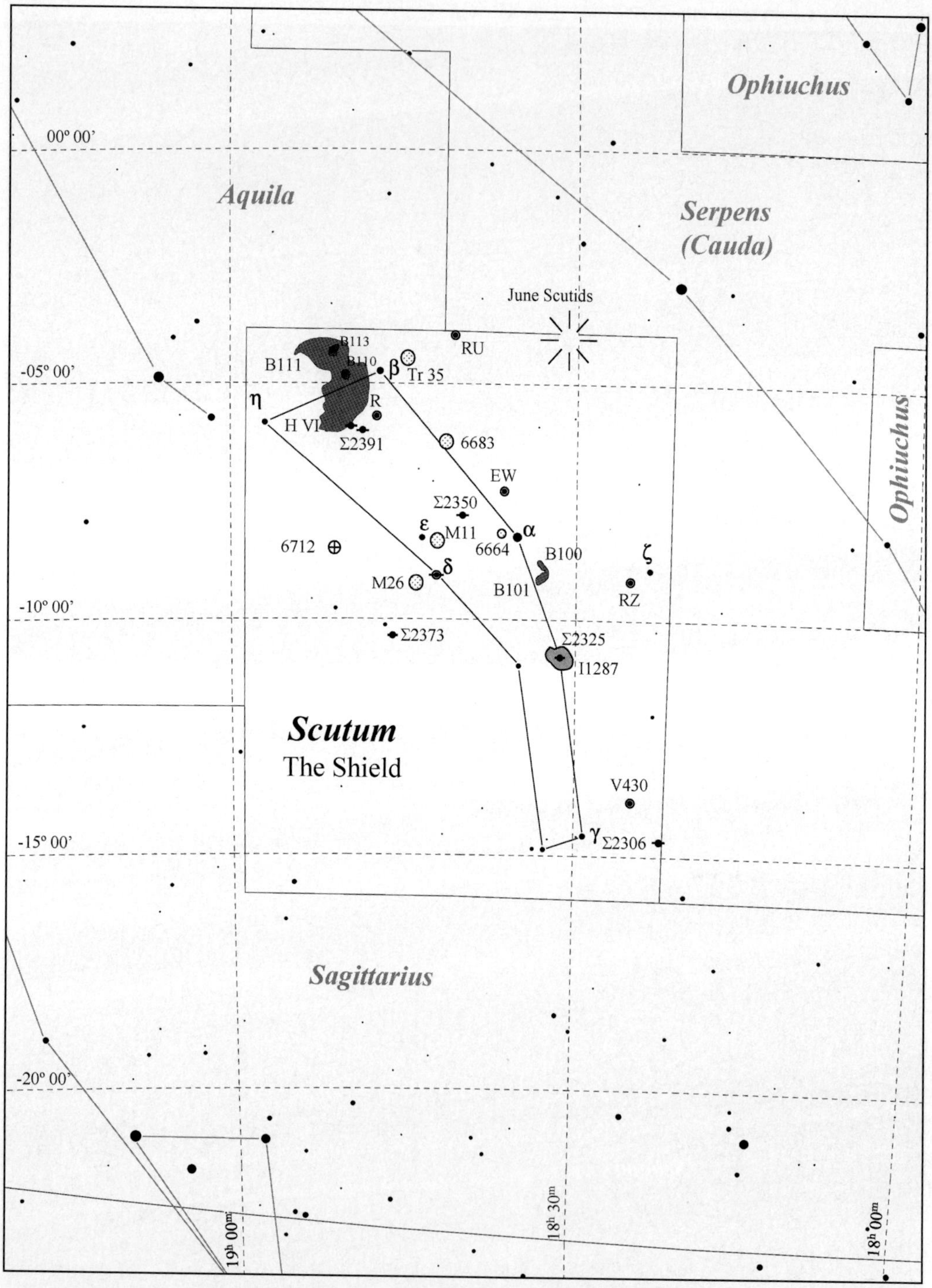

Ophiuchus

Aquila

Serpens
(Cauda)

Ophiuchus

June Scutids

00° 00'

-05° 00'

B113
B111   B110   β Tr 35
η   R
H VI   Σ2391   6683
                    EW
              Σ2350
          ε M11   α
6712 ⊕   6664   B100   ζ
M26   δ   B101
              RZ
      Σ2373   Σ2325
                I1287

-10° 00'

RU

Scutum
The Shield

V430

-15° 00'   γ Σ2306

Sagittarius

-20° 00'

19ʰ 00ᵐ   18ʰ 30ᵐ   18ʰ 00ᵐ

Fig. 17.9. Scutum details.

Fig. 17.10. M26 (NGC 6694), open cluster in Scutum. Astrophoto by Mark Komsa.

Fig. 17.11. M11 (NGC 6705), open cluster in Scutum. Astrophoto by Rob Gendler.

SEXTANS (seks-tanz)	**the Sextant**
SEXTANTIS (seks-**tan**-tis)	**9:00 P.M. CULMINATION:** Apr. 9
Sex	**AREA:** 314 sq. deg., 47th in size

In Hevelius' time, there were few commercial sources for precise instruments, so most astronomers (and other scientists) either made their own instruments, or commissioned their manufacture to their own specifications. One of Hevelius' most useful instruments lost in the fire was his astronomical sextant. When he was creating the new constellations to fill in the celestial voids, Hevelius commemorated his sextant with the constellation of Sextans Uraniae, The Astronomical Sextant.

TO FIND: Although Sextans is a faint constellation, its location is easy to find because its brightest star α Sextantis is only 12° due south of the first magnitude star <u>Regulus</u> in Leo, the Lion (see Figs. 6.3 and 8.1). The figure of the sextant is tilted so that one side runs almost exactly east-west with α Sextantis near its middle. β Sextantis lies 5.6° east of α, and marks the intersection of the horizontal arm and the top of the arc of the sextant. 2.1° south (actually at PA 186°) of β is δ Sextantis which is at the end of the sextant's movable arm. Finally, 8° south-southeast of α is ε Sextantis, which lies at the intersection of the arc and the other fixed arm of the sextant. γ Sextantis, the only other easily visible star in Sextans lies 6.2° west of ε, but does not form any part of the figure. Finally, the pivot point of the movable arm has an extremely faint star (SAO 117960, m6.34) to define its position, but it is barely visible to the unaided eye. Sextans is another constellation requiring good eyes and good skies to see. Good luck! (Figs. 6.1, 8.2, 17.12 and 17.13).

KEY STAR NAMES: None

Magnitudes and spectral types of principal stars:

Bayer desig.	Mag. (m_v)	Spec. type	Bayer desig.	Mag. (m_v)	Spec. type	Bayer desig.	Mag. (m_v)	Spec. type
α*8	4.48	A0 III	β	5.08var	B6 V	γ	5.07	A2 V
δ	5.19	B9.5 V	ε	5.25	F2 III			

Fig. 17.12a. The stars of Sextans and adjacent constellations.

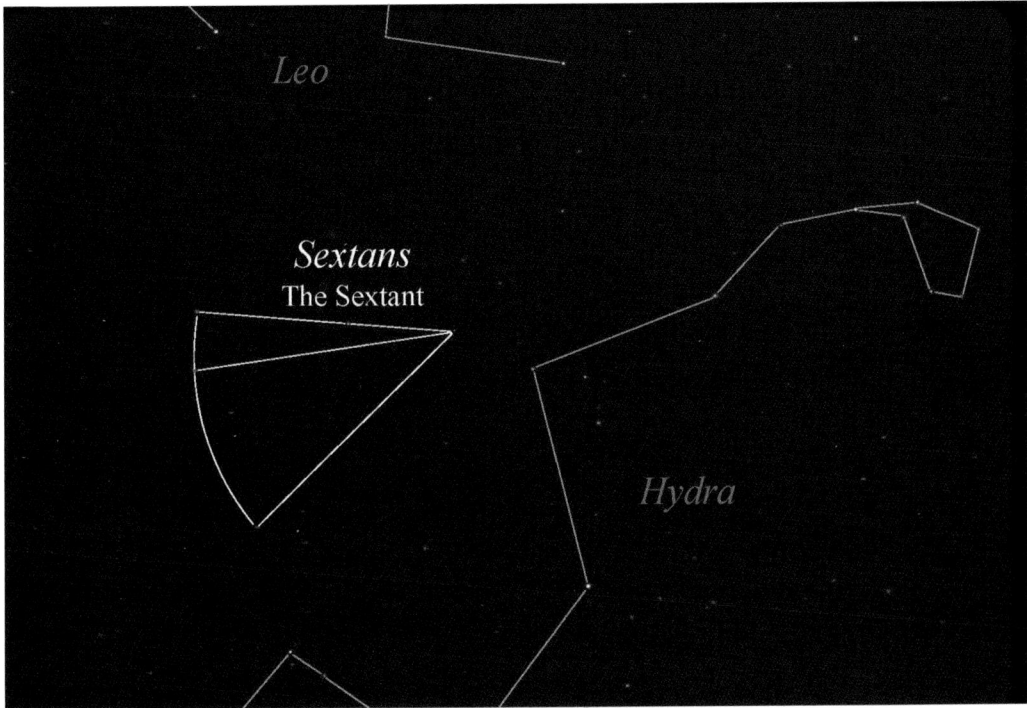

Fig. 17.12b. The figures of Sextans and adjacent constellations.

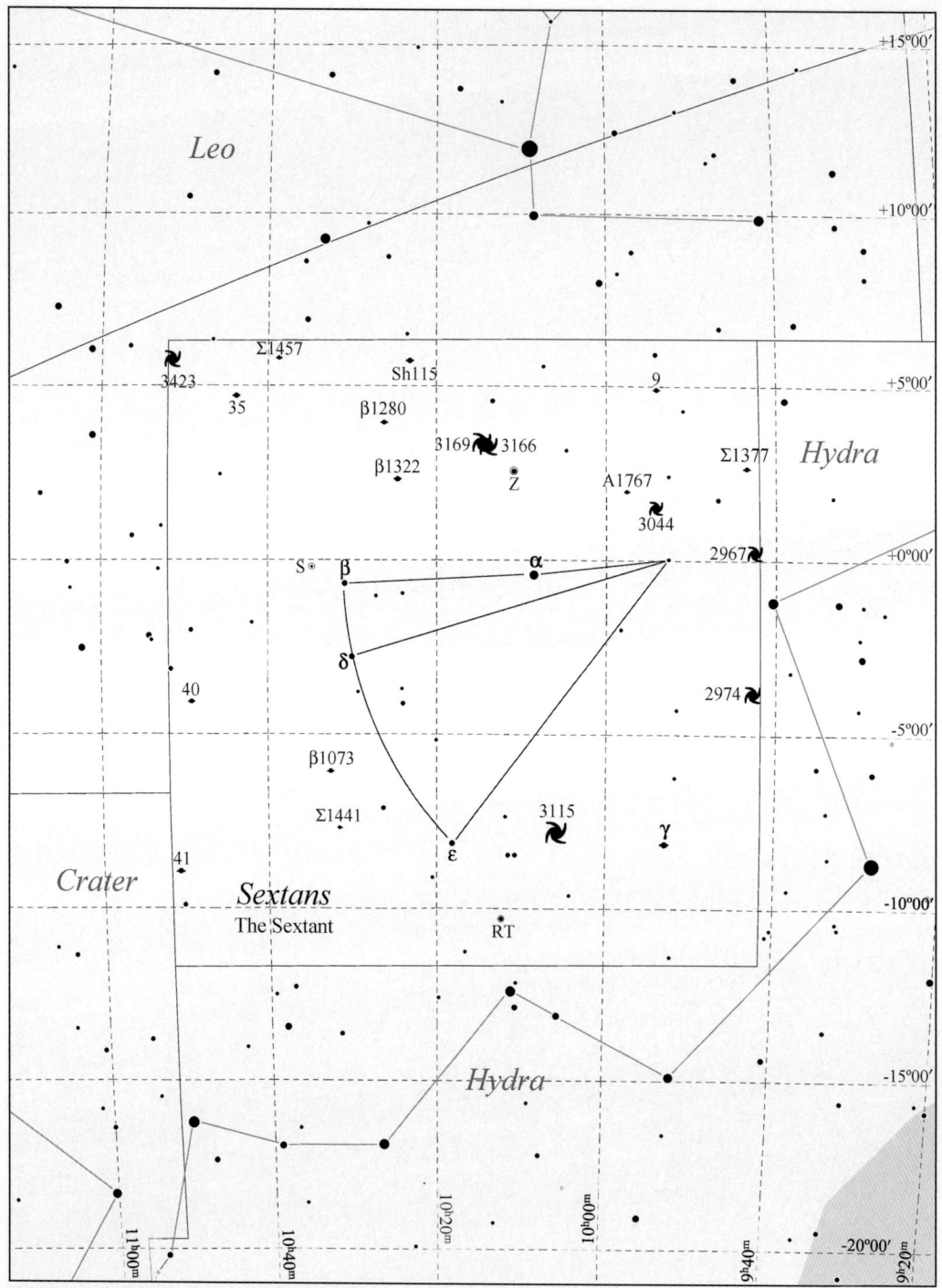

Fig. 17.13. Sextans details.

Table 17.5. Selected telescopic objects in Sextans.

Multiple stars:

Designation	R.A.	Decl.	Type	m_1	m_2	Sep. (")	PA (°)	Colors	Date/Period	Aper. Recm.	Rating	Comments
Σ 1377	09h 43.5m	+02° 38'	A–B	7.5	10.5	4.2	137	yW, B	2000	3/4"	****	Pretty triplet
8, γ Sextantis	09h 52.5m	+08° 06'	A–B	5.1	5.6	0.58	049	–, –	**77.55**	16/17"	*****	a.k.a. h 4256-AC
			A–C	5.1	12.3	37.6	333	–, –	1999	4/5"	***	
9 Sextantis	09h 54.1m	+04° 57'	A–B	6.9	8.4	52.5	287	rO, B	2004	7×50	*****	Lovely colors. a.k.a. S605
A 1767	09h 57.7m	–01° 57'	A–B	6.9	10.3	1.9	023	Y, –	1991	5/6"	****	
Sh 155	10h 23.2m	+05° 42'	A–B	6.5	9.1	63.2	011	Yw, –	2003	7×50	****	
β 1322	10h 24.2m	+02° 22'	A–B	6.3	12.7	12.5	301	Yw, –	2000	5/6"	***	
h 2530			A–C	6.3	6.7	202.4	064	Yw, –	2002	5/6"	****	
β 1280	10h 26.2m	+03° 56'	A–C	6.7	9.4	116.5	192	W, –	2002	15×80	****	
			A–D	6.7	11.3	101.8	130	W, W	2000	3/4"	****	
			B–C	9.6	11.5	0.5	031	–, –	1994	19/20"	******	
Σ1441	10h 31.0m	+07° 38'	A–B	6.5	8.8	2.8	167	O, Y	2003	3/4"	******	Attractive colored pair for small telescopes
β 1073	10h 32.5m	+06° 05'	A–C	6.5	10.1	62.2	314	O, –	2003	15×80	***	Close, slightly unequal pair
Σ1457	10h 38.7m	+05° 44'	A–B	6.9	11.4	3.4	047	oY, –	1940	3/4"	*****	Easy pair for small telescopes.
35 Sextantis	10h 43.3m	+04° 45'	A–B	7.7	8.2	1.8	332	yW, –	2005	5/6"	*****	
			A–B	6.2	7.1	6.7	241	O, yO	2005	2/3"	******	Nice colors. a.k.a. STF1464-AB
40 Sextantis	10h 49.3m	+04° 01'	A–B	7.1	7.8	2.4	015	W, W	2005	4/5"	****	Close pair for small telescopes. a.k.a. STF1476
41 Sextantis	10h 50.3m	+08° 54'	A–B	5.8	11.7	27.9	307	W, W	1999	15×80	****	a.k.a. h838-AB

Variable stars:

Designation	R.A.	Decl.	Type	Range (m.)	Period (days)	F (f_f/f_l)	Spectral range	Aper. Recm.	Rating	Comments
Z Sextantis	10h 11.0m	+02° 33'	Lb	6.7 9.6	–	–	M4–M5III	7×50	****	1° 50' ESE (PA 111°) of 13 Sextantis (m6.43).#
RT Sextantis	10h 12.3m	–10° 19'	SRb	7.9 9.0	96	–	M6	7×50	****	1° 56' S (PA 170°) of 18 Sextantis (m 5.64). #
S Sextantis	10h 34.9m	–00° 21'	M	8.2 13.7	264.9	0.50	M2e–M5e	10/12"	****	1° 12' S (PA 076°) of β Sextantis (m 5.03). ##

(continued)

Table 17.5. (continued)

Star clusters:

Designation	R.A.	Decl.	Type	m_v	Size (')	Brtst. Star	Dist. (ly)	dia. (ly)	Number of Stars	Aper. Recm.	Rating	Comments
None	00^h 00.0^m	+00° 00'										

Nebulae:

Designation	R.A.	Decl.	Type	m_v	Size (')	Brtst./Cent. star	Dist. (ly)	dia. (ly)	Aper. Recm.	Rating	Comments
None											

Galaxies:

Designation	R.A.	Decl.	Type	m_v	Size (')	Dist. (ly)	dia. (ly)	Lum. (suns)	Aper. Recm.	Rating	Comments
NGC 2967	09^h 42.1^m	+00° 20'	SAB	11.6	2.8 × 2.8	101M	42K	21G	8/10"	***	Pretty large, pretty faint, very gradually brighter toward center
NGC 2974	09^h 42.6^m	–03° 42'	SA	10.9	3.5 × 2.4	93M	87K	28G	8/10"	***	Fairly small, bright, brighter in middle. m10 star at SE edge
NGC 3044	09^h 53.7^m	–01° 35'	Scd	11.9	4.4 × 0.9	67M	61K	12G	8/10"	***	Very large, very faint, very much elongated at PA 122°
C53, NGC 3115	10^h 05.2^m	+07° 43'	S0	8.9	8.1 × 2.8	22M	39K	7.2G	8/10"	****	"Spindle galaxy." Large, very bright, very elongated in PA 046° *9
NGC 3166	10^h 13.8^m	+03° 26'	SAB	10.4	4.6 × 2.6	72M	86K	12G	8/10"	***	Bright, pretty small, some brightening toward center. NGC 3169 is in same FOV
NGC 3169	10^h 14.2^m	+03° 28'	SA	10.2	5.0 × 2.8	64M	84K	23G	8/10"	***	Bright, pretty large, elongated, gradually brighter toward center. Near twin of NGC 3166
NGC 3423	10^h 51.2^m	+05° 50'	SA	11.1	4.0 × 3.4	36M	40K	4.6G	8/10"	***	Very large, faint, very gradually brighter toward center

Meteor showers:

Designation	R.A.	Decl.	Period (yr)	Duration	Max Date	ZHR (max)	Comet/ **Asteroid**	First Obs.	Vel. (mi/ **km**/sec)	Rating	Comments
Sextantis	10ʰ 12ᵐ	-02°	4 or 5 yr	09/24 10/09	10/02	-	**2005 UD?**	1957	–	Daylight	Distinct from, but related to, Geminid shower, Discovered by A. A. Weiss (Adelaide)

Other interesting objects:

Designation	R.A.	Decl.	Type	mᵥ	Size (')	Dist. (ly)	dia. (ly)	Lum. (suns)	Aper. Recm.	Rating	Comments
None											

Fig. 17.14. NGC 2967, spiral galaxy in Sextans. Astrophoto by Mark Komsa.

Vulpecula (vul-peck-you-luh)	**The Fox**
Vulpeculae (vul-pec-you-lie)	**9:00 P.M. Culmination:** Sep. 9
Vul	**AREA:** 268 sq. deg., 55th in size

Originally called Vulpecula et Anser (The Fox and the Goose), Vulpecula was based on one of the Aesop's fables instead of classical mythology. In this fable, the fox caught an unwary goose by the neck, and was carrying it off in his mouth. However, his captive was not the "silly goose" the fox expected, and began to explain to the fox that he was resigned to his fate, and did not resent the fox for wanting to eat him because that was the nature of foxes and geese. The goose warned the fox that before he consumed his meal, he should be careful to offer thanks to the gods for his good fortune, lest the gods punish him for impiety. Convinced of the wisdom of the goose's remarks, the fox raised his paws and opened his mouth to praise the gods. Of course, the goose escaped and flew away. In this case, the "sly" fox was tricked out of his meal by the "dull" goose.

According to botanist Allan Foster, a more modern addition to the tale explains the white patch on the Canada goose as the scar left by the fox's teeth.

Apparently, the name Hevelius gave to this constellation, like the goose, was too big a mouthful to handle, and the name was shortened in later catalogs to just Vulpecula (the Fox). A similar fate befell some other modern constellations, including Scutum Sobieski also introduced by Hevelius, and Columba Nohae which was introduced by Plancius.

TO FIND: Vulpecula contains only very faint stars. The brightest star, α, is about 3.3° south of Albireo, and nearly 1½ magnitudes fainter. The constellation figure shown on the chart is very difficult to see, and certainly requires very dark skies to discern. Hevelius must have had very sharp eyes (or a very good imagination). The first step to finding Vulpecula is to study Fig. 11.1, "The Summer Triangle and Lyra Finder." Very carefully note the position of Vulpecula relative to the three bright stars of the Summer Triangle. Vulpecula lies athwart the two longer legs of the triangle, with the asterism forming the goose almost exactly midway between Vega and Altair, and the asterism forming the fox on the longest leg about 40% of the distance from Altair to Deneb. Also note that the small and faint, but distinctive constellations of Sagitta (the Arrow) and Delphinus (the Dolphin) are directly south of Vulpecula. Sagitta lies just south of the Fox's head and forequarters while Delphinus lies just south of the Fox's hindquarters and tail. With good skies, good eyesight, and patience, you should be able to see most of the stars defining both the Fox and the Goose. Good Luck! (Figs. 9.1, 9.3, 12.1–12.3 and 17.15).

KEY STAR NAMES:

Name	Pronunciation	Source
α Anser	**ann**-sir	Anser (Latin), goose

Magnitudes and spectral types of principal stars:

Bayer desig.	Mag. (m_v)	Spec. type
α	4.44	K3 III

Stars of magnitude 5.5 or brighter which have Flamsteed, but no Bayer designations:

Flam-steed	Mag. (m_v)	Spec. type	Flam-steed	Mag. (m_v)	Spec. type	Flam-steed	Mag. (m_v)	Spec. type
1	4.76	G8 III	2	5.46	B0.5 IV	3	5.22	B6 III
4	5.14	K0 III	9	5.00	B8 IIIn	10	5.50	G8 III
12	4.90	B2.5 V	13	4.57	B9.5 V	15	4.66	A4 III
16	5.23	F2 III	17	5.08	B3 V	21	5.19	A7 IVn
22	5.18	G2 Ib	23	4.50	K3 III	24	5.30	G8 III
25	5.50	B8 IIIn	28	5.06	B5 IV	29	4.81	A0 V
30	4.92	K2 III	31	4.56	G8 III	32	5.03	K2 III
33	5.30	K4 III	35	5.39	B0.5 IV*[10]			

Stars of magnitude 5.5 or brighter which have other catalog designations:

Draper Number	Mag. (m_v)	Spec. type
HD 192685	4.79	B3 V

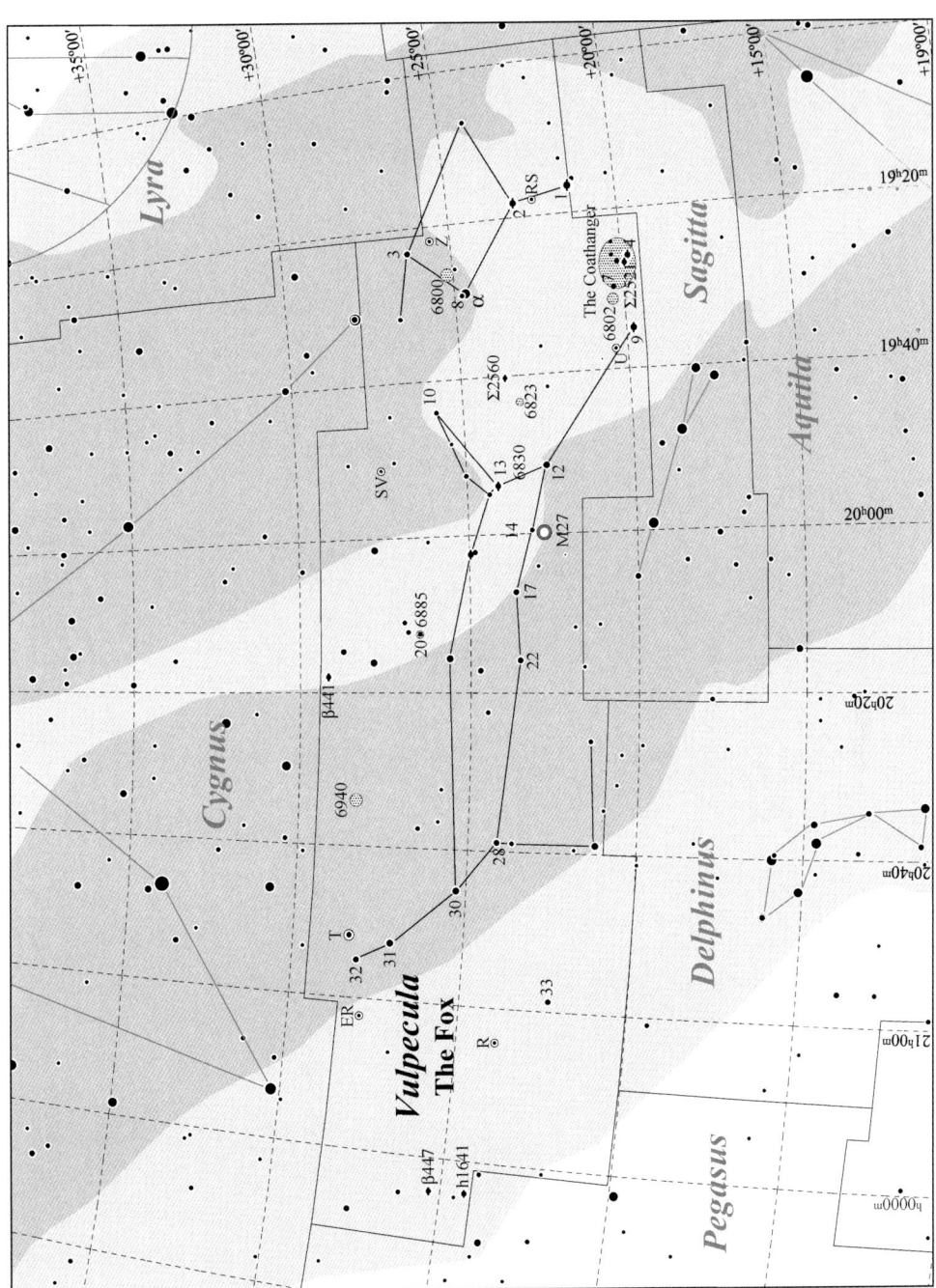

Fig. 17.15. Vulpecula details.

Table 17.6. Selected telescopic objects in Vulpecula.

Multiple stars:

Designation	R.A.	Decl.	Type	m₁	m₂	Sep. (")	PA (°)	Colors	Date/ Period	Aper. Recm.	Rating	Comments
1 Vulpeculae	19ʰ 16.2ᵐ	+21° 23'	A–B	4.8	11.6	38.7	"012	bW, –	1997	15×80	****	a.k.a. h 2862AB
			A–C	4.8	12.8	44.3	157	bW, –	1997	5/6"	****	a.k.a. h 2862AC
2 Vulpeculae	19ʰ 17.7ᵐ	+23° 00'	A–B	5.4	8.8	1.6	131	bW, –	2001	6/7"	****	a.k.a. B 248
			A–C	5.4	11.0	51.1	"119	bW, –	1997	3/4"	****	
4 Vulpeculae	19ʰ 25.5ᵐ	+19° 47'	A–B	5.2	10.0	16.2	093	pY, B	1984	15×80	****	a.k.a. h 2871AB. Brightest star in the "Coat Hanger" asterism
Σ 2521	19ʰ 26.5ᵐ	+19° 54'	A–C	5.2	11.7	51.6	209	pY, B	1997	3/4"	****	a.k.a. h 2871AC
			A–a	5.8	14.5	25.0	071	O, –	1934	11/12"	**	
			Aa–B	5.8	10.5	27.9	033	O, B	1997	15×80	****	
			Aa–C	5.8	10.5	74.1	325	O, –	2002	15×80	****	
			Aa–D	5.8	10.6	152.1	063	O, –	2002	15×80	****	
6 α-8 Vulpeculae	19ʰ 28.7ᵐ	+24° 40'	A–B	4.6	5.9	424.5	030	O, bW	2002	Eye	*****	Binoculars or small telescope needed to show color
9 Vulpeculae	19ʰ 34.6ᵐ	+19° 46'	A–B	5.0	13.4	8.0	030	W, –	1968	7/8"	***	a.k.a. β1130AB
			A–C	5.0	12.5	108.2	318	W, –	1923	4/5"	***	a.k.a. β1130AC
Σ2560	19ʰ 40.7ᵐ	+23° 42'	A–B	6.6	10.5	14.2	298	W, –	2003	2/3"	****	
			A–C	6.6	8.3	118.5	066	W, –	2002	7×50	****	
13 Vulpeculae	19ʰ 53.5ᵐ	+24° 05'	A–B	4.6	7.4	1.3	245	W, –	2003	7/8"	****	a.k.a. Dju 4
16 Vulpeculae	20ʰ 02.0ᵐ	+24° 55'	A–B	5.8	6.2	0.8	124	dY, dY	2003	12/13"	*****	a.k.a. OΣ395
β 441	20ʰ 17.5ᵐ	+29° 09'	A–a	6.4	0.4	–	220	dY, –	1989	–	**	
			Aa–B	6.4	10.9	5.8	067	dY, –	1958	3/4"	****	
h 1641	21ʰ 24.0ᵐ	+24° 16'	A–B	5.7	10.5	53.9	309	yW, –	1923	15×80	****	
β 447	21ʰ 24.1ᵐ	+25° 18'	A–B	6.2	12.2	10.3	311	W, –	1997	4/5"	****	
			A–C	6.2	11.7	58.0	192	W, –	1997	3/4"	****	
			A–D	6.2	11.8	56.6	082	W, –	1960	3/4"	****	
			A–E	6.2	11.0	65.8	116	W, –	1997	3/4"	****	
			A–F	6.2	10.7	83.5	218	W, –	1997	3.4"	****	

Variable stars:

Designation	R.A.	Decl.	Type	Range (m$_v$)		Period (days)	F (f$_r$/f$_f$)	Spectral Range	Aper. Recm.	Rating	Comments
RS Vulpeculae	19^h 17.8^m	+22° 26'	EA/SD	6.8	7.8	4.478	0.14	B4V + A2IV	6×30	****	0°35' S (PA 181 °) of 2 Vulpeculae (m5.46).#
Z Vulpeculae	19^h 21.6^m	+25° 34'	EA/SD	7.3	8.9	2.455	0.18	B4V + A3 III	7×50	****	0°44' SSW (PA 202°) of 3 Vulpeculae (m5.22).#
U Vulpeculae	19^h 34.6^m	+20° 20'	5 Cep	6.7	7.5	7.991	0.33	F6Iab-G2	6×30	****	0°44' NE (PA 040°) of 9 Vulpecuale (m5.00).#
SV Vulpeculae	19^h 51.5^m	+27° 28'	5 Cep	6.7	7.8	45.012	0.23	F7Iab-K0Iab	6×30	****	2°9' W (PA 263°) of 15 Vulpeculae (m4.66).#
T Vulpeculae	20^h 51.5^m	+28° 16'	5 Cep	5.4	6.1	4.435	0.32	F5Ib-G0Ib	6×30	****	0° 22' NNW (PA 335°) of 31 Vulpeculae (m4.66).#
ER Vulpeculae	21^h 02.4^m	+27° 48'	EW/DW/RS	7.3	7.5	0.698	–	G0V + G5V	6×30	****	0° 42' WNW (PA 286°) of 32 Vulpeculae (m5.03).#
R Vulpeculae	21^h 04.4^m	+23° 49'	M	7.0	14.3	136.73	0.49	M3e-M7e	14/16"	****	2° 3' NE (PA 043°) of 37 Vulpeculae (m5.30).##

Star clusters:

Designation	R.A.	Decl.	Type	m$_v$	Size (')	Brst Star	Dist. (ly)	dia. (ly)	Number of Stars	Aper. Recm.	Rating	Comments
NGC 6800	19^h 27.0^m	+25° 06'	Open	–	5	10.0	–	–	20	6/8"	***	
NGC 6802	19^h 30.6^m	+20° 16'	Open	8.8	5	14.0	3.3K	4.8	201	6/8"	***	
NGC 6823	19^h 43.1^m	+23° 18'	Open	7.1	12	8.8	8.8		30	10/12"	***	Surrounded by faint emission nebula, NGC 6820. a.k.a. H9
NGC 6830	19^h 51.0^m	+23° 06'	Open	7.9	12	9.9	5.5K	19	20	10/12"	***	The brightest star, Vulpeculae 20, is near the center of the cluster
NGC 6885	20^h 12.0^m	+26° 29'	Open	8.1	7	5.9	1.9K	3.9	30	6/8"	***	
NGC 6940	20^h 34.0^m	+28° 18'	Open	6.3	30	9.3	2.6K	23	80	4/6"	***	

(continued)

Table 17.6. (continued)

Nebulae:

Designation	R.A.	Decl.	Type	m_v	Size (')	Brtst./ Cent. star	Dist. (ly)	dia. (ly)	Aper. Recm.	Rating	Comments
M27.NGC 6853	19^h 59.6^m	+22° 43'	Plan- etary	7.3	5.8	13.8	815	bl. gr, rd	4/6"	***	Dumbell Nebula. Spectacular in photos and large scopes. Colors shown only in photos

Galaxies:

Designation	R.A.	Decl.	Type	m_v	Size (')	Dist. (ly)	dia. (ly)	Lum. (suns)	Aper. Recm.	Rating	Comments
None											

Meteor showers:

Designation	R.A.	Decl.	Period (yr)	Duration	Max Date.	ZHR (max)	Comet/ Asteroid	First Obs.	Vel. (mi/ km/sec)	Rating	Comments
None	00^h 00^m	+00°									

Other interesting objects:

Designation	R.A.	Decl.	Type	m_v	Size (')	Brtst. Star	Dist. (ly)	dia. (ly)	Number of Stars	Aper. Recm.	Rating	Comments
Collinder 399	19^h 25.4^m	+20° 11'	Asterism	3.6	60	5.2	–	–	10	7×50	****	a.k.a. Brocchi's cluster and "The Coathanger" asterism, chance alignment of stars *11

Fig. 17.16. M27 (NGC 6853), planetary nebula, "Dumbell Nebula" in Vulpecula. Astrophoto by Warren Keller.

Fig. 17.17. NGC 6820-6823, nebulosity with imbedded open cluster in Vulpecula. Astrophoto by Steve Pastor.

NOTES:

1. Flamsteed assigned numbers to only 16 stars in Lacerta, with α and β being 7 and 3, respectively. None of the lettered or numbered stars in Lacerta had their constellations or designations changed by the IAU's official definition of the constellations.

2. BL Lacertae was first thought to be an unusual variable star, but was later found to be a distant galaxy with an active nucleus or a quasar with the jet pointed toward the earth. This made the extremely bright jet so bright that the faint galaxy surrounding it was difficult to distinguish.

3. o Leonis Minoris is also known as 46 Leonis Minoris. If you are using Norton's Star Atlas and Reference Handbook, or some other atlases, you may notice a coincidence. Two stars, less than 1° apart, are both labeled "46." The brighter one is 46 LMi, and the fainter one (0° 52′ SE of 46 LMi) is 46 UMa.

4. In his revision of various catalogs, including that of Flamsteed, Francis Bailey's catalog revision assigned only one Greek letter designation in Leo Minor, β.

5. 46 Leo Minoris is the brightest star in the constellation, and nearly half a magnitude brighter than β. Since Bailey died before he could review the galley proofs, he may have assigned it the designation α only to have it accidently omitted. In 1801, Johann Bode published an extensive catalog in which he extended Bayer's designations for several constellations. In Leo Minor, he assigned several lower case English letters, but only that of "o" has survived.

6. M26 is a rather inconspicuous cluster, with no bright stars. There is a darker area near the center which is visible in larger scopes with averted vision.

7. M11 is a very rich open cluster, with an estimated population of nearly 3,000 stars. About 500 are brighter than magnitude 14, and thus accessible to moderate sized amateur telescopes. A hypothetical observer at the center of M11 would have a night sky (if you call it that) with several hundred stars of first magnitude or brighter. M11 was discovered by Gottfried Kirch in 1681, and resolved into stars in 1733 by William Derham. Messier apparently independently discovered it in 1764. Messier acknowledged that his was not an original discovery. It received its name because it contains an open "V" of about 15–20 nearly equally bright stars which resemble the formation taken by migrating ducks.

8. John Flamsteed assigned numbers to 41 stars in Sextans. Benjamin Gould later assigned Greek letters to the five brightest stars with 15, 30, 8, 29, and 22, respectively, becoming α, β, γ, δ, and ε. A measurement error by Flamsteed resulted in a single star being recorded as both 28 and 29 Sextantis. The number 28 is no longer used in Sextans.

9. NGC 3115 is actually brighter than several Messier objects and can be seen in 7×50 binoculars. 11×80 binoculars reveal its spindle shape.

10. The remaining numbered stars in Vulpecula are fainter than magnitude 6.5, and not included in this list.

11. The "Coat Hanger" consists of 4, 5, and 7 Vulpeculae, plus a line of 5 fainter stars going directly west of 7 Vulpeculae, and two stars north of 4 Vulpeculae and south of 5 Vulpeculae. In an inverting telescope, it appears right-side up.

CHAPTER EIGHTEEN

Lacaille: Abbe and Astronomer

ANTLIA	CAELUM	CIRCINUS	FORNAX	HOROLOGIUM
The Air Pump	The Burin	The Compasses	The Furnace	The Clock

MENSA	MICROSCOPIUM	NORMA	OCTANS	PICTOR
The Mountain	The Microscope	The Square	The Octant	The Easel

PYXIS	RETICULUM	SCULPTOR	TELESCOPIUM
The Compass	The Net	The Sculptor	The Telescope

Nicolas Louis Lacaille (sometimes written La Caille or LaCaille) (1713–1763) was an eighteenth century French astronomer who observed, measured, and cataloged much of the heavens, including many of the fuzzy celestial objects later cataloged by Messier. In fact, Lacaille discovered three of the "Messier Objects" (M55, M69, and M83). He spent 2 years (1751–1753) observing and cataloging over 9,000 southern stars from Cape Town, South Africa. During this time, he cataloged one of the nearest stars, Lacaille 9352 (#19 in distance at 11.77 ly), and made measurements of many double stars, including the first accurate measurements of α Centauri, the nearest star system.

During his stay in South Africa, Lacaille made some significant additions and changes to the list of constellations. In 1752, he sent a report to the French Academy of Science which included 13 new constellations that he had created from "unformed" stars, or stars which lay between the recognized constellations, and were not officially part of any constellation. Most of these represented instruments or equipment related

P. Simpson, *Guidebook to the Constellations: Telescopic Sights, Tales, and Myths,*
Patrick Moore's Practical Astronomy Series, DOI 10.1007/978-1-4419-6941-5_18,
© Springer Science+Business Media, LLC 2012

to the technology of his time. To make these new constellations, he also borrowed some stars from other constellations when these stars did not form any significant part of the generally recognized figure of these constellations. In the case of Reticulum, he apparently "borrowed" the entire constellation. Isaac Habrecht of Strassburg had called the main stars of this area Rhombus, but this constellation had not been widely recognized, and Lacaille renamed it Reticulum Rhomboidalis after a type of telescope reticule he used to measure the positions of the stars.

The constellations introduced in Lacaille's report were:

Original name	Modern name
Antlia Pneumatica (The Air Pump)	Antlia (The Pump)
Apparatus Sculptoris (The Sculptor's Workshop)	Sculptor (The Sculptor)
Caela Sculptoris (The Sculptor's burins or graving tools)	Caelum (The Graving Tool or Chisel)
Circinus (The drafting Compasses)	Circinus (The Compasses)
Equuleus Pictoris (The Painter's Easel)	Pictor (The Easel)
Fornax Chemica (The Chemist's Furnace)	Fornax (The Furnace)
Horologium Oscillatorium (The Pendulum Clock)	Horologium (The Clock)
Microscopium (The Microscope)	Microscopium (The Microscope)
Mons Mensae (The Table Mountain or Mesa)	Mensa (The Table)
Norma et Regula (The Carpenter's Square and Rule)	Norma (The Square)
Octans (The Octant of Hadley)	Octans (The Octant)
Pyxis (The Mariner's Compass)	Pyxis (The Compass)
Reticulum Rhomboidalis (The Rhomboidal Reticule)	Reticulum (The Net)
Telescopium (The Telescope)	Telescopium (The Telescope)

Lacaille created another set of constellations from the ancient constellation of Argo Navis (The Argonauts' Ship). The original constellation covered a huge portion of the southern hemisphere sky – about 1,800–2,000 square degrees, about 4 or 5% of the entire sky. Today's largest constellation is Hydra which covers just over 3% of the sky. Lacaille thought Argo Navis was unwieldy and in a 1763 posthumous publication of his work, it was broken up into four constellations: Carina (The Keel), Puppis (The Stern or Poop Deck), Vela (The Sail), and Malus (The Mast). Malus was the only one of these constellations which was not generally accepted, and thus did not survive until the present time. Most of its stars were incorporated into Vela.

In his new constellations, Lacaille continued the Bayer scheme of assigning Greek letters to the stars in each constellation in rough order of brightness. However, when he broke up Argo Navis, he kept the already assigned designations of the stars. This left each of these constellations with an incomplete set of letters. Because of this, Carina has an α, β, and ε, but no γ or δ, while Vela has γ and δ, but no α, β, or ε, and the designations in Puppis start with ζ, the sixth letter of the Greek alphabet.

Antlia (ant-lee-ah),	The Air Pump
Antliae (ant-lee-eye)	9:00 P.M. Culmination: Apr. 11
Ant	AREA: 239 sq. deg., 62nd in size

Lacaille's atlas showed "la Machine Pneumatique" (Antlia) as a table with a bell jar on it, and a cylindrical pump extending below. This drawing strongly resembled an air pump invented by fellow Frenchman Denis Papin. Some later atlases show a pump with a handle similar to old-fashioned well pumps. Lacaille's own later atlas changed the name to the Latin "Antlia Pneumatica," which was subsequently shortened to just Antlia.

Lacaille's figure overlay several stars, but the stars were in more-or-less random places on the figure. There is no obvious shape of a pump shown by Antlia's stars, and the brightest stars form only an irregular quadrilateral which is shown on the charts in this book.

Not only small, Antlia is very faint, with only about a dozen stars visible to the unaided eye. However, it is rich in galaxies, although only one is brighter than about 11th magnitude. If you have a moderate-sized telescope and enjoy the challenge, there are rich pickings in Antlia, with many cases of multiple galaxies in a single field of view.

TO FIND: See Fig. 8.1, Finder Chart for Corvus, Crater, and Hydra and Fig. 8.2, The Stars of Corvus, Crater, and Hydra for the location of Antlia. There are no bright stars close enough to Antlia to provide an easy guide to locating it. Antlia lies south of the middle part of Hydra, but it is probably easier to use the fact that it is south of the "Sickle" of Leo. Alpha Antliae is 43° south of Regulus. It is not due south, but slightly east of that line. In fact, the line from η Leonis to Regulus can be extended 43° to almost exactly the position of α Antliae. From α, you will have to carefully follow a star chart to identify the rest of Antlia. The western boundary of Antlia lies almost due south of Alphard (α Hydrae), and the eastern boundary lies due south of the midpoint between α (Alkes) and β Crateris (Figs. 8.1, 8.2, 16.1, 18.1 and 18.2).

NAMED STARS: None.

Magnitudes and spectral types of principal stars:

Bayer desig.	Mag. (m_v)	Spec. type	Bayer desig.	Mag. (m_v)	Spec. type	Bayer desig.	Mag. (m_v)	Spec. type
α	42.8	K4 III	β	–	–*1	γ	–	–*2
δ	5.57	B9/9.5 V	ε	4.51	K3 III	ζ¹	5.75	A0 0
ζ²	5.91	A9 IV	η	5.23	A8 IV	θ	45.78	A7 V
ι	4.60	K0 III	U	5.50v	C6.3			

Stars of magnitude 5.5 or brighter which have only catalog designations:

Draper desig.	Mag. (m_v)	Spec. type	Draper desig.	Mag. (m_v)	Spec. type	Draper desig.	Mag. (m_v)	Spec. type
HD 90132	5.34	A8 V	HD 96146	5.43	A0 V	HD 82205	5.49	K2 III

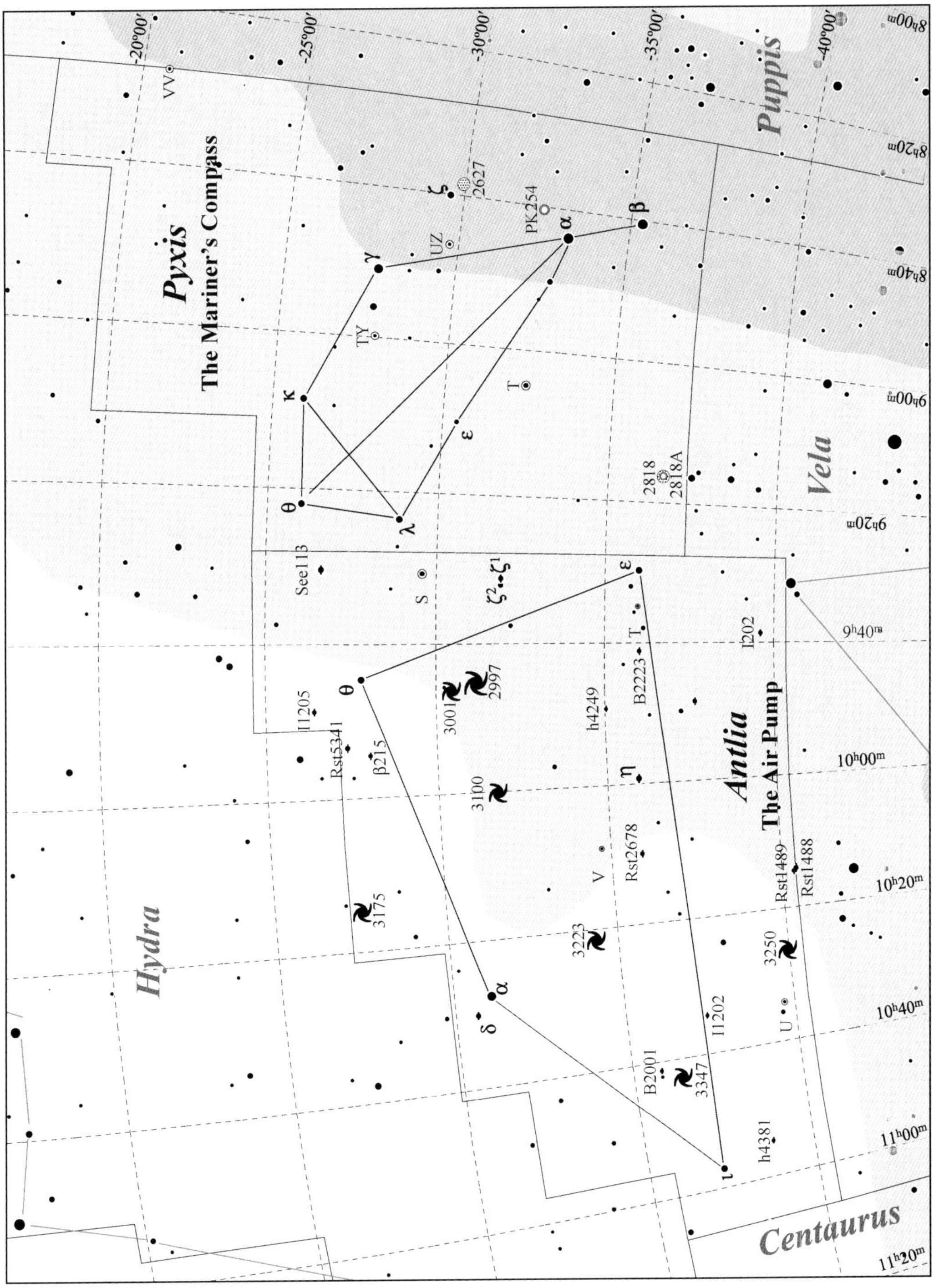

Fig. 18.1. Antlia and Pyxis details.

Table 18.1. Selected telescopic objects in Antlia.

Multiple stars:

Designation	R.A.	Decl.	Type	m₁	m₂	Sep. (")	PA (°)	Colors	Date/Period	Aper. Recm.	Rating	Comments
See113	09ʰ 29.9ᵐ	−26° 35′	A–B	5.5	14.3	11.4	132	O, ?	1999	10/11″	**	Challenging object. For larger scopes
ζ¹ Antliae	09ʰ 30.8ᵐ	−31° 53′	A–B	6.2	6.8	8.1	212	pY, pY	1999	2/3″	****	Nice object for small scopes, a.k.a. Dun 78
I202	09ʰ 38.7ᵐ	−39° 37′	A–B	6.9	8.9	**1.21**	**177**	yW, ?	**110.4**	19/20″	****	Passed min. of 0.08″ in 1999, will reach max. of 1.36″ in 2022
B2223	09ʰ 41.2ᵐ	−36° 02′	A–B	7.3	13.0	11.0	038	Y, ?	1930	6/7″	***	Faint, but pretty nearly identical pair. 2°14′ WNW (PA 292°) of η Ant (m5.23)
Jsp351	09ʰ 48.5ᵐ	−37° 38′	A–B	7.0	12.5	2.3	073	oY, ?	1945	4/5″	***	Close and somewhat difficult
h4249	09ʰ 48.8ᵐ	−35° 01′	A–B	8.0	8.1	4.3	123	W, W	1999	2/3″	****	Similar to above, but a little easier
I205	09ʰ 48.8ᵐ	−26° 25′	A–B	6.7	10.8	1.1	332	yW, ?	2001	8/9″	****	One more step up the ease of separation scale
Rst 5341	09ʰ 53.0ᵐ	−27° 18′	A–B	6.3	10.5	1.3	006	Y, ?	1966	7/8″	***	
β215	09ʰ 54.1ᵐ	−28° 00′	A–B	7.2	9.3	1.8	346	bW, ?	1991	5/3″	***	
η Antliae	09ʰ 58.9ᵐ	−35° 53′	A–B	5.2	11.3	31.4	319	yW, ?	1999	3/4″	***	Brightness difference makes this somewhat challenging. a.k.a. h4271
Rst 2678	10ʰ 09.6ᵐ	−35° 51′	A–B	6.2	10.4	2.3	283	yW, ?	1964	4/5″	***	
Rst 1488	10ʰ 13.8ᵐ	−40° 21′	A–B	5.9	13.0	4.9	103	oY, ?	1945	6/7″	**	
Rst 1489	10ʰ 13.9ᵐ	−40° 19′	A–B	6.4	13.3	12.3	213	O, ?	1945	10/11″	**	For large scopes. Slightly easier than See113 *3
δ Antliae	10ʰ 29.6ᵐ	−30° 36′	A–B	5.6	9.8	10.9	226	pY, W	1999	2/3″	****	Nice object for small scopes. a.k.a. H 50
I2001	10ʰ 34.3ᵐ	−37° 23′	A–B	6.9	10.8	3.9	129	W, ?	1991	3/4″	***	
B2001	10ʰ 40.9ᵐ	−35° 44′	A–B	6.6	9.1	0.6	048	Y, ?	1991	16/17″	***	Very challenging object. Requires excellent seeing
h4381	10ʰ 54.6ᵐ	−38° 45′	A–B	7.0	8.5	25.8	042	bW, W	1999	10×50	****	

Variable stars:

Designation	R.A.	Decl.	Type	Range (mᵥ)		Period (days)	F (fᵣ/f₁)	Spectral range	Aper. Recm.	Rating	Comments
S Antliae	09ʰ 32.3ᵐ	−28° 38′	WE/KE	6.4	6.9	0.648	–	A9Vn	6×30	*****	Prototype of its class of variables. 2° 00′ W (PA 085°) *4 of λ Pyxidis (m7.41).#
T Antliae	09ʰ 33.8ᵐ	−36° 47′	δ Cep	8.9	9.8	5.89	0.22	F6Iab	10×80	****	D ~ 4,000 ly, L ~ 400 ☉. 1°09′ SE (PA 126°) of ε Antliae (m4.71).#
V Antliae	10ʰ 09.2ᵐ	−34° 47′	M	9.2	>12.5	302.76	–	M7IIIe	>6/8″	****	D ~ 4,500 ly, L ~ 200 ☉.##
U Antliae	10ʰ 35.2ᵐ	−39° 34′	Lb	8.1	9.7	–	–	C5,3 (Nb)	10×80	*****	D ~ 8,201Y, L ~ 300 ☉.#

Star clusters:

Designation	R.A.	Decl.	Type	m_v	Size (')	Brtst. Star	Dist. (ly)	dia. (ly)	Number of Stars	Aper. Recm.	Rating	Comments
None												

Nebulae:

Designation	R.A.	Decl.	Type	m_v	Size (')	Brtst./ Cent. star	Dist. (ly)	dia. (ly)	Lum. (suns)	Aper. Recm.	Rating	Comments
None												

Galaxies:

Designation	R.A.	Decl.	Type	m_v	Size (')	Dist. (ly)	dia. (ly)	Lum. (suns)	Aper. Recm.	Rating	Comments
NGC 2997	09h 45.6m	−31° 11'	SA	9.3	10.0×6.3	13.8M	135K	8.9G	8/10"	****	Moderately faint E–W halo with a small, slight central brightening
NGC 3001	09h 46.3m	−30° 26'	SB	11.4	3.1×2.1	106M	108K	22G	8/10"	***	Faint flow, featureless except for slight brightening toward center
NGC 3100	10h 00.7m	−31° 40'	SB	11.2	3.5×2.1	—	—	—	8/10"	***	Faint NNW–SSE halo, bright center. Fainter NGC 3095 and 3108 in FOV
NGC 3175	10h 14.7m	−28° 52'	SB	11.3	5.0×1.3	101M	58K	10.0G	8/10"	****	Faint lenticular NE–SW glow, slightly brighter in middle
NGC 3223	10h 21.6m	−34° 16'	SB	11.0	4.0×2.6	107M	131K	76G	8/10"	***	Faint NW/SE halo, moderate brightening toward center
NGC 3250	10h 26.5m	−39° 57'	SB	11.5	2.8×2.1	122M	67K	50G	8/10"	***	Moderately faint round halo, gradually brightening toward center
NGC 3347	10h 42.8m	−36° 22'	SB	11.4	3.7×1.9	126M	132K	33G	8/10"	***	Faint N–S halo, bright stellar nucleus

Meteor showers:

Designation	R.A.	Decl.	Period (yr)	Comet/ Asteroid	ZHR (max)	Max Date	Duration	Vel. (mi/ km/sec)	First Obs.	Aper. Recm.	Rating	Comments
None												

Other interesting objects:

Designation	R.A.	Decl.	Type	m_v	Size (')	Dist. (ly)	dia. (ly)	Lum. (suns)	Aper. Recm.	Rating	Comments
None										*	

Fig. 18.2a. NGC 2997, spiral galaxy in Antlia. Astrophoto by Rob Gendler.

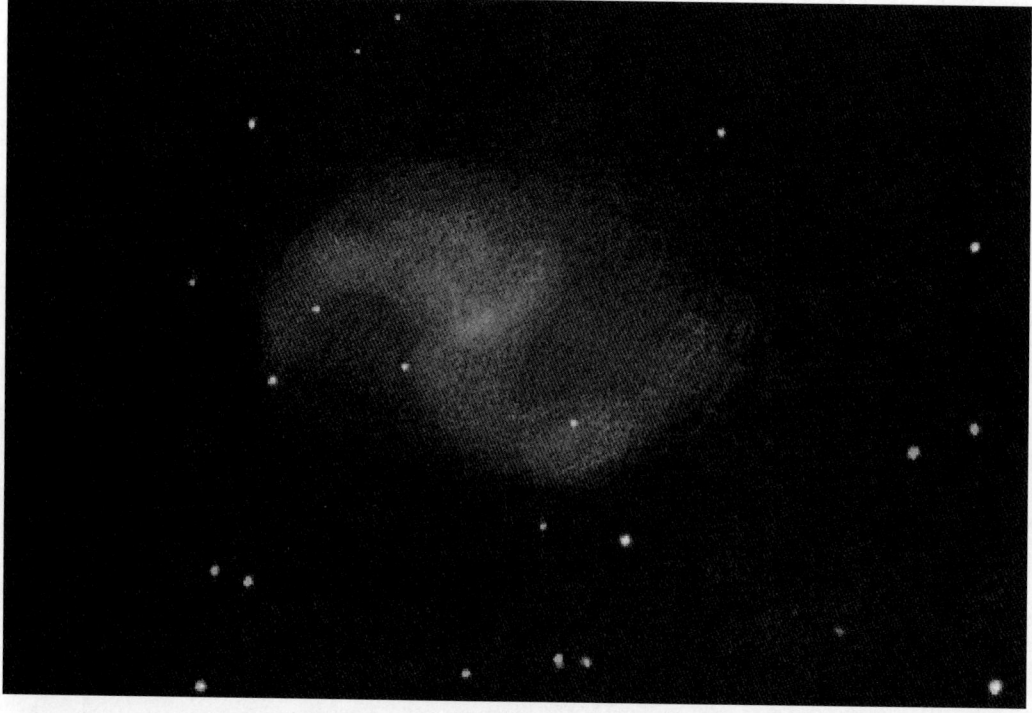

Fig. 18.2b. NGC 2997, spiral galaxy in Antlia. Sketch by Nicolas Biver.

PYXIS (pik-siss),	The Mariner's Compass
Pyxidis (pik-si-diss)	9:00 P.M. Culmination: Mar. 21
Pyx	AREA: 221 sq. deg., 65th in size

Even though its stars are somewhat brighter than those of Antlia, Pyxis is not much more distinctive. Lacaille must have had a great imagination to see an elaborate mariner's compass here, but his atlas shows an eight-pointed compass rose floating over a circular dial, all enclosed in a square box. Because there was no Latin word for compass, Lacaille gave the constellation the name of "Pyxis Nautica" which means "The Nautical Box," but called it the nautical compass.

Although some ancient authors included the stars of Pyxis in the constellation of Argo Navis, these faint stars were never prominent, and by the eighteenth century, they were generally not shown as part of any constellation. Lacaille's own earlier star charts clearly show that he did not include them in Argo or any other constellation.

TO FIND: Pyxis lies just west of Antlia and east of the northern part of Puppis. See Fig. 8.1, Finder Chart for Corvus, Crater, and Hydra and Fig. 8.2, The Stars of Corvus, Crater, and Hydra for the location of Pyxis. A line drawn from the two southernmost of the bright stars in Canis Major, ε (Adhara), and η (Aludra) and extended 20° will end in the center of Pyxis, between α and γ Pyxidis. Together with β Pyxidis, these stars form an arrow pointing toward Alphard (α Hydrae). Again, you will need to consult a star chart to locate the remaining faint stars in Pyxis (Fig. 8.1, 8.2 and 18.1).

NAMED STARS: None.

Magnitudes and spectral types of principal stars:

Bayer desig.	Mag. (m_v)	Spec. type	Bayer desig.	Mag. (m_v)	Spec. type	Bayer desig.	Mag. (m_v)	Spec. type
α	3.68	B15 III	β	3.97	G5 II/III	γ	4.02	K3 III
δ	4.87	A3 IV	ε	5.59	A4 IV	ζ	4.86	G5 III
η	5.24	A0 V	θ	4.71	M0 III*5	κ	4.62	K4/K5 III*6
λ	4.71	G8 III						

Stars of magnitude 5.5 or brighter which have only catalog designations:

Draper desig.	Mag. (m_v)	Spec. type	Draper desig.	Mag. (m_v)	Spec. type	Draper desig.	Mag. (m_v)	Spec. type
HD 73752	5.05	G3/G5V	HD 75605	5.19	G8 III	HD 72310	5.42	B9.5 IV/V

Table 18.2. Selected telescopic objects in Pyxis.

Multiple stars:

Designation	R.A.	Decl.	Type	m₁	m₂	Sep. (")	PA (°)	Colors	Date/ Period	Aper. Recm.	Rating	Comments
Gli 96	08ʰ 28.0ᵐ	−35° 06'	Aα–B	5.7	9.7	25.1	144	bW, –	1999	2/3"	****	A–α is too close (0.10") for amateur scopes
β205	08ʰ 33.1ᵐ	−24° 36'	A–B	6.2	6.3	0.6	293	W, –	142.5	16/17"	***	
h4115	08ʰ 37.5ᵐ	−33° 45'	A–B	6.5	11.8	22.3	159	yW, –	1999	3/4"	****	
			A–C	6.5	12.5	48.8	022	yW, –	1999	4/5"	***	
			A–D	6.5	13.1	25.6	195	yW, –		6/7"		
η Pyxidis	08ʰ 37.9ᵐ	−26° 15'	A–B	5.3	13	16.7	097	W, –	1943	6/7"	****	a.k.a. Rst 2571
β207	08ʰ 38.7ᵐ	−19° 44'	A–B	6.6	9.2	4.4	102	O, W	2003	2/3"	****	
β208	08ʰ 39.1ᵐ	−22° 40'	A–B	5.4	6.8	1.0	042	dY, –	123	10/11"	***	
I314	08ʰ 39.4ᵐ	−36° 36'	A–B	6.4	7.9	0.5	233	yW, –	1981	19/20"	**	
ζ Pyxidis	08ʰ 39.7ᵐ	−29° 34'	A–B	5.0	9.6	52.4	61	Y, pGr	1999	2/3"	****	a.k.a. h4120
β Pyxidis	08ʰ 40.1ᵐ	−35° 18'	A–B	4.0	12.5	12.5	119	dY, –	1999	4/5"	****	a.k.a. Rst 4894
HdO 204	08ʰ 45.2ᵐ	−32° 14'	A–B	6.9	12.0	10.0	310	yW, –	1900	3/4"	***	
			A–C	6.9	12.0	13.0	275	yW, –	1900	3/4"	***	
Gli 101	08ʰ 47.4ᵐ	−34° 36'	A–BC	6.7	10.6	46.7	133	O, W	1999	2/3"	****	
			B–C		11.7	1.9	272	W, –	1990	5/6"	****	
h4144	08ʰ 50.4ᵐ	−35° 55'	A–B	7.0	9.0	2.4	315	bW, –	1991	4/5"	****	a.k.a. I491
κ Pyxidis	09ʰ 08.0ᵐ	−25° 52'	A–B	4.6	10.1	2.1	263	O, –	1911	4/5"	****	
ε Pyxidis	09ʰ 09.9ᵐ	−30° 22'	A–BC	5.6	9.9	17.7	147	pY, pY	1999	15×80	****	At 0.2", B–C is too close for amateur instruments
			A–D	5.6	13.5	38.1	344	pY, ~	1999	7/8"	***	

Variable stars:

Designation	R.A.	Decl.	Type	Range (mᵥ)	Period (days)	F (f₁/f₂)	Spectral Range	Aper. Recm.	Rating	Comments
VV Pyxidis	08ʰ 27.6ᵐ	−20° 51'	EA/DM	6.57 – 7.05	4.596	–	A1V+A1V	6×30	****	1° 30' E (PA 97°) of ζ Pyxidis (m4.01)
UZ Pyxidis	08ʰ 46.6ᵐ	−29° 44'	SRb	6.99 – 7.47	100	–	C5.5j(R8)	6×30	****	56' E (PA 98°) of δ Pyxidis (m4.87)
TY Pyxidis	08ʰ 59.7ᵐ	−27° 49'	EA/D/RS	6.85 – 7.50	3.200	–	G5+G5	6×30	****	
T Pyxidis	09ʰ 04.7ᵐ	−32° 23'	Nr	6.3 – 15.8	–	–	Pec (Nova)	Large	*****	Recurrent nova

Star clusters:

Designation	R.A.	Decl.	Type	m$_v$	Size (')	Brst. Star	Dist. (ly)	dia. (ly)	Number of Stars	Aper. Recm.	Rating	Comments
NGC 2627	08^h 37.3^m	−29° 57'	Open	8.4	9	11	–	–	80	8/10"	***	Moderately rich, 40" SW of ζ Pyxidis. Some nebulosity can be seen in 7×50
NGC 2818	09^h 16.0^m	−36° 37'	Open	8.2	8	11.3	24	10,500	300	8/10"	***	Faint constellation, 3–4 dozen stars 11m or fainter. PN NGC 2118A on W edge. Detectable in 7×50

Nebulae:

Designation	R.A.	Decl.	Type	m$_v$	Size (')	Brst./Cent. star	Dist. (ly)	dia. (ly)	Aper. Recm.	Rating	Comments
PK254	08^h 40.7^m	−32° 23'	Planetary	11.0	9"	–	13.9	–	8/10"	***	Slightly oval disk, brighter toward center. Greenish. a.k.a. Min 3–6
NGC 2818A	09^h 16.0^m	−36° 37'	Planetary	11.6	60"	–	19.5p	10K	16/18"	*	Irregular disk, traces of ring structure

Galaxies:

Designation	R.A.	Decl.	Type	m$_v$	Size (')	Dist. (ly)	dia. (ly)	Lum. (suns)	Aper. Recm.	Rating	Comments
NGC 2613	08^h 33.4^m	−22° 58'	Sb	10.4	6.8×1.7	63M	–	–	4/6"	****	Fairly bright, elongated ESE-WNW. Concentrated. oval core, star-like nucleus

Meteor showers:

Designation	R.A.	Decl.	Period (yr)	Duration	Max Date	ZHR (max)	Comet/Asteroid	First Obs.	Vel. (mi/km/sec)	Rating	Comments
None											

Other interesting objects:

Designation	R.A.	Decl.	Type	m$_v$	Size (')	Dist. (ly)	dia. (ly)	Lum. (suns)	Aper. Recm.	Rating	Comments
None											

Fig. 18.3. NGC 2613, spiral galaxy in Pyxis. Astrophoto by Rob Gendler.

Caelum (see-lum)	The Burin (engraving tool)
Caeli (see-lee)	9:00 P.M. Culmination: Jan. 15
Cae	AREA: 125 sq. deg., 81th in size

Caelum has the second simplest figure of all the constellations. It is just a crossed pair of lines representing two burins (pointed engraving tools or chisels). Only Dorado, with a single line representing the fish's backbone is simpler. It is also one of the star poorest, with only four stars as bright as 5th magnitude.

TO FIND: For northern hemisphere observers: Caelum is completely visible from latitudes south of 41°, but parts may be too near the horizon for viewing. Caelum is very difficult to identify because it is very faint, and it will require careful comparison of the sky with star charts to recognize. Look first for a very sparsely populated region about 20° WNW of <u>Canopus</u>. α Cae (m4.5) lies about 5° S of β Cae (m5.1). This pair marks one of the two burins (scribers) of Lacaille's original constellation. γ Cae (m6.3) lies about 5° ENE of β Cae and 8° NE of α Cae. δ Cae (m5.1) lies about 3.5° SW of α. A line drawn from γ to α and then to δ defines the second burin. Another aid to finding this faint constellation is the use of three naked-eye pairs of stars in Eridanus and Horologium. A line drawn from v^1 to v^2 Eri and extended SE will intersect with a similar line drawn from v^4 to v^3 Eri at a point nearly between γ and β Cae. Also, a line from δ to α Hor and extended about 5° ends near δ Cae. Although these pointer stars are also faint, the pairing makes them relatively easy to identify. See Figs. 10.9 and 18.16 to see the stars of Caelum relative to the surrounding stars of Carina, Columba and Eridanus (Figs. 10.9, 16.1, 18.4 and 18.16).

KEY STAR NAMES: None.

Magnitudes and spectral types of principal stars:

Bayer desig.	Mag. (m_v)	Spec. type	Bayer desig.	Mag. (m_v)	Spec. type	Bayer desig.	Mag. (m_v)	Spec. type
α	4.44	F2 V	β	5.04	F3 V	$γ^1$	4.55	K2 III[*7]
δ	5.07	B2 IV/V[*8]	ν	6.06	F1 III/IV[*9]			

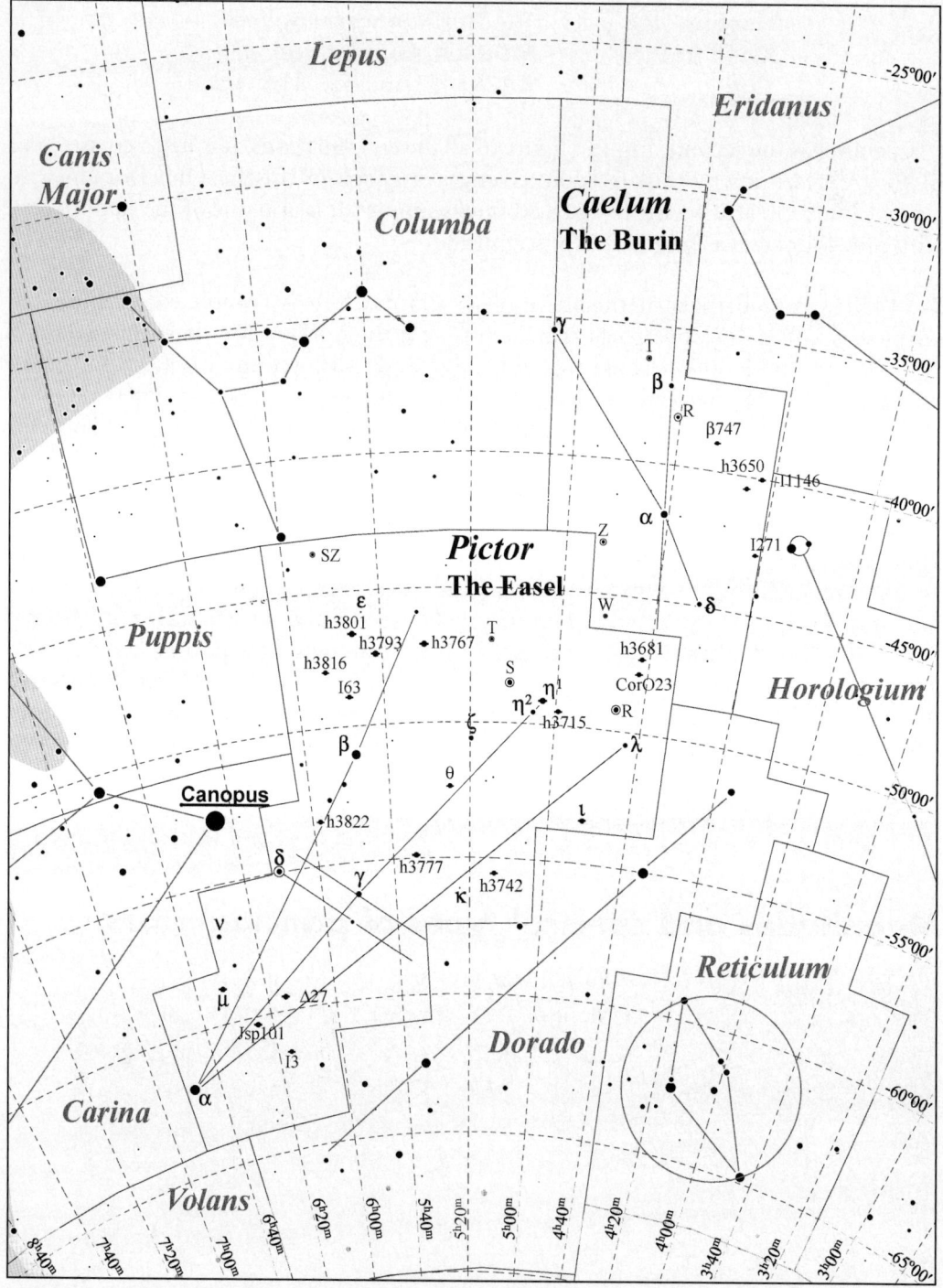

Fig. 18.4. Caelum and Pictor details.

Table 18.3. Selected telescopic objects in Caelum.

Multiple stars:

Designation	R.A.	Decl.	Type	m_1	m_2	Sep. (")	PA (°)	Colors	Date/Period	Aper. Recm.	Rating	Comments
h3650	04^h 26.6^m	-40° 32'	A–B	7.0	8.2	2.9	189	bW, W	1993	3/4"	****	Perhaps the best double in Caelum
I271	04^h 21.7^m	-42° 47'	A–B	7.8	10.4	2.5	143	oY, –	1993	4/5"	****	
I1146	04^h 24.2^m	-40° 03'	A–B	7.3	11.7	6.2	042	bW, –	1998	3/4"	****	
β747	04^h 33.0^m	-38° 17'	A–B	7.9	9.8	3.2	222	W, –	1993	3/4"	****	
h3674	04^h 38.9^m	-37° 20'	A–B	8.5	12.5	28.5	208	oY, –	1998	4/5"	***	
α Caeli	04^h 40.6^m	-41° 52'	A–B	4.5	12.5	6.3	120	yW, –	1933	4/5"	***	a.k.a. Hdo 190
v Caeli	04^h 50.3^m	-41° 19'	A–B	6.1	10.4	13.0	258	W, –	1998	2/3"	****	a.k.a. h3697
h3695	04^h 50.5^m	-38° 34'	A–B	7.3	13.0	32.3	055	yW, –	1998	6/7"	***	
Sec41	04^h 52.6^m	-30° 40'	A–B	7.8	12.1	9.3	122	W, –	1998	3/4"	****	
λ^1 Caeli	05^h 04.4^m	-35° 29'	A–B	4.7	8.2	3.2	305	O, –	2001	3/4"	****	a.k.a. Jc 9

Variable stars:

Designation	R.A.	Decl.	Type	Range (m_v)		Period (days)	F (f_1/f_2)	Spectral Range	Aper. Recm.	Rating	Comments
R Caeli	04^h 40.5^m	-38° 14'	M	6.7	13.7	390.95	0.41	M6e	10/12"	*****	Very red. 3° 32' ESE (PA 105°) of δ Caeli (m5.07.##
T Caeli	04^h 47.3^m	-36° 13'	SR	9.0	**10.8**	156	–	C6, 4 (N4)	3/4"	****	Carbon star (sp. cl. N4). 1° 32' NE (PA 049°) of β Caeli (m5.04).#
W Caeli	04^h 50.6^m	-45° 52'	EA	10	**10.8**	6.979	–	G0	3/4"	***	3° 32' ESE (PA 105°) of δ Caeli. A challenge for southern observers.#
Z Caeli	04^h 52.6^m	-43° 04'	SR	7.84	7.99	52	–	M2III	7×50	**	2° 31″ E (PA119°) of β Caeli

Star clusters:

Designation	R.A.	Decl.	Type	Size (')	Brtst. Star	m_v	Dist. (ly)	dia. (ly)	Number of Stars	Aper. Recm.	Rating	Comments
None												

(continued)

Table 18.3. (continued)

Nebulae:

Designation	R.A.	Decl.	Type	m_v	Size (')	Brst./**Cent.** star	Dist. (ly)	dia. (ly)	Aper. Recm.	Rating	Comments
None											

Galaxies:

Designation	R.A.	Decl.	Type	m_v	Size (')	Dist. (ly)	dia. (ly)	Lum. (suns)	Aper. Recm.	Rating	Comments
NGC 1679	04^h 50.0^m	−31° 59'	Sc	11.5	3.0×1.5	38M	33K		14/16"	***	Somewhat irregular, patchy arms with many knots, a few foreground stars

Meteor showers:

Designation	R.A.	Decl.	Period (yr)	Duration	Max Date	ZHR (max)	Comet/**Asteroid**	First Obs.	Vel. (mi/**km**/sec)	Rating	Comments
None											

Other interesting objects:

Designation	R.A.	Decl.	Type	m_v	Mass (suns)	Dia. (suns)	Dist. (ly)	Planet Mass (J)	Dist. (AU)	Period (days)	Eccentricity	Comments
None												

Pictor (pik-tor)	**The (Painter's) Easel**
Pictoris (pik-**tor**-iss)	**9:00 P.M. Culmination:** Jan. 31
Pic	**AREA:** 247 sq. deg., 59th in size

Lacaille first called this constellation le Chevalet et la Palette, "the Easel and the Palette." This was later Latinized to Equuleus Pictoris, and finally shortened to just Pictor, the Easel. Note that both the French and Latin words for easel literally mean "little horse" similar to the English "sawhorse" for a device to support something while it is being worked upon. The figure of Pictor is a simple three-legged artist's easel.

TO FIND: Although the stars of Pictor are relatively faint, the constellation is conveniently close to Canopus (α Carinae), the second brightest star in the entire sky. Pictor's four brightest stars can all be located by starting with Canopus. α Pictoris lies 9.8° south–southeast (PA 163°) from Canopus and β Pictoris lies only 5.9° west–northwest (PA 282°) of Canopus. Together, α and β define the top part of one leg of the celestial easel. Extending the α–β line another 4.8° takes you to a faint star (HD 36553) which defines the end of that leg. Returning to Canopus and going 6° to the southwest (PA 232°) will take you to γ PIctoris. A line from α to γ pictoris marks the upper part of the middle leg of the easel. Extending this line by 9.4° takes you to the faint pair η^1 and η^2 Pictoris at the end of that leg. The final leg goes from α to λ, but at that distance, it would be difficult to distinguish 5th magnitude λ Pictoris from the other faint stars in that area. An easier way to find λ is to start with the η^1 and η^2 pair and go 3.4° to the west south–west (PA 251°). In addition to marking the middle of the second tripod leg, γ Pictoris also marks the middle of the horizontal bar which supports the artist's canvas in use. This bar points directly to Canopus, and δ Pictoris, about midway between Canopus and γ Pictoris marks the eastern end of the bar. Although the stars are faint, the nearness to brilliant Canopus and the fact that the middle leg of the easel is nearly parallel to the body of Dorado, and about halfway between Canopus and the four stars forming the straight-line center of Dorado gives you an advantage in finding it. See Figs. 10.9, 16.1 and 18.16 to find the stars of Pictor relative to Canopus and the constellations of Dorado and Caelum.

KEY STAR NAMES: None.

Magnitudes and spectral types of principal stars:

Bayer desig.	Mag. (m_v)	Spec. type	Bayer desig.	Mag. (m_v)	Spec. type	Bayer desig.	Mag. (m_v)	Spec. type
α	3.24	A7 IV	β	3.85	A3 V	γ	4.50	K1 III
δ	4.72	B0.5 IV	ε	6.38	F0 III	ζ	5.44	F7 III/IV
η[1]	5.37	F2 V	η[2]	5.05	M2 IIIv*[10]	θ	6.26	A0 V
ι	6.42	F0 IV	κ	6.10	B8/9 V	λ	5.01	K0/1 III
μ	5.69	B9 V	ν	5.60	Am*[11]			

Stars of magnitude 5.5 or brighter which have only catalog designations:

Draper desig.	Mag. (m_v)	Spec. type	Draper desig.	Mag. (m_v)	Spec. type	Draper desig.	Mag. (m_v)	Spec. type
HD 42540	5.04	K2/3 III	HD 39640	5/16	G8 III	HD 40292	5.29	F0 V
HD 38871	5.31	K0/1 III	HD 36553	5.46	G3 IV			

Table 18.4. Selected telescopic objects in Pictor.

Multiple stars:

Designation	R.A.	Decl.	Type	m₁	m₂	Sep. (″)	PA (°)	Colors	Date/ **Period**	Aper. Recm.	Rating	Comments
h3681	04ʰ 41.8ᵐ	−47° 16′	A–B	6.8	10.6	49.6	246	Y, –	1999	15×80	****	
CorO23	04ʰ 41.9ᵐ	−47° 50′	A–B	7.5	9.6	3.7	228	Y, –	2002	2/3″	****	
ι Pictoris	04ʰ 50.9ᵐ	−53° 28′	A–B	5.6	6.2	12.6	059	yW, pY	2002	2/3″	*****	Showcase pair, a.k.a. Δ18
h3715	04ʰ 59.5ᵐ	−49° 27′	A–B	7.2	9.1	9.9	112	Y, Y	1999	2/3″	****	
η¹ Pictoris	05ʰ 02.8ᵐ	−49° 09′	A–B	5.4	10.4	14.0	198	yW, –	1949	2/3″	****	a.k.a. Rst20
h3742	05ʰ 13.5ᵐ	−55° 34′	A–B	7.1	12.7	25.9	295	Y, –	2000	5/6″	***	
θ Pictoris	05ʰ 24.8ᵐ	−52° 19′	AB–C	6.9	7.2	**0.20**	**263**	W, W	**191**	–	****	A–B separation is increasing to a max. of 0.46″ in 2078. a.k.a. Δ20
h3767	05ʰ 30.2ᵐ	−47° 05′	A–C	6.3	6.7	38.1	289	W, W	2002	6×30	****	Easy, fairly well matched pair
			A–BC	5.5	11.1	25.9	257	dY, –	1933	3/4″	***	
			B–C	11.7	12.7	0.6	85	–, –	1979	16/17″	***	a.k.a. Rst136
			A–D	5.5	6.7	197.8	271	Y, Y	2000	6×30	****	a.k.a. A21-AD
h3777	05ʰ 33.7ᵐ	−54° 54′	A–B	6.5	10.0	58.6	342	yW, –	1999	2/3″	****	
			B–C	10.2	12.2	11.1	105	Y, Y	1999	4/5″	****	
h3793	05ʰ 41.5ᵐ	−48° 15′	A–B	7.1	10.4	12.7	135	W, –	1999	2/3″	****	
h3801	05ʰ 46.5ᵐ	−46° 36′	A–B	5.3	12.5	40.0	195	oY, –	1999	4/5″	***	
163	05ʰ 48.2ᵐ	−48° 55′	A–B	7.3	8.6	1.0	358	W, –	1999	9/10″	****	
			A–C	7.3	13.0	32.8	208	W, –	1999	6/7″	***	
h3816	05ʰ 53.2ᵐ	−47° 57′	A–B	6.6	11.6	22.9	180	oY, –	1913	3/4″	****	
h3822	05ʰ 57.2ᵐ	−53° 26′	A–B	6.6	7.6	55.6	303	O, Y	2000	7×50	****	
			B–C	7.6	13	19.3	117	oY, –	2000	6/7″	****	
13	06ʰ 12.5ᵐ	−61° 28′	A–B	7.1	7.6	1.0	005	W, –	1991	09/10	****	a.k.a. LDS157
Δ27	06ʰ 16.3ᵐ	−59° 13′	A–B	6.5	7.6	34.4	233	oY, B	1999	7×50	*****	A somewhat fainter Albireo for the southern hemisphere
Jsp	06ʰ 22.5ᵐ	−60° 13′	A–B	6.5	14.2	40.4	310	B, –	1999	10/11″	**	
			A–B	6.5	10.6	16.2	050	Y, Y	1977	15×80	****	
μ Pictoris	06ʰ 32.0ᵐ	−58° 45′	A–B	5.6	9.3	2.5	230	W, –	1991	3/4″	****	a.k.a. h3874

Variable stars:

Designation	R.A.	Decl.	Type	Range (m)		Period (days)	F (f_r/f_l)	Spectral Range	Aper. Recm.	Rating	Comments
R Pictoris	04ʰ 46.2ᵐ	−49° 15′	SR	6.4	10.1	170.9	–	M1Ie–M4IIe	10×80	****	1° 21′ NNE (PA 24°) of λ Pictoris (M5.31)
S Pictoris	05ʰ 11.0ᵐ	−48° 30′	M	6.5	14.0	428.0	0.36	M6.5e–M8e	12/14″	****	1° 29′ ENE (PA 65°) of η¹ Pictoris (m5.73)

(continued)

Table 18.4. (continued)

Variable stars:

Designation	R.A.	Decl.	Type	Range (m$_v$)	Period (days)	F (f$_r$/f$_f$)	Spectral Range	Aper. Recm.	Rating	Comments
T Pictoris	05^h 15.1^m	−46° 55'	M	7.9	200.58	0.49	M6III	14/16"	****	1° 44' NNE (PA 24°) of S Pictoris
SZ Pictoris	05^h 53.5^m	−43° 33'	RS	8.0	2.441	–	G8V	7×50	****	1° 17' NNW (PA 234°) of λ Colombae (m3.96)
δ Pictoris	06^h 10.3^m	−54° 58'	EB/D	4.7	1.672	0.18	B3III+O9V	Eye	****	A hot blue giant and hotter blue main sequence, estimated distance ~3,000 ly, Lum 8,000

Star clusters:

Designation	R.A.	Decl.	Type	m$_v$	Size (')	Brtst. Star	Dist. (ly)	dia. (ly)	Number of Stars	Aper. Recm.	Rating	Comments
None												

Nebulae:

Designation	R.A.	Decl.	Type	m$_v$	Size (')	Brtst./ Cent. star	Dist. (ly)	dia. (ly)	Aper. Recm.	Rating	Comments
None											

Galaxies:

Designation	R.A.	Decl.	Type	m$_v$	Size (')	Dist. (ly)	dia. (ly)	Lum. (suns)	Aper. Recm.	Rating	Comments
NGC 1705	04^h 54.2^m	−53° 22'	S0p	12.0	1.9×1.4	17M	9.4K	36.5M	6/8"	***	

Meteor showers:

Designation	R.A.	Decl.	Period (yr)	Duration	Max Date	ZHR (max)	Comet/ **Asteroid**	First Obs.	Vel. (mi/ **km**/sec)	Rating	Comments
None											

Other interesting objects:

Designation	R.A.	Decl.	Type	m$_v$	Size (')	dia. (ly)	Dist. (ly)	Lum. (suns)	Aper. Recm.	Rating	Comments
Kapteyn's Star	05^h 11.3^m	−44° 59'	Hi PM	8.8			12.7		7×50	***	Second highest proper motion star, 8.70 arcsec/year in PA 131° *12
β Pictoris	05^h 57.0^m	−51° 04'	Disk star	3.8			6.2		Eye	***	A disk of dust and ice has been detected surrounding β Pictoris *13

Circinus (sir-**sin**-us)	**The (Drafting) Compass**
Circini (sir-**sin**-ee)	**9:00 P.M. Culmination:** June 15
Cir	**AREA:** 93 sq. deg., 85th in size

One of the smallest and simplest of the constellations, Circinus represents a drafting compass or divider. The figure of Circinus first appeared on Lacaille's planisphere published in 1754, but it was not named until 2 years later when Lacaille labeled it "le Compas." In 1763, the name was Latinized to "Circinus."

TO FIND: For northern hemisphere observers: Circinus is completely visible from south of latitude 30°, but parts may be too near the horizon for viewing. Circinus would be somewhat difficult to identify if it were in most parts of the sky because most of its stars are very faint, with only α Cir being brighter than 4th magnitude. However, two of the brightest stars in the southern sky, α and β Centauri, are very close by and point to the middle of Circinus. α Cir lies only 4° south of α Centauri, and is thus very easy to find and identify, even though it is only magnitude 3.2. β and γ Cir lie slightly more than 5° ENE of α Centauri (Figs. 6.26, 16.1 and 18.4).

KEY STAR NAMES: None.

Magnitudes and spectral types of principal stars:

Bayer desig.	Mag. (m_v)	Spec. type	Bayer desig.	Mag. (m_v)	Spec. type	Bayer desig.	Mag. (m_v)	Spec. type
α	3.18	F1 V	β	4.07	A3 V	γ	4.48	B5 III
δ	5.04	O8.5 V	ε	4.85	K2.5 III	ζ	6.09	B3 V
η	5.16	G8 III	θ	5.08	B4 V*[14]			

Stars of magnitude 5.5 or brighter which have only catalog designations:

Designation	Mag. (m_v)	Spec. type	Designation	Mag. (m_v)	Spec. type	Designation	Mag. (m_v)	Spec. type
HD131342	5.18	K1 III	HD129422	5.36	A7 V	HD 40292	5.29	F0 V

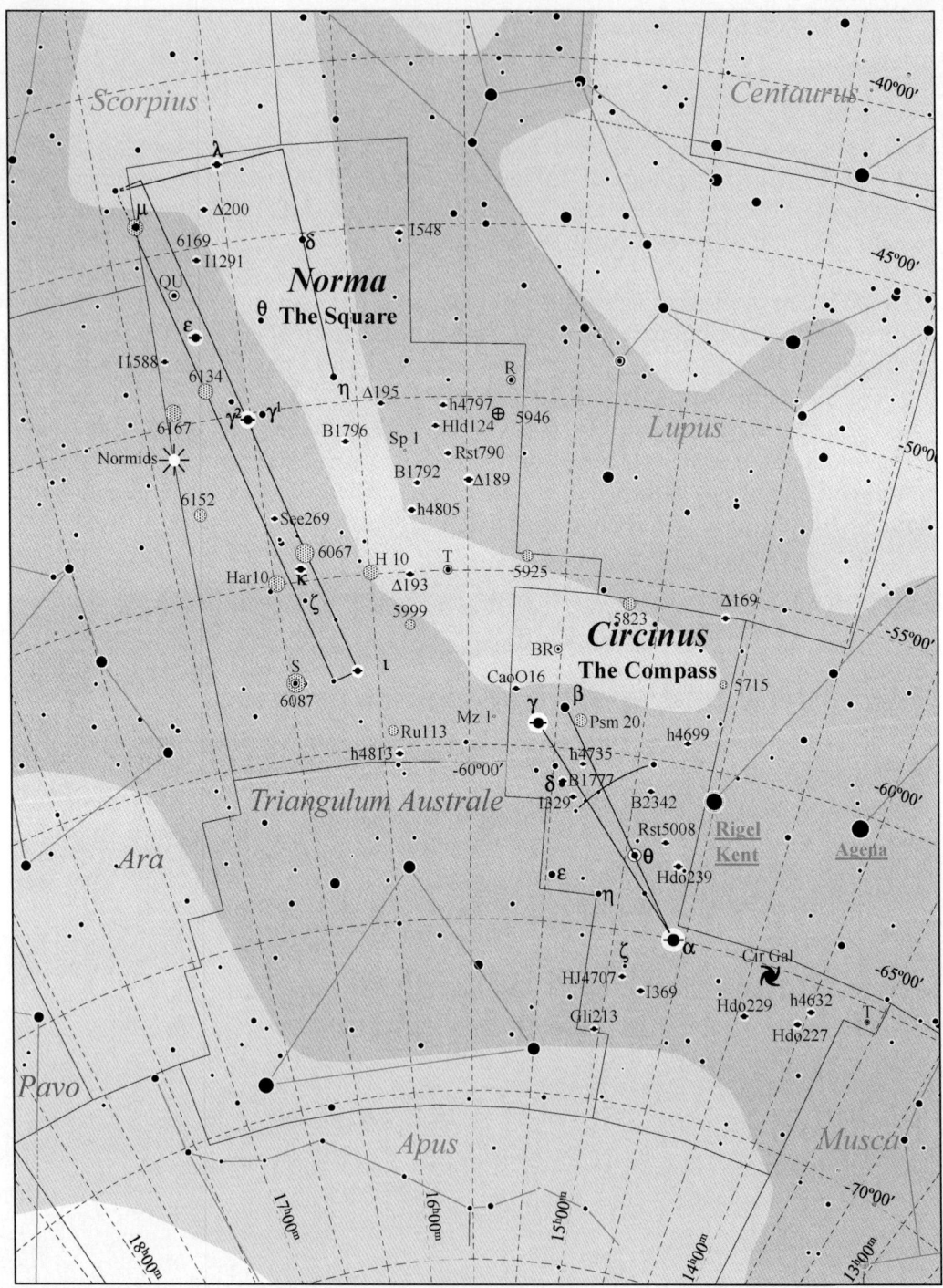

Fig. 18.5. Circinus and Norma details.

Table 18.5. Selected telescopic objects in Circinus.

Multiple stars:

Designation	R.A.	Decl.	Type	m_1	m_2	Sep. (")	PA (°)	Colors	Date/Period	Aper. Recm.	Rating	Comments
h4632	13h 58.5m	−65° 48'	A–B	6.4c	9.5	6.5	014	oY, ?	1991	2/3"	****	
Hdo227	14h 00.9m	−66° 16'	A–B	6.0	12.0	41.5	182	yW, ?	2000	3/4"	***	
Hdo229	14h 16.7m	−66° 35'	A–B	5.8	13.5	23.9	313	bW, ?	2000	7/8"	***	
α Circini	14h 42.5m	−64° 59'	A–B	3.2	8.5	15.1	227	Y, R	2002	15×80	****	a.k.a. Δ166
Δ169	14h 45.2m	−55° 36'	A–B	6.1	7.6	69.2	105	B, O	1999	7×50	****	
Hdo239	14h 45.3m	−62° 53'	A–B	5.4	10.8	34.7	045	yW, ?	2000	3/4"	****	
I369	14h 48.7m	−66° 36'	A–B	5.9	9.0	46.6	034	bW, W	2000	7×50	****	a.k.a. Jrn 16-AD
	14h 48.7m	−66° 36'	A–D	5.9	7.7	87.8	323	bW, ?	2000	7×50	****	
	14h 48.7m	−66° 36'	B–C	9.0	11.0	2.7	241	W, ?	1929	4/5"	****	
h4699	14h 49.2m	−59° 24'	AB–C	6.8	9.1	37.0	125	O, W	1999	2/3"	****	At 0.1", AB (FIN298-AB) is too close for amateur instruments
Rst5008	14h 50.0m	−62° 20'	A–B	7.4	12.3	2.8	114	W, ?	1966	4/5"	***	
HJ 4707	14h 54.2m	−66° 25'	A–B	7.5	8.1	**1.07**	**273**	Y, ?	**346**	8/9"	****	26' S of ζ Cir. Secondary min. of 0.53" in 1975, opening to primary max. of 2.192 in 2107
B2342	14h 56.0m	−60° 54'	A–B	6.8	12.0	5.9	089	dY, ?	2000	3/4"	***	
Gli213	15h 01.3m	−67° 59'	A–B	7.1	9.3	5.2	333	bW, ?	2000	2/3"	****	
h4735	15h 12.8m	−60° 24'	A–B	7.6	10.5	7.4	031	yW, ?	2000	2/3"	****	
I329	15h 14.0m	−61° 21'	A–B	6.7	7.7	0.9	339	bW, ?	1996	10/11"	****	
B1777	15h 16.6m	−60° 54'	A–B	5.8	8.6	1.3	164	bW, ?	1991	7/8"	****	
I428-AC	15h 16.6m	−60° 54'	A–C	5.7	11.6	20.0	315	bW, ?	2000	3/4"	****	
γ Circini	15h 23.4m	−59° 19'	A–B	5.1	5.5	**0.82**	**347**	bW, O	**269.9**	11/12"	*****	Secondary min. of 0.79" in 1995, opening to 1.52" in 2155. a.k.a. h4757
CapO16	15h 29.5m	−58° 21'	A–B	7.0	8.0	2.4	032	W, ?	1991	4/5"	****	

(continued)

Table 18.5. (continued)

Variable stars:

Designation	R.A.	Decl.	Type	Range (m_v)	Period (days)	F (f_r/f_f)	Spectral Range	Aper. Recm.	Rating	Comments
T Circini	13h 43.4m	−65° 28′	EA/SD	**9.3 10.6**	3.298	0.13	B6-B8, II-III	3/4″	****	2° 21′ SW (PA 219°) of three Cent. (m4.71).# *15
θ Circini	14h 56.7m	−62° 47′	γ Cas	5.0 5.4	–	–	B3Vne	Eye	***	Rapidly rotating star losing mass at equator, irregular variation
BR Circini	15h 56.7m	−57° 10′	Irr	7.2 11.4	–	–	A3II-III	4/6″	****	1° 41′ NNE (PA 015°) of β Cir (m4.07). a.k.a. Cir X-I, an X-ray source

Star clusters:

Designation	R.A.	Decl.	Type	m_v	Size (′)	Brtst. Star	Dist. (ly)	dia. (ly)	Number of Stars	Aper. Recm.	Rating	Comments
NGC 5715	14h 43.7m	−57° 34′	Open	9.8	7.0	11.0	–	–	30	3/4″	***	~80 stars visible in 12″. Total luminosity ~650 suns
NGC 5823	15h 05.7m	−55° 36′	Open	7.9	12.0	9.7	3,400	4	100	7×50	****	
Pismis 20	15h 15.4m	−59° 04′	Open	7.8	4.5	8.2	8,270	11	12	7×50	****	~12 stars visible in 12″. Total luminosity ~4,100 suns

Nebulae:

Designation	R.A.	Decl.	Type	m_v	Size (′)	Brtst./**Cent. star**	Dist. (ly)	dia. (ly)	Lum. (suns)	Aper. Recm.	Rating	Comments
NGC 5315	13h 53.9m	−66° 31′	Planetary	9.8	6″		1,600	11.4	9,000	10/12″	****	Stellar at LP, high power needed to distinguish disk

Galaxies:

Designation	R.A.	Decl.	Type	m_v	Size (′)	Dist. (ly)	dia. (ly)	Lum. (suns)	Aper. Recm.	Rating	Comments

Circinus Galaxy	R.A.	Decl.	Type	m$_v$	Size (')	Dist. (ly)	dia. (ly)	Lum. (suns)	Aper. Recm.	Rating	Comments
	14^h 13.2^m	−65° 20'	SA	10.6	6.9×3.0	13M	26K	78M	8/10"	****	a.k.a. ESO 97-G13. Seyfert Galaxy. Stellar nucleus, many foreground stars

Meteor showers:

Designation	R.A.	Decl.	Period (yr)	Duration	Max Date	ZHR (max)	Comet/**Asteroid**	First Obs.	Vel. (mi/ **km**/sec)	Rating	Comments
None											

Other interesting objects:

Designation	R.A.	Decl.	Type	m$_v$	Size (')	Dist. (ly)	dia. (ly)	Lum. (suns)	Aper. Recm.	Rating	Comments
None											

Norma (nor-muh)	The (Carpenter's) Square
Normae (nor-mye)	9:00 P.M. Culmination: July 4
Nor	AREA: 165 sq. deg., 74th in size

Lacaille introduced this constellation as "l'Equerre et la Regle," The Carpenter's Square and Rule. This was first Latinized to Norma Et Regula, The Square and Rule, and then shortened to just Norma, the Square.

TO FIND: To see λ, the northernmost star in Norma, northern hemisphere observers must be south of 47° latitude, and to see the entire constellation, they must be south of 30°. Although Norma is relatively faint, with only α reaching 4th magnitude, and only 15 other stars reaching 5th magnitude, it is not difficult to find. Note from the charts that the line of stars forming the "rule" is fairly straight and fairly regular, with the distances from μ to ε, ε to γ, γ to κ, and κ to ι¹ and ι² averaging about 3.5°, and that the rule lies about midway between ε Scorpii and β Trianguli Australis, and points to both. For northern observers, start with ε Scorpii and look about 10° south–southwest to find μ Normae. You can then follow the line of stars above until you reach ι Normae at the southern end of the rule. Southern observers can reverse this process by starting with Triangulum Australe and extending the line from γ to β Trianguli Australis about 6° to locate ι¹ and ι², then follow the chain of stars to μ Normae at the north end. The square accompanying the rule is fainter and less well defined, consisting only of λ, δ, and η Normae (Figs. 6.38, 16.1 and 18.5).

KEY STAR NAMES: None.

Magnitudes and spectral types of principal stars:

Bayer desig.	Mag. (m$_v$)	Spec. type	Bayer desig.	Mag. (m$_v$)	Spec. type	Bayer desig.	Mag. (m$_v$)	Spec. type
α	–	–*16	β	–	–*17	γ¹	4.97	F9 Ia
γ²	4.01	G8 III	δ	4.73	Am	ε	4.46	B4 V
ζ	0.00	O0 0	η	4.65	G8 III	θ	5.13	B8 V
ι¹	4.63	A7 IV	ι²	5.57	B9 V	κ	4.95	G4 III
λ	5.44	A3 V	μ	4.86	B0 Ia			

Other stars magnitude 5.5 or brighter which have only catalog designations:

Designation	Mag. (m$_v$)	Spec. type	Designation	Mag. (m$_v$)	Spec. type
HD 147152	5.32	B6 IV	HD 148379	5.35	B1.5 Ia
HD 139129	5.43	B9 V	HD 146003	5.45	M2 IV

Table 18.6. Selected telescopic objects in Norma.

Multiple stars:

Designation	R.A.	Decl.	Type	m_1	m_2	Sep. (")	PA (°)	Colors	Date/ Period	Aper. Recm.	Rating	Comments
Δ189	15ʰ 38.8ᵐ	−52° 22′	A–B	5.4	10.7	53.0	280	W,?	1999	15×80	***	
			B–C	10.0	10.7	4.9	023	?,?	1999	3/4″	***	
Rst790	15ʰ 43.6ᵐ	−51° 38′	A–B	6.7	12.9	2.5	041	oY,?	1947	5/6″	***	
h4797	15ʰ 44.2ᵐ	−0° 13′	A–B	6.9	10.6	21.3	257	yW,?	1999	3/4″	****	
Hld 124	15ʰ 45.0ᵐ	−50° 47′	A–B	6.6	8.5	2.2	195	W,?	1996	4/5″	****	
B1792	15ʰ 48.8ᵐ	−52° 26′	A–B	6.2	10.3	1.2	288	oY,?	1991	8.9″	****	
h4805	15ʰ 50.1ᵐ	−53° 12′	A–B	5.8	11.8	28.5	119	W,?	1999	3/4″	****	
I548	15ʰ 50.3ᵐ	−45° 24′	A–BC	6.2	11.0	3.3	188	yW,?	1976	3/4″	****	
			B–C	11.6	11.9	0.3	329	W,?	1976	–	**	
Δ193	15ʰ 51.1ᵐ	−55° 03′	A–B	5.8	9.1	16.5	012	W,O	1999	15×80	****	
Δ195	15ʰ 54.8ᵐ	−50° 20′	A–B	6.8	12.0	5.0	010	Y,W	1999	3/4″	***	
			A–C	7.4	11.5	26.5	299	W,?	1999	3/4″	****	
h4813	15ʰ 55.5ᵐ	−60° 11′	A–B	5.9	8.4	4.4	100	Y,W	2000	2/3″	****	
B1796	16ʰ 02.1ᵐ	−51° 07′	A–B	6.4	14.0	5.2	257	bW,?	1933	9/10″	***	
			A–C	6.4	11.2	11.3	296	bW,?	2000	3/4″	*****	
τ Normae	16ʰ 03.5ᵐ	−57° 47′	A–B	5.2	5.8	0.40	307	Y,Y	26.93 years	–	*****	a.k.a. h4825-AB×C
			AB–C	4.6	8.0	10.9	243	Y,oR	2000.0	2/3″	****	
κ Normae	16ʰ 13.5ᵐ	−54° 38′	A–B	4.9	13.0	14.2	214	dY,?	1998	6/7″	***	a.k.a. Hdo254
See269	16ʰ 17.4ᵐ	−53° 05′	A–B	6.3	13.0	22.7	015	yW,?	2000	6/7″	***	
λ Normae	16ʰ 19.3ᵐ	−42° 40′	A–B	5.8	6.9	0.37	106	W,?	1999	–	**	a.k.a. See271
γ² Normae	16ʰ 19.9ᵐ	−50° 09′	A–B	4.1	10.5	47.7	018	dY,?	1999	2/3″	****	a.k.a. h4841
Δ200	16ʰ 22.5ᵐ	−43° 55′	A–B	6.0	9.7	39.2	195	Y,?	1999	15×80	****	
I1291	16ʰ 24.9ᵐ	−45° 20′	A–B	6.3	12.5	4.6	033	W,?	1999	4/5″	***	
ε Normae	16ʰ 27.2ᵐ	−47° 33′	A–B	4.5	6.1	22.8	335	pY,pB	2002	15×80	*****	Showcase pair, a.k.a. h4853
I1588	16ʰ 33.9ᵐ	−48° 06′	A–B	6.7	12.6	2.9	270	B,?	1936	5/6″	***	
			A–C	6.7	13.2	35.5	359	B,?	1934	6/7″	***	

Variable stars:

Designation	R.A.	Decl.	Type	Range (m)		Period (days)	F (f_1/f_2)	Spectral Range	Aper. Recm.	Rating	Comments
R Normae	15ʰ 36.0ᵐ	−49° 30′	M	6.5	13.9	507.50	–	M3e–M6II	12/14″	****	
T Normae	15ʰ 44.1ᵐ	−54° 59′	M	6.2	13.6	240.7	0.39	M3e–M6e	10/12″	****	
S Normae	16ʰ 18.9ᵐ	−57° 54′	δ Cep	6.1	6.8	9.754	0.48	F8–G0Ib	6×30	****	In NGC 6087 (see below)
QU Normae	16ʰ 29.7ᵐ	−46° 15′	γ Cas	5.3	5.4	–	–	B1.5Iape	Eye	****	1° 23′ NNE [PA 018°] of ∈ Normae (m5.29)

(continued)

Table 18.6. (continued)

Star clusters:

Designation	R.A.	Decl.	Type	m_v	Size (')	Brst. Star	Dist. (ly)	dia. (ly)	Number of Stars	Aper. Recm.	Rating	Comments
NGC 5925	15ʰ 27.7ᵐ	−54° 31'	Open	8.4	15	–	–	–	120	7×50	****	Faint, but rich. Interesting circlet, plus other lines and curves of stars
NGC 5946	15ʰ 35.5ᵐ	−50° 40'	Globular	8.4	3	–	34K	148	–	7×50	****	
NGC 5999	15ʰ 52.2ᵐ	−56° 28'	Open	9.0	3	12.0	–	–	50	8/10"	***	
Ru 113	15ʰ 57.2ᵐ	−59° 28'	Open	–	25	9.0	–	–	20	7×50	****	Large, poorly defined, ~4° N of β Triangulum Australis
NGC 6067	16ʰ 13.2ᵐ	−54° 13'	Open	5.6	15	10.0	5.6K	24	315	7×50	*****	Slightly N of κ Normae. Beautiful, overlooked object
NGC 6087	16ʰ 18.9ᵐ	−57° 54'	Open	5.4	15	8.0	2.9K	4	349	7×50	****	Another striking, but often overlooked southern summer object
Harvard 10	16ʰ 19.9ᵐ	−54° 59'	Open	6.9p	25	11.0	4.3K	31	31	7×50	***	Yet another overlooked object, ~10 stars visible in 7×50 binoculars
NGC 6134	16ʰ 27.7ᵐ	−49° 09'	Open	7.2	6	11.0	2.1K	4	179	7×50	***	Rich field for low power telescope. ~6 stars visible in 7×50 binoculars
NGC 6152	16ʰ 32.7ᵐ	−52° 37'	Open	8.1p	25	11.0	3.4K	25	70	7×50	***	
NGC 6169	16ʰ 34.1ᵐ	−44° 03'	Open	6.6p	12	12.0	3.6K	13	40	7×50	***	μ Normae Cluster
NGC 6167	16ʰ 34.4ᵐ	−49° 36'	Open	6.7	7	7.4	1.9K	4	218	7×50	***	

Nebulae:

Designation	R.A.	Decl.	Type	m_v	Size (')	Brst./Cent. star	Dist. (ly)	dia. (ly)	Aper. Recm.	Rating	Comments
Mz 1	15ʰ 34.3ᵐ	−59° 09'	Planetary	12.1	26"	–	–	–	8/10"	***	Ring, with anomalies
Sp 1	15ʰ 51.7ᵐ	−51° 31'	Planetary	13.6	76"	14.0	–	–	10/12"	***	Large, but very faint ring, use low power and nebular filter

Galaxies:

Designation	R.A.	Decl.	Type	m$_v$	Size (')	Dist. (ly)	dia. (ly)	Lum. (suns)	Aper. Recm.	Rating	Comments
None											

Meteor showers:

Designation	R.A.	Decl.	Period (yr)	Duration	Max Date	ZHR (max)	Comet/ **Asteroid**	First Obs.	Vel. (mi **km**/sec)	Rating	Comments
γ Normids	16^h 36^m	−51°	Annual	Feb. 25	Mar. 22	03/13	8	Unknown	1929	35/56	II	First reported by Ronald A. McIntosh (Auckland, NZ)

Other interesting objects:

Designation	R.A.	Decl.	Type	m$_v$	Size (')	Dist. (ly)	dia. (ly)	Lum. (suns)	Aper. Recm.	Rating	Comments
None											

Fig. 18.6. NGC 6067 (C89), open cluster in Norma. Astrophoto by Christopher J. Picking.

Fornax (**for**-naks)	The (Chemist's) Furnace
Fornacis (for-**nay**-siss)	9:00 P.M. Culmination: Dec. 18
For	AREA: 398 sq. deg., 41st in size

In most cases, Lacaille formed his new constellations with mainly stars which formed no part of the earlier figure. However, for Fornax, he appears to have stolen an entire "U" bend and shortened the winding river of Eridanus significantly. Perhaps, he felt justified in removing a few stars which were fainter than the main course of the long river. In any case, what is left of the southern end of Eridanus is now considerably straighter than the northern parts.

Lacaille first called it "le Fourneau," but later latinized it to "Fornax Chemica." Bode later renamed it the "Apparatus Chemicus," but the name did not stick, and it was finally shortened to just Fornax in the twentieth century.

Lacaille's version of Fornax showed a rectangular iron box with an opening on one side to feed the fire, and what appears to be a distillation setup on the flat top of the furnace. The figure shown here represents only the top surface of the furnace since there are not enough stars to suggest any greater detail.

TO FIND: See Fig. 11.1a, The stars of Eridanus and adjacent constellations, Fig. 11.1b, and Fig. 11.2, Eridanus Finder chart. The celestial river Eridanus starts just west of <u>Rigel</u> (β Orionis) and winds westward for about 35° before turning first south, then east. The first part of this eastward section is formed by a nearly 20° long chain of fourth and fifth magnitude stars (τ^1 to τ^9). The eastern half of Fornax lies directly south of the western part (τ^1 to τ^7) of this star chain. α Fornacis lies 5.8° south–southeast of τ^3 and 11.25° south–southwest of τ^4. It also lies directly between τ^4 and θ Eridani (Acamar), somewhat closer to the former. β Fornacis is almost precisely 6° southwest of α. These are the only two-fourth magnitude stars in Fornacis, and these mark the northern and southern edges of the furnace top. ω Fornacis is 5.3° northwest of β, and the pair defining the western edge of the rectangle, μ and ν Fornacis, lie about 5° south west and 6.5° west–southwest of ω. The final star of the rectangle, δ Fornacis, lies 7° east–southeast of α (Figs. 11.7 and 18.7).

KEY STAR NAMES: None.

Magnitudes and spectral types of principal stars:

Bayer desig.	Mag. (m$_v$)	Spec. type	Bayer desig.	Mag. (m$_v$)	Spec. type	Bayer desig.	Mag. (m$_v$)	Spec. type
α	3.80	F8 V	β	4.45	G8 III	γ^1	6.14	K0 III
γ^2	5.39	A1 V	δ	4.99	B5 III	ε	5.88	G8/K0 V
ζ	5.69	F3 V	η^1	6.51	K0/1 III	η^2	5.92	K0 III
η^3	5.48	K2 III	θ	6.21	A1 V*[18]	ι^1	5.74	G8/K0 III
ι^2	5.84	F5 V	κ	5.19	G2 V	λ^1	5.91	K0/1 III
λ^2	5.78	G1 V	μ	5.27	A0 V	ν	4.68	B9.5p
ξ	–	–*[19]	o	6.19	A5 V*[18]	π	5.34	G8 III
ρ	5.52	G6 III	σ	5.91	A2 V	τ	6.01	A0 V
υ	8.89	A3*[18]	ϕ	5.13	A2/3 V	χ^1	6.39	A1 IV
χ^2	5.71	K2 III	χ^3	6.49	A1 V	ψ	5.93	F5 V
ω	4.96	B9 V						

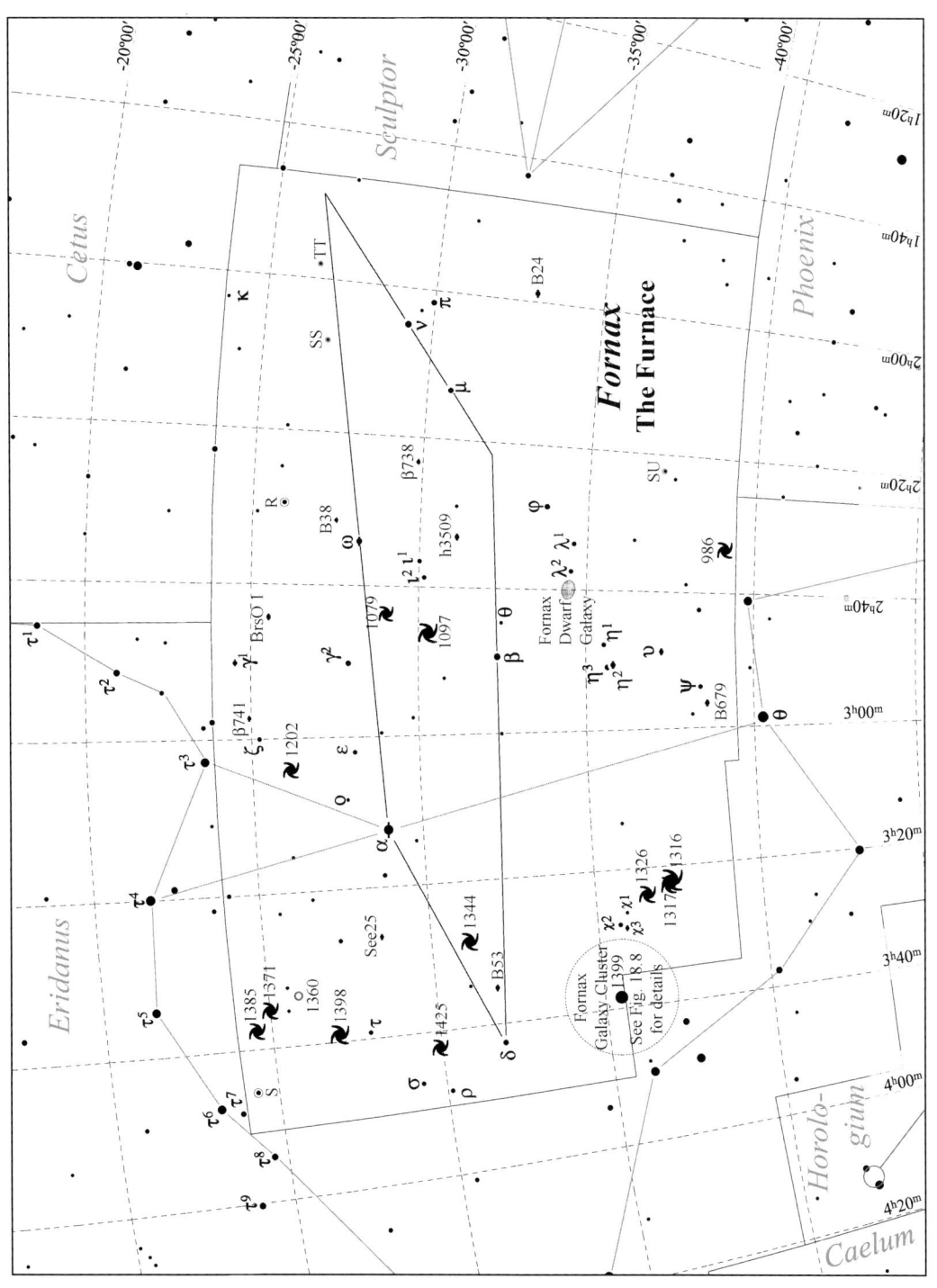

Fig. 18.7. Fornax details.

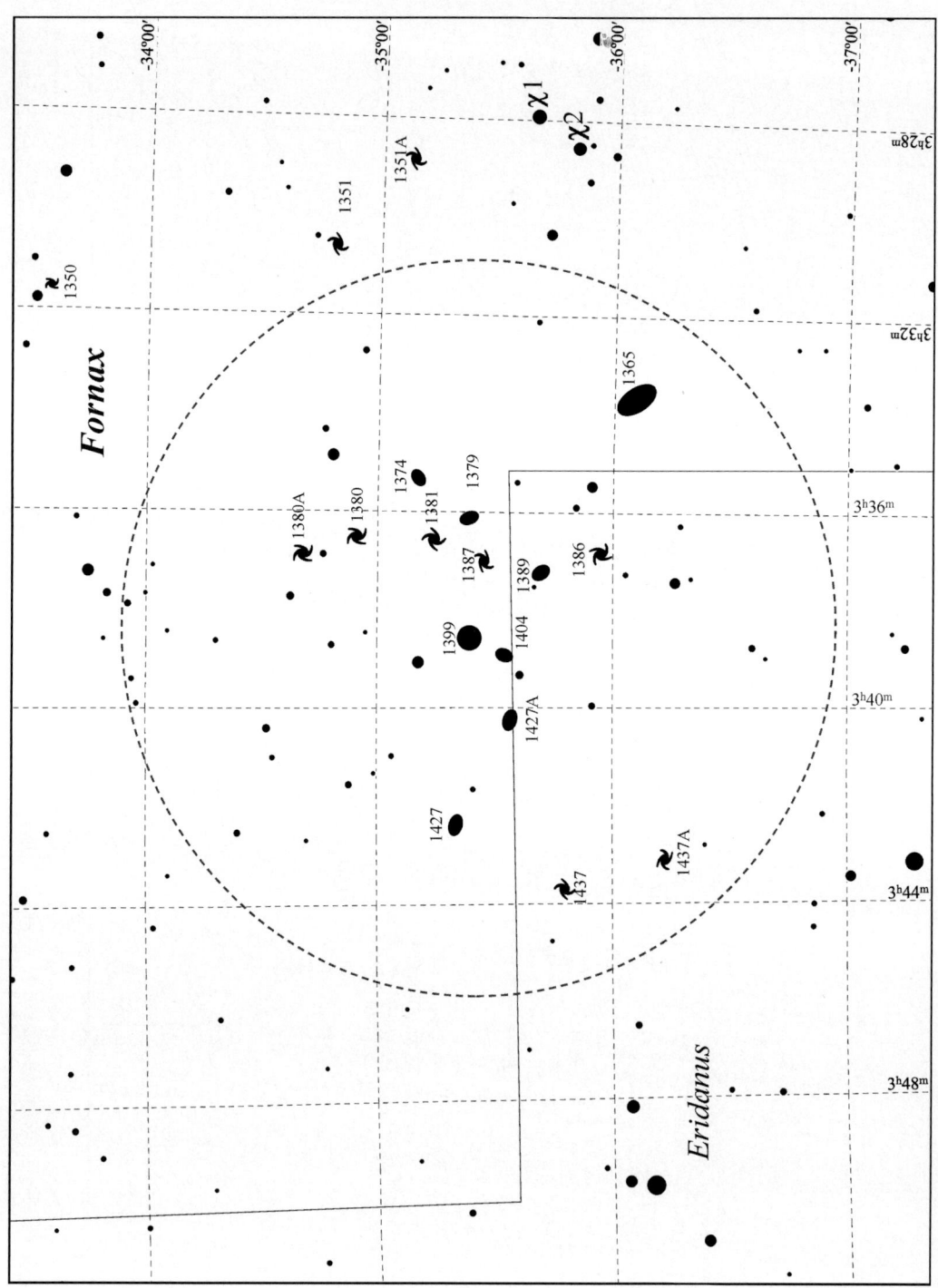

Fig. 18.8. Fornax Galaxy Cluster details.

Table 18.7. Selected telescopic objects in Fornax.

Multiple stars:

Designation	R.A.	Decl.	Type	m₁	m₂	Sep. (")	PA (°)	Colors	Date/ Period	Aper. Recm.	Rating	Comments
B 24	01ʰ 58.4ᵐ	−33° 03'	A–B	6.4	13.3	6.1	122	oY, ?	1927	6/7"	***	Passed min. of 0.6" in 1960, widening to max. of 3.35" in 2170
β 738	02ʰ 23.3ᵐ	−29° 52'	A–B	7.6	8.0	**1.83**	**212**	Y, ?	560	5/6"	****	
B 38	02ʰ 31.6ᵐ	−27° 30'	A–B	7.7	10.3	1.5	259	W, ?	1996	6/7"	****	a.k.a. h3506
ω Fornacis	02ʰ 33.8ᵐ	−28° 14'	A–B	5.0	7.7	10.6	245	oW, W	2002	2/3"	****	Showcase pair
h3509	02ʰ 34.2ᵐ	−31° 31'	A–a	8.1	8.9	0.2	290	yW, ?	1991	–	*	Too close for amateur telescopes
BrsO 1	02ʰ 44.2ᵐ	−25° 30'	Aa–B	7.5	11.5	23.4	059	yW, ?	1998	15×80	****	A has a faint companion at ~0.1" distance
			Aa–B	7.0	8.5	15.0	192	Y, ?	1998	15×80	****	
υ Fornacis	02ʰ 48.6ᵐ	−37° 24'	A–B	7.0	8.1	5.3	145	Y, pB?	2000	5/6"	*****	a.k.a. h3532
γ¹ Fornacis	02ʰ 49.8ᵐ	−24° 34'	A–B	6.1	12.7	11.9	145	Y, ?	1998	5/6"	***	a.k.a. β 877-AB
			A–C	6.1	10.7	37.8	133	Y, ?	1998	15×80	****	a.k.a. h2161
η² Fornacis	02ʰ 50.2ᵐ	−35° 51'	A–B	6.0	10.0	4.9	018	yW, ?	1991	2/3"	****	a.k.a. h3536
B 679	02ʰ 55.9ᵐ	−38° 57'	A–B	7.1	12.0	2.2	056	oY, ?	1938	4/5"	****	
β 741	02ʰ 57.2ᵐ	−24° 57'	A–B	8.1	8.2	**0.35**	**350**	Y, ?	137	–	***	Closing to min. of .06" in 2012, then opening to max. of 1.342" in 2037
α Fornacis	03ʰ 12.1ᵐ	−28° 59'	A–B	4.0	7.2	**5.23**	**300**	pY, ?	269	2/3"	*****	Showcase pair. Min. of 0.20" in 1951, max. of 5.627" in 2014
Sec 25	03ʰ 26.8ᵐ	−28° 34'	A–B	7.7	11.7	10.3	9.9	oY, ?	1957	3/4"	****	
χ³ Fornacis	03ʰ 28.2ᵐ	−35° 51'	A–B	6.5	10.1	6.5	247	W, ?	1998	2/3"	****	a.k.a. I 58
B 53	03ʰ 34.6ᵐ	−31° 52'	A–B	6.6	9.9	1.4	227	yO, ?	1991	6/7"	****	

Variable stars:

Designation	R.A.	Decl.	Type	Range (m)		Period (days)	F (f/f₁)	Spectral Range	Aper. Recm.	Rating	Comments
TT Fornacis	01ʰ 58.0ᵐ	−26° 29'	SRb	9.4	10.2	100?	–	M2	10×80	***	3° 10' NNW (PA 332°) of ν For (m.4.68).#
SS Fornacis	02ʰ 07.9ᵐ	−26° 52'	RRab	9.5	10.6	0.495	0.14	A3–G0	2/3"	****	2° 33' NNE (PA 017°) of ν For (m4.68).#
SU Fornacis	02ʰ 21.6ᵐ	−37° 13'	EA/SD	9.5	11.0	2.435	0.12	A2	3/4"	****	3° 39' SSW (PA 201°) of φ For (m5.13).#
R Fornacis	02ʰ 29.3ᵐ	−26° 04'	M	7.5	13.0	388.73	0.52	C4, 3e(Ne)	8/10"	****	2° 22' NNW (PA 334°) of ω. For (m4.96). Deep red color.##
S Fornacis	03ʰ 46.2ᵐ	−24° 24'	Const?	5.6	8.5	–	–	F8	7×50	***	1° 09' S (PA 187°) of τ⁶ Eri (m4.22). May not be variable. Bears watching.#

(continued)

Table 18.7. (continued)

Star clusters:

Designation	R.A.	Decl.	Type	Size (')	m_v	Brtst. Star	Dist. (ly)	dia. (ly)	Number of Stars	Aper. Recm.	Rating	Comments
NGC 1049	02h 39.7m	−34° 17'	Globular	0.4	12.6		2M	2.9M	18.4	8/10"	*	Brightest of 6 globular clusters in the Fornax Dwarf Galaxy

Nebulae:

Designation	R.A.	Decl.	Type	Size (')	m_v	Brtst./ **Cent.** star	Dist. (ly)	dia. (ly)	Aper. Recm.	Rating	Comments
NGC 1360	03h 33.3m	−25° 51'	Planetary	6.5	9.4	10.98	–	Green	4/6"	*****	One of the 20 brightest planetary nebulae. Can be glimpsed in 2" scope

Galaxies:

Designation	R.A.	Decl.	Type	m_v	Size (')	Dist. (ly)	dia. (ly)	Lum. (suns)	Aper. Recm.	Rating	Comments
NGC 986	02h 33.6m	−39° 02'	SB(rS)	10.9	3.8×1.9	85M	94K	25G	4/5"	***	Pretty bright, pretty large, elongated, very bright nucleus, with dark lanes
Fornax Dwarf Galaxy	02h 40.0m	−34° 27'	E	8.1	60×48	2M	35K	180M	–	*	a.k.a. ESO 356-G4 *20
NGC 1079	02h 43.7m	−29° 00'	SAB(rs)a	11.3	3.2×2.1	94M	85K	21G	4/5"	***	Bright, pretty large, elongated, small bright nucleus. 6" shows both nucleus and halo
NGC 1097	02h 46.3m	−30° 16'	SB	9.2	10.5×6.3	53M	162K	47G	3/4"	****	Very bright, large, elongated
NGC 1201	03h 04.1m	−26° 04'	SAB	10.8	3.5×1.9	71M	72K	19G	2/3"	***	Bright, small, bright diffuse nucleus
NGC 1316	03h 22.7m	−37° 12'	SAB(s)	8.2	13.5×9.3	71M	279K	210G	15×80	****	Large, slightly elongated, diffuse center, X-ray and radio source, can be glimpsed in 7×50
NGC 1317	03h 22.8m	−37° 06'	SAB(rs)	10.8	3.5×3.0	83M	85K	26G	2/3"	***	Pretty bright, pretty small, extremely bright nucleus
NGC 1326	03h 23.9m	−36° 28'	SB	10.4	4.4×3.2	51M	62K	14G	2/3"	***	Pretty small, very bright nucleus
NGC 1344	03h 28.3m	−31° 04'	SAB(rs)	10.2	4.7×3.0	47M	64K	15G	2/3"	***	Quite bright, pretty large, irreg. round, bright elongated nucleus
NGC 1350	03h 31.1m	−33° 38'	SAB®	10.3	6.2×3.2	72M	130K	31G	2/3"	****	Bright, large, very elongated, pretty bright arms, stellar nucleus

Object	R.A.	Decl.	Type	m_v	Size (')	Dist. (ly)				Rating	Comments
NGC 1365	03h 33.6m	−36° 08'	SB(s)	9.3	8.9×6.5	65M	170K	64G	2/3"	****	Very bright, very large, quite elongated, extreme bright nucleus. Remarkable barred galaxy
NGC 1371	03h 35.0m	−24° 56'	SB®	10.6	4.9×3.4	61M	87K	17G	3/4"	***	Pretty bright, pretty large, small bright nucleus
NGC 1387	03h 36.0m	−35° 31'	SAB(s)	10.8	3.1×2.8	48M	43K	2.7G	3/4"	***	Very bright, pretty large, round, very bright elongated nucleus
NGC 1385	03h 37.5m	−24° 30'	SB(s)	10.7	3.6×2.4	60M	63K	15G	3/4"	***	Pretty bright, pretty small, round. pretty arms
NGC 1399	03h 38.5m	−35° 27'		8.8	6.0	56M	98K	75G	3/4"	***	Very bright, pretty large, bright nucleus
NGC 1398	03h 38.9m	−26° 54'	SB®	9.5	7.1×5.2	57M	118K	41G	2/3"	*****	Quite bright, quite large, round, large bright nucleus. Internal ring w bar
NGC 1404	03h 38.9m	−35° 35'	E1	9.7	4.8×3.9	77M	108K	63G	2/3"	***	Very bright, pretty large, roundish, small bright diffuse nucleus
NGC 1425	03h 42.2m	−29° 54'	SA(rs)	10.8	5.4×2.2	65M	102K	16G	3/4"	*****	Faint, pretty large, irregularly round, very small very bright nucleus
NGC 1427	03h 42.3m	−35° 25'	SA0	10.9	3.4×2.5	62M	61K	13G	3/4"	***	Pretty faint, small, roundish. Small, bright nucleus

Meteor showers:

Designation	R.A.	Decl.	Period (yr)	Duration	Max Date	ZHR (max)	Comet/ Asteroid	First Obs.	Vel. (mi/ km/sec)	Rating	Comments
None											

Other interesting objects:

Designation	R.A.	Decl.	Type	m_v	Size (')	Dist. (ly)	dia. (ly)	Number of Galaxies	Aper. Recm.	Rating	Comments	
The Fornax Galaxy Cluster	03h 38m	−35° 27'	Gal. Clust.	8.12	>10°	65M	>10M	>60	–	***	The Fornax cluster appears to be centered around NGC 1399	*21

Fig. 18.9a. NGC 1097, barred spiral galaxy in Fornax. Astrophoto by Mark Komsa.

Fig. 18.9b. NGC 1097, lines point to supernova SN2003b. Drawing by Eiji Kato.

Fig. 18.10a. Barred spiral galaxy in Fornax. Astropot by Mark Komsa.

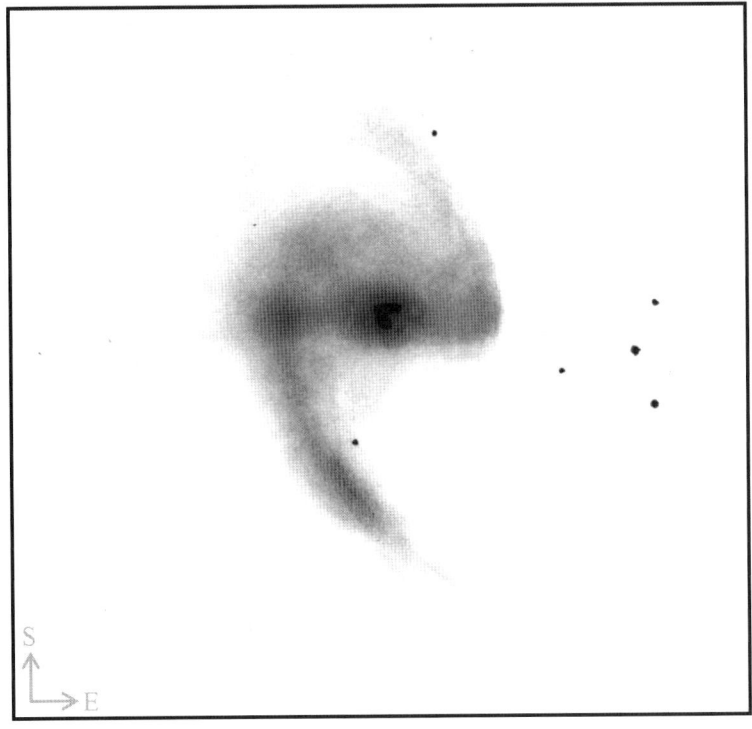

Fig. 18.10b. NGC 1365. Sketch by Eiji Kato.

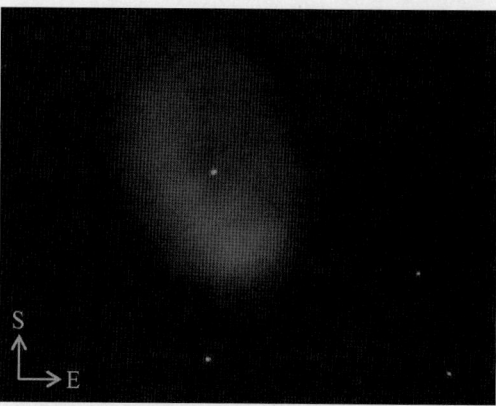

Fig. 18.11. NGC 1960, planetary nebula in Fornax. Sketch by Eiji Kato.

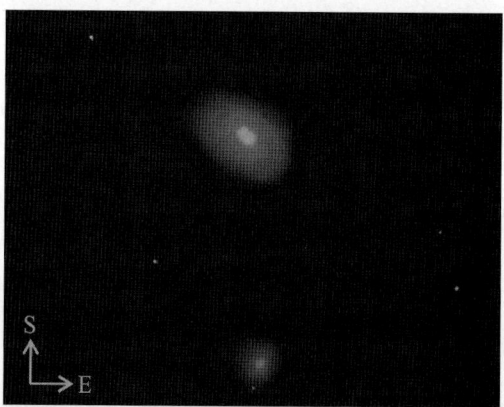

Fig. 18.12. NGC 1316 (top) and 1317, spiral galaxies in Norma. Sketch by Eiji Kato.

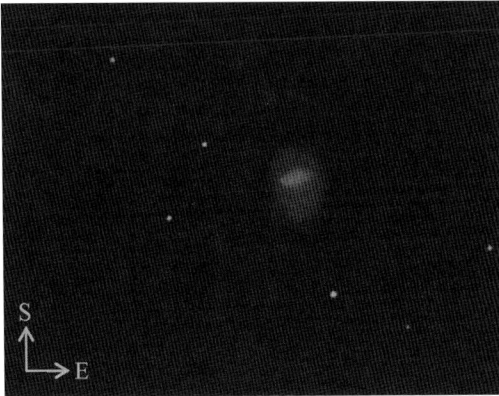

Fig. 18.13. NGC 1385, spiral galaxy in Fornax. Sketch by Eiji Kato.

Fig. 18.14. NGC 1386, spiral galaxy in Fornax. Sketch by Eiji Kato.

Fig. 18.15. NGC 1398, spiral galaxy in Fornax. Sketch by Eiji Kato.

Horologium (hor-uh-**low**-jih-um)	The (Pendulum) Clock
Horologii (hor-owe-**low**-jih-ee)	9:00 P.M. Culmination: Dec. 26
Hor	AREA: 249 sq. deg., 58th in size

As with his other southern constellations, Lacaille commemorated an important scientific tool in Horologium. His 1754 planisphere showed a pendulum clock, but named it only l'Horloge, the clock. This was later Latinized to "Horologium Oscillatorium," or the oscillating pendulum clock, and finally shortened to just Horologium. Lacaille may have intended to honor Galileo, who had long been interested in properties of the pendulum, and first described the pendulum clock to a companion in 1541. Since Galileo was already blind by that time, the drawings made from his description were inadequate for building a working clock. The first working pendulum clock was constructed from descriptions and drawings published in 1657 by Christian Huygens.

TO FIND: See Fig. 11.1a, The stars of Eridanus and adjacent constellations, Fig. 11.1b, Eridanus and adjacent constellations, and Fig. 11.3, Eridanus detail (southern portion), for the location of Horologium. α and δ Horologii lie about 5° due west of α Caeli (which marks the crossing point of the two Burins). The pair is easy to recognize because they are separated by only 40′, close enough to attract attention, but still very easy to separate. α and δ Horologii define the pendulum bob (weight), and a dim trio of stars southwest of the pair trace out the bottom 11° of the pendulum was viewed from the southern hemisphere. Continuing this direction another 11° will put you in the middle of a group of five faint stars which define the rectangular box holding the clock face and mechanism. Horologium is one of those constellations requiring sharp eyes, dark skies, and good luck to recognize (Figs. 16.1, 18.16 and 18.17).

KEY STAR NAMES: None.

Magnitudes and spectral types of principal stars:

Bayer desig.	Mag. (m$_v$)	Spec. type	Bayer desig.	Mag. (m$_v$)	Spec. type	Bayer desig.	Mag. (m$_v$)	Spec. type
α	3.85	K1 III	β	4.98	A5 III	γ	5.73	G8 III/IV
δ	4.93	A9 V	ε	6.78	B9 V[22]	ζ	5.21	F4 IV
η	5.30	A6 V	θ	6.15	A2 V[22]	ι	5.40	G3 IV
κ	–	–[23]	λ	5.36	F2 III	μ	5.12	F0 IV[24]
ν	5.25	A2 V[24]						

Stars of magnitude 5.5 or brighter which have only catalog designations:

Designation	Mag. (m$_v$)	Spec. type
HD 27588	5.33	K2 III

Fig. 18.16a. The stars of Reticulum and parts of adjacent constellations, including Horologium. Astrophoto by Christopher J. Picking.

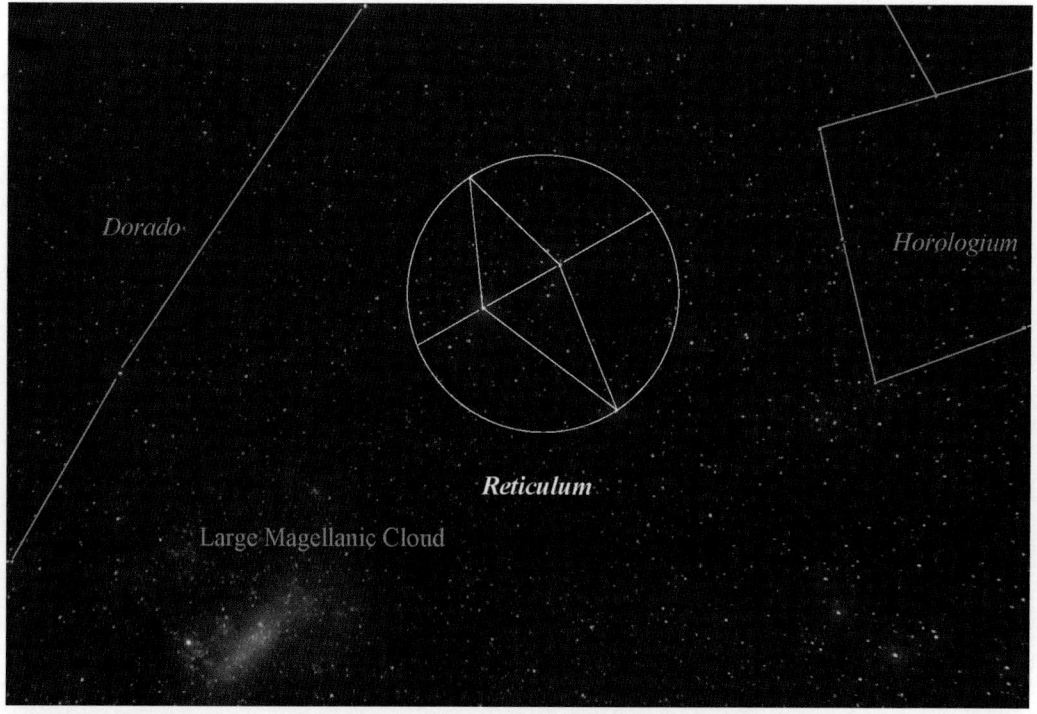

Fig. 18.16b. The figures of Reticulum and part of Horologium and adjacent constellations.

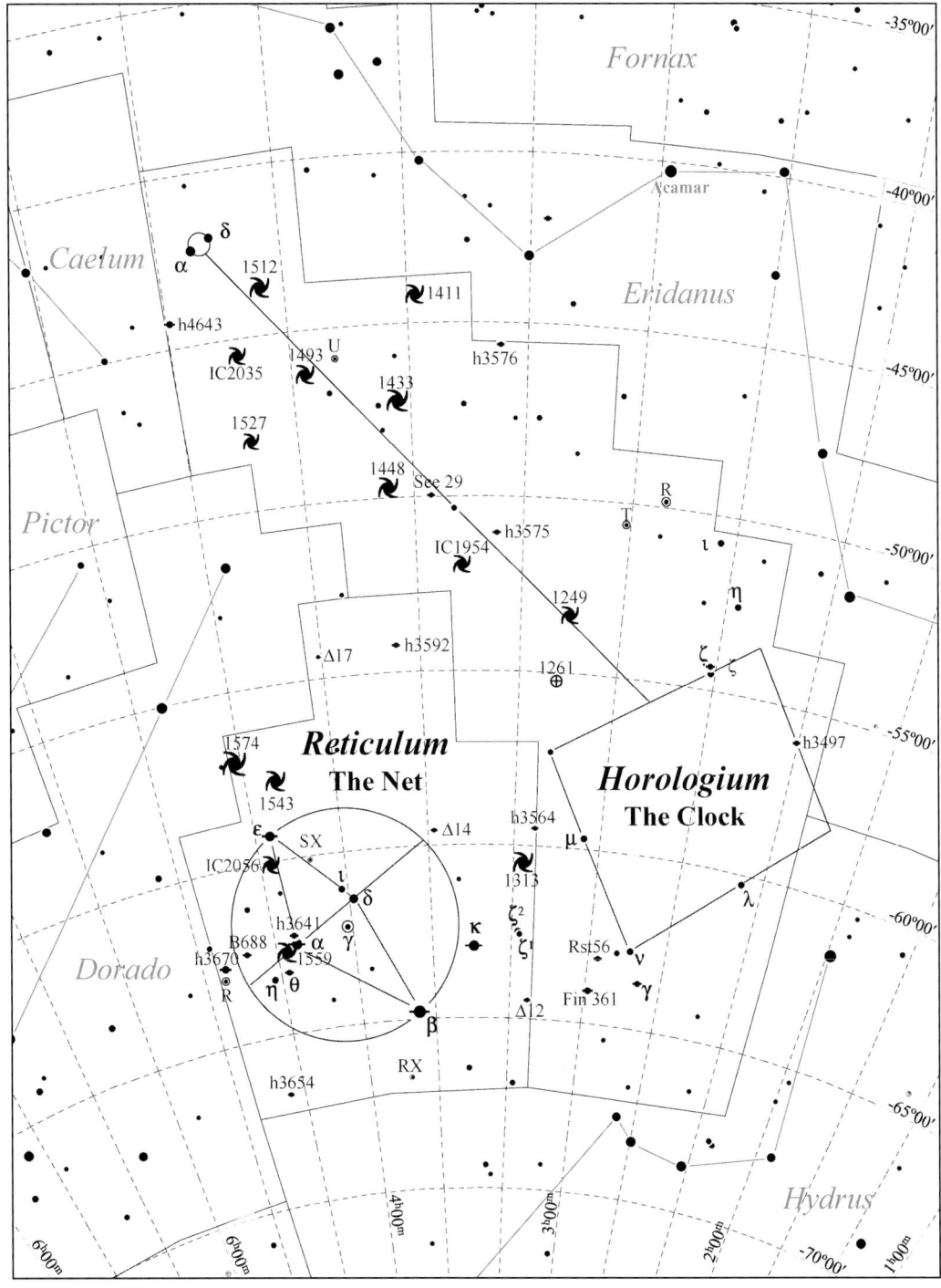

Fig. 18.17. Horologium and Reticulum details.

Table 18.8. Selected telescopic objects in Horologium.

Multiple stars:

Designation	R.A.	Decl.	Type	m₁	m₂	Sep. (″)	PA (°)	Colors	Date/Period	Aper. Recm.	Rating	Comments
h3497	02ʰ 19.9ᵐ	−55° 57′	A–B	6.0	10.6	36.5	083	O, ?	2000	2/3″	****	
γ Horologii	02ʰ 45.5ᵐ	−63° 42′	A–B	5.7	13.0	19.2	080	dY, ?	1999	6/7″	***	a.k.a. Hdo306
Rst56	02ʰ 57.0ᵐ	−63° 02′	A–B	7.5	13.0	7.5	143	dY, ?	1945	6/7″	***	
Fin 361	03ʰ 10.1ᵐ	−63° 55′	A–a	7.4	7.6	0.1	288	W, ?	1991	–	*	A–a is too close for amateur instruments
h3559			Aa–B	6.7	10.1	42.9	041	W, –	1999	15×80	****	
h3575	03ʰ 24.5ᵐ	−51° 04′	A–B	6.7	10.2	35.8	048	W, ?	1999	10×80	****	
h3576	03ʰ 24.6ᵐ	−45° 40′	A–B	7.3	8.8	2.9	341	pY, ?	1993	3/4″	*****	
See29	03ʰ 36.9ᵐ	−49° 57′	A–B	7.5	11.8	18.3	262	Y, ?	1999	3/4″	***	
h3643	04ʰ 19.3ᵐ	−44° 16′	A–B	5.5	8.6	70.4	115	dY, ?	1999	7×50	****	

Variable stars:

Designation	R.A.	Decl.	Type	Range (mᵥ)	Period (days)	F (fᵢ/fₜ)	Spectral Range	Aper. Recm.	Rating	Comments
R Horologii	02ʰ 53.8ᵐ	−49° 53′	M	6.0 – 13.0	407.6	0.40	M7IIIe	8/10″		2° 01′ ENE (PA 064°) of ι Hor (m5.40).##
T Horologii	03ʰ 00.9ᵐ	−50° 38′	M	7.2 – 13.7	217.66	0.46	M5Ie	10/12″	****	2° 54′ E (PA 089°) of ι Hor (m5.40).##
U Horologii	03ʰ 52.6ᵐ	−45° 50′	M	7.6 – 15.1	348.4	–	M6IIIe	20/22″	****	5° 12′ SW (PA 225°) of α Hor (m3.85).##

Star clusters:

Designation	R.A.	Decl.	Type	Size (′)	Brtst. Star	Dist. (ly)	dia. (ly)	Number of Stars	Aper. Recm.	Rating	Comments
NGC 1261	03ʰ 12.3ᵐ	−55° 13′	Globular	6.8	13.5	–	55	94K	15×80	****	Bright, large, round. High concentration of stars, partially resolved. a.k.a. C87

Nebulae:

Designation	R.A.	Decl.	Type	m$_v$	Size (')	Brst./Cent. star	Dist. (ly)	dia. (ly)	Aper. Recm.	Rating	Comments
None											

Galaxies:

Designation	R.A.	Decl.	Type	m$_v$	Size (')	Dist. (ly)	dia. (ly)	Lum. (suns)	Aper. Recm.	Rating	Comments
NGC 1249	03^h 10.0^m	−53° 20'	SBc	11.5	4.8×2.3	36M	50K	2.6G	4/6"	***	
IC 1954	03^h 31.5^m	−51° 54'	SBb	11.4	3.1×1.6	40M	36K	3.5G	4/6"	***	
NGC 1411	03^h 38.7^m	−44° 06'	S0	11.1	2.3×1.7	39M	29K	4.4G	4/6"	***	
NGC 1433	03^h 42.0^m	−47° 13'	SBa	9.8	6.8×6.0	62M	123K	37G	3/4"	****	Very bright nucleus
NGC 1448	03^h 44.5^m	−44° 39'	Sc	10.7	8.1×1.8	44M	104K	8.1G	3/4"	***	
NGC 1493	03^h 57.5^m	−46° 13'	SBc	11.2	3.8	36M	40K	3.4G	4/6"	***	
NGC 1512	04^h 03.9^m	−43° 21'	SBa	10.3	8.9×5.6	24M	62K	3.5G	3/4"	***	Bright nucleus
NGC 1527	04^h 08.4^m	−47° 54'	S0	10.7	4.1×1.7	37M	44K	5.7G	4/5"	***	
IC 2035	04^h 09.0^m	−45° 31'	SB0	11.5	1.2×0.9	55M	19K	6.1G	5/6"	***	

Meteor showers:

Designation	R.A.	Decl.	Period (yr)	Duration	Max Date	ZHR (max)	Comet/Asteroid	First Obs.	Vel. (mi/km/sec)	Rating	Comments
None											

Other interesting objects:

Designation	R.A.	Decl.	Type	m$_v$	Size (')	Dist. (ly)	dia. (ly)	Lum. (suns)	Aper. Recm.	Rating	Comments
None											

Table 18.9. Selected telescopic objects in Reticulum.

Multiple stars:

Designation	R.A.	Decl.	Type	m_1	m_2	Sep. (")	PA (°)	Colors	Date/Period	Aper. Recm.	Rating	Comments
h3564	03h 14.9m	−59° 31'	A–B	7.0	12.0	32.8	279	yO, ?	1999	3/4"	***	Obvious in very sparse star field
Δ12	03h 15.2m	−64° 27'	A–B	6.7	9.0	19.1	104	Y, ?	1999	15×80	****	Nearly matched naked eye/binocular double, a.k.a. ALBI Aa–B *26
ζ¹, ζ² Reticuli	03h 18.2m	−62° 30'	Aa–B	5.3	5.6	309.3	217	Y, Y	1991	6×30	****	
κ Reticuli	03h 29.4m	−62° 56'	A–B	4.8	10.8	54.4	127	Y, ?	1999	15×80	****	a.k.a. h3580
Δ14	03h 38.2m	−59° 47'	A–B	7.0	8.3	57.3	272	Y, ?	1999	7×50	****	
β Reticuli	03h 44.2m	−46° 48'	A–B	4.0	7.9	80.0	095	oY, ?	1991	7×50	****	a.k.a. LDS104
h3592	03h 44.6m	−54° 16'	A–B	6.5	9.3	5.0	017	O, B	1993	2/3"	****	
Δ17	04h 00.8m	−54° 22'	A–B	7.7	8.2	64.6	142	W, yW	1999	7×50	****	a.k.a. Δ17AB
			B–C	8.2	11.7	3.5	079	yW, ?	1991	3/4"	***	a.k.a. I269BC
			B–D	8.2	12.8	27.6	204	yW, ?	1999	5/6"	***	a.k.a. I269BD
α Reticuli	04h 14.4m	−62° 28'	A–B	3.4	12.0	48.5	355	dY, ?	1998	3/4"	***	a.k.a. h3638
h3641	04h 16.5m	−62° 11'	A–B	5.5	11.0	12.7	216	oY, ?	1998	3/4"	****	α Reticuli is in the same FOV
ε Reticuli	04h 16.5m	−59° 18'	A–B	4.4	12.5	13.5	036	oY, ?	1064	4/5"	***	a.k.a. Jsp 56
θ Reticuli	04h 17.7m	−63° 15'	A–B	6.0	7.7	4.3	003	Y, dY	2000	2/3"	****	a.k.a. Rmk3
h3654	04h 23.9m	−66° 44'	A–B	7.0	12.7	18.6	130	W, yW	2000	5/6"	***	
B688	04h 27.8m	−62° 31'	A–B	5.8	12.5	9.9	126	dY, ?	1998	4/5"	***	
h3670	04h 33.6m	−62° 49'	A–B	5.9	9.3	31.8	100	dY, B	1998	15×80	*****	Very nice color

Variable stars:

Designation	R.A.	Decl.	Type	Range (m_v)		Period (days)	F (f_r/f_f)	Spectral Range	Aper. Recm.	Rating	Comments
RX Reticuli	03h 47.7m	−66° 42'	SRd	9.1	11.2	–		G8–K0 II–IIIep	3/4"	***	1° 55' S (PA 169°) of ι Reticuli (m3.84)
γ Reticuli	03h 47.7m	−62° 10'	SR	4.42	4.64	25	–	M4III	Eye	****	50' NE (PA 50°) of ι Reticuli (m4.97)
SX Reticuli	04h 07.3m	−60° 27'	SRb	9.24	9.80	100		M4–5III	7×50	***	
R Reticuli	04h 33.5m	−63° 02'	M	6.5	14.2	278.46	0.48	M4e–M7.5e	14/16"	****	1° 22' ENE (PA 74°) of η Reticuli (m5.24)

Star clusters:

Designation	R.A.	Decl.	Type	m_v	Size (')	Brtst. Star	Dist. (ly)	dia. (ly)	Number of Stars	Aper. Recm.	Rating	Comments
None												

Nebulae:

Designation	R.A.	Decl.	Type	m_v	Size (')	Brtst./ Cent. star	Dist. (ly)	dia. (ly)	Aper. Recm.	Rating	Comments
None											

Galaxies:

Designation	R.A.	Decl.	Type	m_v	Size (')	Dist. (ly)	dia. (ly)	Lum. (suns)	Aper. Recm.	Rating	Comments
NGC 1313	03h 18.2m	−60° 30'	SBd	8.8	8.8×6.8	10M	26K	2.4G	4/6"	****	
NGC 1543	04h 12.8m	−57° 44'	S0	10.6	3.0×1.3	51M	46K	12G	6/8"	***	
IC 2056	04h 16.4m	−60° 12'	SB0p	11.6	1.9×1.6	83M	46K	13G	8/10"	***	
NGC 1559	04h 17.6m	−62° 47'	SBc	10.6	3.5×2.0	47M	48K	10G	6/8"	***	
NGC 1574	04h 22.0m	−56° 58'	S0	10.4	4.1×3.7	20M	24K	2.2G	6/8"	***	

Meteor showers:

Designation	R.A.	Decl.	Period (yr)	Duration	Max Date	ZHR (max)	Comet/ **Asteroid**	First Obs.	Vel. (mi/ **km**/sec)	Rating	Comments
None											

Other interesting objects:

Name/ Designation	R.A.	Decl.	Type	m_v	Mass (suns)	Dia. (suns)	Dist. (ly)	Planet Mass (J)	Dist. (AU)	Period (days)	Eccentricity	Comments
ε Reticuli	03h 39.4m	−59° 18'	Exo-planets	4.44	1.49	–	59.4	1.56	1.27	428.1	0.06	Sep. 17, 2002, 100th announced exo-planet discovery. a.k.a. HD 216435, SAO113044

Reticulum (reh-**tik**-yoo-lum),	**The Net (Reticule)**	
Reticuli (reh-**tik**-yoo-lee),	**9:00 P.M. Culmination:** Jan. 3	
Ret	**AREA:** 114 sq. deg., 82nd in size	

The figure of Reticulum is that of a transparent circular disk with a horizontal line and a vertical diamond etched upon it. This type of reticle could be rotated until the vertical line ran through the centers of two objects in the field of gave the position angle (PA), a second rotation until two parallel lines of the diamond ran through the centers of the two objects allowed the distance between the objects to be calculated from the amount of rotation. Lacaille use a reticule similar to this for determining the positions of stars and other objects which he observed.

TO FIND: See Fig. 11.1a, The stars of Eridanus and adjacent constellations, Fig. 11.1b, Eridanus and adjacent constellations, and Fig. 11.3, Eridanus detail (southern portion), for the location of Reticulum. α Ret (m3.3) is the brightest star in Reticulum and can be found 2.5° south of the midpoint of a line drawn between <u>Canopus</u> and <u>Acherner</u>, which are above the pole and about 40° apart on a nearly horizontal line when Reticulum is at its highest. The rhomboidal pattern of Lacaille's constellation is formed by α, β, δ, and ε Ret, with γ near its center.

Reticulum is circumpolar for most of the southern hemisphere and no part of it is visible to northern hemisphere observers, north of 37° latitude. Those south of 23° can see all the constellation, but most of it will be low in the south. Like Caelum and Mensa, Reticulum is a faint constellation and difficult to identify, but somewhat easier because of its slightly brighter stars and distinctive pattern (Figs. 16.1, 18.16 and 18.17).

KEY STAR NAMES: None

Magnitudes and spectral types of principal stars:

Bayer desig.	Mag. (m$_v$)	Spec. type	Bayer desig.	Mag. (m$_v$)	Spec. type	Bayer desig.	Mag. (m$_v$)	Spec. type
α	3.33	G7 III	β	3.84	K0 IV	γ	4.48	M4 III
δ	4.56	M2 III	ε	4.44	K2 IV	ζ^1	5.11	G2 V
ζ^2	5.24	G1 V	η	5.24	G7 III	θ	5.88	B9 III/IV*[25]
ι	4.97	K4 III	κ	4.71	F5 IV/V			

Stars of magnitude 5.5 or brighter which have only catalog designations:

Desig-nation	Mag. (m$_v$)	Spec. type
HD 27304	5.45	K0 III

Fig. 18.18. NGC 1559, spiral galaxy in Reticulum. Drawing by Eiji Kato.

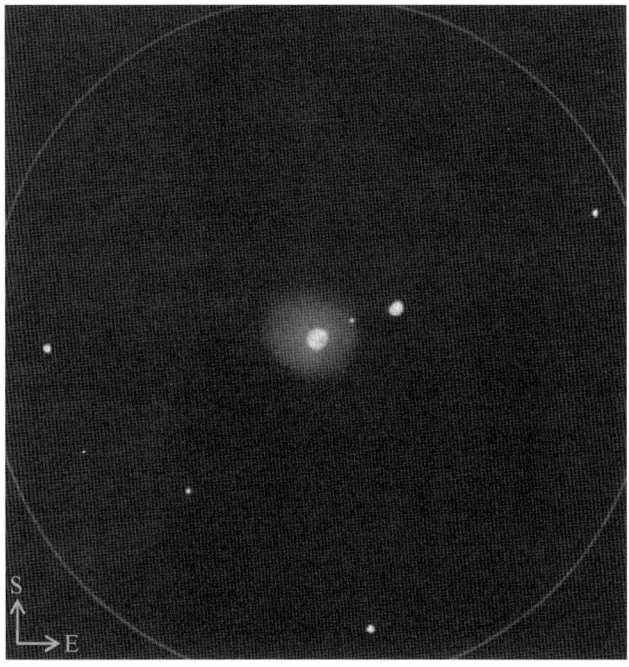

Fig. 18.19. NGC 1574, spiral galaxy in Reticulum. Drawing by Eiji Kato.

Mensa (men-suh)	The Table (Mountain)
Mensae (men-sigh)	9:00 P.M. Culmination: Jan. 29
Men	AREA: 153 sq. deg., 75th in size

Lacaille named this constellation after Table Mountain, the mesa near the Cape of Good Hope, where he conducted many of his observations. He first called it "Montagne de la Table," then latinized it to Mons Mensae. The name was later shortened to just Mensa, the Table. There is an appropriateness to Mons Mensae, because when it is highest in the southern hemisphere sky, the Large Magellanic Cloud appears to rest on its top, just as the real Table Mountain is often capped with a low-lying cloud.

TO FIND: While it is not difficult to find the location of Mensa because the southern part of the Large Magellanic Cloud occupies the northern part of the constellation, it is difficult to pick out the pattern defined by Lacaille. Mensa is one of the faintest of the constellations, with the brightest star (α) being only magnitude 5.08, and a total of only 23 stars of magnitude 6.5 or brighter. In addition, there are no easy-to-see patterns formed by the stars. The most practical way to recognize Mensa is to find one or two of the stars and then use a chart to pick out the rest one-by-one. You can start this process in one of the two ways. First, locate η¹ and η² Doradus, the pair of stars at the extreme ESE end of the straight line described in how to find Dorado. Just 3° due south of this pair is ν Doradus (m5.05), and 6° further south is α Mensae. Another method is to start with the bright star Miaplacidus (β Car, m1.67) and move ~8° to the WSW to find ζ Volantis (m3.93). Continuing that line for about 1.5 times its length (~12°) will bring you to β Mensae (m5.3). Another clue to recognizing β Mensae is that it lies at the southwestern edge of the Large Magellanic Cloud. Once you have found these two stars, you should be able to identify the remaining visible stars in Mensa. A pair of 6 × 30 binoculars will help in the last process, but be careful to not let the fainter stars confuse you. This is a difficult constellation to find and you must have very dark skies. Good luck! (Fig. 18.1 and 18.20).

KEY STAR NAMES: None.

Magnitudes and spectral types of principal stars:

Bayer desig.	Mag. (m_v)	Spec. type	Bayer desig.	Mag. (m_v)	Spec. type	Bayer desig.	Mag. (m_v)	Spec. type
α	5.08	G6 V	β	5.31	G8 III	γ	5.19	K2 III
δ	5.69	K2/3 III	ε	5.54	K2/3 III	ζ	5.61	A5 III
η	5.47	K4 III	θ	5.45	B9.5 V	ι	6.04	B8 III
κ	5.47	B9 V	λ	6.54	K0 III	μ	5.53	B9 IV
ν	5.78	F0/2 III	ξ	5.84	G8 III			

Fig. 18.20a. The stars of Octans and adjacent constellations. Astrophoto by Christopher J. Picking.

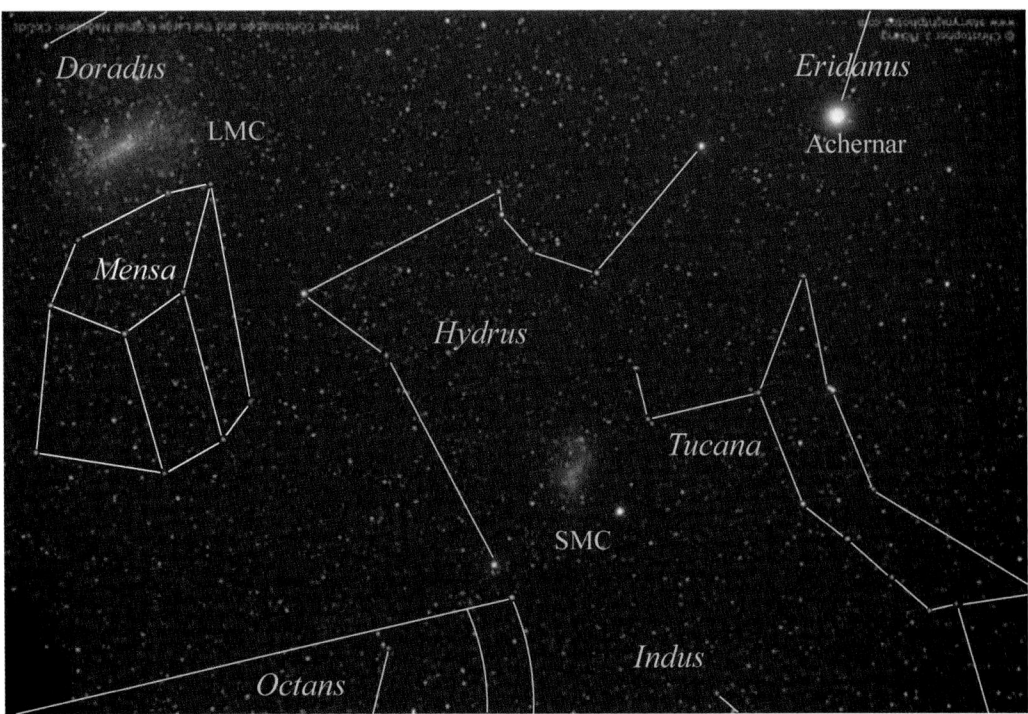

Fig. 18.20b. The figures of Octans and adjacent constellations, with south celestial pole (SCP) indicated.

Table 18.10. Selected telescopic objects in Mensa.

Multiple stars:

Designation	R.A.	Decl.	Type	m_1	m_2	Sep. (″)	PA (°)	Colors	Date/Period	Aper. Recm.	Rating	Comments
h3607	03h 37.4m	−80° 51′	A–B	8.5	9.1	35.8	128	Y, O	2000	7×50	******	More nearly equal in brightness than most colored doubles
h3612	03h 42.4m	−80° 01′	A–B	8.3	9.8	19.3	164	Y, –	2000	2/3″	****	
h3692	04h 22.9m	−82° 54′	A–B	6.8	12.5	48.6	184	W, –	199	4/5″	***	
h3741	05h 00.2m	−78° 18′	A–B	6.5	10.7	49.5	111	O, R	1999	15×80	****	
Hrg2	05h 01.2m	−74° 20′	A–B	7.4	8.0	0.8	169	W, –	1991	12/13″	****	a.k.a. h3795
γ Mensae	05h 31.9m	−76° 20′	A–B	5.2	11.1	48.2	143	O, –	1999	3/4″	****	At 0.1″, A–a is too close
h3911	06h 47.5m	−76° 51′	Aa–B	7.1	10.0	21.9	048	Y, –	1998	15×80	****	for amateur scopes
Rst2448	06h 47.7m	−81° 04′	A–B	7.2	12.0	3.8	163	O, –	2035	3/4″	***	
h3932	06h 54.1m	−77° 47′	A–B	7.7	9.8	8.4	286	W, –	1998	2/3″	****	
h3996	07h 11.7m	−84° 28′	A–B	7.5	11.9	16.2	260	bW, bW	2000	3/4″	****	

Variable stars:

Designation	R.A.	Decl.	Type	Range (m_v)		Period (days)	F (f_r/f_t)	Spectral Range	Aper. Recm.	Rating	Comments
U Mensae	04h 09.6m	−83° 51′	SRa	8.0	10.9	407.3	–	Me	3/4″	****	30′ SW (PA 235°) of ν Mensae
TZ Mensae	05h 30.2m	−84° 47′	Ea	6.19	6.87	8.56898	0.04	A1 III+B9V	6×30	****	2° 28′ NNW (PA 335°) of ζ Mensae

Star clusters:

Designation	R.A.	Decl.	Type	m_v	Size (′)	Brtst. Star	Dist. (ly)	dia. (ly)	Number of Stars	Aper. Recm.	Rating	Comments
NGC 1841	04h 45.4m	−84° 00′	Globular	14.1	2.4	–	37K	26	740K	14/16″	**	Fairly large, irregularly round, brighter toward center. A large scope is needed
NGC 1754	04h 54.4m	−70° 27′	Globular	12.0	1.6	–	180K	85	–	8/10″	***	In LMC
NGC 2019	05h 31.9m	−70° 10′	Globular	10.9	1.0	–	180K	50	–	4/6″	****	In LMC

Nebulae:

Designation	R.A.	Decl.	Type	m_v	Size (')	Brst./ **Cent.** star	Dist. (ly)	dia. (ly)	Aper. Recm.	Rating	Comments
None											

Galaxies:

Designation	R.A.	Decl.	Type	m_v	Size (')	Dist. (ly)	dia. (ly)	Lum. (suns)	Aper. Recm.	Rating	Comments
Large Magellanic Cloud	05h 00m	–69° 00'	Irr.	1.0	~6°	180K	50K	36M	Eye	*****	Nearest of the medium-sized galaxies *27

Meteor showers:

Designation	R.A.	Decl.	Period (yr)	Duration	Max Date	ZHR (max)	Comet/ **Asteroid**	First Obs.	Vel. (mi/ **km**/sec)	Rating	Comments
None											

Other interesting objects:

Designation	R.A.	Decl.	Type	m_v	Size (')	Dist. (ly)	dia. (ly)	Lum. (suns)	Aper. Recm.	Rating	Comments
None											

Octans	The Octant
Octantis	9:00 P.M. Culmination: Circumpolar
Oct	AREA: 291 sq. deg., 50th in size

Lacaille paid tribute to the explorers who sailed the southern seas to find new lands and the routes which led to countries with desirable trade products by putting two of their most important navigational instruments in the skies as Pyxis, the Mariner's Compass, and Octans, the Octant. Like the better known sextant, the octant has a graduated arc (one-eighth of a circle), and a movable arm bearing a small telescope or tube through which stars could be sighted to measure their altitude above the horizon. By using the altitude of two stars and the time of the sightings, it is possible to determine the location of a ship or a point on land.

TO FIND: Unfortunately, Octans is an extremely faint constellation, with only three stars reaching 4th magnitude. It also has no easily recognized patterns of stars to help you find a starting point for learning the constellation pattern. A good time to find Octans is about mid-October, when ν Octantis, the brightest star in the constellation, is at its highest above the south celestial pole about 9:00 PM. One way to begin is to find β Hydri, the southernmost star in the snake's tail, and the brightest star in the constellation, but only by the insignificant amount of 0.04 magnitudes. Less than 1.5° west of β Hydri, you will find φ Octantis, about two magnitudes fainter, but easy to identify. About 7.5° farther west is ν Octantis, the brightest star in its constellation. ν Octantis is flanked by α and ψ Octantis, in a pattern that mimics Altair and its sidekicks some 90° north, but about 3 magnitudes fainter than Altair and its companions. ν Octantis defines not only the arc of the octant, but also the location of the movable arm used to measure angles. Continuing the arc formed by θ, ψ, ν, and α Octantis about 10° farther west leads to φ Octantis, which completes the arc of the octant. η Octantis is about 18° from the stars in the octant's arc and marks the pivot point of the octant's arm. It may be the hardest star in the pattern to find. At only magnitude 6.19, it is barely visible to most people, even with the best skies. η Octantis lies on a line between ν Octantis and γ Chameleontis, and about three-quarters of the distance from the former to the latter. Octans is very faint, and very difficult to pick out, so do not be disappointed if you have difficulty (Figs. 16.1, 18.20 and 18.21).

KEY STAR NAMES: None.

Magnitudes and spectral types of principal stars:

Bayer desig.	Mag. (m_v)	Spec. type	Bayer desig.	Mag. (m_v)	Spec. type	Bayer desig.	Mag. (m_v)	Spec. type
α	5.13	F4 III	β	4.13	A9 IV/V	γ^1	5.10	G7 III[*28]
γ^2	5.72	K0 III	γ^3	5.29	K1/2 III	δ	4.31	K2 III
ε	5.09	M6 III	ζ	5.43	F0 III	η	6.19	A1 V
θ	4.78	K2 III	ι	5.45	K0 III	κ	5.56	A2m
λ	5.27	G8/K0 III	μ^1	5.99	F4 III/IV	μ^2	6.51	G1 V
ν	3.73	K0 III	ξ	5.32	B6 IV	o	7.21	B9.5 V[*29]
π^1	5.65	G8/K0 III	π^2	5.65	G8 Ib	ρ	5.57	A2 V
σ	5.45	F0 III[*30]	τ	5.50	K2 III	υ	5.76	K0 III
φ	5.47	A0 V	χ	5.29	K3 III	ψ	5.49	F3 III
ω	5.88	B9.5 V						

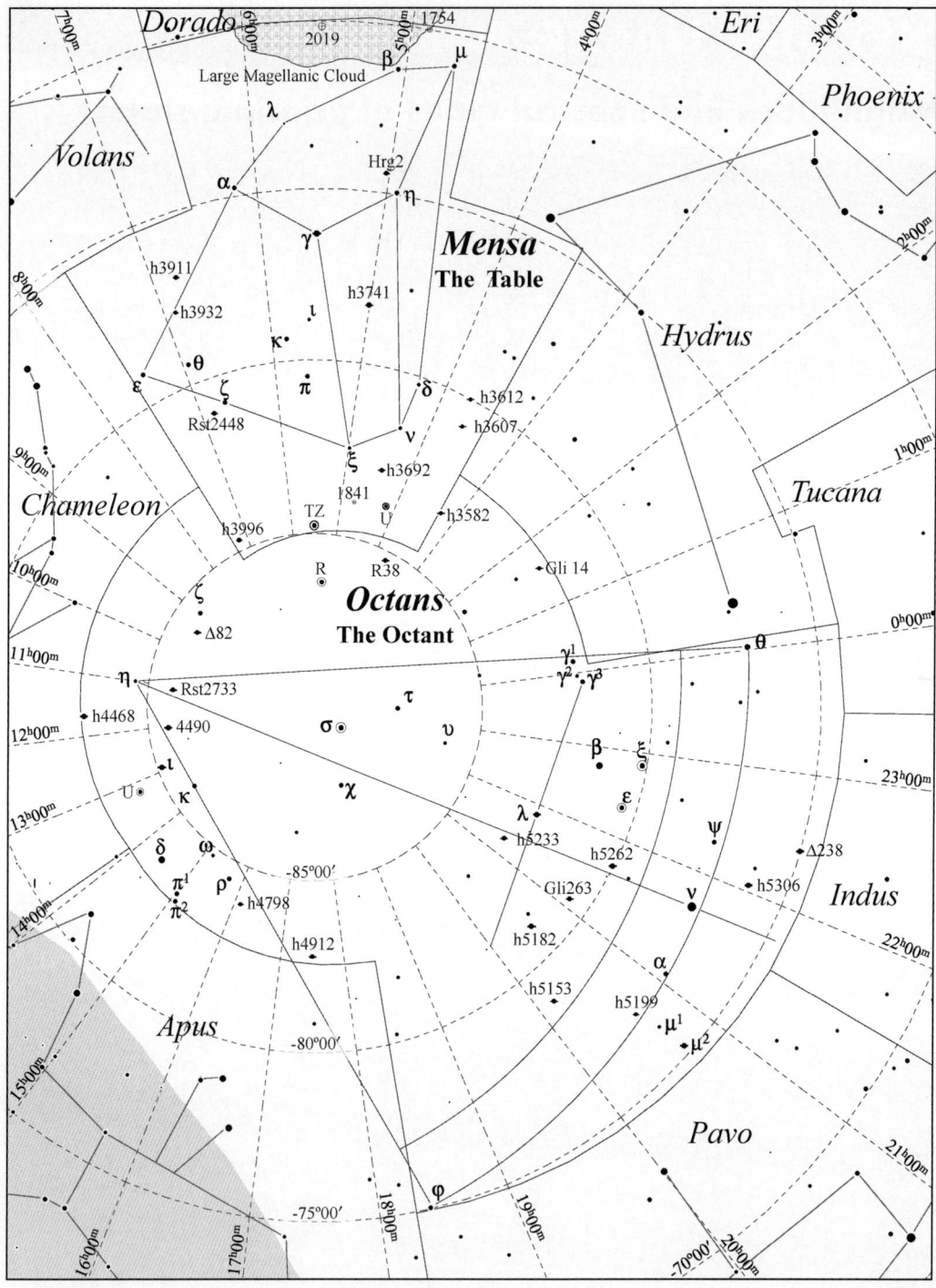

Fig. 18.21. Mensa and Octans details.

Table 18.11. Selected telescopic objects in Octans.

Multiple stars:

Designation	R.A.	Decl.	Type	m_1	m_2	Sep. (")	PA (°)	Colors	Date/Period	Aper. Recm.	Rating	Comments
Gli 14	01h 37.4m	−82° 17'	A-B	7.6	8.4	5.4	054	Y, Y	2000	2/3"	****	
h3582	03h 09.9m	−85° 32'	A-B	7.7	10.9	19.4	301	yW, ?	1999	3/4"	****	
R38	03h 42.5m	−85° 16'	A-B	6.6	8.1	1.9	252	W, ?	1993	5/6"	****	
Δ82	09h 33.3m	−86° 01'	A-B	7.1	7.6	15.4	275	yW, yW	2000	2/3"	****	
Rst2733	11h 06.2m	−85° 45'	A-B	7.2	13.0	4.0	080	W, ?	1975	6/7"	***	
h4468	11h 41.0m	−83° 06'	A-B	6.3	11.4	25.3	140	oY, ?	2000	3/4"	****	
h4490	12h 02.3m	−85° 38'	A-B	6.2	9.0	24.8	146	O, ?	2000	10×50	****	
ι Octantis	12h 55.0m	−85° 07'	A-B	5.9	6.9	0.7	239	O, ?	1991	13/14"	****	a.k.a. Rst 2819
h4798	16h 10.3m	−84° 14'	A-B	7.7	11.2	20.4	130	oY, ?	2000	3/4"	****	
h4912	17h 28.7m	−82° 47'	A-B	6.6	10.7	24.8	123	W, ?	2000	3/4"	****	
h5153	20h 10.3m	−79° 07'	A-B	7.4	12.8	46.3	124	dY, ?	2000	5/6"	***	
h5182	20h 33.3m	−80° 58'	A-B	5.9	10.8	26.1	359	Y, ?	1991	15×80	****	
h5199	20h 38.1m	−76° 54'	A-B	7.9	12.9	27.1	210	dY, O	2000	5/6"	***	Nice color combination
μ² Octantis	20h 41.7m	−75° 21'	A-B	6.5	7.1	16.8	020	Y, Y	2002	15×80	****	a.k.a. Δ232
Gli 263	21h 06.7m	−80° 42'	A-B	7.3	9.6	4.6	245	Y, ?	2000	3/4"	****	
h5233	21h 15.9m	−83° 16'	A-B	7.7	11.7	11.7	269	oY, ?	2000	3/4"	***	
h5262	21h 33.4m	−80° 02'	A-B	6.5	11.5	24.4	092	W, ?	2000	15×80	****	
λ Octantis	21h 50.9m	−82° 43'	A-B	5.6	7.3	3.5	063	dY, O	2002	2/3"	****	a.k.a. h5233
H5306	22h 03.1m	−76° 07'	A-B	6.0	10.6	34.7	071	yW, ?	1999	15×80	****	
Δ238	22h 25.9m	−75° 01'	A-B	6.2	8.9	20.8	080	dY, Y	1999	15×80	****	

Variable stars:

Designation	R.A.	Decl.	Type	Range (m)		Period (days)	F (f_i/f_j)	Spectral Range	Aper. Recm.	Rating	Comments
R Octantis	05h 26.1m	−86° 23'	M	6.4	13.2	409.39	0.44	M5.5e	8/10"	****	
U Octantis	13h 24.5m	+59° 23'	M	7.0	14.1	308.44	0.47	M4e-M6(II-III)e	12/14"	****	1° 8' NE(PA048°) of ι Octantis (m5.45).##
σ Octantis	21h 08.7m	−88° 57'	δ Sctb	5.4	5.5	0.097	–	F0IV	Eye	****	By coincidence, both pole stars are small amplitude variables
Octantis	22h 20.0m	−80° 26'	SRb	4.6	5.3	55	–	M6III	Eye	*****	ξ, ε, and β form a nearly equilateral triangle.#
ξ Octantis	22h 50.4m	−80° 07'	SRb	4.6	5.3	–	–	B6IV	Eye	****	1° 6' N (PA 008°) of β Octantis (m4.13)

(continued)

Table 18.11. (continued)

Star clusters:

Designation	R.A.	Decl.	Type	m_v	Size (')	Brst. Star	Dist. (ly)	dia. (ly)	Number of Stars	Aper. Recm.	Rating	Comments
None												

Nebulae:

Designation	R.A.	Decl.	Type	m_v	Size (')	Brst./ Cent. star	Dist. (ly)	dia. (ly)	Aper. Recm.	Rating	Comments
None											

Galaxies:

Designation	R.A.	Decl.	Type	m_v	Size (')	Dist. (ly)	dia. (ly)	Lum. (suns)	Aper. Recm.	Rating	Comments
None											

Meteor showers:

Designation	R.A.	Decl.	Period (yr)	Duration	Max Date	ZHR (max)	Comet/ Asteroid	First Obs.	Vel. (mi/ km/sec)	Rating	Comments
None											

Other interesting objects:

Designation	R.A.	Decl.	Type	m_v	M_v	Dist. (ly)	dia. (ly)	Lum. (suns)	Aper. Recm.	Rating	Comments
σ Octantis	21^h 08.8^m	−88° 57'	F0 III	5.45	0.83	270		40	Eye/6×30		a.k.a. Polaris Australis. ~1° 03' from S. Cel. Pole in 2000. Distance is increasing

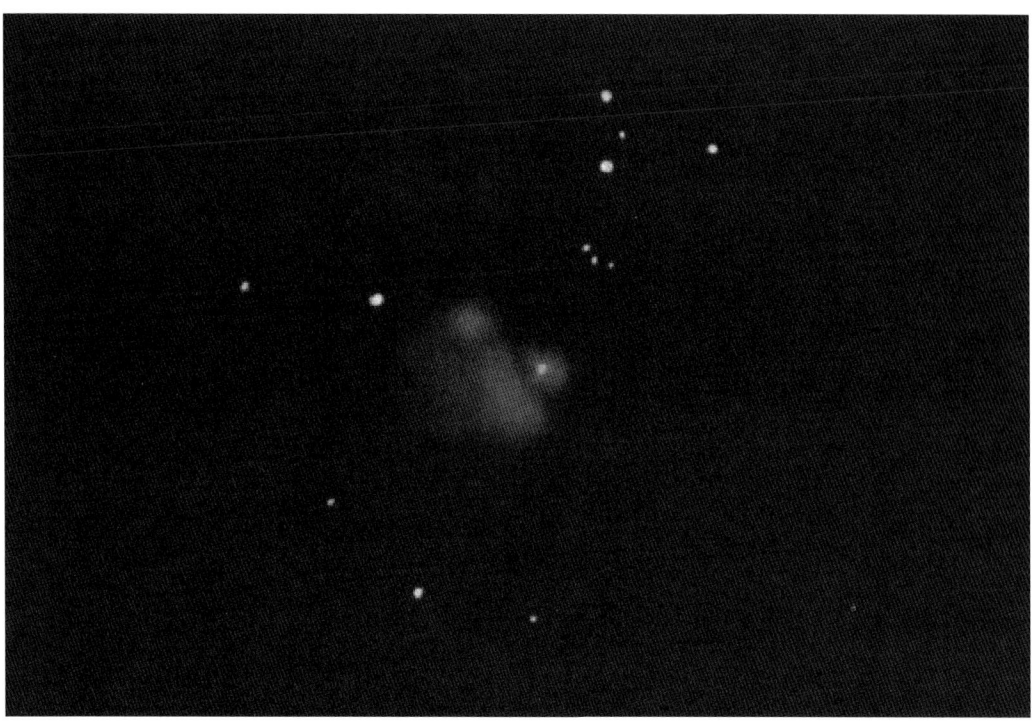

Fig. 18.22. NGC 6438 and 6438A, colliding galaxies in Octans. Drawing by Eiji Kato.

Fig. 18.23. NGC 7098, barred spiral galaxy in Octans. Drawing by Eiji Kato.

Microscopium (my-krow-**skope**-ee-um)	**The Microscope**
Microscopii (my-krow-**skow**-pee-ee)	**9:00 P.M. Culmination:** Sep. 19
Mic	**AREA:** 210 sq. deg., 66th in size

The figure used by Lacaille to portray this new constellation was that of the most common microscope used in his time. This usually consisted of a table or box-like stand with the optical tube attached to a vertical post mounted on the top of the table or stand. The image portrayed here is very similar to Lacaille's. Many of today's microscopes are essentially a miniaturized version of this form with enhancements like joints to make it more comfortable to use, and a light source to improve the brightness and contrast of the magnified image formed.

TO FIND: Although the stars of Microscopium are faint, its location is relatively easy to find since it is bordered on the east, north, and west by Piscis Austrinus, Capricornus, and Sagittarius, all moderately bright constellations. However, it is still hard to trace out the figure because of both the faintness of the stars and the lack of a sufficient number of stars to define the key points of a recognizable figure. One way to visualize Microscopium is to start with ω Capricorni, the southwesternmost hoof of the goat. Slightly less than 7.9° almost due south of this star is α Microscopii, and just 0.7° northeast of α is β Microscopii. These two stars define the upper part of the post which supports the optical tube. As unlettered magnitude 5.2 star just 4° south of α defines the middle of the western side of the post, while a similar 5.3 magnitude 6.6° south of β defines not only the eastern side of the support post, but also the intersection of the post with the supporting surface. The optical tube is also outlined by four stars. First, γ Microscopii, magnitude 4.7, the brightest star in the constellation, defines the western side of the optical tube, while an unnumbered star of magnitude 5.2 just 1.1° east of γ defines the eastern side. Magnitude 5.7 δ Microscopii is just 2.2° north of the last pair defines the top of the optical tube, and magnitude 5.3 ζ Microscopii is 6.3° south of γ. The support top is defined by θ^1 and the above star in the western side of the support post as the southeast and northwest corners of the table or box top. Two more faint stars define the eastern and western sides of the table top (Figs. 16.1 and 18.24).

KEY STAR NAMES: None.

Magnitudes and spectral types of principal stars:

Bayer desig.	Mag. (m_v)	Spec. type	Bayer desig.	Mag. (m_v)	Spec. type	Bayer desig.	Mag. (m_v)	Spec. type
α	4.89	G8 III	β	6.06	A1 IV	γ	4.67	G8 III[*31]
δ	5.69	K0/1 III	ε	4.71	A0 V[*31]	ζ	5.32	F3 V
η	5.55	K3 III	θ¹	4.80	A2p[*32]	θ²	5.76	A0 III[*32]
ι	5.11	F1 IV[*33]	μ¹	5.99	F1 III/IV	μ²	6.64	G0 V
ν	5.12	K0 III[*34]	2	5.20	K2 III	3	5.41	K3 III

Other stars magnitude 5.5 or brighter which have only catalog designations:

Desig-nation	Mag. (m_v)	Spec. type	Desig-nation	Mag. (m_v)	Spec. type
HD 201772	5.32	F3 V	HD 198716	5.34	K2 III
HD 196737	5.47	K1 III	HD 197630	5.48	B8/9 V
HD 204018	5.50	Am			

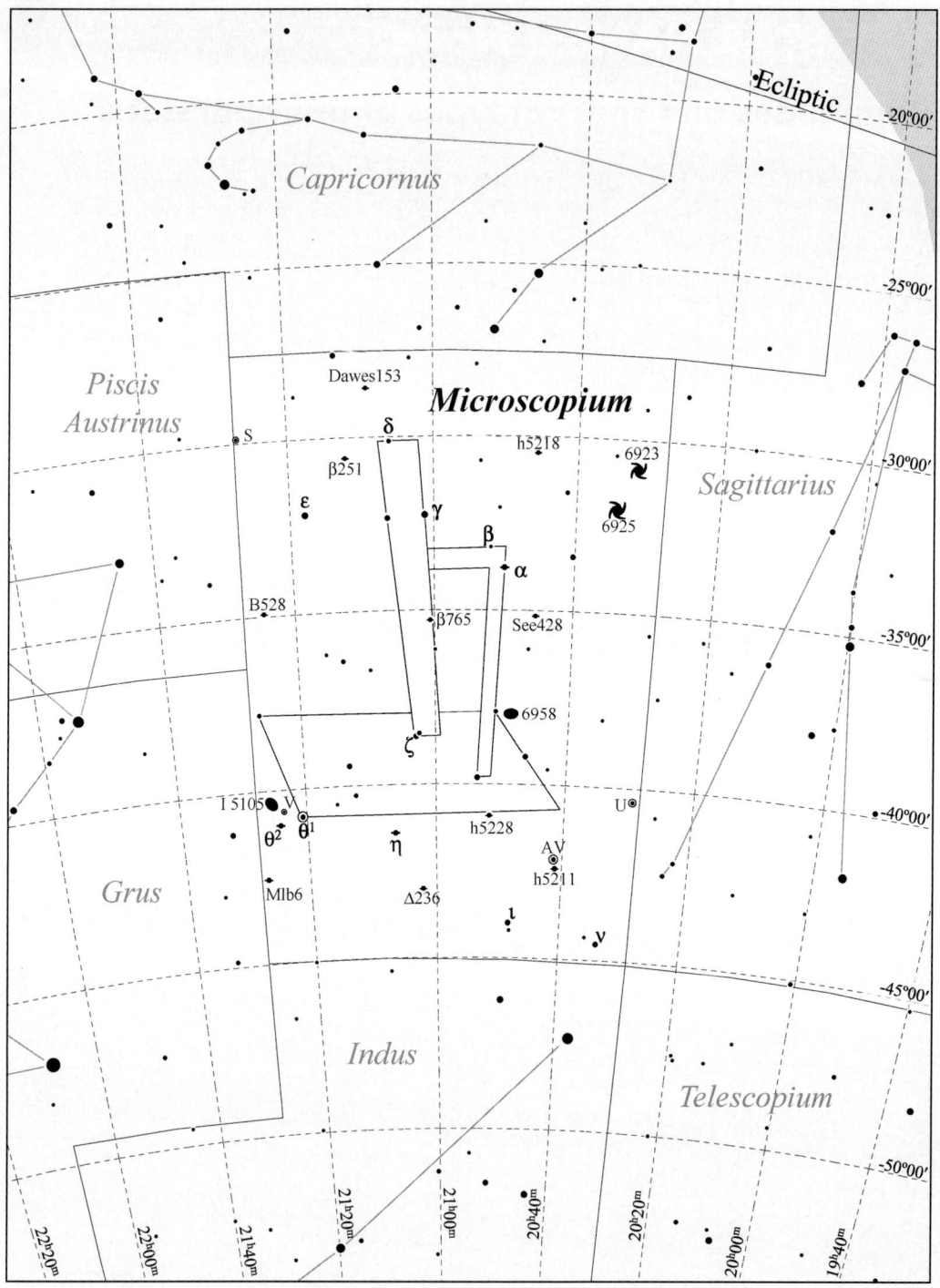

Fig. 18.24. Microscopium details.

Table 18.12. Selected telescopic objects in Microscopium.

Multiple stars:

Designation	R.A.	Decl.	Type	m₁	m₂	Sep. (")	PA (°)	Colors	Date/Period	Aper. Recm.	Rating	Comments
h5211	20ʰ 40.9ᵐ	-42° 24'	A–B	6.5	10.4	20.1	300	Y, ?	1999	15×80	****	
			A–B	10.4	12.8	8.9	244	?, ?	1999	5/6"	***	
See428	20ʰ 45.2ᵐ	-35° 10'	A–B	6.9	10.9	3.9	186	pY, ?	1998	3/4"	*****	
H5218	20ʰ 45.4ᵐ	-33° 29'	A–B	6.6	11.0	9.7	191	yW, ?	1998	3/4"	*****	Beautiful
α Microscopii	20ʰ 50.0ᵐ	-33° 47'	A–B	4.9	10.0	20.4	166	Y, W	1999	15×80	****	a.k.a. h5224
h5228	20ʰ 51.7ᵐ	-40° 54'	A–B	7.4	9.3	3.2	104	dY, ?	2000	3/4"	****	
β765	20ʰ 00.7ᵐ	-35° 18'	A–B	6.9	10.5	1.0	095	oY, ?	1991	9/10"	***	
Δ236	21ʰ 02.2ᵐ	-43° 00'	A–B	6.7	7.0	57.5	073	Y, Y	1999	6×30	*****	Very nice, slightly unequal pair
η Microscopii	21ʰ 06.4ᵐ	-41° 23'	A–B	5.8	7.9	128.3	088	pY, pY	1999	7×50	****	a.k.a. See437-AC
Dawes153	21ʰ 09.0ᵐ	-28° 29'	A–B	7.0	11.7	3.5	280	Y, ?	1977	3/4"	****	
Δ251	21ʰ 12.1ᵐ	-30° 35'	A–B	7.4	9.4	2.1	232	pY, ?	1991	4/5"	****	
θ² Microscopii	21ʰ 24.4ᵐ	-41° 00'	A–B	6.2	6.9	0.3	210	W, ?	1999	–	*	a.k.a. β766-AB. Too close for amatur instruments
			A–C	6.2	10.5	79.1	065	W, ?	1999	–15×80	****	a.k.a. β766-AC
B528	21ʰ 24.6ᵐ	-34° 58'	A–B	7.0	13.0	6.8	170	oY, ?	1933	6/7"	***	
MlbO6	21ʰ 27.0ᵐ	-42° 33'	A–B	5.6	8.2	2.9	150	dY, W	1999	3/4"	****	a.k.a. Burnham 767

Variable stars:

Designation	R.A.	Decl.	Type	Range (m$_v$)	Period (days)	F (f$_r$/f$_l$)	Spectral Range	Aper. Recm.	Rating	Comments
U Microscopii	20ʰ 29.6ᵐ	-40° 24'	M	7.0 14.4	334.29	0.39	M5e-M7e	14/16"	***	2° 04' NE (PA 039°) of κ1 Sgr (m5.60).##
AV Microscopii	20ʰ 41.8ᵐ	-42° 07'	Lc	6.3 6.4	–	–	MeII	6×30	****	2° 15' NNW (PA 324°) of ι Mic (m5.11). Small variations, photometric measurements only
θ¹ Microscopii	21ʰ 21.8ᵐ	-40° 47'	αCVn	4.8 4.9	2.122	–	A2p (Cr, Eu, Sr, Mg)	Eye	****	Small variations, photometric measurements only
S Microscopii	21ʰ 27.1ᵐ	-29° 50'	M	7.8 14.8	209.68	0.43	M3e-M5.5	18/20"	***	2° 59' NE (PA 040°) of ε Mic (m4.71).##
V Microscopii	21ʰ 23.8ᵐ	-40° 42'	M	9.4 14.0	381.15	–	M3e-M6e	14/16"	***	0° 35' E (PA 080°) of θ¹ Mic (m4.76).##

(continued)

Table 18.12. (continued)

Star clusters:

Designation	R.A.	Decl.	Type	m$_v$	Size (')	Brtst. Star	Dist. (ly)	dia. (ly)	Number of Stars	Aper. Recm.	Rating	Comments
None												

Nebulae:

Designation	R.A.	Decl.	Type	m$_v$	Size (')	Brtst./ **Cent.** star	Dist. (ly)	dia. (ly)	Aper. Recm.	Rating	Comments
None											

Galaxies:

Designation	R.A.	Decl.	Type	m$_v$	Size (')	Dist. (ly)	dia. (ly)	Lum. (suns)	Aper. Recm.	Rating	Comments
NGC 6923	20^h 31.6^m	–30° 50'	SB+	12.0	2.9 × 1.4	120M	101K	18G	6/8"	***	
NGC 6825	20^h 34.3^m	–31° 59'	Sb	11.3	4.4 × 1.2	110M	141K	29G	4/6"	****	
NGC 6958	20^h 48.7^m	–38° 00'	E1	11.3	2.2 × 1.7	120M	77K	35G	4/6"	****	
I 5105	21^h 24.4^m	–40° 32'	E4	11.4	2.8 × 1.7	230M	187K	116G	4/6"	****	

Meteor showers:

Designation	R.A.	Decl.	Period (yr)	Duration	Max Date	ZHR (max)	Comet/ **Asteroid**	First Obs.	Vel.(mi/ **km**/sec)	Rating	Comments
None											

Other interesting objects:

Designation	R.A.	Decl.	Type	m$_v$	M$_v$	Mass (suns)	Dist. (ly)	Lum (suns)	Effective Temp (K)	Aper. Recm.	Rating	Comments
Lacaille 8760	21^h 17.3^m	–38° 52'	Red dwarf	6.67	8.69	0.6	12.87	0.028	3,340	6×30	****	Brightest red dwarf. 2° 48' E (PA 096°) of ζ Mic (m5.32) *35

TELESCOPIUM (tel-es-KOH-pee-um)	**The Telescope**
TELESCOPII (tel-es-**KOH**-pee-ee)	**9:00 P.M. culmination:** Aug. 25
Tel	**AREA:** 252 sq. deg., 57th in size

A small group of mostly faint stars, this area of the sky was unformed (not part of any constellation) until 1756 when Nicholas Louis LaCaille published his atlas including this and other new constellations. Like many of LaCaille's other new constellations, Telescopium represents one of the scientific instruments of his time, the refractor telescope. In LaCaille's time, there were no official boundaries to constellations, and constellations were often shown overlapping other constellations. LaCaille's Telescopium represented one of the aerial telescopes used before the advent of achromatic lenses. To obtain both high magnification and freedom from color fringes around bright objects, these telescopes were extremely long (some over 60 yards!). LaCaille's original celestial telescope was over twice the length of our modern one, extending from the northern edge of Pavo to the southern edge of Ophiuchus, and crossing the modern constellations of Telescopium, Corona Australis, and Scorpius. In addition, the supporting pole extended from northern Pavo into southern Sagittarius.

While all of Telescopium is technically visible from any place south of 33° north latitude, the fact that it contains only 17 stars brighter than magnitude 5.5 (the brightest one being magnitude 3.5), combined with atmospheric absorption near the horizon makes it impractical to view north of about 25° N. This means good viewing is not possible in Europe, and in the USA it is possible to view only from Hawaii and extreme southern Florida or Texas.

TO FIND: α Telscopii (m3.5) and the naked-eye double δ^1 and δ^2 (m5.0 and 5.1) can be found about 5° south of the western tip of the oval of Corona Australis. ε Telescopii (m4.5) lies just under 3° due west of α, and the pair form the objective of the refracting telescope. Also, λ, κ, and ι^1 Scorpii, the three brightest stars in the eastern part of the Scorpion's tail, form a 4° long northwest to southeast arrow pointing down the telescope's tube. Following the direction of this arrow leads your eye between α and ε Telescopii, less than 10° away. Continuing this direction for another 17° leads you down a broad path of about a dozen faint stars which define the telescope's tube. This path of stars ends about 5° west of Peacock (m1.9), the brightest star in this barren region of the sky. Since the star Peacock represents the head of the celestial peacock, you might visualize Pavo as turning its head toward the telescope's eyepiece to look at the Scorpion's tail (Figs. 16.1, 18.25 and 18.26).

NAMED STARS: None.

Magnitudes and spectral types of principal stars:

Bayer desig.	Mag. (m$_v$)	Spec. type	Bayer desig.	Mag. (m$_v$)	Spec. type	Bayer desig.	Mag. (m$_v$)	Spec. type
α	3.49	B3 IV	β	–	_*36	γ	–	_*37
δ¹	4.92	B6 IV	δ²	5.07	B3 III	ε	4.52	G5 III
ζ	4.10	G8/K0 III	η	5.03	A0 Vn	θ	–	_*38
ι	4.88	G9 V	κ	5.18	G8/K0 III	λ	4.85	A0 V
μ	6.29	F5 V	ν	5.33	A9 Vn	ξ	4.93	M1 II
o	–	_*39	π	–	_*39	ρ	5.17	F7 V
σ	–	_*40	τ	5.69	K2 III*41	υ	–	_*42

Stars of magnitude 5.5 or brighter which have only catalog designations:

Designation	Mag. (m$_v$)	Spec. type	Designation	Mag. (m$_v$)	Spec. type
HD 167128	5.36	B3 IIIpe	HD 179886	5.38	K3 III
HD 169405	5.44	K0/K1 III			

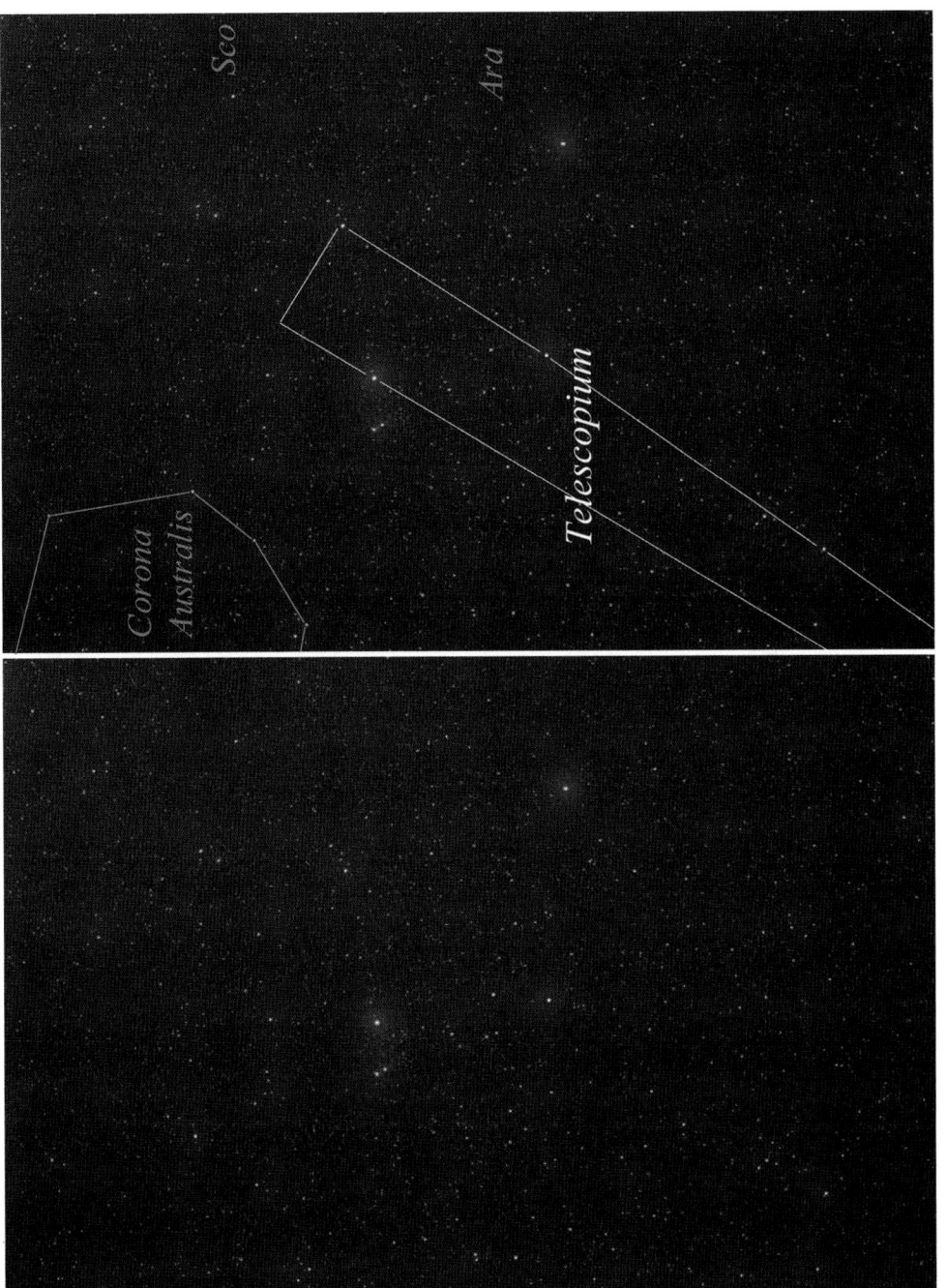

Fig. 18.25. **(a)** The stars of Telescopium and adjacent constellations. Photo by Christopher J. Picking. **(b)** The figures of Telescopium and adjacent constellations. About one-third of the telescope tube is outside the picture.

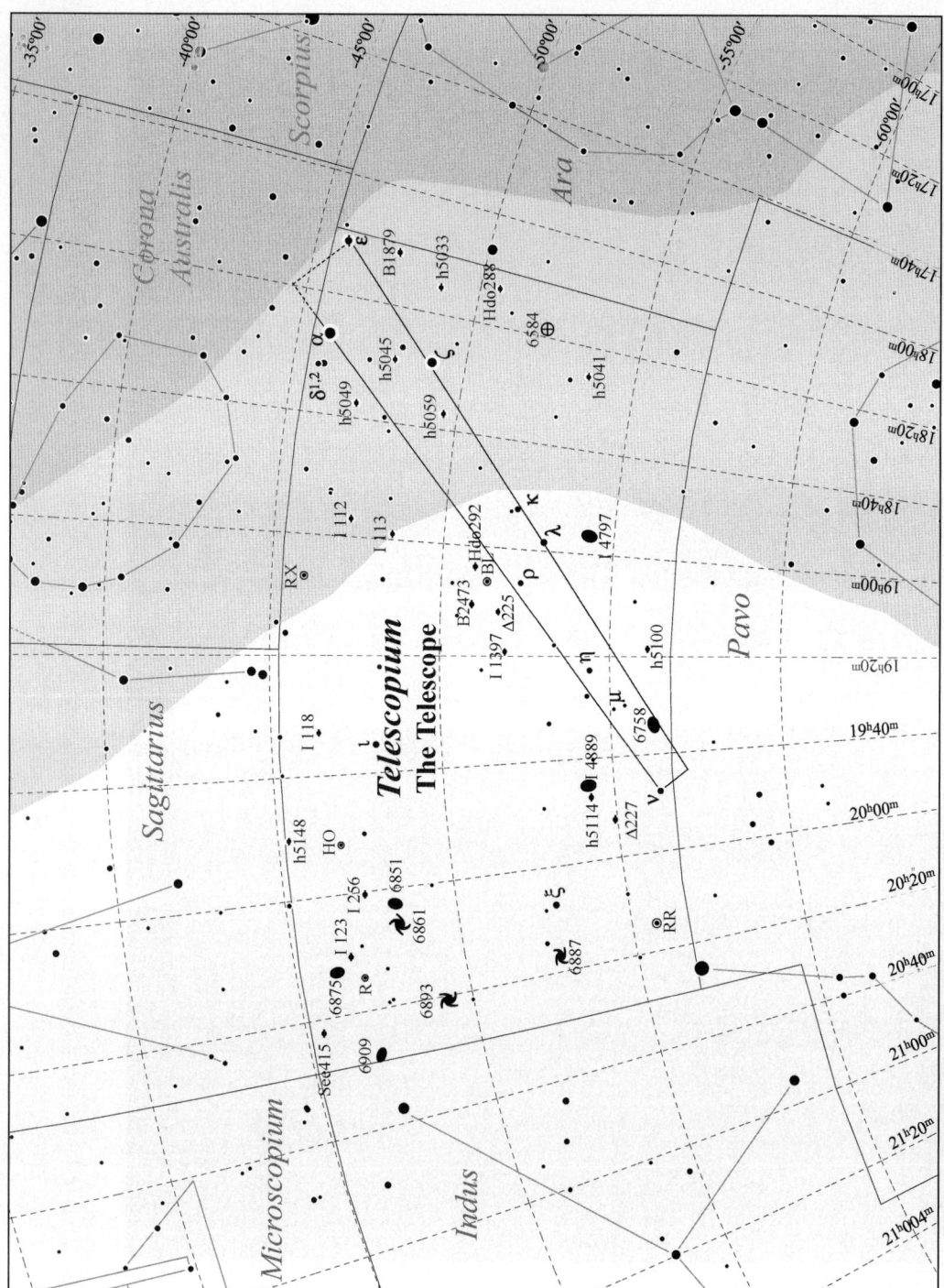

Fig. 18.26. Telescopium details.

Table 18.13. Selected telescopic objects in Telescopium.

Multiple stars:

Designation	R.A.	Decl.	Type	m₁	m₂	Sep. (")	PA (°)	Colors	Date/ Period	Aper. Recm.	Rating	Comments
B 1879	18h 11.1m	−47° 31'	A-B	6.1	11.5	1.7	266	oY, ?	1934	6/7"	****	
ε Telescopii	18h 11.2m	−46° 57'	A-B	4.5	13	20.7	223	oY, ?	1905	6/7"	****	
Hdo288	18h 12.9m	−50° 34'	A-B	6.6	9.94	1.6	132	yW, ?	1991	6/7"	****	
h5033	18h 15.4m	−48° 51'	A-B	6.7	9.8	17.2	114	yW, ?	1999	15×80	****	
			A-C	6.7	12.2	16.0	010	yW, ?	1999	4/5"	***	
			A-D	6.7	10.7	25.5	005	yW, ?	1999	15×80	****	
h5041	18h 25.8m	−53° 38'	A-B	7.3	9.2	3.0	260	yW, ?	1999	3/4"	****	
h5045	18h 30.9m	−48° 01'	A-B	6.7	9.7	8.1	021	yW, ?	2000	2/3"	****	
δ¹ + δ²	18h 31.8m	−52° 13'	A-B	4.9	5.1	592	017	bW, bW	2000	Eye	*****	Can be seen without optical aid by individuals with excellent eyesight. Unusual chance alignment, δ² is 320 ly more distant than δ¹
h5049	18h 37.4m	−47° 04'	A-B	6.9	11.1	19	248	yW, ?	2000	3/4"	****	
h5059	18h 47.4m	−49° 38'	A-B	6.9	12.3	25	241	W, ?	1999	4/5"	***	
			A-C	6.9	12.1	30	201	W, ?	1999	3/4"	***	
I112	18h 54.0m	−47° 16'	A-B	7.1	9.1	2	190	yW, ?	1991	5/6"	****	a.k.a. Δ224
			A-B	7.1	7.3	87	062	W, ?	1998	7×50	***	
I113	18h 58.9m	−48° 30'	A-B	6.8	10.7	3.2	227	O, ?	1991	3/4"	****	
Hdo292	19h 04.0m	−51° 01'	A-B	5.9	13.5	10.8	040	oY, ?	1926	7/8"	***	
Δ225	19h 12.4m	−51° 48'	A-B	7.2	8.4	70.1	250	O, ?	1999	8×50	****	
			A-B	8.4	11.0	26.8	088	?, ?	1976	3/4"	****	
B2473	19h 12.8m	−50° 29'	A-B	6.2	11.0	7.2	320	dY, ?	1991	3/4"	****	
h5100	19h 18.7m	−56° 09'	A-B	6.8	11.5	21.6	142	bW, ?	1999	15×80	****	
I1397	19h 19.6m	−51° 34'	A-B	6.6	10.6	1.5	338	W, ?	1928	6/7"	****	
h5114	19h 27.8m	−54° 20'	A-a	5.9	11.0	1.5	260	O, yW	1857	6/7"	*****	3/4° East of η Tel
			A-B	5.9	8.2	75.1	238	O, ?	1999	15×80	****	
			A-D	5.9	10.9	72.9	094	O, ?	1999	15×80	****	
			B-C	8.3	12.8	8.3	195	O, ?	1932	5/4"	****	
I118	19h 32.9m	−46° 47'	A-B	7.4	9.4	1.1	138	oY, ?	1991	8/9"	***	
h5148	19h 50.2m	−45° 23'	A-B	7.2	12.0	13.8	319	oY, ?	1999	3/4"	***	
Δ227	19h 52.6m	−54° 58'	A-B	5.8	6.4	22.5	147	Y, pG?	2002	15×80	*****	Unusual apparent color effect
I256	20h 01.4m	−47° 24'	A-B	7.1	9.6	0.8	189	yW, ?	1991	12/13"	****	
I123	20h 10.7m	−46° 44'	A-B	6.8	11.2	6.4	200	oY, ?	1999	3/4"	****	
See415	20h 20.9m	−45° 33'	A-B	7.3	12.0	11.1	095	Y, ?	1999	3/4"	****	

(continued)

Table 18.13. (continued)

Variable stars:

Designation	R.A.	Decl.	Type	Range (m.)	Period (days)	F (f₊/f₋)	Spectral Range	Aper. Recm.	Rating	Comments
BL Telescopii	19ʰ 06.6ᵐ	–51° 25′	+ SR	7.1 9.4	778.6	0.10	F5Iab/B+M	7×50	*****	55′ N (PA 002°) of ρ Tel (m5.71).#
RX Telescopii	19ʰ 07.0ᵐ	–45° 58′	Lc	7.0 7.5	–	–	M3Iab	6×30	****	Luminous, rapidly evolving red supergiant.#
HO Telescopii	19ʰ 52.0ᵐ	–46° 52′	EA/D	8.3 8.7	1.613	0.15	A7III(m)	7×30	****	Algol type, stars completely separate.#
RR Telescopii	20ʰ 04.5ᵐ	–55° 44′	Nc	6.5 16.5	–	–	pec	7×50	*****	Normally periodic, 387 days~m 12–16
R Telescopii	20ʰ 14.7ᵐ	–46° 59′	M	7.6 14.8	461.88	0.43	M5IIe–M7e	18/20″	****	Mira type pulsating variable.##

*43

Star clusters:

Designation	R.A.	Decl.	Type	m_v	Size (′)	Brst. Star	Dist. (ly)	dia. (ly)	Number of Stars	Aper. Recm.	Rating	Comments
NGC 6584	18ʰ 18.6ᵐ	–52° 13′	Globular	7.9	6.6	13.5	31K	70	–	10×80	****	Resolved in larger scopes

Nebulae:

Designation	R.A.	Decl.	Type	m_v	Size (′)	Brst./Cent. star	Dist. (ly)	dia. (ly)	Aper. Recm.	Rating	Comments
None											

Galaxies:

Designation	R.A.	Decl.	Type	m_v	Size (′)	Dist. (ly)	dia. (ly)	Lum. (suns)	Aper. Recm.	Rating	Comments
I 4797	18ʰ 56.5ᵐ	–54° 18′	E5	11.3	2.7×1.3	110M	86K	29G	6/8″	****	
NGC 6758	19ʰ 13.9ᵐ	–56° 19′	E1	11.5	2.3×1.7	140M	94K	39G	6/8″	****	
I 4889	19ʰ 45.3ᵐ	–54° 20′	E5	11.3	3.0×1.9	100M	87K	24G	8/10″	***	
NGC 6851	20ʰ 03.6ᵐ	–48° 17′	E4	11.8	1.9×1.5	130M	72K	26G	8/10″	***	
NGC 6861	20ʰ 07.3ᵐ	–48° 22′	S0	11.1	2.9×1.9	120M	101K	42G	6/8″	****	
NGC 6875	20ʰ 13.2ᵐ	–46° 10′	E6	11.9	2.3×1.3	130M	87K	23G	8/10″	***	

Designation	R.A.	Decl.	Type	m_v	Size (')	Dist. (ly)	dia. (ly)	Lum. (suns)	Aper. Recm.	Rating	Comments
NGC 6887	20ʰ 17.3ᵐ	–52° 48'	Sb	12.0	3.2 × 1.3	120M	112K	18G	8/10"	***	
NGC 6893	20ʰ 20.8ᵐ	–48° 14'	S(B)0	11.7	2.7 × 1.9	120M	94K	24G	8/10"	***	
NGC 6909	20ʰ 27.6ᵐ	–47° 02'	E6	11.8	2.3 × 1.1	110M	74K	18G	6/8"	****	

Meteor showers:

Designation	R.A.	Decl.	Period (yr)	Duration	Max Date	ZHR (max)	Comet/ Asteroid	First Obs.	Vel. (mi/ km/sec)	Rating	Comments
None											

Other interesting objects:

Designation	R.A.	Decl.	Type	m_v	Size (')	Dist. (ly)	dia. (ly)	Lum. (suns)	Aper. Recm.	Rating	Comments
None											

SCULPTOR (skulp-tore)	**The Sculptor's Workshop**
SCULPTORIS (skulp-**tore**-iss)	**9:00 P.M. Culmination:** Nov. 11
Scl	**AREA:** 475 sq. deg., 36th in size

There is no definite myth or story associated with the constellation of Sculptor, but given the emphasis on classical subjects in Lacaille's time, he may have been influenced by the mythological story of Pygmalion and Galatea. A sculptor from Crete, Pygmalion, fell in love with the goddess Aphrodite, who rebuffed him. Pygmalion remained so infatuated with the goddess that he sculpted a statue of her in either ivory or white stone. His love for Aphrodite made him strive for perfection in the sculpture to the point where he fell in love with the statue itself. At the next festival of Aphrodite, Pygmalion went to her temple to make a sacrifice and prayed to her to give him a woman as beautiful as the statue. When he returned to his studio, the statue had become a living woman with the complexion of the ivory or white stone. Pygmalion named the beautiful woman Galatea (from "the whiteness of milk"). They soon married and had one son named Paphos.

The South Galactic Pole lies in the north-eastern part of Sculptor, just 0° 37′ south–southwest (position angle 209°) from globular cluster NGC 288. When we look in that direction, our line-of-sight is along the shortest path out of our galaxy, and thus there are very few stars and little gas and dust in that direction. Because of this, Sculptor has one of the lowest average brightnesses of any constellation. The integrated brightness of Sculptor is only about a single sixth magnitude star per 6 sq. deg. of area. This makes it only about 1/30th as bright as the constellation Crux, which lies in the Milky Way. On the other hand, this direction gives us a nearly perfect view of intergalactic space and Sculptor is rich in galaxies.

Lacaille's original drawing of the Sculptor's workshop showed a block of stone (probably marble) on which rested a mallet and two chisels. Next to the block was a three-legged table holding a finished bust wearing a crown fern. Obviously, Lacaille's imagination (and artistry) far outstrips that of the author, who was able to construct only the rectangular stone out of the available stars.

TO FIND: See Figs. 11.1a, b and 18.16a, b for the general location of Sculptor. All the stars of Sculptor are quite faint, magnitude 4.3 or fainter, so it is a difficult constellation to see. Also, there is no way to make these stars show any figure resembling a sculptor or his work shop, so there is little reason to pick out the individual stars except to use them as guides to nearby telescopic objects. However, they can be found as follows. Most of the stars of Sculptor are contained in a triangle formed by three relatively bright stars: Fomalhaut (α Piscis Austrinus, m1.16), Diphda (β Ceti, m2.04), and Ankaa (α Phoenicis, m2.39). α Sculptoris lies 12° SSE (PA 164°) of Diphda. Although it is faint, there is no equal to it within 10° of its location. At the western end of Sculptoris, γ Scl and β Scl lie 5°23′ and 11°00′ SE of Fomalhaut. Although the region around γ and β Scl is brighter than that around α Scl, each is still the brightest star within an 8° diameter circle around its position, so they should also not be difficult to pick out. The remaining stars in Sculptor are 5th magnitude or fainter, but the principal ones can be located by using the two approximate parallelograms and the near diamond shape they form with α, β, and γ Scl (Figs. 3.1, 3.24, 16.1 and 18.27).

KEY STAR NAMES: None.

Magnitudes and spectral types of principal stars:

Bayer desig.	Mag. (m_v)	Spec. type	Bayer desig.	Mag. (m_v)	Spec. type	Bayer desig.	Mag. (m_v)	Spec. type
α	4.30	B7 IIIp	β	4.38	B9.5 IV	γ	4.41	K1 III
δ	4.59	A0 V	ε	5.29	F2 IV	ζ	5.04	B4 V
η	4.86	M2.5 III	θ	5.24	F4 V	ι	5.18	K0 III
κ¹	5.42	F3 V	κ²	5.41	K2 III	λ¹	6.05	B9.5V
λ²	5.90	K1 III	μ	5.30	K0 III	ν	6.69	B9 V*44
χ	5.59	K0 III*45	ξ	5.59	K1 III	o	–	–*46
π	5.25	K1 III	ρ	6.65	A9 V*47	σ	5.50	A2 V
τ	5.69	F2 V	υ	–	–*48			

Stars of magnitude 5.5 or brighter which have only catalog designations:

Designation	Mag. (m_v)	Spec. type
HD 9525	5.49	K1/K2 III

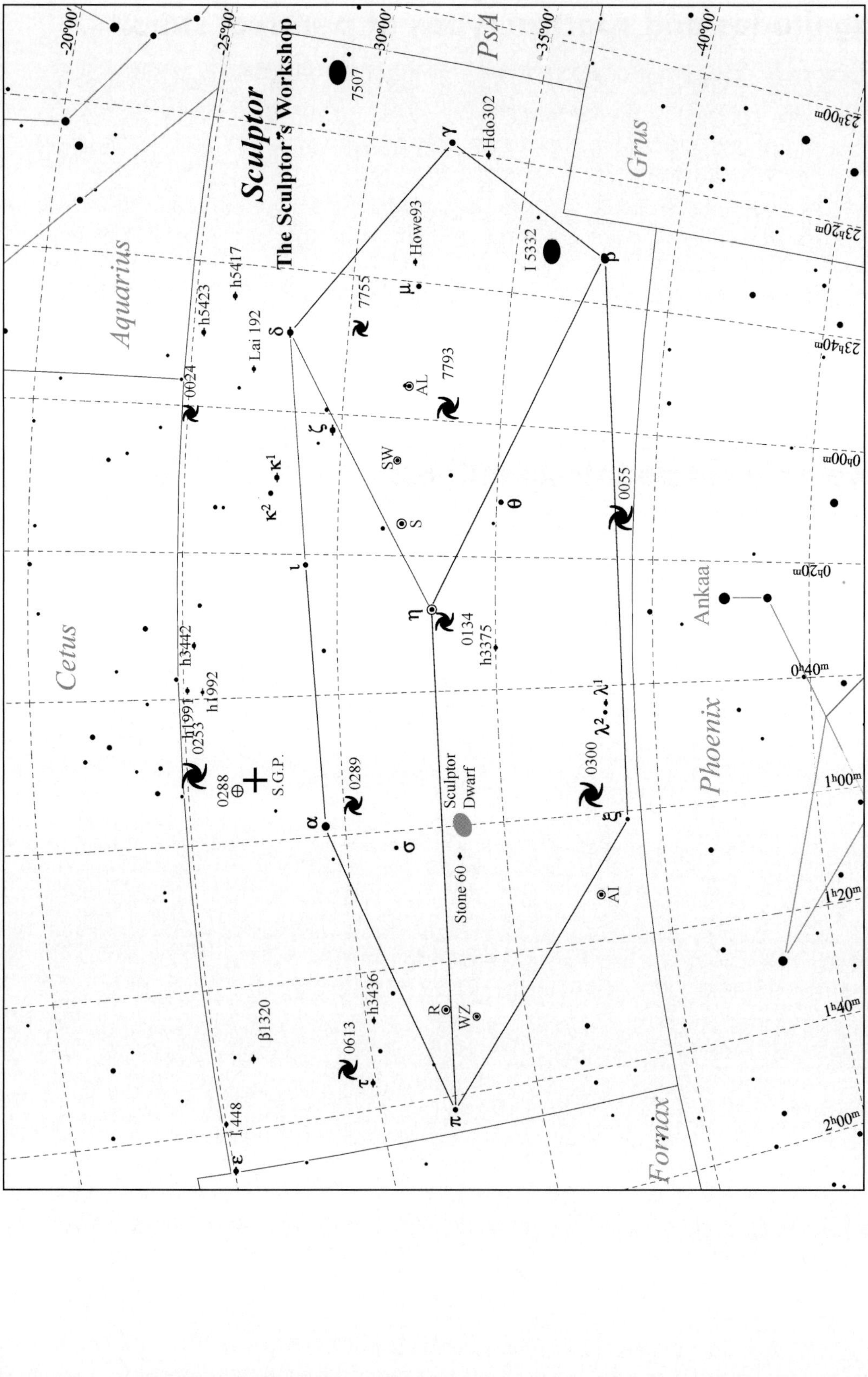

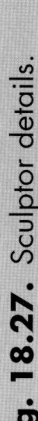

Fig. 18.27. Sculptor details.

Table 18.14. Selected telescopic objects in Sculptor.

Multiple stars:

Designation	R.A.	Decl.	Type	m_1	m_2	Sep. (")	PA (°)	Colors	Date/ **Period**	Aper. Recm.	Rating	Comments
Hdo 302	23h 19.7m	−33° 42′	A–B	6.4	12.5	41.0	306	oY, ?	1999	4/5″	*****	Very nice
Howe 93	23h 37.1m	−31° 52′	A–B	6.7	9.9	5.5	251	oY, ?	1991	2/3″	****	
H5417	23h 44.5m	−26° 15′	A–B	6.3	9.4	8.1	319	Y, ?	2002	2/3″	****	Easy
δ Sculptoris	23h 48.9m	−28° 08′	A–B	4.6	11.6	3.5	250	O, ?	1964	3/4″	***	a.k.a. β 1013
			A–C	4.6	9.4	74.7	296	O, ?	1998	15×80	****	a.k.a. h3216
h5423	23h 49.8m	−25° 20′	A–B	6.4	11.6	13.7	307	W, ?	1998	3/4″	****	
Lal 192	23h 54.4m	−27° 03′	A–B	6.7	7.4	6.7	271	W, ?	2006	2/3″	******	Attractively close, not too unequal
ζ Sculptoris	00h 02.3m	−29° 43′	A–B	5.0	13.0	3.0	330	bW, ?	1927	6/7″	****	a.k.a. B 631. Significant brightness difference. ~1,600:1
κ¹ Sculptoris	00h 09.4m	−27° 59′	A–B	6.1	6.2	1.3	260	Y, ?	2006	7/8″	*****	a.k.a. β 391. Nearly perfectly matched close pair
h3442	00h 32.7m	−25° 22′	A–B	6.9	11.0	18.3	186	oY, ?	1998	2/3″	****	
h3375	00h 33.7m	−35° 00′	A–B	6.6	8.5	4.5	169	Y, Y	1998	2/3″	****	
h1991	00h 38.8m	−25° 06′	A–B	6.6	9.7	47.1	095	dY, ?	1998	15×80	******	Attractive, fairly easy
h1992	00h 38.9m	−25° 36′	A–B	7.8	8.9	45.9	248	yW, ?	1998	7×50	****	
λ¹ Sculptoris	00h 42.7m	−38° 28′	A–B	6.6	7.0	0.7	322	bW, bW	1996	13/14″	***	a.k.a. Hdo 182. Pair is slowly separating
Stone 60	01h 04.5m	−33° 32′	A–B	6.6	10.2	8.5	218	Y, Y	1998	2/3″	****	
h3436	01h 27.1m	−30° 14′	A–B	6.9	9.6	9.7	128	oY, ?	1998	2/3″	****	
β 1230	01h 30.4m	−26° 12′	A–B	6.0	11.0	3.0	226	oY, ?	1959	4/5″	****	
τ Sculptoris	01h 36.1m	−29° 54′	A–B	6.0	7.4	**2.38**	**342**	Y, Y	**1,875.50**	4/5″	****	Minimum separation of 0.87″ in 1962, opening to 5.68″ in 2694. a.k.a. h3447
l448	01h 38.8m	−25° 01′	A–B	6.7	12.3	20.2	350	W, ?	1999	4/5″	****	
∈ Sculptoris	01h 45.6m	−25° 03′	A–B	5.4	8.5	**4.65**	**020**	Y, Y	**1,192.20**	2/3″	******	Showcase pair. Perfectly circular apparent orbit, separation is constant, angle changes

Variable stars:

Designation	R.A.	Decl.	Type	Range (m_v)	Period (days)	F (f_r/f_t)	Spectral Range	Aper. Recm.	Rating	Comments
AL Sculptoris	23h 55.3m	−31° 55′	EA/DM	6.1 6.3	2.445	–	B6V+A0V	6×30	****	
SW Sculptoris	00h 06.2m	−32° 49′	SRc	7.3 9.3	146	–	M1Ib-IIe–M4IIIe	7×50	****	

(continued)

Table 18.14. (continued)

Variable stars:

Designation	R.A.	Decl.	Type	Range (m$_v$)		Period (days)	F (f$_r$/f$_l$)	Spectral Range	Aper. Recm.	Rating	Comments
S Sculptoris	00^h 15.4^m	−32° 03′	M	5.5	13.6	362.57	0.18	M3e–M9e(Te)	10/12″	****	Range too small for eye – photoelectric or CCD measurements only *49
η Sculptoris	00^h 27.9^m	−33° 00′	Lb	4.8	4.9	–	–	M4IIIa	Eye	***	
Al Sculptoris	01^h 12.7^m	−37° 51′	δ Sct	5.9	6.0	0.05	–	A7III	6×30	***	Range too small for eye – photoelectric or CCD measurements only
R Sculptoris	01^h 27.0^m	−32° 33′	SRb	6.0	9.1	370	–	C6.5ea(Np)	7×50	****	Carbon star, extremely red.##
WZ Sculptoris	01^h 28.7^m	−33° 46′	δ Sct	6.5	6.6	0.090	–	F01V	6×30	***	Range too small for eye – photoelectric or CCD measurements only

Star clusters:

Designation	R.A.	Decl.	Type	m$_v$	Size (′)	Brtst. Star	Dist. (ly)	dia. (ly)	Number of Stars	Aper. Recm.	Rating	Comments
NGC 288	00^h 52.8^m	−26° 35′	Globular	8.1	13	12.6	29K	110	–	6×30	****	Large, bright, slightly elongated, sparse concentration of stars. <1° from S. Galactic Pole. Resolved in 4/6″

Nebulae:

Designation	R.A.	Decl.	Type	m$_v$	Size (′)	Brtst./Cent. star	Dist (ly)	dia. (ly)	Aper. Recm.	Rating	Comments
None											

Galaxies:

Designation	R.A.	Decl.	Type	m$_v$	Size (′)	Dist. (ly)	dia. (ly)	Lum. (suns)	Aper. Recm.	Rating	Comments
NGC 7507	23^h 12.1^m	−28° 32′	E0	10.6	4.3 × 1.5	73M	91K	24G	4/6″	***	Small, bright, round
I5332	23^h 34.5^m	−36° 06′	E0	10.3	8.9 × 8.2	31M	80K	5.8G	8/10″	**	Very large, extremely faint, with a small, bright nucleus
NGC 7755	23^h 47.9^m	−30° 31′	Sc	11.4	3.6 × 2.8	130M	136K	37G	6/8″	***	Large, bright, round. Faint companion galaxy
NGC 7793	23^h 57.8^m	−32° 35′	SA	9.2	10.5 × 6.2	8M	27K	1.4G	4/6″	****	Large, faint, elongated. Patchy halo, small, bright nucleus

Designation	R.A.	Decl.	Type	m_v	Size	Dist. (ly)	dia. (ly)	Lum. (suns)	Aper. Recm.	Rating	Comments
NGC 24	00ʰ 09.9ᵐ	−24° 58′	Sbc	11.3	6.3×1.3	26M	48K	1.6G	6/8″	*****	Large, faint, elongated, patchy
NGC 55	00ʰ 15.1ᵐ	−39° 13′	SB	8.1	31×6.3	4M	36K	735M	2/3″	*****	Second brightest galaxy in Sculptor Group
NGC 134	00ʰ 30.4ᵐ	−33° 15′	S(B)b	10.4	8.0×1.7	67M	156K	25G	4/6″	*****	Beautiful nearly edge-on galaxy. NGC 131 is in same medium power FOV. Great in 10″
NGC 253	00ʰ 47.6ᵐ	−25° 17′	Scp	7.6	30×6.9	11M	96K	8.8G	7×50	*****	Brightest member of Sculptor Group. Remarkable. Complex dust lanes
NGC 289	00ʰ 52.7ᵐ	−31° 12′	Sb	10.6	5.0×4.2	78M	113K	28G	6/8″	***	Very bright, large, slightly elongated, four arms, small, bright nucleus
NGC 300	00ʰ 54.9ᵐ	−37° 41′	SBmp	8.1	20×13	7M	41K	2.2G	3/4″	***	Face-on, very bright, very large, elongated. Small, bright nucleus. Low surface brightness
NGC 613	01ʰ 34.3ᵐ	−29° 25′	S(B)b	10.0	5.2×2.6	64M	97K	33G	6/8″	****	Very nice in 10″

Meteor showers:

Designation	R.A.	Decl.	Period (yr)	Duration	Max Date	ZHR (max)	Comet/Asteroid	First Obs.	Vel. (mi/km/sec)	Rating	Comments
None											

Other interesting objects:

Designation	R.A.	Decl.	Type	m_v	Size (′)	Dist. (ly)	dia. (ly)	Lum. (suns)	Aper. Recm.	Rating	Comments
South Galactic Pole	00ʰ 51.4ᵐ	−27° 08′	Point	–	–	–	–	–	–	–	
Sculptor Dwarf	01ʰ 00.1ᵐ	−33° 43′	dE	10.1	40×30	280K	3,360	610K	–	**	Dwarf spheroidal satellite of MW Galaxy. Very sparse, prob. visible only in photographs

Fig. 18.28. NGC 7793, spiral galaxy in Sculptor. Astrophoto by Rob Gendler.

Fig. 18.29. NGC 55, spiral galaxy in Sculptor. Astrophoto by Rob Gendler.

Fig. 18.30. NGC 134, spiral galaxy in Sculptor. Drawing by Eiji Kato.

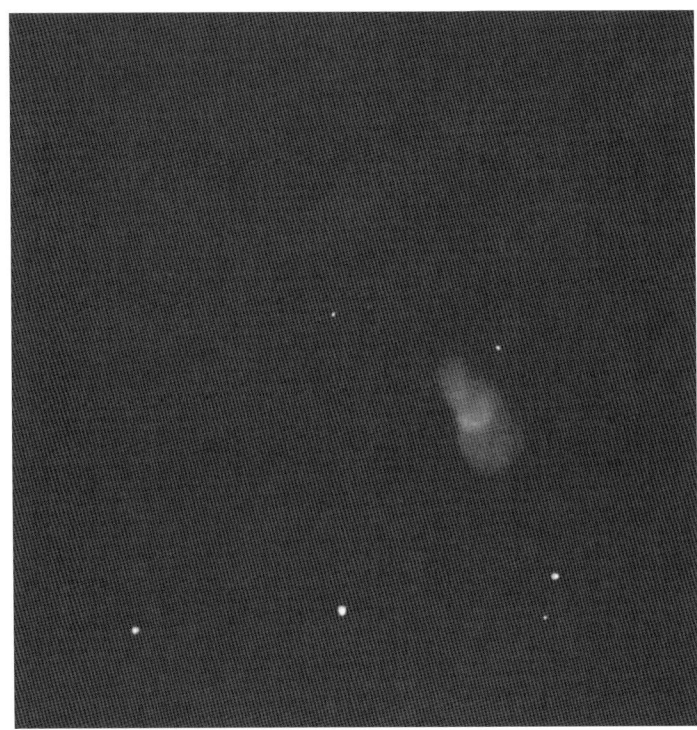

Fig. 18.31. NGC 150, spiral galaxy in Sculptor. Drawing by Eiji Kato.

Fig. 18.32a. NGC 253, skpiral galaxy in Sculptor. Astrophoto by Mark Komsa.

Fig. 18.32b. NGC 253, drawing by Eiji Kato.

Fig. 18.33. NGC 300, spiral galaxy in Sculptor. Astrophoto by Rob Gendler.

NOTES:

1. The star originally labeled β Antliae was later moved to Hydra and left unlettered by Baily.
2. γ Antliae was moved to Hydra by the IAU redefinition of the constellations.
3. The double star Rst 1489 requires larger amateur telescopes for resolution. A magnitude 6.0 star 3′55″ away makes this group similar to ε Lyrae for the unaided eye. A magnitude 7.8 star is also close by.
4. S Antliae variables have short periods and low amplitudes with a fast rise, a slower decline, and a flat-bottomed light curve.
5. Lacaille did not assign any letters beyond θ in Pyxis.
6. Lacaille thought that κ was a 6th magnitude star and did not assign it a letter. Gould judged it more appropriately and assigned it the letter κ.
7. Lacaille saw two stars through his telescope, but only one was visible to the unaided eye, so he assigned only one letter to γ Caeli. Some later observers assigned γ¹ and γ² to the components, so some atlases and catalogs may differ in the designation of this star.
8. Lacaille did not assign any letters beyond δ in Caelum.
9. Bode extended the lettering to ρ Caeli, but only ν has survived.
10. Lacaille gave the single letter designation η Pictoris to two stars which were close together.
11. Lacaille did not use letters past ν in Pictor.
12. Kapteyn's star has the second highest proper motion in the sky, 8.70″/year at p.a. 131° (SE). It covers a distance equal to the moon's diameter in ~200 years. Its rapid motion is due to a combination of being fairly close (22nd closest star) and its extremely high space motion of 175 mi/s.
13. The β Pictoris disk is about twice the diameter of our solar system and may be coalescing to form planets orbiting β Pictoris. Located 6 W, and 1.5 N of Canopus, and being the brightest star of a short chain, β Pictoris is relatively easy to find.
14. Lacaille used only the letters from α to θ in Circinus.
15. T Circini is one of the most interesting eclipsing binaries in the southern sky. It is similar to Algol, except that the stars are highly evolved, and so close together that they are distorted into teardrop shapes. As a result, the light curve is not only very deep, 1.5 magnitudes, but it is also very rounded, without a long flat light output between eclipses.
16. The star labeled α Normae by Lacaille is now N Scorpii.
17. Lacaille included β Normae on his maps of the southern sky, but accidentally left it out of his catalog. This star is now known as H Scorpii.
18. Lacaille did not use the letter ξ in Fornax.
19. Lacaille assigned the letters θ, o, and υ, but ξ some catalogs dropped some or all of these letters because of their faintness, so many atlases and catalogs omit these letters.
20. Although the Fornax Dwarf Galaxy is less than 15° from the center of the Fornax Galaxy Cluster, the two are not related. The Dwarf Galaxy is a member of the Local Group of Galaxies, and a satellite of the Milky Way Galaxy, and only 500,000 light years distant. The Fornax Dwarf is very inconspicuous because the brightest stars are only 19th magnitude and are spread thinly, with no central concentration.
21. At a distance of 65 million light years, the Fornax Cluster of galaxies is the second largest cluster within 100,000,000 light years of our Milky Way Galaxy. With about 60 large galaxies and a few hundred dwarf galaxies, it is considerably smaller than the Virgo Cluster of approximately 150 large galaxies and over 1,000 dwarf galaxies. Being only about 30° from the south galactic pole, The Fornax Cluster is well away from the star and dust clouds of the Milky Way, so we get a clear view of its horde of galaxies. A nearby cluster of galaxies, the Eridanus Cluster, is a near-twin of the Fornax cluster, being about the same size, but with somewhat fewer members. In fact, the two clusters are often called Fornax I and Fornax II, and considered as essentially one cluster.
22. Lacaille assigned the ε and θ designations to these two stars. Baily discarded them because of the stars' dimness.
23. In his 1763 catalog, Lacaille assigned the letter κ to a 6th magnitude star, but no reasonable star was found at the position he gave, and none of the prominent southern observers could positively identify any star as being the one Lacaille intended.
24. Lacaille cataloged these two stars, but did not assign letters to them. Gould assigned the letters μ and ν because they were brighter than several other stars which Lcaille did letter.

25. θ Reticuli is a double star whose components are near enough in brightness that features of both can be seen in the combined spectrum.

26. Slightly fainter than ε^1 and ε^2 but almost 50% farther apart. High proper motion, ~1.5″/year

27. The Large Magellanic Cloud was long thought to be a satellite of the Milky Way Galaxy, with a mass about 10% of our own galaxy. New measurements of the velocity of the LMC in 2008 showed that it was moving too fast to be orbiting the Milky Way unless the Milky Way had a mass of about four trillion suns, or twice the estimated mass. New measurements of the Milky Way mass gave a new estimate of about three trillion suns, high enough to make it possible that with errors in both measurements, the LMC might be a satellite of the Milky Way. The LMC is close enough that many interesting nebulae and other objects are visible in medium to large amateur telescopes. About 25% of the LMC lies in Mensa, with the rest in Dorado.

28. Lacaille assigned the letter γ to this group of three stars. The superscripts were added later.

29. Charles Rumker noticed that it had been the pole star in Lacaille's time and gave it the designation of o. However, Lacaille had made an error in the coordinates and thought that σ was closer to the pole and called it the pole star.

30. σ Octantis is currently the closest visible star (for most people) to the southern pole and is thus also known as Polaris Australis.

31. Flamsteed designated γ and ε Microscopii in Piscis Austrinus at numbers 1 and 4. Gould considered them to be in Microscopium and his judgment was upheld by the IAU in 1930.

32. Lacaille assigned θ to both these stars. The superscripts were assigned later.

33. Lacaille did not assign ι or any subsequent letters in Microscopium.

34. Lacaille mistakenly gave the designation of ν to two stars in Indus, and Gould transferred one of them to Microscopium.

35. Although Lacaille 8760 is not the closest red dwarf, its distance of 12.87 ly combined with its relatively high luminosity make it the brightest red dwarf in our skies. Proxima Centauri and 24 other red dwarf stars are closer to us. Lacaille 8760 has a mass of 0.60 suns, a radius of 0.66 suns, but a luminosity of only 0.028 suns. It also has an age of 4.6 billion years, almost identical to our sun. Because of its closeness and relative similarity to our sun, it has been the target of intensive searches for accompanying planets, but none have yet been found. Like most red dwarf stars, Lacaille 8760 is a flare star, also known as AX Microscopii. It experiences about one flare per day. However, these flares, which would make a less luminous star to increase its total brightness significantly, cause only an approximately 0.1 magnitude increase in its brightness, barely detectable to the eye.

36. Lacaille appropriated η Sagittarii into Telescopium, but it was later restored to Sagittarius.

37. Lacaille's γ Telescopii was later transferred to Scorpius and given the designation "G".

38. Lacaille renamed Bayer's d Ophuchii as θ Telescopii, but it was later restored to Ophiuchius and renumbered as 45 Ophiuchii.

39. Lacaille did not use either o or π in Telescopium.

40. Lacaille's σ Telescopii is now an unlettered star in Corona Australis.

41. Gould considered τ Telescopii too faint to merit a letter designation, and it now does not appear in most catalogs and atlases.

42. Lacaille did not assign any letters beyond τ in Telescopium.

43. Although RR Telescopii is normally a periodic variable, in 1944, it had a nova-like rise to magnitude 7 and stayed bright until 1950, when it began fading back to its normal brightness.

44. Lacaille assigned the letter ν to this star, but Baily dropped it because of the star's faintness.

45. Lacaille cataloged this star, but did not assign a letter to it. Gould assigned it the letter χ, which had not been used by Lacaille.

46. Lacaille did not use the letter o in Sculptor.

47. The star Lacaille called ρ Sculptoris is now an unlettered star in Cetus.

48. Lacaille did not use υ or subsequent letters in Sculptor.

49. Carbon stars, such as R Sculptoris, are red giants which have converted nearly all their hydrogen into helium and much of the helium into carbon and oxygen. Most stars in this stage of their lives produce more oxygen than carbon, but a few do the reverse. In the latter stars, all the oxygen is consumed in producing carbon monoxide, and the excess carbon can form compounds like CH, C_2, and CN. The metallic oxides, such as TiO and ZrO, often found in red giants are no longer present. This combination of compounds adsorbs almost all the blue light emitted and the carbon stars are thus extremely red compared to other stars of the same temperature.

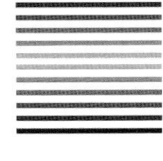

Glossary of Terms

Altitude   The angular distance of a celestial object above the horizon, usually expressed in degrees.

Asterism   A group of stars forming a recognizable pattern. Asterisms are not formally recognized by the IAU. Asterisms may contain stars from one or more constellations.

Azimuth   The angular distance of an object, from the north point, usually measured in degrees from north to east, i.e., an object due east of the observer has an azimuth of 90°, one due west 270°, etc.

Bayer letters   A Greek lower case letter assigned by Johannes Bayer in his Uranometria of 1603. He usually divided the stars into groups by their magnitudes, e.g., 1, 2, 3, etc. and gave all the first magnitude stars' designations of α, β, γ, etc. until all stars of that magnitude were lettered, then continued with the second and fainter magnitude stars. In some cases, Bayer followed a sequence of designations based on locations in the constellation rather than a strictly brightness sequence. Since Bayer's time, the magnitudes of the stars have been precisely defined by a mathematical formula, and the modern magnitudes sometimes differ from the older estimates, making Bayer's letters seem out of order.

Celestial Equator   The line traced around the celestial sphere by the Earth's extended equatorial plane.

Constellation   An officially designated area of the sky containing all the celestial objects within that area.

P. Simpson, *Guidebook to the Constellations: Telescopic Sights, Tales, and Myths,*
Patrick Moore's Practical Astronomy Series, DOI 10.1007/978-1-4419-6941-5,
© Springer Science+Business Media, LLC 2012

Culmination   A celestial object is said to culminate or reach culmination when it passes the meridian. At that point, it has reached its highest altitude above the southern (or northern for southern hemisphere observers) horizon. See Transit for an exception.

Declination (dec.)   The celestial equivalent of latitude on the Earth. Both are customarily measured in degrees, with the latitude of the equator (either celestial or terrestrial) being 0°, and the north and south poles being +90° and −90°, respectively.

Ecliptic   The line tracing the annual apparent path of the Sun around the sky. The intersection of the ecliptic plane with the sky. So named because solar and lunar eclipses occur only when the moon is very close to this line.

Ecliptic plane   The plane of the Earth's orbit around the Sun.

First Point of Aries   The point where the apparent Sun crosses the celestial equator each spring. It is the zero point of the celestial coordinate system.

Flamsteed numbers   John Flamsteed, the British Astronomer Royal from 1675 to 1719, listed all the stars visible to the unaided eye and gave numbers to the stars of each constellation in strict order of their right ascension (R.A.) regardless of their brightness.

Galactic cluster   See "open cluster."

Globular cluster   A cluster consisting of such a large number of stars that gravity forces it into a nearly spherical shape. Globular clusters are randomly distributed throughout a large spherical region centered on the host galaxy's center. The globular clusters in the Milky Way Galaxy are generally much older and more evolved than the galactic clusters.

Ind-Arabic (indigeneous Arabic)   The Arabic language as used by the average Arabic person in the medieval period.

IAU   The International Astronomical Union. An international organization of professional astronomers which has many groups of specialties within astronomy and organizes commissions to investigate specific areas and report or make recommendations for standardization of terminology and other topics to the entire IAU. In 1930, one of these commissions created the first official boundaries of the constellations consisting of a series of straight lines of either constant right ascension or declination. Until that time, there had been no official definition of the constellation boundaries, and some stars had been included in more than one constellation and some in no constellation at all. Under the new definition, every star was assigned to one and only one constellation.

Lunar Mansion   A region of the sky through the moon passed periodically, analogous to zodical signs through which the Sun passes.

Meridian   The great circle passing through both the celestial pole and the zenith pole and extending from the northern to the southern horizon is called the meridian.

Nebula   Any diffuse celestial object. Examples include planetary nebulae, emission nebulae, dark nebulae, and reflection nebula. Before telescopes were able to resolve galaxies into individual stars, they were included in the class of nebula, but now they constitute their own separate class.

Open cluster   A loose assemblage of stars. Open clusters generally comprise 10–10,000 stars, are loosely held together by gravity, and usually found near the galactic plane (the plane of the Milky Way). Also called "galactic clusters."

Precession of the equinoxes   Gravitational forces from the other bodies in the solar system cause the axis of the Earth's rotation to slowly precess or move. This motion is analogous to the motion of a spinning top, but much slower, requiring nearly 26,000 years for one cycle.

Psuedo-Eratsthenes et al.   In many cases, there are documents which were written in the style of a particular classical author, and perhaps meant so as to seem that they were authentic, but were actually probably not. In these cases, historians often designate the unknown authors putting "Psuedo-" before the purported author's name.

Right Ascension (R.A.)   The celestial equivalent of terrestrial longitude. While longitude is measured in degrees east or west from Greenwich, England, Right Ascension is measured in hours, minutes, and seconds of time. The Right Ascension of any celestial object is the amount of time between the passage of the First Point of Aries across the meridian and the passage of the celestial object.

Sci-Arabic (scientific Arabic)   The Arabic language used by Arabic astronomers and scientists of the medieval period.

Transit   A celestial object crosses, or transits, the meridian halfway between its rising and setting. For circumpolar objects only, there can be a second meridian crossing when the object crosses the meridian below the celestial pole. These are called the upper and lower transits.

Wolf-Rayet star   Often called WR stars, these are massive (>20 solar masses), hot (25,000–50,000 K), very evolved stars. They are believed to be approaching the supernova stage. They also have extremely rapid mass loss through stellar winds which typically exceed the solar wind by 10,000,000 times. They usually show bright (emission) lines of hydrogen, helium, carbon, oxygen, and silicon in their spectra.

Abbreviations and Symbols

1. General

AAS   American Astronomical Society
BAA   British Astronomical Society
IAU   International Astronomical Union
NGC   New General Catalog
RFT   Rich(est) Field Telescope, a telescope designed to show the largest number of stars for its aperture

2. Double Star Discoverer Codes

β   S. W. Burnham (1838–1921)
βpm   S. W. Burnham's proper motion catalog, 1913
Δ   J. Dunlop (1798–1848)
δ   B. H. Dawson (1890–1960)
Σ   F. G. Wilhelm Struve (1793–1864)
ΣI   W. Struve, first supplement
Σ II   W. Struve, second supplement
A   R. G. Aitkin (1864–1951)
AGC   Alvan G. Clark (1832–1897)
Alden   H. L. Alden (1890–1964)
B   Williem H. Van den Bos (1896–1974)
Bla   A. Blazil

BrsO	Brisbane Observatory, Australia
CapO	Cape Observatory, South Africa
Chr	Center for High Resolution Astronomy
CorO	Cordoba Observatory, Argentina
Cou	Paul Couteau (1923–?)
Dawes	W. R. Dawes (1799–1868)
Es	T. E. H. Espin (1858–1934)
Fin	W. S. Finsen (1903–1979)
Frk	W. S. Franks
Gale	W. F. Gale (1865–1945)
Gli	J. M. Gilliss (1811–1865)
H	William Herschel (1739–1822)
H 1–6	William Herschel, difficulty classes from 1 to 6, as assigned in 1782–1784 catalogs
H N	William Herschel's 1821 catalog
h	John Herschel (1792–1871)
HdO	Harvard Observatory, USA and other locations
HDS	Hipparchos Double Star
Ho	G. W. Hough (1836–1909)
Howe	H. A. Howe (1858–1926)
Hu	William Hussey (1864–1926)
Jc	W. S. Jacob
Jrn	A. Jorissen
Jsp	Morris K. Jessup (1900–1959)
Ku	F. Küstner (----)
Kui	Gerard P. Kuiper (1905–1973)
Lal	F. de Lalande (1732–1807)
Lam	J. von Lamont (1805–1879)
LDS	W. J. Luyten (1899–1994), proper motion survey, 1941
OΣ	Otto Struve (1819–1905), Pulkovo catalog
OΣΣ	Otto Struve, Pulkovo catalog supplement
OΣΣA	Otto Struve, Pulkovo catalog supplement, Appendix List
R	H. C. Russell
Roe	E. D. Roe
Ru	R. A. Rossiter (1896–1977)
S	James South (1785–1867)
S,h	James South and J. Herschel, joint catalog, 1824
Schj	S. C. F. C. Schjellerup (----)
See	T. J. J. See (1866–1962)
Stein	J. Stein (1821–1951)
Stone	Ormond Stone
Wal	A. Wallenquist
Wirtz	Carl W. Wirtz (1876–1939)
WNO	U.S. Naval Observatory, Washington, DC

3. Variable Stars

#	Variable star shown in the AAVSO Variable Star Atlas
##	Variable star which also has an AAVSO comparison chart available
$	Variable star for which a BAA comparison chart is available

4. Star Clusters

Cr	Per Arne Collinder
Ha	Harvard Observatory
M	Charles Messier
Mel	Philibert J. Mellote
Ru	Ruprecht
Ste	Stephenson
Tr	Robert J. Trumpler

5. Nebulae

Abell	George O. Abell (planetary nebula)
B	Edward. E. Barnard (dark nebula)
Be	Bernes (dark nebula)
Gum	Colin S. Gum (bright nebula)
Jn	Jones (planetary nebula)
Mz	D. N. Menzel (planetary nebula)
PK	Lubos Perek and Lubos Kohouteck (planetary nebula)
Sa	Aage Sandqvist
Sh	Stewart Sharpless (bright nebula)
Sp	Shapley, Harold

6. Galaxies

A	George O. Abell (galaxy clusters)
Abell	George O. Abell (galaxy clusters)
Arp	Halton C. Arp (peculiar galaxies)

7. Other Objects

C	Caldwell-Moore, Patrick Alfred (deep-sky objects of all types)
GJ	W. von Gliese and H. Jahreiss (nearby stars)
Gl	W. von Gliese (nearby stars)
M	Charles Messier (deep-sky objects of all types)

8. Variable Star Types

Cep	Long-period cepheids, 1 to ~50–70 days, 0.1–2 mag., high lum. (G-K)
Cδ	Long-period classical, disk component of MW, open clusters (δ Cep)
CW	Long-period cepheids, spherical component of MW, 1.5–2 mag. fainter than Cδ, W Vir
L	Slow irreg., little or no trace of periodicity
Lb	Slowly varying irreg. var., late spec classes, giants (K, M, C, and S) (CO Cyg)
Lc	Irreg. var. supergiants of late spec classes, TZ Cas
M	Mira type, >2.5 mag. var., fairly reg., 80–1,000 days (Me, Ce, and Se), o Cet
SR	SemiReg variables, giants/supergiants, periodicity interrupted by variations
SRa	Giants of late spec class (M, C, and S), <2.5 m comparatively regular variation, with strong changes, Z Aqr
SRb	Giants of late spec class (M, C, and S), poor periodicity, RR CrB, AF Cyg
SRc	Supergiants of late spec class, μ Cep, RS Cnc
SRd	Giants/supergiants of spec class F, G, and K, S Vul, UU Her, AG Aur
RR var.	RR Lyrae (short-period cepheids), cluster variables, 0.05–1.2 days, 1–2 m max.
RRab	Sharply asymmetric light curve (steep asc branch)
RRc	Almost symmetric light curve, ~0.3 day, per >0.21 day
RRs	Supergiants, period <0.21 day, lum. 2–3 m fainter than RRc, SX Phe
RV	RV Tauri with constant mean brightness
#	The star is shown in the AAVSO Variable Star Atlas
##	An AAVSO chart featuring this star is available
S	This star is included in the BAA variable star list

9. General Star Catalogs

HD	Henry Draper
HIP	Hipparchos (satellite)
HR	Henry Russell
PPM	Positions and Proper Motion catalog
SAO	Smithsonian Astrophysical Observatory

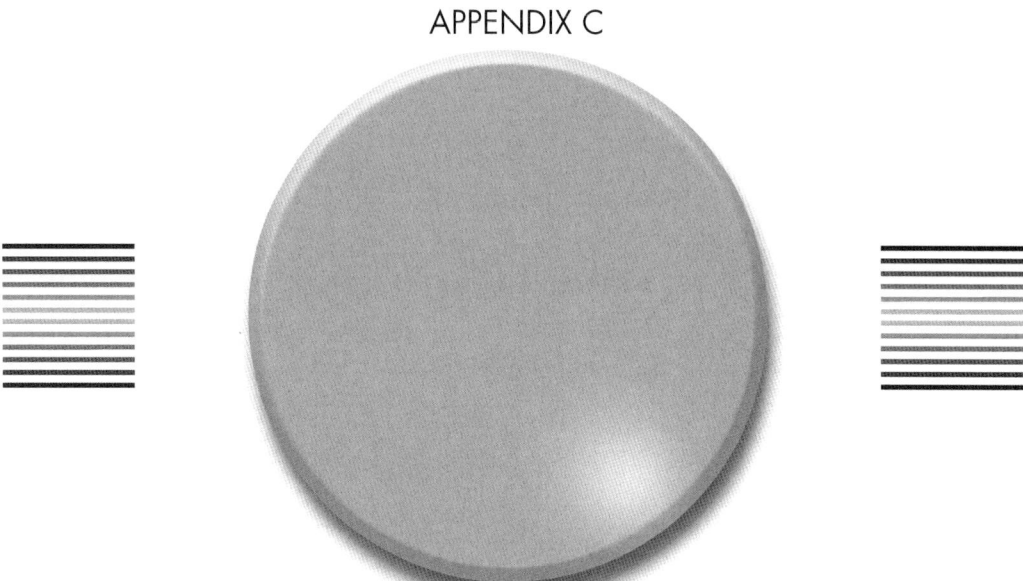

Mythological Names with Pronunciations

1. Characters and Creatures

Abas (**ahb**-us): King of Argos, decreed that twin sons Acrisius and Proëtus were to rule in alternate years.

Acastus (ah-**kas**-tus): A son of Pelias, ruled Iolcus after Pelias' death.

Achilles (a-**kill**-eez): Son of Peleus and Thetis, greatest Greek hero of the Trojan war.

Acrisius (ah-**kriss**-ih-us): Son of Abas, refused to give up throne after his term expired, expelled brother Proëtus. Fathered Danaë who was foretold to father a son who would kill Acrisius, locked Danaë and infant Perseus into covered boat, and set adrift to die.

Aeëtes (ee-**ee**-teez): Son of Helius, king of Colchis, father of two daughters Medea (a sorceress) and Chalciope (a.k.a. Iophossa).

Aeson (**ee**-son): Son of Cretheus and Tyro, father of Jason, rejuvenated by Medea.

Agamemnon (ag-uh-**mem**-non): Son of Atreus, King of Mycenae.

Alcides (al-**sigh**-deez): The son of Zeus and Alcmenes, born Alcides, renamed Heracles.

Alcmenes (alk-**mee**-neez): Granddaughter of Perseus, daughter of Electryon and Anaxo, wife of Amphitryon, mother of Heracles.

Alcimede (al-**sim**-ih-dee): Queen of Iolcus, mother of Jason.

Amalthea (**am**-al-**THEE**-ah): The goat which nursed the infant Zeus and was rewarded with a place in the sky.

Amazon (**ah**-mah-zon): One of the warlike women who inhabited the northern limits of the ancient Greek world.

Amphitryon (am-**fit**-rye-ohn): Son of Perseus, husband of Alcmeme, father of Iphicles (the half brother of Heracles).

Amycus (**am**-ih-kus): Son of Poseidon, belligerent boxer, killed by Polydeuces.

Androgeus (an-**droh**-jih-us): Son of Minos and Phasiphaë, brother of Ariadne, Deucalion and Phaedra, killed in the hunt for the Marathonian bull.

Andromeda (an-**drom**-uh-duh): Daughter of King Cepheus and Queen Cassiopeia, wife of Perseus, mother of Perses, founder of Persia.

Aphrodite (af-roe-**die**-tee): Goddess of Love, born from sea foam which carried Uranus' genitals, married to Haephaestus, mother of Aeneas, Deimus, Harmonia, Hermaphroditus, and Priapus.

Apollo (uh-**pol**-oh): Son of Zeus and Leto, God of fine arts, medicine, music, and poetry, a.k.a Phoebus Apollo or Phoebus (usually known as the Sun god).

Apsyrtus (ap-**sir**-tus): Son of Aeëtes, brother of Medea, killed by her to aid Jason.

Arcas (**are**-kas): Illegitimate son of Zeus by Callisto. Changed into constellation of Ursa Minor.

Ares (**ahr**-ees): Greek god of war, equivalent to Roman Mars.

Argus (**ahr**-gus): Designer and builder of the Argo, also son of Phrixus and Chalciope named after the first.

Argonaut (**ar**-guh-nawt): Any of the persons who participated in the voyage of the ship Argo.

Ariadne (ar-ih-**ad**-nee): Daughter of Minos and Phasiphaë, helped Theseus escape, stranded, found, and married Dionysus.

Arion (ah-**rye**-ahn): A Greek poet from Methymna (ca. 700 BC) who was robbed and thrown overboard by the crew he hired to take him back to Corinth. Rescued by a dolphin.

Artemis (**ar**-tuh-miss): Daughter of Zeus and Themis, virgin goddess of the moon and the hunt, sister of Apollo.

Ascalaphus (ass-**kal**-uh-fuss): Gardner for Hades, reported seeing Persephone eating six pomegranate seeds, forcing her to spend 6 months of each year in Hades.

Asclepius (ass-**klee**-pih-us): Son of Apollo, minor god of healing and medicine, mentor of several Greek heros, immortalized in sky as the constellation Ophiuchus.

Asterion (as-**tee**-rih-on): (1) One of the Argonauts, (2) A king of Crete who married Europa after she was abandoned there by Zeus, and (3) An alternate name for Chara (β Canis Venaticorum).

Astraea [sometimes spelled Astraia] (ass-**tree**-ah): Daughter of Zeus and Themis, goddess of justice, last immortal to leave the Earth for Olympus.

Astraia: See Astraea.

Athamas (**ath**-a-mas): Husband of Nephele and Ino, father of Phrixus and Helle.

Athene (uh-**thee**-nah): Greek goddess of wisdom, patroness of Athens.

Atlas (**at**-las): A Titan, the son of Iapetus and Clymene, father of the Pleiades by Pleione. Condemned by Zeus to hold up the heavens for siding with his brother Titans.

Augeias (au-**jee**-as): King of Elis (an area in the Peloponnese).

Cadmus (**kad**-mus): Theban king, brother of Europa, husband of Harmonia.

Calaïs (**kay**-lah-iss): Winged son of Boreas and Orithyia, brother of Zetes.

Calliope (kuh-**lie**-uh-pee): Daughter of Zeus and Mnemosyne, muse of epic and heroic poetry.

Callisto (kuh-**liss**-toe): Beautiful daughter of Lycaon, the king of Arcadia. Seduced by Zeus, she was the mother of Arcas. She was placed in the sky as Ursa Major.

Canopus (kuh-**noh**-pus): Pilot for Menelaus on voyage to Troy and of the Argo for Jason.

Cassiopeia (kass-ih-oh-**pee**-uh): Queen of Æthiopia, wife of Cepheus, who boasted that she and daughter Andromeda were more beautiful than the sea nymphs.

Castor (**kas**-ter, **kas**-tore): Son of Tyndareus, brother of Clytemnestra, half brother of Polydeuces and Helen. One of the Dioscuri (twins).

Cepheus (**see**-fee-us): King of Æthiopia, husband of Cassiopeia, father of Andromeda, brother of Phineus and/or Agenor.

Cerberus (**ser**-buh-rus): The three-headed dog which stood watch at the gates of Hades.

Cetus (**see**-tus): Sea monster sent by Poseidon to ravage the coast of Æthiopia.

Chalciope (kal-**sigh**-oh-pee): Daughter of Aeëtes, wife of Phrixus.

Chalybes (**kal**-ih-beez): Iron-working people inhabiting the southern coast of the Black Sea. Considered by the Greeks to be hostile savages.

Chara (**kar**-ah or **kay**-rah): One of the hunting dogs of Boötes.

Charon (**kar**-un, **kay**-run): Ferryman who carried the dead across the river Styx.

Chiron or Cheiron (**kye**-ron): Immortal king of the centaurs who gave up his mortality.

Clytemnestra (kleye-tem-**ness**-truh): Daughter of Tyndareus and Leda, sister of Castor, half sister of Polydeuces and Helen.

Coronis (kuh-**roe**-nis): Daughter of Phlegyas, king of the Lapiths, unwilling lover of Apollo, unfaithful to him, and killed by him, mother of Asclepius by Apollo.

Corybantes (kor-ih-**ban**-tez): Male attendants of Cybele, who celebrated with armed dances in which they noisily clashed their shields and spears.

Cretheus (**kree**-the-us, **kree**-thus): Founder and first king of Iolcus, husband of Tyro, father of Aeson, Pheres, and Amythaon, and possibly daughter Hippolyte.

Cronus or Kronus (**kroh**-nuhs); A son of Ge (Earth) and Uranus (sky), ruler of the Titans, father of Zeus and several other Olympians.

Curetes (koo-**ree**-teez): Cretan spirits who danced around the mouth of the cave where the infant Zeus was hidden, and clashed their spears and shields to mask the infant's crying.

Cybeles (**sib**-ih-lee): A Phrygian mother-goddess, equivalent to the Greek Rhea, mother of Zeus.

Cyclopes (sigh-**kloh**-pez, **kyk**-lopes): Sons of Ge (Earth) and Uranus (sky), born after the Titans. One-eyed giants, they were released from their prison in Tartarus by Zeus to help the Olympians in their war with Uranus.

Cytisorus (si-tis-**oh**-rus): Son of Phrixus and Chalciope.

Cyzicus (**siz**-ih-kus): Eponymous King of Cyzicus on the Asiatic shore of the Propontis (Sea of Marmara).

Daedalus (**ded**-uh-lus, **dee**-duh-luss): Brilliant inventor and artisan, servant of Minos, builder of the Layrinth, father of Icarius.

Dagon (**day**-gon): Fish god of the Philistines, often portrayed as half man, half fish.

Danaë (**dan**-ay-ee): Daughter of Acrisius and Aganippe, locked up by Acrisius, mother of Perseus by Zeus.

Decerto (**dee**-ker-toe): Syrian goddess similar to Aphrodite. Sometimes portrayed as a fish goddess with the head of a woman and body of a fish.

Delphyne (dell-**fie**-nee): Dragon-woman, half woman, half serpent. She guarded the severed tendons of Zeus during his battle with the Titans.

Demeter (dee-**mee**-ter): Daughter of Cronus and Rhea, goddess of harvest and fertility, sister of Hades, Hera, Hestia, Poseidon, and Zeus, mother of Persephone by Zeus.

Demodice (dee-**mod**-ih-see): Wife of Athamas, falsely accused her nephew Phrixus of attempted rape, leading to his condemnation and near execution.

Dictys (**dik**-tiss): Poor fisherman, brother of king Polydectes, found Danaë and Perseus, protected Danaë in Perseus' absence, became king after Perseus killed Polydectes.

Dionysus (die-oh-**nigh**-sus): Greek god of grapes and wine.

Ea (**ee**-ah): Babylonian god with role similar to that of Cronus. Sometimes called Enki.

Eleusis (el-**you**-sis): King of country by the same name, entertained Demeter, but misinterpreted her attempt to immortalize his son, attacked her, and was killed by the goddess.

Enki (**en**-kee): Sumerian god of wisdom, alternate name of Ea.

Epimetheus (epp-ih-**mee**-thoos): A Titan who helped the Olympians in their war with Cronus and his followers. Also, creator of the animals and humans.

Erichthonius (er-ik-**thoe**-nigh-us): Son of Haephaestus, and born with the same handicap as Haephaestus.

Erigone (ee-**rig**-uh-nee): Daughter of Icarius, who hanged herself after finding the body of her father.

Eros (**err**-ros): One of the earliest gods, god of love. Later, the son of Aphrodite, who made unsuspecting people fall in love.

Euryale (yoo-**rye**-uh-lee): One of the Gorgons, sister of Medusa and Stheno.

Europa (yoo-**roe**-puh): Daughter of Agenor and Telephassa, sister of Cadmus, kidnapped and seduced by Zeus, mother of Minos.

Eurydice (yoo-**rid**-ih-see): Beloved wife of Orpheus, fled unwanted suitor, stepped on snake and was bitten. Orpheus attempted to rescue her from Hades, but failed.

Eurythemis (you-**rih**-theme-is): Wife of King Thestius, mother of Leda.

Eurytheus (you-**riss**-thoos): King of Mycenae, master of Heracles during his 12 labors of atonement.

Gaea [or Ge] (**jee**-ah): Goddess of the Earth, sometimes called Gaia (**gay**-ah).

Galatea (agl-uh-**tee**-uh): Woman created from statue sculpted by Pygmalion.

Ganymede (gan-ih-**mee**-dee): Handsome youth abducted to become Zeus' companion and cup-bearer of the Olympians.

Ge (jee): See Gaea.

Glaucus or Glaucus (**glaw**-kus): Mortal fisherman who ate an unknown herb and turned into a minor sea god.

Hades (**hay**-deez): Son of Cronus and Rhea, brother of Demeter, Hera, Hestia, Poseidon, and Zeus, lord of the Underworld, kidnapped Persephone to be his wife.

Harmonia (har-**moan**-nih-ah): Daughter of Ares and Aphrodite, given to Cadmus as a wife by Zeus.

Harpies (**har**-peez): Bird-like monsters with the heads of women, sometimes called "snatchers," who were blamed for various disappearances.

Hebe (**hee**-bee): Olympian goddess, daughter of Zeus and Hera, Cup bearer of the Olympians, married Heracles after his ascension to Olympus.

Helen (**hell**-en): Daughter of Zeus and Leda, most beautiful woman in the world, married to Menelaus but kidnapped by (or eloped with) Paris, causing the Trojan war.

Helios, Helius (**hee**-lih-us): Son of Hyperion and Theia, brother of Selene and Eos, God of the Sun, father of Phaethon and Aeëtes.

Helle (**hel**-ee): Daughter of Athamas and Nephele, sister of Phrixus, eponym for the Hellespont.

Hephaestus (hee-**fehs**-tus): Crippled Olympian who forged armor, chariots, weapons, and art works for the Olympians.

Hera (**hee**-rah, **hair**-uh): Daughter of Cronus and Rhea, jealous wife of Zeus, and queen of Olympia.

Heracles (**hehr**-uh-kleez): Greek hero, son of Zeus and Alcmena, elevated to be a god. Roman name: Hercules (**her**-cue-leez).

Hermes (**her**-meez): The precocious herald and messenger of the Olympians.

Hesperides (hes-**per**-ih-deez): The daughters of Atlas and Hesperis, guardians of the golden apples Hera received as a wedding gift when she married Zeus.

Hippodamia [or Hippodameia] (hip-uh-duh-**my**-uh): Daughter of King Oenomaus of Pisa, wife of Pelops.

Hydra (**high**-drah): The mythological serpentine beast with nine heads, one of which was immortal. When one head was cut off, two grew back in place. The beast was slain by Heracles, who burned each stump to prevent new heads from growing.

Hylas (**high**-las): Kidnapped by Heracles to become his squire and lover, Argonaut, pulled into a spring by a nymph to become her lover.

Icarius (eye-**ker**-ih-us): Athenian herdsman/grape grower taught to make wine by Dionysus, murdered by friends who thought he had poisoned them with the wine.

Ida (**eye**-dah): A nymph of Crete, who went to Phrygia where a famous mountain was named after her.

Idas (**eye**-duss): Son of Aphareus, twin brother of Lynceus, cousin of Castor and Polydeuces.

Ino (**eye**-noh): Daughter of Cadmus and Harmonia, sister of semele, caretaker for infant Dionysus, wiver of Athamas.

Iolaos or Iolaus (eye-oh-**lay**-uhs): Heracles' cousin and chariot driver.

Iphicles (**if**-eh-kleez): Half brother of Heracles.

Iris (**eye**-riss): Daughter of Titan Thaumas and Oceanid Electra, messenger for Hera, sister of the Harpies.

Ischys (**iss**-kiss): Son of Elatus, lover of Coronis, slain by Apollo.

Jason (**jay**-son): Son of Aeson and Alcimede, leader of the quest for the golden fleece, captain of the Argo.

Ladon (**lay**-don): Dragon which guarded the tree of golden apples in the garden of the Hesperides before Hera placed it in the heavens to separate Ursa Major and Ursa Minor.

Leda (**lee**-duh): Daughter of Thestius, wife of Tyndareus, mother of Castor, Polydeuces, Helen, and Clytemnestra.

Lycaon (lie-**kay**-on): King of Arcadia, father of Callisto.

Lycurgus (**lie**-kur-gus): King of Edonian tribe of Thrace, drove Dionysus and followers out of Thrace, driven mad by Dionysus, killed by the Edonians.

Lynceus (**lynn**-suse): Son of Aphareus, twin brother of Idas, cousin of Castor and Polydeuces. Gifted with extraordinary eyesight.

Lyncus (**lin**-cuss): King of Scythia, changed into a lynx by Demeter for trying to murder Triptolemus.

Maenad (**mee**-nad): Female follower of Dionysus who engaged in orgiastic revels with demigods and humans.

Maera (**mee**-rah): Female dog owned by Icarius, extremely loyal, led Erigone to body of Icarius.

Medea (mee-**dee**-ah): Daughter of King Aeetes of Colchis. She helped Jason seize the Golden Fleece and became his wife.

Medusa (muh-**doo**-suh, -sah): One of the three Gorgons, immortal, whose gaze turned all living creatures to stone. Killed by Perseus.

Megara (**meg**-uh-rah): Princess, married Heracles, and killed by him during a bout of insanity.

Melas (**mee**-las): Son of Phrixus and Chalciope.

Minos (**my**-noss): Son of Zeus and Europa, King of Crete, husband of Pasiphaë, father of Androgeus, Aridne, Deucalion, and Phaedra.

Minotaur (**min**-uh-tor): Half human, half bull son of Pasiphaë and a sacred bull sent by Poseidon for a sacrifice.

Muse (myooz): Any of the nine daughters of Zeus and Mnemosyne, goddesses who inspired artists. Individually, they were: Calliope, Clio, Erato, Euterpe, Melpomene, Polynomnia, Terpsichore, Thaleia, and Urania.

Myrtilus (**mer**-tih-luss): Son of Hermes, fastest of charioteers, bribed by Pelops to sabotage the chariot of King Oenomaus.

Neleus (**nee**-lus, **nee**-leh-us): Son of Poseidon and Tyro, king of Pylos, father of Nestor.

Nephele (**nef**-eh-lee): "Cloud" woman created by Zeus to fool a suitor of Hera, mother of Centaurus, wife of Athamas, mother of Phrixus and Helle.

Niobe (**nigh**-oh-bee): Daughter of Tantalus, wife of King Amphion of Thebes, mother of seven sons and seven daughters who were killed by Apollo and Artemis after she boasted that she had better children than their mother, Leto.

Nymph (nimf): A semidevine maiden associated with trees, rivers, and other aspects of nature.

Oannes (oh-**ahn**-nes): Babylonian half man, half fish figure who may have been a god or a foreign priest wearing a fish-like costume. Sometimes called Ea.

Oenomaus (ee-nom-**ay**-us): King of Pisa, father of Hippodamia, killed in crash of sabotaged chariot.

Olympians (oh-**lim**-pih-ans): The 12 major Greek gods, led by Zeus, who lived on Mt. Olympus.

Oncaea (ohn-**kee**-ah): Mother of Arion by Poseidon.

Orion (oh-**rye**-on): Generally considered the son of Poseidon and Euryale parentage varies according to the specific source, giantic hunter, became companion of Artemis and Leto.

Orpheus (**or**-fih-us): Son of Apollo and Calliope, devoted husband to Eurydice, famed musician, journeyed to Hades in an attempt to free his deceased wife, killed by angry Maenads.

Pan (pan): Amorous god, son of either Hermes or Zeus. Born with the body of a man and the legs of a goat. His enormously loud yell induced panic in opponents, causing them to flee.

Paphos (**pay**-fos): The daughter (or in some versions, the son) of Pygmalion and Galatea.

Pasiphaë (pah-**sif**-ay-ee): Daughter of Helius and Perse, wife of Minos, lover of sacred bull, mother of Androgeus, Aridne, Deucalion, and Phaedra by Minos, mother of Minotaur by the bull.

Pegasus (**peg**-uh-sus): Winged horse which sprang from blood shed by Medusa.

Pelias (**pee**-lih-as): Son of Poseidon and Tyro, usurped throne of Iolcus from Aeson, killed by Medea.

Pelops (**pee**-lops): Son of Tantalus and Dione, brother of Niobe and Broteas, killed and served in stew to Zeus by father, fathered Hippodameia.

Perseus (**per**-sih-us, **per**-syoos): Hero who was the son of Zeus and Danaë, killed Medusa, killed the sea monster, and rescued and married Andromeda. Father of Perses.

Persephone (per-**sef**-uh-nee): Daughter of Zeus and Demeter, Wife of Hades, goddess of the underworld, mother of Dionysus and the Erinyes. Her story of periodic return was one of the central themes of Greek mythology.

Periander (pehr-ih-**an**-der): A real king who ruled Corinth from 625 to 585 BC.

Phaeton (**fay**-uh-thon): Son of Helius and Clymene, borrowed the Sun chariot and lost control, killed by Zeus to prevent destruction of the world.

Phanothea (**fan**-oh-the-uh): Wife of Icarius, mother of Erigone, became a priestess of Apollo after the death of her husband.

Philyra (**fill**-ih-ruh): Niece of Poseidon, mother by him of Chiron, the first centaur.

Phineus (**fin**-ee-us, **fin**-us): Seer blinded by Zeus for revealing the future to humans, plagued by the Harpies, rescued by the Argonauts.

Phoenix (**fee**-niks): Mythological bird of the Arabian desert. Only one phoenix existed at a time, but it lived 500 years, then died in a funeral pyre, only to rise again for another 500 years.

Pholus (**foe**-luhs): Centaur who entertained Heracles and became a constellation.

Phrixus, Phryxus (**frik**-sus): Son of Athamas and Nephele, sister of Helle, saved from sacrifice by appearance of a golden-fleeced ram.

Phrontis (**fron**-tis): Son of Phrixus and Chalciope.

Polydectes (pol-ih-dek-teez): Son of Cadmus and Harmonia, King of Thebes, established Thebes as the center of worship of Dionysus.

Polydeuces (pol-ih-**duu**-seez) (Pollux to the Romans).

Poseidon (puh-**sigh**-don): Son of Cronus and Rhea, God of the sea, horses, and earthquakes, husband of Amphitrite, father of Theseus, father of the golden-fleeced ram by Theophane.

Prometheus (pro-**mee**-thih-uhs): One of the Titans who helped the Olympians in their war against Chronus and his followers. Creator and benefactor of the human race.

Pygmalion (pig-**may**-lih-on): Cretian sculptor who fell in love with the statue he created.

Rhea (**ree**-uh): One of the 12 Titans, wife of Cronus, mother of some of the Olympians.

Semele (**sem**-eh-lee): Daughter of Cadmus and Harmonia. She conceived a child by Zeus, and was persuaded by a disguised Hera to demand that her lover appear in his full glory. He did, and Semele was killed by the sight.

Semiramis (sem-ih-**rah**-miss): Ancient Babylonian queen who was reputed to be the daughter of Decerto.

Stellio (ste-**lih**-oh): A young boy who laughed at Demeter was turned into a lizard.

Stheno (**sthee**-noh): Gorgon, sister of Euryale and Medusa.

Stymphalian Birds (stim-**fay**-lih-an): Giant birds with bronze claws which inhabited the region around Stymphalus (city and lake) in north-eastern Arcadia. Killed by Heracles poisoned arrows.

Talus (**tay**-lus): Brazen giant sent by Hera to guard the island of Crete by pelting incoming ships with giant boulders, killed by the Argonauts.

Tantalus (**tan**-tuh-luhs): Son of Zeus and Pluto (a daughter of Cronus), King of Sipylus. Served human flesh to Zeus, punished by eternally having food snatched from his reach.

Themis (**thee**-miss): One of 12 Titans, daughter of Uranus and Gaia, mother of the Horae (Seasons) and the Moirai (Fates) by Zeus.

Theseus (**thee**-sih-us, **thee**-syoos): Son of Poseidon and Aethra, husband of Phaedra, King of Athens, one of the greatest Greek heroes.

Thestius (**thess**-tih-us): Son of Ares and Demonice, father of Leda.

Tiphys (**tie**-fis): Helmsman of the Argo, guided it through the Clashing Rocks and other dangers before being killed in a battle.

Titan (**tie**-ten): Any of the six sons (Coeus, Crius, Cronus, Hyperion, Iapetos, and Oceanus) and six daughters (Mnemosyne, Phoebe, Thea, Theia, Themis, and Thetis) of Uranus and Gaia. Also, the descendants of the 12 original titans.

Triptolemus (trip-**tol**-uh-muss): Son of Eleusis. Demeter's attempt to immortalize him was broken up by his father and failed.

Tros (trose): King and eponym of Troy, father of Ganymede.

Tyndareus (ten-**day**-ree-us): King of Lacedaemon, husband of Leda, father of Caster and Clytemnestra, raised from death by Asclepius.

Typhon (**tie**-fon): Gigantic son of Gaea, born especially to revenge the defeat of the Titans. He was taller than the mountains, and had hands with a hundred burning snake heads instead of fingers.

Tyro (tie-roh): Loved by Poseidon, wife of Cretheus, mother of Neleus and Pelias by Poseidon.

Zetes (**zee**-teez): Winged son of Boreas and Orithyia, brother of Calaïs.

Zeus (zuse, zoos): Leader of the Olympian Gods.

Stellar Spectral Classes

The Harvard system of stellar classification arranges stars into a one-dimensional temperature sequence designated by the capital letters O, B, A, F, G, K, and M. These classes are subdivided into ten subclasses of 0, 1, 2...9, except for class M, in which the largest number used is usually 8. Occasionally, these numbers are appended with "0.5" to indicate a classification about midway between two integer subclasses, e.g., B0.5. Three additional classes are often used for later (in the sequence) stars to distinguish those with different chemical compositions. These are R, N, and S. R and N are subdivisions of an earlier class C (for carbon), which is also sometimes still used. A rare class of extremely hot stars is called "WR" in honor of the discoverers, Charles Wolf and Georges Rayet.

The major spectral classes can easily be remembered by using the mnemonic, "Oh, Be A Fine Girl, Kiss Me Right Now, Sweetheart."

Originally, there were several more letters arranged alphabetically. This arrangement was based primarily on the strength of the hydrogen absorption lines in the spectra, with lines of additional elements used to distinguish between classes with about the same strength of hydrogen lines. It was later found that the present arrangement represented a unidimensional ordering of the stars from the hottest to the coolest. Because of the relationship between temperature and the distribution of wavelengths in the light emitted from radiating bodies, this sequence of classes is also a sequence of star colors ranging from blue (O) to blue-white (B), white (A), yellow-white (F), yellow (G), orange-yellow (K), and orange-red (M). The R, N, and S classes are generally a deeper red than class M. When this order of classes was first determined, most astronomers believed that all stars started out large, hot, and blue, and that they slowly cooled, shrank, and reddened until they faded from sight. Because of this belief, the hotter classes, i.e., O, B, and A are sometimes called "early" classes, and the cooler classes, i.e., G, K, and M are called "late classes."

In addition to the temperature, other factors cause difference in the spectra of stars. One of these is luminosity or size. The luminosity classes are designated by Roman numerals following the spectral classes, with a "0" added to the beginning to accommodate the previously unknown hypergiants:

0. Hypergiants
I. Supergiants

  Ia-0. Extremely luminous supergiants
  Ia.   More luminous supergiants
  Iab. Intermediate luminosity supergiants
  Ib.   Less luminous supergiants

II. Bright giants

  IIa.   More luminous bright giants
  IIab. Intermediate luminosity bright giants
  IIb.   Less luminous bright giants

III. Giants
  IIIa. More luminous giants
  IIIab. Intermediate luminosity giants
  IIIb. Less luminous giants

IV. Subgiants
  IVa. More luminous subgiants
  IVb. Less luminous subgiants

V. Main sequence stars (dwarfs)
  Va.   More luminous dwarfs
  Vab. Intermediate luminosity dwarfs
  Vb.   Less luminous dwarfs

VI. Subdwarfs (rarely used)
  VII. White dwarfs (rarely used)

Lower case letters are appended to indicate other effects apparent from the stellar spectra:

e = emission lines present
f = some "O" type stars with emission lines
p = a peculiar spectrum
n = broad spectral lines
s = sharp spectral lines
k = lines due to interstellar gas
m = stars with strong lines of metals

When a star appears to be intermediate between temperature or luminosity categories, a "–" or "/" is used between the two classes, e.g., G9/K0 II/III.

A final group of stars is that of the white dwarfs. These are very old stars which have burned all their nuclear fuel and collapsed into extremely small, dense stars. These stars can be approximately the size of the Earth, but have masses similar to that of the Sun, indicating densities up to 1,000,000 times that of water. The white dwarfs are designated by a capital "D", followed by A, B, O, Q, or Z to indicate spectral differences:

DA – hydrogen lines dominate
DB – neutral helium lines dominate
DO – ionized helium lines dominate
DQ – carbon lines present
DZ – metal lines dominate
DX or DXP – unidentified features, probably due to a strong magnetic field

More detail can be found at http://en.wikipedia.org/wiki/Stellar_classification, or in most college freshman level astronomy textbooks.

Helpful Books and References

1. Instructional Observing Guides

Double and Multiple Stars and How to Observe Them, James Mullaney, Springer-Verlag London Limited, 2005.

Observing Variable Stars, Gerry A. Good, Springer-Verlag London Limited, 2003.

Star Clusters and How to Observe Them, Mark Allison, Springer-Verlag London Limited, 2006.

Nebulae and How to Observe Them, Steven R. Coe, Springer-Verlag London Limited, 2007.

Galaxies and How to Observe Them, Woldfang Steinicke and Richard Jakiel, Springer-Verlag London Limited, 2007.

2. Other Observing Guides

The Night Sky Observer's Guide, George Robert Kepple and Glen W. Sanner, Willman-Bell, Inc., Vol. 1 Autumn and Winter, Vol. 2, Spring and Summer, 1998, Vol. 3, The Southern Skies, 2008. An extended list of all types of night sky objects, with detailed descriptions of appearances of deep-sky objects in various sizes of telescopes.

The Messier Objects, Stephen James O'Meara, Sky Publishing Corporation/Cambridge University Press, 1998. Excellent coverage of Charles Messier's famous 110 deep-sky objects, with extensive descriptions.

Observing the Caldwell Objects, David Ratledge, Springer-Verlag London Limited, 2000. A handy guide to Patrick Alfred Caldwell Moore's extension of Messier's list. Each object is presented on a pair of facing pages which include a photograph and catalog data on one side, and a finder chart, visual and physical description on the facing page.

The Caldwell Objects, Stephen James O'Meara, Sky Publishing Corporation/Cambridge University Press, 2002. A thorough coverage of Patrick Moore's extension of Messier's list.

Binocular Astronomy, 2nd ed., Craig Crossen and Wil Tirion, Willman-Bell, Inc., 1992–2008. Includes Bright Star Atlas. Good coverage and instructions for binocular users.

Double Stars for Small Telescopes, Sissy Haas, Sky Publishing Corporation, Cambridge, MA, USA, 2006. Over 2,100 double and multiple stars for small telescopes.

3. Star Atlases

Norton's Star Atlas and Reference Handbook, edited by Ian Ridpath, 18th ed., Longman Group UK Limited, 1989. Classic star atlas and handbook for beginners. Shows all the stars visible to the unaided eye.

Uranometria 2000.0, Deep Sky Atlas, Tirion, Rappaport, Remaklus, Vols. 1 and 2, Willman-Bell, Inc. Richmond, VA, USA, 1987–2001. An atlas for more advanced observers. Plots 280,035 stars to magnitude 9.75 and over 30,000 nonstellar objects. Also includes a magnitude 6.5 atlas to correlate the charts to sky as visible to the unaided eye.

Uranometria 2000.0, Deep Sky Field Guide, Cragin and Bonanno, Vol. 3. Lists data on all the nonstellar objects plotted in Vols. 1 and 2 by the chart on which they appear.

4. Constellations

The Cambridge Guide to the Constellation, Michael E. Bakich, Cambridge University Press, New York, NY, USA, 1995. Lists data on all 88 constellations, with lists of the stars and constellations sorted by various parameters.

Star Lore – Myths, Legends and Facts, William Tyler Olcott, Dover Publications, Mineola, NY, USA, 2004. Republication of a classic work on the mythology of the constellations.

Field Book of the Skies, William Tyler Olcott, G. P. Putnam's Sons, New York, NY, USA, 1936, 1954. Another republication of a classic work. Emphasis is on observations with naked eye, binoculars, and small telescopes.

5. Web Sites

A. Double Stars:

http://www.virtualcolony.com/sac/star-search-form.html. A search form which uses the Sacramento Astronomy Club (Arizona, USA) dataase allow the observer to define search parameters to select double stars of interest for each constellation.

http://ad.usno.navy.mil/wds/wds.html. A listing of the Washington Double Star Catalog of over 98,000 double and multiple stars. Maintained and updated frequently by the U.S. Naval Observatory.

http://www.dibonsmith.com/orbits.htm. Plots of the apparent orbital motion of 150 binary stars observable with amateur telescopes.

B. Variable Stars:

http://www.aavso.org/. The home page of the American Association of Variable Star Observers, the leading variable star organization. Extensive information about variable stars, including descriptions of variable star types and charts for individual variable stars.

http://www.aavso.org/vsx/index.php?view=search.top. A form for entering data to search for information about individual variable stars.

http://britastro.org/baa/content/view/44/82/. The variable star section of the British Astronomical Association. This site has some variable star charts not available through the AAVSO.

C. Star Clusters:

http://www.univie.ac.at/webda/ocl_list.html. List of about 1,200 Open Clusters in the Milky Way Galaxy.

http://www.nightskyinfo.com/open_clusters/. List of Open Clusters with data and pictures of many.

http://www.seds.org/messier/xtra/supp/mw_gc.html. List of all 158 known Globular Clusters associated with the Milky Way Galaxy.

D. Nebulae:

http://www.noao.edu/outreach/aop/observers/pn.html. Over 70 images of beautiful examples of planetary nebula.

http://www.blackskies.org/links.html. Links to Web sites containing information about dark-sky objects of all types, including bright and dark nebulae.

E. Galaxies:

http://en.wikipedia.org/wiki/List_of_nearest_galaxies. A list of 115 of the nearest galaxies, including type, distance, group membership, and pictures of many.

http://en.wikipedia.org/wiki/List_of_galaxies. Lists of many types of galaxies, with data for most, and pictures of some.

http://www.dmoz.org/Science/Astronomy/Galaxies/. Links to Web sites including many different types of galaxies and galaxy clusters.

F. Stars, General:

http://en.wikipedia.org/wiki/List_of_stars_by_constellation. Sortable lists of stars with Bayer and Flamsteed designations as well as Henry Draper (HD) and Hipparchos (HIP) catalog numbers, coordinates, distances, apparent, and absolute magnitudes and common (and not-so-common) names.

http://stellar-database.com/. A Web site which allows you to search for information on stars by common names as well as by Bayer designation and several catalog numbers.

http://www.solstation.com/stars.htm. Links to data on dozens of special interest stars.

http://simbad.u-strasbg.fr/simbad/sim-fid. A cross reference to star numbers for over 50 scientific catalogs.

G. Constellations:

http://stars.astro.illinois.edu/sow/const.html. A guide to the constellations, including charts, and origin stories. By Jim Kaler, a retired astronomy professor.

http://www.dibonsmith.com/cvn_con.htm. A guide to the constellations, including origins, charts, and lists of a few telescopic objects in each constellation. By Richard Dibon-Smith.

http://chandra.harvard.edu/photo/constellations/. A Guide to the constellations, including photos of the classical figures represented by each constellation.

http://www.windows.ucar.edu/tour/link=/mythology/stars.html. Classical, Native American, Hindu, and other mythology of some constellations.

http://www.slivoski.com/astronomy/androm.htm. Photographs of constellations, with outlines.

Contributors

Nicolas Biver was born in 1969 and became an amateur astronomer at the age of 15, when he started drawing celestial objects as seen with a 60 mm refractor. He has continued observing and drawing planets and deep-sky wonders while progressing to an 8 home-made Newtonian in 1989, a 10 in 1996, and a 16 in 2003. He became a professional astronomer after defending his Ph.D. thesis on radio observations of comets in 1997 at Paris-Meudon Observatory. He then spent 3 years as a Post-doc in Hawaii (Institute for Astronomy, University of Hawaii, Honolulu) in 1997–2000, then back in Meudon for 1 year, and 15 months at the European Space Agency in 2001–2002 working on the Rosetta mission. He now has a permanent (National Center for Scientific Research) position at Paris-Meudon Observatory, where he is still working on radio (mm to sub-mm) astronomy and comets. More of Nicolas' drawings and astrophotography can be seen at: http://www.usr2.obspm.fr/~biver/.

Rob Gendler is a physician whose interest in astronomy dates back to his childhood in New York City, where frequent visits to Hayden Planetarium with its magnificent astrophotographs inspired him. He was unable to pursue his astronomical interests until 1993, when he moved to a dark cul-de-sac in Connecticut. After taking an introductory astronomy course, he bought a pair of binoculars and spent a year and a half learning the night sky. A 10 Dobsonian and another year led to the purchase of a 10 Schmidt-Cassegrain and a CCD camera, which started his astro-imaging career. With more and more light pollution making his Connecticut imaging less suitable, Rob started imaging remotely at New Mexico Skies, and later at their affiliated facility in Western Australia. Rob is one of the best known amateur astrophotographers and regarded as one of the best in the world. He devotes at least one full night and often several to taking images of each single object and considerable time to combining and processing to create each of his astrophotos.

Rob has published two books and several articles on astronomy, astrophotography, and image processing. His spectacular images and more information about his work can be seen at http://www.robgendlerastropics.com/.

Stuart J. Goosen: In 1999, Stuart bought a 12 Meade SCT for his daughter's Xmas present, and a few week's later came the "Daddy, I'd like to take pictures" request, and so began his astrophotography journey.

After Stuart retired from Intel in 2006, where he was the Director of the Core Software Division, and responsible for ensuring the optimization of both the Windows and Linux software platforms for Intel architecture, he and his wife Gail moved full time to the New Mexico Skies Astronomers Enclave in Mayhill, New Mexico.

He currently has 2 Paramount ME's with RCOS 12.5 and TOS 130 telescopes and STL-11000, ST-10XME, and Canon XTi cameras in his roll-off roof observatory. Having plenty of free time while waiting for his imaging sessions to complete, Stuart and his wife now spend many hours playing World of Warcraft, where they run the "Astronomers Guild."

Andy Homeyer is the retired owner of a high-tolerance machine shop and design facility. Most of his work was for companies that were working under government contracts. Included was a fair amount of design work. He was fortunate in starting his career with Chester Silvernail of the Regulus Corporation where he did design and custom fabrication of advanced spectrographic systems, Schlieren systems, and various other types of custom designed optical systems. Andy has been active in the amateur astronomy world off and on for the past 30+ years. Among his products, the Flip Mirror units and Motorized Filter Wheels have been custom designed and built for Off Axis Guiders. Probably his best known product has been the cradle systems for mounting larger scopes, C-11 thru Meade16 SCT's, to the Paramount and AP 1200 mounts. He has also designed and built various custom cradles to hold up to three refractors. These cradles have been so well received that they are now distributed all over the world.

Some of the telescopes Andy has owned and used include: a 90+ mm John Mellish refractor, a truly spectacular Vixen 4 in. Flourite refractor, a C-11, a 5 in. Astro-Physics refractor, and his favorite, a 7 in. Astro-Physics refractor. All but two of the images on his Web page were taken with the 7 in. A-P. Andy's equipment and pictures can be seen on his Web site at: http://www.intint.com/andy/index.html.

Eiji Kato operates the Twinstar Guesthouse in Queensland, Australia. From his small observatory attached to the B&B, he introduces the night sky to his guests on every clear night. He uses a 18 (46 cm) reflector for observing and sketching. He is an astronomy enthusiast of 40 years. More examples of Eiji's astronomical drawings are available at:

http://www.geocities.com/eijikato2001/constellationlist.htm, and more about his guest house can be seen at:

http://www.twinstarguesthouse.com/night_sky_tour.html.

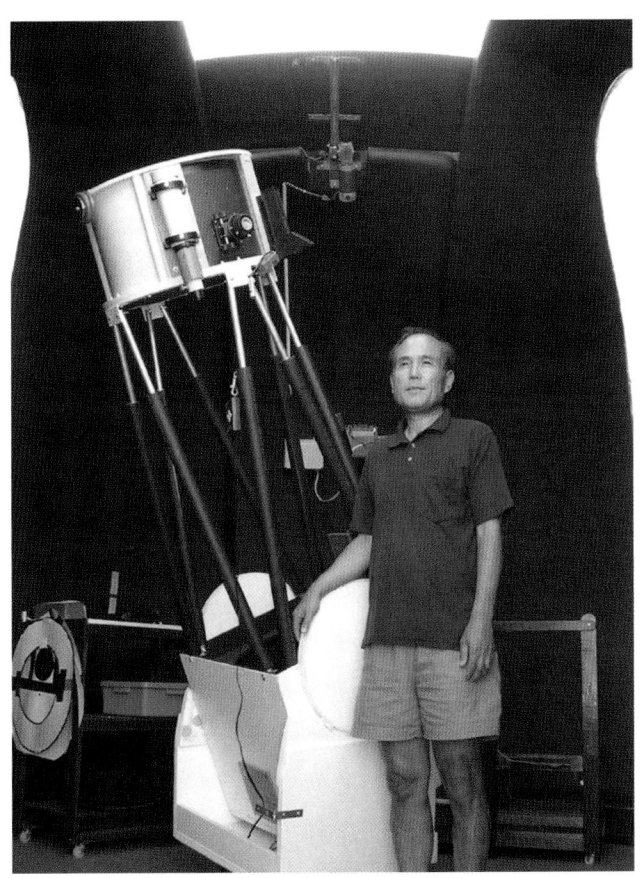

Warren A. Keller was deeply and lastingly affected when, as a child, he saw the movie *2001, A Space Odyssey*. Then, upon receiving Fred Hoyle's <u>Astronomy</u> as a gift, he says, "The cover had the glossiest color photo of M27 – the Dumbbell Nebula – in the blackest velvet sky. I knew from that moment I must someday photograph the heavens' wonders." The results of that compulsion are displayed at: http://www.BillionsandBillions.com.

Warren describes his approach to astrophotography as "Artistic by nature, it's less about Cosmology and more about the hunt for the myriad of beautiful shapes and color throughout the universe."

Warren also has the ability to effectively teach the essentials of astro-imaging to others, and his tutorial business IP4AP (http://www.ip4ap.com) was named a Sky & Telescope Magazine "Hot Product." Warren is proud to have been published in *Sky & Telescope, Astronomy, Amateur Astronomy, AstroPhoto Insight*, and many places on the Web, including *NASA's APOD* and *Universe Today*.

Mark Komsa grew up in the "perpetually cloudy skies of north-eastern Ohio." In spite of this handicap, Mark nursed an intense interest in astronomy for nearly 40 years before he was able to move to New Mexico to work at the National Solar Observatory as an instrument engineer. Along with other tasks, he integrates cameras and other instruments with the Dunn Solar Telescope. Mark became interested in astro-imaging about 10 years ago and has been developing his techniques since that time. Mark says that he feels very fortunate to be able to image from the pristine skies of a remote mountain top at 9,300 ft elevation in south-central New Mexico. His equipment includes a 16 Ritchey-Crétien (shown), a TMB 203 apo-chromatic refractor, an Astro-Physics 140 mm EDF refractor, and an STL-11000 CCD camera.

Steve Pastor. "My interest in astronomy began in 1956 when I was nine. My father bought me a 3.5-in. Sky Scope reflector and with my first view of Saturn, like many of us, I knew I was hooked! I progressed to a 2.4 mm refractor (Monolux), an 80 mm refractor (R. Hutchinson Co., Metuchen, NJ), and about 10 years later to a 10-in. reflector (Cave Optical Co.). After graduating high school in 1965, I obtained a B.A. Degree in chemistry from Rutgers University, New Brunswick, NJ. After college I served 2 years in the military, which included a year in Vietnam (1st Lt. Infantry, US Army). Upon return to civilian life, I worked for a period of 10 years for National Starch and Chemical Corporation in the Synthetic Polymer Research Group during which time I obtained an M.S. degree in chemistry. I then returned to Rutgers University and obtained my Ph.D. in chemistry. I then obtained a position in Ciba-Geigy Corporation in New York and worked in a diverse number of areas including organophosphorus chemistry as well as a 2-year assignment in the Central Research Department of Ciba-Geigy (Basel, Switzerland) where I worked on asymmetric synthesis using transition-metal catalysts. I also held the position of an adjunct Professor of chemistry at Pace University, Pleasantville, NY. A benefit of being involved in basic research was being allowed to publish many of my findings in the chemical literature."

"My interest in both 'classic telescopes' as well as modern telescopes has continued through the years – and being somewhat of a collector – I still have most of my scopes including my 3.5-in. Sky Scope. My primary visual scopes (visual observing is my favorite activity) are a TeleVue 140 refractor, a 6-in. f/12 Astro-Physics refractor, a Takahashi Mewlon 300 Cassegrain, and a 20-in. Obsession (in my case, what better word than 'obsession'). For photography, I use a Takahashi FSQ-106ED, a Takahashi Epsilon 180, or an Astro-Physics 178 mm f/9 refractor, usually with a Canon DSLR. I currently live at 7,300 ft in the Sacramento Mountains of New Mexico and couldn't think of a better place to live and pursue my hobby of astronomy."

Jeremy Perez is a graphic artist from Flagstaff, Arizona where he takes advantage of dark, high altitude skies. He sketches to get the most from his observations, and as a means to convey the captivating views to others. Jeremy contributes sketches to astronomy publications, writes a monthly sketching column for Astronomy Now magazine, and is a co-author of Astronomical Sketching: A Step-by-Step Introduction. He maintains a Web site http://www.perezmedia.net/beltofvenus, where you will find astronomical sketching resources, observing templates, and links to sketch galleries from observers around the world.

Christopher J. Picking: Born and raised in the Wairarapa region of New Zealand, I have always been interested in the night sky. After joining the Phoenix Astronomical Society in 1998, I became a keen astrophotographer and built a small roll-off roof observatory in my backyard. I regularly give talks about various aspects of astrophotography and was also involved in the construction of Stonehenge-Aotearoa.

Chris says, "I live in New Zealand at latitude 41 degrees south. This allows me to photograph all of the spectacular objects visible in southern hemisphere skies. Objects such as the Large and Small Magellanic Clouds, Eta Carina nebula and globular cluster Omega Centauri, to name a few, can be viewed. I enjoy capturing wide field images particularly of the southern night sky as well as the occasional Aurora Australis. Many of my images have been used in books, magazines, multimedia presentations, television, exhibitions, and various publications world wide."

Gian Michele Ratto, who prefers to be called "Gimmi," is a tenured senior scientist at the Institute of Neuro-physiology CNR in Pisa, Italy. His hobbies include exploring ice caves and astrophotography.

Some of his astrophotography can be viewed at: http://www.collectingphotons.com.

Gimmi says, "By day, when I am not processing some astrophotograph, I work as a biophysicist engaged in some reasonably interesting neurobiology study. Yet again my main tool is an imaging platform consisting in a scanning microscope powered by a pretty serious pulsed infrared laser. I landed in Pisa after several years spent in Berkeley and Cambridge: here I have got a country house with an observatory."

David Richards is an Amateur astronomer observing from Aberdeenshire, UK. He is shown by his roll-off roof observatory which houses an 8 LX200 and SBIG ST-7E CCD camera. He does imaging and observing of a wide range of astronomical objects. David is a member of the BAA and the Aberdeen Astronomical Society. David says, "I've been interested in stars and planets since I was about 9 or 10 years old. I particularly remember a couple of events – staying up to watch the television coverage of the Moon Landings and being given a small 2 telescope for a birthday present. Although the scope was rather limited in what it could see, and had no mechanism to track across the sky, it left with me a desire to one day get myself a proper telescope. Many years later (1995 in fact), I finally got that 'proper' telescope – a Meade 8 in. LX200." David's sketches and images can be seen at his Web site: http://www.richweb.f9.co.uk/astro/.

Richard Weatherston lives near Sarnia in Southern Ontario not far from the border with Michigan. He has been a member of the Royal Astronomical Society of Canada since 1995. He is primarily a visual observer and although Sarnia has relatively dark skies, Richard enjoys traveling to places where there is no light pollution. He does his observing with his 10 in. Schmidt Cassegrain and binoculars. Richard does not yet use a computerized telescope as he likes the challenge of tracking down distant objects.

Index

OTHER OBJECTS:

NGC 2422, See M47
NGC 2437, See M46
NGC 2440, 459, 462
NGC 2442, 652, 653, 701
NGC 2447, See M93
NGC 2477, 459
NGC 2516, 451, 453, 473
NGC 2547, 469, 471
NGC 2548, See M48
NGC 2613, 761, 762
NGC 2632, See M44
NGC 2682, See M67
NGC 2859, 719, 723
NGC 2903, 234, 240
NGC 2997, 757, 758
NGC 3031, See M81
NGC 3034, See M82
NGC 3109, 386, 391
NGC 3242, 386
NGC 3293, 451, 453
NGC 3351, See M95
NGC 3368, See M96
NGC 3379, See M105
NGC 3384, 235
NGC 3521, 235
NGC 3532, 451, 454, 474
NGC 3587, See M97
NGC 3623, See M65
NGC 3627, See M66
NGC 3628, 235
NGC 3656, See M108
NGC 3699, 269, 271, 273
NGC 3992, See M109
NGC 4038–39, 375
NGC 4192, See M98
NGC 4236, 28, 29, 31
NGC 4244, 313, 317
NGC 4254, See M99
NGC 4258, See M106
NGC 4274, 255
NGC 4303, See M61
NGC 4321, See M100
NGC 4369, See M90
NGC 4374, See M84
NGC 4382, See M85
NGC 4406, See M86
NGC 4472, See M49
NGC 4486, See M87
NGC 4501, See M88
NGC 4552, See M89
NGC 4558, See M91
NGC 4559, 256
NGC 4565, 256
NGC 4579, See M58
NGC 4590, See M68
NGC 4594, See M104
NGC 4621, See M59
NGC 4631, 314, 317, 363
NGC 4649, See M60
NGC 4666, 333

NGC 4706, 269, 274
NGC 4709, 274
NGC 4725, 256, 260
NGC 4736, See M94
NGC 4762, 333
NGC 4826, See M64
NGC 4945, 269, 275
NGC 5024, See M53
NGC 5033, 314, 318
NGC 5128, 270, 272, 293
NGC 5139, See ω Centauri
NGC 5189, 637, 638
NGC 5194, See M51
NGC 5236, See M83
NGC 5253, 270, 275, 293
NGC 5272, See M3
NGC 5457, See M101
NGC 5643, 282, 283
NGC 5746, 333
NGC 5824, 268, 280, 281, 283
NGC 5866, See M102
NGC 5904, See M5
NGC 5905, 28, 29
NGC 5907, 28, 29, 32
NGC 5908, 28, 29, 31
NGC 5963, 30, 32
NGC 5965, 30
NGC 5981, 30
NGC 5982, 30
NGC 5985, 30
NGC 6093, See M80
NGC 6121, See M4
NGC 6171, See M107
NGC 6205, See M13
NGC 6218, See M12
NGC, See M10
NGC 6266, See M62
NGC 6273, See M19
NGC 6333, See M9
NGC 6341, See M92
NGC 6402, See M14
NGC 6405, See M6
NGC 6475, See M7
NGC 6494, See M23
NGC 6503, 30, 33
NGC 6514, See M20
NGC 6523, See M8
NGC 6531, See M21
NGC 6543, 28, 29
NGC 6603, See M24
NGC 6611, See M16
NGC 6613, See M18
NGC 6618, See M17
NGC 6626, See M28
NGC 6637, See M69
NGC 6656, See M22
NGC 6681, See M70
NGC 6694, See M26
NGC 6705, See M11
NGC 6715, See M54

Printed in the United States of America